Intelligent Control in Energy Systems

Intelligent Control in Energy Systems

Special Issue Editor

Anastasios Dounis

MDPI • Basel • Beijing • Wuhan • Barcelona • Belgrade

Special Issue Editor
Anastasios Dounis
University of West Attica
Greece

Editorial Office
MDPI
St. Alban-Anlage 66
4052 Basel, Switzerland

This is a reprint of articles from the Special Issue published online in the open access journal *Energies* (ISSN 1996-1073) from 2018 to 2019 (available at: https://www.mdpi.com/journal/energies/special_issues/intelligent_control)

For citation purposes, cite each article independently as indicated on the article page online and as indicated below:

LastName, A.A.; LastName, B.B.; LastName, C.C. Article Title. *Journal Name* **Year**, *Article Number*, Page Range.

ISBN 978-3-03921-415-0 (Pbk)
ISBN 978-3-03921-416-7 (PDF)

Contents

About the Special Issue Editor

Anastasios Dounis received a B.Sc degree in physics from the University of Patras (Greece) in 1983, an M.Sc. degree in electronic automation from the National and Kapodistrian University of Athens (Greece) in 1988, and a Ph.D. degree in fuzzy control systems in buildings from the Technical University of Crete in 1993.

From 1993 to 1995, he was an Adjunct Professor in the Department of Electronic Engineering, Hellenic Air Force Academy, Greece. From 2000 to 2002, he was a Postdoc at the Hellenic State Scholarship Foundation.

He is currently a Professor in Industrial Design and Production Engineering at the University of West Attica, Greece. His research encompasses many aspects of smart energy microgrids and intelligent buildings. He is author or co-author of over 90 published papers in international journals and conferences. His current research interests include computational intelligence, intelligent control, evolutionary computation, multi-agent systems, and machine learning.

He participated in funded research projects including Archimedes I and II, Excellence I, and supporting researchers with emphasis on young researchers—cycle B 2019. Prof. Dounis is currently a guest editor for the Special Issue "Intelligent Buildings and Home Energy Management in a Smart Grid Environment," in *IEEE Transactions on Smart Grid*. Simultaneously, he is also the guest editor for *Algorithms'* Special Issues "Algorithms for PID Controller" and "Algorithms for PID Controller 2019;" for *Energies* Special Issues "Intelligent Control in Energy Systems," "Intelligent Control in Energy Systems II," and "Intelligent Decentralized Energy Management in Microgrids;" and for the *Applied Sciences* special issue "Artificial Intelligence for Smart Buildings." He is a reviewer for a number of international journals such as *Applied Soft Computing, IEEE Transactions on Fuzzy Systems, Energy and Buildings, Solar Energy,* and *IEEE Transactions on Smart Grid.*

Editorial

Special Issue "Intelligent Control in Energy Systems"

Anastasios Dounis

Department of Industrial Design and Production Engineering, University of West Attica, 12244 Egaleo, Greece; aidounis@umiwa.gr

Received: 3 July 2019; Accepted: 2 August 2019; Published: 5 August 2019

Abstract: The editor of this special issue on "Intelligent Control in Energy Systems" have made an attempt to publish a book containing original technical articles addressing various elements of intelligent control in energy systems. The response to our call had 60 submissions, of which 27 were published submissions and 33 were rejections. This book contains 27 technical articles and one editorial. All have been written by authors from 15 countries (China, Netherlands, Spain, Tunisia, United States of America, Korea, Brazil, Egypt, Denmark, Indonesia, Oman, Canada, Algeria, Mexico, and Czech Republic), which elaborated several aspects of intelligent control in energy systems. It covers a broad range of topics including fuzzy PID in automotive fuel cell and MPPT tracking, neural network for fuel cell control and dynamic optimization of energy management, adaptive control on power systems, hierarchical Petri Nets in microgrid management, model predictive control for electric vehicle battery and frequency regulation in HVAC systems, deep learning for power consumption forecasting, decision tree for wind systems, risk analysis for demand side management, finite state automata for HVAC control, robust μ-synthesis for microgrid, and neuro-fuzzy systems in energy storage.

Keywords: intelligent control; artificial intelligence; energy management system; smart micro-grid; energy systems; intelligent buildings; forecasting; multi-agent control; optimization

---

1. Introduction

Energy systems (ES) are a complex and constantly evolving research area. Since energy systems are multi-layered and distributed, there is a growing interest in integrating heterogeneous entities (energy sources, energy storage, micro-grids, grid networks, buildings, electrical vehicles, etc.) into distribution systems. The challenge in handling the vast volume of information is the requirement of the use of modern efficient management control strategies such as intelligent control technologies.

Intelligent control (IC) describes a class of control techniques that use various artificial intelligence techniques such as neural network control, Bayesian control, fuzzy logic control, neuro-fuzzy control, evolutionary computation, machine learning, and intelligent agents. IC systems are very useful when no mathematical model is available a priori. IC is inspired by the intelligence and genetics of living beings.

IC, communications infrastructure, and wireless networking play an important role in a smart grid network in achieving reliable, efficient, secure, distribution, cost-effective generation and consumption. IC on energy storage devices provide reliability and economic impacts on the energy systems.

Buildings consume a large portion of the world's energy and they are a source of greenhouse gas emissions. The concept of sustainable and zero energy buildings is emerging as an important area for the smart micro-grid initiative. In addition, effective energy management is becoming more feasible using the innovative smart micro-grid technologies and IC. These changes have resulted in an environment of high complexity, uncertainty, and imprecision. The IC can play a remarkable and vital role in handling a significant part of this high uncertainty and nonlinearity by providing new smart solutions for a more efficient and reliable operation of ESs.

This Special Issue is focused on to bring together innovative developments and emerging synergetic technologies in the fields of intelligent control and energy systems. The particular topics of interest in the original call for papers included, but were not limited to:

- Energy management and IC in energy micro-grids
- ESs modeling and IC
- IC and optimization for zero energy buildings
- Evolutionary control in ESs
- IC in hybrid ESs of isolated areas
- Fuzzy logic control in ESs
- Intelligent multiagent control systems in ESs
- Artificial neural networks for control in ESs
- IC of holonic ESs
- IC in energy storage systems
- IC in sustainable smart ESs
- Fault diagnosis and IC in ESs
- Chaos control in ESs
- Bayesian control in renewable energy systems
- Neuro-fuzzy control in ESs
- Machine learning in ESs
- IC in distributed electrical energy generation system
- IC in smart grid network
- IC and ESs stability
- IC and demand side forecasting in ESs
- IC and uncertainty analysis of ESs.

2. Brief Overview of the Contributions to This Special Issue

This Special Issue focused on bringing together innovative developments and synergies in the fields of intelligent control and energy systems. Therefore, a variety of topics were presented:

In [1], AbouOmar, Zhang and Su presented a fractional order fuzzy PID controller for a Proton Exchange Membrane Fuel Cell (PEMFC) air feed system, in order to achieve maximum power point tracking for the PEMFC stack, used the Neural Network Algorithm (NNA), a new metaheuristic optimization algorithm inspired by the structure and operation of ANNs to optimize the controller. A detailed simulation on MATLAB/SIMULINK environment proved the efficiency of the proposed controller over other types of controllers. The NNA optimized FOFPID controller achieved a better set point tracking and disturbance rejection with minimal fluctuations around the set value, with better transient response and minimum time domain performance indices. It was also shown that the system had satisfactory robustness against the considered parameter uncertainty range. In the future, the modification of the original NNA algorithm for improving its convergence with applications to PEMFC control using new control schemes may be expected.

Shen, Tang, Yi et al. [2] proposed an online switching methodology to relieve voltage violations based on a three-stage strategy which includes screening, ranking, and detailed analysis and assessment stages for high speed and accuracy. This online methodology rapidly identified effective candidate lines, ranking the effective candidates performing detailed analysis of the top ranked candidates, and supplied a set of solution for the power system, balanced speed and accuracy for online applications by combining linear and nonlinear methods. A significant feature of this methodology is that it can provide a set of high-quality solutions, allowing the operators to select a preferred solution. The results showed that this scheme led to promising results that could provide single-line switching as well as multiple-line switching and also, in comparison to other methods has more advantages in accuracy and speed, but also requires less CPU time, especially in a large-scale system.

A control architecture aimed at performing frequency regulation with renewable hybrid power plants comprised of a wind farm, a solar photovoltaic and a battery storage system while considering the causes that lead to inertia loss in worldwide grids, was the focus in [3]. The proposed architecture considered the latest regulations and recommendations published by ENTSO-E when implementing the fast frequency response and the frequency containment reserve, the first two stages of frequency control. The proposed system architecture was tested against two scenarios studying the same event: One according to the ENTSO-E recommendations, and ROCOF (Rate of Change of Frequency). The results showed a great improvement of the system's frequency behavior. Additionally, the dynamic response of the generators in the system was smoother in scenario II, satisfying the concerns of wind turbine manufacturers about mechanical stresses and premature aging due to frequency support provisions. The impact of the proposed architecture was the speed in event identification, which gave the plant, the time to react to a fault. ROCOF was proposed in event identification, since it allowed the identifications of fast excursions before the frequency could reduce its value and thus eased the system necessary reactions.

Kong, Ma, Zhao et al. [4] proposed an online estimation method for the operating error of electric meters which used the recursive least squares (RLS) and introduced a double parameter method with dynamic forgetting factors λ_a and λ_b to track the meter parameters changes in real time. The case analysis results showed that the estimated performance of the proposed approach was better than other estimation methods, with a false detection rate below 2%. By this proposed method, the parameters of the electric meter error and line loss could simultaneously be estimated and thus, the accuracy of the online estimation of electric meter errors could be improved. Finally, the proposed method was based on the elimination of abnormal data such as light load data and null data. The reduction of these effects in the process of the data processing and algorithm solving need further study.

In [5], three intelligent control systems (fuzzy logic, artificial neural network (ANN) and adaptive neuro-fuzzy inference system (ANFIS)) were used to carry out a strategy of controlling the air discharge of a small scale compressed air energy storage (SS-CAES) prototype to produce a constant voltage according to the user set point. The purpose was to simplify the control of the SS-CAES, so that it could be integrated with a grid based on a constant voltage reference. The experimental evaluation used two scenarios, demonstrating that ANN had the best performance in both of them since it had less iteration than the other controls resulting to a fast response. A high overshoot was observed in the second scenario due to the effect of high pressure when the load was still installed.

Choi and Cho [6], proposed an advanced continuous voltage control method that implemented multiple-point control to ensure peak power system performance. The utilization of generators to regulate the pilot point voltage of a control area was common in most of the control schemes. The influence of adjacent areas in a meshed power system made the exact control of a single pilot point difficult. In the proposed method, multiple pilot points were accessed to mitigate the effects of the neighboring area. The Multiple Continuous Voltage Control (MCVC) algorithm demonstrated effectiveness in regulating the voltages at a group of pilot points to remain around set-point values while dealing with the evolution of those voltages separately. Simulation with realistic data obtained from the Korean power system demonstrated the feasibility of this control scheme for reducing the severity of mutual interactions between adjacent zones. The same dynamic simulations were also used to study how the MCVC could return the system to stability from more severe conditions.

An adaptive damping control strategy of a wind integrated power system was analysed in [7], by tracking the variation of system operating points and updating the controller's functions to achieve a robust damping control effect. This occurred in three steps, including the division of the space of the operating point into operating subspaces by the even interval of wind power outputs, the pre-design of the coordinated Power System Stabilizers (PSSs) for each operating subspace, and the formation of a classification tree by training the distances to the hyperplanes and the regression tree was used to identify the subspaces with the help of on line measurement from PMUs. The proposed adaptive control demonstrated robustness to stochastic varying operating conditions and showed good performance in the case of multiple wind farms connected at different buses.

In [8], a repetitive controller used in a SEPIC PFC converter was designed using a third-order model approximation-instead of fifth-order for reasons of simplicity in order to reduce the input current distortion and the stability of the controller was verified with an error transfer function. The proposed controller was then validated by simulation in 100 W to 800 W load conditions in buck and boost mode. The THDs of the input current were significantly decreased in both modes. The experimental result also showed that the controller based on simplified model was well designed.

A simple strategy for controlling an interleaved boost converter that was used to reduce the current fluctuations in proton exchange membrane fuel cells was presented by Barhoumi, Ben Belgacem, Khiareddine et al. in [9], which had a high impact on the fuel cell lifetime. A neural network controller was employed in order to keep the output voltage at the desired reference value under the fluctuations of the fuel flow rate, the fuel supply pressure and temperature. The proposed converter has reduced the ripples of load voltage to less than 2 V. The simulation results indicated that the controller demonstrated robustness and efficiency of the converter to regulate the load voltage as well as minimize the voltage ripples. It also showed that using limited number of tests allows one to develop efficient ANN controllers for the regulation of the load voltage.

Tchoketch Kebir, Larbes, Ilinca et al. in [10] proposed a control method that offered high performance to get a maximum power output tracking by using a fuzzy logic approach which entailed a maximum speed of power achievement, a good stability, and high robustness. A fuzzy controller was used, based on a special choice of a combination of inputs and outputs. The choice of inputs and outputs as well as fuzzy rules, was based on the principles of mathematical analysis of the derived functions for the purpose of finding the optimum. It was also proved that using the simplest possible fuzzy model by using only 3 sets of linguistic variables to decompose the membership functions of the inputs and the outputs of the fuzzy controller could achieve the best results and answers for a PV system. A comparison of the fuzzy controller model with conventional perturb and observe controllers' models proved higher efficiency in maximum power point tracking for the fuzzy logic controller and in maximum power tracking time delay, stability, and robustness in all cases. The fuzzy logic algorithm was a robust and efficient algorithm which worked at the optimal point without oscillations and with a good transient behavior.

In [11], a solution in load balancing issue in urban μgrids with the use of hierarchical Petri Nets (PNs) in phase-load balance method was presented. The new system design composed of combined algorithms, called Load-Balance Control (LBC) system contributing to the load amount identification to transfer between feeders, and with the single-phase consumer unit selection to the switch operation of load balance procedure. The hierarchical PN was used to represent and validate the workflow of each inner algorithm of the control system for LV and for the upper hierarchical levels as Microgrid Central Control (MGCC) and the legacy LV grid as well. Both networks were tested through dynamic simulation, verifying reliable and reliable dynamic performance in both, free of conflicts, stops, and deadlock. The attainability of all its states also verified identifying that were both limited and safe networks. The identification of two inviolable workflows in both networks guarantees the efficient execution of the load transfer algorithm and its evaluation of each fuzzy inference rule used to identify load transfer. This provided an efficient and reliable load-balancing algorithm that ensured a single and admissible load-balancing solution to the integrated control flow as well as a unique and admissible inference rule to the load transfer. Furthermore, the combined algorithm of the LBC system was also tested by dynamic simulation which presented load imbalance between its phases, showing the identification of the load transfer amount in each phase, the limits of variation of load in relation to the discrete states of consumption in each phase, the future consumption matrix and the future load consumption states. A second application of the LBC system was also tested demonstrating the efficiency of the proposed system.

A study for the design of an active equilibrium control strategy based on model predictive control (MPC) for series battery packs is presented in [12]. Bidirectional active equalization was modelled and analyzed, and the MPC algorithm was applied to the established state space equation. The optimization problem that minimized the equilibrium time transformed to a linear programming form in each cycle. The solution of the linear programming problem gave a group of control optimal solutions and the series equalization problem was decoupled. The dynamic adjustment of the equalization current shortens the equalization time. The experimental result indicated that the equilibrium time was reduced by 31%. The main idea of this method was that the balance current was adjustable. One drawback of computation process of local optimal solutions was that it was a time-consuming process, so in the future other optimization algorithms must be tried to reduce the computation time, increasing the efficiency further.

In [13], atmospheric pollution and Total Suspended Particle emissions control was analysed; the development of a non-linear model for TSP emissions estimation from an industrial boiler based on a one-layer neural network was reported. The model used expansion polynomial basic functions combined with an orthogonal least square and model structure approach and required five independent boiler variables for TSP emissions estimation. The experimental results demonstrated that orthogonal least square algorithms were a great tool that provided extra information about internal model behaviors. It also showed that finite expansion polynomial basic functions could be implemented with one-layer ANN and agreed with the universal approximation theorem. The precision of the PBF network was excellent in predicting TSP emissions and this methodology can be replicated for other pollutants emitted into the atmosphere such as NO_x and CO emissions etc.

A hybrid deep learning neural network framework that combined Convolutional Neural Network (CNN) with Long Short Term Memory (LSTM) with a multi-step forecasting strategy was proposed in [14] in order to fill the research gaps that existed in power consumption forecasting problems and were considered as disadvantages in practical applications of LSTM: The prediction's accuracy and the shortness of the forecasting time. The proposed framework was tried against some of the known existing approaches, such as ARIMA, persistent model, SVR, and LSTM alone. Additionally, a k-step power consumption forecasting strategy was demonstrated in order to promote the proposed framework for real world application usage. The results obtained based on five real world households using MAPE as the error metric and the CNN-LSTM framework with multistep forecasting strategy outperforms the conventional methods.

The planning of an Integrated Energy Micro-Grid (IEMG), formed by connecting multiple regions' Integrated Energy Systems (IES) was the subject in [15]. Compared with isolated IES, an IEMG, could further improve the reliability, flexibility, cleanliness and the economy of a regional energy supply. An IEMG planning model was presented with distributed photovoltaic developed by Mixed Integer Linear Programming (MILP). First, the capacity construction configuration of the energy production equipment by known electricity, heating and cooling loads was determined. Second, an operational cost analysis of heating, cooling, transitional and extreme load scenarios was conducted, in order to improve the feasibility of the planning results. As the main investors, the model takes the district energy suppliers, and the optimized capacity configuration could meet the overall energy demand of the region in different scenarios and, at the same time, give the construction and operation cost of different sub-regions. A case study was given to improve the validity of the model and, according to the results of the model calculation, the proposed model could be seen as a theoretical reference for the planning of multi-district IES (an IEMG).

Su, Dong and Shen [16] introduced an improved adaptive backstepping method based on error compensation (ABEC), a method which considered the damping coefficients. Then, an improved adaptive backstepping sliding mode variable control based on error compensation method (ABSMVCEC) was introduced. This method stabilized the system more quickly and with a κ-class function addition to the selection of the intermediate virtual variable function, the convergence of the system speeded up. The nonlinear controller for the Static Var Compensator (SVR) system simulated with the two above mentioned methods and the method of adaptive backstepping. The proposed methods had better performance. Additionally, the ABSMVCEC method was more effective in improving the transient stability.

A static switch in order to feed a High-Speed Train (HST) through the Neutral Section (NS) was proposed by Canales, Aizpuru, Iraola et al. [17]. An NS operation was analyzed to identify impacts within the electric system and solution requirements were developed. A low-scale prototype switch was used to experimentally validate the solution which was based on Thyristor technology and the medium-voltage AC static switch was designed. The final tests took place on the Cordoba-Malaga High-Speed Railway. The thyristor demonstrated the ability to feed high-speed trains in neutral sections, avoiding the electrical transient of connection and disconnection of traction transformers and failure of the train's main breaker. The AC switch solution was suitable for conventional railways where the train's speed can be very low and there is a risk of stopping in a neutral section without electrical power.

Blaauwbroek, Nguyen and Slootweg [18] were presenting a time horizon three-phase grid supportive demand side management methodology for low voltage networks by using a universal interface established between the DSM application and the network's operator monitoring and network analysis tools. By using time horizon predictions of the system stated that the probability of operational limit violations was identified. Contributing with a probabilistic approach by presenting a Monte Carlo as well as a Neural Network based approach, they reduced the probability of geographical dependent operation limit violations to acceptable levels. Numerical simulations showed that NN-based approach offered a significant benefit over the PPF based approach in terms of computational complexity. Moreover, from the findings of the proposed approach, it seems that the research could be extended in several directions.

The development of an innovative solar monitoring system was presented in [19]. The system aimed at measuring the main parameters and characteristics of solar plants; collecting, diagnosing and processing data. The system communicated with the inverters, electrometers, metrological equipment and additional components of the PV arrays using special data collecting technologies. This monitoring system contributed in quality management of plants and provided data for scientific purposes; it helped to identify and eliminate installation errors and contributed to the continuous operation of the PV arrays by providing information to the staff about the potential error. The increased number of input lines and secured communications were some of the benefits of the system. The high frequency of data saving allowed a higher accuracy of the mathematical models. Moreover, a significant advantage was the capability to collect additional data from other power plant components.

The improvement of the efficiency of HVAC systems was the target of the study in [20], by providing accurate occupancy prediction to the HVAC control in order to ensure that HVAC is not run needlessly when a room or a zone is unoccupied. Simple but effective algorithms were proposed to predict occupancy, along with an algorithm for automatically assigning temperature set-points. The three techniques for occupancy prediction were carried out by utilizing past occupancy observations. The methods employed were Expectation Maximization (EM), Finite State Automata (FSA) reconstructed by a General Systems Problem Solver (GSPS) and an alternative stochastic model based on uncertain basis functions. All three methods achieved more than 70% accuracy in experimental studies. Along with a Model Predictive Control (MPC) algorithm to assign temperature set point based on occupancy information, the paper delivered a novel end to end solution.

In [21], the mechanical vibration characteristics of a dry-type transformer winding were studied. A vibration simulation model of a dry type transformer was established based on actual short circuit experimental conditions of an SCB10-1000/10 dry type transformer in which the vibration signal at the surface was measured. The model has been developed using COMSOL Multiphysics software. A Multiphysics coupling simulation of the circuit, magnetic field and solid mechanics of the transformer was performed on the model. After the validation, the model used to develop simulation models of winding failures, such as looseness, deformation or insulation failure. The results then were used as a basis for analyzing and detect the mechanical state of transformer windings.

Xu, Mao et al. [22] investigating a dynamic optimization energy management strategy called Hybrid Electric Vehicle Based on Compound Structured Permanent Magnet Motor (CSPM-HEV) which had obvious advantages on power density, heat dissipation efficiency, torque performance and energy transmission efficiency. The topology and working principle of CSPM-HEV were described and an analysis of its operating mode and corresponding energy flow laws were also analyzed. Back Propagation neural network was employed for the real time energy management of CSPM-HEV, solving the problem of complex algorithms and poor real time performance. It was shown that the instantaneous optimal control of the vehicle target achieved, along with a real time energy management strategy based on a BP neural network and the instantaneous optimal control of CSPM-HEV which were also fulfilled. As a future research topic, a global optimization algorithm is expected to improve the fuel economy of CSPM-HEV.

Bhuiyan and Lee [23] proposed a position control method for a low cost exhaust which had a recirculation (EGR) valve system for automotive applications, that could be applied under the high difference friction mechanical system and overcome the restrictions that common position control systems with the conventional P-PI linear controller faced because of the large differences in static and Coulomb friction resulting in position and current vibrations. The mathematical analysis showed that the proposed system could achieve the proper control performance with errors within the acceptance boundaries.

Two control strategies that allowed HVAC systems in commercial and residential buildings to provide frequency regulation services were investigated in [24]. The first was based on model predictive control acting on a variable air volume HVAC system. The second strategy was rule-based control acting on an aggregate of on/off HVAC systems considering the hardware constraints. The first strategy could be applied in large commercial buildings and the second to residential and small to medium size commercial buildings. The second strategy provided the required flexibility for ancillary services to the grid with little impact on indoor environments. The presented strategies demonstrated a novelty: to use HVAC loads as ancillary service to the grid. A study for the design of a decentralized framework for the rule-based control strategy would be expected in the future.

In [25], a preventive control methodology to increase the capacity of voltage sag recovery (Fault Ride Through Capability-FRTC) of a doubly-fed induction generator connected in an electrical network was presented. The methodology was based in decision trees (DT) technique and assisted with monitoring and support for security and preventive control, ensuring that wind systems remain connected to the system even after the occurrence of disturbances in the electric system. The presented methodology was tested using the IEEE 39 bus system, which was modified by the insertion of doubly fed induction generators. The presented results verified that the wind power system voltage and the reactive power of synchronous generator contribute to the systems operation security and to the continuity of electricity supply from a wind turbine after the occurrence of a disturbance in the electrical network. It was also possible to verify that active power and voltage contribute to the continuity and lack of wind system shutdown. Furthermore, it was shown that the decision tree classified the system's operational state with goof accuracy and indicated the way to maintain the electrical system dynamic security for each topology. The use of the optimization tool may guarantee optimal operating conditions. Conclusively, the presented methodology consisted of a DT based support tool which could be directly integrated into operation centers.

Li, Wang and Xiao [26] investigated the secondary load frequency controller of the power systems with renewable energies taking into account internal parameter perturbations and stochastic disturbances induced by the integration of renewable energies and the power unbalance caused between the supply and the demand side. The robust μ-synthesis approach was used for load frequency control in a microgrid power system. A load frequency control state space model with uncertainty was established. The results showed that ultracapacitors could enhance the frequency stability of microgrid power systems with μ-synthesis controller which demonstrated robustness and better nominal performance than the H∞ controller and could greatly improve the load frequency stability of a μgrid power system.

In [27], a detection method for the internal short circuit of a Lithium-Ion battery pack was proposed by estimating the resistance with the whole terminal voltages and the load currents of the pack. The open circuit voltage of a faulted cell in the pack was extracted to reflect the self-discharge phenomenon obviously yielding accurate estimates of the resistance. The proposed algorithm was verified for various soft ISCr fault conditions such as diverse magnitudes of true R_{ISCr} and two load current profiles in both the simulation and the experiment. Additionally, through estimating the R_{ISCr} from the normal battery pack and analyzing it, it was checked that estimated resistances in the various scenarios were reliable. With the proposed algorithm, it was possible to estimate accurately the R_{ISCr} and the soft ISCr in the battery pack could be calculated using the R_{ISCr} as the fault index. The error of estimated resistance did not exceed 31.2% in the experiment, enabling the battery management system to detect the internal short circuit early.

The above-mentioned articles that constitute this book critically reviewed various intelligent control technologies in energy systems and provided systematic solutions for the readers to easily understand the concepts used and outcomes produced. The editor believes that this book will be useful to many researchers and industries working on intelligent control in energy systems. The editor of the book would like to record their sincere thanks and acknowledgements to all the contributors of the articles and the continuous support they received from the Energies journal editorial staff team, without whose dedication it would have not been possible to publish this book.

Funding: This research received no external funding.

Conflicts of Interest: The author declares no conflict of interest.

References

1. AbouOmar, M.; Zhang, H.; Su, Y. Fractional Order Fuzzy PID Control of Automotive PEM Fuel Cell Air Feed System Using Neural Network Optimization Algorithm. *Energies* **2019**, *12*, 1435. [CrossRef]
2. Shen, Z.; Tang, Y.; Yi, J.; Chen, C.; Zhao, B.; Zhang, G. Corrective Control by Line Switching for Relieving Voltage Violations Based on A Three-Stage Methodology. *Energies* **2019**, *12*, 1206. [CrossRef]
3. Vázquez Pombo, D.; Iov, F.; Stroe, D. A Novel Control Architecture for Hybrid Power Plants to Provide Coordinated Frequency Reserves. *Energies* **2019**, *12*, 919. [CrossRef]
4. Kong, X.; Ma, Y.; Zhao, X.; Li, Y.; Teng, Y. A Recursive Least Squares Method with Double-Parameter for Online Estimation of Electric Meter Errors. *Energies* **2019**, *12*, 805. [CrossRef]
5. Soenoko, R.; Wahyudi, S.; Siswanto, E. Comparison of Intelligence Control Systems for Voltage Controlling on Small Scale Compressed Air Energy Storage. *Energies* **2019**, *12*, 803. [CrossRef]
6. Choi, Y.; Cho, Y. Multiple-Point Voltage Control to Minimize Interaction Effects in Power Systems. *Energies* **2019**, *12*, 274. [CrossRef]
7. Deng, J.; Suo, J.; Yang, J.; Peng, S.; Chi, F.; Wang, T. Adaptive Damping Control Strategy of Wind Integrated Power System. *Energies* **2019**, *12*, 135. [CrossRef]
8. Kim, J.; Han, S.; Cho, W.; Cho, Y.; Koh, H. Design and Analysis of a Repetitive Current Controller for a Single-Phase Bridgeless SEPIC PFC Converter. *Energies* **2019**, *12*, 131. [CrossRef]
9. Barhoumi, E.; Ben Belgacem, I.; Khiareddine, A.; Zghaibeh, M.; Tlili, I. A Neural Network-Based Four Phases Interleaved Boost Converter for Fuel Cell System Applications. *Energies* **2018**, *11*, 3423. [CrossRef]
10. Tchoketch Kebir, G.; Larbes, C.; Ilinca, A.; Obeidi, T.; Tchoketch Kebir, S. Study of the Intelligent Behavior of a Maximum Photovoltaic Energy Tracking Fuzzy Controller. *Energies* **2018**, *11*, 3263. [CrossRef]
11. Sicchar, J.; Da Costa, C.; Silva, J.; Oliveira, R.; Oliveira, W. A Load-Balance System Design of Microgrid Cluster Based on Hierarchical Petri Nets. *Energies* **2018**, *11*, 3245. [CrossRef]
12. Song, S.; Xiao, F.; Peng, S.; Song, C.; Shao, Y. A High-Efficiency Bidirectional Active Balance for Electric Vehicle Battery Packs Based on Model Predictive Control. *Energies* **2018**, *11*, 3220. [CrossRef]

13. Ronquillo-Lomeli, G.; Herrera-Ruiz, G.; Ríos-Moreno, J.; Ramirez-Maya, I.; Trejo-Perea, M. Total Suspended Particle Emissions Modelling in an Industrial Boiler. *Energies* **2018**, *11*, 3097. [CrossRef]

14. Yan, K.; Wang, X.; Du, Y.; Jin, N.; Huang, H.; Zhou, H. Multi-Step Short-Term Power Consumption Forecasting with a Hybrid Deep Learning Strategy. *Energies* **2018**, *11*, 3089. [CrossRef]

15. Huang, H.; Liang, D.; Tong, Z. Integrated Energy Micro-Grid Planning Using Electricity, Heating and Cooling Demands. *Energies* **2018**, *11*, 2810. [CrossRef]

16. Su, Q.; Dong, F.; Shen, X. Improved Adaptive Backstepping Sliding Mode Control of Static Var Compensator. *Energies* **2018**, *11*, 2750. [CrossRef]

17. Canales, J.; Aizpuru, I.; Iraola, U.; Barrena, J.; Barrenetxea, M. Medium-Voltage AC Static Switch Solution to Feed Neutral Section in a High-Speed Railway System. *Energies* **2018**, *11*, 2740. [CrossRef]

18. Blaauwbroek, N.; Nguyen, P.; Slootweg, H. Data-Driven Risk Analysis for Probabilistic Three-Phase Grid-Supportive Demand Side Management. *Energies* **2018**, *11*, 2514. [CrossRef]

19. Beránek, V.; Olšan, T.; Libra, M.; Poulek, V.; Sedláček, J.; Dang, M.; Tyukhov, I. New Monitoring System for Photovoltaic Power Plants' Management. *Energies* **2018**, *11*, 2495. [CrossRef]

20. Dong, J.; Winstead, C.; Nutaro, J.; Kuruganti, T. Occupancy-Based HVAC Control with Short-Term Occupancy Prediction Algorithms for Energy-Efficient Buildings. *Energies* **2018**, *11*, 2427. [CrossRef]

21. Duan, X.; Zhao, T.; Liu, J.; Zhang, L.; Zou, L. Analysis of Winding Vibration Characteristics of Power Transformers Based on the Finite-Element Method. *Energies* **2018**, *11*, 2404. [CrossRef]

22. Xu, Q.; Mao, Y.; Zhao, M.; Cui, S. A Hybrid Electric Vehicle Dynamic Optimization Energy Management Strategy Based on a Compound-Structured Permanent-Magnet Motor. *Energies* **2018**, *11*, 2212. [CrossRef]

23. Bhuiyan, H.; Lee, J. Low Cost Position Controller for Exhaust Gas Recirculation Valve System. *Energies* **2018**, *11*, 2171. [CrossRef]

24. Olama, M.; Kuruganti, T.; Nutaro, J.; Dong, J. Coordination and Control of Building HVAC Systems to Provide Frequency Regulation to the Electric Grid. *Energies* **2018**, *11*, 1852. [CrossRef]

25. Vieira, D.; Nunes, M.; Bezerra, U. Decision Tree-Based Preventive Control Applications to Enhance Fault Ride Through Capability of Doubly-Fed Induction Generator in Power Systems. *Energies* **2018**, *11*, 1760. [CrossRef]

26. Li, H.; Wang, X.; Xiao, J. Differential Evolution-Based Load Frequency Robust Control for Micro-Grids with Energy Storage Systems. *Energies* **2018**, *11*, 1686. [CrossRef]

27. Seo, M.; Goh, T.; Park, M.; Kim, S. Detection Method for Soft Internal Short Circuit in Lithium-Ion Battery Pack by Extracting Open Circuit Voltage of Faulted Cell. *Energies* **2018**, *11*, 1669. [CrossRef]

Article

Fractional Order Fuzzy PID Control of Automotive PEM Fuel Cell Air Feed System Using Neural Network Optimization Algorithm

Mahmoud S. AbouOmar [1,2,*], Hua-Jun Zhang [1,*] and Yi-Xin Su [1,*]

[1] School of Automation, Wuhan University of Technology, Wuhan 430070, China
[2] Industrial Electronics and Control Engineering Department, Faculty of Electronic Engineering, Menoufia University, Menouf 32952, Egypt
* Correspondence: mahmoud_samy_09@el-eng.menofia.edu.eg (M.S.A.); zhanghj@whut.edu.cn (H.-J.Z.); suyixin@whut.edu.cn (Y.-X.S.)

Received: 14 March 2019; Accepted: 9 April 2019; Published: 14 April 2019

Abstract: The air feeding system is one of the most important systems in the proton exchange membrane fuel cell (PEMFC) stack, which has a great impact on the stack performance. The main control objective is to design an optimal controller for the air feeding system to regulate oxygen excess at the required level to prevent oxygen starvation and obtain the maximum net power output from the PEMFC stack at different disturbance conditions. This paper proposes a fractional order fuzzy PID controller as an efficient controller for the PEMFC air feed system. The proposed controller was then employed to achieve maximum power point tracking for the PEMFC stack. The proposed controller was optimized using the neural network algorithm (NNA), which is a new metaheuristic optimization algorithm inspired by the structure and operations of the artificial neural networks (ANNs). This paper is the first application of the fractional order fuzzy PID controller to the PEMFC air feed system. The NNA algorithm was also applied for the first time for the optimization of the controllers tested in this paper. Simulation results showed the effectiveness of the proposed controller by improving the transient response providing a better set point tracking and disturbance rejection with better time domain performance indices. Sensitivity analyses were carried-out to test the robustness of the proposed controller under different uncertainty conditions. Simulation results showed that the proposed controller had good robustness against parameter uncertainty in the system.

Keywords: fractional order fuzzy PID controller; neural network algorithm; PEM fuel cell; MPPT operation; sensitivity analysis

---

1. Introduction

In recent years, fuel cells gained a lot of interest as one of the most promising renewable energy sources because of its high efficiency, flexibility and sustainability. Fuel cells produce electricity via electrochemical reactions between hydrogen and oxygen. The byproducts of the electrochemical reactions are only water and heat so fuel cells are considered as clean energy sources. The most common type of fuel cells is the proton exchange membrane (PEM) fuel cell. Proton exchange membrane fuel cells (PEMFCs) are used in vehicular applications because of its high electrical efficiency [1].

A PEMFC stack works as an autonomous energy source in automotive systems where the compressor motor of the air feeding system is powered by the PEMFC stack acting as an auxiliary load. The air feeding system is one of the most important systems in the PEMFC stack that has a great impact on the stack performance because it consumes up to 30% of the fuel cell power [2,3]. The role of the air feeding system is to regulate the oxygen excess ratio also known as stoichiometry at it is a predefined value using compressed air to prevent both oxygen starvation and oxygen saturation and to obtain the maximum net power output from the PEMFC stack. Oxygen starvation occurs at a sudden increase in the fuel cell stack current causing damage of the fuel cell membrane and the catalyst layer leading to fuel cell damage. On the contrary, oxygen saturation, which means a high level of oxygen availability, increases the power consumption of the air compressor resulting in a reduction of the net power output of the fuel cell. The PEMFC air feeding system is a complex nonlinear multi-input multi-output (MIMO) system that may include parameter uncertainty, so an efficient controller is required for the precise regulation of the oxygen excess ratio at different disturbance levels.

For the PEMFC air feeding system, several control techniques have been investigated in the literature such as feedforward control [4–7], LQR/LQG control [8], feedforward plus PI feedback control [6], sliding mode control (SMC) [9], adaptive sliding mode observer based control [10], adaptive control [11], model predictive control (MPC) [12], time delay control (TDC) with static feedforward [13] and disturbance-observer-based control [2]. Recently, soft computing techniques gained a lot of interest for the control of the PEMFC air feeding system. A B-spline neuro controller (B-SNN) was proposed in [14], PID feedback control with a single-input single-output (SISO) fuzzy feedforward controller [15] and hybrid fuzzy-PID controller [16].

A fuzzy logic controller (FLC) is widely accepted as an efficient controller, which is capable of controlling system without knowledge of its underlying dynamics and without using extensive mathematical analysis. Applications of FLCs in the literature witness that FLC is very efficient for nonlinear and uncertain systems [17]. However, the design of FLC is difficult because it involves several parameters without a distinct method for tuning. The design parameters for FLC are input/output scaling factors, membership function parameters and the rule base. Several heuristic methods have been proposed for the design and tuning of FLCs usually involving trial and error methods. The use of metaheuristic optimization techniques is an efficient method for tuning FLC, which proved efficient for different applications in literature [18,19].

Fractional order controllers are a generalization of standard controllers by using fractional order calculus where the order of the differentiators or integrators is a fractional number rather than an integer number used in standard controllers. The use of fractional operators increases the degree of freedom of the controller allowing it to generate outputs, which cannot be generated using integer order operators. A fractional order PID (FOPID) controller was proposed by Podlubny [20] where it demonstrated better control performance compared to the standard integer order PID controller. As a result, Fractional order PID controller gained a lot of interest in different control applications [21–25].

The application of fractional order operators has been extended to be used with fuzzy logic controllers where it was firstly proposed by Das et al. in [26]. Results demonstrated the superiority of the fractional order fuzzy PID compared to the standard Fuzzy PID controller. As a result, the fractional order fuzzy PID (FOFPID) controller gained a lot of interest and it is considered as an active and promising research area for different control applications [27–30].

This paper proposes a fractional order fuzzy PID (FOFPID) controller as an efficient controller for the PEMFC air feeding system. The proposed controller is optimized using the neural network algorithm (NNA). NNA is a new metaheuristic optimization algorithm developed by Sadollah et al. [31]. Sadollah et al. concluded that the artificial neural networks (ANNs) could be modeled and used as a metaheuristic optimization algorithm for handling optimization problems. NNA was developed based on the structure and the operators of the artificial neural networks (ANNs) for solving optimization problems [31]. NNA is one of the parameter free metaheuristic optimization algorithms where it does not require the user to fine-tune any algorithm parameters.

In this paper, the proposed fractional order fuzzy PID (FOFPID) controller is optimized using the NNA, where the NNA is used to optimize the input and output scaling factors, the membership function parameters of the controller inputs as well as the order of the fractional order differentiator and integrator.

The main contributions of this paper are:

- A new application for the FOFPID and FOPID controllers is proposed to apply in the PEMFC air feed control to improve performance and robustness.
- This paper employs a direct discretization approach using an Al-Alawi operator for the first time to implement fractional order fuzzy PID controllers rather than indirect discretization approach based on Oustaloup's recursive approximation.
- This paper is the first application of the NNA algorithm in controller design applications.
- The proposed NNA optimized FOFPID controller is tested for a constant set value for the oxygen excess ratio as well as the maximum power point operation by tracking a time varying set value for the oxygen excess ratio.
- Sensitivity analyses are performed to test the robustness of the proposed controller under various uncertainty conditions.

2. PEMFC Model

A PEM fuel cell stack consists mainly of three subsystems which are: (i) A hydrogen supply subsystem that feeds the anode side with hydrogen, (ii) an air feed subsystem that feeds the cathode by oxygen from the air and (iii) a humidification and thermal management subsystem that regulates the humidity and the temperature of the fuel cells, respectively. The main components of a PEMFC stack system are shown in Figure 1.

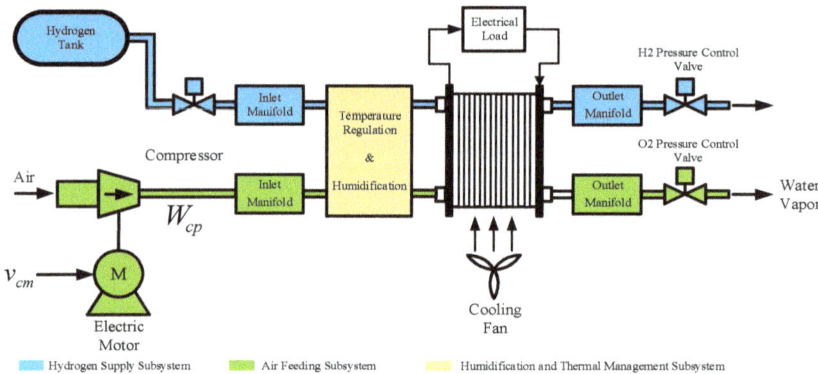

Figure 1. The main components of a PEMFC stack system.

The air feeding subsystem has a great impact on the PEMFC stack performance because it consumes up to 30% of the fuel cell power [2,3]. The air feed subsystem consists of an electromechanical air compressor, which maintains the required oxygen pressure and mass flow rate in the cathode of PEMFC [2].

2.1. Air Feed System Model for PEMFC

There are several models for the PEMFC air feed system. Pukrushpan et al. introduced an accurate 9th order model for the air feed system in [4,5,32]. A reduced order model of Pukrushpan's model was introduced by Suh in [33] where the 9th order model of Pukrushpan et al. was reduced into a 4th order model while preserving the dynamic behavior. Some assumptions have been assumed for PEMFC model reduction: The hydrogen subsystem dynamics are neglected by assuming perfect hydrogen supply control. Humidity and temperature variations are neglected by assuming perfect humidity and temperature control. The DC motor dynamics are neglected due to its small time constant compared to the mechanical system [33,34]. This model has been widely accepted by researchers for the design of the air feed system controller [16,35,36].

According to Suh's model, the PEMFC air feed system equations are expressed as follows:

$$\dot{x} = f(x) + g_u u + g_\omega \omega \tag{1}$$

with a state vector $x = [x_1, x_2, x_3, x_4]^T$, where $x_1 = P_{O_2}$ is the partial pressure of oxygen in the cathode, $x_2 = P_{N_2}$ is the partial pressure of nitrogen in the cathode, $x_3 = \omega_{cp}$ is the angular velocity of the compressor, $x_4 = P_{sm}$ is the pressure of the supply manifold, $u = v_{cm}$ is the compressor-motor voltage as the control input and $\omega = I_{st}$ is the PEMFC stack current representing the measurable disturbance to the system. The components of $f(x)$ are [16,35,37]:

$$f_1(x) = c_1(x_4 - \chi) - \frac{c_3 x_1 \alpha(x_1, x_2)}{c_4 x_1 + c_5 x_2 + c_6} \tag{2}$$

$$f_2(x) = c_8(x_4 - \chi) - \frac{c_3 x_2 \alpha(x_1, x_2)}{c_4 x_1 + c_5 x_2 + c_6} \tag{3}$$

$$f_3(x) = -c_9 x_3 - \frac{c_{10}}{x_3}\left(\left(\frac{x_4}{c_{11}}\right)^{c_{12}} - 1\right)\psi_3(x_3, x_4) \tag{4}$$

$$f_4(x) = c_{14}\left(1 + c_{15}\left(\left(\frac{x_4}{c_{11}}\right)^{c_{12}} - 1\right)\right)\cdot\left(\psi_3(x_3, x_4) - c_{16}(x_4 - \chi)\right) \tag{5}$$

where $\chi = x_1 + x_2 + c_2$ is the cathode pressure (P_{ca}) and $\alpha(x_1, x_2)$ is the total flow rate at the cathode outlet, which is given by:

$$\alpha(x_1, x_2) = \begin{cases} c_{17}\chi\left(\frac{c_{11}}{\chi}\right)^{c_{18}}\cdot\left(1 - \left(\frac{c_{11}}{\chi}\right)^{c_{12}}\right)^{0.5} & for \ \frac{c_{11}}{\chi} > c_{19} \\ c_{20}\chi & for \ \frac{c_{11}}{\chi} \leq c_{19} \end{cases} \tag{6}$$

The input vectors g_u and g_ω are given by:

$$g_u = [0\ 0\ c_{13}\ 0]^T \tag{7}$$

$$g_\omega = [-c_7\ 0\ 0\ 0]^T \tag{8}$$

The constants c_i, $i = 1, 2, \ldots, 24$ depend on the physical parameters of the PEMFC stack. The definition of these constants is given in Table 1 [16,35]. The values of the model parameters are shown in the Appendix A in Table A1 [16].

The measurement outputs vector is:

$$y = \begin{bmatrix} y_1 \\ y_2 \\ y_3 \\ y_4 \end{bmatrix} = \begin{bmatrix} \psi_1(x_1, x_2) \\ x_4 \\ \psi_3(x_3, x_4) \\ x_3 \end{bmatrix} \tag{9}$$

where $y_1 = \psi_1(x_1, x_2)$ is the stack voltage (V_{st}) given by:

$$V_{st} = n v_{FC} \tag{10}$$

where v_{FC} is the voltage of a single fuel cell and n is the number of fuel cells in the stack. The voltage of a single fuel cell is defined by:

$$v_{FC} = E - v_{act} - v_{ohm} - v_{conc} \tag{11}$$

with E as the open circuit voltage and v_{act}, v_{ohm} and v_{conc} are the activation, ohmic and concentration overvoltages, respectively. For more details about $\psi_1(x_1, x_2)$, the reader can refer to [4,5,7]. $y_3 =$

$\psi_3(x_3, x_4)$ is the airflow rate inside the compressor (W_{cp}) also known as the compressor flow map. It is approximated as follows [16,35]:

$$\psi_3^{aproximated} = \frac{y_3^{max} x_3}{x_3^{max}}\left(1 - exp\left(\frac{-r_c\left(s_c + \frac{x_3^2}{q_c} - x_4\right)}{s_c + \frac{x_3^2}{q_c} - x_4^{min}}\right)\right) \tag{12}$$

where, $r_c = 15$, $s_c = 10^5$ Pa and $q_c = 462.25$ rad$^2/\left(\text{s}^2\text{Pa}\right)$.

Table 1. PEMFC model constants [a].

PEMFC Model Constants	
$C_1 = \frac{RT_{st}K_{ca,in}}{M_{O_2}V_{ca}}\left(\frac{x_{O_2,atm}}{1+\omega_{atm}}\right)$	$C_{14} = \frac{RT_{atm}\gamma}{M_{a,atm}V_{sm}}$
$C_2 = P_{sat}$	$C_{15} = \frac{1}{\eta_{cp}}$
$C_3 = \frac{RT_{st}}{V_{ca}}$	$C_{16} = K_{ca,in}$
$C_4 = M_{O_2}$	$C_{17} = \frac{C_D A_T}{(RT_{st})^{0.5}}\left(\frac{2\gamma}{\gamma-1}\right)^{0.5}$
$C_5 = M_{N_2}$	$C_{18} = \frac{1}{\gamma}$
$C_6 = M_v P_{sat}$	$C_{19} = \left(\frac{2}{\gamma+1}\right)^{\frac{\gamma}{\gamma-1}}$
$C_7 = \frac{RT_{st}n}{4FV_{ca}}$	$C_{20} = \frac{C_D A_T}{\sqrt{RT_{st}}}\gamma^{0.5}\left(\frac{2}{\gamma+1}\right)^{\frac{\gamma+1}{2\gamma-2}}$
$C_8 = \frac{RT_{st}K_{ca,in}}{M_{N_2}V_{ca}}\left(\frac{1-x_{O_2,atm}}{1+\omega_{atm}}\right)$	$C_{21} = \frac{1}{R_{cm}}$
$C_9 = \frac{\eta_{cm}k_t k_v}{J_{cp}R_{cm}}$	$C_{22} = k_v$
$C_{10} = \frac{C_p T_{atm}}{J_{cp}\eta_{cp}}$	$C_{23} = K_{ca,in}\left(\frac{x_{O_2,atm}}{1+\omega_{atm}}\right)$
$C_{11} = P_{atm}$	$C_{24} = \frac{nM_{O_2}}{4F}$
$C_{12} = \frac{\gamma-1}{\gamma}$	$x_{O_2,atm} = \frac{y_{O_2,atm}M_{O_2}}{M_{a,atm}}$
$C_{13} = \frac{\eta_{cm}k_t}{J_{cp}R_{cm}}$	$\omega_{atm} = \frac{M_v}{M_{a,atm}}\frac{\phi_{atm}P_{sat}}{P_{atm}-\phi_{atm}P_{sat}}$

[a]: Adopted with permission from Reference [16] Copyright (2017) Elsevier.

2.2. Control Objective

The performance variables vector for the PEMFC stack system is defined by:

$$z = \begin{bmatrix} z_1 \\ z_2 \end{bmatrix} = \begin{bmatrix} P_{net} \\ \lambda_{O_2} \end{bmatrix} \tag{13}$$

where $z_1 = P_{net}$ is the net power output of the PEMFC stack and $z_2 = \lambda_{O_2}$ is the oxygen-excess ratio.

$$z_1 = y_1\omega - c_{21}u(u - c_{22}x_3) \tag{14}$$

$$z_2 = \frac{c_{23}}{c_{24}\omega}(x_4 - \chi) \tag{15}$$

Oxygen starvation occurs when the value of z_2 falls below 1, i.e., ($z_2 < 1$). Hence, the oxygen excess ratio z_2 must be regulated at a certain point that prevents oxygen starvation at different disturbance conditions. For hydrogen/air fuel cells, $z_2^{ref} = 2$ has been proposed as an optimal value [15,16,36]. Although, keeping the oxygen excess ratio at $z_2^{ref} = 2$ can avoid oxygen starvation, it cannot guarantee the maximum net power output from the fuel cell stack. The z_1/z_2 performance curve for different stack currents from 100 A to 300 A is shown in Figure 2.

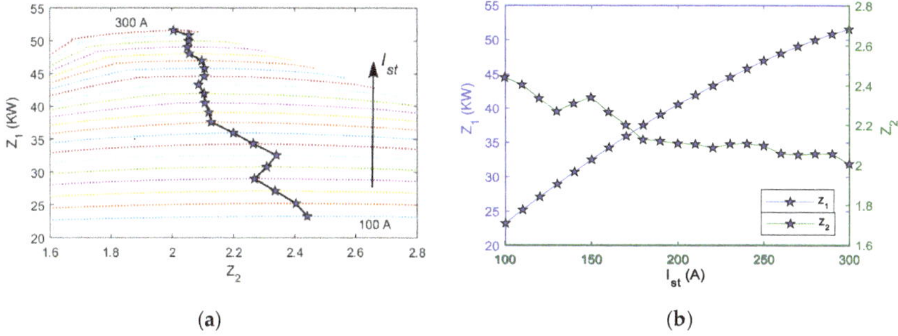

Figure 2. The z_1/z_2 performance curves for the PEMFC stack at different levels of disturbance. (**a**) Maximum power point ($z_1{}^\star$, $z_2{}^\star$) for different levels of disturbance (I_{st}). (**b**) $z_1{}^\star$, $z_2{}^\star$ as a function of the disturbance (I_{st}).

Figure 2 show that the optimal operating point ($z_1{}^\star$, $z_2{}^\star$) depend on the stack current I_{st}, meaning that for different values of the stack current I_{st}, there exists an optimal operating point ($z_1{}^\star$, $z_2{}^\star$) between $z_2 = 2$ and $z_2 = 2.5$ where the maximum net power output is achieved. The optimal values $z_1{}^\star$ and $z_2{}^\star$ are functions of the stack current I_{st} and are given by:

$$z_2{}^\star = \varphi_1(\omega) \tag{16}$$

$$z_1{}^\star = \varphi_2(\omega) \tag{17}$$

where $\varphi_1(\omega)$ and $\varphi_2(\omega)$ are approximated from the z_1/z_2 performance curve given in Figure 2. $\varphi_1(\omega)$ is obtained using shape preserving interpolation while $\varphi_2(\omega)$ is a quadratic function with parameters obtained using the least squares method.

Hence, to obtain the maximum power output from the stack, Z_2^{ref} must be determined based on the stack current I_{st} as follows:

$$z_2{}^{ref} = z_2{}^\star \tag{18}$$

$$z_1{}^{ref} = z_1{}^\star \tag{19}$$

The control objective is to design an optimal controller for the oxygen excess ratio z_2 to regulate it at the required level to prevent oxygen starvation and obtain the maximum net power output z_1 from the PEMFC stack at different disturbance conditions.

3. Air Feeding System Controller Design

The PEMFC air feeding system is a highly nonlinear MIMO system so an efficient controller is required for achieving the control objectives. This paper proposes a fractional order fuzzy PID controller as an efficient candidate for solving the PEMFC air feeding control problem. The proposed control scheme is shown in Figure 3. Fuzzy control simplifies the controller design procedures especially for complex nonlinear systems because FLCs apply the control actions in human-like thinking rather than a complex mathematical design [38]. The hybridization of fractional order operators for integration and differentiation with a fuzzy PID controller increases the degrees of freedom of the fuzzy controller allowing it to produce outputs, which cannot be produced with an integer order fuzzy controller.

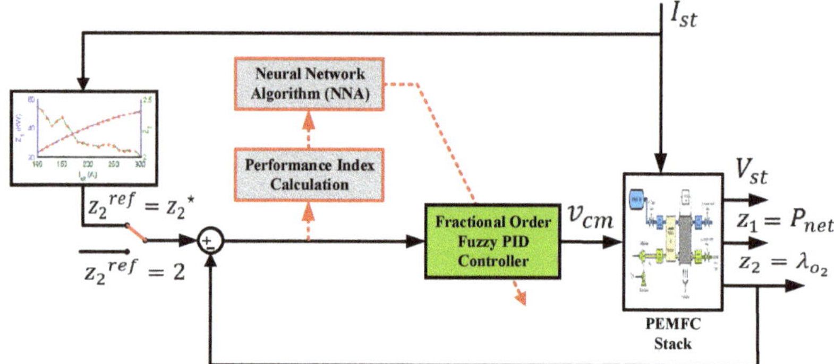

Figure 3. The proposed NNA optimized fractional order Fuzzy PID control scheme for the PEMFC stack.

3.1. Fractional-Order Operators and Its Discretization

Among the several definitions, the most common definitions for fractional order operators (differentiator/integrator) are the Grünwald-Letnikov (G-L) definition and Riemann-Liouville (R-L) definition [39].

The Grünwald-Letnikov (G-L) definition is given by:

$$_aD_t^r f(t) = \lim_{h \to 0} \frac{1}{h^r} \sum_{i=0}^{[(t-a)/h]} (-1)^i \binom{r}{i} f(t - ih) \tag{20}$$

where the time domain operator D^r is equivalent to the frequency domain operator S^r, $r \in [-1, 1]$. A positive value of r implies a fractional order differentiator while a negative value of r implies a fractional order integrator.

To obtain digital implementation of a fractional order controller (FOC), two discretization methods can be used: Direct discretization and indirect discretization [39]. Indirect discretization methods are two-step methods, where, the first step is to perform a frequency-domain approximation in a continuous time domain such as the Oustaloup's band-limited rational approximation, the second step is to discretize the obtained fit s-transfer function. Several frequency-domain approximations can be used but the stable minimum-phase discretization cannot be guaranteed [39]. Direct discretization methods are used to obtain the discrete approximation transfer function directly.

Generally, direct discretization of the fractional-order differentiator/integrator $S^{\pm r}$, $(r \in R)$, can be carried out using the generating function $S = \omega(z^{-1})$. The generating function used and its expansion determine the form and the coefficients of the approximation [39]. Direct discretization methods include the direct power series expansion (PSE) of the Euler operator, continuous fractional expansion (CFE) of the Tustin operator and the numerical integration-based method [39].

In this paper, the direct discretization approach was used to obtain a discrete approximation of the fractional order operator $S^{\pm r}$, $(0 < r < 1)$, in the infinite impulse response (IIR) form of discretization using the Al-Alaoui operator, which is a mixed scheme of the Euler and Tustin Operators. The Al-Alaoui operator as a generating function is given by:

$$\omega(z^{-1}) = \frac{8}{7T} \frac{1 - z^{-1}}{1 + z^{-1}/7} \tag{21}$$

where T is the sampling interval.

The discretized fractional-order operator is given by:

$$D^{\pm r}(z) = \left(\omega(z^{-1})\right)^{\pm r} = \left(\frac{8}{7T} \frac{1 - z^{-1}}{1 + z^{-1}/7}\right)^{\pm r} \tag{22}$$

Equation (22) is a rational discrete-time transfer function of infinite orders. CFE is an efficient way to approximate Equation (22) with a finite order rational one [39]. The resulting discrete transfer function approximating a fractional-order operator can be expressed as:

$$
D^{\pm r}(z) \approx \left(\tfrac{8}{7T}\right)^{\pm r} CFE\left\{\left(\tfrac{1-z^{-1}}{1+\tfrac{z^{-1}}{7}}\right)^{\pm r}\right\}_{p,q} = \left(\tfrac{8}{7T}\right)^{\pm r}\tfrac{P_p\left(z^{-1}\right)}{Q_q\left(z^{-1}\right)}
$$
$$
= \left(\tfrac{8}{7T}\right)^{\pm r}\tfrac{P_p\left(z^{-1}\right)}{Q_q\left(z^{-1}\right)} = \left(\tfrac{8}{7T}\right)^{\pm r}\tfrac{p_0+p_1 z^{-1}+p_2 z^{-2}+...+p_m z^{-p}}{q_0+q_1 z^{-1}+q_2 z^{-2}+...+q_n z^{-q}}
\tag{23}
$$

where $CFE\{u\}$ denotes the continued fraction expansion of u; p and q are the order of approximation. Normally, it could be set $p = q = n$. The discretization of S^r result is an infinite impulse response (IIR) form. An approximate rational function can be obtained by truncation.

The continued fractions expansion (CFE) of any well-behaved function $G\left(z^{-1}\right)$ is given by:

$$
G\left(z^{-1}\right) \simeq a_0\left(z^{-1}\right) + \cfrac{b_1\left(z^{-1}\right)}{a_1\left(z^{-1}\right) + \cfrac{b_2\left(z^{-1}\right)}{a_2\left(z^{-1}\right)+\cfrac{b_3\left(z^{-1}\right)}{a_3\left(z^{-1}\right)+...}}}
\tag{24}
$$

where the coefficients a_i and b_i are either constants or rational functions of the variable z^{-1}.

The advantage of using the direct discretization method with the Al-Alaoui operator as a generating function is that it always gives discrete transfer functions with stable minimum phase characteristics, which is not always guaranteed when using the indirect discretization approach. The other advantage is that there is only one tuning knob [40,41].

The transfer function of the fractional order PID controller ($PI^\lambda D^\mu$) is given by:

$$
G_{FOPID}(s) = K_p + \frac{K_i}{s^\lambda} + K_d s^\mu
\tag{25}
$$

where K_p, K_i, K_d are proportional, integral and derivative gains respectively. μ and λ are positive numbers that represent the order of differentiation and integration [30]. The control signal in the time domain representation given by:

$$
u_{FOPID}(k) = K_p e(k) + K_i D^{-\lambda} e(k) + K_d D^\mu e(k)
\tag{26}
$$

3.2. Fractional Order Fuzzy PID Controller

A fuzzy logic PID controller consists basically of a fuzzy PI and a fuzzy PD controller connected in parallel [18]. Hybridization of fractional order operators with a fuzzy controller is achieved by replacing the integer order differentiator and integrator at the input and the output of the FLC by a fractional order operator [26]. The use of fractional order operators adds extra degrees of freedom for tuning.

The structure of two inputs fractional order fuzzy PID controller with its tunable parameters is shown in Figure 4 where G_E and G_{DE} are the input scaling factors while α and β are output scaling factors. D^μ is a fractional order differentiator with non-integer order μ while $D^{-\lambda}$ is a fractional order integrator with a non-integer order λ. Integer order fuzzy PID controller can be obtained easily from a fractional order fuzzy PID controller by setting the order of the differentiator and integrator in Figure 4 to an integer value, i.e., $\mu = 1$, $\lambda = 1$. However, the use of fractional order operators increases the degrees of freedom (DOF) of the fuzzy controller allowing it to generate output values that cannot be generated using an integer order fuzzy controller.

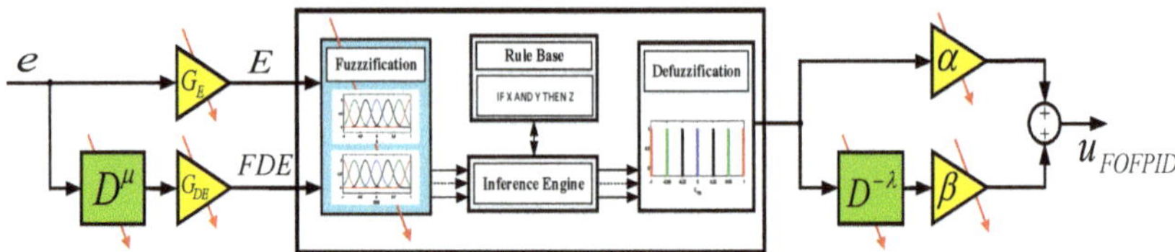

Figure 4. Fractional order fuzzy PID controller with tunable parameters.

The input scaling factors G_E, G_{DE} perform a scaling or normalization of the inputs from the real values into a normalized universe of discourse $[-1,1]$, while the output scaling factors α, β perform an inverse scaling or denormalization of the fuzzy controller output into applicable values suitable for the system. The performance of the fuzzy logic PID controller depends strongly on the values of these scaling factors [19]. Scaling factors has a global effect on the performance of the fuzzy controller. Therefore, these scaling factors must be properly tuned to achieve the desired system performance. Optimization algorithms represent an efficient tool for tuning the scaling factors of fuzzy controllers [19].

The inputs of the fractional order fuzzy PID controller are the error $e(k)$ and the fractional derivative of error $D^{\mu}(z)e(k)$, which are scaled using the input scaling factors G_E and G_{DE} respectively into $E(k)$ and $D^{\mu}E(k)$. The output is the control signal u_{FOFPID} which is scaled using the output scaling factors α and β where:

$$e(k) = z_2^{ref} - z_2(k) \tag{27}$$

$$fde(k) = D^{\mu}(z)e(k) = \left(\omega\left(z^{-1}\right)\right)^{\mu}e(k) \tag{28}$$

$$E(k) = G_E.e(k) \tag{29}$$

$$FDE(k) = D^{\mu}(z)E(k) = G_{DE}.D^{\mu}(z)e(k) \tag{30}$$

$$u_{FOFPID}(k) = \alpha.u_{FIS}(k) + \beta.D^{-\lambda}(z)u_{FIS}(k) = \alpha.u_{FIS}(k) + \beta.\left(\omega\left(z^{-1}\right)\right)^{-\lambda}u_{FIS}(k) \tag{31}$$

where

$$u_{FIS}(k) = f_{Fuzzy}(E(k), D^{\mu}(z)E(k)) \tag{32}$$

where $D = \omega\left(z^{-1}\right)$ is the generating function for the Al-Alaoui operator and $\left(\omega\left(z^{-1}\right)\right)^{\mu}$ and $\left(\omega\left(z^{-1}\right)\right)^{-\lambda}$ are discrete transfer functions approximating the fractional-order differentiator and integrator respectively obtained using the Al-Alaoui operator. f_{Fuzzy} is a nonlinear function represent the fuzzy reasoning.

In this paper, seven membership functions (MFs) namely NB, NM, NS, Z, PS, PM and PB are used for the inputs E, $D^{\mu}E$ and the output u_{FIS}. Gaussian MFs are selected for the input variables. The Gaussian MF is defined by:

$$g(x;C,\sigma) = exp\left(\frac{-(x-C)^2}{2\sigma^2}\right) \tag{33}$$

where C is the mean of the membership function and σ is the standard deviation.

In this work, for computational efficiency, a zero-order Takagi-Sugeno-Kang (TSK) fuzzy inference is used, where the output of each rule is simply a constant or a singleton. The type and the parameters of the membership functions used affect the performance of the fuzzy controller. An optimization algorithm has been used for tuning the parameters of the membership functions [19,42]. The inputs and output membership functions for the fractional order fuzzy PID controller with its design parameters are shown in Figure 5.

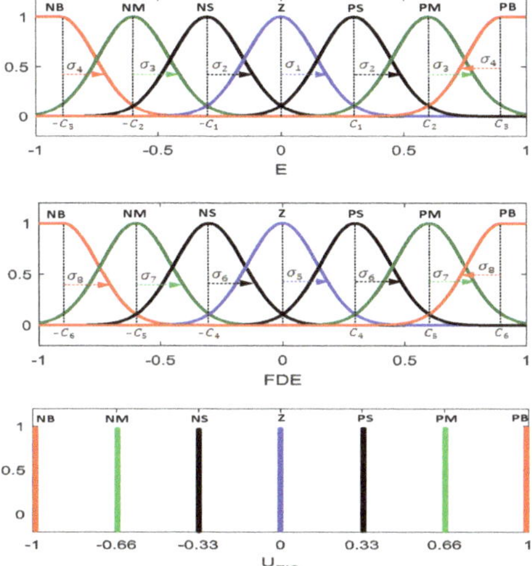

Figure 5. Input and output MFs for the fractional order fuzzy PID controller with its design parameters.

The *i*th fuzzy rule used for the fractional order fuzzy PID controller has the following form:

$$\text{Rule } R_i : \text{ IF E is } A_i \text{ AND } D^\mu E \text{ is } B_i \text{ THEN } u_{FIS} \text{ is } y_i.$$

where A_i and B_i are Gaussian fuzzy sets, while y_i is a singleton. The complete rule base of the fractional order fuzzy PID controller with 49 rules is given in Table 2. This rule base has been selected according to [43].

Table 2. Fractional order fuzzy PID controller rule base [43].

u_{FIS}		E						
		NB	**NM**	**NS**	**Z**	**PS**	**PM**	**PB**
	NB	NB	NB	NB	NB	NM	NS	Z
	NM	NB	NB	NB	NM	NS	Z	PS
	NS	NB	NB	NM	NS	Z	PS	PM
$D^\mu E$	**Z**	NB	NM	NS	Z	PS	PM	PB
	PS	NM	NS	Z	PS	PM	PB	PB
	PM	NS	Z	PS	PM	PB	PB	PB
	PB	Z	PS	PM	PB	PB	PB	PB

4. Optimization Tool

4.1. Neural Network Algorithm (NNA)

Artificial Neural Networks (ANNs) map the input data to the target data through an iterative update of the weights w_{ij} of the ANNs to reduce the mean square error between the predicted output and the target output. The neural network algorithm (NNA) is based on the concepts and the structure of the ANNs to generate new solutions where the best searching agent in the population is considered as the target and the procedures of the algorithm tries to make all the searching agents follow that target solution [31].

NNA is a population-based algorithm where it starts with an initial population of randomly generated solutions within the search space. Each individual or searching agent in the population is called a "pattern solution", each pattern solution is a vector of $1 \times D$ representing the input data of the NNA. *Pattern Solution$_i$* $= [x_{i1}, x_{i2}, x_{i3}, \ldots \ldots, x_{iD}]$.

To start the NNA optimization algorithm, a pattern solution matrix X with size $N_{pop} \times D$ is randomly generated between the lower and upper bounds of the search space. The population of pattern solution X is given by:

$$X = \begin{bmatrix} X_1 \\ \vdots \\ X_i \\ \vdots \\ X_{N_{pop}} \end{bmatrix} = \begin{bmatrix} x_{11} & x_{12} & \cdots & x_{1D} \\ \vdots & \vdots & \vdots & \vdots \\ x_{i1} & x_{i2} & \cdots & x_{iD} \\ \vdots & \vdots & \vdots & \vdots \\ x_{N_{pop}1} & x_{N_{pop}2} & \cdots & x_{N_{pop}D} \end{bmatrix} \tag{34}$$

where

$$x_{ij} = LB_j + rand\left(UB_j - LB_j\right), \; i = 1,2,\ldots,N_{pop}, \; j = 1,2,\ldots,D \tag{35}$$

where LB and UB are $1 \times D$ vectors representing the lower and upper bounds of the search space.

Like ANNs, in NNA each pattern solution X_i will have its corresponding weight W_i where $W_i = \left[w_{i1}, \, w_{i2}, \ldots, w_{iN_{pop}}\right]^T$. The weights array W is given by:

$$W = \left[W_1, \ldots, W_i, \ldots, W_{N_{pop}}\right] = \begin{bmatrix} w_{11} & \cdots & w_{i1} & \cdots & w_{N_{pop}1} \\ w_{12} & \cdots & w_{i2} & \cdots & w_{N_{pop}2} \\ \vdots & & \vdots & \vdots & \vdots \\ w_{1N_{pop}} & \cdots & w_{iN_{pop}} & \cdots & w_{N_{pop}N_{pop}} \end{bmatrix} \tag{36}$$

where W is a square matrix $\left(N_{pop} \times N_{pop}\right)$ of uniformly distributed random numbers between 0 and 1. The weight of the pattern solution is involved in the generation of a new candidate solution.

In NNA, the initial weights are random numbers and its value is updated as the iteration number increases according to the calculated error of the network. The weight values are constrained such that the summation of the weights for any pattern solution should not exceed one, defined mathematically as follows:

$$w_{ij} \in U(0,1), \; i,j = 1,2,3,\ldots,N_{pop} \tag{37}$$

$$\sum_{j=1}^{N_{pop}} w_{ij} = 1, \; i = 1,2,3,\ldots, N_{pop} \tag{38}$$

These constraints for weight values are used to control the bias of movement and the generation of new pattern solutions. Without this constraint, the algorithm will be stuck in a local optimum solution [31].

The fitness C_i of each pattern solution is computed by the evaluation of the objective function f_{obj} using the corresponding pattern solution X_i.

$$C_i = f_{obj}(X_i) = f_{obj}(x_{i1}, x_{i2}, \ldots, x_{iD}), \; i = 1,2,\ldots, N_{pop} \tag{39}$$

where f_{obj} is the objective function.

After the fitness calculation for all pattern solutions, the pattern solution with the best fitness is considered as the target solution with a target position X^{Target}, target fitness F^{Target} and target weight W^{Target}. The NNA models an ANN with N_{pop} inputs each input of D dimension(s) and only one target output X^{Target} [31].

Inspired by the weight summation technique used in ANNs, the new pattern solution is generated as follows:

$$\vec{X}_j^{New}(k+1) = \sum_{i=1}^{N_{pop}} w_{ij}(k)\cdot\vec{X}_i(k), \ j = 1,2,3,\dots,N_{pop} \tag{40}$$

$$\vec{X}_i(k+1) = \vec{X}_i(k) + \vec{X}_i^{New}(k+1), \ i = 1,2,3,\dots,N_{pop} \tag{41}$$

where k is an iteration index.

After the new pattern solutions are generated from the previous population, the weight matrix is updated as well using the following equation:

$$\vec{W}_i^{Updated}(k+1) = \vec{W}_i(k) + 2\cdot rand\cdot\left(\vec{W}^{Target}(k) - \vec{W}_i(k)\right), \ i = 1,2,3,\dots,N_{pop} \tag{42}$$

where the constraints (37) and (38) must be satisfied during the optimization process.

For better exploration of the search space, a bias operator is used in the NNA algorithm. The bias operator is used to modify a certain percentage of the pattern solutions generated in the new population $\vec{X}_i(k+1)$ as well as the updated weight matrix $W_i^{Updated}(k+1)$. The bias operator prevents the algorithm from premature convergence by modifying a certain number of individuals in the population to explore other places in the search space, which has not been visited by the population. For more details about the bias strategy, the reader can refer to reference [31].

A modification factor β_{NNA} is used to determine the percentage of the pattern solutions to be modified using the bias operator. The initial value of β_{NNA} is set to 1 meaning that all individuals in the population are biased. The value of β_{NNA} will be adaptively reduced at each iteration using any possible reduction technique such as follows:

$$\beta_{NNA}(k+1) = 1-\left(\frac{k}{Max\_iteration}\right), \ k = 1,2,3,\dots, \ Max\_iteration \tag{43}$$

$$\beta_{NNA}(k+1) = \beta_{NNA}(k)\cdot\alpha_{NNA}, \ k = 1,2,3,\dots, \ Max\_iteration \tag{44}$$

where α_{NNA} is a positive number smaller than 1 originally selected as 0.99.

The reduction of the modification factor β_{NNA} is made to enhance the exploitation of the algorithm as the iterations increase by allowing the algorithm to search for the optimum solution near to the target solution especially at the final iterations.

Unlike ANNs, in NNA the transfer function operator is used to generate better-quality solutions. The transfer function operator (TF) is defined by the following equation:

$$\vec{X}_i^*(k+1) = TF\left(\vec{X}_i(k+1)\right) = \vec{X}_i(k+1) + 2\cdot rand\cdot\left(\vec{X}^{Target}(k) - \vec{X}_i(k+1)\right), i = 1,2,3,\dots,N_{pop} \tag{45}$$

Using the transfer function operator, the ith updated pattern solution $\vec{X}_i(k+1)$ is transferred from its current position to a new updated position $\vec{X}_i^*(k+1)$ towards the target pattern solution $\vec{X}^{Target}(k)$.

In NNA, at early iterations the bias operator has more chances to generate a new pattern solution meaning that more possible opportunities for discovering unvisited pattern solutions as well as using new weight values. As the iteration number increases, the chance of applying the bias operator decreases while the transfer function (TF) operator will have more chance enhancing the exploitation of the NNA especially at the final iterations.

NNA is considered as a dynamic optimization model because the generation of a new updated solution does not depend only on the previous value of that solution but also depends on all the population described mathematically as follows:

$$\vec{X}_i(k+1) = f\left(\vec{X}_i(k), X(k)\right), \; i = 1,2,3,\ldots,N_{pop} \tag{46}$$

where $\vec{X}_i(k+1)$ and $\vec{X}_i(k)$ are the next and current locations of the ith pattern solution respectively.

4.2. Formulation of FOFPID Controller Design as an Optimization Problem

In this paper, the neural network algorithm (NNA) was used to optimize the fractional order fuzzy PID controller. NNA was used to obtain the optimal or suboptimal value of the four scaling factors $\{G_E, G_{DE}, \alpha, \beta\}$, membership functions parameters for the two inputs E and $D^\mu E$ $\{C_1, C_2,\ldots, C_6, \sigma_1, \sigma_2,\ldots, \sigma_8\}$ as well as the order of the fractional order operators $\{\mu, \lambda\}$. Each candidate pattern solution must contain these parameters of the FOFPID controllers as follows:

$$\vec{X}_i = \{G_{Ei}, G_{DEi}, \alpha_i, \beta_i, C_{1i}, C_{2i},\ldots, C_{6i}, \sigma_{1i}, \sigma_{2i},\ldots, \sigma_{8i}, \mu_i, \lambda_i\} \tag{47}$$

Gaussian membership functions are used for the inputs of the FOFPID controller. The Gaussian membership function is characterized by two parameters, which are the center C_i and the standard deviation σ_i. In this paper, a technique for encoding the membership functions using the minimum number of parameters is used, where, the peer positive and negative membership functions have the same value of the mean C_i, but with the opposite sign, and have the same standard deviation σ_i as shown in Figure 5. This approach of encoding reduces the total number of the membership functions' parameters to be optimized to half, reducing the dimension of the optimization problem leading to a reduction of the computational cost. The total problem dimension is 20. The encoding of the controller parameters into a pattern solution is given in Figure 6.

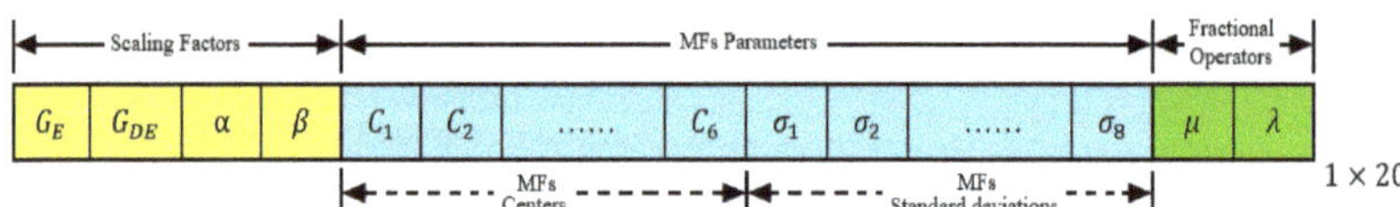

Figure 6. The encoding of FOFPID controller parameters into a pattern solution.

The formulation of FOFPID controller design as an optimization problem is described as follows:
Minimize

$$J = f_{obj}(G_E, G_{DE}, \alpha, \beta, C_1, C_2,\ldots, C_6, \sigma_1, \sigma_2,\ldots, \sigma_8, \mu, \lambda) \tag{48}$$

Such that,

$$G_{Emin} \leq G_E \leq G_{Emax} \tag{49}$$

$$G_{DEmin} \leq G_{DE} \leq G_{DEmax} \tag{50}$$

$$\alpha_{min} \leq \alpha \leq \alpha_{max} \tag{51}$$

$$\beta_{min} \leq \beta \leq \beta_{max} \tag{52}$$

$$0 < C_i \leq 1 \tag{53}$$

$$\sigma_{imin} \leq \sigma_i \leq \sigma_{imax} \tag{54}$$

$$0 \leq \mu, \lambda \leq 1 \tag{55}$$

With the constrains:

$$0 < C_1 < C_2 < C_3 \leq 1 \tag{56}$$

$$0 < C_4 < C_5 < C_6 \leq 1 \tag{57}$$

where,

$$J = ITAE = \int t \cdot |e| \, dt \tag{58}$$

is the integral of the time weighted squared error, e is the error signal and t is the time.

The detailed procedures for using NNA for the optimization of the FOFPID controller are described in Figure 7.

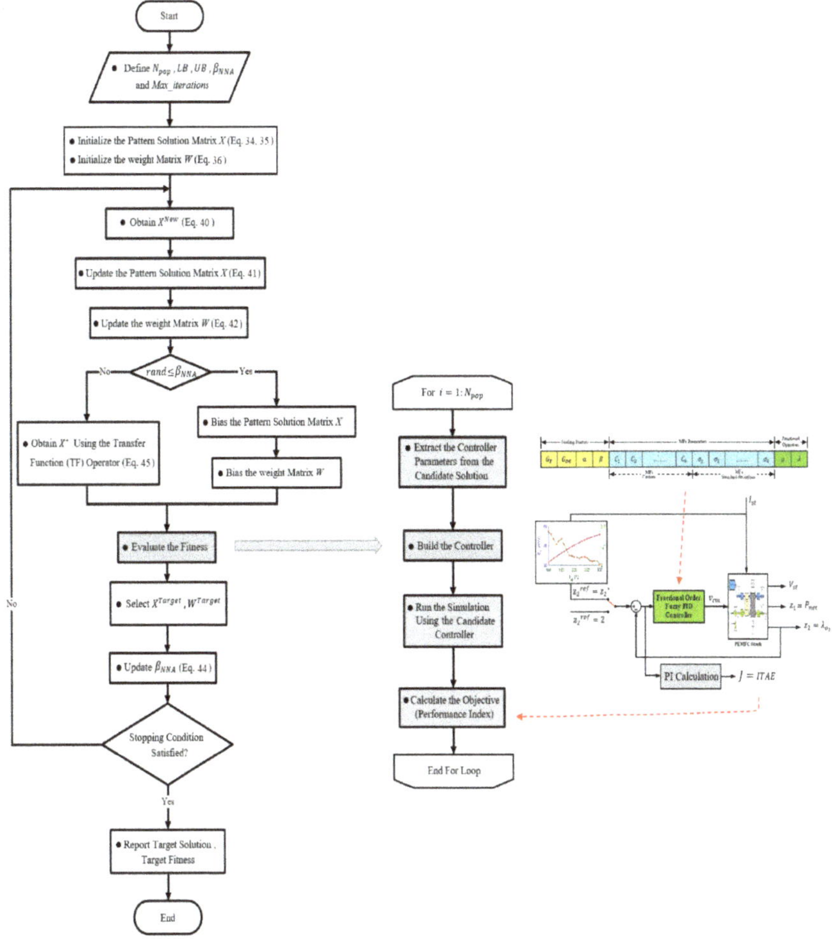

Figure 7. The procedures of FOFPID controller optimization using the NNA algorithm.

The optimized membership functions for both inputs of the FOFPID controller are shown in Figure 8. The optimal values for μ, λ are: $\mu = 0.8644$ and $\lambda = 1$. Using the Al-Alawi operator, the truncated 5th order discrete transfer functions approximating $D^{0.8644}$ and D^{-1} with a sampling interval $T = 0.001$ s are:

$$D^{\mu} = D^{0.8644} = \left(\frac{8}{7T}\frac{1-z^{-1}}{1+z^{-1}/7}\right)^{0.8644} = \frac{2.537\times10^6 - 6.69\times10^6\ z^{-1} + 6.163\times10^6\ z^{-2} - 2.264\times10^6\ z^{-3} + 2.483\times10^5\ z^{-4} + 6291\ z^{-5}}{5770 - 9515\ z^{-1} + 4245\ z^{-2} - 160.6\ z^{-3} - 121.6\ z^{-4} + z^{-5}} \tag{59}$$

$$D^{-\lambda} = D^{-1} = \left(\frac{8}{7T}\frac{1-z^{-1}}{1+z^{-1}/7}\right)^{-1} = \frac{0.7478 - 1.175\ z^{-1} + 0.4782\ z^{-2} - 0.001453\ z^{-3} - 0.01458\ z^{-4} - 0.000125\ z^{-5}}{854.6 - 2320\ z^{-1} + 2221\ z^{-2} - 865.4\ z^{-3} + 108.6\ z^{-4} + z^{-5}} \tag{60}$$

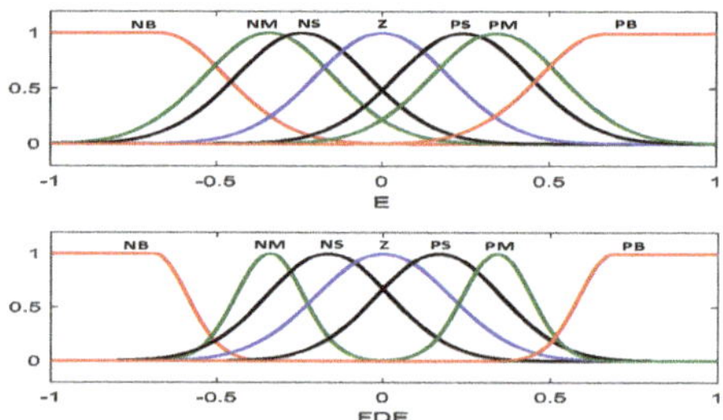

Figure 8. The optimized input MFs for the FOFPID controller.

5. Simulation Results and Discussion

To verify the performance, the efficiency and the robustness of the proposed controller (NNA optimized fractional order fuzzy PID controller), detailed simulations using a MATLAB/SIMULINK environment were carried-out and analyzed. The performance verification was divided into three tasks. The first task was to test the controller for constant set point mode with $z_2^{ref} = 2$. The second task was to test the controller for maximum power point operation mode with $z_2^{ref} = z_2^{\star}$, where $z_2^{\star} \in [2, 2.5]$ is a function of the stack current. The third task was to test the robustness of the proposed controller against parameter uncertainty in the PEMFC stack system using the sensitivity analyses. For the validation of the simulation results, this work uses the same numerical values of the model parameters as well as the same profile of the disturbance used in a recent paper (reference [16]). Moreover, the simulation results are compared and verified to that of reference [16].

5.1. The First Task (Tracking Constant z_2^{ref})

In this task, the controller is tested by applying different values of the disturbance I_{st}, which cover the whole range of the operation of the PEMFC stack while keeping the oxygen excess ratio at a constant set point value $z_2^{ref} = 2$. The profile of the disturbance, i.e., the PEMFC stack current I_{st} is shown in Figure 9.

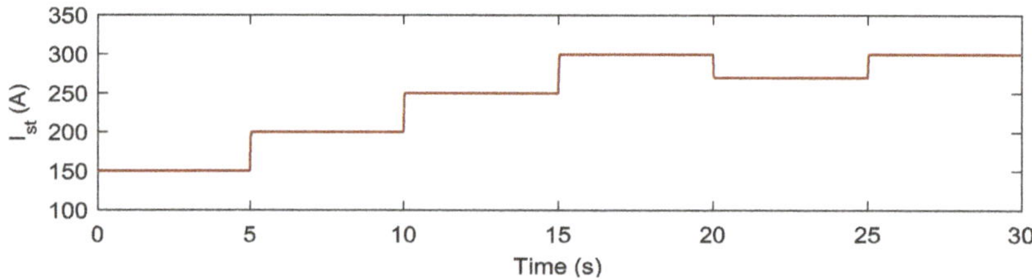

Figure 9. The disturbance ω (I_{st}).

The PEMFC performance using four different controllers, which are the NNA optimized PID controller (NNA PID), NNA optimized fractional order PID controller (NNA FOPID), NNA optimized fuzzy PID controller (NNA FPID) and NNA optimized fractional order fuzzy PID controller (NNA

FOFPID) is shown in Figure 10. Simulation results showed that a sudden increase in the stack current I_{st} representing the disturbance to the system resulted in a sudden reduction of the oxygen excess ratio z_2. The proposed controller (NNA FOFPID) recovered from the disturbance effect faster than other controllers achieving the least settling time, rise time and fluctuations around the set point. At time = 20 s, a sudden reduction in the stack current caused a sudden increase in the oxygen excess ratio z_2. The proposed controller (NNA FOFPID) recovered from the disturbance effect faster than other controllers achieving the least settling time and fluctuations around the set point.

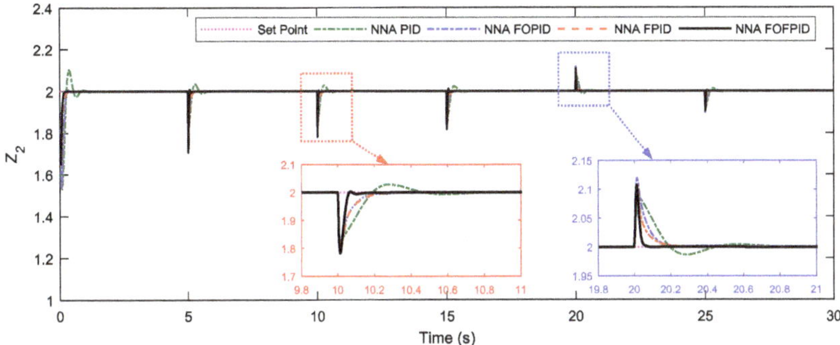

Figure 10. The response of z_2 using four NNA optimized controllers for task 1 ($z_2^{ref} = 2$).

Simulation results showed that, the proposed NNA optimized fractional order fuzzy PID controller (NNA FOFPID) significantly improved the transient response of the PEMFC air feeding system by reducing the settling time and fluctuations around the set point compared to other controllers.

Simulation results showed that the NNA FOPID controller could outperform the NNA PID controller. However, it could not outperform the NNA FPID controller. Simulation results showed that the proposed NNA FOFPID could outperform all the other three types of controllers achieving a better performance.

The variation of the stack voltage V_{st} and the net power output z_1 of the PEMFC stack using the four controllers is shown in Figures 11 and 12, respectively. It could be noticed that a sudden increase in the PEMFC stack current resulted in a sudden reduction in the oxygen excess ratio z_2 reducing the stack voltage V_{st}. Although, the reduction of the PEMFC stack current, at time = 20 s, resulted in an increase in the oxygen excess ratio z_2 increasing the stack voltage V_{st}, the net power output z_1 of the stack was reduced because of the increased power consumption of the compressor motor. The compressor motor voltage v_{cm} using the proposed NNA optimized fractional order fuzzy PID (NNA FOFPID) controller is shown in Figure 13.

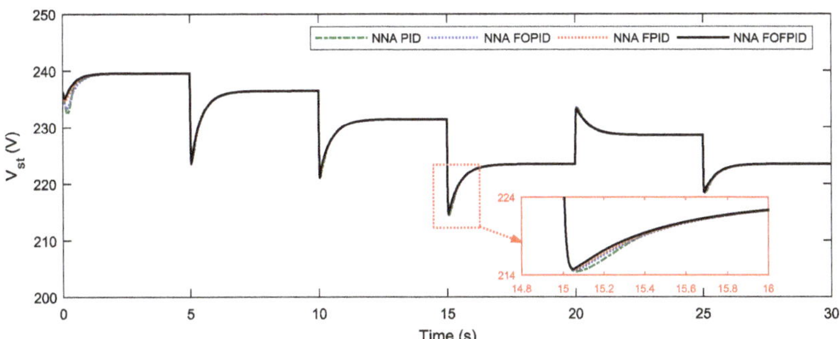

Figure 11. The stack voltage variation using the four NNA optimized controllers for task 1 ($z_2^{ref} = 2$).

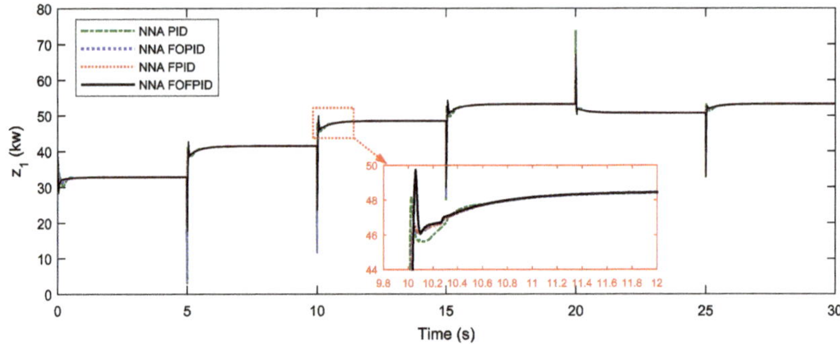

Figure 12. The net power output (z_1) using the four NNA optimized controllers for task 1 ($z_2^{ref} = 2$).

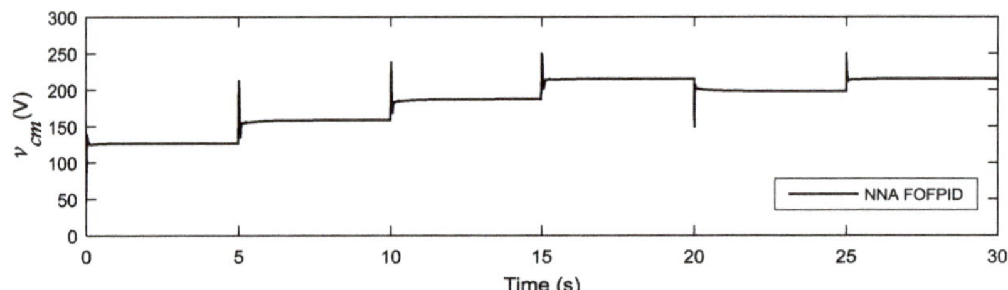

Figure 13. The compressor motor voltage (v_{cm}) using the proposed controller for task 1 ($z_2^{ref} = 2$).

A performance comparison based on the time domain performance indices is given in Table 3. Results showed that, NNA optimized controllers could outperform the controllers presented in reference [16]. The proposed NNA optimized fractional order fuzzy PID controller (NNA FOFPID) was superior and achieved the best time domain performance indices.

Table 3. Performance indices using different controllers for task 1 ($z_2^{ref} = 2$).

Controller	ISE	IAE	ITSE	ITAE
PID [16]	0.0627	0.2903	NA	2.2741
FLC [16]	0.5045	1.1047	NA	8.0201
HFPID [16]	0.0249	0.1005	NA	0.6781
NNA PID	0.03711	0.1995	0.1032	1.356
NNA FOPID	0.02652	0.1261	0.07036	0.8443
NNA FPID	0.014	0.09013	0.06015	0.6539
NNA FOFPID (proposed)	0.009186	0.05291	0.04193	0.3639

5.2. The Second Task (MPPT)

In this task, the proposed NNA FOFPID controller was tested for the maximum power point operation for the PEMFC stack by tracking a time-varying set-value $z_2^{ref} = z_2^\star$, where the set-value $z_2^\star$ is a function of the stack current I_{st} to obtain the maximum net power output P_{net} from the PEMFC stack as described in Section 2. The same profile of the disturbance used in task 1 was used in the maximum power point tracking mode (MPPT) task.

The PEMFC performance using the proposed controller and the other controllers for the maximum power point tracking mode (MPPT) is shown in Figure 14. Simulation results showed that the proposed NNA optimized fractional order fuzzy PID controller (NNA FOFPID) could outperform the other controllers achieving a better set point tracking with the least settling time and minimal fluctuations around the time-varying set value for both positive and negative set point changes achieving a better transient response. Results showed that the proposed NNA optimized fractional order PID controller (NNA FOPID) was better in both set point tracking and the disturbance rejection.

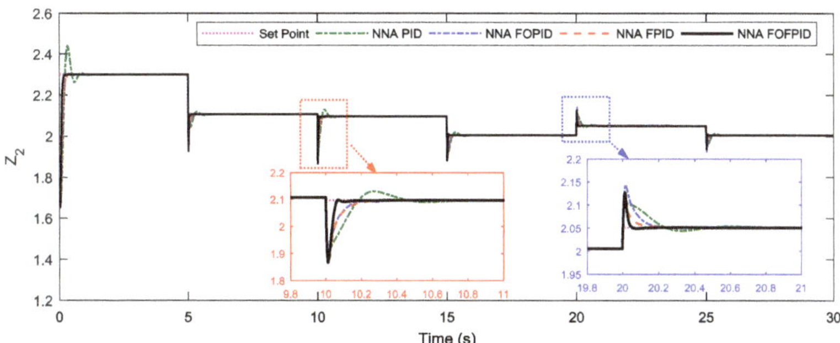

Figure 14. The response of z_2 using four NNA optimized controllers for task 2 (MPPT).

The variation of the stack voltage V_{st} in the MPPT operation mode using the four controllers is shown in Figure 15. By comparing Figures 11 and 15, it could be noticed that the stack voltage V_{st} in the case of the MPPT operation mode was larger than that in case of the constant set point operation mode. The net power output z_1 of the PEMFC stack is shown in Figure 16. Simulation results showed that using a time-varying set-value $z_2{}^{ref} = z_2{}^\star$, the net power output of the PEMFC stack was maximized. The compressor motor voltage v_{cm} in the MPPT operation mode using the NNA optimized fractional order fuzzy PID controller (NNA FOFPID) is shown in Figure 17.

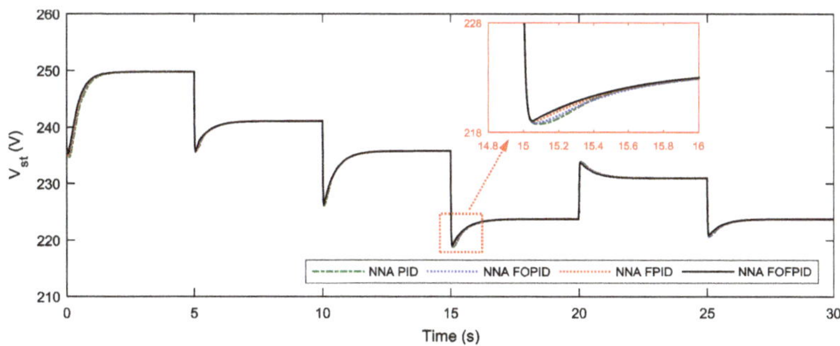

Figure 15. The stack voltage variation using the four NNA optimized controllers for task 2 (MPPT).

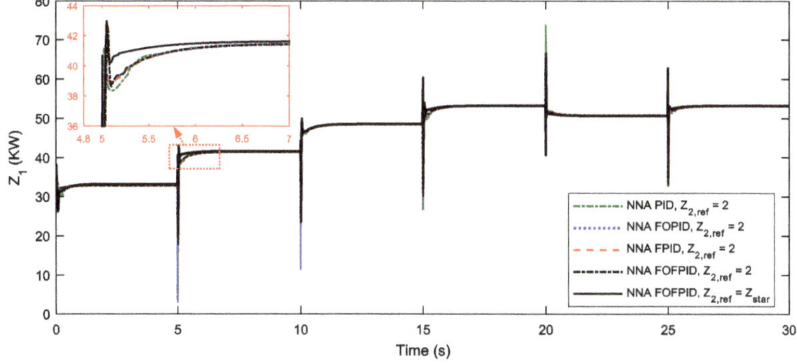

Figure 16. The net power output comparison for constant $z_2^{ref} = 2$ and maximum power point operation $z_2^{ref} = z_2^\star$.

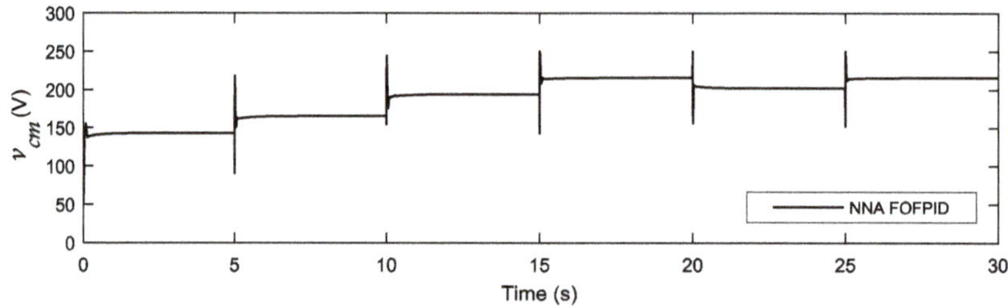

Figure 17. The compressor motor voltage (v_{cm}) using the proposed controller for task 2 (MPPT).

A performance comparison based on the time domain performance indices for tracking a time-varying set-value $z_2^{ref} = z_2^\star$ is given in Table 4. The proposed NNA optimized fractional order fuzzy PID (NNA FOFPID) controller was superior and achieved the best time domain performance indices.

Table 4. Performance indices using different controllers for task 2 (MPPT).

Controller	ISE	IAE	ITSE	ITAE
HFPID [16]	NA	NA	NA	NA
NNA PID	0.04931	0.1938	0.062	1.034
NNA FOPID	0.03104	0.1238	0.04675	0.7024
NNA FPID	0.03671	0.1127	0.03346	0.4368
NNA FOFPID (proposed)	0.02459	0.0701	0.02513	0.2619

5.3. The Third Task (Sensitivity Analysis)

Sensitivity analyses were carried-out for testing the robustness of the proposed NNA optimized FOFPID controller against system parameters changes. The system parameters were varied independently by ±25% of their nominal values without changing the optimized parameter of the proposed NNA FOFPID controller. The time domain performance indices (ISE, IAE, ITSE and ITAE) for the nominal PEMFC air feeding system as well as the perturbed systems are shown in Table 5. The performance of the system with the different considered parameter uncertainty is shown in Figure 18.

Table 5. Sensitivity analysis for the PEMFC air feeding system with the proposed NNA FOFPID.

Parameter	% Change	ISE	IAE	ITSE	ITAE
Nominal	0	0.02459	0.0701	0.02513	0.2619
J_{cp}	+25%	0.0239	0.07184	0.03055	0.3069
	−25%	0.0263	0.07017	0.01969	0.2255
R_{cm}	+25%	0.0256	0.07463	0.03141	0.3158
	−25%	0.0239	0.06669	0.01997	0.225
V_{sm}	+25%	0.03594	0.08698	0.03287	0.3187
	−25%	0.01389	0.05485	0.01815	0.2263
k_t	+25%	0.02459	0.0701	0.02513	0.2619
	−25%	0.02459	0.0701	0.02513	0.2619
k_v	+25%	0.03929	0.1005	0.04467	0.519
	−25%	0.01774	0.05987	0.02576	0.2671
T_{atm}	+25%	0.0313	0.07813	0.02625	0.2707
	−25%	0.01966	0.06367	0.02441	0.2571

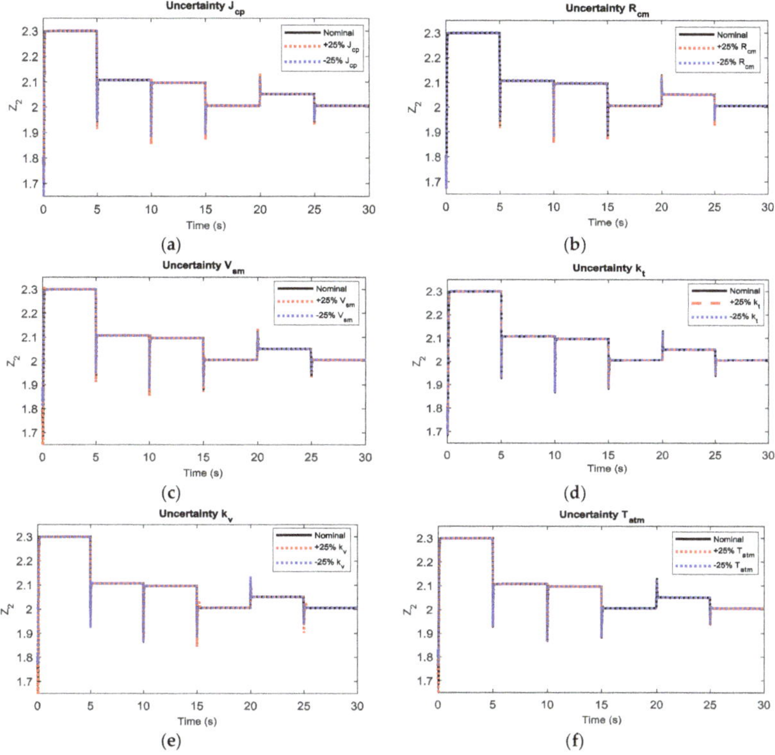

Figure 18. Sensitivity analyses for the PEMFC air feeding system with the proposed NNA optimized FOFPID controller for 25% uncertainty in different PEMFC stack parameters. (**a**) Uncertainty in J_{cp}. (**b**) Uncertainty in R_{cm}. (**c**) Uncertainty in V_{sm}. (**d**) Uncertainty in k_t. (**e**) Uncertainty in k_v. (**f**) Uncertainty in T_{atm}.

The results of Table 5 and Figure 18 showed that applying ±25% uncertainties in J_{cp}, R_{cm}, V_{sm}, k_t, k_v and T_{atm} caused the time domain performance indices, overshoots, undershoots and settling time to deviate from their nominal values. However, these deviations were slight within an acceptable range and the system was dynamically stable. Sensitivity analyses showed that the PEMFC air feeding system with an NNA optimized FOFPID controller had satisfactory robustness against the considered parameter uncertainty range. It can be concluded that the NNA optimized FOFPID controller parameters obtained with the nominal system parameters can be used without retuning or resetting even the system parameters change in a considerable range.

6. Conclusions

In this paper, a fractional order fuzzy PID controller was proposed as an efficient controller for the PEMFC air feeding system. The proposed controller was optimized using the neural network algorithm (NNA). NNA was used to obtain the optimal value of the controller scaling factors and the order of the fractional differentiator and integrator as well as the optimal parameters of the input membership functions. Detailed simulation using a MATLAB/SIMULINK environment was carried-out to test the performance of the proposed NNA optimized FOFPID controller for different modes of operation of the PEMFC stack. Simulation results proved the efficiency and the superiority of the proposed NNA optimized FOFPID controller over other types of controllers. The proposed controller achieved a better set point tracking and disturbance rejection with minimal fluctuations around the set value with better transient response and minimum time domain performance indices. Sensitivity analyses were carried-out to test the robustness of the proposed controller against parameter uncertainty in the PEMFC air feeding system. Future research will concentrate on modifying the original NNA algorithm for improving its convergence with applications to PEMFC control using new control schemes.

Author Contributions: M.S.A. has proposed, conceptualization, simulations, analysis and written the manuscript. H.-J.Z. and Y.-X.S. validated the main idea and results, supervision and checked the whole manuscript. All authors organized and refined the manuscript in present form.

Funding: The APC was funded by Wuhan University of Technology.

Acknowledgments: M.S. AbouOmar acknowledges the Chinese Scholarship Council (CSC) and the Egyptian Government for the financial support. The authors would like to acknowledge the reviewers for their time and valuable comments regarding a high-quality paper.

Conflicts of Interest: The authors declare no conflict of interest.

Appendix A

Table A1. System Parameters for Simulation [b].

Fuel Cells (FC) Data			
n	Number of cells in fuel cell stack	381	–
V_{ca}	The volume of the cathode	0.01	m^3
C_D	Throttle discharge coefficient for the cathode outlet	0.0124	–
A_T	Cathode outlet throttle area	0.00175	m^2
V_{sm}	Supply manifold volume	0.02	m^3
T_{st}	Fuel cell temperature	353.15	K
$K_{ca,in}$	Cathode inlet orifice constant	0.3629×10^{-5}	$Kg\ s^{-1}Pa^{-1}$
Air & Steam Properties			
γ	Ratio of specific heats of air	1.4	–
M_{N_2}	Nitrogen molar mass	28×10^{-3}	$Kg\ mol^{-1}$
M_{O_2}	Oxygen molar mass	32×10^{-3}	$Kg\ mol^{-1}$
M_v	Vapor molar mass	18.02×10^{-3}	$Kg\ mol^{-1}$
$M_{a,atm}$	Air molar mass	28.97×10^{-3}	$Kg\ mol^{-1}$
T_{atm}	Atmospheric temperature	298.15	K
P_{atm}	Atmospheric pressure	1.01325×10^5	Pa
C_p	Specific heat of air at constant pressure	1004	$J\ Kg^{-1}\ K^{-1}$
ϕ_{atm}	Average relative humidity of the ambient air	0.5	–
$y_{O_2,atm}$	Oxygen mole fraction	0.21	–
Electrochemistry			
F	Faraday constant	96.487	$C\ mol^{-1}$
R	Universal gas constant	8.31451	$J\ mol^{-1}\ K^{-1}$
Compressor (CP)			
η_{cp}	Compressor efficiency	80%	–
J_{cp}	Compressor inertia	5×10^{-5}	$Kg\ m^2$
Compressor Motor (CM)			
R_{cm}	Compressor motor resistance	0.82	Ω
k_t	Motor constant	0.0225	$N\ m\ A^{-1}$
k_v	Motor constant	0.0153	$V\ rad^{-1}\ s$
η_{cm}	Motor mechanical efficiency	98%	–

[b]: Adopted with permission from reference [16] Copyright (2017) Elsevier.

References

1. Pathapati, P.R.; Xue, X.; Tang, J. A new dynamic model for predicting transient phenomena in a PEM fuel cell system. *Renew. Energy* **2005**, *30*, 1–22. [CrossRef]
2. Liu, J.; Gao, Y.; Su, X.; Wack, M.; Wu, L. Disturbance-observer-based control for air management of PEM fuel cell systems via sliding mode technique. *IEEE Trans. Control Syst. Technol.* **2018**, *99*, 1–10. [CrossRef]
3. Vahidi, A.; Stefanopoulou, A.; Peng, H. Current management in a hybrid fuel cell power system: A model-predictive control approach. *IEEE Trans. Control Syst. Technol.* **2006**, *14*, 1047–1057. [CrossRef]
4. Pukrushpan, J.T.; Peng, H.; Stefanopoulou, A.G. Control-Oriented Modeling and Analysis for Automotive Fuel Cell Systems. *J. Dyn. Syst. Meas. Control* **2004**, *126*, 14–25. [CrossRef]
5. Pukrushpan, J.T.; Stefanopoulou, A.G.; Huei, P. Modeling and Control for PEM Fuel Cell Stack System. In Proceedings of the American Control Conference, Anchorage, AK, USA, 8–10 May 2002; Volume 4, pp. 3117–3122.
6. Pukrushpan, J.T.; Stefanopoulou, A.G.; Huei, P. Control of fuel cell breathing. *IEEE Control Syst.* **2004**, *24*, 30–46.
7. Grujicic, M.; Chittajallu, K.M.; Law, E.H.; Pukrushpan, J.T. Model-based control strategies in the dynamic interaction of air supply and fuel cell. *Proc. Inst. Mech. Eng. Part A J. Power Energy* **2004**, *218*, 487–499. [CrossRef]
8. Niknezhadi, A.; Allué-Fantova, M.; Kunusch, C.; Ocampo-Martínez, C. Design and implementation of LQR/LQG strategies for oxygen stoichiometry control in PEM fuel cells based systems. *J. Power Sour.* **2011**, *196*, 4277–4282. [CrossRef]
9. Talj, R.; Hilairet, M.; Ortega, R. Second order sliding mode control of the moto-compressor of a PEM fuel cell air feeding system, with experimental validation. In Proceedings of the 35th Annual Conference of IEEE Industrial Electronics, Porto, Portugal, 3–5 November 2009; pp. 2790–2795.
10. Laghrouche, S.; Liu, J.; Ahmed, F.S.; Harmouche, M.; Wack, M. Adaptive second-order sliding mode observer-based fault reconstruction for PEM fuel cell air-feed system. *IEEE Trans. Control Syst. Technol.* **2015**, *23*, 1098–1109. [CrossRef]
11. Han, J.; Yu, S.; Yi, S. Adaptive control for robust air flow management in an automotive fuel cell system. *Appl. Energy* **2017**, *190*, 73–83. [CrossRef]
12. Gruber, J.K.; Doll, M.; Bordons, C. Design and experimental validation of a constrained MPC for the air feed of a fuel cell. *Control Eng. Pract.* **2009**, *17*, 874–885. [CrossRef]
13. Wang, Y.-X.; Xuan, D.-J.; Kim, Y.-B. Design and experimental implementation of time delay control for air supply in a polymer electrolyte membrane fuel cell system. *Int. J. Hydrogen Energy* **2013**, *38*, 13381–13392. [CrossRef]
14. Sanchez, V.M.; Barbosa, R.; Arriaga, L.G.; Ramirez, J.M. Real time control of air feed system in a PEM fuel cell by means of an adaptive neural-network. *Int. J. Hydrogen Energy* **2014**, *39*, 16750–16762. [CrossRef]
15. Beirami, H.; Shabestari, A.Z.; Zerafat, M.M. Optimal PID plus fuzzy controller design for a PEM fuel cell air feed system using the self-adaptive differential evolution algorithm. *Int. J. Hydrogen Energy* **2015**, *40*, 9422–9434. [CrossRef]
16. Baroud, Z.; Benmiloud, M.; Benalia, A.; Ocampo-Martinez, C. Novel hybrid fuzzy-PID control scheme for air supply in PEM fuel-cell-based systems. *Int. J. Hydrogen Energy* **2017**, *42*, 10435–10447. [CrossRef]
17. Robles Algarín, C.; Taborda Giraldo, J.; Rodríguez Álvarez, O. Fuzzy logic based MPPT controller for a PV system. *Energies* **2017**, *10*, 2036. [CrossRef]
18. Sahu, B.K.; Pati, S.; Mohanty, P.K.; Panda, S. Teaching–learning based optimization algorithm based fuzzy-PID controller for automatic generation control of multi-area power system. *Appl. Soft Comput.* **2015**, *27*, 240–249. [CrossRef]
19. Omar, M.S.A.; Khedr, T.Y.; Zalam, B.A.A. Particle swarm optimization of fuzzy supervisory controller for nonlinear position control system. In Proceedings of the 8th International Conference on Computer Engineering & Systems (ICCES), Cairo, Egypt, 26–28 November 2013; pp. 138–145.
20. Podlubny, I. Fractional-order systems and $PI^\lambda D^\mu$ controller. *IEEE Trans. Autom. Control* **1999**, *44*, 208–214. [CrossRef]
21. Alomoush, M.I. Load frequency control and automatic generation control using fractional-order controllers. *Electr. Eng.* **2010**, *91*, 357–368. [CrossRef]

22. Debbarma, S.; Saikia, L.C.; Sinha, N. AGC of a multi-area thermal system under deregulated environment using a non-integer controller. *Electr. Power Syst. Res.* **2013**, *95*, 175–183. [CrossRef]

23. Sondhi, S.; Hote, Y.V. Management, Fractional order PID controller for load frequency control. *Energy Convers.* **2014**, *85*, 343–353. [CrossRef]

24. Taher, S.A.; Fini, M.H.; Aliabadi, S.F. Fractional order PID controller design for LFC in electric power systems using imperialist competitive algorithm. *Ain Shams Eng. J.* **2014**, *5*, 121–135. [CrossRef]

25. Pan, I.; Das, S. Fractional-order load-frequency control of interconnected power systems using chaotic multi-objective optimization. *Appl. Soft Comput.* **2015**, *29*, 328–344. [CrossRef]

26. Das, S.; Pan, I.; Das, S.; Gupta, A. A novel fractional order fuzzy PID controller and its optimal time domain tuning based on integral performance indices. *Eng. Appl. Artif. Intell.* **2012**, *25*, 430–442. [CrossRef]

27. Pan, I.; Das, S. Fractional order fuzzy control of hybrid power system with renewable generation using chaotic PSO. *ISA Trans.* **2016**, *62*, 19–29. [CrossRef] [PubMed]

28. Kumar, A.; Kumar, V. Communications, Hybridized ABC-GA optimized fractional order fuzzy pre-compensated FOPID control design for 2-DOF robot manipulator. *AEU Int. J. Electron.* **2017**, *79*, 219–233. [CrossRef]

29. Mahto, T.; Mukherjee, V. Distribution, Fractional order fuzzy PID controller for wind energy-based hybrid power system using quasi-oppositional harmony search algorithm. *IET Gener. Transm.* **2017**, *11*, 3299–3309. [CrossRef]

30. Arya, Y.; Kumar, N. BFOA-scaled fractional order fuzzy PID controller applied to AGC of multi-area multi-source electric power generating systems. *Swarm Evol. Comput.* **2017**, *32*, 202–218. [CrossRef]

31. Sadollah, A.; Sayyaadi, H.; Yadav, A. A dynamic metaheuristic optimization model inspired by biological nervous systems: Neural network algorithm. *Appl. Soft Comput.* **2018**, *71*, 747–782. [CrossRef]

32. Pukrushpan, J.T. *Modeling and Control of Fuel Cell Systems and Fuel Processors*; University of Michigan: Ann Arbor, MI, USA, 2003.

33. Suh, K.W. Modeling, analysis and control of fuel cell hybrid power systems. Ph.D. Thesis, University of Michigan, Ann Arbor, MI, USA, January 2006.

34. Deng, H.; Li, Q.; Cui, Y.; Zhu, Y.; Chen, W. Nonlinear controller design based on cascade adaptive sliding mode control for PEM fuel cell air supply systems. *Int. J. Hydrogen Energy* **2018**. [CrossRef]

35. Gruber, J.K.; Bordons, C.; Dorado, F. Nonlinear control of the air feed of a fuel cell. In Proceedings of the American Control Conference, Seattle, WA, USA, 11–13 June 2008; pp. 1121–1126.

36. Talj, R.J.; Hissel, D.; Ortega, R.; Becherif, M.; Hilairet, M. Experimental Validation of a PEM Fuel-Cell Reduced-Order Model and a Moto-Compressor Higher Order Sliding-Mode Control. *IEEE Trans. Ind. Electron.* **2010**, *57*, 1906–1913. [CrossRef]

37. Talj, R.; Ortega, R.; Hilairet, M. A controller tuning methodology for the air supply system of a PEM fuel-cell system with guaranteed stability properties. *Int. J. Control* **2009**, *82*, 1706–1719. [CrossRef]

38. Liu, J.; Gao, Y.; Geng, S.; Wu, L. Nonlinear control of variable speed wind turbines via fuzzy techniques. *IEEE Access* **2017**, *5*, 27–34. [CrossRef]

39. Chen, Y.; Vinagre, B.M.; Podlubny, I. Continued fraction expansion approaches to discretizing fractional order derivatives—An expository review. *Nonlinear Dyn.* **2004**, *38*, 155–170. [CrossRef]

40. Chen, Y.Q.; Moore, K.L. Discretization schemes for fractional-order differentiators and integrators. *IEEE Trans. Circuits Syst. Fundam. Theory Appl.* **2002**, *49*, 363–367. [CrossRef]

41. Petrás, I. Fractional derivatives, fractional integrals, and fractional differential equations in Matlab. In *Engineering Education and Research Using MATLAB*; InTech: London, UK, 2011.

42. Cheng, Y.-S.; Liu, Y.-H.; Hesse, H.; Naumann, M.; Truong, C.; Jossen, A. A PSO-Optimized Fuzzy Logic Control-Based Charging Method for Individual Household Battery Storage Systems within a Community. *Energies* **2018**, *11*, 469. [CrossRef]

43. Sharma, R.; Rana, K.; Kumar, V. Performance analysis of fractional order fuzzy PID controllers applied to a robotic manipulator. *Expert Syst. Appl.* **2014**, *41*, 4274–4289. [CrossRef]

Article

Corrective Control by Line Switching for Relieving Voltage Violations Based on A Three-Stage Methodology

Zhengwei Shen [1,*], Yong Tang [2], Jun Yi [2], Changsheng Chen [2], Bing Zhao [2] and Guangru Zhang [3]

[1] School of Electrical and Information Engineering, Tianjin University, Tianjin 300072, China
[2] State Key Laboratory of Power Grid Security and Energy Conservation, China Electric Power Research Institute, Haidian District, Beijing 100192, China; tangyong@epri.sgcc.com.cn (Y.T.); yjun@epri.sgcc.com.cn (J.Y.); ccs1989@163.com (C.C.); zhaobing@epri.sgcc.com.cn (B.Z.)
[3] Department of Power Supply Technology Center, State Grid Gansu Electric Power Research Institute, Lanzhou 730070, Gansu Province, China; guangruzhang@163.com
* Correspondence: shenzhengwei@epri.sgcc.com.cn; Tel.: +86-155-1014-5550

Received: 13 February 2019; Accepted: 26 March 2019; Published: 28 March 2019

Abstract: An online line switching methodology to relieve voltage violations is proposed. This novel online methodology is based on a three-stage strategy, including screening, ranking, and detailed analysis and assessment stages for high speed (online application) and accuracy. The proposed online methodology performs the tasks of rapidly identifying effective candidate lines, ranking the effective candidates, performing detailed analysis of the top ranked candidates, and supplying a set of solutions for the power system. The post-switching power systems, after executing the proposed line switching action, meet the operational and engineering constraints. The results provided by the exact Alternating Current (AC) power flow are used as a benchmark to compare the speed and accuracy of the proposed three-stage methodology. One feature of the methodology is that it can provide a set of high-quality switching solutions from which operators may choose a preferred solution. The effectiveness of the proposed online line switching methodology in providing single-line switching and multiple-line switching solutions to relieve voltage violations is evaluated on the IEEE 39-bus and 2746-bus power system. The CPU time of the proposed methodology compared with that under AC power flow constitutes a speed-up of 9905.32% on a 2746-bus power system, showing good potential for online application in a large-scale power system.

Keywords: line switching; voltage violations; three-stage

---

1. Introduction

It is widely known that the modern power grid is a large-scale and extremely complex interconnected network. Fulfilling the demand for electric power is essential from economic, protective, and societal standpoints [1,2]. Unfortunately, it is not easy to keep the grid running at a stable point all the time: voltage variation problems seriously affect the stable operation of the system.

Line switching is a cost-effective measure to improve the operational stability of power systems. There are several instances where line switching is employed for corrective applications by the industry today. One of the line switching operations mentioned in the Pennsylvania-New Jersey-Maryland Interconnection (PJM) transmission operations manual is described below: Loadings on the Sunnyside–Warner–Torrey 138 kV for the loss of the S. Canton–Torrey 138 kV can be controlled by opening the S.E. Canton–Sunnyside 138 kV line at Sunnyside via supervisory control. Contingency loadings need to be watched on the SE Canton–Canton Central 138 kV and S. Canton–Torrey 138 kV circuits when this procedure is implemented [3].

Hedman [4] and Rolim [5] presented the development and applications of line switching. Compared with other methods, line switching has more advantages in terms of cost reduction, fast speed, and accuracy improvement [6,7]. Relevant work has illustrated the effectiveness of line switching in relieving line overloads [8–15], reducing transmission losses and generation cost [16–20], assisting in load recovery, improving voltage profiles, relieving system congestion, and enhancing system reliability [21–26].

Guo [27] proposed that for extra-high-voltage power systems, line switching can be used to relieve voltage violations at low-load periods. In order to avoid computational intractability, a basic mixed-integer nonlinear program formula is transformed into a mixed-integer linear program in this reference. Based on fast decoupled power flow with finite iterations, a new algorithm was proposed by Shao [28] to find the best line and bus bar switching action to relieve overloads and voltage violations caused by system faults. Although the algorithm developed in references [27,28] can relieve the problem of voltage violation, it is very time-consuming and difficult to implement in a practical power system.

However, the effectiveness of line switching depends on the selection of lines to be switched off. It is well recognized that linear methods are usually satisfactory in speed but not in accuracy, while nonlinear methods are usually accurate but can be slow [29]. In other words, by simplifying the size of the network that linear methods perform linear algebra operations, the speed is improved but the accuracy is reduced. In order to further improve the speed and accuracy, a novel online line switching methodology is proposed in this paper to relieve voltage violations. Instead of dealing with the combinatorial character of optimal transmission switching (OTS), the proposed methodology combines linear and nonlinear methods to achieve the goal of online application. The proposed methodology employs a three-stage strategy combining linear and nonlinear methods: (i) a screening stage using a linear method, (ii) a ranking stage using a PQ decoupled method, and (iii) a detailed analysis stage which utilizes AC power flow.

The task of the screening stage is to quickly select effective lines from the list of credible candidate lines that can relieve voltage violations after switching. The effective candidates selected by the screening stage are ranked in order in the ranking stage. The detailed analysis stage performs a detailed evaluation of the several top-ranked lines of stage (ii) and provides multiple high-quality solutions that can relieve voltage violations, while the post-switching system satisfies operational and engineering constraints.

The innovation of the proposed methodology in this paper is mainly reflected in the following aspects:

(1) It can find the "best" line switching scheme to relieve bus voltage violations for the power system.
(2) It can provide a variety of high-quality line switching schemes for multiple-line switching, from which the system operator can select a "desired" one.
(3) It shows fast speed, which means that it is suitable for determining switching lines of large-scale power systems in an online environment.

The effectiveness of the proposed online switching methodology is evaluated on the IEEE 39-bus and 2746-bus power systems.

2. Problem Formulations

We consider a comprehensive power system quasi-steady-state model of the following general form:

$$0 = f(x) \tag{1}$$

where x is the vector of state variables.

The proposed line switching problem can be generically expressed as

$$\min_{\mathcal{N} \in N} Num(\mathcal{N}) \tag{2}$$

subject to

$$f_{\mathcal{N}}(x) = 0 \tag{3}$$

$$|S_{ij}| = (P_{ij}^2 + Q_{ij}^2)^{1/2} \leq S_{ij}^{max} \quad i, j \in 1, 2, 3 \cdots, n; \tag{4}$$

$$V_i^{min} < V_i < V_i^{max} \quad i \in 1, 2, 3 \cdots, n; \tag{5}$$

$$Num(\mathcal{N} - \mathcal{N}_{base}) \leq m \tag{6}$$

where Equation (2) is the minimum number of switched lines required for relieving voltage violations; $\mathcal{N}$ represents the new topology after the lines are switched out; S_{ij}, P_{ij}, and Q_{ij} represent the apparent, active, and reactive power flows, respectively, of branch i–j; V_i^{min} and V_i^{max} are the minimum and maximum voltage magnitudes at bus i; and S_{ij}^{max} is the maximum apparent power flow of line i–j. The constraints given by Equations (4) and (5) limit the line flows and voltage violation. $Num(\cdot)$ indicates the number of line switching actions needed to change the network topology $\mathcal{N}_{base}$ to the new network topology $\mathcal{N}$, $\mathcal{N}_{base}$ denotes the network topology of the base case power system, and m is the upper bound of the number of switching lines allowed.

In this paper, we use the proposed three-stage methodology to relieve voltage violations. Equation (2) is used to search the network topology such that the voltage violation is relieved with a minimum number of lines switched out within the boundaries of the constraints of the power flow equation (Equation (3)), operational and engineering constraints (Equations (4) and (5)), and the upper limit of the number of switched lines (Equation (6)).

3. Solution Methodology

The proposed methodology employs a three-stage strategy that contains screening, ranking, and detailed analysis and assessment stages. The solution methodology used in each stage is presented as follows: A sensitivity-based method was used to increase the speed of screening to achieve the goal of Stage 1. Stage 2 is based on the PQ decoupled method to achieve fast and accurate ranking goals, while Stage 3 utilizes AC power flow to assess the switching solutions for the post-switching power systems. The architecture of the proposed methodology is shown in Figure 1.

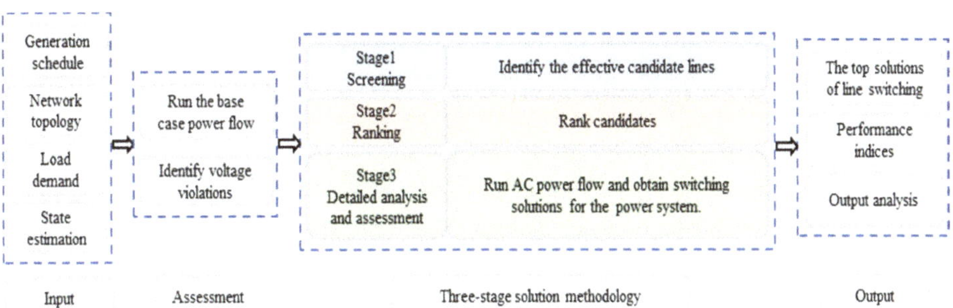

Figure 1. Architecture of the proposed three-stage solution methodology.

3.1. Stage 1: Screening

The task of this stage is to identify candidate lines whose disconnection may relieve voltage violations. In the screening stage, we use a sensitivity method to rapidly estimate the voltage variations on bus i of the power system due to the switching-out action of each candidate [30].

Assume that the network has n buses and b branches and that we can relieve a voltage violation of bus i by switching line k–m. Let the base case system impedance be expressed as

$$\mathbf{Z} = \begin{bmatrix} Z_{11} & \cdots\cdots & Z_{1n} \\ \vdots & & \vdots \\ Z_{n1} & \cdots\cdots & Z_{nn} \end{bmatrix} \tag{7}$$

where Z_{ij} is the impedance between buses i and j.

The current system operating state and the power flow of the post-switching power system will be changed when the lines are switched out. Once the line k–m has been switched out, its impact on the voltage variation can be described as

$$\Delta \mathbf{V} = \mathbf{Z}^{post} \Delta \mathbf{I} \tag{8}$$

where $\Delta \mathbf{V}$, $\Delta \mathbf{I}$ denote the other branches' voltage and current variations, respectively, due to switching out k–m. After line k–m has been switched out, the system impedance matrix $\mathbf{Z}^{post}$ can be expressed as

$$\mathbf{Z}^{post} = \begin{bmatrix} Z_{11}^{post} & \cdots & Z_{1k}^{post} & \cdots & Z_{1m}^{post} & \cdots & Z_{1n}^{post} \\ \vdots & & \vdots & & \vdots & & \vdots \\ Z_{k1}^{post} & \cdots & Z_{kk}^{post} & \cdots & Z_{km}^{post} & \cdots & Z_{kn}^{post} \\ \vdots & & \vdots & & \vdots & & \vdots \\ Z_{m1}^{post} & \cdots & Z_{mk}^{post} & \cdots & Z_{mm}^{post} & \cdots & Z_{mn}^{post} \\ \vdots & & \vdots & & \vdots & & \vdots \\ Z_{n1}^{post} & \cdots & Z_{nk}^{post} & \cdots & Z_{nm}^{post} & \cdots & Z_{nn}^{post} \end{bmatrix} \tag{9}$$

Then, the voltage variations with lines out of service can be obtained as follows:

$$\Delta V_{i,km} = Z_{ik}^{post} \Delta I_k + Z_{im}^{post} \Delta I_m = (Z_{ik}^{post} - Z_{im}^{post}) I_{km} \tag{10}$$

Using the branch-adding method [31], Z_{ik}^{post} and Z_{im}^{post} are described by the following equations:

$$Z_{ik}^{post} = Z_{ik} + \frac{1}{z_{km} - Z_{kk} - Z_{mm} + 2Z_{km}} (Z_{ik} - Z_{im})(Z_{kk} - Z_{mm}) \tag{11}$$

$$Z_{im}^{post} = Z_{im} + \frac{1}{z_{km} - Z_{kk} - Z_{mm} + 2Z_{km}} (Z_{ik} - Z_{im})(Z_{km} - Z_{mm}) \tag{12}$$

where z_{km} is the impedance of branch k–m and Z_{kk}, Z_{mm}, Z_{km}, Z_{im}, and Z_{ik} are elements of $\mathbf{Z}$.

Then, Equation (10) can be rewritten as

$$\Delta V_{i,km} = (Z_{ik} - Z_{im}) \frac{z_{km}}{z_{km} - Z_{kk} - Z_{mm} + 2Z_{km}} I_{km} \tag{13}$$

where I_{km} is the current from k to m in the base case power system.

Consider that the reactance in the transmission line is much larger than the resistance; in this paper, we replace the impedance in the above formula with reactance. Then, the voltage variations ΔV_{i-km} on bus i caused by switching line k–m can be obtained by the following:

$$\Delta V_{i-km} = (X_{ik} - X_{im}) \frac{x_{km}}{x_{km} - X_{kk} - X_{mm} + 2X_{km}} I_{km} \tag{14}$$

where X_{ik}, X_{im}, X_{kk}, X_{mm}, and X_{km} are the corresponding elements in the reactance matrix $\mathbf{X}$, and x_{km} represents the reactance of branch k–m.

We define the impact factor β_{i-km} as the voltage variations on bus i caused by switching line k–m. with unit current. Then we have

$$\beta_{i-km} = (X_{ik} - X_{im}) \frac{x_{km}}{x_{km} - X_{kk} - X_{mm} + 2X_{km}} \tag{15}$$

The screening stage uses the sensitivity equation (Equation (15)) to rapidly identify effective candidate lines. All of the candidates with $\beta_{i-km} > \varepsilon$ or $\beta_{i-km} < \varepsilon$ (where ε is a pre-defined value) are captured and sent to Stage 2 for further ranking.

3.2. Stage 2: Ranking

In this stage, the effective candidate lines selected in Stage 1 are ranked. The PQ decoupling method is used to compute the voltage of the violation bus with each effective candidate switched out, and the effectiveness of each candidate is ranked based on the calculated voltage variations of the post-switching power system. PQ decoupling is a variation of the Newton–Raphson method that exploits the approximate decoupling of active and reactive flows in well-behaved power networks and additionally fixes the value of the Jacobian matrix during the iteration in order to avoid costly matrix decompositions [32]; it is also referred to as fixed-slope, decoupled Newton–Raphson. Within the algorithm, the Jacobian matrix gets inverted only once and is simplified to form two separate matrices of **P** and **Q**. This simplification splits the Jacobian matrix into two small matrices, which means that the PQ decoupling method can return the answer within seconds, whereas the Newton–Raphson method takes much longer.

Figures 2 and 3 show the transmission line π-equivalent model and reactive power flow model [33]:

$$Q_{km} = (B_{km} - B_{cap})V_k^2 + G_{km}V_kV_m \sin(\theta_k - \theta_m) - B_{km}V_kV_m \cos(\theta_k - \theta_m) \tag{16}$$

$$Q_{mk} = (B_{km} - B_{cap})V_m^2 + G_{km}V_kV_m \sin(\theta_m - \theta_k) - B_{km}V_kV_m \cos(\theta_m - \theta_k) \tag{17}$$

$$Q_{km} = \frac{(B_{km} - B_{cap})(V_k^2 - V_m^2)}{2} + G_{km}V_kV_m(\theta_k - \theta_m) \tag{18}$$

$$Q_{k,loss} = \frac{1}{2}(B_{km} - B_{cap})(V_k^2 + V_m^2) - B_{km}V_kV_m \tag{19}$$

where Q_{km} and Q_{mk} denote the reactive power flow from bus k to bus m and bus m to bus k, respectively; B_{km} and G_{km} are the imaginary and real parts, respectively, of the reactance for branch k–m; B_{cap} is the admittance to ground of branch k–m; and θ_k and θ_m are the voltage phase angles of bus k and bus m, respectively.

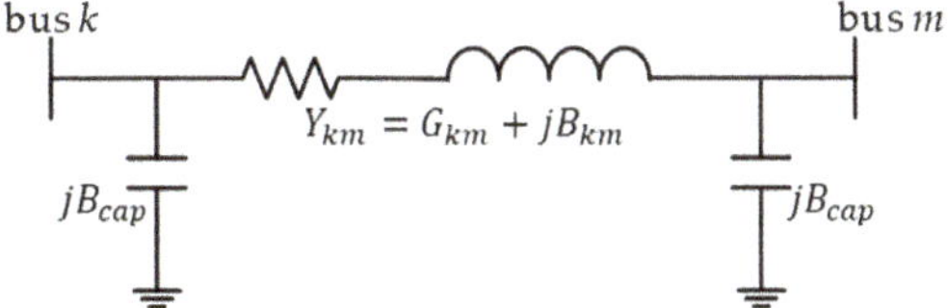

Figure 2. Transmission line π-equivalent model.

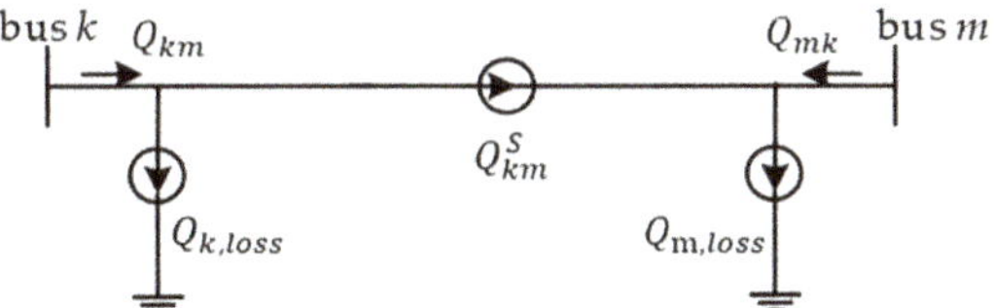

Figure 3. Transmission line reactive power flow model.

Figure 4 shows the pre-switch power system. We have

$$Q_k = Q_{k,\mu} + Q_{km}^S + Q_{k,loss} \tag{20}$$

$$Q_m = Q_{m,\vartheta} + Q_{mk}^S + Q_{m,loss} \tag{21}$$

where buses μ is the set of buses connected to bus k excluding bus m and buses ϑ is the set of buses connected to bus m excluding bus k; $Q_{k,\mu}$ and $Q_{m,\vartheta}$ are the reactive power flows in lines connecting bus k and bus m to buses μ and buses ϑ, respectively; Q_{km}^S represents the reactive power flow of the transmission part; and $Q_{k,loss}$, $Q_{m,loss}$ are the reactive power flows of the loss part.

As shown in Figure 5, with line k–m out of service, we have

$$Q_k = Q\prime_{k,\mu} \tag{22}$$

$$Q_m = Q\prime_{m,\vartheta} \tag{23}$$

where $Q\prime_{k,\mu}$ and $Q\prime_{m,\vartheta}$ are the reactive power flows in lines connecting bus k and bus m to buses μ and buses ϑ, respectively, after line k–m is taken out of service.

Assume that bus k and bus m still connect, as shown in Figure 6, then the simulated line part of reactive power flow from bus k to bus m ($Q\prime_{km}^S$) and the simulated loss part of reactive power flow in bus k and bus m ($Q\prime_{k,loss}$) can be described as

$$Q\prime_{km}^S = \Delta Q_{k1} = -\Delta Q_{m1} = -Q\prime_{mk}^S \tag{24}$$

$$Q\prime_{k,loss} = \Delta Q_{k2} = \Delta Q_{m2} = Q\prime_{m,loss} \tag{25}$$

where ΔQ_{k1}, ΔQ_{m1} are the injected reactive powers with values equal to $Q\prime_{km}^S$ and $Q\prime_{mk}^S$ and ΔQ_{k2}, ΔQ_{m2} are the injected reactive powers with values equal to $Q\prime_{k,loss}$ and $Q\prime_{m,loss}$, respectively.

Then

$$Q_k + \Delta Q_{k1} + \Delta Q_{k2} = Q\prime_{k,\mu} + Q_{km}^S + Q_{k,loss} \tag{26}$$

$$Q_m + \Delta Q_{m1} + \Delta Q_{m2} = Q\prime_{m,\vartheta} + Q_{mk}^S + Q_{m,loss} \tag{27}$$

Then, the voltage variations of bus i can be rewritten as

$$\Delta V_i = \alpha_{i-km} Q_{km}^S + \delta_{i-km} Q_{k,loss} \tag{28}$$

In the equation above,

$$\begin{cases} \alpha_{i-km} = \dfrac{X\prime_{ik} - X\prime_{im}}{1 - a(X\prime_{kk} - X\prime_{km}) - b(X\prime_{mk} - X\prime_{mm})} \\ \delta_{i-km} = \dfrac{X_{ik} + X_{im}}{1 - g(X\prime_{kk} + X\prime_{km}) - h(X\prime_{mk} + X\prime_{mm})} \end{cases} \tag{29}$$

$$\begin{cases} a = V_k(B_{km} - B_{cap}) + G_{km} V_m(\theta_k - \theta_m) \\ b = -V_m(B_{km} - B_{cap}) + G_{km} V_k(\theta_k - \theta_m) \\ g = V_k(B_{km} - B_{cap}) + B_{km} V_m \\ h = V_m(B_{km} - B_{cap}) + B_{km} V_k \end{cases} \tag{30}$$

where $X\prime_{ik}$, $X\prime_{im}$, $X\prime_{kk}$, $X\prime_{km}$, $X\prime_{mk}$, and $X\prime_{mm}$ are the corresponding elements in the reactance matrix $X\prime$ and $X\prime$ is the inverse matrix of the coefficient matrix of the PQ decoupling method [31].

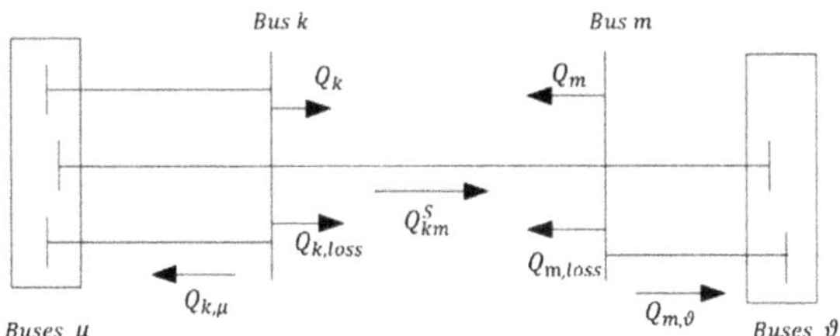

Figure 4. Pre-switch state of the power system.

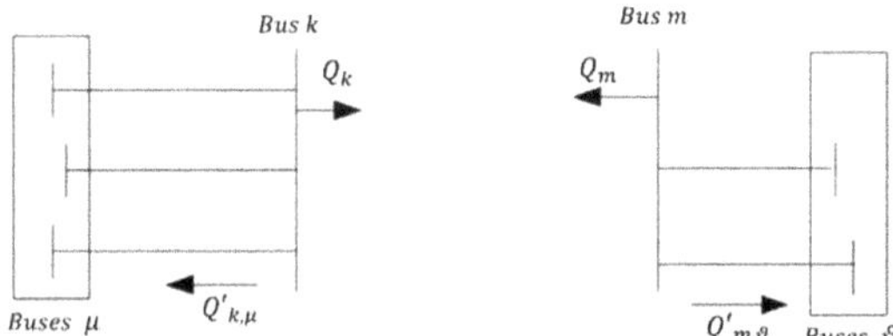

Figure 5. Post-switch state of the power system.

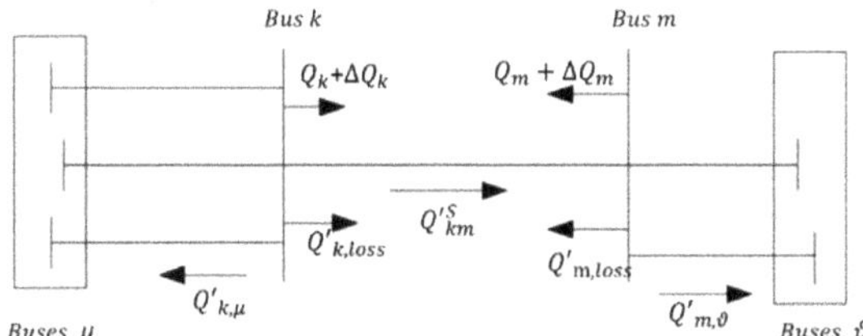

Figure 6. Simulated state of the power system.

We calculate the alleviation contribution ΔV_i of each candidate line out of service on the violated bus using Equation (28) and rank them in order.

3.3. Stage 3: Detailed Analysis and Assessment

To perform a detailed analysis of the several top candidates ranked at the ranking stage, the AC power flow is employed to compute the exact post-switching bus voltage. Based on the exact calculation of the AC power flow, the optimal network topology of the post-switching power system and the needed action of line switching are assessed.

We define the performance index NAM as follows:

$$NAM = min\left\{ \frac{V_{i\_max} - V_i^N}{V_{i\_max}}, \frac{V_i^N - V_{i\_min}}{V_{i\_min}} \right\} * 100\% \quad i \in 1, 2, 3 \cdots, n; \tag{31}$$

where V_{i_max} and V_{i_min} are the maximum and minimum voltage magnitudes of bus i, and V_i^N is the actual voltage magnitude of bus i with line k–m switched out. By using Equation (31), the line switching solutions list is assessed.

4. The Overall Solution Methodology

A step-by-step description of the proposed three-stage methodology for online applications is summarized in the following steps and shown in Figure 7.

Step 1: Input the online data, including the generation schedule, load demands, state estimation, network topology, and candidate lines for online line switching action.

Step 2: Run the AC power flow according to the given operating state. If voltage violations exist, go to Step 3; otherwise, stop and output the base case assessment results.

Step 3: If all candidate line combinations have been checked, stop and output "no solution found"; otherwise, go to Step 4.

Step 4: Apply the sensitivity formula (Equation (15)) to each candidate line.

Step 5: If effective candidate lines are found, then send them to Step 6 for ranking. Otherwise, go to Step 3.

Step 6: Apply Equation (28) to calculate the alleviation contribution ΔV_i of each line from Step 5 and rank them in order. Send the top candidate lines to Step 7 for detailed analysis.

Step 7: Apply AC power flow to compute the post-switching bus voltage corresponding to each top-ranked line switching and assess the line switching solutions for the power system.

Step 8: Rank the line switching solutions in order using NAM.

Step 9: Output the ordered effective line switching solutions and analysis; if no effective line is found, go to Step 3.

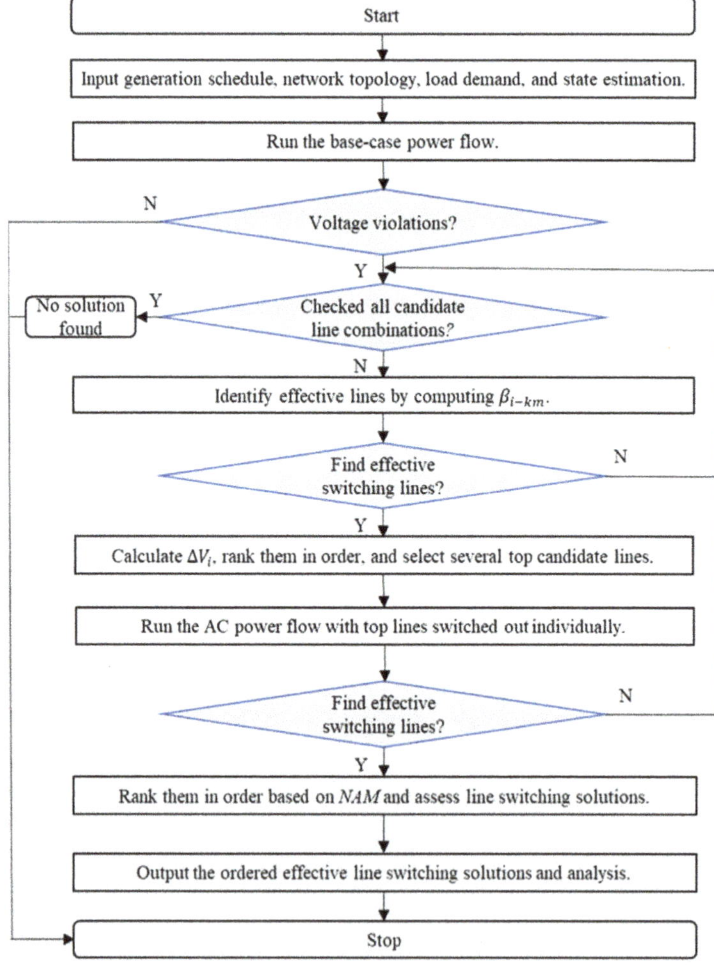

Figure 7. Flow chart of the proposed methodology.

5. Numerical Schemes

The proposed online line switching methodology was applied to the IEEE 39-bus and 2746-bus power systems to validate its effectiveness and accuracy. The proposed methodology was implemented in MATPOWER 6.0 on a ThinkPad PC with Intel Core 2.50 GHz i5-7200U CPU and 8 GB of memory. The results provided by the exact AC power flow were used as a benchmark to compare the speed and accuracy of the proposed three-stage methodology.

5.1. Single-Line Switching

The IEEE 39-bus system has 46 branches and 10 generators, and the active power of total loads is 6254.2 MW. The maximum and minimum voltage magnitudes of bus 26 are 1.0494 p.u. and 0.94 p.u., respectively; those of the other buses are 0.94~1.060 p.u. The power flow was run at current operating conditions and a voltage violation was found on bus 26.

By applying the proposed methodology, several solutions were assessed to relieve the voltage violation of bus 26. The screening, ranking, and detailed analysis results are shown in Table 1, and the CPU time for this example is displayed in Table 2. The voltages of bus 26 for the pre-switching and post-switching power systems are shown in Figure 8. We made the following observations from the results:

- *Stage 1*: By using β_{i-km}, 20 effective candidate lines were identified from 45 candidate lines.
- *Stage 2*: The ΔV_i of each line (20 effective candidates from Stage 1) was calculated to select the top seven lines and rank them in order: lines 26–29, 26–28, 26–27, 2–3, 28–29, 16–21, and 21–22.
- *Stage 3*: The AC power flow was used to check for any voltage violation at the current operating point with the top seven lines switched out individually. With lines 26–27 and 16–21 switched out individually, we found that there were still voltage violations on bus 26 of 1.0740 p.u. and 1.0497 p.u., respectively. Thus, the high-quality line switching solutions found to relieve voltage violation of bus 26 were lines 28–29,26–29, 26–28, 21–22, and 2–3. With each of these top five lines switched out, the voltage magnitudes on bus 26 were 1.0326 p.u. (NAM = 1.6009), 1.0366 p.u. (NAM = 1.2197), 1.0404 p.u. (NAM = 0.8576), 1.0414 p.u. (NAM = 0.7623), and 1.0416 p.u. (NAM = 0.7433).

Table 1. Result of single lines switched off.

Stage 1 Screening			Stage 2 Ranking			Stage 3 Detailed Analysis and Assessment		
Effective Candidates			Highly Ranked Candidates	ΔV_{26}	Error	Top Candidates	NAM	V_{26}/p.u.
2–3	5–6	8–9	26–29	−0.0109	−0.0019	28–29	1.6009	1.0326
3–4	7–8	21–22	26–28	−0.0062	−0.0028	26–29	1.2197	1.0366
4–14	6–7	16–21	26–27	−0.0022	0.0268	26–28	0.8576	1.0404
6–11	16–24	28–29	2–3	−0.0010	−0.0068	21–22	0.7623	1.0414
17–18	22–23	4–5	28–29	−0.0008	−0.0160	2–3	0.7433	1.0416
10–11	26–27	26–29	16–21	−0.0007	0.0010			
5–8	26–28	/	21–22	−0.0006	−0.0074	/		

The error is the difference between the actual voltage variation and the calculated value ΔV_{26}.

To evaluate the speed and accuracy of the proposed methodology, all 45 candidate lines were switched out individually and then lines 28–29, 26–29, 26–28, 21–22, and 2–3 were selected to relieve the voltage violation by using AC power flow. With the five lines switched out individually, the voltages on bus 26 were 1.0326 p.u., 1.0366 p.u., 1.0404 p.u., 1.0414 p.u., and 1.0416 p.u., respectively. This is consistent with the solutions assessed by the proposed methodology. It is noteworthy that the scheme given in this example is locally optimal when evaluated using full AC power flow.

The total CPU time of the proposed methodology in this study was 0.1054 s, whereas full AC power flow takes 0.9766 s. Compared with AC power flow, the CPU time speed-up given by the method in this study is 826.57%, as shown in Table 2.

We then compared the speed and accuracy of the proposed methodology with Shao's method in [28]. Lines 28–29, 26–29, and 26–28 are given to relieve the violation on bus 26 by using the method in [28], and the CPU time is 0.3816 s. From the results we can see that compared with the method in [28], the proposed methodology can provide more effective solutions and the speed is faster.

As can be seen from Figure 8, the line switching solutions obtained using the proposed methodology relieved the voltage violation on bus 26 in this study.

Table 2. CPU time required for the example (seconds).

The Proposed Methodology				AC Power Flow	Speed-Up
Stage 1	Stage 2	Stage 3	Total		
0.0007	0.0026	0.1021	0.1054	0.9766	826.57%

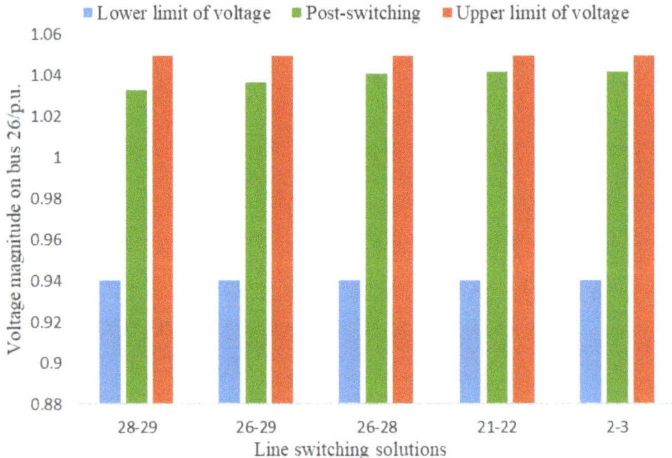

Figure 8. The voltage of bus 26 for the pre-switching and post-switching systems.

The above results illustrate the effectiveness of the proposed methodology in relieving voltage violations by switching out single lines compared with AC power flow. This study also shows the accuracy and fast speed of the proposed methodology compared with Shao's method in [28].

It is worth noting that considering the existence of errors in stage 1 and 2, the solutions provided in this paper may omit some solutions, but the proposed method can still provide a set of high-quality schemes to relieve voltage violations. All high-quality solutions are given in this study.

5.2. Multiple-Line Switching

The IEEE 39-bus system has 46 branches; the maximum and minimum voltage magnitudes of bus 26 are 1.0494 p.u. and 0.94 p.u., respectively, and those of the other buses are 0.94~1.060 p.u. The power flow was run at current operating conditions and a voltage violation was found on bus 26: V_{26} = 1.0526 p.u. A single line switched off cannot effectively relieve this voltage violation, so we increased the number of switching lines by 1 and utilized the proposed methodology to provide a set of multiple-line switching solutions. In this study, we set the number of switching lines at 2. The obtained solutions are summarized in Table 3 and the CPU times are displayed in Table 4. The voltages of bus 26 for the pre-switching and post-switching systems are shown in Figure 9.

Table 3. Results of multiple lines switched off.

Stage 1 Screening			Stage 2 Ranking			Stage 3 Detailed Analysis and Assessment		
Effective Candidates			Highly Ranked Candidates	ΔV_{26}	Error	Top Candidates	*NAM*	V_{26}/p.u.
2–3	5–6	8–9	28–29, 21–22	−0.0317	−0.0025	28–29, 21–22	2.9541	1.0184
3–4	7–8	21–22	2–3, 28–29	−0.0315	−0.0021	2–3, 28–29	2.8969	1.0190
4–14	6–7	16–21	26–29, 21–22	−0.0272	−0.0021	26–29, 21–22	2.4871	1.0233
6–11	16–24	28–29	26–28, 21–22	−0.0216	−0.0037	26–29, 2–3	2.4490	1.0237
17–18	22–23	4–5	26–28, 2–3	−0.0197	−0.0053	26–28, 21–22	2.1060	1.0273
10–11	26–27	26–29	26–29, 2–3	−0.0181	−0.0108	26–28, 2–3	2.0774	1.0276
5–8	26–28	/	2–3, 21–22	−0.0159	−0.0086	2–3, 21–22	2.0297	1.0281

The error is the difference between the actual voltage variation and the calculated value ΔV_{26}.

We made the following observations from the results:

Twenty-one candidates were identified from 45 candidate lines at Stage 1 and sent to Stage 2 for ranking. The top seven single lines were identified: lines 26–29, 26–28, 26–27, 2–3, 28–29, 16–21, and 21–22, as in Section 5.1. We combined the top seven single switching lines in pairs and calculated the ΔV_i of each candidate solution. Then, the seven most highly ranked multiple-line candidates were captured and sent to Stage 3 for detailed analysis and assessment: lines 28–29 and 21–22, lines 2–3 and 28–29, lines 26–29 and 21–22, lines 26–28 and 21–22, lines 26–28 and 2–3, lines 26–29 and 2–3, and lines 2–3 and 21–22.

Lines 28–29 and 21–22, lines 2–3 and 28–29, lines 26–29 and 21–22, lines 26–29 and 2–3, lines 26–28 and 21–22, lines 26–28 and 2–3, and lines 2–3 and 21–22 were assessed to relieve the voltage violation of bus 26 for the current power system by using AC power flow at Stage 3. For the successful line switching solutions, the voltage violations on bus 26 were 1.0184 p.u. (*NAM* = 2.9541), 1.0190 p.u. (*NAM* = 2.8969), 1.0233 p.u. (*NAM* = 2.4871), 1.0237 p.u. (*NAM* = 2.4490), 1.0273 p.u. (*NAM* = 2.1060), 1.0276 p.u. (*NAM* = 2.0774), and 1.0281 p.u. (*NAM* = 2.0297) in the current power system, respectively.

To verify the effectiveness of the obtained solutions, we performed an exhaustive search with all 45 candidate lines combined in pairs and then switched out. Switching solutions including lines 28–29 and 21–22, lines 2–3 and 28–29, lines 26–29 and 21–22, lines 26–29 and 2–3, lines 26–28 and 21–22, lines 26–28 and 2–3, and lines 2–3 and 21–22 were obtained to relieve the voltage violation by using AC power flow. It can be clearly seen that the effective line switching solutions obtained by the proposed methodology are the same as those in the AC power results.

The CPU time of the proposed methodology in this case was 0.1372 s, while the exhaustive search based on AC power flow took 7.3152 s. The speed-up was 5231.78%. Figure 9 shows the effectiveness of the proposed methodology for relieving the voltage violation of bus 26.

Similarly, lines 2–3 and 28–29, lines 26–28 and 2–3, lines 26–29 and 2–3, and lines 2–3 and 21–22 are provided by using the method in [28], and the CPU time is 0.8293s. Compared with the method in [28], the proposed method shows more advantages in accuracy and speed.

This study shows that the proposed methodology can provide several high-quality multiple-line switching solutions to relieve voltage violations. The CPU time test verifies the fast speed of the three-stage methodology.

Table 4. CPU time required for the example (seconds).

The Proposed Methodology				AC Power Flow	Speed-Up
Stage 1	Stage 2	Stage 3	Total		
0.0007	0.0047	0.1318	0.1372	7.3152	5231.78%

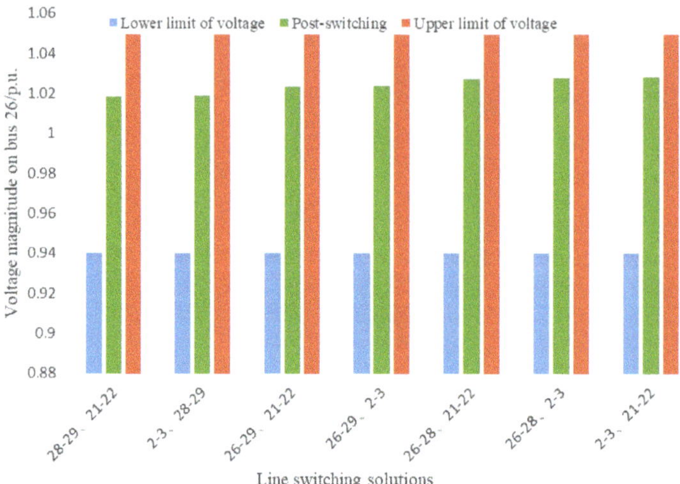

Figure 9. The voltage of bus 26 for the pre-switching and post-switching systems.

5.3. The 2746-Bus System

The proposed online methodology for line switching was evaluated on a 2746-bus power system containing 3514 transmission lines; the voltage limit on bus 249 was 0.94~1.06 p.u. and the actual voltage magnitude of the base case power system on bus 249 was 1.0829 p.u., meaning the existence of a voltage violation on bus 249.

The line switching solutions obtained after applying the proposed methodology are summarized in Table 5. The voltages of bus 249 for the pre-switching and post-switching systems are shown in Figure 10. The output from each stage is summarized as follows:

- *Stage 1:* There were 79 candidates identified from 2836 candidate lines. Due to space limitations, Table 5 displays 21 effective candidates.
- *Stage 2:* The 79 candidates were ranked, and the top seven candidates are selected for detailed analysis and assessment to be performed at Stage 3: lines 17–3, 249–3, 474–210, 474–248, 471–210, 249–247, and 374–247.
- *Stage 3:* For each top candidate line, AC power flow was performed to assess the effectiveness of each candidate. Consequently, lines 17–3, 249–3, and 474–248 were assessed to be most effective for relieving the voltage violation in the power system.

Table 5. Result of lines switched off.

Stage 1 Screening			Stage 2 Ranking	Stage 3 Detailed Analysis and Assessment		
Effective Candidates			**Highly Ranked Candidates**	**Top Candidates**	*NAM*	V_{249}/p.u.
7–8	350–287	249–247	17–3	17–3	3.2128	0.9702
7–17	2588–2460	471–437	249–3	249–3	1.9623	1.0392
17–3	287–218	474–210	474–210	474–248	1.1226	1.0481
249–3	370–286	2714–2604	474–248			
25–192	374–247	2460–2714	471–210			
383–370	474–248	513–278	249–247		/	
374–270	471–210	553–299	374–247			

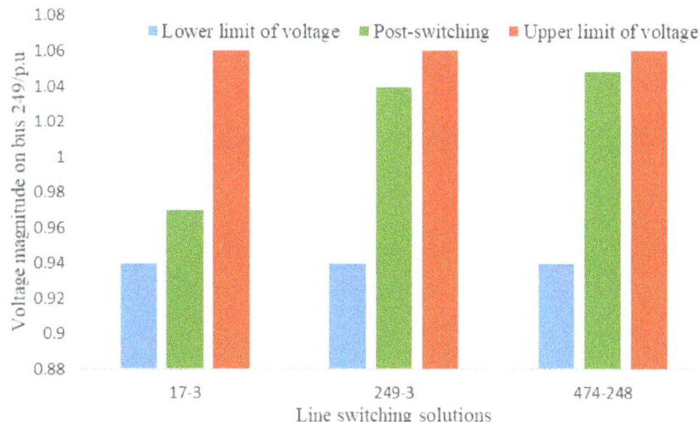

Figure 10. The voltages of bus 249 for the pre-switching and post-switching systems.

To evaluate the speed and accuracy of the proposed methodology, all 2836 candidate lines were switched out individually, and lines 17–3, 249–3, and 474–248 were found to relieve the voltage violation by using AC power flow. This is consistent with the solutions obtained by the proposed methodology.

As shown in Table 6, the total CPU time of the proposed methodology in this study was 2.3093 s, whereas the exhaustive search based on AC power flow took 231.0528 s—an improvement in speed by the proposed methodology of 9905.32%.

Table 6. CPU time required for the example (seconds).

The Proposed Methodology				AC Power Flow	Speed-Up
Stage 1	Stage 2	Stage 3	Total		
0.1969	0.8592	1.2532	2.3093	231.0528	9905.32%

Figure 10 shows that all the line switching solutions provided by the proposed methodology relieved the voltage violation on bus 249.

Simulation studies on the 2746-bus system showed that the proposed methodology is able to effectively solve the problems of voltage violations, and that the computation time is also satisfactory for online application in large-scale systems.

6. Conclusions

This paper proposed a novel online methodology of line switching for relieving voltage violations. The proposed methodology employs a three-stage strategy that contains screening, ranking, and detailed analysis and assessment stages. The proposed methodology balances speed and accuracy for online applications by combining linear and nonlinear methods to relieve voltage violations.

One distinguishing feature of the proposed methodology is that it can provide a set of high-quality solutions from which operators may select a preferred solution. Numerical schemes and methods were developed and implemented for each stage of the proposed methodology. It was evaluated on the IEEE 39-bus and 2746-bus power systems with promising results. The results provided by exact AC power flow were used as a benchmark to compare the speed and accuracy of the proposed three-stage methodology.

The results showed that the proposed methodology can provide single-line switching as well as multiple-line switching to relieve voltage violations. Compared with the method in [28], the proposed method shows more advantages in accuracy and speed. In addition, compared with AC power flow, the three-stage methodology requires less CPU time, especially in a large-scale system. A numerical study was conducted on the 2746-bus power system and revealed the fast speed (a speed-up of

9905.32% over AC power flow) and effectiveness of the proposed methodology when applied to large-scale systems, showing good potential for online applications.

Author Contributions: Conceptualization, Z.S.; Methodology, Z.S.; Software, Z.S.; Validation, Z.S.; Writing—Original Draft Preparation, Z.S.; Writing—Review and Editing, Z.S.; Supervision, Y.T.; Data Curation, J.Y., B.Z., C.C., G.Z.

Funding: This research was funded by National Key Research and Development Plan of China (2016YFB0900602); Science and Technology Project of State Grid Corporation of China (XT71-16-032).

Acknowledgments: The authors would like to thank National Key Research and Development Plan of China (2016YFB0900602) and Science and Technology Project of State Grid Corporation of China (XT71-16-032) for its financial support.

Conflicts of Interest: The authors declare no conflict of interest.

References

1. Song, F.; Wang, Y.; Yan, H.; Zhou, X.; Niu, Z. Increasing the utilization of transmission lines capacity by quasi-dynamic thermal ratings. *Energy* **2019**, *12*, 792. [CrossRef]
2. The, J.; Ooi, C.A.; Cheng, Y.H. Composite reliability evaluation of load demand side management and dynamic thermal rating systems. *Energy* **2018**, *11*, 466. [CrossRef]
3. PJM Manual 03: Transmission Operations, Revision: 54. 2018. Available online: http://www.pjm.com/~{}/media/documents/manuals/m03.ashx (accessed on 27 March 2019).
4. Hedman, W.K.; Oren, S.S.; O'Neil, P.R. A Review of Transmission Switching and Network Topology Optimization. In Proceedings of the IEEE Power and Energy Society General Meeting, Detroit, MI, USA, 24–29 July 2011; pp. 1–7.
5. Rolim, J.G.; Machado, L.J.B. A study of the use of corrective switching in transmission systems. *IEEE Trans. Power Syst.* **1999**, *14*, 336–341. [CrossRef]
6. Wang, L.; Chiang, H.D. Toward online bus-bar splitting for increasing load margins to static stability limit. *IEEE Trans. Power Syst.* **2017**, *32*, 3715–3725. [CrossRef]
7. Henneaux, P.; Kirschen, D.S. Probabilistic security analysis of optimal transmission switching. *IEEE Trans. Power Syst.* **2016**, *31*, 508–517. [CrossRef]
8. Li, M.; Luh, P.B.; Michel, L.D.; Zhao, Q.; Luo, X. Corrective line switching with security constraints for the base and contingency cases. *IEEE Trans. Power Syst.* **2008**, *23*, 125–133. [CrossRef]
9. Shao, W.; Vittal, V. Corrective switching algorithm for relieving overloads and voltage violations. *IEEE Trans. Power Syst.* **2005**, *20*, 1877–1885. [CrossRef]
10. Shao, W.; Vittal, V. BIP-based OPF for line and bus-bar switching to relieve overloads and voltage violations. In Proceedings of the IEEE Power and Energy Society General Meeting, Atlanta, GA, USA, 29 October–1 November 2006; pp. 2090–2095.
11. Liu, W.L.; Chiang, H.D. Toward On-Line Line Switching Method for Relieving Overloads in Power Systems. In Proceedings of the IEEE Power and Energy Society General Meeting, Denver, CO, USA, 26–30 July 2015; pp. 1–5.
12. Mazi, A.A.; Wollenberg, B.F.; Hesse, M.H. Corrective control of power system flows by line and bus-bar switching. *IEEE Trans. Power Syst.* **1986**, *1*, 258–264. [CrossRef]
13. Bacher, R.; Glavitsch, H. Network topology optimization with security constraints. *IEEE Trans. Power Syst.* **1986**, *1*, 103–111. [CrossRef]
14. Khanabadi, M.; Ghasemi, H.; Doostizadeh, M. Optimal transmission switching considering voltage security and N–1 contingency analysis. *IEEE Trans. Power Syst.* **2013**, *28*, 542–550. [CrossRef]
15. Wrubel, J.N.; Rapcienski, P.S.; Lee, K.L.; Gisin, B.S.; Woodzell, G.W. Practical experience with corrective switching algorithm for on-line applications. *IEEE Trans. Power Syst.* **1996**, *11*, 415–421. [CrossRef]
16. Hedman, K.W.; O'Neil, R.P.; Fisher, E.B.; Oren, S.S. Optimal transmission switching-sensitivity analysis and extensions. *IEEE Trans. Power Syst.* **2008**, *23*, 1469–1479. [CrossRef]
17. Hou, L.R.; Chian, H.D. Toward Online Line Switching Method for Reducing Transmission Loss in Power Systems. In Proceedings of the IEEE Power and Energy Society General Meeting, Boston, MA, USA, 17–21 July 2016; pp. 1–5.

18. Hedman, K.W.; O'Neil, R.P.; Fisher, E.B.; Oren, S.S. Optimal transmission switching with contingency analysis. *IEEE Trans. Power Syst.* **2009**, *24*, 1577–1578. [CrossRef]

19. Escobedo, A.R.; Moreno-Centeno, E.; Hedman, K.W. Topology control for load shed recovery. *IEEE Trans. Power Syst.* **2014**, *29*, 908–916. [CrossRef]

20. Fisher, E.B.; O'Neil, R.P.; Ferris, M.C. Optimal transmission switching. *IEEE Trans. Power Syst.* **2008**, *23*, 1346–1355. [CrossRef]

21. Mak, T.W.K.; Hentenryck, P.V.; Hiskens, I.A. A Nonlinear Optimization Model for Transient Stable Line Switching. In Proceedings of the American Control Conference (ACC), Seattle, WA, USA, 24–26 May 2017; pp. 2085–2092.

22. Sahraei-Ardakani, M.; Li, X.; Balasubramanian, P.; Hedman, K.W.; Abdi-Khorsand, M. Real-time contingency analysis with transmission switching on real power system data. *IEEE Trans. Power Syst.* **2016**, *31*, 2501–2502. [CrossRef]

23. Khodaei, A.; Shahidehpour, M. Transmission switching in security-constrained unit commitment. *IEEE Trans. Power Syst.* **2010**, *24*, 1937–1945. [CrossRef]

24. Li, C.; Chiang, H.D.; Du, Z. Investigation of an effective strategy for computing small-signal security margins. *IEEE Trans. Power Syst.* **2018**, *33*, 5437–5445. [CrossRef]

25. Wang, L.; Chiang, H.D. Toward online line switching for increasing load margins to static stability limit. *IEEE Trans. Power Syst.* **2015**, *31*, 1744–1751. [CrossRef]

26. Liu, C.; Wang, J.; Ostrowski, J. Static switching security in multi-period transmission switching. *IEEE Trans. Power Syst.* **2012**, *27*, 1850–1858. [CrossRef]

27. Guo, W.M.; Wei, Q.; Liu, G.J.; Yang, M.; Zhang, X.K. Transmission Switching to Relieve Voltage Violations in Low Load Period. In Proceedings of the IEEE Region, Lyngby, Denmark, 6–9 October 2013; pp. 1–4.

28. Shao, W.; Vittal, V. A New Algorithm for Relieving Overloads and Voltage Violations by Transmission Line and Bus-Bar Switching. In Proceedings of the IEEE PES Power Systems and Exposition, New York, NY, USA, 10–13 October 2004; Volume 1, pp. 322–327.

29. Fu, B.; Ouyang, C.X.; Li, C.S.; Wang, J.W.; Gui, E. An improved mixed integer linear programming approach based on symmetry diminishing for unit commitment of hybrid power system. *Energy* **2019**, *12*, 833. [CrossRef]

30. El-Abiad, A.H.; Stagg, G.W. Automatic evaluation of power system performance-effects of line and transformer outages. *AIEE Trans. Power Appar. Syst.* **1963**, *81*, 712–715. [CrossRef]

31. Chen, M.; Shi, D.; Li, Y.; Zhu, L.; Liu, H. Research on Branches Group Based Method for Adding Mutual Inductance Branches to Y-Matrix and Z-Matrix. In Proceedings of the IEEE PES Power Systems and Exposition, Boston, MA, USA, 17–21 July 2014; pp. 1–5.

32. Stott, B.; Alsac, O. Fast decoupled load flow. *IEEE Trans. Power Appar. Syst.* **1974**, *PAS-93*, 859–869. [CrossRef]

33. Lee, C.Y.; Chen, N. Distribution factors of reactive power flow in transmission line and transformer outage studies. *IEEE Trans. Power Syst.* **1992**, *7*, 194–200. [CrossRef]

 *energies*

Article

A Novel Control Architecture for Hybrid Power Plants to Provide Coordinated Frequency Reserves

Daniel Vázquez Pombo *, Florin Iov and Daniel-Ioan Stroe

Department of Energy Technology, Aalborg University, 9220 Aalborg, Denmark; fi@et.aau.dk (F.I.); dis@et.aau.dk (D.-I.S.)
* Correspondence: dvp@et.aau.dk; Tel.: +45-5222-2202

Received: 25 January 2019; Accepted: 4 March 2019; Published: 9 March 2019

Abstract: The inertia reduction suffered by worldwide power grids, along with the upcoming necessity of providing frequency regulation with renewable sources, motivates the present work. This paper focuses on developing a control architecture aimed to perform frequency regulation with renewable hybrid power plants comprised of a wind farm, solar photovoltaic, and a battery storage system. The proposed control architecture considers the latest regulations and recommendations published by ENTSO-E when implementing the first two stages of frequency control, namely the fast frequency response and the frequency containment reserve. Additionally, special attention is paid to the coordination among sub-plants inside the hybrid plant and also between different plants in the grid. The system's performance is tested after the sudden disconnection of a large generation unit (N-1 contingency rules). Thus, the outcome of this study is a control strategy that enables a hybrid power plant to provide frequency support in a system with reduced inertia, a large share of renewable energy, and power electronics-interfaced generation. Finally, it is worth mentioning that the model has been developed in discrete time, using relevant sampling times according to industrial practice.

Keywords: hybrid power plant; control architecture; coordination of reserves; frequency support; frequency control dead band; fast frequency response; frequency containment reserve

---

1. Introduction

During the last decades, penetration rates of renewable energy sources (RES) generation have steadily increased due to environmental concerns and positive market stances [1]. This trend is expected to continue during the following decades, partly motivated by international regulations [2,3]. In fact, some countries already present high shares of RES generation in their power systems, e.g., 43.4% of the Danish electricity consumption was covered by wind power in 2017 [4]. This renewable energy transition is positive and must be continued [3]. However, most of these generation units do not use traditional synchronous generators (SG), and thus are unable to provide inertia to the grid since all these units are not synchronously coupled with the grid. In addition, as RES-based generation increases, traditional power plants equipped with SG are gradually phased-out; ergo, new challenges arise in the power system, like loss of inertia, volatile frequency, and unwanted disconnection of distributed generation units [5].

In this context, even though frequency stability has, traditionally, been a simple task, nowadays it is becoming an increasingly complex activity due to the fluctuations of both generation and demand. European regulatory agencies, such as ENTSO-E, have already started giving the first guidelines and rules in order to ensure the correct operation of the electric system [6,7]. Such documents maintain the traditional structure with three stages of frequency control, acknowledging the fact that inertial response from SG is no longer sufficient, and opening the possibility to include smart control strategies in the new generation units in order to compensate such a lack. However, as will be presented in the next section, these new guidelines and rules are not yet clearly defined and stablished, as discussed in ref [8]. This new set of regulation not only has come to complete the normative of some countries were frequency control with wind power is not considered—like in Regelleistungand (association of German Transmission System Operators (TSOs)) [9]—but also modifies others like the most recent one from ENERGINET (Denmark), as will be presented in Section 7.1. [10]

On the other hand, industry has recently called attention to increasing the full load hours of wind farms (WF), which are defined according to the loading of the transformer in the plant. This value oscillates around 35% and 45% for onshore and offshore, respectively [11]. Briefly, if additional generation units are added to the system, the under-utilization of the plant will be reduced. Different manufacturers [12,13] are considering over-planting and the inclusion of a solar photovoltaic plant (PVP) in order to increase the production rate and complement it with a battery energy storage system (BESS). Such a plant will be capable of providing power smoothing, production loss minimization, and sudden power injections to help in the frequency regulation. Subsequently, a plant combining these three elements is referred to in this paper as a hybrid power plant (HyPP) and based on the benchmark model presented in ref [8].

Regarding frequency support with RES, there are barely any relevant works presented in academia, especially when considering HyPPs. In work such as refs [14,15], a combination of WF and PVP is used to provide an inertial response while others like refs [16,17] also include storage like flywheels. However, in all these studies, only the WF and the storage (if present) respond to the frequency excursion; none of them use a BESS, the model is always in continuous time (Laplace), and time delays accounting for event identification, measurements, or communications are dismissed. On the other hand, there are a few examples of site-tests made by industry, e.g., ref [18], a HyPP similar the one considered in this research that will finish its construction in 2019 in Australia. Then in refs [19,20], two combinations of WF and BESS in Denmark are also examples of how the Hybrid technology presents increasing interest to companies. However, the frequency control capabilities are still quite limited.

The main objective of this research is to evaluate the ability of HyPP to participate in the regulation of the system frequency by following current standards and system operation grid codes. Furthermore, a control strategy for the coordinated provision of frequency reserves in a system with a high share of RES and very low inertia is presented. In that respect, the model accounts for event identification and communication delays, and it uses current operational limitations of typical industrial controllers. It is expected that the HyPP's combined response will be a sufficient and effective solution that is able to substitute conventional generating units due to the promising results presented in ref [8]. Although, in that work, only the first stage of frequency control was considered. Finally, it is worth mentioning that the proposed control algorithm including the hybrid power plant and external grid model has been developed for real-time hardware-in-the-loop (RT-HIL) studies. However, the results presented in this paper were obtained during the off-line verification stage according to the model-based design approach. The RT-HIL testing of the proposed control strategy is currently ongoing and will be presented in future publications.

The structure of the paper is as follows: A brief background review of frequency behavior is presented in Section 2, while current frequency regulation requirements and standards are covered in Section 3. Then, the system modelling, along with the HyPP concept, are covered in Section 4. Subsequently, in Section 5, the control architecture for the provision of reserves and its coordination is presented; whereas in Section 6, the design of every control stage is presented. Thereafter, in Section 7, the evaluation of the architecture and the model is addressed after defining relevant scenarios. Finally, the main conclusions of the study are stated in Section 8, and new research paths available for future work are highlighted.

2. Background in Frequency Behavior

Traditionally, Equation (1) has been used to define the simplified frequency behavior of any power system [21]. Such an equation expresses how a generation–demand imbalance causes a frequency variation, given that the speed or rate of that change is inversely proportional to the inertia and the size of the system:

$$ROCOF = \frac{P_G - P_L}{2HS} f_n \tag{1}$$

where $ROCOF$, P_G, P_L, H, S, and f_n stand for Rate of Change of Frequency [Hz/s or $\frac{\partial f}{\partial t}$], total generated power [W], total consumed power [W], time constant related to the grid's inertia [s], system's total installed power [VA], and nominal frequency [Hz], respectively. $ROCOF$, which is defined as the time derivative of the frequency, has been historically dismissed in frequency control due to its low relevance in systems with high inertia. However, nowadays, due to the loss of such inertia, its use is becoming increasingly relevant. For the purposes of this study, the inertia constant has been set to 3 s, since it is a standard value presented by systems with high shares of RES, like Denmark [21].

3. Relevant Frequency Regulation Codes

The transmission system operator (TSO) is the agent in charge of ensuring the frequency stability of any network. At the European level, ENTSO-E elaborates and distributes a baseline of regulations to be followed by the countries belonging to such organizations. However, national standards can further develop regulations on top of those proposed by ENTSO-E. Nevertheless, recently, such regulations have been subjected to revision due to the changes experienced in power system behavior and operation resulting from the increasing RES and power electronic interfaces integration [6,7,22]. Subsequently, this paper uses up-to-date standards and regulations during the model development stage.

During steady state, the frequency oscillates in the vicinity of the nominal frequency value (50 ± 0.2 Hz); then; after the occurrence of an event, the nominal value is lost, and the three stages of frequency regulation start, namely: fast frequency response (FFR), frequency containment reserve (FCR), and frequency restoration reserve (FRR). Both the traditional and modern regulations acknowledge these three stages; however, there are certain differences and uncertainties that must be acknowledged. Additionally, the modern regulation opens the possibility of including a 4th stage called replacement reserves (RR). A typical frequency response to be expected after a down frequency event is presented in Figure 1. The pink line on the image illustrates the concept of a second dip, which is not yet covered in any regulation, but nevertheless, as will be explained, it represents a factor of major importance.

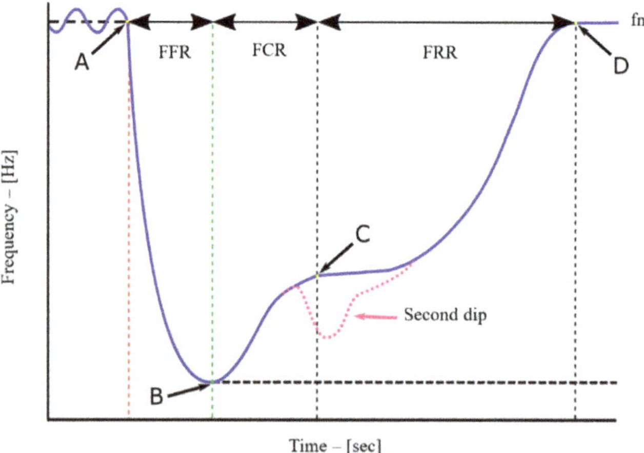

Figure 1. Typical frequency response curve.

It should be mentioned how dips are caused by a generation vs. demand mismatch, which leads to discontinuities in the frequency recovery. In fact, systems with low inertia are especially vulnerable against such mismatches, which are usually caused by aggressive frequency response approaches.

3.1. First Stage: Inertial Response–Fast Frequency Response

This stage starts in the instant of the event detection and finalizes once the frequency reaches its minimum value. Such a point is known as 'Nadir', with a corresponding frequency of f_{nadir}. Although there is no consensus regarding its duration, it is given mainly by the overall inertia present in the system, and thus it has a maximum from 2 to 5 s. It can be considered short, especially when compared with the other stages. As a reference, and due to the results presented in ref [8], it is considered to last from 0.5 to 2 s in this work.

The main difference between inertial response (IR) and FFR is that IR is provided exclusively by SG and in a natural, uncontrollably manner [12]. This means that, due to the physics ruling synchronism, SG naturally reacts in order to keep balance between generation and demand, and thus, stopping the frequency from changing. On the other hand, FFR is a controllable non-spontaneous reaction of the generators in a grid [12], which is, in short, a control-driven sudden power injection aiming to stop the frequency from continuing to modify its value. There are certain techniques to achieve this, like virtual synchronous machines, synthetic inertia, or the inclusion of spinning reserves [12].

In ref [7], FFR is acknowledged; however, there is neither a time length definition nor a specific approach to be followed in order to perform it. All the considerations taken regarding FFR in this paper are based on the work presented in ref [8].

3.2. Second Stage: Primary Frequency Regulation–Frequency Containment Reserve

This stage starts once the frequency stops dropping after the event (Nadir point); however, its end is not clearly defined in the regulations. In ref [7], it is simply stated that this stage finishes once the frequency has been partially restored, meaning that the frequency value is close to the nominal, but there is still an offset or error. The units participating in this stage are required to be able to provide full power injection during a certain period (around 30 min), although this time can be less if the third stage is activated prior to that. Again, there is no consensus regarding its time length; however, it is in the scale of several minutes [7,23,24].

The main differences between primary frequency regulation (PFR) and FCR is that PFR only considers the control action to be performed by the governors of different plants. Usually, PFR is performed only by one plant in the system in order to avoid the hunting effect. On the other hand, FCR is based on local frequency measurements. Basically, this stage is approached in both regulations as a droop control with a certain deadband in order to avoid over-actuation of the control system [7,23,24].

In ref [7], the procedure to be followed in order to estimate the FCR needs for a certain grid is stated. It also gives recommendations related to the droop characteristic and deadband to be implemented. Additionally, it states several time constraints: First, the FCR must start 3 to 5 s after the event is triggered and be fully activated in less than 30 s. Finally, it should be able to provide its maximum power capacity for at least 15 min. Lastly, the end of this stage is defined by this 15 min limitation or the activation of the third stage, whichever occurs first.

It is worth mentioning that the 15 min rule does not allow the renewable generation plants to participate in the regulatory market due to their dependency and uncertainty on meteorological conditions. However, the improvement of the short term meteorological forecast or the modification of this regulation might eliminate such limitations. Additionally, the inclusion of a storage system as in the case of the HyPP may already solve this challenge.

3.3. Third Stage: Secondary Frequency Response–Frequency Restoration Reserve

This stage starts after the second stage reaches steady state or after the time limitation; and finalizes once the frequency is restored to its nominal value or marginally close to it. Again, there is no consensus regarding its time length, but it is usually in the range of tens of minutes. It should be stated that this stage falls beyond the scope of this research, but it will be addressed in future publications.

There are no differences worth mentioning between secondary frequency response (SFR) and FRR as according to refs [7,23,24].

3.4. Fourth Stage: Tertiary Frequency Response–Replacement Reserves

This stage does not have clear start and ending points and, in both regulations, is considered to be optional. Thus, it is usually neglected in research. In the case of the tertiary frequency response (TFR), it consists of a set-point change in the conventional power plants based on an economic dispatch and unit commitment algorithms. While in the case of RR, it accounts for load variations occurring during the event clearance.

Again, this stage falls beyond the scope of this research, but will be addressed in future publications.

3.5. Final Considerations

The new regulations try to capture new technological realties present in power systems like the inclusion of renewable energy, inertia provision, etc. However, they seem like an ongoing work, especially due to the amount of amendments released during 2018 by ENTSO-E. Now, while these regulations are still being defined and established, is the moment to carefully analyze and review them.

In Table 1, a comparison of both terminologies is presented along with the main objective of each stage.

Table 1. Summary of the frequency regulation processes.

Stage	Similar To	Main Objective
FFR	IR	Stop frequency variation
FCR	PFR	Approximate frequency to nominal value
FRR	SFR	Restore nominal frequency
RR	TFR	Final support

A second or subsequent dip is an additional frequency reduction that occurs during the restoration process, caused by a non-smooth recovery of the frequency. The frequency reduction of the second and subsequent dips is always of smaller amplitude than the Nadir. However, a grid's stability is threatened even more, since frequency protections are triggered unnecessarily and thus will activate load-shedding schemes. The main reason for the protection to needlessly trigger is that first and successive dips are detected as a single fault with a comparatively long duration. Currently, most of the research regarding frequency restoration does not acknowledge the importance the second dip, as discussed in ref [8]. Although this second dip is not covered or defined in any standard yet, TSOs have raised their concerns in public talks (conferences, etc.).

4. System Modeling

In this section, the electric power system model is presented, including the description of the SG's governors and loads included in the system. In a different subsection, the HyPP's balance of plant is introduced.

4.1. Electric Power System

The considered grid—topology presented in Figure 2—represents an equivalent UK topology. It is based on the standard IEEE 12-bus system and adapted for wind power integration studies [25]. The system has four differentiated areas led by thermal power plants. Area 1 presents a large thermal power generation and a combination of residential and industrial loads. Meanwhile, rural loads dominate area 2, where there generation is also present. Subsequently, area 3 constitutes a heavily industrial load center with reduced thermal generation. Lastly, area 4 is the one marked in ref [25] as the connection point of wind farms for integration studies; thus, the HyPP is considered to be connected on Bus 5.

This network is suitable for the purposes of this research due to its simplicity caused by the fact of being an island and therefore avoiding the additional complexity of continental connections (limited size), which ultimately allows for easy implementation of frequency studies. However, the sizes of the conventional units have been altered in order to create a high HyPP penetration scenario. Table 2 summarizes the considered sizes.

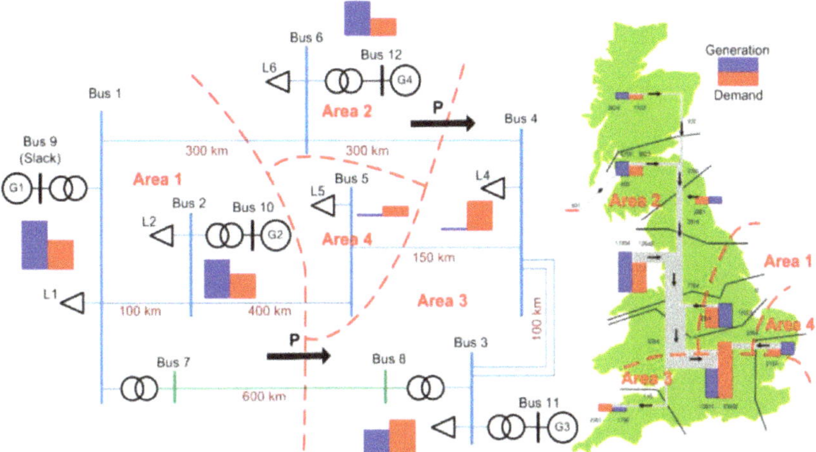

Figure 2. IEEE 12-bus system topology [25].

Table 2. Summary of generation units' sizes.

Generator	Size [MW]	Ratio [%]
SG: G01	750	21.75
SG: G02	640	18.56
SG: G03	384	11.14
SG: G04	474	13.75
HyPP	1200	34.8

The role of the SG is to represent the IR and primary frequency response as it is implemented nowadays. Those units respond to frequency deviations by means of governors, which are an extensively covered topic in literature [26,27]. It is worth mentioning that all the SG's governors have been modeled as the F10 type (specifications can be found in ref [28]). On the other hand, the demand is aggregated and represents a mixture of different load characteristics, e.g., residential, rural, urban, industrial, agricultural, etc. Thus, half of this load is considered pure resistive loads while the rest are frequency dependent; that is, their active power demand is based on the grid's frequency [26]. In order to account for those changes, all the loads are modeled as frequency-dependent, which is done by following Equation (2):

$$P_l = P_{lo}(1 + D \cdot \Delta f) \tag{2}$$

where P_l, P_{lo}, D, and Δf represent the total load, the non-frequency dependent part of the load, the load-damping constant, and the frequency deviation, respectively. In this work, D is assigned to be 1%, meaning that every 1% change in frequency would cause 1% change in the system load, whereas the standard value of the D constant is between 1% and 2% according to ref [29].

4.2. Hybrid Power Plant

As mentioned in Section 1, the HyPP concept is born from pursuing the idea of increasing the full production hours of WFs by adding additional turbines, a PVP, and a BESS. Thus, the included HyPP, which is based on the 100 MW benchmark model presented in ref [8], is compounded by three different sub-plants: WF, PVP, and BESS of sizes 100, 31.5, and 28 MW, respectively. The sizing, configuration and intra-plant loss estimation is extensively covered in ref [8]. However, it is important to note how the losses at nominal power and voltage in the PCC can be accounted for as a gain K_{loss} of 0.9921, 0.9658, and 1 for the WF, the PVP, and the BESS, respectively. This value was obtained in ref [8] after performing an extensive sensitivity analysis of the losses as a function of voltage level, active power, reactive power, short-circuit ratio, and R/X.

The balance of the plant is presented in Figure 3 where POC and PCC stand for point of connection and point of common coupling, respectively. It is assumed that in those points, there are measurement devices installed, i.e., grid meters. Lastly, Table 3 states the sizes of the HyPP both in the benchmark reference model and the implemented size in the tested scenarios of this research. As aforementioned, the benchmark model was 100 MW in ref [8]. Therefore, in this research, 12 of these plants are considered to be connected in Bus 5 of Figure 2, for a total implemented size of 1200 MW.

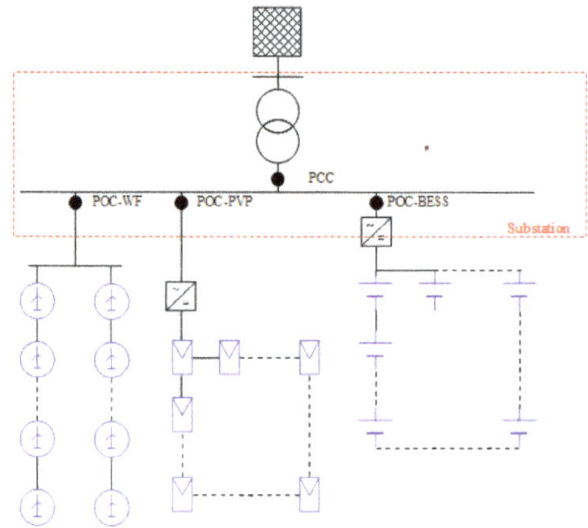

Figure 3. HyPP balance of plant [8].

Table 3. Sizing of the HyPP in the benchmark model and in the implemented (scaled up) system.

Plant	Benchmark Size [MW]	Implemented Size [MW]
WF	100	1200
PVP	31.5	378
BESS	28	336
HyPP	100	1200

5. Control Architecture

In this section, a control architecture suitable to implement FFR, FCR with coordination of reserves is proposed. The objective of this approach is to allow the performance of frequency control while taking into consideration technical limitations like, capacity of commercial controllers, event identification and transmission of data (telecommunications).

The base line of this architecture is the proposed in [8] and it is presented in Figure 4.

The operational process is as follows; in normal operation, the TSO establishes a certain production set-point for the HyPP, which constitutes the generation reference. The Dispatch function will then divide the reference according to the available power of each sub-plant and to certain operational priorities (i.e., produce with the WF instead of discharging the BESS). Subsequently, each sub-plant controller will again perform a distribution of the reference among the individual assets (i.e., individual turbines). Then, after accounting for the internal losses by using a meter connected to the POC; the production of the three sub-plants is added, thus, obtaining the resulting performance in the PCC. The production of the HyPP is then fed into the grid as presented in Section 3.1. Then, by using the meter at the PCC, the frequency and, subsequently, ROCOF signals can be obtained. Both frequency and ROCOF, are fed back to the frequency controller; which will remain inactive until an event is detected.

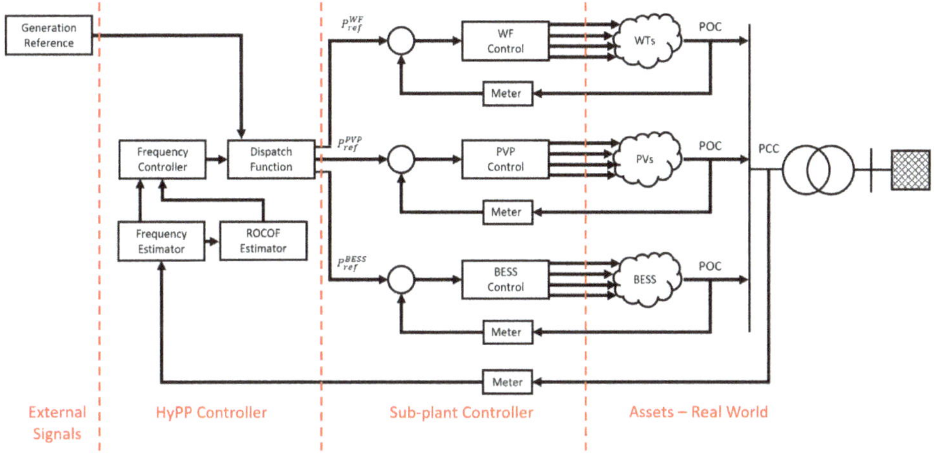

Figure 4. Proposed control structure of a hybrid power plant (HyPP; based on ref [8]).

According to the standards, an event is defined when the frequency falls beyond 50 ± 0.2 Hz, as defined in ref [7]. However, in the proposed method, an event can also be defined by a *ROCOF* value of ± 0.2 Hz/s. In this way, the frequency contingency strategies are activated faster in case of a sudden event thanks to the *ROCOF* and also in low *ROCOF* events due to the frequency. Additionally, the over-activation of the frequency control is still avoided.

After an event is detected, a flag is raised, thus altering the dispatch's operation and starting the frequency controller. Briefly, this controller uses frequency and *ROCOF* inputs to modify the operational set-point of the HyPP, thus coordinating the different frequency recovery stages. Regarding the operation of the Dispatch after the event is detected, FFR starts, with the objective to slow down the frequency excursion by combining the three sub-plants and dividing the effort as much as possible in order to provide a fast and harmless response. Then, FCR starts once the Nadir is reached, a point where *ROCOF* is 0. It should be stated that, according to the standards, FCR actions should start as soon as possible and in less than 2 s after the event identification, and thus being added on top of FFR as traditionally was with IR and PFR [7]. Subsequently, the HyPP combines the three sub-plants in order to approximate the frequency to its nominal value. Then, the BESS alone will perform the FCR actions since it is easier to control and it can act fast, avoiding behaviors that might threaten the frequency recovery or the plant's lifetime (e.g., over-oscillations, vibrations, etc.). Since it is also possible to know the available energy and power available in the BESS from its state-of-charge, the minimum time of operation can be ensured. On the other hand, in the power system block, the SG reacts uncontrollably to the frequency excursion with IR and then PFR, in accordance with the standards.

In order to estimate the available power of the HyPP at PCC, power requirements from the grid–P_{ref} and external parameters influencing sub-plant production—wind speed, temperature, irradiance—are considered. Power requirements from the grid are based on requirements established from the system operator (SO) and those related to frequency regulation and the provision of FFR and

FCR. On the sub-plant level, the input is a reference power from the control structure comprising of a 'Dispatch' block, which coordinates the different sub-plants. The output of each sub-plant is based on the input reference power requirement, external parameters, and unit dynamics. The overall output of the HyPP is then fed into the PCC.

6. Control Design

In this section, the design of the frequency controllers is addressed, starting with the evaluation criteria and continuing with each control stage implemented in the HyPP. It is worth mentioning that the developed model has been built by following industrial standards and common practices in order to resemble a real system, i.e., including sampling times for various blocks and subsystems, communication delays, etc. Therefore, even though the controllers were first designed in continuous time (Laplace), they were discretized using backward Euler due to its simplicity [30] and usefulness [8].

6.1. Success Criteria (Objectives)

The minimum performance requirements are established by different grid codes, regulations, and technical limitations of the involved elements [6,7,23]:

- Sub-plant's settling time must be less than 10 s for a 0.1 p.u. variation.
- Sub-plant's steady state error must be less than 2% without overshoot.
- Frequency must not drop below 49.2 Hz, a point were load-shedding protocols are activated.
- After FCR activation, frequency should be recovered to a value closer to the nominal than to the Nadir.

6.2. Plant Controllers

The selected controllers are PI, which were designed by replacing the slowest pole of the sub-plant's transfer function with a pole in the origin. This approach yields a closed loop transfer function with a response similar to the sub-plant. Equation (3) presents the transfer function of the controller where G_{PI}, K_{PI}, T_{PI}, and s stand for the controller's transfer function, gain, time constant [s], and the Laplace operator respectively:

$$G_{PI} = K_{PI} \, \frac{T_{PI}s + 1}{T_{PI}\, s} \tag{3}$$

T_{PI} has the same value as the sub-plant's time constant, while K_{PI} is obtained as presented in Equation (4), where K_{Loss} represents the intra-plant losses as presented in Section 4.2. Finally, Table 4 presents the values of the implemented PI controllers' parameters.

$$K_{PI} = \frac{T_{PI}}{K_{Loss}} \tag{4}$$

The Root locus and Bode plot of the closed loop of every plant after the addition of a PI controller were analyzed in order to study the gain range within stable operation. Figure 5 presents both diagrams for the WF's controller where G_{ol}, G_{cl}, and PI stand for open-loop, closed-loop, and controller's plants. However, a detail design for PV and BESS sub-plants as well as more details can be found in ref [8]. It can be seen how the system is stable for any gain, and the gain margin is also infinite, since the phase never crosses 180° and the phase margin corresponds to 89.1°, resulting in a stable system as was expected from the root locus. Subsequently, Figure 6 presents the step response of the WF, where the signals corresponding to the reference, the controller, and the plant behavior are shown.

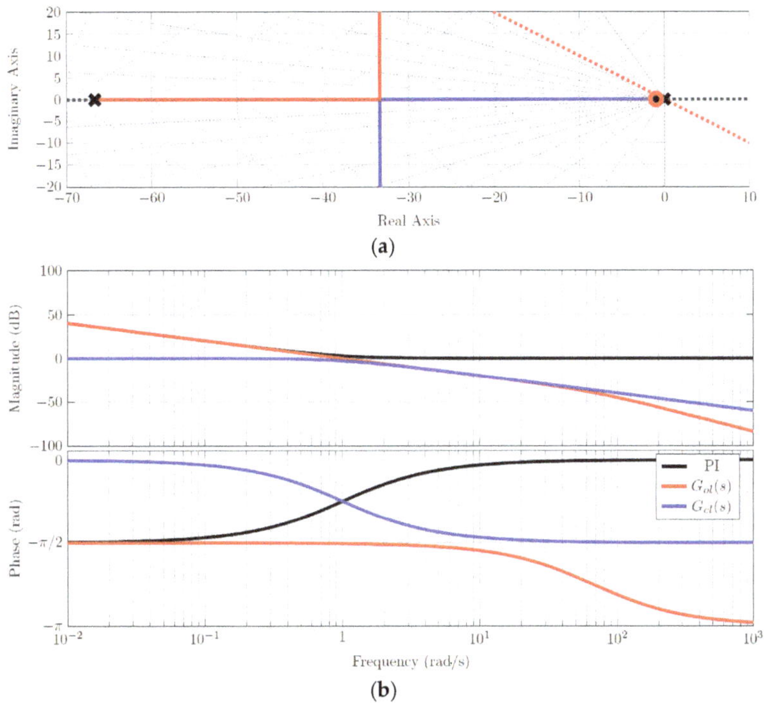

Figure 5. Diagrams for the WF's controller (**a**) Root locus and (**b**) bode plot of WF's controller.

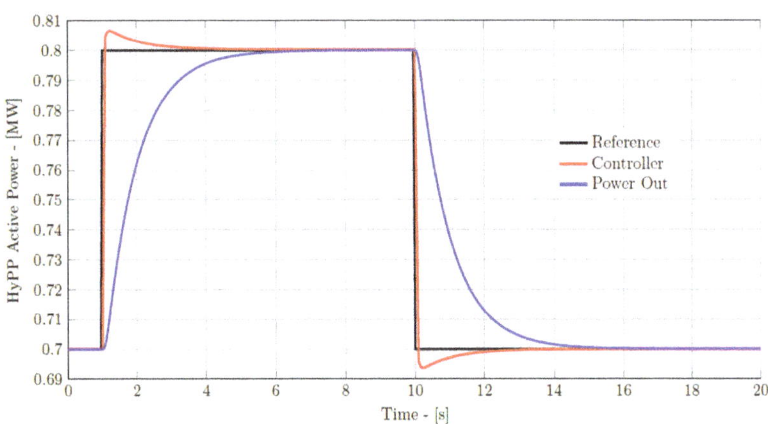

Figure 6. Step response of WF after controller inclusion.

Table 4. Implemented PI-controller parameters [8].

$K_{PI}{}^{WF}$	$T_{PI}{}^{WF}$ [s]	$K_{PI}{}^{PVP}$	$T_{PI}{}^{PVP}$ [s]	$K_{PI}{}^{BESS}$	$T_{PI}{}^{BESS}$ [s]
1.008	1	0.3106	0.3	0.005	0.005

6.3. FFR Controller

Due to the non-existence of consensus regarding the topology of FFR. However, as explained in [8], droop and derivative (df/dt) control have been used for years in automatic frequency control.

Thus, a combination of them has been implemented; being the main novelties implementing it in a HyPP, use discrete modelling and real-time testing. Both control stages act in a similar way, once the deadband is cleared they apply a constant gain, which will be positive or negative depending on the event, to the reference signal; frequency for the droop and *ROCOF* for the derivative. The composition of the FFR controller is presented in Figure 7, while a summary of the implemented parameters is shown in Table 5 [8].

Table 5. Implemented PI-controller parameters.

Parameter	Unit	Value	Notes
Deadband Droop	Hz	0.5	Reduce activity of the control
Droop Constant	%	5	Slope of the Droop control
Deadband df/dt	Hz/s	0.2	Reduce activity of the control
T_{LPF}	s	0.025	Low-pass filter's time constant
H_{HyPP}	-	5	Gain of the derivative control

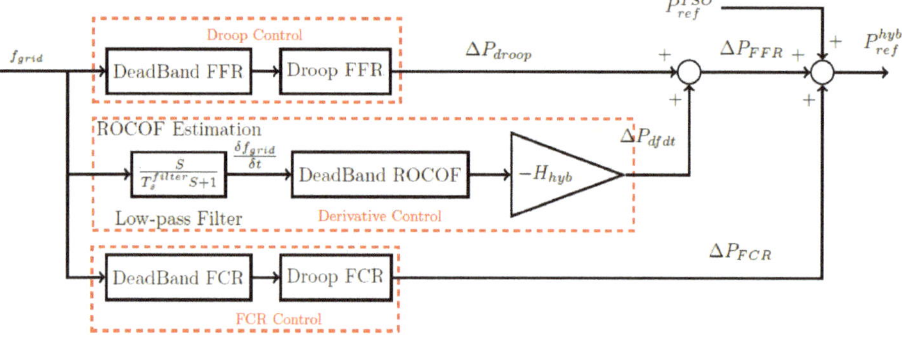

Figure 7. Block diagram of the frequency control.

6.4. FCR Controller

In Figure 7, the FCR control is also present. It should be stated that there is no strict rule regarding the appropriate droop constant to be implemented, although ref [7] recommends values between 2% and 12%, whereas it states that values between 3 and 5 are implemented in practice. Thus, during the FCR stage, a value of 5% was chosen.

7. Case Study

The selected scenarios evaluate the HyPP's participation in frequency control. As aforementioned, the considered system presents a low inertia due to its high RES penetration. The considered event is a N-1 Contingency, which is defined in ref [7] as the loss of a single generation unit or a transmission component. In this case, SG G02 is tripped, causing a sudden considerable generation loss. After the event is detected, the system reacts to it, first by stopping the frequency drop and, subsequently, bringing it back to stable values close to the nominal. These are the purposes of FFR and FCR, respectively. The system is also considered to be in a steady state with a nominal frequency during the initialization, and the demand does not change throughout the simulation. Although, due to the frequency dependency of part of the load, as explained in Section 4.1, its effective value will change according to the frequency value. Therefore, the only variations are related to the active power production of the plants, which is caused by the frequency controller and demand due to their correlation with frequency. Table 6 presents the steady-state operative point of the system, and it should be noted how the last column refers to the operational point of the plant related to its size, which can be checked in Table 3.

Table 6. Steady-state operative points.

Unit Name	Steady-State Power Output [MW]	Steady-State Share in Total Generation [%]	Steady-State Power Output [%]
HyPP	600	40.0	50.0
G01	300	20.0	40.0
G02	200	13.3	31.3
G03	100	6.7	26.0
G04	300	20.0	63.3

7.1. Scenario Definition

The selected scenarios represent the same system, event, and response strategy, but a different detection and deadband. In Scenario I, ENTSO-E grid codes are fully complied with, which means that the event is detected when the frequency drops below 49.8 Hz, and the controller starts to act once its value is 49.5 Hz. In Scenario II, the event is detected either when the frequency or *ROCOF* fall below 49.8 Hz or -0.2 Hz/s, respectively, and the controller starts acting immediately after the detection. However, in both scenarios, the rest of the system is kept unaltered. Table 7 shows the major differences between the considered scenarios.

It should be mentioned that a deadband for the frequency controllers was necessary in traditional power systems with high inertia, since the uncontrollable synchronous response will dampen and correct all the small excursions. In this way, over-actuation of the controllers was avoided. However, in modern, low inertia systems, this deadband is not useful anymore, since due to the low inertia, the response of the system will be extremely limited. That, combined with higher levels of *ROCOF*, makes it crucial to act fast. In the past, the frequency response was only activated after large excursions; however, in future scenarios with virtually no inertia, FFR has to be activated continuously in order to counteract small imbalances in generation and demand as the traditional IR does. Subsequently, in Scenario II, no deadband is set for the frequency in order to smoothly compensate with the FFR while the *ROCOF* deadband accounts for the identification of major excursions. Due to lack of scientific literature covering the topic, the value of ± 0.2 Hz/s has been obtained after studying the response of the system during normal operation and after events triggered. Values up to ± 0.1 Hz/s could be found if small load variations were inserted for the load, and thus the selected value provides a wide error margin, which is sufficient for the purposes of this research.

The objective pursued with these two scenarios is to highlight the importance of reviewing the recommendations provided by ENTSO-E in ref [7].

Table 7. Differences between Scenarios.

Difference	Scenario I	Scenario II
Event Detection	50 ± 0.2 Hz	Frequency $\notin (50 \pm 0.2$ Hz$)$ ROCOF $\notin (\pm 0.2$ Hz/s$)$
Frequency Deadband	50 ± 0.5 Hz	none
ROCOF Deadband	none	± 0.2 Hz/s

7.2. Tests Results

In Figure 8, the results of both Scenarios I and II are presented in different columns. The frequency response of the system is presented in Figure 8a,e where the Nadir is highlighted with a vertical green line. Subsequently, in Figure 8b,f, the active power demand of the system is presented and has a shape is similar to the frequency due to the dependency presented in Section 4.1. Thereafter, the generation provided by the HyPP is presented in Figure 8c,g, which is divided into each sub-plants' actuation as well as the overall result. It should be noted how negative values of production in the BESS case represent a charging process. Finally, Figure 8d,h present the actuation of the governors, in which the disconnection of Gov2 is clearly shown. Note, how in these pictures, the legend of the left axis, which corresponds to the signals from individual governors, is only present in Figure 8d while the right axis legend, which corresponds to the overall response from the governors, is only present in Figure 8h.

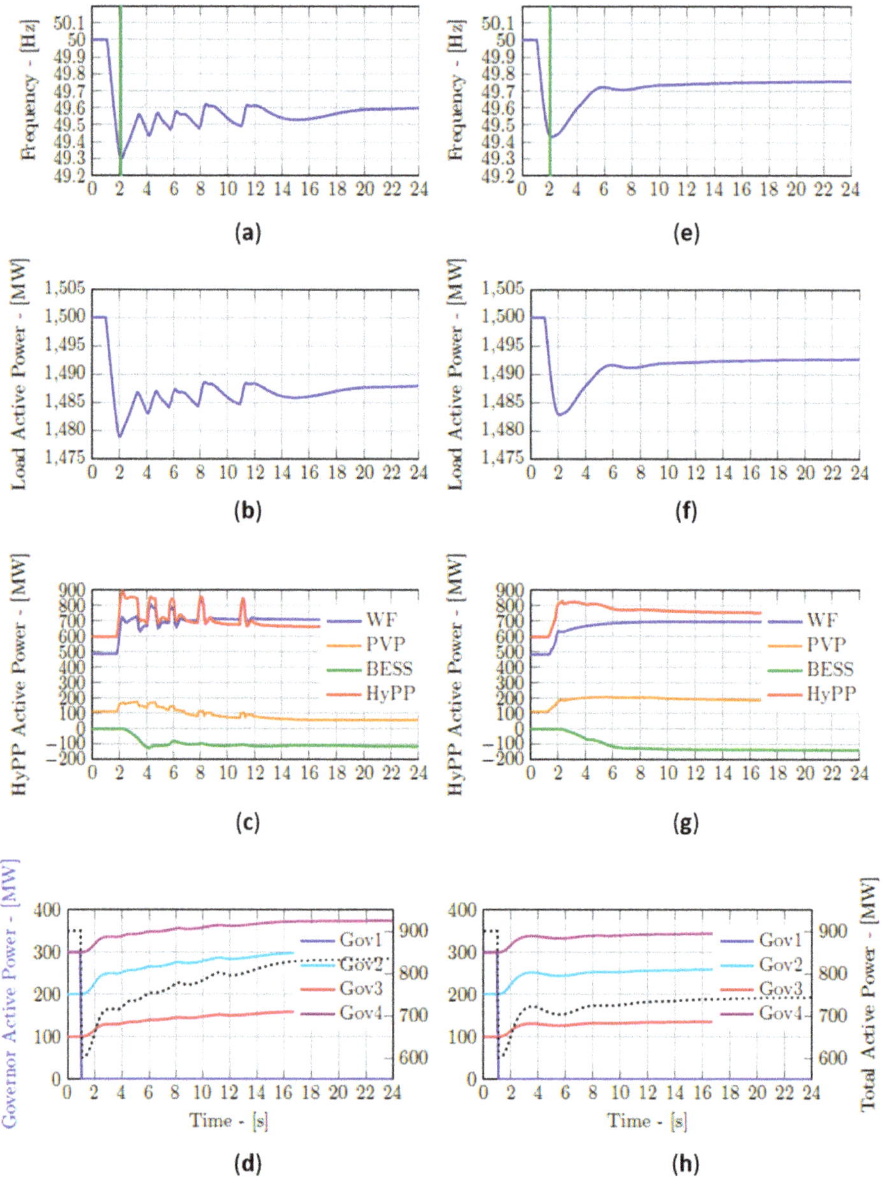

Figure 8. Results of the test scenario. Scenario I (**a–d**); Scenario II (**e–h**).

In both scenarios, the SG and HyPP react to the N-1 contingency—ones based on a natural phenomenon and others by applying a control approach after the event is identified and the deadbands surpassed. It should be stated that the event identification time delay is accounted for along with the discrete actuation of modern controllers.

7.3. Discussion

In Figure 8, is clear that the change in the reference signal and deadband level improves the behavior of the whole system's response. A comparative of the frequency behavior is presented in Table 8, where the Nadir is shown to be reached faster but at a lower frequency for Case I. This is because the HyPP starts reacting after the frequency drops below 49.5 Hz, while in Case II, it does this immediately after the detection of the event, which is, again, faster due to using *ROCOF*. Thus, the reaction can be smoother. Additionally, the steady-state value reached by the FCR reaches stability faster in Scenario II, but the frequency value is also closer to the nominal. Furthermore, when comparing the dynamic responses in Figure 8c–g, it can be seen that the stresses suffered by the HyPP are reduced in Scenario II. This is important, since wind turbine manufacturers are concerned about the mechanical stresses suffered by WF providing frequency support. The PVP is also almost completely curtailed in Scenario I, while it increases its production in the second scenario, making the system more efficient. Finally, in Figure 8d–h, the response of the governors is also smoothed out and reduced. Finally, Figure 9 has been included in order to highlight the most important differences between both scenarios. Figure 9a presents how, in Scenario II, the frequency response is not only smoother but also recovers values closer to the nominal after the event, while in Figure 9b, the active power response in the PCC highlights how the power is also injected in a smoother manner into the power system, which improves the overall response of the grid and thus avoiding over-oscillations.

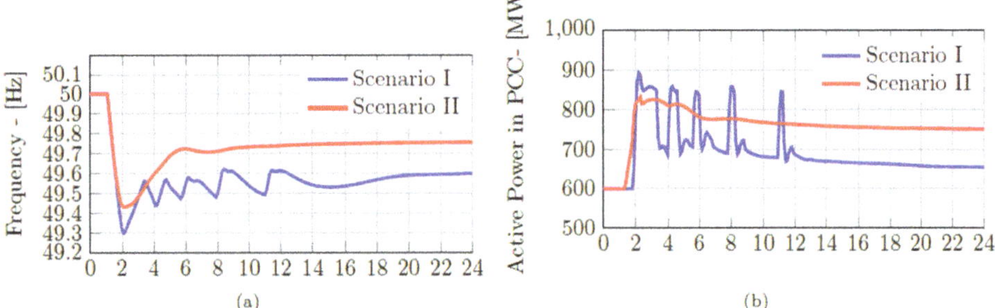

Figure 9. Results of the test scenario. (**a**) Frequency; (**b**) active power at the point of common coupling (PCC).

Table 8. Overview of the simulation results.

Parameter	Case I	Case II
Nadir frequency [Hz]	49.29	49.43
Nadir time [s]	1.1	1.2
Frequency steady-state [Hz]	49.61	49.76
Time to reach steady-state [s]	27	20

8. Sensitivity Analysis

Since all the work has been developed as a simulation, it is important to assess the influence of modifying different parameters on their behavior. In Table 9, a summary of the parameters subjected to the sensitivity analysis can be found. It is worth mentioning how all the studies have been conducted on top of Scenario II.

Table 9. Parameter summary of the sensitivity analysis.

Parameter	Values		Notes
	Original Scenarios	**Tested**	
H	3 s	2 s 5 s	The inertia constant of the grid depends of the size of the synchronous units connected to it
FCR delay	3 s	1 s 5 s	This value corresponds to the time delay in the activation of the FCR stage after reaching the Nadir. Ref [7] specifies that it should start in less than 3 s.

The results of this analysis are presented in Figure 10, where it can be seen how reducing and increasing the delay for FCR actuation improves and worsens the system's response, respectively. Similar behavior also occurs after the modification of H, where larger parameters imply a higher stability of the grid, and thus higher stiffness.

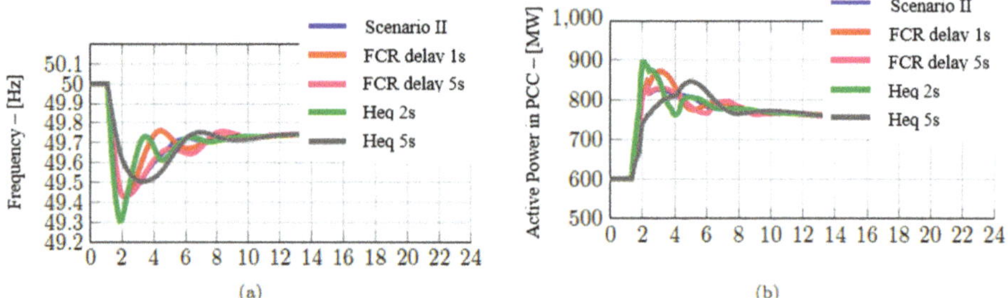

Figure 10. Results of the test scenario. (**a**) Frequency; (**b**) active power at the PCC.

9. Conclusions

In this paper, the importance of frequency provision has been addressed along with the causes that lead to inertia loss in worldwide grids. Subsequently, the relevant background related to frequency behavior has been introduced and associated to current regulations and recommendations from ENTSO-E, with focus on ongoing and questionable components. Since those regulations are still in the development stage, it is a matter of utter importance to assess whether they are targeting appropriate objectives with the correct methods. This is exactly the aim of this research—to evaluate if there are open paths for improving the frequency behavior, taking into account possibilities that so far appear to be dismissed by regulating agencies.

In Section 4, the system modeling is briefly presented, since most of it is extensively described in ref [8]. Thereafter, Section 5 presented the proposed architecture of the system, while the design of the different control stages was covered in Section 6. Then, in Section 7, two different scenarios were defined, both studying the same event—an N-1 contingency—but with a different event detection approach. Scenario I implements a frequency monitoring detection technique, as ENTSO-E recommends, with a considerably large deadband for the controller, while in Scenario II, *ROCOF* is the signal triggering the activation of the frequency controller. Consequently, the results show how the frequency behavior of the system is greatly improved. Even though the Nadir is reached later in the proposed method, the frequency reached is 8.5% higher, and the steady state error improved by 61.53 % in Scenario II. Additionally, the dynamic response of the generators in the system is smoother in Scenario II, satisfying one of the greatest concerns for wind turbine manufacturers: mechanical stresses and premature aging due to frequency support provisions. Finally, Section 8 presented a brief sensitivity analysis on top of Scenario II, pointing out the effects of modifying model parameters such as H.

The impact of the proposed architecture is the speed in event identification. The sooner an event is detected as a fault, the sooner the plant will react. On the other hand, some oscillations are always present in any system working within normal operation, which makes the objective of the existent deadbands to reduce over-actuation. However, the value of the nominal frequency has traditionally been used as a deadband. This paper proposes the use of *ROCOF* in event identification, since it allows the identification of fast excursions (like the ones appearing after an N-1 contingency) before the frequency can reduce its value and thus easing the system necessary reactions. Finally, it is worth mentioning that since the model has been developed for RT-HIL studies, future publications will cover the testing of this model in such frameworks, including industrial controllers for HyPPs.

Author Contributions: Conceptualization, D.V.P.; investigation, D.V.P.; supervision, F.I. and D.-I.S.; validation, D.V.P.; writing—original draft, D.V.P.; writing—review and editing, D.V.P.

Funding: This research received no external funding.

Acknowledgments: The authors would like to thank the Department of Energy Technology at Aalborg University for the support in the development of this paper.

Conflicts of Interest: The authors declare no conflict of interest.

References

1. Ren21 Renewables 2017 Global Status Report. Available online: http://www.ren21.net/wp-content/uploads/2017/06/17-8399_GSR_2017_Full_Report_0621_Opt.pdf (accessed on 6 February 2018).
2. Delbeke, J.; Vis, P. EU Climate Policy Explained. Available online: https://ec.europa.eu/clima/sites/clima/files/eu_climate_policy_explained_en.pdf (accessed on 8 February 2018).
3. Plessmann, G.; Blechinger, P. How to meet EU GHG emission reduction targets? A model based decarbonization pathway for Europe's electricity supply system until 2050. *Energy Strategy Rev.* **2017**, *15*, 19–32. [CrossRef]
4. Energinet Environmental Report 2017. Available online: https://en.energinet.dk/About-our-reports/Reports/Environmental-Report-2017 (accessed on 7 February 2018).
5. National Grid. *System Operability Framework 2016*; National Grid: London, UK, 2016.
6. ENTSO-E. *High Penetration of Power Electronic Interfaced Power Sources*; Tech Report; ENTSO-E: Brussels, Belgium, 2017.
7. ENTSO-E. *Supporting Document for the Network Code on Load-Frequency Control and Reserves*; ENTSO-E: Brussels, Belgium, 2013.
8. Vázquez Pombo, D. *Coordinated Frequency and Active Power Control of Hybrid Power Plants—An Approach to Fast Frequency Response*; Aalborg University: Aalborg, Denmark, 2018.
9. Regelleistung, Internetplattform zur Vergabe von Regelleistung. 50 Hertz, Amprion, Tennet, Transnet BW. Available online: https://www.regelleistung.net/ext/ (accessed on 26 February 2019).
10. Technical Regulation 3.2.5 for Wind Power Plants with a Power Output above 11 kW, Signal List. Available online: https://en.energinet.dk/Electricity/Rules-and-Regulations/Regulations-for-grid-connection (accessed on 26 February 2019).
11. Chabot, B. Onshore and Offshore Wind Power Capacity Factors: How Much They Differ Now and in the Future. 2013. Available online: https://www.erneuerbareenergien.de/files/smfiledata/3/1/7/2/7/1/V2BC37NhCFWindDK.pdf (accessed on 4 March 2019).
12. Website, S.G. Hybrid Power Solutions. 2018. Available online: https://goo.gl/gZKNBZ (accessed on 4 March 2019).
13. Energy, G.E.R. Hybrid Power. 2018. Available online: https://www.gerenewableenergy.com/hybrid (accessed on 4 March 2019).
14. Liu, Y.; You, S.; Liu, Y. Study of Wind and PV Frequency Control in US Power Grids—EI and TI Case Studies. *IEEE Power Energy Technol. Syst. J.* **2017**, *4*, 65–73. [CrossRef]
15. Hoke, A.F.; Shirazi, M.; Chakraborty, S. Rapid active power control of photovoltaic systems for grid frequency support. *IEEE J. Emerg. Sel. Top. Power Electron.* **2017**, *5*, 1154–1163. [CrossRef]
16. Thatte, A.A.; Zhang, F.; Xie, L. Coordination of wind farms and flywheels for energy balancing and frequency regulation. In Proceedings of the Power and Energy Society General Meeting, Detroit, MI, USA, 24–29 July 2011; pp. 1–7.

17. Wu, Z.; Gao, D. Coordinated Control Strategy of Battery Energy Storage System and PMSG-WTG to Enhance System Frequency Regulation Capability. *IEEE Trans. Sustain. Energy* **2017**, *8*, 1330–1343. [CrossRef]

18. Petersen, L.; Hesselbæk, B.; Martinez, A.; Borsotti-Andruszkiewicz, R.M.; Tarnowski, G.C.; Steggel, N.; Osmond, D. Vestas Power Plant Solutions Integrating Wind, Solar PV and Energy Storage. In Proceedings of the 3rd International Hybrid Power Systems Workshop, Energynautics, Tenerife, Spain, 8–9 May 2018.

19. Nielsen, T.; McMullin, D.; Lenz, B.; Gamboa, D. Towards 100% Renewables in the Faroe Islands: Wind and Energy Storage Integration. In Proceedings of the 3rd International Hybrid Power Systems Workshop, Energynautics, Tenerife, Spain, 8–9 May 2018.

20. *Energi Danmark A/S, Vestas Wind Systems A/S, BESS Project Smart Grid Ready Battery Energy Storage System for future grid Final Report ForskEL project No. 10739*; Danish Technological Institute: Taastrup, Denmark, 2017.

21. Rezkalla, M.; Pertl, M.; Marinelli, M. Electric power system inertia: Requirements, challenges and solutions. *Electr. Eng.* **2018**, *100*, 2677–2693. [CrossRef]

22. Anderson, P.M.; Mirheydar, M. A low-order system frequency response model. *IEEE Trans. Power Syst.* **1990**, *5*, 720–729. [CrossRef]

23. ENTSO-E European Power System 2040–Completing the Map–Technical Appendix. Available online: https://docstore.entsoe.eu/Documents/TYNDPdocuments/TYNDP2018/System_NeedReport.pdf (accessed on 4 March 2019).

24. Policy 1: Load-Frequency Control and Performance. Available online: https://www.entsoe.eu/fileadmin/user_upload/_library/publications/entsoe/Operation_Handbook/Policy_1_final.pdf (accessed on 16 February 2018).

25. Adamczyk, A.; Altin, M.; Goksu, O.; Teodorescu, R.; Iov, F. Generic 12-bus test system for wind power integration studies. In Proceedings of the 2013 15th European Conference on Power Electronics and Applications, Lille, France, 2–6 September 2013; pp. 1–6.

26. Machowski, J.; Bialek, J.W.; Bumby, J.R. *Power System Dynamics: Stability and Control*; John Wiley & Sons: Hoboken, NJ, USA, 2008.

27. IEEE Task Force on Turbine-Governor Modeling. *Dynamic Models for Turbine-Governors in Power System Studies*; IEEE: Piscataway, NJ, USA, 2013.

28. Anderson, P.M.; Fouad, A.A. *Power System Control and Stability*, 2nd ed.; Wiley-IEEE Press: Hoboken, NJ, USA, 2003.

29. Kundur, P.; Balu, N.J.; Lauby, M.G. *Power System Stability and Control*; McGraw-Hill: New York, NY, USA, 1994.

30. Comanescu, M. Influence of the discretization method on the integration accuracy of observers with continuous feedback. In Proceedings of the 2011 IEEE International Symposium on Industrial Electronics, Gdansk, Poland, 27–30 June 2011; pp. 625–630. [CrossRef]

Article

A Recursive Least Squares Method with Double-Parameter for Online Estimation of Electric Meter Errors

Xiangyu Kong [1],*, Yuying Ma [1], Xin Zhao [1], Ye Li [2] and Yongxing Teng [2]

[1] Key Laboratory of Smart Grid of Ministry of Education, Tianjin University, Tianjin 300072, China; shinyying@tju.edu.cn (Y.M.); 18832356846@sina.cn (X.Z.)

[2] Tianjin Electric Power Research Institute, State Grid Tianjin Electric Power Company, Tianjin 300384, China; ibly@163.com (Y.L.); yuxi092@126.com (Y.T.)

* Correspondence: eekongxy@tju.edu.cn; Tel.: +86-022-27405477

Received: 30 January 2019; Accepted: 25 February 2019; Published: 28 February 2019

Abstract: In view of the existing verification methods of electric meters, there are problems such as high maintenance cost, poor accuracy, and difficulty in full coverage, etc. Starting from the perspective of analyzing the large-scale measured data collected by user-side electric meters, an online estimation method for the operating error of electric meters was proposed, which uses the recursive least squares (RLS) and introduces a double-parameter method with dynamic forgetting factors λ_a and λ_b to track the meter parameters changes in real time. Firstly, the obtained measured data are preprocessed, and the abnormal data such as null data and light load data are eliminated by an appropriate clustering method, so as to screen out the measured data of the similar operational states of each user. Then equations relating the head electric meter in the substation and each users' electric meter and line loss based on the law of conservation of electric energy are established. Afterwards, the recursive least squares algorithm with double-parameter is used to estimate the parameters of line loss and the electric meter error. Finally, the effects of double dynamic forgetting factors, double constant forgetting factors and single forgetting factor on the accuracy of estimated error of electric meter are discussed. Through the program-controlled load simulation system, the proposed method is verified with higher accuracy and practicality.

Keywords: electric meter; error estimation; line loss; RLS; double forgetting factors

---

1. Introduction

With the construction and development of smart grids, the power industry has entered an era of big data. Electric meter is an important part of acquiring big data which have received wide attention. According to reports [1], before and after the World Metrology Day on May 20, 2018, the State Grid Corporation of China has installed more than 457 million electric meters, covering the 99.57% of the user service area. Facing the huge amounts of electric meters with complex application sites, how to improve electric meters' self-diagnosis and verification, and improve their status evaluation and analysis capabilities, has become the focus of power grid companies. At present, the main way for power companies to verify the accuracy of electric meters is to use professionals that regularly carry instruments and equipment to the site for periodic sampling inspection [2,3]. The existing calibration mode has some disadvantages such as high working intensity and long calibration cycle and low managed efficiency, so it is difficult to meet the requirements of the maintenance and replacement of electric meter status. The measured results are directly related to grid security and whether the trade settlement between the two parties is fair and reasonable. However, the global power industry has not found yet a practical theory and technology that can accurately measure and monitor the operating errors of electric meters in real time. Therefore, in order to realize the change of the electric meter from the periodic replacement to the state replacement and to judge the operational error's status of electric meter, it is imperative to find an efficient and accurate method for estimating online the operational errors of electric meters.

With the development of smart grids and especially the popularization of Advanced Metering Infrastructure (AMI), power companies have acquired large-scale measured data. In recent years, some achievements have been made in the application of large-scale data measured from electric meters. In the research on anti-theft methods, the measured data of electric meters are used in gray correlation analysis [4,5], support vector machine and local anomaly factor algorithm [6] and estimated of loss in distribution line [7]. Through these methods, the identification of the substation and phase sequence, as well as the detection of the power stealing of the user substation, can effectively realize the location of abnormal power users. In addition, researchers have used measured data of electric meters to build a power consumption prediction model based on artificial neural networks, aiming at achieving accurate demand for ecasting in the smart grid system as well as acquiring power consumption profiles for demand response purposes [8]. With the use of smart meters in smart cities, a data-driven probabilistic peak demand estimation framework using fine-grained smart meter data and sociodemographic data of the consumers was proposed in [9], which drives fundamental electricity consumption across different categories.

In the field of analysis of electric meter operational errors, there are relatively few studies in academic and industrial fields. In [10], Korhonen used automatic meter reading data to deduce a calculation method for the operating errors of electric meters. The applicability of the method depends on the user's power-consumption level, the number of user's meters and other factors and the method will not be able to accurately estimate the errors of an electric meter if the loss between the head meter and the users' meters is large. Guo [11] proposed a method to solve the errors of electric meter by the generalized law of energy conservation, but the method of decomposing the reading matrix into the upper and lower triangular matrix is verbose and matrix is prone to failure. Due to factors such as data size and quality, data in some unit measurement periods can't satisfy the requirements of independence and orthogonality, and the method lacks real-time performance. In [12] researchers described a method of online smart meter error calibration using meter reading data. The approach of data analysis using sum meter reading and branch meter reading in a tree topology grid was studied. An algorithm based on the combination of K-means clustering and regularization theory is proposed to evaluate smart meter errors precisely. The proposed method has a better solution, but there may still be small solution errors caused by random factors such as the inaccurate estimation of energy losses. A mathematical model of On-Off-Key (OOK) dynamic load current was built in [13], a mathematical model of dynamic load energy sequences is proposed and three dynamic load power modes are defined: transient, short-term and long-term, based on which, an algorithm for measuring the dynamic errors of electric meter was proposed. Results indicate that the dynamic errors of electric meters are closely related to the dynamic load power mode of driving and the characteristics of dynamic errors are quite different among different kinds of electric meter. In [14], an error verification device for harmonic electric meters was proposed, that can output different amplitudes, different phases, different frequencies of the voltage and current through a power source, then transform the data detected by the measured harmonic meters and from a reference standard error calculator, it can verify the metering errors of harmonic meters quickly. In [15], an estimation method for electric meter errors based on a parameter degradation model was proposed. The comprehensive influence of various error factors such as temperature, humidity and load under actual working conditions are considered, but the error parameters of electric meter can only be estimated in a short time and real-time estimated tracking can't be achieved. In [16], the adaptive variable weight method is introduced in the fuzzy analytic hierarchy process and the state evaluation model of the meter is established to solve the problem that the weight of the index remains unchanged, which ensures the influence of each index on the measurement error of the electric meter is dynamically reflected. Aiming at solving the electric meter failure prediction problem and based on historical failure data of electric meters in some regions, a smart meter fault identification model is proposed based on a C5.0 algorithm in [17] which shows the accuracy of the failure prediction model for smart meters is higher, achieving good prediction effects. Reference [18] adopts a stratified sampling method to sample the electric meters in typical areas with high humidity and heat and severe cold that have been running for one year and a sampling life tolerance model is established. Considering the effect of temperature and humidity in errors, based on the bathtub curve, a Weibull lifetime distribution model is established. In addition, some related electric power workers have analyzed the factors affecting the accuracy of electric meter measurement from the perspectives such as communication transmission harmonic characteristics, ambient temperature, operating load size, and data sampling method [19–22].

In summary, although some research results have been obtained in the application of electric meter measured data and online estimation of operational errors, there are still some shortcomings in estimating accuracy and real-time performance. Based on the existing research, an online estimation method for solving the operational error of electric meters by using the recursive least squares algorithm with double-parameter is proposed. First, the measured data in similar operational states of each power user is selected. Then the recursive least squares algorithm with double-parameter is applied to calculate the operational error parameter of the electric meter and the line loss parameter of the low-voltage substation simultaneously. As explained throughout this paper, the key contributions of this paper can be summarized as follows:

(1) Compared with the existing methods, the proposed method realizes online estimation of line loss parameters and electric meter error parameters simultaneously by using the double-parameter recursive least squares algorithm with double dynamic for getting factors.

(2) In addition, due to the introduction of double dynamic forgetting factors in the recursive least-squares algorithm, the flexible correction ability of the new data to the double-parameter estimation is ensured. The forgetting factors are adjusted according to the frequency of measured data collection system, which enhances the real-time performance and the parameter changes can be better tracked, so a large number of electric meters can be checked online.

The remainder of this paper is organized as follows: Section 2 introduces the overall estimation framework of the solution of the operational error and the method of processing the data that will be useful for obtaining the measured data to satisfy the requirements of the estimation model. Section 3 constructs the theoretical model for estimating the operational error of electric meters, and the error parameter is calculated based on the double-parameter recursive least squares algorithm. Section 4 presents the evaluation metrics and the methods to be compared in the case studies. Finally, the conclusions are presented in Section 5.

2. Acquisition and Processing of Information of Electric Meters

2.1. Implementing Scheme for Online Estimation of Electric Meter Error

The online estimated method of electric meter operational error mainly includes four steps: firstly, information, which includes profile information from power marketing information systems, (this system is composed of people, computers and computer programs, for power market decision makers to collect, select, analyze, evaluate and distribute the marketing information timely and accurately) and measured data from electric meters (user-side and substation) should be acquired. Then the acquired data is preprocessed by clustering to get rid of the abnormal measured data such as null data and light load data. Next, an online estimation model for the operational errors of electric meters is established and the electric meter errors based on the proposed double-parameter RLS algorithm are calculated. Finally, we simulate real data of electric meters by using a program-controlled load simulation system (the whole simulation system can simulate the power consumption of real substations, and support the adjustment of different power consumption conditions, different line losses and different error conditions) and using Mean Absolute Percent Error (MAPE) and Root Mean

Square Error (RMSE) as the evaluation metrics for checking the accuracy of the estimated error result of the electric meter. The whole process is shown in Figure 1.

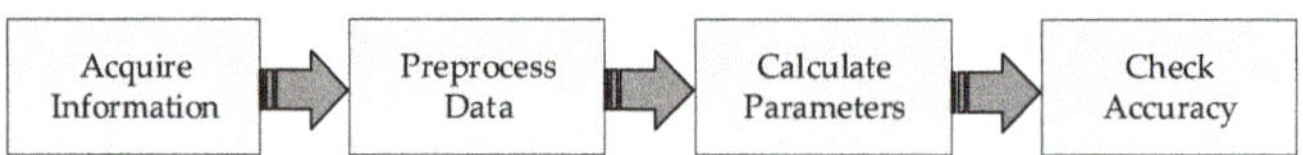

Figure 1. Online estimation process of electric meter errors.

2.2. Required Electric Meter Information

The information needed for online estimation of electric meter errors is as follows:

(1) The profile information contains five types:

- User's information such as user number, name, address, category of power consumption, etc.
- Electric meter's information such as electric meter's number, name, type, address, current transformer (CT)/phase voltage transformer (PT) ratio, etc.
- Metering point record information of the user information collection system: metering point identification, number, name, address, classification, nature, main purpose type, metering point side, voltage level, etc.
- Electrical parameters of the head meter in the low-voltage substation, such as voltage, average power factor, active energies, reactive energies, etc.
- Main electrical characteristic parameters related to the load, such as load rate, operating load type, quality and proportion of the power consumption, terminal voltage of users.

(2) Measured data of electric meters which is collected by electric meters data acquisition system based on AMI.

Though system schemes may be slightly different for different regions, a typical scheme is shown in Figure 2. The distribution transformer station as a unit forms the acquisition system, and the concentrator is installed under the common distribution transformer, which realizes the power consumption information collection of all users under the substation with a communication system, such as power line carrier (PLC), RS485 or micro-power wireless. At the same time, it gathers calibrated meter data of the distribution transformer to realize the collection of the distribution transformer. According to the overall design scheme of the state grid corporation in China, the users' collection structure of electricity information is a small-scale centralized collection, with a unified upload to the master station [23].

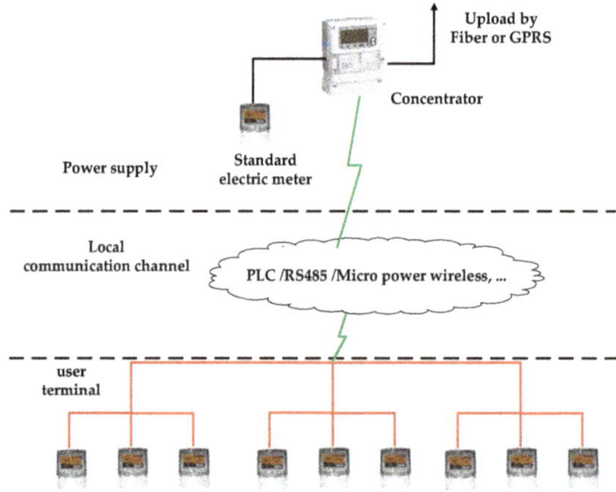

Figure 2. Electric meter data acquisition physical architecture based on AMI.

2.3. Preprocessing Data Measured by Electric Meter

The operational error of the electric meter generally refers to the relative error between the measured value and the actual value collected during the operation, which can be deduced by the following formula[24]:

$$\zeta = \frac{z_{\text{real}} - z}{z},\tag{1}$$

where ζ is the measurement error of the electric meter; z_{real} is the actual electric energy value in the unit measured period, which can be understood as the actual electricity consumption (true value) through the electric meter; z is the increment (view value) of the electric meter reading during the unit measured period.

In order to meet the requirements of the error estimation model established and method proposed, which will be elaborated in Section 3, it is important to ensure the operational error parameters of the electric meter and line loss all be close to a certain value during the unit measurement period.

However, in reality, the difference of the load level in the user terminal will change the current on the distribution line, the voltage at the delivery end of the substation side, and the voltage on the user side of the power supply, which causes the line loss rate of the distribution area not being constant and changing as the users' load levels fluctuate.

In addition, according to the working principle of the electric meter, on the one hand, the current and voltage fluctuation of line load will affect current and voltage of the sampling circuit of electric meters and the operation of the computing chip, such as the value of the load current and voltage cause the sampling circuit power consumption and heating changes and the harmonics of the load current and voltage affect the frequency characteristics of the sampling circuit [25]. These factors discussed above will affect the measured value of the electric meter. From this we can conclude that the errors generated by the electric meter in the work are not constant values, and the operational error of the electric meter becomes larger as the power factor decreases and increases as the relative value of the voltage and current amplitude changes. On the other hand, it is called light load when the operational load current is below 5%~10% of the rated current. The light load affects the creep performance of the electric meter and the working state of the current transformer, so that the accuracy of the electric meter is greatly affected and the measured data error is higher and it is necessary to remove the data measured under light load conditions.

Therefore, in order to reduce the influence of the fluctuation of the load rate on the measurement error of the electric meter and line loss rate, it is necessary to ensure that the meter error and the line loss rate in the unit measured period have a certain value. It is also indispensable to cluster the collected electric meter measured data for the purpose of obtaining the measured value of the electric meter in a similar operating state.

Thence, preprocessing the measured data collected by the electric meter combined with the improved fuzzy clustering algorithm [26] in the process of online estimation of operational error, in which the abnormal measured data such as null data and light load data are rejected. The clustering process for selecting the measured data in the similar operational states of each power user is shown in Figure 3.

We sort the preprocessed data in the measured chronological order to form a head meter and every user meters' matrix of the measured data, which are used as input variable for recursive least squares with double-parameter method.

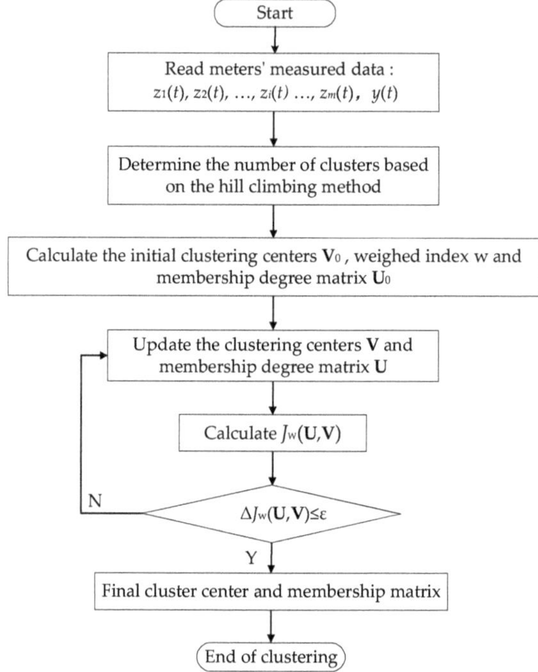

Figure 3. Flow chart of our improved fuzzy clustering algorithm clustering for measured data.

3. The Solution Method of Electric Meter Operational Error

3.1. Establishing an Estimated Model of Operational Error

The typical power distribution topology is shown in Figure 4. A head electric meter is installed under a concentrator which is connected to a common distribution transformer. In the situation shown in the figure, the number of user meters is m. We consider the fact that the head electric meter under each distribution transformer has been calibrated, so the head electric meter is assumed standard. In addition, there is line loss in the line topology of the power distribution area. Based on the law of conservation of energy, the reading of the head electric meter in the low-voltage substation is equal to the sum of the true values of the electric meters of each user plus any line loss of the power distribution area during the unit measured period (line loss refers to the loss of electric energy in the form of thermal energy in the process of transmission, substation, distribution and marketing from power plant to power user, generally referred to as active loss. The term w_{loss} mentioned in (2) refers to the entire process of power loss from the main transformer to the user electric meter). For any unit measured period, the electric meters reading in the station have the following relationship:

$$y_0(t) = \sum_{i=1}^{m} y_i(t) + w_{\mathrm{loss}}(t), \tag{2}$$

where $y_0(t)$ is the power supplied by the head standard meter during the tth measured period; $w_{\mathrm{loss}}(t)$ is power loss between the head meter and each users' meter during the tth measured period; m is total number of users' electric meters in the substation; $y_i(t)$ is the real power consumption of the ith user during the tth measured period, according to Equation (1), using $y_i(t)$ instead of z_{real}, the relationship between actual electricity consumption (true value) through the electric meters and the increment (view value) of the electric meter reading during the *t*th measured period as follows:

$$y_i(t) = z_i(t)(1 + \zeta_i(t)), \tag{3}$$

where, $z_i(t)$ is power consumption of the *i*th user measured by electric meter during the *t*th measured period; $\zeta_i(t)$ is measurement error of the *i*th electric meter.

The specific calculation process of $w_{\text{loss}}(t)$ is as follows [27]:

$$w_{loss}(t) = \frac{E(t) \cdot \Delta U(t)}{100} \times K_p(t), \tag{4}$$

where $E(\text{MW·h})$ is the active power of the head electric meter in the low-voltage substation; ΔU is the loss rate of voltage in low voltage lines (the voltage loss is the difference between the amplitude of the voltage at the beginning and the ending, and its value is approximately equal to the vertical component of the voltage drop when the voltage phase difference between the two ends is very small); K_p is the ratio of percentage of power loss to percentage of voltage loss which is related to the load power factor and the selected phase angle difference γ between current and voltage of the head electric meter.

ΔU can be defined as follows:

$$\Delta U(t) = \frac{U_1(t) - U_2(t)}{U_1(t)} \times 100\%, \tag{5}$$

where U_1 (kV) is the outlet voltage of the distribution side, usually taking the average value of three-phase electricity; U_2 (kV) is the lowest point voltage on the user-side, if the low voltage is a single phase load, several low voltage averages must be measured.

K_p can be written as follows:

$$K_p(t) = \frac{1 + (tg\gamma(t))^2}{1 + (\frac{X}{R}(t))tg\gamma(t)}, \tag{6}$$

where X/R is the ratio of wire reactance to resistance; γ is the phase angle difference between current and voltage of the head electric meter, which is the power factor angle and $tg\gamma(t)$ can be written as follows:

$$tg\gamma(t) = \frac{Q(t)}{E(t)}, \tag{7}$$

where, Q (Mvar·h) is reactive power of the head electric meter in the substation.

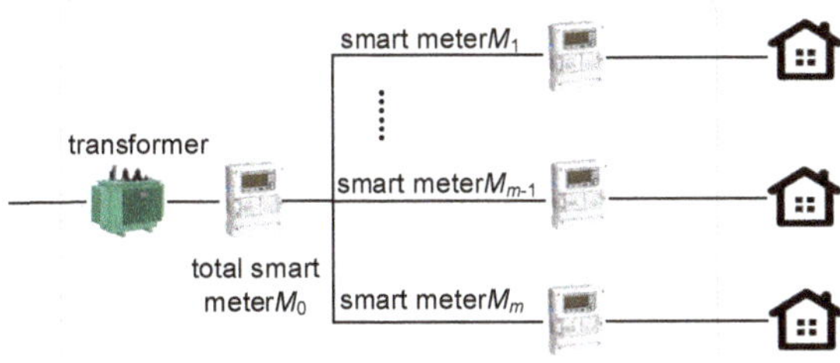

Figure 4. Topology of typical power distribution area.

3.2. Parameter Estimation of Meter Error and Line Loss

According to Equations (2), (4) and (6), using $\varphi(t)$ and $\hat{\theta}_b(t)$ replace $\frac{E(t)\cdot\Delta U(t)}{100}$ and $K_p(t)$ respectively, we obtain the recursive least squares with double-parameter equation as follows:

$$y_0(t) = \mathbf{Z}^{\mathrm{T}}(t)\hat{\mathbf{\Theta}}_a(t) + \varphi(t)\hat{\theta}_b(t),\tag{8}$$

where, $\mathbf{Z}^{\mathrm{T}}(t) = [z_1(t), z_2(t), \cdots, z_m(t)]$ is the matrix of measured data of each user meter during the tth measured period; $\hat{\mathbf{\Theta}}_a(t) = \left[\hat{\theta}_{a1}(t), \hat{\theta}_{a2}(t), \cdots, \hat{\theta}_{am}(t)\right]^T$ is the matrix of error parameter to be estimated in each user's meter, and define the electric meter operational error $\hat{\zeta}_i(t)$ as follows:

$$\hat{\zeta}_i(t) = \hat{\theta}_{ai}(t) - 1,\tag{9}$$

Through direct estimation of unknown parameters $\hat{\mathbf{\Theta}}_a(t) = \left[\hat{\theta}_{a1}(t), \hat{\theta}_{a2}(t), \cdots, \hat{\theta}_{am}(t)\right]^T$ and $\hat{\theta}_b(t)$ by recursive least squares with double-parameter algorithm. The electric meter operational error $\hat{\zeta}_i(t)$ and the ratio of wire reactance to resistance X/R can be indirectly obtained.

3.3. A Recursive Least Square Scheme with Double-Parameter

When researching the measurement error of the electric meter and the loss of electric energy of the distribution line in the low-voltage distribution area, two limitations in the RLS algorithm based on a single forgetting factor are noticed [10]:

(1)   The electric meter error parameter changes with the line loss parameter at the same rate.
(2)   In the formulation of the loss-function defined in the RLS algorithm of a single forgetting factor and the resulting recursive scheme in the subsequent formula, the error due to all parameters is classified as a single scalar term.

Therefore, the algorithm has no way to realize whether the error is caused by one or more parameters. As a result, if there is drift in one of the parameters, then all parameters that cause the estimated overshoot or undershoot will be corrected in the same order. If the drift continues for a while, it may cause the overall performance of the estimate to deteriorate and may even cause the so-called estimate to be tightened or amplified, so the goal is to conceptually 'separate' the errors caused by each parameter and then apply a suitable forgetting factor for each parameter.

Therefore, an estimated method is proposed which is based on a double forgetting factor. The recursive least squares with double parameter which can not only use the real-time measured information to modify the estimated result repeatedly but also can adapt to the situation where the different parameters change speed in the multi-parameter estimated is different. The method can simultaneously estimate the operational error of the Electric meter and line loss. The electric meter operational error estimated model established above shows that there are two unknown parameters $\hat{\mathbf{\Theta}}_a(t)$ and $\hat{\theta}_b(t)$ need to be estimated, so two forgetting factors λ_a and λ_b are introduced. And the residual cost function defining this estimated model is as follows:

$$J(\hat{\mathbf{\Theta}}_a(t), \hat{\theta}_b(t), t) = \frac{1}{2}\sum_{j=1}^{t}\lambda_a^{t-j}\left(y_0(j) - \mathbf{Z}^{\mathrm{T}}(j)\mathbf{\Theta}_a(t) - \varphi(j)\theta_b(j)\right)^2 +$$
$$\frac{1}{2}\sum_{j=1}^{t}\lambda_b^{t-j}\left(y_0(j) - \mathbf{Z}^{\mathrm{T}}(j)\mathbf{\Theta}_a(j) - \varphi(j)\hat{\theta}_b(t)\right)^2\tag{10}$$

With this definition for the residual cost function, the first term on the right side of equation (10) only represents the error of the step t due to first parameter estimate, $\hat{\mathbf{\Theta}}_a(t)$ and the second term corresponds to the second parameter estimate, $\hat{\theta}_b(t)$. Now, each of these errors can be discounted by an exclusive forgetting factor. Notice that $\mathbf{\Theta}_a(j)$ and $\theta_b(j)$ are unknown, and we will later replace them with their estimates $\hat{\Theta}_a(t)$ and $\hat{\theta}_b(t)$. The swapping between the estimated and the actual parameters allows us to formulate the proposed modification to the classical Least Squares (LS) method with for getting factors.

Here, λ_a and λ_b are forgetting factors for the first and second parameters, respectively. Incorporating multiple forgetting factors provides more degrees of freedom for tuning the estimator and, as a result, parameters with different rates of variation could possibly be tracked more accurately. The optimal estimates are those that minimize the loss function and are obtained as follows [28]:

$$\frac{\partial J}{\partial \hat{\Theta}_a(t)} = 0 \Rightarrow \sum_{j=1}^{t} \lambda_a^{t-j}\left(-Z^T(j)\right)\left(y_0(j) - Z^T(j)\hat{\Theta}_a(t) - \varphi(j)\theta_b(j)\right) = 0, \tag{11}$$

Rearranging Equation (11), $\hat{\Theta}_a(t)$ is found to be:

$$\hat{\Theta}_a(t) = \left(\sum_{j=1}^{t} \lambda_a^{t-j}Z^T(j)Z(j)\right)^{-1}\left(\sum_{j=1}^{t} \lambda_a^{t-j}(y_0(j) - \varphi(j)\theta_b(j))\right). \tag{12}$$

Similarly, $\hat{\theta}_b(t)$ will be:

$$\hat{\theta}_b(t) = \left(\sum_{j=1}^{t} \lambda_b^{t-j}(\varphi(j))^2\right)^{-1}\left(\sum_{j=1}^{t} \lambda_b^{t-j}\left(y_0(j) - Z^T(j)\Theta_a(j)\right)\right), \tag{13}$$

For real time estimated, a recursive form is required. Using the analogy that is available between Equations (12), (13) and the classical form (8), the recursive form can be readily deduced:

$$\hat{\Theta}_a(t) = \hat{\Theta}_a(t-1) + K_a(t)\left(y_0(t) - Z^T(t)\hat{\Theta}_a(t-1) - \varphi(t)\theta_b(t)\right), \tag{14}$$

where:

$$K_a(t) = P_a(t-1)Z^T(t)\left(\lambda_a + Z(t)P_a(t-1)Z^T(t)\right)^{-1}, \tag{15}$$

$$P_a(t) = (I - K_a(t)Z(t))P_a(t-1)\frac{1}{\lambda_a}, \tag{16}$$

and similarly:

$$\hat{\theta}_b(t) = \hat{\theta}_b(t-1) + K_b(t)\left(y_0(t) - Z^T(t)\Theta_a(t) - \varphi(t)\hat{\theta}_b(t-1)\right), \tag{17}$$

where:

$$K_b(t) = P_b(t-1)\varphi(t)\left(\lambda_b + \varphi^T(t)P_b(t-1)\varphi(t)\right)^{-1}, \tag{18}$$

$$P_b(t) = \left(I - K_b(t)\varphi^T(t)\right)P_b(t-1)\frac{1}{\lambda_b}. \tag{19}$$

In the two aforementioned equations, $\Theta_a(j)$ and $\theta_b(j)$ are unknown, so we replace them with their estimates, $\hat{\Theta}_a(t)$ and $\hat{\theta}_b(t)$, as is customary in similar situations, such as the 'separation principle' in optimal control. The substitution is also justified when the actual and the estimated values are very close to each other or within the algorithm region of convergence. A convergence proof, or conditions for convergence of the algorithm under this assumption, remains open for future research. Upon substitution for $\Theta_a(j)$, $\theta_b(j)$ and rearranging equations (14) and (17), we obtain:

$$\hat{\Theta}_a(t) + K_a(t)\varphi(t)\hat{\theta}_b(t) = \hat{\Theta}_a(t-1) + K_a(t)\left(y_0(t) - Z^T(t)\hat{\Theta}_a(t-1)\right), \tag{20}$$

$$K_b(t)Z^T(t)\hat{\Theta}_a(t) + \hat{\theta}_b(t) = \hat{\theta}_b(t-1) + K_b(t)(y_0(t) - \varphi(t)\hat{\theta}_b(t-1)), \tag{21}$$

for which the solution is:

$$\begin{bmatrix} \hat{\Theta}_{\mathbf{a}}(t) \\ \hat{\theta}_{\mathbf{b}}(t) \end{bmatrix} = \begin{bmatrix} 1 & \mathbf{K}_{\mathbf{a}}(t)\varphi(t) \\ \mathbf{K}_b(t)\mathbf{Z}^{\mathrm{T}}(t) & 1 \end{bmatrix}^{-1} \begin{bmatrix} \hat{\Theta}_{\mathbf{a}}(t-1) + \mathbf{K}_{\mathbf{a}}(t)\left(y_0(t) - \mathbf{Z}^{\mathrm{T}}(t)\hat{\Theta}_{\mathbf{a}}(t-1)\right) \\ \hat{\theta}_b(t-1) + \mathbf{K}_b(t)\left(y_0(t) - \varphi(t)\hat{\theta}_{\mathbf{b}}(t-1)\right) \end{bmatrix}, \quad (22)$$

Using the fact that $\mathbf{P}_{\mathbf{a}}(t)$ and $\mathrm{P}_b(t)$ are always positive, it can be proved that the determinant of the matrix is always non-zero and therefor e the inverse always exists. With some more mathematical manipulations, Equation (22) can be written as follows:

$$\hat{\Theta}(t) = \hat{\Theta}(t-1) + \mathbf{K}(t)\left(y_0(t) - \mathbf{\Phi}^{\mathrm{T}}(t)\hat{\Theta}(t-1)\right), \quad (23)$$

where, $\mathbf{K}(t)$ is defined as follows:

$$\mathbf{K}(t) = \frac{1}{1 + \frac{\mathbf{P}_{\mathbf{a}}(t-1)\left(\mathbf{Z}^{\mathrm{T}}(t)\right)^2}{\lambda_a} + \frac{\mathrm{P}_b(t-1)(\varphi(t))^2}{\lambda_b}} \begin{bmatrix} \frac{\mathbf{P}_{\mathbf{a}}(t-1)\mathbf{Z}^{\mathrm{T}}(t)}{\lambda_a} \\ \frac{\mathrm{P}_b(t-1)\varphi(t)}{\lambda_b} \end{bmatrix}, \quad (24)$$

where, λ_a and λ_b are the forgetting factors corresponding to the two parameters $\zeta_i(t)$ and X/R to be estimated, respectively, and the value range is [0, 1]. Considering that $\zeta_i(t)$ is a fast variable, in order to ensure that the estimated result of $\zeta_i(t)$ has better tracking performance, it should happen that $\lambda_a > \lambda_b$.

3.4. Real-Time Adjustment of For getting Factors

The abovementioned parameter estimation method uses the constant forgetting factor, which can only be used in slow time-varying systems. However, in the actual problem of electric meter error estimation, the dynamic characteristics of the system do not always change according to the same law, and sometimes change rapidly. Sometimes the change is very slow, and sometimes there is a mutation. If a small forgetting factor is selected according to the fast change of the parameter, the information obtained from the data is less when the parameter changes slowly which will cause the parameter estimation error to increase exponentially, which is very sensitive to interference. If we choose a large forgetting factor based on the slow change of parameters, it can memorize data that is far away and it will be insensitive to sudden changes in the system parameters. Therefore, if a constant forgetting factor is chosen, satisfactory results cannot be obtained.

According to the characteristics of the estimation parameters required, appropriate automatic adjustment methods are selected for the forgetting factors λ_a and λ_b, respectively. The specific algorithm is as follows [29,30]:

For getting factor λ_a corresponding to the electric meter operational error estimated parameter:

$$\lambda_a(t) = 1 - (1 - \frac{\mathbf{Z}(t)\mathbf{P}_{\mathbf{a}}(t-1)\mathbf{Z}^{\mathrm{T}}(t)}{1 + \mathbf{Z}(t)\mathbf{P}_{\mathbf{a}}(t-1)\mathbf{Z}^{\mathrm{T}}(t)})\frac{e^2(t)}{R}, \quad (25)$$

where, $0 < \lambda_a(t) < 1$, $R \in (0,1)$ is the observed noise variance, in this paper, its value is 0.5, $e^2(t)$ is the variance of the estimated value.

For getting factor λ_b corresponding to the line loss estimated parameter:

$$\lambda_b = R\left[(1 + (\hat{\theta}_b(t) - \hat{\theta}_b(t-1))\right]^{-1}, \quad (26)$$

4. Case Analysis

4.1. Accuracy of Error Estimated under Different λ

In order to verify the validity of the proposed method and the necessity of adopting the dynamic double forgetting factors method, we select a typical electricity low-voltage substation of urban residential users as an analysis case. The research area contains a head electric meter and three hundred user-side electric meters. The actual measured electric meter data from January 2016 to December 2018 in one substation are analyzed. The collection frequency of electricity consumption of ordinary residential users is 24 h. We eliminate abnormal data such as light load data and null data by appropriate clustering methods and obtaining the measured data preprocessed as the analysis samples. The operational error recursive estimated curves of some electric meters solved by the proposed method are shown in Figure 5. The y-axis represents the error rate of the electric meter (error rate is the percentage form of $\hat{\zeta}_i(t)$, which represents the extent that the measured value of the electric meter deviates from the true value of the electric meter; when the error rate exceeds 2%, the electric meter is defined as an error-over meter), x-axis represents the frequency of data acquisition, the output frequency of the error analysis result is as the same as the frequency of the measured data acquisition, which is generated once a day.

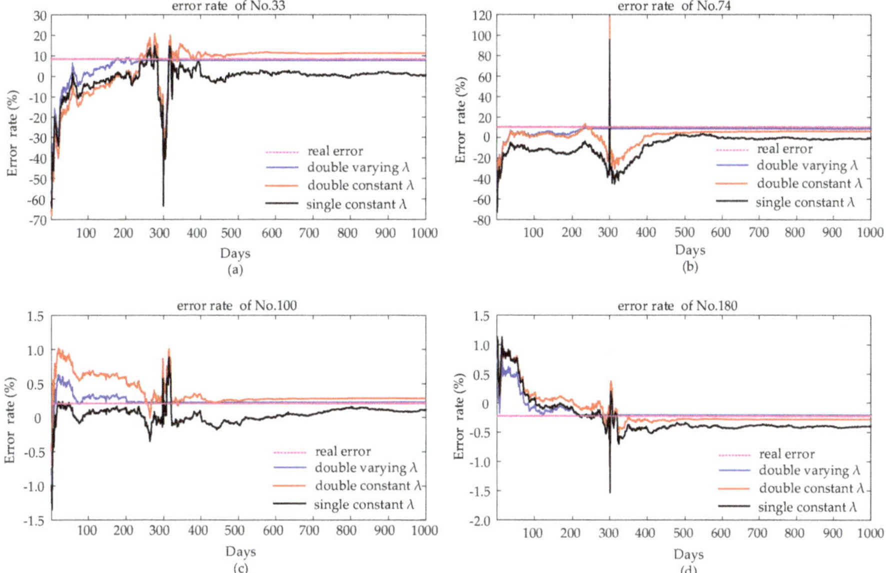

Figure 5. The operational error recursive estimation curves: (**a**) represents the error rate recursive estimation curve of No.33 electric meter; (**b**) represents the error rate recursive estimation curve of No.74 electric meter; (**c**) represents the error rate recursive estimation curve of No.100 electric meter; (**d**) represents the error rate recursive estimation curve of No.180 electric meter.

It can be seen from Figure 5 that the estimated consequence of the double dynamic forgetting factors recursive least squares algorithm is shown by the blue line which is closest to the real error of the electric meter, and the convergence speed of the algorithm is faster than the other two algorithms; The estimated consequence of the double constant forgetting factors algorithm is shown by the red line. The estimated accuracy is not as high as the double dynamic forgetting factors algorithm. Because different estimated parameters change at different rates, the constant forgetting factors can't be made according to the change rule of the parameters and the adjustment also causes the convergence speed of the algorithm to be slower. The estimated effect of the single forgetting factor algorithm is shown by the black line. As can be seen from the figure, the calculation accuracy of the single forgetting factor is the lowest because the line loss rate is regarded as a certain value so that the total power consumption of each user's electric meter in the unit measured period generates a large error, thereby causing a large interference to the estimated error of the user electric meter and resulting in greatly reduced accuracy of the estimated operational error of the electric meter. It also leads to recursive estimated curves are unstable with the circumstances of large noise.

Therefore, the forgetting factor should be automatically adjusted as the dynamic characteristics of the estimated parameters change. When the system parameter changes are abrupt, the small forgetting factor should be automatically selected to improve the sensitivity of the identification. When the system parameters change slowly, a large forgetting factor should be automatically selected in order to improve the recognition accuracy based on the memory length.

4.2. Distribution of Meter Error Rate under Different λ

In order to verify the effectiveness of the proposed method and analyze the influence of forgetting factors in different situations on the estimated value of the meter errors we use the typical urban residential station mentioned in Section 4.1 as the research object. The measurement data collected by 300 electric meters in this area are used as the input variable of the algorithm, and the simulation results are shown in the following figure.

Figure 6 shows the online estimated results of the operational errors of all the user electric meters under a station. From Figure 6, we can conclude that the estimated results based on the double dynamic forgetting factors recursive least squares algorithm have the highest accuracy, the double forgetting factors is the second, and the single forgetting factor is the worst. Therefore, the double dynamic forgetting factor are introduced by simultaneously estimating the operating error parameters of the electric meter and line loss parameters for ideal real-time online estimation of the operational error of large-scale user electric meters under a power substation can be obtained.

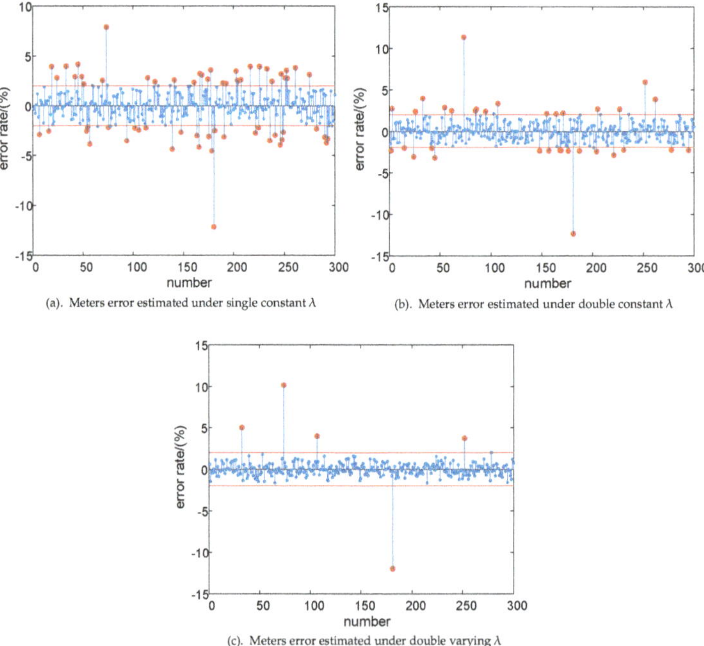

Figure 6. Result of operational error of 300 meters under different states of λ.

4.3. Estimation of Line Loss Rate

There are many factors affecting grid power loss. For distribution grids below 380/220 V, the users' power load has a significant impact on the line loss rate of the power distribution area. The user electrical load is the most important power consumption factor of the power supply system. Therefore, the power load determines the losses of the power supply system. The numerical distribution of the load and the spatial distribution of each load point affect the loss of the multi-branch line. The load curve reflects the variation of the load within a certain time interval. It is inevitable to increase the equipment capacity to meet the safe and stable power supply if the electric load curve fluctuates greatly, which resulting in an increase in line loss.

For three types of electricity users in industrial, commercial, and urban residents, choosing their typical distribution area as the analysis object to calculate the network loss rate respectively, and still selecting the measured data of electric meter from January 2016 to December 2018 as data source.

Figure 7 shows the estimated results of the line loss rate of different type power distribution stations. The black curve represents the type of the substation occupied by industrial users. Since industrial users are in high-load power state all the year round and the load curve fluctuates greatly, the result that line loss rate of this type distribution area is the largest, which is between 3.3% and 3.8%; The blue curve represents the type of commercial-based area, the electrical load characteristics are similar to industrial stations, but the overall electricity consumption is lower than that of the industry, which is between 2.4% and 3.1%; The red curve represents the type of the power distribution stations are main occupied by ordinary residents. When compared with industrial users and commercial users, we could get the result that the residential electricity load is the lowest, but resulting in a large fluctuation in the line loss rate of the distribution area due to the influence such as season, holidays and other factors have created the large fluctuations in the power load. The line loss rate ranging from 1.5% to 2.6%.

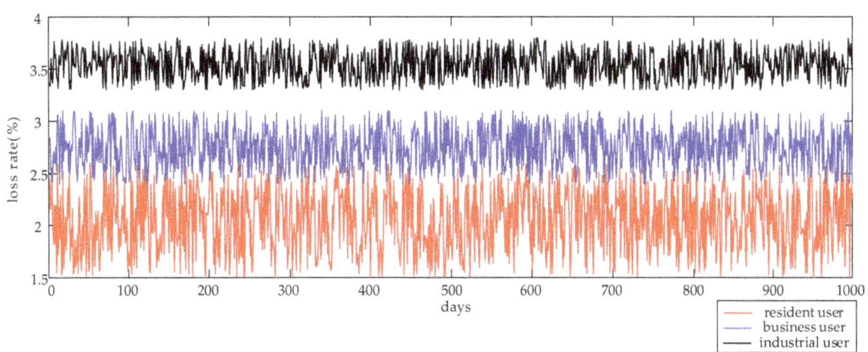

Figure 7. The estimated line loss rate results.

4.4. Error Detection Rate Analysis

In order to further analyze the accuracy of the error rate of electric meters estimated by the proposed method, the actual operation state of the typical station user can be fully simulated in the laboratory environment through the programmable control load simulation system of the substation. The entire simulation system can simulate the power of the real substation and support the adjustment such as different power usage conditions, different line losses, and different electric meter error conditions. Therefore, the power consumption and line loss status of four types which include typical industrial users, commercial users, urban residents and rural residents are simulated respectively by using this simulation system and the real error of the electric meter are obtained.

In practice, the power consumption levels of different industries are quite different. The power consumption of industrial areas is the highest, followed by commercial areas. Compared with rural areas, in developed cities, residents consume more electricity. Therefore, the simulation system is used to simulate the power consumption of four type users: industrial, commercial, rural residents and urban residents. In the simulation process, it is guaranteed that all factors except the power consumption of different types of users are the same, such as the frequency of data collection and the number of users. In the simulation, set the data collect frequency to 1 h, and the number of user's electric meters is p.

Then MAPE and RMSE are used as the index for evaluation. In the online estimation of electric meter error, the smaller MAPE and RMSE values indicate the higher accuracy of the estimated error parameters. MAPE and RMSE can be expressed as:

$$MAPE = \frac{1}{p}\sum_{i=1}^{p}\frac{\left|\hat{\zeta}_i - \zeta_i\right|}{\zeta_i} \times 100\%,\tag{27}$$

$$RMSE = \sqrt{\frac{1}{p}\sum_{i=1}^{p}(\hat{\zeta}_i - \zeta_i)^2},\tag{28}$$

where, p is the total number of users in the simulation; $\hat{\zeta}_i$ and ζ_i are the estimated value and actual value of the measurement error of the ith electric meter respectively.

As shown in Figure 8, under the situation that multiple simulation experiments are perfor med for each type of user, the missed detection rate and over-detection rate of the error estimation method proposed in this paper are analyzed. Through statistical calculation and analysis, the missed detection rate and over-detection rate of industrial users and commercial users are lower than for ordinary residents and the overall rate of missed detection is lower than the over-detection rate. The detection rate is below 1%. It can be inferred that the proposed method is applicable to the real-time online estimation of the operational errors of electric meters with high precision, and can also realize the estimation of the operational error for large-scale electric meters.

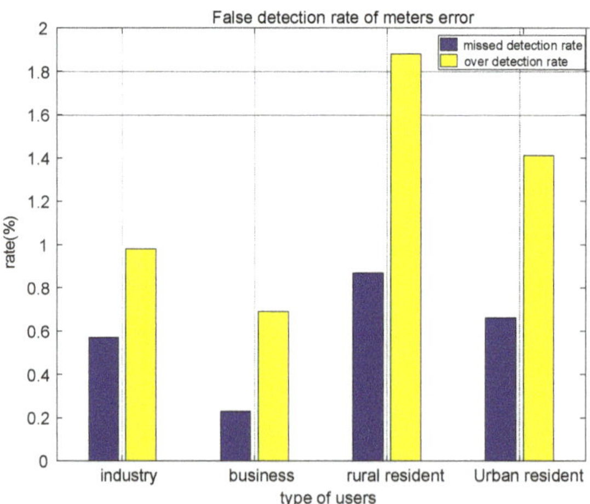

Figure 8. The false detection rate of electric meter errors.

5. Conclusions

In order to study how to estimate online the errors of electric meters, this paper proposes a double-parameter recursive least squares estimation method, and a double-varying forgetting factor strategy that is in line with the development trend of AMI. The case analysis results show that the estimated performance of the double-varying forgetting factor is better than other estimation methods, such as the double-constant forgetting factor and single constant forgetting factor, and its false detection rate is below 2%. Moreover, the proposed method can simultaneously estimate the parameters of the electric meter error and line loss, and improve the accuracy of the online estimation of electric meter errors. In addition, the estimation method proposed is based on the elimination of abnormal data such as light load data and null data, although how to reduce the above discussed effects in the process of the data processing and algorithm solving needs further study.

Author Contributions: Conceptualization, X.K. and Y.M.; methodology, Y.M.; software, Y.M.; validation, Y.M.; for mal analysis, Y.M.; investigation, X.Z.; resources, X.K. and Y.L. and Y.T.; data curation, X.K. and X.Z.; writing—original draft preparation, Y.M.; writing—review and editing, X.K. and Y.M.; supervision, X.K.; project administration, X.K. and Y.L. and Y.T.

Funding: This research was funded by the National Natural Science Foundation of China under Grant. No. 51877145.

Acknowledgments: The authors would like to acknowledge the support in part by Electric Power Research Institute of Tianjin Power Grid, and the National Key Research and Development Program of China (2017YFB0902900 and 2017YFB0902902).

Conflicts of Interest: The author declares no conflict of interest. The funders had no role in the design of the study; in the collection, analyses, or interpretation of data; in the writing of the manuscript; and in the decision to publish the results.

Abbreviations

RLS   Recursive Least Squares
AMI   Advanced Metering Infrastructure
CT   Current Transfor mer
PT   Phase voltage Transfor mers
MAPE   Mean Absolute Percent Error
RMSE   Root Mean Square Error
PLC   Power Line Carrier

References

1. Review of the Development of the Electricity Meter Industry in 2018 and the Outlook for 2019 (1st Episode). Available online: https://www.sohu.com/a/282141253_732907 (accessed on 16 December 2018).
2. Huang, M.; Shi, Y.; Lin, N.; Lin, Z.; Liu, Y.; He, Y. Common faults and treatment methods in the verification process of electric meter verification device. *Electr. Eng.* **2018**, *2*, 89–91.
3. Hong, S.; Zhang, H.; Ding, Q.; Zhang, H.; Peng, H. Study on verification method during single-phase electric energy meter automatic verification system. *Jilin Electr. Power.* **2017**, *45*, 16–18.
4. Yang, N.; Li, H.; Yuan, J.; Li, S.; Wang, X. Medium and long-term load grey prediction method considering grey correlation analysis. *Electr. Power Syst. Autom.* **2018**, *30*, 108–114.
5. Pan, M.; Tian, S.; Wu, B.; Ye, J. Research on detection and tamper detection method based on Electric meter data. *Wisdom Power* **2017**, *45*, 80–84.
6. Ding, L. Research on Anti-Stealing Technology Based on Data Mining in Electricity Information Collection System. Master's Thesis, Chongqing University of Posts and Telecommunications, Chongqing, China, 2017.
7. Nikovski, D.; Wang, Z.; Esenther, A.; Sun, H.; Sugiura, K.; Muso, T.; Tsuru, K. Smart meter data analysis for power theft detection. In Proceedings of the International Workshop on Machine Learning and Data Mining in Pattern Recognition, New York, NY, USA, 19–25 July 2013; Springer: Berlin/Heidelberg, Germany, 2013; pp. 379–389.
8. Lee, J.; Kim, Y.C.; Park, G.L. An Analysis of Smart Meter Readings Using Artificial Neural Networks. In Proceedings of the International Conference on Hybrid Information Technology, Daejeon, Korea, 23–25 August 2012; pp. 182–188.
9. Sun, M.; Wang, Y.; Goran, S.; Kang, C. Probabilistic Peak Load Estimation in Smart Cities Using Smart Meter Data. *IEEE Trans. Ind. Electron.* **2019**, *66*, 1608–1618. [CrossRef]

10. Korhonen, A. Verification of Energy Meters Using Automatic Meter Reading Data. Master's Thesis, AALTO University, Espoo, Finland, 2012; pp. 12–20.

11. Guo, J. Line Measured Key Technology for Smart Grid AMI and User Electricity Data Mining Research. Ph.D. Thesis, Tianjin University, Tianjin, China, 2011.

12. Liu, F.; He, Q.; Hu, S.; Wang, L.; Jia, Z. Estimation of Smart Meters Errors Using Meter Reading Data. In Proceedings of the 2018 Conference on Precision Electromagnetic Measurements (CPEM), Paris, France, 8–13 July 2018.

13. Wang, X.; Wen, L.; Jia, X.; Wang, L.; Wang, Q.; Yuan, R.; Zhou, L. OOK-driven dynamic error measurement of smart energy meter. *Electr. Power Autom. Equip.* **2014**, *9*, 143–147.

14. Dong, X.; Li, Z.; Dai, Y.; Wang, Q.; Zhang, Z.; Wang, P.; Chen, Z.; Wang, T.; Xu, Z. Research on error verification of harmonic smart electricity meter. In Proceedings of the 2018 3rd International Conference on Mechanical, Control and Computer Engineering (ICMCCE), Huhhot, China, 14–16 September 2018; pp. 286–290.

15. Yang, J.; Jiang, P.; Chen, G.; Yuan, T.; Xue, F. Application of C5.0 algorithm in failure prediction of smart meters. In Proceedings of the 13th International Computer Conference on Wavelet Active Media Technology and Information Processing, Chengdu, China, 16–18 December 2016.

16. Zhang, P.; Zhang, B.; Chen, S. Research on residual life prediction method of running intelligent electricity meters based on stratified sampling. *Electr. Meas. Instrum.* **2019**, *32*, 1–5.

17. Zhang, Z.; Gong, H.; Li, C.; Wang, Z.; Xiang, X. Research on estimating method for the smart electric energy meter's error based on parameter degradation model. *Mater. Sci. Eng.* **2018**, *366*, 012065. [CrossRef]

18. Cheng, Y.; Du, J.; Zhou, F.; Xiao, J.; Yang, H. Application of variable weight fuzzy analytic hierarchy process in evaluation of electric energy meter. In Proceedings of the IEEE 2nd Advanced Information Technology, Electronic & Automation Control Conference, Chongqing, China, 25–26 March 2017.

19. Peng, X.; Wang, L.; Jia, Z.; Wang, X. Evaluation of measurement error of digital electric energy meter based on Monte Carlo method. *Electr. Meas. Instrum.* **2017**, *54*, 100–105.

20. Yin, X.; Liu, H.; Yin, C.; Man, J. The error model of the Electric meter under the influence of low temperature. In Proceedings of the International Conference in Communications, Signal Processing, and Systems, Harbin, China, 14–17 July 2017; pp. 1909–1915.

21. Yip, S.C.; Wong, K.S.; Hew, W.P.; Gan, M.T.; Phan, C.W.; Tan, S.W. Detection of energy theft and defective Electric meters in smart grids using linear regression. *Int. J. Electr. Power Energy Syst.* **2017**, *91*, 230–240. [CrossRef]

22. Yao, L.; Hu, Y.; Wu, J.; Liu, Y. Error Analysis and Compensation for Non-Integer-Period Sampling in Energy Metering of Smart Substation. *Zhejiang Electr. Power* **2015**, *34*, 1–5.

23. Dong, Y. Design and Implementation of Power User Information Collection System for Power Users in Rong Cheng Power Supply Company. Master's Thesis, Shandong University, Shandong, China, 2016.

24. Calibration Specification of Digital. Available online: http://www.biaozhuns.com/archives/20180128/show-178130-86-1.html (accessed on 28 December 2018).

25. Deng, J. Error analysis and processing of electronic energy meter. *Technol. Enterp.* **2013**, *5*, 319.

26. Kong, X.; Hu, Q.; Dong, X.; Zeng, Y.; Wu, Z. Load data identification and repair method based on improved fuzzy C-means clustering. *Autom. Electr. Power Syst.* **2017**, *41*, 90–95.

27. Tang, H. Research on Line Loss Analysis and Loss Reduction Measures in Low Voltage Substation. Master's Thesis, Southeast University, Nanjin, China, 2017.

28. Vahidi, A.; Stefanopoulou, A.; Peng, H. Recursive least squares with for getting for online estimated of vehicle mass and road grade: Theory and experiments. *Veh. Syst. Dyn.* **2005**, *43*, 31–55. [CrossRef]

29. Aguayo, M.; Bellido, L.; Lentisco, C.M.; Pastor, E. DASH Adaptation Algorithm based on Adaptive For getting Factor estimation. *IEEE Trans. Multimed.* **2018**, *20*, 1224–1232. [CrossRef]

30. Liu, C.; Sun, X.; Li, H. A Limited Memory Identification Algorithm with For getting Factor. *Electro-Opt. Control* **2006**, *13*, 48–49.

Article

Comparison of Intelligence Control Systems for Voltage Controlling on Small Scale Compressed Air Energy Storage

Widjonarko [1,*], Rudy Soenoko [2], Slamet Wahyudi [2] and Eko Siswanto [2]

[1] Energy Conversion, Electrical Engineering Departement, Universitas Jember, Jalan Kalimantan No. 37, Jember, East Java 68121, Indonesia
[2] Energy Conversion, Mechanical Engineering Departement, Universitas Brawijaya, Jalan M.T. Haryono No. 167, Malang, East Java 65145, Indonesia; rudysoen@ub.ac.id (R.S.); slamet_w72@ub.ac.id (S.W.); eko_s112@ub.ac.id (E.S.)
* Correspondence: widjonarko.teknik@unej.ac.id; Tel.: +62-813-5822-3843

Received: 3 January 2019; Accepted: 18 February 2019; Published: 28 February 2019

Abstract: This study presents the strategy of controlling the air discharge in the prototype of small scale compressed air energy storage (SS-CAES) to produce a constant voltage according to the user set point. The purpose of this study is to simplify the control of the SS-CAES, so that it can be integrated with a grid based on a constant voltage reference. The control strategy in this study is carried out by controlling the opening of the air valve combined with a servo motor using three intelligence control systems (fuzzy logic, artificial neural network (ANN), and adaptive neuro-fuzzy inference system (ANFIS)). The testing scenario of this system will be carried out using two scenes, including changing the voltage set point and by switching the load. The results that were obtained indicate that ANN has the best results, with an average settling time of 2.05S in the first test scenario and 6.65S in the second test scenario.

Keywords: ANFIS; artificial neural network; fuzzy; small scale compressed air energy storage (SS-CAES); voltage controlling

---

1. Introduction

The development of a combination of renewable energy technologies and energy storage is the most rapidly developing research topic at this time [1,2]. Problems related to the use of non-renewable energy which is still high [3] and becomes the world's main problem (especially in climate change [4]) can be solved [5,6] by using a combination of this technologies. In some applications for renewable energy use, this energy source is not used as the main support for an area's load [7,8]. However, this energy source is more widely used as a support for overcoming peak loads at certain times [7,9–15]. The reason is that, in some renewable energy sources, it is still very dependent on weather conditions, such as the use of photovoltaics (PV) [16–19], which will only produce energy during the day. Given this problem, the existence of storage technology that will store energy when the energy not in use, such as batteries, is vital [20]. However, batteries still have problems with environmental aspects because of toxic waste [21,22] and they can explode due to excessive heat [23]. Therefore, some researchers have begun to shift a lot on the topic of developing energy storage technology that is more environmentally friendly, has no degradation over time, such as batteries, and is relatively inexpensive on an energy base [1]. One of the technologies chosen is Compressed Air Energy Storage (CAES), or on a small scale known as SS (Small Scale)-CAES [3,9,24]. This technology is considered to be capable of overcoming environmental problems, because the energy source used is atmosphere gas [9,25–28], and it does not require large space, as on a large scale (CAES) [29].

To be able to help the grid in maintaining supply at peak loads, the combination of renewable energy technology and energy storage must be synchronized [30] with the network, so that energy can be transferred and not cause the grid to be damaged [31]. One of them is by controlling several parameters that are contained in the system the formed, such as frequency or voltage [32–34]. Information regarding several studies focus on controlling parameters in the SS-CAES, and can be found in the paper that has been published [35], one of which was carried out by Martinez [36,37]. In the Martinez study, he simulated the control of air valves in the SS-CAES to supply power according to the grid requirements. The simulation of SS-CAES formed was using an AC-PMSG generator (Permanent magnet synchronous generator). In that study, the generator is connected to several converters before being connected to the grid (including AC-DC converters and inverters). The control carried out is to control the pneumatic valve that is operated in the open-close mode to get the appropriate pressure in achieving the desired power. In another study that was conducted by Maia [38], the SS-CAES prototypes were made using a three-phase generator. The prototype made is observed without controlling the parameters that were contained in the SS-CAES. In another study that was conducted by Kokaew, V. [39–41], it controlled the rotational speed rotation parameters in the SS-CAES prototype. Fighting the mechanical torque that arises because the air passing through the airmotor with electric torque is regulated using a Buck converter carries out speed regulation. The purpose of this control is to get the speed referenced in achieving maximum power transfer or known as MPPT (Maximum power point tracking). However, from several previous studies, no research discusses how SS-CAES can produce a constant voltage to be indirectly integrated into the grid [5,12]. Whereas, in some concepts that have been put forward by several researchers, such as Vongmanee [42], Lemofouet [43], and Martinez [37], to be able to integrate the SS-CAES system into the grid, it must be combined with an inverter. In the general concept, the inverter requires a specific DC input voltage to operate by the voltage parameters on the grid. Therefore, to facilitate integrated systems, energy sources must adjust the inverter's working voltage or it has a stable voltage [44–46], and the energy can be transferred to the grid.

However, in detailed research, Martinez has published his research [36,37] to integrate his system. The strategy that was used by Martinez in his simulation is to control the air pressure using pneumatic valves to reach the required power. However, in the results of his study, the power had a high and low effect because of the pneumatic open-close mechanism. This results happened, because the air pressure that entered the turbine (air motor) is controlled by an open-close mechanism, so that the power also has the same characteristics [36,37]. Because of this phenomenon, the inverter will have to done two jobs. The first is to stabilize the voltage and the second is to synchronize with the phase on the grid. These multiple actions cause the control system that is used to be more complicated.

To be able to simplify the control and eliminate the effects that are caused by the previous study, the changes will be made in this study. There are two strategies used in this study to solve that problem. The first is to remove the high-low effect in previous studies, that is replacing the pneumatic valve control with a combination of valve and servo motor. By using this way, the airflow rate will have smoother air transfer and will also make the produced voltage smoother, so the high-low effects can be eliminated. This could be happened because the control concept using a servo motor works by adjusting the direction of the rotation rather than an open-close mechanism such as in pneumatics system. The second is to replace the control reference to a voltage reference. Thus, the integration between SS-CAES and the grid will be much easier, since the inverter will only adjust the network phase rather than doing two actions (voltage and phase synchronization according to the previous research). Since the controlled system has a high workload, the system is very susceptible to parameter changes and input disturbances. Those issues will be a big problem if a conventional model is used to control the system, therefore artificial intelligence (AI) is chosen. By using AI, the system will be able to work with robust controls and can adapt to non-linear systems [47]. Some AI systems that will be used in the system must have these criteria, there are Fuzzy Logic, Artificial Neural Network, and adaptive neuro-fuzzy inference system (ANFIS). In this study, an experiment will be done to compare the use of these three AIs to control the SS-CAES prototype with 60W that has been built by the researcher to reach the desired set point voltage of the user. The purpose of comparing the performance of the three AIs is to be able to find out the most appropriate control system to solve this problem. In this study, the system has been tested with two scenarios for testing the settling time variable by implementing a microcontroller that was programmed with three artificial intelligence, as a control device.

2. Small Scale Compressed Air Energy Storage Design

2.1. Prototype Design

The SS-CAES system prototype block diagram that can be schematically seen in Figures 1 and 2 is a picture of a prototype made. Several SS-CAES components, including air tanks, air valves (which combine with continuous servo motors), air motors, and DC generators form the system [38] (Figure 2a). In this study, several sensors were installed to retrieve the response from the control system. Some of these parameters are voltage, current, air pressure passing through air-motor, and speed sensor. Data associated with these parameters will be saved in the data logger to observe the effect of changes in the control process that was carried out.

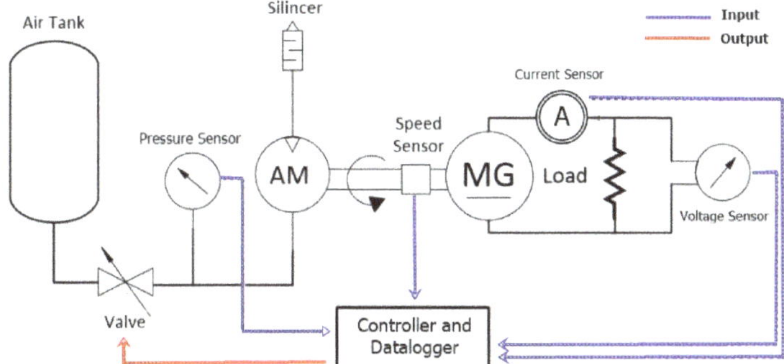

Figure 1. Block diagram of prototype small scale compressed air energy storage (SS-CAES).

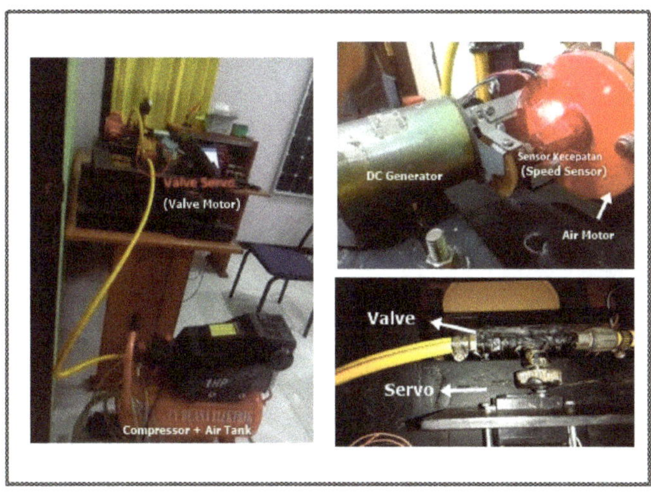

(a)

Figure 2. *Cont.*

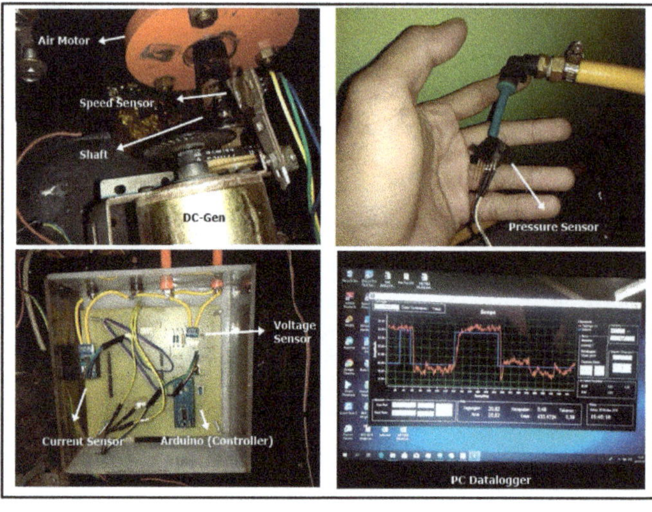

(b)

Figure 2. (a) The main components of prototype SS-CAES; and, (b) The installed sensor, controller on prototype SS-CAES and PC data logger.

In this study, there are four types of sensors that are used in the prototype. The first one is the voltage sensor that is made using the principle of the voltage divider; the second is the speed sensor that uses a hall effect sensor. The third is the air pressure sensor using MPX 5010 (Freescale Semiconductor, Inc., Austin, TX, USA), and the fourth is the current sensor using ACS712. All of the sensors used in this test have been calibrated so the value that appears on the sensor is the real value of the measured parameter. This experiment was done using a servo that was controlled and coupled with an air valve; the servo type that was used in this experiment is MG 996R (TowerPro, Singapore City, Singapore). For the controller, the Arduino UNO microcontroller is used, which is connected to the Computer (monitoring, controlling, and data logger function). Since the primary target of this study is to stabilize the voltage to connect an inverter, then the load used is a resistor with a value of 150 Ω. Figures from sensors, controllers, and PC data loggers can be seen in Figure 2b. Details of the control mechanism of this prototype will be explained in the control block section.

2.2. Control Block

There are four sensors installed on the prototype, but only one sensor will be used as a control reference, which is the voltage sensor. Even so, all the data from the sensor will be saved in the data logger. These data were used to analyze the system performance. The reason for using one sensor in this study is because this system implemented a closed-loop control system where one of the inputs is used in the controlling system as a feedback control from the plant [48]. A detailed scheme of this control block can be seen in Figure 3.

Figure 3 shows that there are two parameters that are used as inputs in this control system. The first parameter is the set point voltage, as determined by the user, and the second parameter is the output voltage (from voltage sensor) of the prototype. To fulfill the closed-loop control system, the input is changed to become two control inputs from the system. These two inputs are Error and Delta Error [48,49], where we can use Equations (1) and (2). After the Error and Delta Error value are obtained, the value of the two parameters will be processed in the controller block. The controller that was used in this prototype was the Arduino UNO microcontroller, as described earlier. This microcontroller will be programmed with three different artificial intelligence according to the scenario that will be tested at the plant. The output of this controller is a pulse width modulator (PWM) signal. The PWM signal is used to control the air valve combined with a servo motor.

$$e(t) = \text{Set Point} - \text{Actual Voltage} \tag{1}$$

$$de = e(t) - e(t-1) \tag{2}$$

where e is Error, de is DeltaError, $e(t)$ is Error at time t, and $e(t-1)$ Error in time $t-1$ or time before.

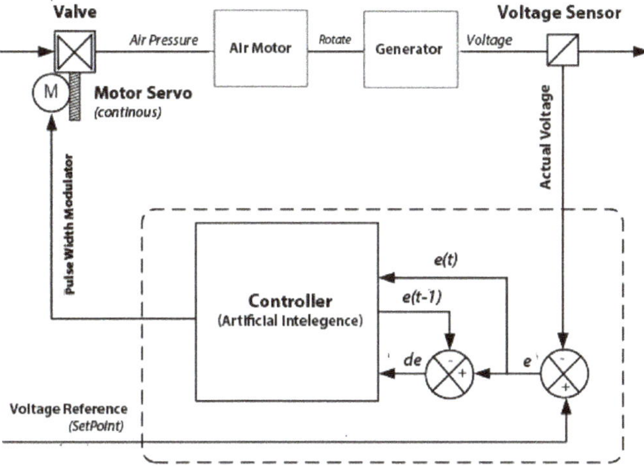

Figure 3. Control block of prototype SS-CAES.

To achieve a constant voltage, the key is to control the airflow entering the system. This condition can be achieved because of the characteristics of the SS-CAES where the voltage generated is directly proportional to the speed of the air motor [40,50]. The faster that the generator changes its speed, the higher the voltage will be generated. Because the rotation speed of an air motor needs to be controlled, it is done by controlling the rate of airflow through the air motor. Therefore, the key to control the voltage is to adjust the airspeed by controlling the valve.

The valve that was used in the prototype is a combination of air valves and a continuous servo. The width of the airline on the valve can be adjusted to control the rate of the airflow, this can be done by changing the servo rotation. However, in continuous servo control, it differs slightly from the general servo. In the continuous servo, the rotation control is not based on the desired angle but rather is based on the direction of rotation (rotating clockwise or counterclockwise). From the servo that was installed in this study, to widen the valve opening, the servo must be controlled clockwise by giving a PWM value >100. Whereas to reduce valve openings, the servo must be controlled so that it rotates counterclockwise, that is by providing a PWM value <100. To stop the rotating servo, PWM = 100 is given. It should also be noted that, the higher the PWM value of the neutral value when the servo stops (PWM = 100), the faster the servo rotation goes in that direction. For example, when PWM = 105, the servo will move quickly in a clockwise direction with different speeds with PWM = 101, and so does the opposite direction. As for the AI, the control output value is the number of actual PWM values with PWM AI output control. For more details, see Equation (3).

$$PWM(t) = PWM(t-1) + PWM(\text{AIOutput}) \tag{3}$$

where the $PWM(t)$ is the actual output of the PWM, $PWM(t-1)$ is PWM value in time $t-1$ or time before, and $PWM(\text{AIOutput})$ is the PWM of the AI process.

3. Intelegence Controller

There are three artificial intelligence systems that are used in this study. The explanation and design of the artificial intelligence system are explained in this subsection.

3.1. Fuzzy Logic Controller

Fuzzy Logic is a rule-based decision-making process that aims to solve problems, where systems are difficult to model or where there is ambiguity [51,52]. Fuzzy logic is determined by logical equations, not from complex differential equations and it comes from thinking that identifies and utilizes obscurity between two extreme lines. Fuzzy logic systems consist of fuzzification, defuzzification, rule base, and inference systems [47]. Fuzzy can work according to the rules that are given by fuzzy designers. By using rules, the relationship between the input that enters the system can be known for its output value. The structure of processing fuzzy logic can be seen in Figure 4.

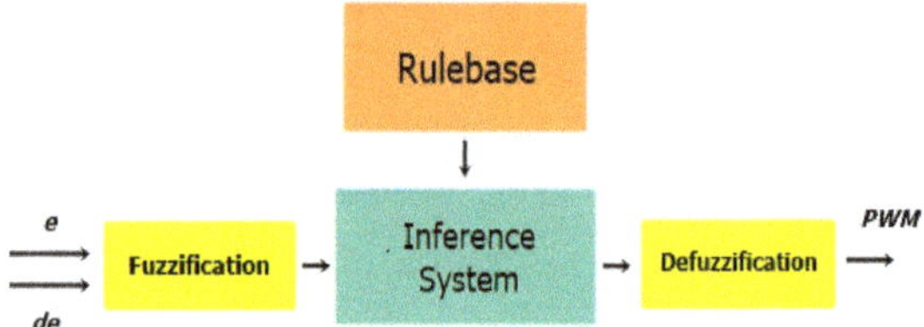

Figure 4. Structure of fuzzy logic controller.

In designing the control system using Fuzzy Logic, the number of membership input and output members is 5, including NB (Negative Big), NS (Negative Small), Zero, PS (Positive Small), and PB (Positive Big). For each membership value, the value of the input (Error and Delta Error) can be seen in Figures 5 and 6. While, for membership, output can be seen in Figure 7.

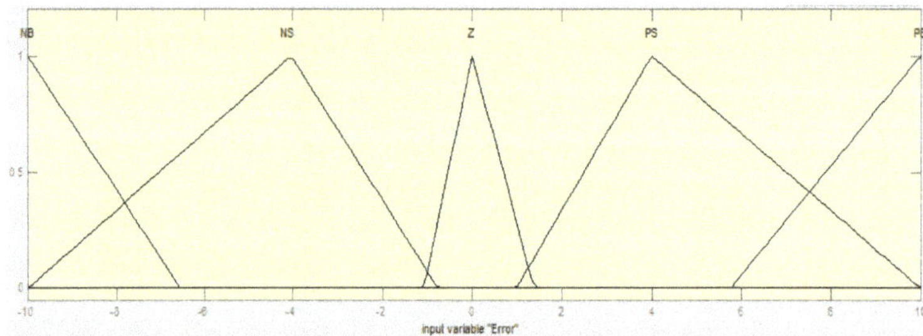

Figure 5. Membership input of error.

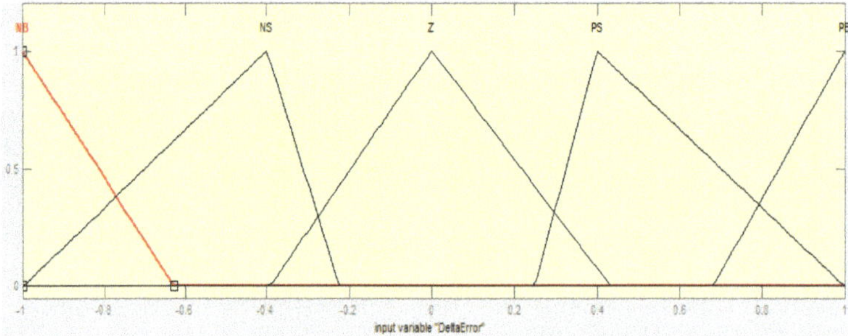

Figure 6. Membership input of delta error.

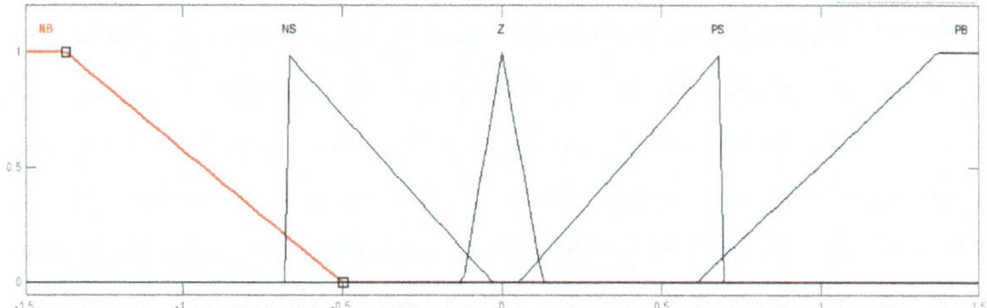

Figure 7. Membership output of pulse width modulator (PWM).

3.2. Artificial Neural Network

Artificial neural network (ANN) is an information processing technique or approach that is inspired by the workings of the biological nervous system, especially in the cells of the human brain in processing information [53,54]. A key element of this technique is the unique and diverse structure of information processing systems for each application. The Neural Network consists of a large number of information processing elements (neurons) that are interconnected and work together to solve certain problems [55]. In this study, the Artificial Neural Network (ANN) was used with the architecture, as shown in Figure 8.

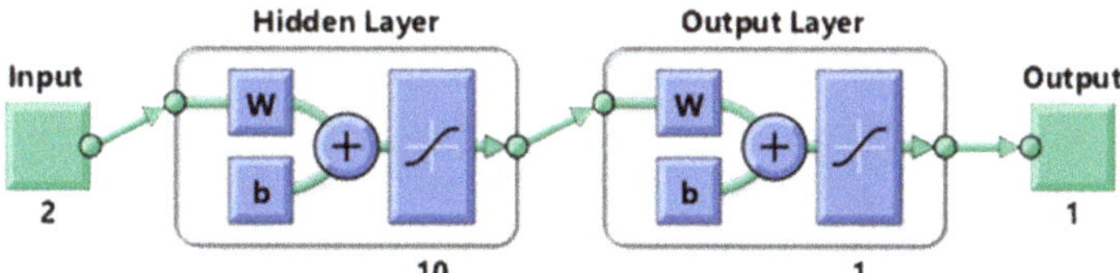

Figure 8. Architecture of artificial neural network used in this study.

The ANN structure that was built by researchers consists of two inputs, namely error and delta error. For the layer used, there are two layers with ten neurons for the hidden layer and one neuron for the output layer with output that is in the value of PWM. The activation function that is used in this structure uses sigmoid activation. The structure of the hidden layer can be seen in Figure 9, while for the whole structure, it can be seen in Figure 10.

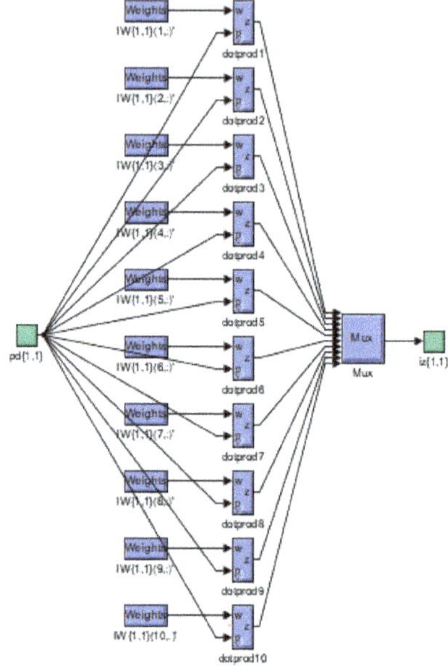

Figure 9. Structure of hidden layer artificial neural network used in this study.

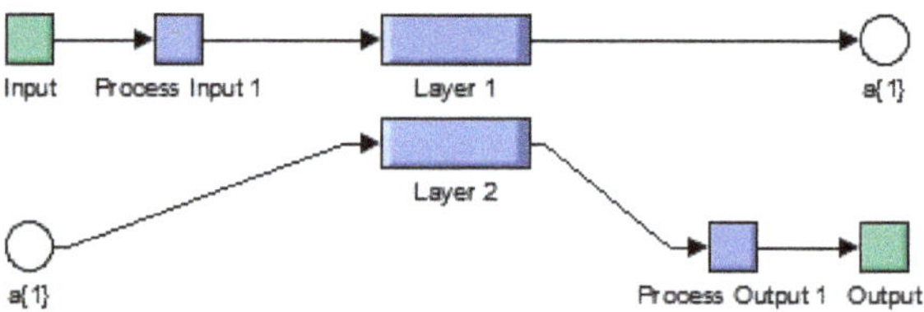

Figure 10. Structure of artificial neural network used in this study.

Artificial neural networks are built using the Levenberg–Marquardt back propagation training algorithm. The training data to create a network is obtained from conventional control data that was applied to the plant. The total data used is $3 \times 816{,}160$ with a portion of training data for 70%, while 15% of the data is for testing and validation. The relationship between training, testing, and network validation formed by ANN has a high correlation coefficient and it can be seen in Figure 11.

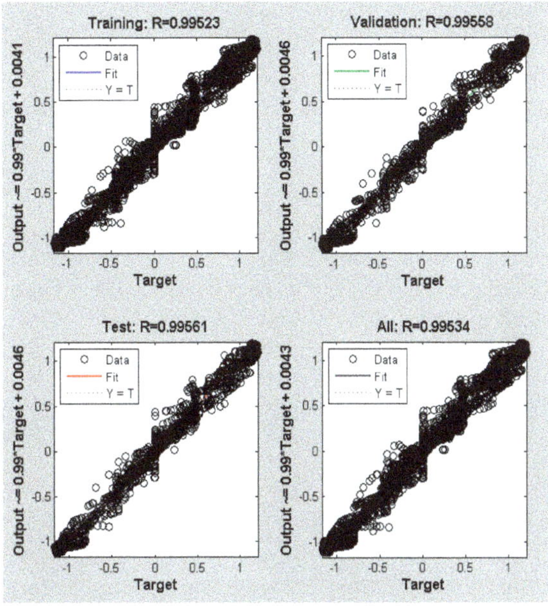

Figure 11. Regression plot of Artificial Neural Network (ANN).

3.3. Adaptive Neuro Fuzzy Inference System

The adaptive neuro-fuzzy inference system (ANFIS) is a method that uses artificial neural networks to implement fuzzy inference systems. The advantage of the fuzzy inference system is that it can translate knowledge from experts in the form of rules, but it usually takes a long time to determine membership functions [47,56]. Therefore, learning techniques from ANN are needed to automate the process, so that it can reduce search time; this causes the ANFIS method to be very well applied in various fields [57].

The ANFIS structure that was used in this study uses two inputs, namely error and delta error. While, from the results of the ANFIS training, nine rules will be applied for the implementation of the prototype. The ANFIS structure in this study can be seen in Figure 12.

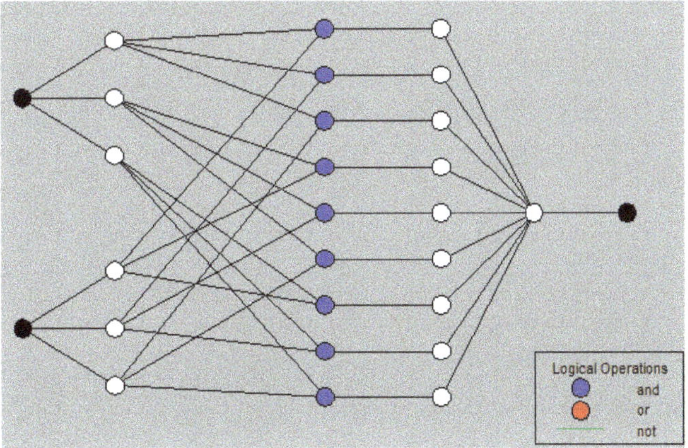

Figure 12. Structure of adaptive neuro-fuzzy inference system (ANFIS).

4. Testing Scenario

This system will be tested in two scenarios with a different AI control program on each test. The first test is to change the set point value. The experiment flowchart for the first scenario can be seen in Figure 13. In this first scenario, the system will be programmed with one of the AI controllers. Subsequently, the prototype will run with the initial set point value of 24 V. After the initial set point is reached, the set point will then be adjusted to 20 V. As long as the system reaches the steady state, all of the installed sensors data, set point values, and system response values (PWM Value) will be saved in the data logger. The purpose of saving these data is to see and analyze the response of data in offline mode (after the system has been tested). Afterwards, after 10 s or more, when the system has reached steady state, the set point value will be reduced. The set point that was previously set at 20 V decreased to 15 V. Subsequently, the cycle data is being saved while the system is running to reach a steady state again. After 10 s or more, after the system reached a steady state, the set point testing model changed. In this section, the set point will be set to higher value. This test starts by using the previous set point, which starts from 15 V, and then the set point is raised to 20 V. As the previous test, all installed sensor data, set point value, and system response will also be saved into the data logger. After 10 s or more, after the system has reached steady state, the set point value will be increased to 24 V. Afterwards, the data saving cycle is taken again while the system running to reach the steady state. After the system reached the steady state point, at the last stage, the system will be set to do final set point jump. The set point jump is from 24 V to 15 V. If we sort the set point test, it will be 24 V, 20 V, 15 V, 20 V, 24 V, 12 V, 24 V, and 15 V. These set point options are based on the voltage that is commonly used by the inverter in the grid application. The first AI Control test has been done for the first scenario. To be able to compare and see the overall AI control response that was used in this study, the first scenario will be repeated three times for three AI controls.

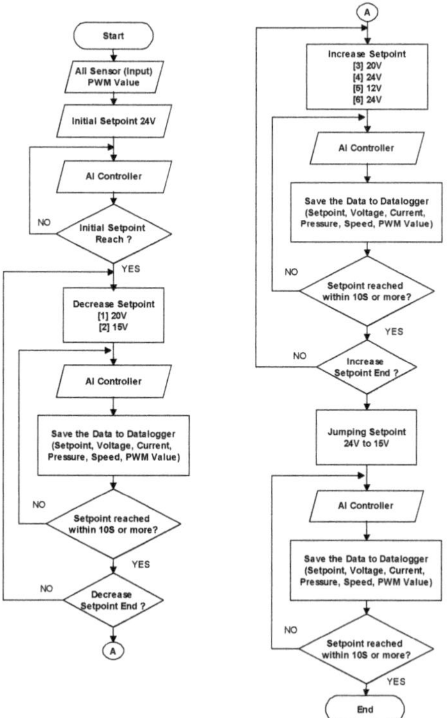

Figure 13. Flowchart Scenario 1.

For the second scenario, the system will be tested by keeping the set point given by the user. In this test, after the AI control is programmed into the microcontroller, the system will run where the system has to maintain a set point value of 20 V. At the initial condition, the 150 Ω load is not connected to the circuit. After the system reaches steady state at 20 V, then the load is connected to the circuit. Due to changes in the load, a change in voltage value occurred. During the process towards the steady state, the entire installed sensor data, voltage set point, and system response (PWM value) will be saved in the data logger. After 10 s or more, the system will reach a steady state condition, then the load will be released afterwards, and the cycle data saving process will be repeated for offline analysis. This process will be repeated three times in one experiment (switch and release load) to obtain the system response. The goal of this test is to determine the reliability of the system. That process is only testing for one cycle in the second scenario. To be able to test and see the overall AI control response that was used in this study, the first scenario will be repeated three times for three AI controls. The flowchart of the overall test in the second scenario is shown in Figure 14.

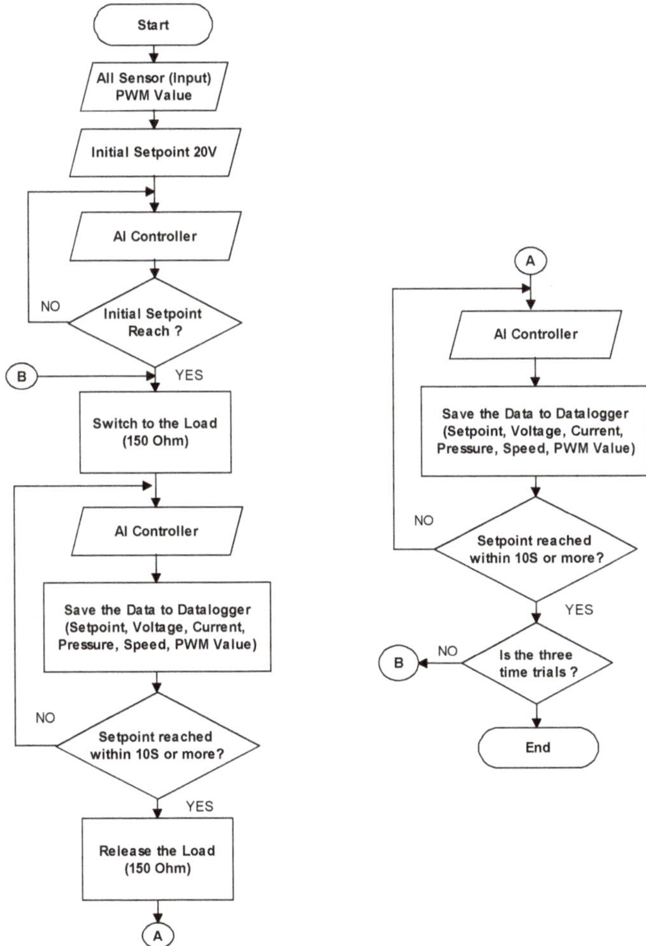

Figure 14. Flowchart Scenario 2.

5. Result and Discussion

In the experiment section, the SS-CAES prototype will run according to the predetermined scenario. The total number of trials is six experiments with different AI systems. The results of these experiments can be seen in Figures 15–26. The results in these figures are obtained data from the data logger that saved while the system was running according to the tested scenario.

The first test uses fuzzy logic. The system is tested with some scenarios and the obtained results can be seen in Figures 15–19. For the first scenario, the set point is changed in an order that already explained in the previous section. From the obtained data, the results shows that, to reach steady-state condition in the first scenario, fuzzy logic has an average settling time of 2.27 s. This result can be seen in Figures 15 and 16, which have been zoomed at 64 to 68 s.

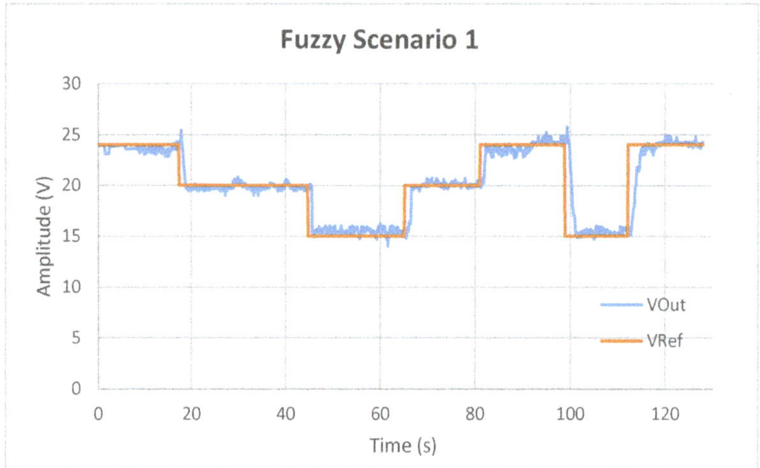

Figure 15. The results of set point changes with fuzzy logic controller.

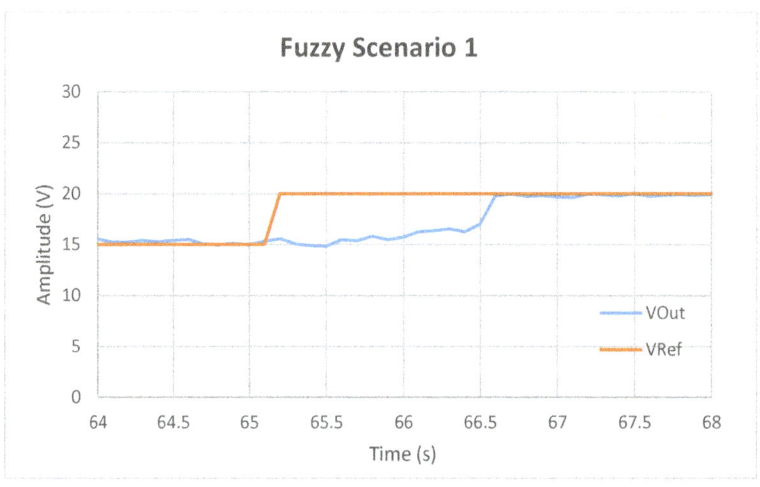

Figure 16. The results of set point changes with fuzzy logic controller on 64 to 68 s.

The results for the second test on the Fuzzy Logic Controller are shown in Figure 17. This scenario testing is done by removing the load three times and connecting the load two times. The effect that is caused by the release of the load is a surge in the output voltage with the highest value of 31.6 V. While the impact caused when adding the load is a reduction in the output voltage with the lowest value of 14.4 V. The average settling time on the results of this test is 8.18 s. For detailed results of displacements transitions, we have presented the zoomed results on a time scale of 85–125 s in Figure 18.

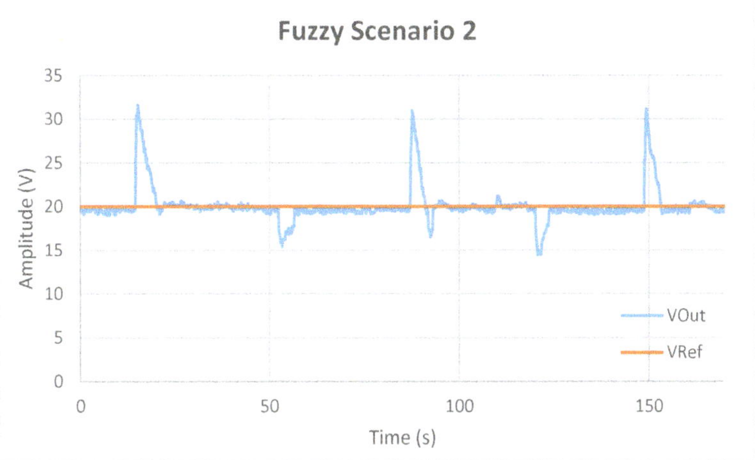

Figure 17. The response results of load changes with fuzzy logic controller.

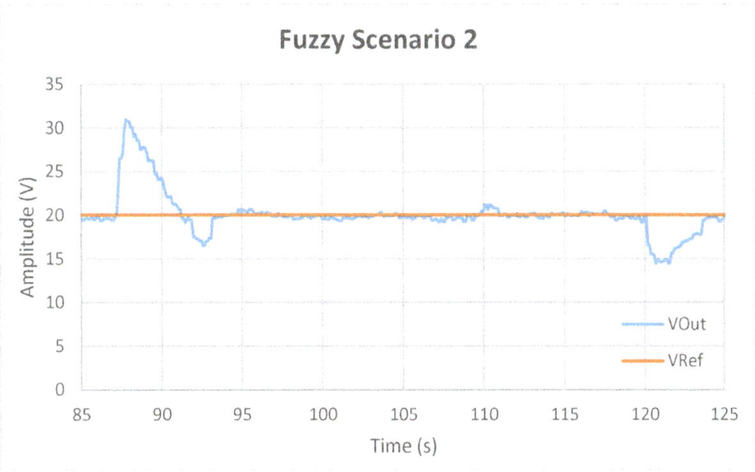

Figure 18. The results of set point changes with fuzzy logic controller on 85 to 125 s.

The next experiment is testing the artificial neural network. The first scenario can be seen in Figure 19. This scenario test resulted in an average settling time of 2.05 s. For a detailed result, we presented data that zoomed at 118–122 s in Figure 20.

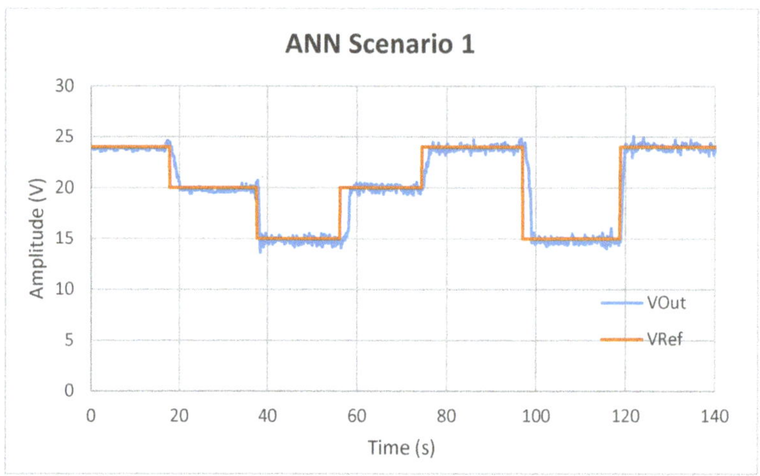

Figure 19. The response results of set point changes with artificial neural network.

Figure 20. The response results of set point changes with artificial neural network controller on 118 to 122 s.

At the second scenario, the highest jump in the output voltage due to the load is released at 37.6 V, and the lowest voltage drop due to the load is connected at 14.7 V. The average settling time is 6.65 s. The results of this test can be seen in Figure 21. The zoomed result at 135–195 s can be seen in Figure 22.

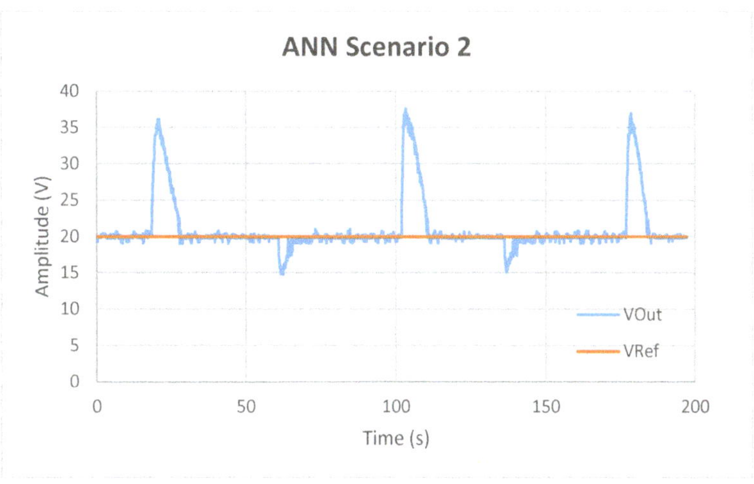

Figure 21. The response results of load changes with artificial neural network controller.

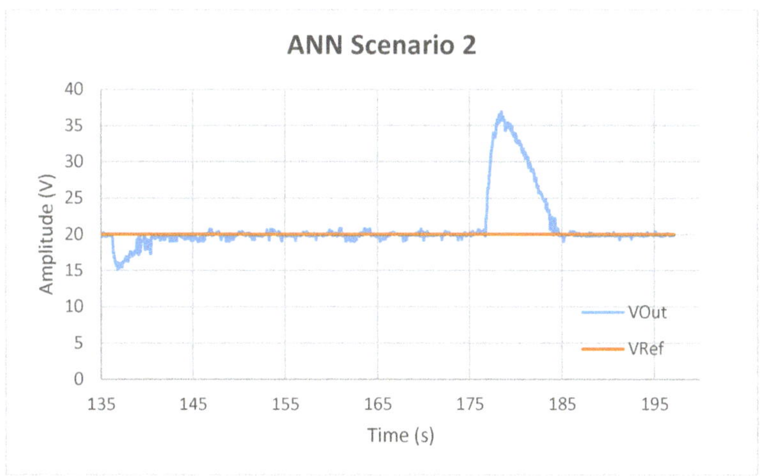

Figure 22. The response results of load changes with artificial neural network controller on 135 to 195 s.

The last experiment used ANFIS and the results can be seen in Figure 23. The first scenario test resulted in an average settling time of 3.49 s. The zoomed result at 56–60 s can be seen in Figure 24.

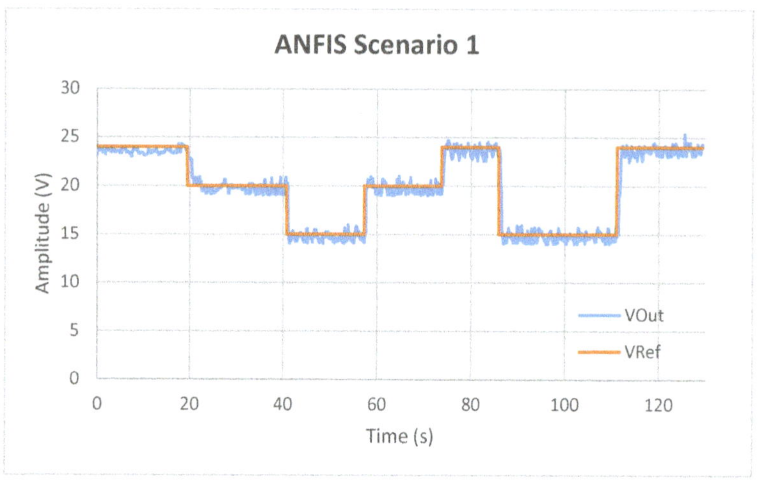

Figure 23. The response results of set point changes with ANFIS controller.

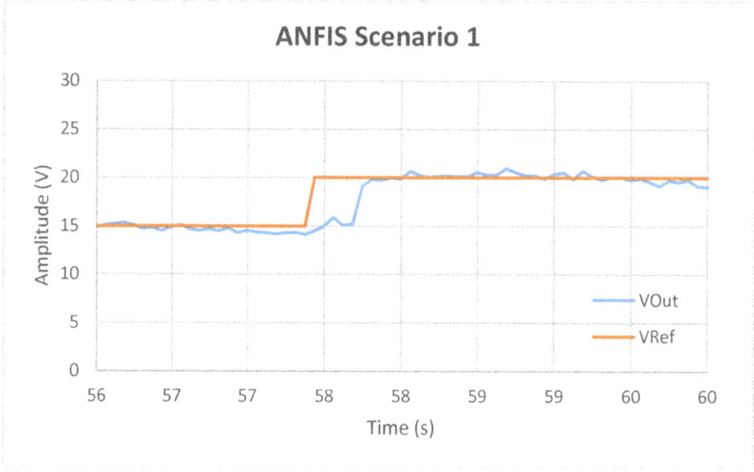

Figure 24. The response results of set point changes with ANFIS Controller on 56 s to 60 s.

For experiments using the second scenario of ANFIS, the system response can be seen in Figure 25. This test resulted in a voltage surge with the highest value of 31.6 V and the lowest voltage at 15.43 V. The average settling time in this test is 8.92 s. Figure 26 shows the results of the second scenario testing zoomed at a 145–210 s time scale.

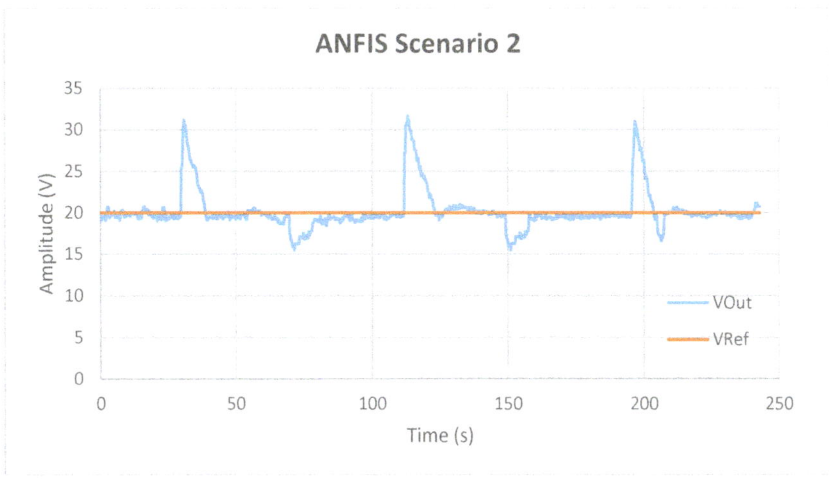

Figure 25. The response results of load changes with ANFIS controller.

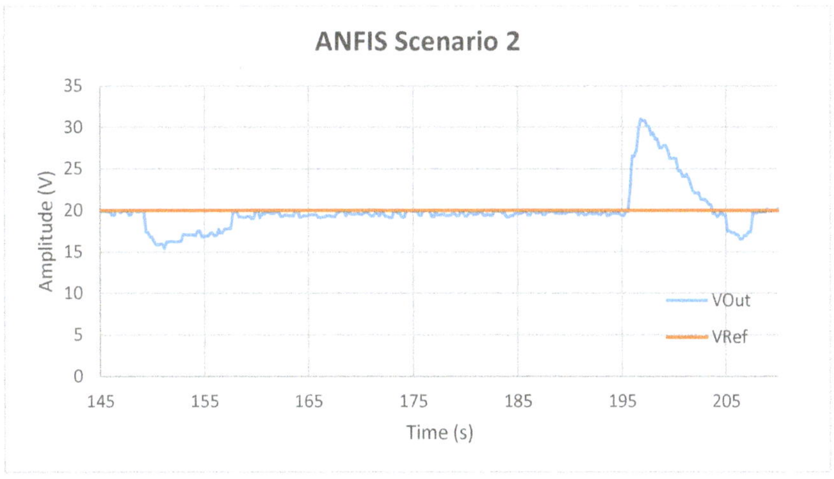

Figure 26. The response results of load changes with ANFIS Controller on 145 s to 210 s.

Those experimental results show that the best result for the first scenario (changing voltage set point) that was achieved by using ANN with an average settling time of 2.05 s. While for the second scenario (maintaining voltage), the best results were also achieved by using ANN with an average settling time of 6.65 s to reach steady state conditions. The time comparison of the results for two scenarios of the three Artificial Intelligence systems can be seen in Figure 27.

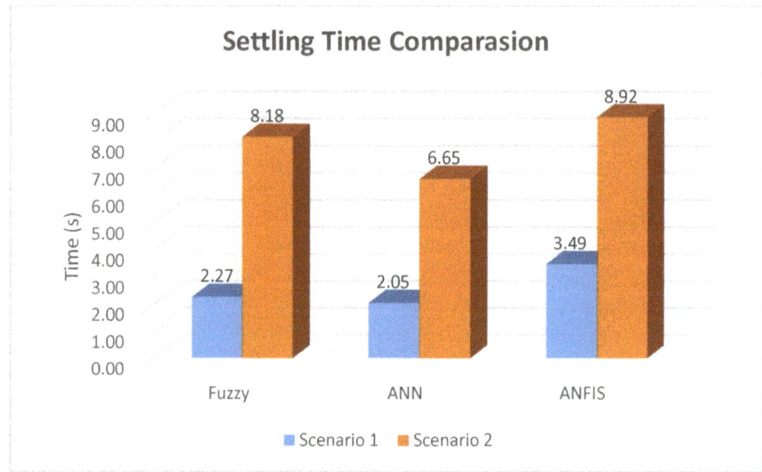

Figure 27. Settling time comparison on three intelligent control systems.

From the acquired results that are presented in Figure 27, the results show that the difference between intelligent control systems has relatively small time differences. By evaluating the first test between Fuzzy and ANN, the time difference is very close to 1 s and, when compared with ANFIS, the time difference is less than 1 s. In the second scenario, the time difference is also around 1–2 s. This difference can be occurred because of the iterative problems in each intelligent control. ANN is superior to other AI because the ANN program has shorter iterations than others, this made ANN more responsive when compared to other intelligent control systems.

If we evaluate the test results in the second scenario, we will find a high overshoot value. The high overshoot value that occurred in the second scenario was caused by the high pressure when the circuits are loaded with a 150 Ω resistor. This overshoot can be seen in Figure 25, which shows a graph of the comparison between voltage and pressure in the ANN test for the second scenario. At the early stage, the load is installed; therefore, the pressure rises to 0.36 bar. Subsequently, the load is released, the voltage does not immediately go down, but it rises for a short period before it starts to drop. This effect happened because the electrical force that opposes the mechanical force suddenly drops due to the load being released, and this caused a high shaft rotation that resulted in high generator rotation and generating high voltage. The voltage decreases corresponding to the shaft rotation that was coupled with the generator. The generator slowly decreases its speed, even though the pressure through the air-motor has been drastically reduced. As shown in Figure 28, the voltage drops slowly, which corresponds to the shaft rotation. In Figure 28, the pressure value has been multiplied by 100 to simplify the analysis process.

The graph shown in Figure 29 is about the comparison of settling time for each cycle of ANN testing Scenario 2. The number shown on each cycle is based on Figure 28. The results show that the longest cycle to normalize towards steady state is the first Cycles 1, 3, and 5 (cycles when the load is released). While Cycles 2 and 4 have a shorter settling time than others. This results shows that the problem of shaft rotation after the load is removed is one of the variables that must be resolved to accelerate the normal process.

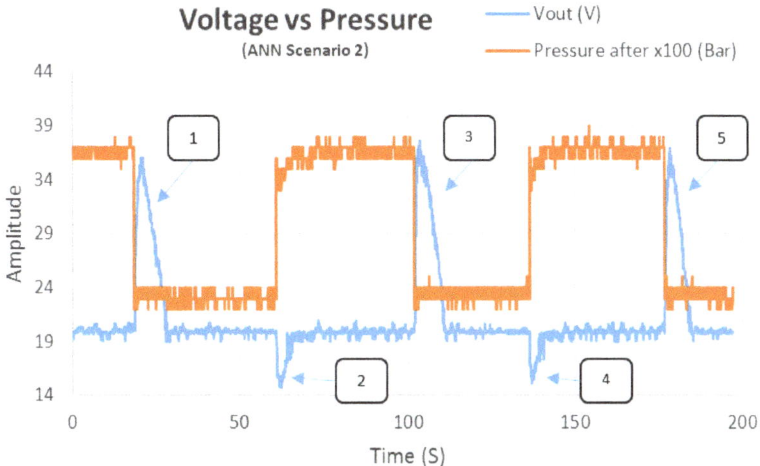

Figure 28. Comparison of voltage and pressure on the ANN scenario.

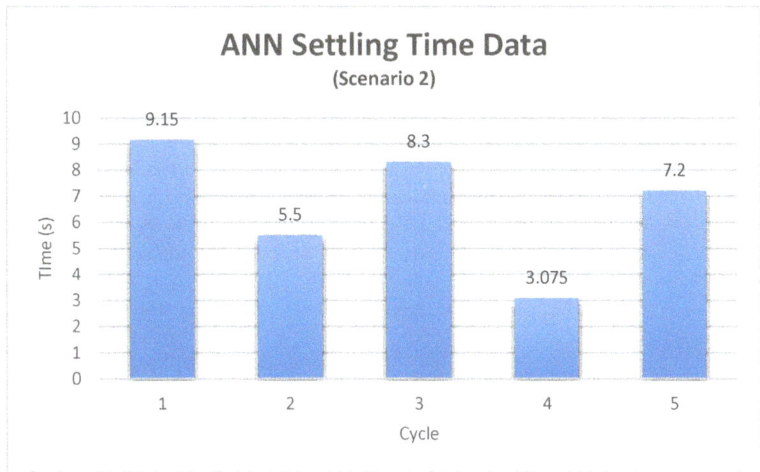

Figure 29. Comparison of settling time on each cycle in ANN scenario.

6. Conclusions

In this paper, an experimental evaluation is done to stabilize the voltage that is generated by the SS-CAES, so that it can be integrated into the grid indirectly. The voltage is controlled by converting air passing through the airmotor using a valve combined with a servo motor. This experiment also applied three artificial intelligences; Fuzzy Logic, ANN, and ANFIS.

This experiment uses two scenarios, the first scenario was done by changing the set point and the best results was obtained by using ANN with the average settling time of 2.05 s. The second scenario was done by connecting the load and specifying which load has the best results, as obtained using ANN 6.65 s. Those results could happen because ANN has less iteration than other intelligent controls, and this made the processing have a fast response. However, in the second scenario, there is a high overshoot value when the load was released. This overshoot is happened due to the effect of high pressure when the load was still installed. Accordingly, when the load was released, the electric torque drops suddenly and it caused the air motor to spin again tight, since the remaining mechanical energy was still high. Overall, the results shows that the system successfully stabilized the voltage smoothly.

Author Contributions: Widjonarko has proposed, conceptualization and written the remaining manuscript. R.S., S.W. and E.S. as validated the main idea and supervision. All authors together organized and refined the manuscript in the present form.

Funding: This research is supported by the LPDP scholarships under grant No. 20161141011767. The author gratefully acknowledges Universitas Brawijaya for the support for this research.

Conflicts of Interest: The author declares no conflict of interest.

Nomenclature

AI	Artificial intelligence
AM	Air motor
b	Bias
$de(t)$	Delta error
$e(t)$	Error
$e(t-1)$	Error in time before
MG	Motor generator
PWM(t)	Pulse Width Modulator (actual output)
PWM($t-1$)	Pulse Width Modulator (actual output in time $t-1$)
PWM(AIOutput)	Pulse Width Modulator from AI output process
V	Voltage, V
V_{Out}	Voltage output, V
V_{Ref}	Voltage reference, V
w	Weight

References

1. Alhamali, A.; Farrag, M.E.; Bevan, G.; Hepburn, D.M. Review of Energy Storage Systems in electric grid and their potential in distribution networks. In Proceedings of the 2016 Eighteenth International Middle East Power Systems Conference (MEPCON), Cairo, Egypt, 27–29 December 2016; pp. 546–551.

2. Ubertini, S.; Facci, A.L.; Andreassi, L. Hybrid Hydrogen and Mechanical Distributed Energy Storage. *Energies* **2017**, *10*, 2035. [CrossRef]

3. Luo, X.; Wang, J.; Dooner, M.; Clarke, J. Overview of current development in electrical energy storage technologies and the application potential in power system operation. *Appl. Energy* **2015**, *137*, 511–536. [CrossRef]

4. Li, W.; Kong, D.; Wu, J. A New Hybrid Model FPA-SVM Considering Cointegration for Particular Matter Concentration Forecasting: A Case Study of Kunming and Yuxi, China. *Comput. Intell. Neurosci.* **2017**, *2017*, 1–11. [CrossRef] [PubMed]

5. Lal, A.; Kumar, R.; Mehta, U. Energy dispatch fuzzy model in hybrid power system. *Int. Energy J.* **2014**, *14*, 133–142.

6. Ellabban, O.; Abu-Rub, H.; Blaabjerg, F. Renewable energy resources: Current status, future prospects and their enabling technology. *Renew. Sustain. Energy Rev.* **2014**, *39*, 748–764. [CrossRef]

7. Lv, S.; He, W.; Zhang, A.; Li, G.; Luo, B.; Liu, X. Modelling and analysis of a novel compressed air energy storage system for trigeneration based on electrical energy peak load shifting. *Energy Convers. Manag.* **2017**, *135*, 394–401. [CrossRef]

8. Mohan, V.; Singh, J.G.; Ongsakul, W.; Madhu, M.N.; Suresh, M.P.R. Economic and network feasible online power management for renewable energy integrated smart microgrid. *Sustain. Energy Grids Netw.* **2016**, *7*, 13–24. [CrossRef]

9. Luo, X.; Wang, J.; Dooner, M.; Clarke, J.; Krupke, C. Overview of Current Development in Compressed Air Energy Storage Technology. *Energy Procedia* **2014**, *62*, 603–611. [CrossRef]

10. Akinyele, D.; Rayudu, R.; Rayudu, R. Review of energy storage technologies for sustainable power networks. *Sustain. Energy Technol. Assess.* **2014**, *8*, 74–91. [CrossRef]

11. Hussain, H.; Javaid, N.; Iqbal, S.; Hasan, Q.; Aurangzeb, K.; Alhussein, M. An Efficient Demand Side Management System with a New Optimized Home Energy Management Controller in Smart Grid. *Energies* **2018**, *11*, 190. [CrossRef]

12. Khamis, A.; Shahrieel, M.; Aras, M.; Nizam, H.; Shah, M.; Zamzuri, M.; Rashid, A. Design and Analysis of Diesel Generator with Battery Storage for Microgrid System. *Int. J. Adv. Eng. Res. Technol.* **2017**, *5*, 26–32.

13. Lim, Y.; Al-Atabi, M.; Williams, R.A. Liquid air as an energy storage: A review. *J. Eng. Sci. Technol.* **2016**, *11*, 496–515.

14. Kassim, A.H.; Miskon, M.T.; Rustam, I.; Mat Zain, M.Y.; Mod Arifin, A.I.; Rizman, Z.I. Optimum storage size for thermal energy storage system. *ARPN J. Eng. Appl. Sci.* **2017**, *12*, 3304–3307.

15. Han, X.; Liao, S.; Ai, X.; Yao, W.; Wen, J. Determining the Minimal Power Capacity of Energy Storage to Accommodate Renewable Generation. *Energies* **2017**, *10*, 468. [CrossRef]

16. Sun, H.; Luo, X.; Wang, J. Feasibility study of a hybrid wind turbine system—Integration with compressed air energy storage. *Appl. Energy* **2015**, *137*, 617–628. [CrossRef]

17. Li, S.; Dai, Y. Design and Simulation Analysis of a Small-Scale Compressed Air Energy Storage System Directly Driven by Vertical Axis Wind Turbine for Isolated Areas. *J. Energy Eng.* **2015**, *141*, 04014032. [CrossRef]

18. Krupke, C.; Wang, J.; Clarke, J.; Luo, X. Modeling and Experimental Study of a Wind Turbine System in Hybrid Connection with Compressed Air Energy Storage. *IEEE Trans. Energy Convers.* **2017**, *32*, 137–145. [CrossRef]

19. Saadat, M.; Shirazi, F.A.; Li, P.Y. Modeling and control of an open accumulator Compressed Air Energy Storage (CAES) system for wind turbines. *Appl. Energy* **2015**, *137*, 603–616. [CrossRef]

20. Hadjipaschalis, I.; Poullikkas, A.; Efthimiou, V. Overview of current and future energy storage technologies for electric power applications. *Renew. Sustain. Energy Rev.* **2009**, *13*, 1513–1522. [CrossRef]

21. Winslow, K.M.; Laux, S.J.; Townsend, T.G. A review on the growing concern and potential management strategies of waste lithium-ion batteries. *Resour. Conserv. Recycl.* **2018**, *129*, 263–277. [CrossRef]

22. Orlins, S.; Guan, D. China's toxic informal e-waste recycling: Local approaches to a global environmental problem. *J. Clean. Prod.* **2016**, *114*, 71–80. [CrossRef]

23. Richa, K.; Babbitt, C.W.; Gaustad, G. Eco-Efficiency Analysis of a Lithium-Ion Battery Waste Hierarchy Inspired by Circular Economy. *J. Ind. Ecol.* **2017**, *21*, 715–730. [CrossRef]

24. Castellani, B.; Morini, E.; Nastasi, B.; Nicolini, A.; Rossi, F. Small-Scale Compressed Air Energy Storage Application for Renewable Energy Integration in a Listed Building. *Energies* **2018**, *11*, 1921. [CrossRef]

25. Chen, H.; Zhang, X.; Liu, J.; Tan, C. Compressed Air Energy Storage. In *Energy Storage—Technologies and Applications*; Zobaa, A.F., Ed.; InTechOpen: London, UK, 2013; pp. 101–112. ISBN 978-953-51-0951-8.

26. Wang, J.; Lu, K.; Ma, L.; Dooner, M.; Miao, S.; Li, J. Overview of Compressed Air Energy Storage and Technology Development. *Energies* **2017**, *10*, 991. [CrossRef]

27. Safaei, H.; Aziz, M.J. Thermodynamic Analysis of Three Compressed Air Energy Storage Systems: Conventional, Adiabatic, and Hydrogen-Fueled. *Energies* **2017**, *10*, 1020. [CrossRef]

28. Salvini, C. CAES Systems Integrated into a Gas-Steam Combined Plant: Design Point Performance Assessment. *Energies* **2018**, *11*, 415. [CrossRef]

29. Chong, S.-H. High Development of a Numerical Approach to Simulate Compressed Air Energy Storage Subjected to Cyclic Internal Pressure. *Energies* **2017**, *10*, 1620. [CrossRef]

30. Jerin, R.A.; Thomas, M.; Palanisamy, K.; Umashankar, S. Enhancing Low Voltage Ride Through Capability in Utility Grid Connected Single Phase Solar Photovoltaic System. *J. Eng. Sci. Technol.* **2018**, *13*, 1016–1033.

31. Gafazi, A.; Cheknane, A.; Merzouk, I.; Seddik, B. Control of A Three Levels AC/DC Converter With Virtual-Flux Direct Power Controlling Method for Grid-Connected Wind Power System Under Grid'S Fault. *J. Eng. Sci. Technol.* **2018**, *13*, 2236–2245.

32. Castillo, A.; Gayme, D.F. Grid-scale energy storage applications in renewable energy integration: A survey. *Energy Convers. Manag.* **2014**, *87*, 885–894. [CrossRef]

33. Phochai, O.; Ongsakul, W.; Mitra, J.; Member, S. Voltage Control Strategies for Grid-Connected Solar PV Systems. In Proceedings of the 2014 International Conference and Utility Exhibition on Green Energy for Sustainable Development, Pattaya, Thailand, 19–21 March 2014; pp. 19–21.

34. Veerasathian, C.; Ongsakul, W.; Mitra, J. Voltage stability assessment of DFIG wind turbine in different control modes. In Proceedings of the 2014 International Conference and Utility Exhibition on Green Energy for Sustainable Development, Pattaya, Thailand, 19–21 March 2014; pp. 1–6.

35. Venkataramani, G.; Parankusam, P.; Ramalingam, V.; Wang, J. A review on compressed air energy storage—A pathway for smart grid and polygeneration. *Renew. Sustain. Energy Rev.* **2016**, *62*, 895–907. [CrossRef]

36. Martinez, M.; Molina, M.G.; Frack, P.F.; Mercado, P.E.; Molina, M. Dynamic Modeling, Simulation and Control of Hybrid Energy Storage System Based on Compressed Air and Supercapacitors. *IEEE Lat. Am. Trans.* **2013**, *11*, 466–472. [CrossRef]

37. Martinez, M.; Molina, M.; Mercado, P.E. Dynamic performance of compressed air energy storage (CAES) plant for applications in power systems. In Proceedings of the 2008 IEEE/PES Transmission and Distribution Conference and Exposition: Latin America, Sao Paulo, Brazil, 8–10 November 2010; pp. 496–503.

38. Maia, T.A.; Barros, J.E.; Filho, B.J.C.; Porto, M.P. Experimental performance of a low cost micro-CAES generation system. *Appl. Energy* **2016**, *182*, 358–364. [CrossRef]

39. Kokaew, V.; Moshrefi-Torbati, M.; Sharkh, S.M. Maximum Efficiency or Power Tracking of Stand-alone Small Scale Compressed Air Energy Storage System. *Energy Procedia* **2013**, *42*, 387–396. [CrossRef]

40. Kokaew, V.; Sharkh, S.; Moshrefi-Torbati, M. A hybrid method for maximum power tracking of a small scale CAES system. In Proceedings of the 2014 9th International Symposium on Communication Systems, Networks & Digital Sign (CSNDSP), Manchester, UK, 23–25 July 2014; pp. 61–66.

41. Kokaew, V.; Sharkh, S.M.; Moshrefi-Torbati, M. Maximum Power Point Tracking of a Small-Scale Compressed Air Energy Storage System. *IEEE Trans. Ind. Electron.* **2016**, *63*, 985–994. [CrossRef]

42. Vongmanee, V.; Monyakul, V. A New Concept of Small-Compressed Air Energy Storage System Integrated with Induction Generator. In Proceedings of the 2008 IEEE International Conference on Sustainable Energy Technologies, Singapore, 24–27 November 2008; pp. 866–871.

43. Lemofouet, S.; Rufer, A. A Hybrid Energy Storage System Based on Compressed Air and Supercapacitors with Maximum Efficiency Point Tracking (MEPT). *IEEE Trans. Ind. Electron.* **2006**, *53*, 1105–1115. [CrossRef]

44. Moradi, M.H.; Eskandari, M.; Hosseinian, S.M. Cooperative control strategy of energy storage systems and micro sources for stabilizing microgrids in different operation modes. *Int. J. Electr. Power Energy Syst.* **2016**, *78*, 390–400. [CrossRef]

45. Tayab, U.B.; Al Humayun, M.A. Operation and control of cascaded H-bridge multilevel inverter with proposed switching angle arrangement techniques. *J. Eng. Sci. Technol.* **2017**, *12*, 3148–3157.

46. Siri, K. Voltage-mode grid-tie inverter with active power factor correction. In Proceedings of the 2015 IEEE 24th International Symposium on Industrial Electronics (ISIE), Buzios, Brazil, 3–5 June 2015; pp. 1133–1139.

47. Kilic, E.; Yilmaz, S.; Ozcalik, H.R.; Sit, S. A comparative analysis of FLC and ANFIS controller for vector controlled induction motor drive. In Proceedings of the 2015 Intl Aegean Conference on Electrical Machines & Power Electronics (ACEMP), 2015 Intl Conference on Optimization of Electrical & Electronic Equipment (OPTIM) & 2015 Intl Symposium on Advanced Electromechanical Motion Systems (ELECTROMOTION), Side, Turkey, 2–4 September 2015; pp. 102–106.

48. Grosenick, L.; Marshel, J.H.; Deisseroth, K. Closed-Loop and Activity-Guided Optogenetic Control. *Neuron* **2015**, *86*, 106–139. [CrossRef] [PubMed]

49. Liu, C.; Wang, J.; Deng, B.; Wei, X.-L.; Yu, H.-T.; Li, H.-Y. Variable universe fuzzy closed-loop control of tremor predominant Parkinsonian state based on parameter estimation. *Neurocomputing* **2015**, *151*, 1507–1518. [CrossRef]

50. Khamis, A.; Badarudin, Z.M.; Ahmad, A.; Ab Rahman, A.; Bakar, N. Development of mini scale compressed air energy storage system. In Proceedings of the 2011 IEEE Conference on Clean Energy and Technology (CET), Kuala Lumpur, Malaysia, 27–29 June 2011; pp. 151–156.

51. Namazov, M. DC motor position control using fuzzy proportional-derivative controllers with different defuzzification methods. *Turk. J. Fuzzy Syst.* **2010**, *1*, 36–54.

52. Agarwal, P.A.P. Brushless Dc Motor Speed Control Using Proportional-Integral and Fuzzy Controller. *IOSRJEEE* **2013**, *5*, 68–78. [CrossRef]

53. Kamble, L.V.; Pangavhane, D.R. Heat transfer studies using artificial neural network—A review. *Int. Energy J.* **2014**, *14*, 25–42.

54. Jafarian, M.; Ranjbar, A. Fuzzy modeling techniques and artificial neural networks to estimate annual energy output of a wind turbine. *Renew. Energy* **2010**, *35*, 2008–2014. [CrossRef]
55. Tino, P.; Benuskova, L.; Sperduti, A. *Artificial Neural Network Models*; Springer Handbooks: Berlin, Germany, 2015; Volume 8, pp. 455–472.
56. Yang, H.; Fu, Y.-T.; Zhang, K.-P.; Li, Z.-Q. Speed tracking control using an ANFIS model for high-speed electric multiple unit. *Control Eng. Pract.* **2014**, *23*, 57–65. [CrossRef]
57. Kassem, Y.; Çamur, H.; Bennur, K.E. Adaptive Neuro-Fuzzy interference System (ANFIS) and Artificial Neural Network (ANN) for predicting the kinematic viscosity and density of biodiesel-petroleum diesel blends. *Am. J. Comput. Sci. Technol.* **2018**, *1*, 8–18.

Article

Multiple-Point Voltage Control to Minimize Interaction Effects in Power Systems

Yun-Hyuk Choi and Yoon-Sung Cho *

School of Electronic and Electrical Engineering, Daegu Catholic University, Gyeongbuk 38430, Korea; yhchoi@cu.ac.kr
* Correspondence: philos@cu.ac.kr; Tel./Fax: +82-53-850-2767

Received: 8 November 2018; Accepted: 9 January 2019; Published: 16 January 2019

Abstract: This paper proposes an advanced continuous voltage control method that implements multiple-point control to ensure peak power system performance. Most control schemes utilize generators to regulate the pilot point voltage of a control area. However, exact control of a single pilot point is difficult because of the influence of adjacent areas in a meshed power system. To address this challenge, the proposed method accesses multiple pilot points to mitigate the effects of the neighboring area. In simulations of the Korean power system, the proposed control scheme offered a considerable improvement in performance when compared with the conventional, currently implemented voltage control system.

Keywords: continuous voltage control; multiple-point control; interaction minimization; pilot point; adjacent areas

---

1. Introduction

The problem of controlling voltage and reactive power in large and complex electric systems requires a great deal of effort on the part of system operators to design and implement sophisticated control schemes. Various devices are used to control these parameters in electrical power systems. Generators are usually equipped with automatic regulators that smooth the voltage variations caused by load fluctuations or failures. Other devices are also installed for this purpose, such as capacitors, reactors, and transformers with load tap changers. Faced with rapid changes in network and operating conditions, electric utilities are increasingly becoming interested in holistic, coherent control systems. These systems are expected to coordinate local facilities for better voltage control, allowing more stable and faster reactions within different regions of the network in case of high voltages and reactive power variations.

Electrical power utilities have always been aware of the need for voltage control facilities in the transmission network, and a wide variety of approaches have been designed and implemented worldwide [1–6]. In Europe, hierarchical control structures are deployed to automatically coordinate reactive power resources to support a constant system voltage. Novel approaches called secondary voltage regulation (SVR) have been tested in France, Italy, Belgium, and Spain, and some of these have already been extended to the national level working on real systems [6]. Hierarchical control systems are organized in a three-level structure: primary, secondary, and tertiary voltage regulation. The primary level controls the terminal voltage of the generator with an automatic voltage regulator. The secondary level, which is based on the pilot point concept, controls the voltage at pilot points by varying the output of generators in each control area of the transmission network. Figure 1 illustrates the concept of secondary-level voltage regulation. Finally, tertiary control computes changes in generator voltage to regulate load voltages on the entire interconnected system.

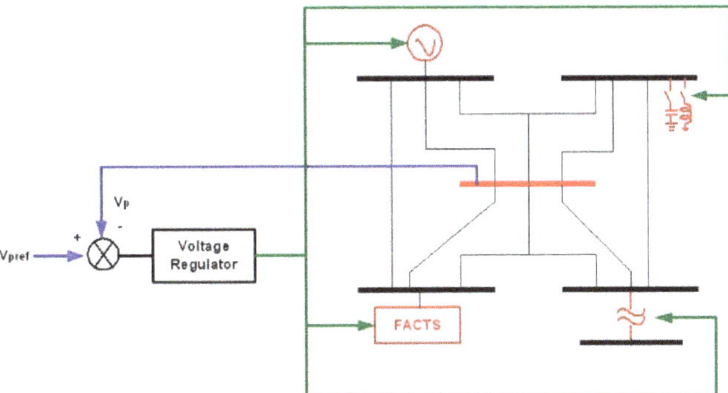

Figure 1. Secondary voltage regulation concept.

Recent studies on closed-loop control have mainly focused on the quality of models in model predictive control implementations [7] and the feedback design of proportional–integral–derivative (PID) control systems [8]. Various control methods are applied to the closed-loop control. While voltage control of an interconnected, large-scale power system is widely recognized as an important problem, the basic formulation of a control scheme is often specific to the utility. Voltage control is typically viewed as a static problem, whose solution is identical to centralized open-loop, optimization-based voltage management. This approach is often referred to as tertiary control, particularly in the literature about European power systems [9,10].

To ensure that the different levels of hierarchical control do not adversely affect each other and to reduce the risks of oscillation or hunting, each hierarchical level has a different time response. At the primary level, control devices, such as generator automatic voltage regulators (AVRs), act locally on rapid voltage variations to keep the local voltages at their reference values. The time constant of these devices is generally in the range of hundreds of milliseconds to seconds. At the secondary level, slower and larger voltage variations in the control area, such as those caused by hourly load changes or contingencies, are fed back to the controller as voltage variations from the reference value of a pilot point. Secondary-level controllers act upon these deviations and update the reference values at the primary level with a time constant of the order of tens of seconds to a few minutes. Finally, at the tertiary level, power system data is used to compute optimal pilot point voltages to ensure economy and security of the power system operations. These computations are achieved by solving optimization problems, either automatically or manually. The time constant for these computations can be tens of minutes.

This paper develops a multiple-point algorithm for continuous voltage control (CVC), which processes multiple voltage points that interact with multiple continuous voltage control (MCVC) in each zone. The CVC is based on the SVR but has several problems such as interaction effects and voltage oscillations. Therefore, the MCVC algorithm can regulate the voltages at pilot points around set-point values while separating the evolutions of those voltages. Operators generally want to modify device operations locally, which is possible when adjusting the set-point voltage of a pilot point. In addition, the proposed control method can prevent voltage oscillations at adjacent pilot points via the existing CVC. The MCVC algorithm addresses the reference voltage at all pilot points. First, the target voltage is regulated for voltage stability within a zone; then, adjustments are made to the target voltages in neighboring zones to smooth the voltage profile of the whole system. The proposed control algorithm was tested using the data of the Korean power system.

2. Multiple-Point Control Algorithm for the CVC

2.1. Principle of the Control Algorithm

Voltage variations in each control zone are represented by the variations at the pilot points. The aim of the multiple-point control algorithm is to hold the voltages at these pilot points at set-point values, such as the CVC. However, the number of generators, and thus the number of control variables, is generally greater than the number of pilot-point output variables. The multiple-point control algorithm improves the CVC so that it can control larger zones compared to the original CVC, which was designed to control smaller zones. With enhanced coordination, the advanced CVC achieves better performance in terms of the interactions between control zones. Consequently, the reactive power generation of the control generator can be minimized, and reactive power reserves can be conserved to cope with any disturbance in the control zone [11,12].

The basic structure of the multiple-point control algorithm with distributed hierarchical control systems is shown in Figure 2. The MCVC is treated as two controllers: a coordination controller governed by the corresponding execution controllers and an individual primary voltage controller for the reactive power dispatcher (RPD). Taken together, these controllers form the multiple-point control system. All the control generators are coordinated for a common objective: to minimize the voltage deviation under normal operating conditions and to maintain an acceptable regional voltage profile in case of system contingencies.

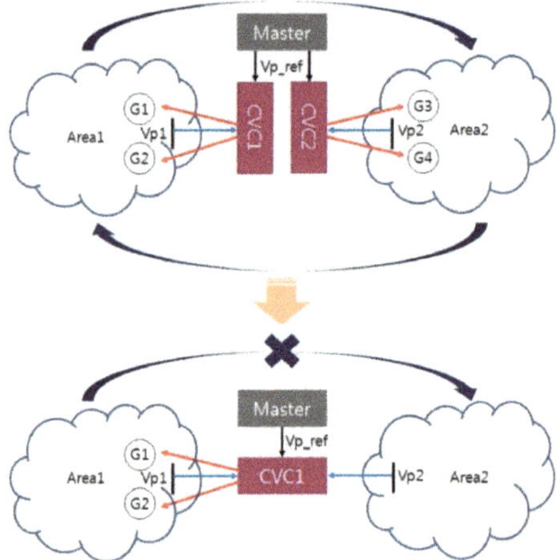

Figure 2. Multiple-point control algorithm concept.

The MCVC algorithm accounts for the existing voltage of pilot points in neighboring zones along with the reference voltage of a pilot point in a single zone, as depicted in Figure 2. The primary objective is to regulate pilot points in a single zone, which was obviously the role of the original CVC. The MCVC also includes the secondary objective of preventing mutual influence between two zones. The Master in Figure 2 gives the voltage reference value. In MCVC, the Master sends the voltage reference to the CVC, and the CVC then compares the voltage from Area 1 and Area 2. The CVC decides the set-point of the generator by comparing the results.

2.2. Operating Mechanism of the Control Algorithm

The control algorithm is composed of four sections: a dead-band, a decision-making section, a proportional integral (PI) controller, and an integral (I) controller. The relationships between these steps are illustrated in Figure 3.

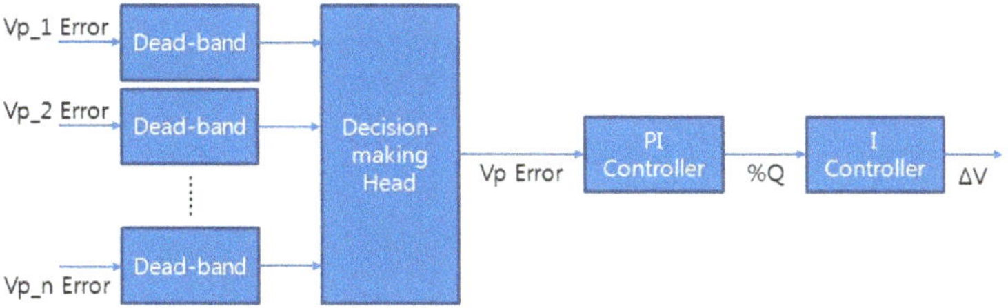

Figure 3. Architecture of the control algorithm.

The dead-band section includes two processes. The first process is a protection logic that enables the controller to avoid abnormal signals that are, for example, very intense or out of sequence. If the input signal is over a standard threshold, the control block changes this output signal to zero. An alarm condition is also included if no sequence is input to the dead-band block. In this case, the algorithm is stopped. The rule of the protection logic is described in Figure 4, and the abnormal signal condition therein is as follows. V_{max} and V_{min} are generically defined as 1.05 (p.u.) and 0.95 (p.u.) in Figure 4.

$$\left| V_p - V_{p\_ref} \right| \geq \varepsilon 1 \tag{1}$$

The second process within the dead-band accounts for sampling error. This prevents control oscillations or unnecessary control interventions. Its concept is similar to that underlying the time integration method [13]. The concept behind the second process is illustrated in Figure 5. E_{max} and E_{min} are 0.01 and -0.01, respectively, in Figure 5. $\varepsilon 1$ and $\varepsilon 2$ are the heuristic values and are set by the system operator. The sampling error condition is as follows:

$$\left| V_p - V_{p\_ref} \right| \leq \varepsilon 2 \tag{2}$$

The decision-making head is the core of the control algorithm. To ensure optimal control, the decision-making head compares the voltage error of a particular pilot point with those of the other pilot points. It selects the most important bus, which needs to be controlled more than the other buses, for every time constant. The procedure for selecting the main pilot point for control is described in Figure 6.

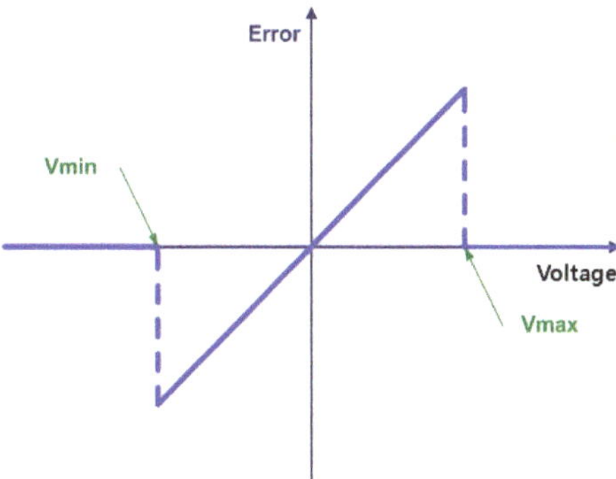

Figure 4. Protection logic rule in the dead-band step.

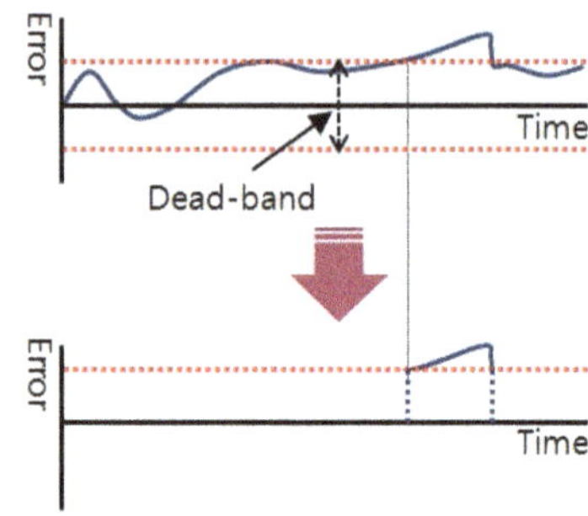

Figure 5. Concept underlying the second process in the dead-band step.

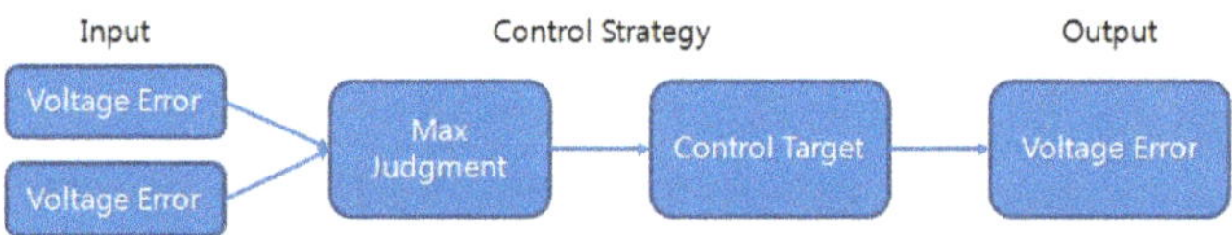

Figure 6. Decision-making principles.

First, the voltage errors of all pilot points are input to the max judgment block. In this mode, the maximum value among all the voltage errors is identified. The bus with this maximum error is in more urgent need of regulation than the other buses. After this mode, the original voltage error is normalized to the maximum value. In the final mode, a logic control switch judges the situation and determines whether to alter the generator output. If the selected voltage error is equal to the maximum value, the controller sends the output to the power plant's CVC; if it is not, this means that one of the other pilot points should be taken as the control target. Equation (3) relates the input errors caused by all pilot points and considers the mutual effects between them.

$$\dot{e}_{r1}(t) = V_{P_{REF1}}(t) - V_{P_1}(t)$$

$$\dot{e}_{r2}(t) = V_{P_2}(t) - V_{P_2}(0) \tag{3}$$

$$\cdots$$

$$\dot{e}_{rn}(t) = V_{P_N}(t) - V_{P_N}(0)$$

where

$\dot{e}_{r1}(t), \ldots, \dot{e}_{rn}(t)$ represent the voltage error;
$V_{P1}(t)$ represents the present voltage of the target pilot point at time t;
$V_{P2}(t), \ldots, V_{PN}(t)$ represent the present voltages of the participating pilot points from the 2nd to the Nth adjacent zone at time t, respectively;
$V_{P\_REF1}(t)$ is the reference voltage of the target pilot point at time t; and
$V_{P2}(0), \ldots, V_{PN}(0)$ are initial voltages of pilot points that are included in the 2nd to the Nth adjacent zone at the initial time, respectively.

The maximum value is selected from this set of vectors as in Equation (4):

$$\dot{e}_r \in \left\{ \dot{e}_r \middle| max\left(|\dot{e}_{r1}|, |\dot{e}_{r2}|, \ldots, |\dot{e}_{rn}|\right) \right\} \tag{4}$$

Using this scheme, the control target is selected by the decision-making head, and the control action is initiated. The PI controller section implements the MCVC. The relevant mathematical equations are as follows:

$$Q_G\%(t) = K_{PC}\dot{e}_r(t) + \int K_{IC}\dot{e}_r(t)dt \tag{5}$$

where

$Q_G\%(t)$ is the reactive power to be generated in each RPD, and K_{PC} and K_{IC} are the proportional and integral gain, respectively, in the MCVC.

The reactive power levels are generated in each RPD and sent to the RPD controller. The PI controller calculates the reactive power level using the difference between the pilot point voltage and the reference value. Then, the I controller adjusts the reference voltage of the AVR using the difference between the calculated reactive power level and the generated reactive power. K_{PC} and K_{IC} are chosen using the sensitivity matrix that relates the pilot points with the control generators. These coefficients must also take time constants into consideration to clearly establish the control hierarchy. The control signals must be arranged to prevent any negative effects resulting from overlapping with the other controllers, such as AVRs. The time constants are generally assumed to be such that AVR (ms) < RPD (5 s) < MCVC (50 s). In addition, $Q_G\%(t)$ has a limit block that corresponds to the generator's available capacity. To respond to the reactive power generation appropriately, the generator's available capacity curve must be known. However, a constant reactive power limit determined from the supervisory control and data acquisition/energy management system (SCADA/EMS) data has been used in this study to determine the upper and lower limits because the actual generator specifications were not available.

The I controller section is called the RPD. The equations involved are as follows:

$$\dot{e}_q(t) = Q_{G_{REF}}(t) - Q_G(t)$$

$$Q_{G_{REF}}(t) = Q_G\%(t)Q_{G_{MIN/MAX}}(t) \tag{6}$$

$$\Delta V_G(t) = \int K_{IR}\dot{e}_q(t)dt$$

where

$Q_G(t)$ represents the reactive power of each generator at time t;
$Q_{G\_REF}(t)$ is the reference reactive power at time t;
$Q_G\%(t)$ is the reactive power to be generated according to the MCVC control signals;
$Q_{G\_MIN/MAX}(t)$ is the lower/upper limit of the reactive power; and
K_{IR} is the integral gain in the RPD.

The RPD controller outputs a control signal to regulate the reactive power output of its own generator, and it also sends a signal to the AVR to change the reference voltage of the terminals at other generators. The amount of reactive power that is to be generated is calculated from $Q_G\%(t)$ in the MCVC controller. The calculated reactive power is then compared with the reactive power that is presently being generated, and the difference is used to adjust the reference voltage of the AVR by the I controller. K_{IR} is calculated with consideration of the time constant of the RPD and the reactance of the step-up transformer of each generator. Figure 7 describes the control blocks for the MCVC and the RPD.

The parameters of the control block can be calculated from the following equations:

$$K_{PC} = \frac{1}{Q_{G\_MAX/MIN} \times X_t}$$
$$K_{IC} = \frac{1 + K_{PC} \times Q_{G\_MAX/MIN} \times X_{eq}}{T_{CS} \times Q_{G\_MAX/MIN} \times X_{eq}} \tag{7}$$
$$K_{IR} = \frac{X_t + X_{eq}}{T_{RS}}$$

where

$$T_{CS} = 50s$$

$$T_{RS} = 5s$$

$$Q_{G\_REF} = \begin{cases} qQ_{G\_MAX} \; for \; 0 < q < 1 \\ -qQ_{G\_MIN} \; for \; -1 < q < 0 \end{cases}$$

As mentioned above, X_t is the reactance of the step-up transformer installed with each generator, and X_{eq} is calculated using the sensitivity matrix of the pilot point and the control generators.

$$\begin{bmatrix} \Delta Q_i \\ \Delta Q_j \end{bmatrix} = \begin{bmatrix} B_{ii} & B_{ij} \\ B_{ji} & B_{jj} \end{bmatrix} \begin{bmatrix} \Delta V_i \\ \Delta V_j \end{bmatrix} \tag{8}$$

$$\Delta Q_i = \left[B_{ii} - B_{ij} B_{jj}^{-1} B_{ji} \right] \Delta V_j \tag{9}$$

where

ΔQ_i and ΔQ_j are the reactive power of the load and the generator, respectively; Bii, Bij, Bji, and Bjj are the transmission line conductance; ΔV_i and ΔV_j are the voltage of the bus, respectively.

The equations for calculating X_{eq} in Equations (8) and (9) are drawn from the Jacobian matrix, which represents the sensitivity of the relationship between the pilot point and the control generators [14]. Table 1 lists the values of the control parameter for the control generators in the Korean power system.

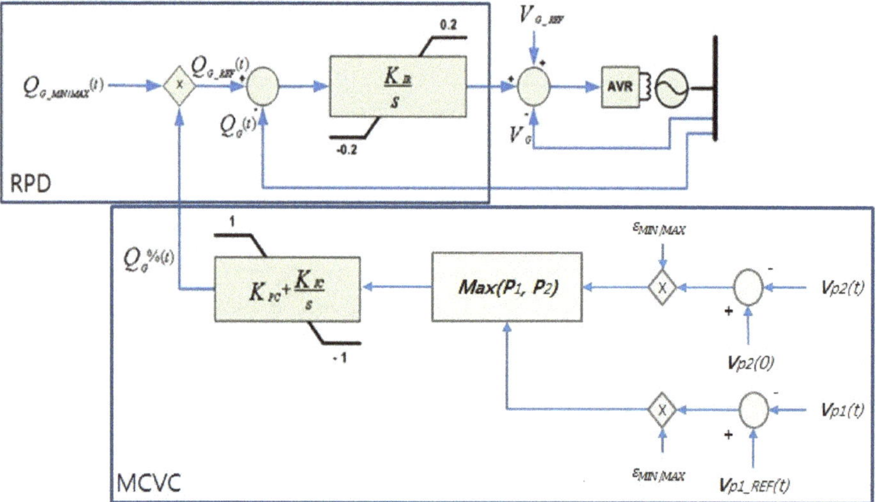

Figure 7. Control block for the multiple continuous voltage control (MCVC) and the reactive power dispatcher (RPD).

Determining the response time of MCVC is the time constants of the PI and I controllers. Therefore, the response time can be reduced by adjusting the time constant. The time constant of the current PI controller is 50 s, and the time constant of the I controller is 5 s.

Table 1. Control parameters of the control generators in Korean power system.

Name	Q_{G_MAX} [MVAR]	Q_{G_MIN} [MVAR]	X_t	X_{eq}	T_{CS} [s]	T_{RS} [s]	K_{PC}	K_{IC}
Yonggwang NP	387.0	−160.0	0.1248	0.0010	50	5	26.7094	67.2009
Tangjin TP	270.0	−164.0	0.2636	0.0010	50	5	22.2875	117.9546
Seoinchon CC	108.0	−70.0	0.0980	2.0175	50	5	25.1953	0.5284
Youngheong TP	384.0	−262.0	0.0980	2.0175	50	5	25.1953	0.5284
Pyongtaek TP	150.0	−89.0	0.4870	2.0175	50	5	55.4970	1.3779

3. Control Scheme for the MCVC

The flowchart of the control scheme for the MCVC, shown in Figure 8, has the following steps:

Step 1. Monitor the pilot points in the control zones of the target power system.

Step 2. Compare the voltage of a target pilot point to its reference voltage.

Step 3. Compare the voltages of pilot points in neighboring zones with their present voltage.

Step 4. Select the control target with the smallest voltage violation among pilot points.

Step 5. Go to step 1 if none of the pilot points have abnormal voltage.

Step 6. According to the control target, determine the requisite reactive power ratios of generators with the MCVC algorithm.

Step 7. According to these reactive power ratios, determine the terminal reference voltages of control generators with the RPD.

Step 8. Repeat the control process from steps 1 to 3.

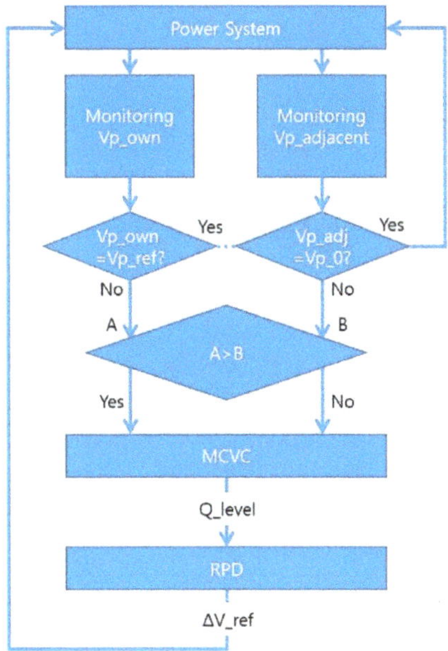

Figure 8. Control scheme of the MCVC.

4. Characteristics of the Korean Power System

4.1. Summary of the Korean Power System

The Korean power system includes about 260 generators and 1400 load buses. Approximately 40% of the total load is concentrated in metropolitan areas, and most of the generators are in nonmetropolitan areas. Furthermore, most of the generation plants in the nonmetropolitan areas have low operating costs. For this reason, a large amount of active power is transmitted from nonmetropolitan to metropolitan areas via interface lines to maximize economy. This transfer of power is defined as interface flow. Any increase in this interface flow, however, may lead to voltage instability due to the lack of reactive power support in the metropolitan areas [15]. Therefore, it is important that the reactive power reserves of power plants in the metropolitan area are defined accurately. The Korean power system is summarized in Table 2.

Table 2. Summary of the Korean power system.

Area	Active Power [MW]		Reactive Power [MVAR]		Number of Installed Generators
	Generations	Loads	Generations	Loads	
Metropolitan	13,779	22,034	3701	10,474	113
Nonmetropolitan	45,089	31,435	10,112	14,222	150

The peak load on the Korean power system is approximately 53,470 MW in the summer. Most plants are in the southwest and southeast regions, while most loads are concentrated in the northern metropolitan area. In addition, switched shunt capacitors and reactors are installed for voltage control at the substations in the north. Figure 9 shows a map of the transmission networks that handle more than 345 kV and the major generating plants in the Korean power system [16].

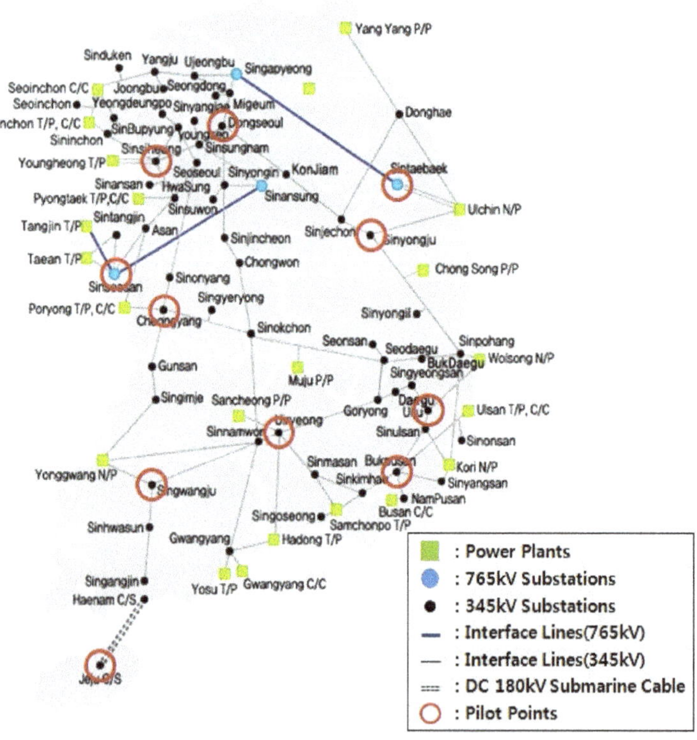

Figure 9. Map of the Korean power system.

4.2. Difficulties of CVC Application in the Korean Power System

The Korean power system is a tightly coupled network, especially in the metropolitan area. Therefore, pilot points in adjacent zones are difficult to isolate from each other. For this reason, the actions of control generators need to be coordinated to allow efficient control of coupled zones.

The transmission lines in the metropolitan area form a ring network, as depicted in Figure 10 [17]. The metropolitan area of the Korean power system has two distinct voltage control zones. The buses feeding the corresponding pilot points are called the Dongseoul and Sinsiheung buses. Each pilot point is closely connected to the other load buses. Therefore, this meshed system responds to the voltage controller as if it is only a single bus. For example, if the voltage at the Dongseoul control target is lowered from an initial value of 1.02 (p.u.) to 1.01 (p.u.), the voltages in the other buses will also fall. The Sinsiheung, Sinbupyeong, Yeongseo, Seoseoul, and Yeongdeungpo pilot points are influenced by this condition. This situation is illustrated in Figure 11.

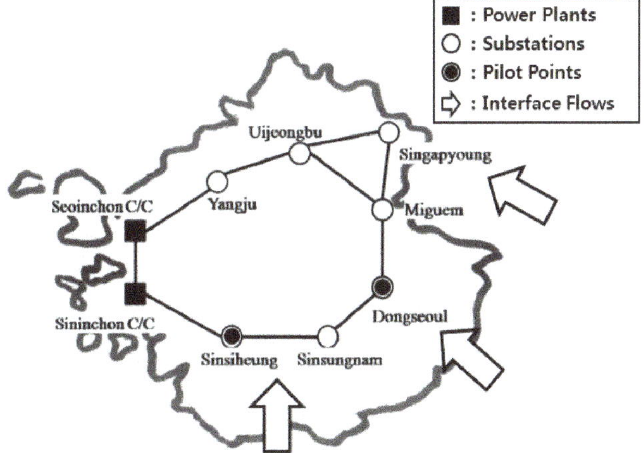

Figure 10. Map of the metropolitan area in the Korean power system.

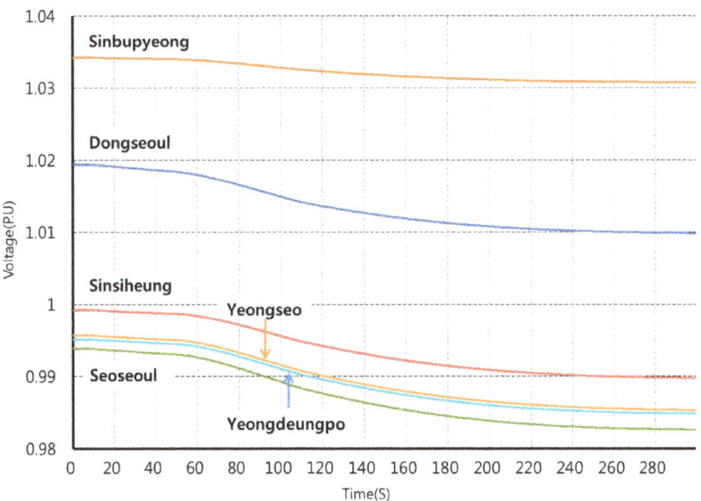

Figure 11. Voltages of pilot points in the metropolitan area.

The meshed power system makes voltage control difficult in some situations. For example, circular var flow and clogging voltage instability may arise. Circular var flow mainly occurs in complicated, large power systems. It results in wasted reactive power resources and can lead to higher line currents as well as unnecessary switching of the reactive power controls [10]. Clogging voltage instability occurs when the network configuration of generators and a particular bus blocks the flow of reactive power to some subregion without expending any reactive power reserves. This means that the needed extra reactive power loads cannot be supplied reliably, or the generators in the network will have an insufficient amount of reactive power in reserve [18–20].

Although the metropolitan area of the Korean power system has many control generators, the control generators in the Dongseoul zone have less than 50 Mvar of reactive capacity. Table 3 compares the reactive power reserves of the Dongseoul and Sinsiheung zones. The Dongseoul zone has much less reactive power in reserve than the Sinsiheung zone. Due to this, the Dongseoul zone faces potential voltage instability. To detect voltage instability and promptly respond to severe system conditions, an effective generator control scheme is needed.

Table 3. Comparison of generator reactive power reserves between pilot points in the metropolitan area of the Korean power system.

Zone	Reactive Power Reserve [Mvar]
Dongseoul	1022
Sinsiheung	5778

5. Simulation Results

To test how the MCVC strategy can manage serious system abnormalities, the results of three cases are summarized in this section. Table 4 lists the cases that were tested.

Table 4. Case summary for simulation.

Case	Description	Purpose of This Case
I	With and without protection logic	Show the usefulness of the protection logic
II	Increase the desired voltage of a pilot point	Test the effectiveness of following a desired voltage at the pilot point
III	Three-phase fault at the interface line	Check the effectiveness of the algorithm
IV	Load increase at the interface line	Comparison between the existing method and the proposed method

5.1. Case (I) Protection Logic Test

After 60 s, an abnormal signal of very low voltage was inserted to the input data. With the protection logic, the control block made the reactive power level zero, so the voltage of the pilot point was not changed (Figure 12). Without the protection logic, the voltage of the pilot point dropped under 0.90 (p.u.). This means that the protection logic is valuable.

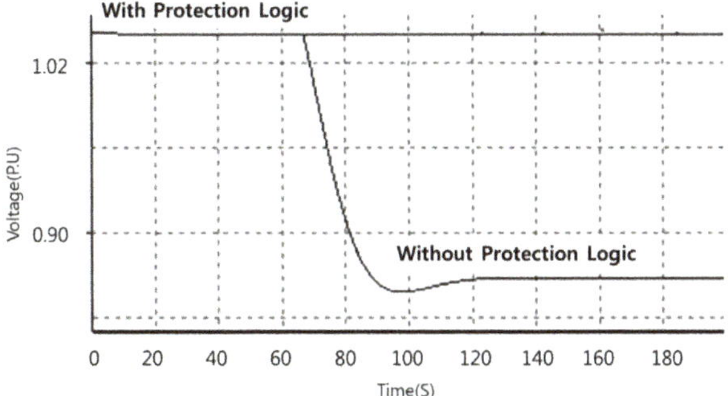

Figure 12. Result of protection logic at pilot point.

5.2. Case (II) Target Voltage Change of a Pilot Point

As shown in Figure 13, the desired voltage of the pilot point at Dongseoul was successfully increased by RPD operations of the MCVC for about 260 s. The desired voltage change was from 1.02 (p.u.) to 1.025 (p.u.) at 50 s. The voltage of the neighboring Sinsiheung pilot point changed only very slightly, as shown in Figure 13. This result indicates that the MCVC reduces the effect of a voltage change on pilot points in the neighboring area.

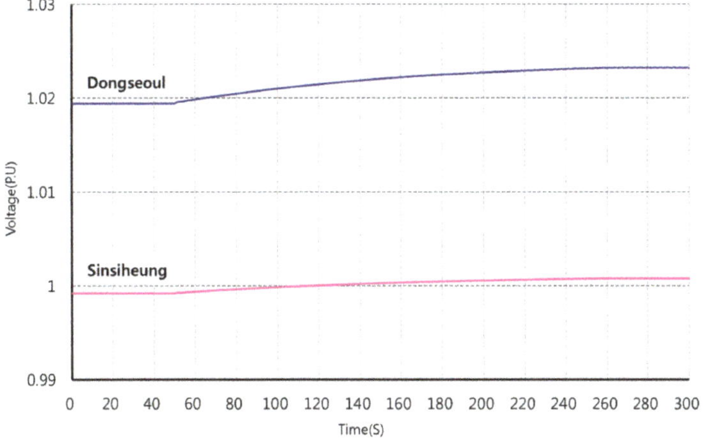

Figure 13. Result of change of target voltage at a pilot point.

5.3. Case (III) Contingency Scenario of an Interface Root

As already mentioned in Section 2, very important interface lines run from rural areas to Korea's metropolitan area. Among them, the Hawseong–Asan line suffers the most severe risk of instability because its carries the most interface flow among the six possible line faults. As shown in Table 5, the margin decreased from 2607.5 MW to 776.1 MW. Therefore, it was considered for an additional case study for a three-phase fault at the Hawseong–Asan line.

Table 5. Interface flow margin of interface line.

Interface Line	No. of Lines	Interface Flow Margins [MW]
Base	-	2607.5
Singapyeong–Sintaebaek	# 2	1761.8
Sinansung–Sinseosan	# 2	1991.6
Hwasung–Asan	# 2	776.1
Seoseoul–Sinonyang	# 2	1492.6
Sinyongin–Sinjincheon	# 2	2444.6
Konjiam–Sinjechon	# 2	1406.8

As the data in Figure 14 shows, a three-phase fault was modeled at 50 s and was cleared after a single cycle. The voltages of the pilot points were returned to the target voltage with the MCVC algorithm applied. The desired voltages of the pilot points are 1.02 (p.u.) at Dongseoul and 1.0 (p.u.) at Sinsiheung. Without the MCVC algorithm, however, the voltages remained lower than their target values after the fault. This means that the proposed control algorithm can effectively handle contingency cases and return the power system to stability within a matter of seconds.

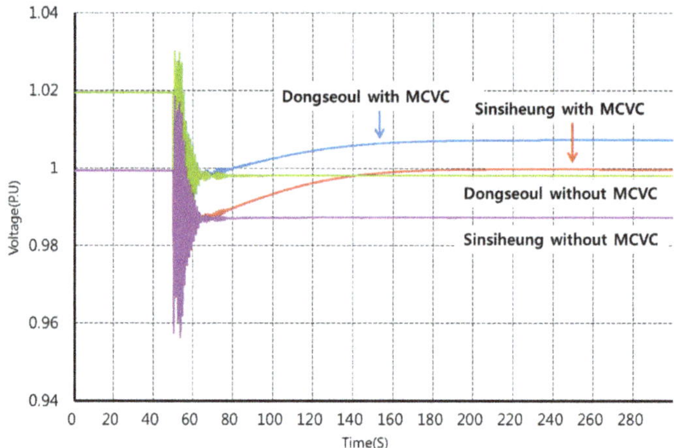

Figure 14. Result of three-phase fault at an interface line.

5.4. Case (IV) Load Increase of the Hwasung–Asan Transmission Line

The control results are described in Figure 15. Figure 15a shows that the MCVC regulated the Dongseoul voltage close to the target voltage and regulated the Sinsiheung voltage close to the initial voltage. The voltages of the Dongseoul and Sinsiheung were stabilized by the MCVC control. This means the Sinsiheung was not influenced by the Dongseoul control. The Dongseoul voltage went to the target voltage due to the CVC control. However, the initial voltage of the Sinsiheung oscillated due to the load increase. Therefore, the voltage control of Dongseoul had a negative influence on the Sinsiheung voltage.

The reactive power control of all generators worked nicely owing to the MCVC control, as shown in Figure 15b. On the other hand, the CVC made unnecessary generations to all generators. Although the CVC of the Sinsiheung zone kept the initial voltage of the Sinsiheung, control generator outputs of the Sinsiheung zone moved in the opposite direction. The Sinsiheung voltage oscillated due to this situation.

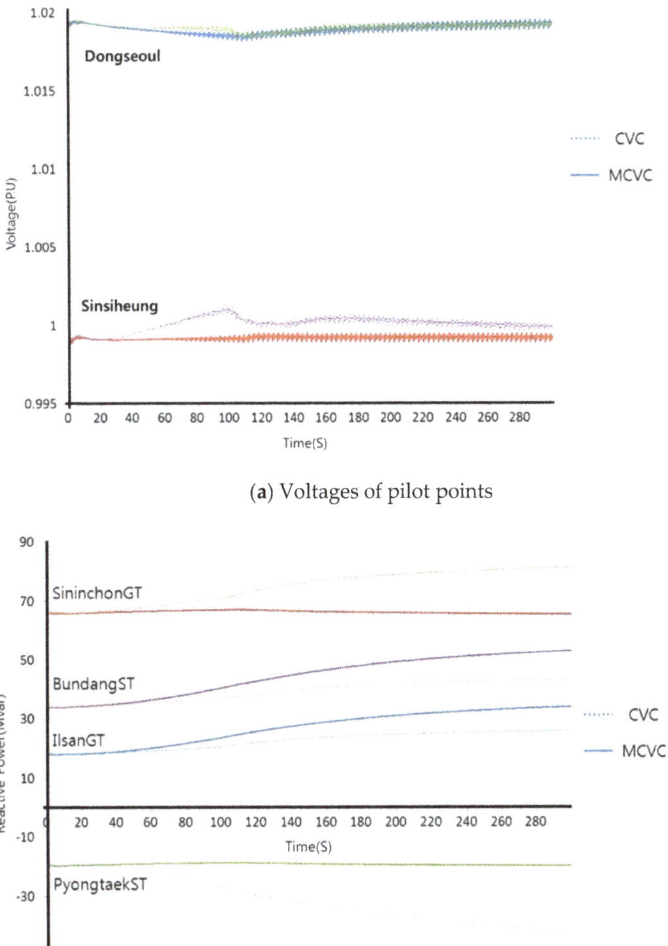

(**a**) Voltages of pilot points

(**b**) Reactive powers of control generators

Figure 15. Result of load increase at the interface line.

6. Conclusions

This paper proposes a multiple-point control algorithm for preventing mutual interference in a power transmission network. The main objective of the control system is to regulate the voltage profiles of the power system in case of emergency system conditions. With the voltage control schemes deployed at present, the control of a pilot point can lead to interference in neighboring zones. In the proposed algorithm, voltage variations in each control zone are represented by the variations at representative pilot points. The MCVC algorithm can regulate the voltages at a group of pilot points to remain around set-point values while dealing with the evolution of those voltages separately. This strategy allows operators to make local modifications to generator output, and these local changes can be applied easily by adjusting the set-point voltage of the pilot point. By selectively regulating the set-point voltage, the proposed control scheme can prevent oscillations of the voltage of neighboring pilot points. The controller accounts for the reference voltages of multiple pilot points. First, the target voltage is regulated to maintain voltage stability in its own zone; then, the neighboring target voltages are adjusted to ensure a smooth voltage profile in neighboring zones.

Simulation tests with realistic data sampled from the Korean power system demonstrate the feasibility of this control scheme for reducing the severity of mutual interactions between adjacent zones. These dynamic simulations were also used to study how the MCVC could return the system to stability from more severe conditions. In future work, coordination between continuous and discrete devices will be included in the control scheme to allow the effective voltage control of a large-scale power system.

Author Contributions: Y.-H.C. put forward to the main idea, guided the experiments and wrote the whole of this paper. Y.-S.C. designed the whole structure of this paper and completed data preprocessing. All authors have read and approved the final manuscript.

Funding: This work was supported by the National Research Foundation of Korea (NRF) grant funded by the Korea government (MSIT) (No. NRF-2017R1C1B5018163). This work was also supported by "Human Resources Program in Energy Technology" of the Korea Institute of Energy Technology Evaluation and Planning (KETEP), granted financial resource from the Ministry of Trade, Industry & Energy, Republic of Korea. (No. 20174010201330).

Conflicts of Interest: The authors declare no conflict of interest.

References

1. Taylor, C.W. *Power System Voltage Stability*; McGraw-Hill Inc.: New York, NY, USA, 1994.
2. Cutsem, T.V.; Vournas, C. *Voltage Stability of Electric Power Systems*; Kluwer Academic Publishers: Dordrecht, The Netherlands, 1998.
3. Kundur, P. *Power System Stability and Control*; McGraw-Hill: New York, NY, USA, 1994.
4. Ilic, M.; Liu, S. *Hierarchical Power System Control*; Springer: Berlin, Germany, 1996.
5. Seo, S.; Choi, Y.H.; Kang, S.; Lee, B.; Shin, J.H.; Kim, T.K. Hybrid control system for managing voltage and reactive power in the JEJU power system. *J. Electr. Eng. Technol.* **2009**, *4*, 429–437. [CrossRef]
6. Corsi, S. *Feasibility Study of the Automatic Voltage Control of the KEPCO Transmission Grid*; Intermediate Report; CESI: Milan, Italy, 2010.
7. Ali, E.; Li, J.; Xie, J.; Joshua, D.I. Closed-loop identification for plants under model predictive control. *Control Eng. Pract.* **2018**, *72*, 206–218.
8. Robert, L.C.; John, F.O. Large feedback control design with limited plant information. *Control Eng. Pract.* **2018**, *72*, 219–229.
9. Kim, T.K.; Choi, Y.H.; Seo, S.S.; Lee, B.J. A study on voltage and reactive power control methodology using integer programming and local subsystem. *Trans. Korean Inst. Electr. Eng. (KIEE)* **2018**, *57*, 543–550.
10. Choi, Y.H.; Lee, B.; Kim, T.K. Optimal shunt compensation for improving voltage stability and transfer capability in metropolitan area of the Korean power system. *J. Electr. Eng. Technol.* **2015**, *10*, 1502–1507. [CrossRef]
11. Kvasov, D.E.; Menniti, D.; Pinnarelli, A.; Sergeyev, Y.D.; Sorrentino, N. Tuning fuzzy power-system stabilizers in multi-machine systems by global optimization algorithms based on efficient domain partitions. *Electr. Power Syst. Res.* **2008**, *78*, 1217–1229. [CrossRef]
12. Sheng, G.; Jiang, X.; Duan, D.; Tu, G. Framework and implementation of secondary voltage regulation strategy based on multi-agent technology. *Int. J. Electr. Power Energy Syst.* **2009**, *31*, 67–77. [CrossRef]
13. Choi, Y.H.; Lee, B.; Kang, S.; Kwon, S.H. Coordinated voltage-reactive power control schemes based on PMU measurement at automated substations. *J. Electr. Eng. Technol.* **2001**, *10*, 1400–1407. [CrossRef]
14. Powell, L. *Power System Load Flow Analysis*; McGraw Hill: New York, NY, USA, 2004; pp. 101–105.
15. Lee, B.; Song, H.; Kwon, S.H.; Jang, G.; Kim, J.H.; Ajjarapu, V. A study on determination of interface flow limits in the KEPCO system using modified continuation power flow (MCPF). *IEEE Trans. Power Syst.* **2002**, *17*, 557–564.
16. Available online: http://www.kpx.or.kr/english/ (accessed on 12 August 2018).
17. Kang, S.G.; Seo, S.; Lee, B.; Chang, B.; Myung, R. Centralized control algorithm for power system performance using FACTS devices in the korean power system. *J. Electr. Eng. Technol.* **2010**, *5*, 353–362. [CrossRef]
18. Capitanescu, F. Assessing reactive power reserves with respect to operating constraints and voltage stability. *IEEE Trans. Power Syst.* **2011**, *26*, 2224–2234. [CrossRef]

19. Choi, Y.H.; Seo, S.; Kang, S.; Lee, B. Justification of Effective Reactive Power Reserves with respect to a Particular Bus using Linear Sensitivity. *IEEE Trans. Power Syst.* **2011**, *26*, 2118–2124. [CrossRef]
20. Choi, Y.H.; Lee, B.; Kim, T.K.; Shin, J.H.; Cho, J.M.; Jung, E.; Kang, B.I. Identification of Voltage Control Areas and Reactive Power Reserves in Power Flow Traceable System for Voltage Stability Assessment. In Proceedings of the System Operation and Control of Electric Power Systems CIGRE Regional Meeting, Recife, Brazil, 3–6 April 2011.

Article

Adaptive Damping Control Strategy of Wind Integrated Power System

Jun Deng [1,*], Jun Suo [1], Jing Yang [2], Shutao Peng [1], Fangde Chi [3] and Tong Wang [2,*]

1   State Grid Shaanxi Electric Power Research Institute, Xi'an 710100, China; powersuo@163.com (J.S.); newpking@126.com (S.P.)
2   School of Electrical & Electronic Engineering, North China Electric Power University, Changping District, Beijing 102206, China; yangjing@ncepu.edu.cn
3   State Grid Shaanxi Electric Power Company, Xi'an 710048, China; chifangde@163.com
*   Correspondence: d.jone@163.com or dengjun@dky.sn.sgcc.com.cn (J.D.); hdwangtong@ncepu.edu.cn (T.W.); Tel.: +86-2989698550 (J.D.); +86-1061771407 (T.W.)

Received: 9 November 2018; Accepted: 10 December 2018; Published: 1 January 2019

Abstract: Random variation of grid-connected wind power can cause stochastic variation of the power system operating point. This paper proposes a new scheme to design an adaptive damping controller by tracking the variation of system operating points and updating the controller's functions to achieve a robust damping control effect. Firstly, the operating space is classified into different modes according to the classification of wind power outputs. Multiple power system stabilizers (PSSs) are then designed. Secondly, the method of optimal classification and regression decision tree (CART) is utilized for classifying subspaces of system operating point and it is proposed that the on-line measurements from wide area measurement system (WAMS) are used for tracking the dynamic behaviors of stochastic drifting point and thus guide the updating of appropriate PSSs be switched on adaptively. A 16-generator-68-bus power system integrated with wind power is presented as a test system to demonstrate that the adaptive control scheme by use of the CART can damp multi-mode oscillations effectively when the wind power output changes.

Keywords: stochastic power system operating point drift; wind integrated power system; power oscillations; adaptive damping control

---

1. Introduction

Traditional power generation is controllable and predictable. The stability and control of the power system is determined for a given set of circumstances [1]. With the rapid growth of wind power resources, the stochastic flotation of wind power has become more and more significant and cannot be neglected [2,3]. In most cases, lack of robustness of the controllers designed for the deterministic operating point when the wind power output changes a lot has become a dominant problem [4]. In a future power system, if the scale of grid-connected wind generation is comparable to the traditional power generation, system stability analysis and control has to consider the random variation features of wind power when the system operating point varies stochastically [5–7].

Therefore, as far as power system stability control is concerned, to guarantee the robust performances in the case of stochastic variation of system operating point as affected by variable integrated wind power, it is important to investigate the design of adaptive damping controller which can accommodate the stochastic variation of the system operating point. Robust control [8] is an effective way to deal with the uncertainties, while when the operating conditions change over a wide range, the robustness might not be guaranteed. Therefore, adaptive control methods such as the fuzzy control method [9], multiple-model method [10,11], Kalman filter method [12], etc., which can track the stochastic behavior of systems have become more emphasized. However, when the operating conditions change a lot, the nonlinearity of power systems is more prominent and the disturbance models more difficult to interpret [13]. Therefore, machine study methods using statistical analysis may be more suitable for this particular application. For example, the classification and regression tree (CART) [14], neural network method [15,16] and so on. Therefore, this paper presents an adaptive control scheme based on CART for a wind integrated power system to accommodate the stochastic variation of wind generation.

The rest part of the paper is organized as follows: firstly, the operating space is classified into different modes according to the classification of wind power outputs. On the basis of the subspaces, multiple PSSs are designed in coordination for mapping rules. Secondly, the CART is formed, trained and tested to identify the subspaces with the help of wide area measurement system (WAMS). The CART determines the selection of appropriate damping control adaptively according to the identification of the subspaces of system operating point. Finally, the results of a 16-generator 68-bus example power system are presented to verify that the proposed adaptive control scheme in the paper can effectively handle the stochastic variation of system operating point and suppress system low-frequency power oscillations when grid-connected wind power varies.

2. The CART-Based Adaptive Damping Control Scheme

2.1. The Formation of Subspaces

The subspaces are formed by dividing the wind power outputs which could cause the changing operating conditions. The lower and higher boundaries correspond to the minimum and maximum wind power outputs as shown in Figure 1a, respectively. The partition method has been used widely in power load prediction as a simple and effective approach. The stochastic wind power outputs are divided into several intervals as shown in Figure 1b, thus the subspaces are formed. All of the subspaces compose the whole operating space and they have the relation: $B_1 \cup B_2 \cup \ldots \cup B_i \cup \ldots \cup B_n \equiv B$. Based on the subspaces, the mapping rules are built for each subspace.

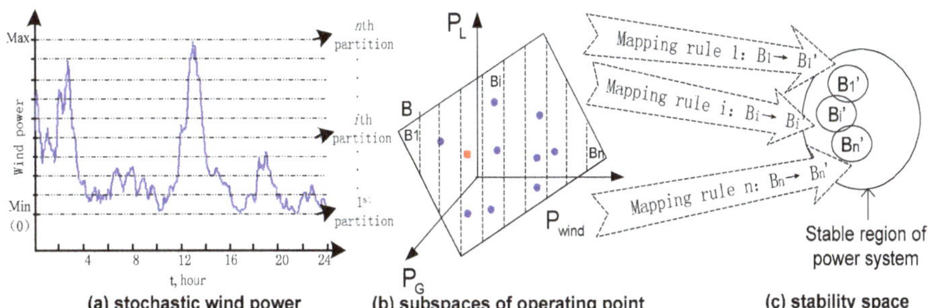

Figure 1. The formation of subspaces.

2.2. Coordinated Design of PSSs

Power system stabilizers (PSSs) have been an effective and economic way to dampen power system low-frequency oscillations for decades. In each operating subspace, multiple PSSs are designed in coordination to ensure the power system operating point to be within the stable region [1]. The participators are used to determine the locations of PSSs to be tuned to achieve effective damping result. The generator speed and electromagnetic power are the input signals of PSS2A as shown in Figure 2. To save the space, details of coordinated design are not presented here. Principle of design is given in reference [17]. $\Delta\omega$ is the generator speed deviation, ΔP is the electromagnetic power deviation. $T_{w1}, T_{w2}, T_{w3}, T_{w4}$ are the time constants of washout blocks. T_1, T_3, T_8 are the lead time constants of phase compensation blocks. T_2, T_4, T_9 are the lag time constants of phase compensation blocks. N is the number of low-pass filters. T_6, T_7 are the time constants of first-order inertia blocks. K_{s1}, K_{s2}, K_{s3} are the gains of PSS.

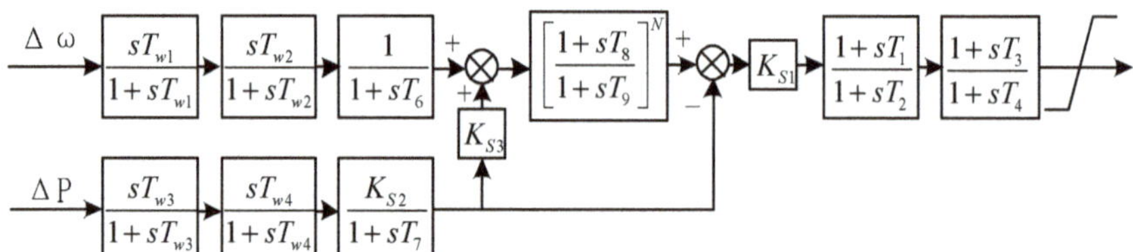

Figure 2. Structure of PSS2A.

2.3. The CART-Based Adaptive Damping Control Scheme

CART is a recursive partitioning method which builds classification and regression trees. Classification trees are built for obtaining the splitting rules of different operating point subspaces whereas regression trees are built for identifying which subspace of an operating point the power system is in. In this way, the coordinated PSSs pre-designed off-line can be switched adaptively using on-line measurements. As long as the wind power output fluctuates, the operating point of the power system will deviate from the initial subspace and randomly move forward to other operating point subspaces or back to the initial subspace. The CART will function to identify the attraction subspace in order for the PSSs to be switched on line.

The CART is structured from the top to the bottom consisting of root node, test nodes and terminal nodes. Each test node or root node corresponds to an optimal splitting rule and a subset of the learning dataset. A terminal node is a pure node which could not be split further. The learning dataset itself corresponds to the root node of the CART. The classification process starts from the top root node, and at each level the subsets will be divided according to the optimal splitting rules. The optimal rules are in the form of "if-then-else" rules. In this paper, each terminal node represents an operating point subspace. The measurements from all operating point subspaces consist of the learning dataset which are the input data of the CART. Reference [18] gives a full introduction of general theory and methods of the CART.

To ensure the accuracy of classification, those measurable and controllable measurements are used as the training data which could characterize the subspaces. Since generators' speed includes the information of power flow routing, topology and the power oscillation modals in a power system, the CART uses the speeds of generators as the learning dataset. However, in most cases, the speeds of generators are not measurable or they are measurable without time tags. Therefore, the generator bus frequencies are employed as the learning dataset in this paper instead of generators' speed. The reason is that the generator bus frequencies, which are the derivation of generators' external bus angle, is an approximation to the generators' speed and the generator bus frequencies can be obtained directly from PMUs.

In a large scale power system, only one measurement could not adequately characterize an operating point subspace. Thus multiple measurements need to be employed for tracking the variation of the power system operating point. Multiple measurements from multiple subspaces make the classification process complex. Therefore, in order to distinguish the features of measurements from different subspaces, the Euclidean distance to the hyperplanes is used as the classification algorithm to process large amounts of measurement data.

Take Figure 3 as an example. The small circles and stars represent the measurements from subspaces α and β, respectively. The horizontal ordinate and vertical ordinate denote the measurements 1 and 2, respectively. Then the dataset from two measurements is expressed in a two dimensional space. In the same way, an n-dimension space is needed when there are n measurements. As shown in Figure 3, a line can distinguish two groups of data and obviously a plane is needed for a three dimensional space. In the large scale power system, with multiple measurements, a hyper-plane is used in this paper to distinguish the subspaces of operating points. In order to explain the algorithm developed in this paper, an example with two measurements is deduced as follows.

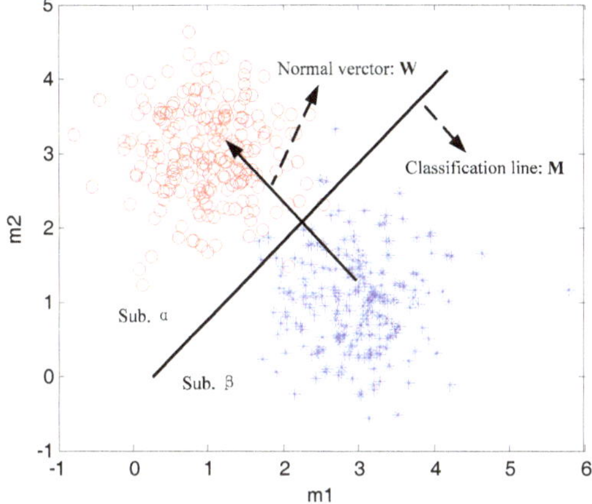

Figure 3. The optimal classification line to classify two groups of data in two dimension space.

Assume a sample from α group has the coordinates value $(x_{\alpha i}, y_{\alpha i})$, where $x_{\alpha i}$ and $y_{\alpha i}$ represent the values of x-axis and y-axis of the i-th sample. The values of x-axis and y-axis of the j-th sample from group β are denoted $(x_{\beta j}, y_{\beta j})$. Then the means of samples from α group could be obtained by:

$$\mu_\alpha = \left(\sum_{i=1}^{n} x_{\alpha i}, \sum_{i=1}^{n} y_{\alpha i} \right) \Big/ n \tag{1}$$

where n represents the number of samples from the α group. The means of samples from the β group could be obtained by:

$$\mu_\beta = \left(\sum_{j=1}^{m} x_{\beta j}, \sum_{j=1}^{m} y_{\beta j} \right) \Big/ m \tag{2}$$

where m represent the number of samples from the β group. Denote Σ_α and Σ_β the covariance of measurements of subspaces α and β, respectively. There should be many lines between those two groups of data. **M** denotes the classification line, and **W** denotes the normal vector of **M**. The classification rules of two classes with different covariance is defined as the ratio of the variance between the classes to the variance within the classes, which is named as FLD index [19]. Mathematically, it is:

$$S = \frac{\left(\mathbf{W}^\mathrm{T}\mu_\alpha - \mathbf{W}^\mathrm{T}\mu_\beta\right)^2}{\mathbf{W}^\mathrm{T}\Sigma_\alpha\mathbf{W} + \mathbf{W}^\mathrm{T}\Sigma_\beta\mathbf{W}}, \tag{3}$$

The FLD index is the best for discriminating two groups of data when the FLD index S is greatest. To achieve maximum value of S, the normal vector **W** is found to be given by:

$$\mathbf{W} = \frac{\mu_\alpha - \mu_\beta}{\Sigma_\alpha + \Sigma_\beta}, \tag{4}$$

The middle point $\mathbf{C}_{mid} = \frac{1}{2}(\mu_\alpha + \mu_\beta)$ should be on this line. With the point on the line and the normal vector **W** identified, the best line is determined. Then the following step is to find the distance from a point (x, y) in the two-dimensional space to the classification line **M** as shown below:

$$d(x,y) = \frac{ax + by + c}{\sqrt{a^2 + b^2}},\tag{5}$$

where the classification line is **M**: $ax + by + c = 0$. In this way, a single dimensional variable (distance) $d(x,y)$ is used as the training data instead of two dimensional data. If $d \geq 0$, the operating point is identified to be inside subspace α, otherwise the point is inside subspace β.

Extend the example above in two-dimension space to the case in a multiple-dimension space. It is easy to understand that the hyper-plane can be used to classify subspaces in the high-dimension space as illustrated as follows.

In an n-dimension space, the normal vector of a hyper-plane can also be calculated by (3). The vector composed of middle points between the means is also on the hyper-plane **N**:

$$\mathbf{N} : \mathbf{W} \cdot (\gamma - \mathbf{C}_{mid}) = 0,\tag{6}$$

where γ represents the position of a point in n-dimension space, i.e., $\gamma = (x, y, z, \cdots)$. The distance from a point $\gamma_i = (x_i, y_i, z_i, \cdots)$ to the hyper-plane can be obtained according to (5):

$$d_i = \frac{\mathbf{W} \cdot (\gamma_i - \mathbf{C}_{mid})}{\|\mathbf{W}\|},\tag{7}$$

In this way, the distance vectors composed of the distance from the points of the subspaces to the hyper-planes are identified. Then they are used as the input variables for the CART to perform the classification process and achieve the splitting rules. After the process of classification, the regression process is performed to identify which subspace the current operating point is in. A new distance vector d' from current operating point γ' to the hyper-planes is obtained and used as inputs to the CART, thus the terminal node which characterizes a subspace is reached. In this way, the CART can track the variation of the system operating point and thus guide the updating of appropriate PSSs into service adaptively.

2.4. Design Procedure of Adaptive Control Scheme

The design process of the adaptive control scheme presented in this paper includes two stages as shown in Figure 4. The first stage is off-line. Firstly, the power system operating space is divided into different operating subspaces according to the partition of the partition of stochastically variable wind power output. Then the mapping rules are set up and the coordinated multiple PSSs are predesigned for each subspace of power system operating points according to the mapping rules, on the other hand, the model of hyper-planes for classifying subspaces are built, thus the optimum CART and the splitting rules. The second stage is on-line, as shown in Figure 5. The wide-area information from PMUs is collected and the suitable bus frequencies are chosen as the original learning dataset. Then, the distance vector to the hyper-planes are calculated for the inputs of CART for regression. The splitting rules are applied for a regression tree and finally the terminal node could demonstrate the operating subspace, thus the switch order according to the output of the CART and the appropriated PSSs are put into service for damping. Hence in the adaptive damping control scheme, the CART plays a centralized role of decision maker and the multiple PSSs play the role of actuators.

The process of computing the parameters of designed PSS is done off-line, and as a matter of fact, the total duration of the proposed method only include the time for collecting data from PMUs, communication lag time, time of classification and regression caused by CART. The data collection from PMUs could be neglected as it is predesigned beforehand. The delays of WAMS communication are less than several hundred milliseconds. The time of classification and regression caused by CART could also be neglected without the consideration of the commutating ability of the computers. Therefore, the total duration of the on-line action is only composed of the communication lag time, so it is could be fast enough as an online method in a power system. In China, the online decision maker has been applied to the South China power grid using the modulation signals to be sent remotely to the DC links and it works well for damping the oscillations [20,21].

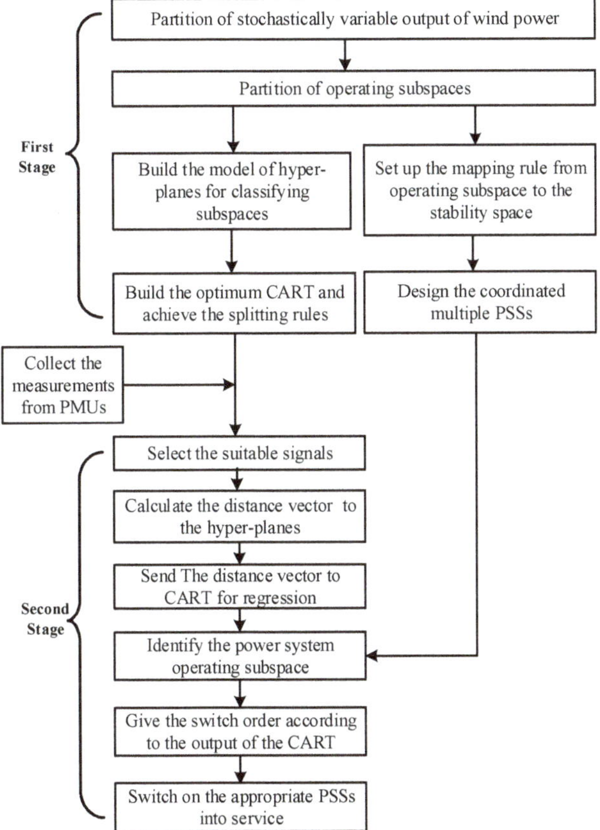

Figure 4. The flowchart of the design process.

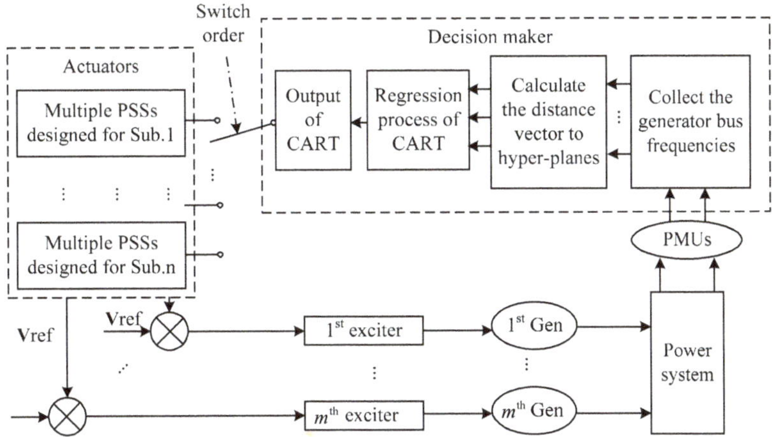

Figure 5. Configuration of the proposed adaptive damping control scheme.

3. Results and Analysis

3.1. Test System

A modified 16 generator, 68 bus system [17] with an integrated wind farm is employed for testing the performance of the proposed adaptive damping control scheme. The configuration of the power system is shown in Figure 6. The wind farm is assumed to be at bus 69. The wind farm is modeled as a dynamic equivalence of doubly-fed wind farms using the method shown in reference [22]. In China, at present the most dominant wind turbine generators are doubly fed Induction Generator (DFIG) units, therefore, in this paper, the dynamic equivalence model of doubly-fed wind farms is applied.

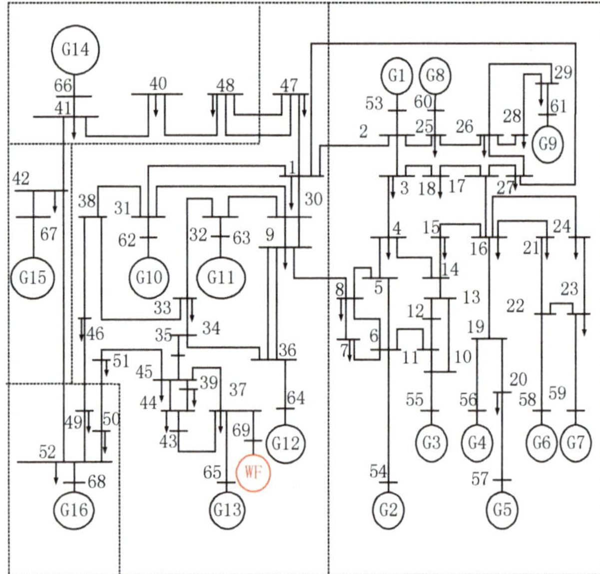

Figure 6. The modified of 16-machine 68-bus system.

The frequency and damping ratio of those modes are given in Figure 7 when the wind power is variable. The stochastically variable output of wind power is divided into 10 subspaces. Each subspace characterizes an operating subspace. The minimum and maximum output of wind power is 0 MW and 4048 MW, respectively. In China, the maximum capacity of the Jiuquan wind power base in Gansu Province is up to 5409.2 MW with a penetration of about 22.3% in 2014 [23]. The large scale of wind power generation has grown rapidly, and ultra-high voltage transmission lines are built for transferring this renewable resource generation powers from the northwest part of the country to the southeast part. Therefore, it is reasonable to consider a large wind farm with high capacity connected to one bus. From Figure 7 it can be seen that with the changing wind power penetration, the damping ratios of different modes changed in a different way. The stochastic fluctuation of wind power affects the damping of system low-frequency power oscillation and can lead to instability in the worst case.

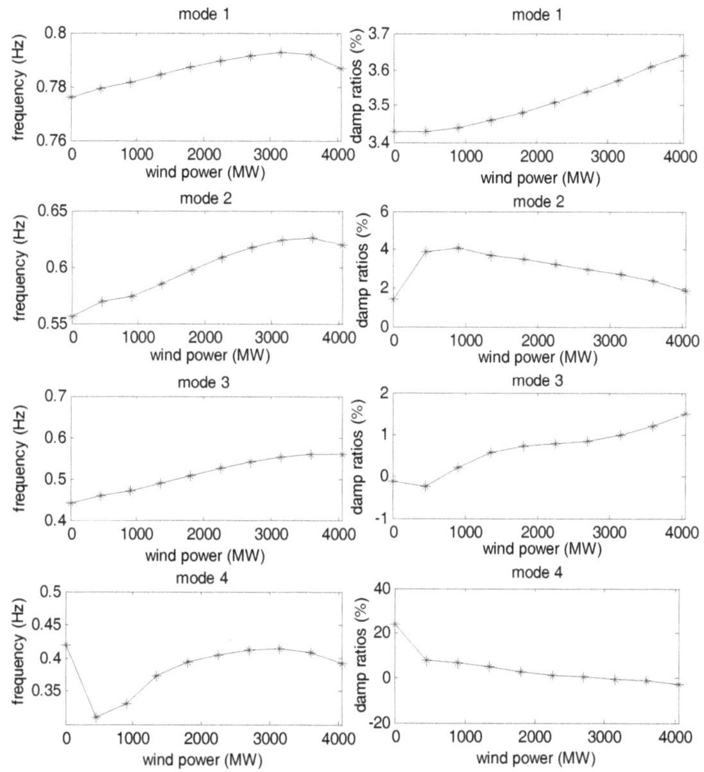

Figure 7. Variation of frequency and damp ratio of oscillation modes with wind power output.

Each generator is installed with a PSS. For each of 10 subspaces of the system operating point, a group of PSSs is selected by use of participation factors and the parameters of the group of PSSs are set in coordination to provide sufficient damping to four oscillation modes. The coordinated PSSs are designed by the traditional method, which includes phase compensator tuning and gain tuning. The compensator tuning is used to provide the required phase compensation for the phase lag between the exciter reference point and the generator air-gap torque. The tuning of an appropriate PSS gain is to ensure that the maximum possible damping can be provided by the PSS for the target modes while limiting the side effect within the acceptable level. The variation from wind power has an impact to the oscillation modes, thus only one set of coordinated PSSs could not provide enough damping for all the subspaces. Therefore, many set of PSSs are designed for many subspaces. The parameters of PSSs designed for one subspace are different from other subspaces. There are 10 subspaces in this paper, thus 10 sets of PSSs. The PSSs combination of each subspace is different from each other as each PSSs combination is designed at its own subspace, respectively.

Once the wind power output changes, the PSSs are switched from one set to another with the guidance of CART. Meanwhile, as long as the parameters of PSSs are coordinated well, a suitable set of parameters of PSSs could provide enough damping for subspaces even if the wind power impact is considered. The results of off-line coordinated design of PSSs are presented in Table 1, and the off-line coordinated design of PSSs is effective and different set of coordinated PSSs could provide satisfying damping to its own subspace.

The system space is divided into *n* subspaces according to two constraints. The first one is that predesigned PSSs in each subspace could give good performance on the boundary of each subspace to guarantee the subspaces could cover the whole system space. The second one is that CART could classify the system space with high accuracy and less complexity to avoid that too large number of subspaces causes the computation burden.

Table 1. Damping ratio and frequency of inter-area modes with PSSs.

Subspace	Wind Power Outputs	Modes	1	2	3	4
1	0	f (Hz)	0.6752	0.6414	0.5621	0.3482
		damp (%)	7.39	14.04	11.95	24.73
2	450	f (Hz)	0.6973	0.6463	0.5746	0.3839
		damp (%)	10.81	13.47	12.94	22.6
3	900	f (Hz)	0.6917	0.6505	0.5895	0.4046
		damp (%)	14.09	12.98	15.14	21.85
4	1350	f (Hz)	0.8942	0.6687	0.6175	0.5589
		damp (%)	6.55	16.13	10.08	40.78
5	1800	f (Hz)	0.8441	0.5556	0.5513	0.4962
		damp (%)	6.35	9.13	32.6	19.94
6	2250	f (Hz)	0.8455	0.5914	0.5811	0.5631
		damp (%)	6.56	15.89	27.81	10.08
7	2700	f (Hz)	0.8469	0.6016	0.5727	0.5400
		damp (%)	6.27	19.28	10.06	55.07
8	3150	f (Hz)	0.8473	0.6255	0.5771	0.5177
		damp (%)	6.45	17.92	10.14	58.6
9	3600	f (Hz)	0.6511	0.6048	0.5452	0.3948
		damp (%)	11.43	1023	18.7	23.18
10	4048	f (Hz)	0.8572	0.6258	0.5738	0.4817
		damp (%)	6.27	14.02	10.03	24.3

The interval of subspaces are [0 0.5*h*], [0.5*h* 1.5*h*], [1.5*h* 2.5*h*], [2.5*h* 3.5*h*], [3.5*h* 4.5*h*], [4.5*h* 5.5*h*], [5.5*h* 6.5*h*], [6.5*h* 7.5*h*], [7.5*h* 8.5*h*], [8.5*h* 4048], in which h = 450. The boundaries of subspaces are 0, 0.5*h*, 1.5*h*, ... , 4048. The lower boundary of one subspace is the upper boundary of other subspace. For example, the 2nd boundary 0.5 h is the upper boundary in the first subspace and lower boundary of the second subspace. In this way, the whole system space could be covered by subspaces. The lower boundary of the first subspace is 0 as it is the minimum output of wind power. The upper boundary of 10th subspace is 4048 as it is the maximum output of wind power. The coordinated PSSs of each subspace are designed at the center point of each subspace, e.g., 0, h = 450, 2h = 900, 3h = 1350, ... , while the PSSs of the first and last subspace are designed at the points of minimum and maximum of wind power outputs, respectively.

Nine boundary points are used as test cases to demonstrate the robustness of coordinated PSSs and the simulation results are shown in Figure 8. Each sub-plot represents one boundary point. Take the second boundary as shown in Figure 8a as an example, the 2nd boundary point is the upper and lower boundaries of subspaces 1 and 2, respectively. As a matter of fact, this boundary point is in both subspace 1 and subspace 2. The dashed curves represent the dynamic response without PSSs.

The blue and red curves represent the dynamic responses with different set of coordinated PSSs, respectively. The blue curve represents the response with the first set of coordinated PSSs applied which is designed for subspace 1. The red curve represents the response with the second set of coordinated PSSs applied which is designed for subspace 2. From this figure, it can be seen that both set of PSSs could give good performance which demonstrate the coordinated PSSs designed at the center point of subspace could provide enough damping at the boundaries points which means the classified subspaces could be able to cover the whole system space. From other sub-plots, the simulation results at other boundary points also demonstrate the robustness of the coordinated PSSs. In this way, different set of PSSs could be switched from one to another adaptively without unsatisfactory control performance.

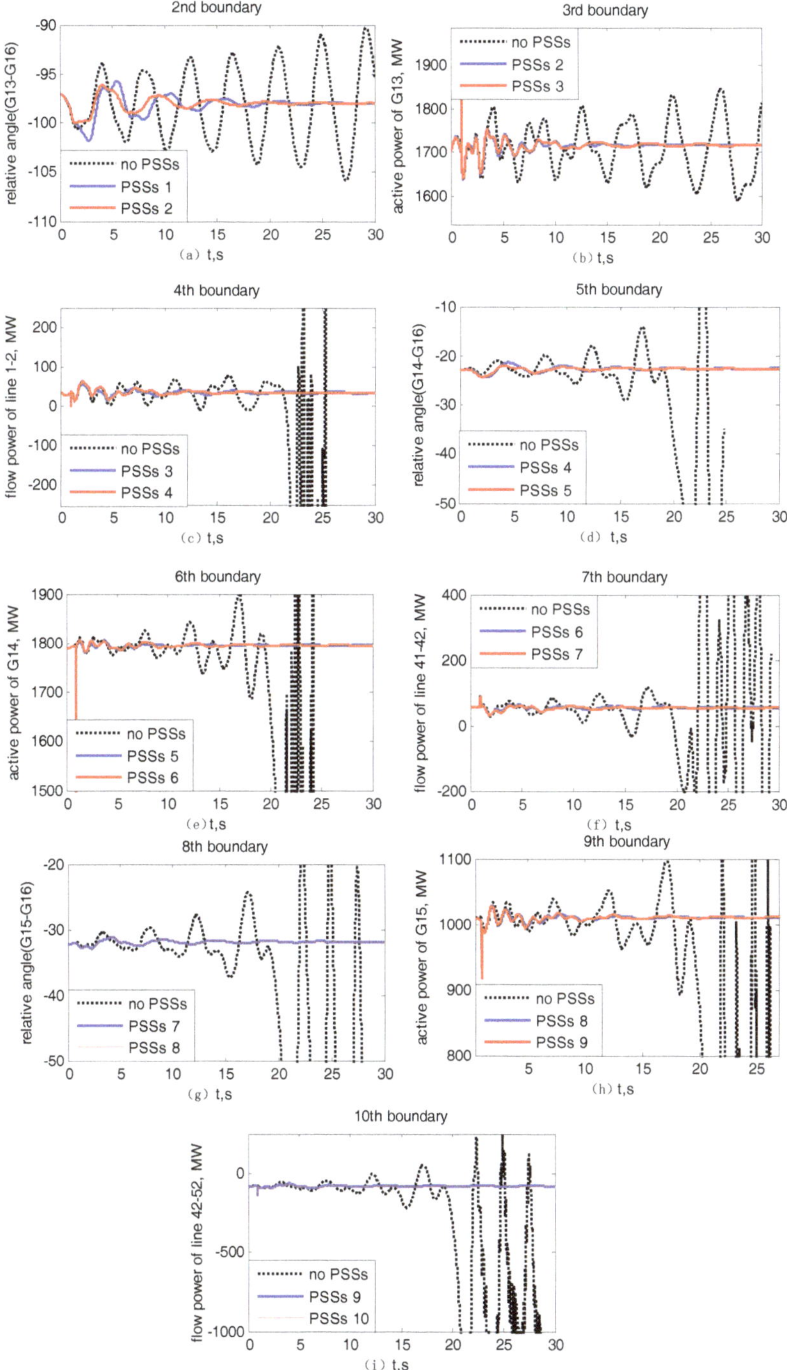

Figure 8. Dynamic responses with different controllers in the case of subspaces boundaries.

It can be seen from Figure 8 that the inter-area power oscillations can be suppressed effectively with the pre-designed multiple PSSs being matched to the correct subspace of system operating point even when the wind generation varies. Hence PSSs pre-designed need to be adaptively adjusted to following the wind power variation. Otherwise poorly damped inter-area oscillations are observed. An adaptive damping control robust to the changes of power system operating point caused by variation of grid-connected wind power need to be able to track the variation of system operating point and to assign correct group of pre-designed PSSs into the service to suppress system inter-area power oscillations. Function of tracking system operating point when wind generation varies is implemented by the CART formed for the test system.

3.2. Formation of the CART

All 10 subspaces are considered to measure the participation factor of each generator. The participation factors with wind power variation are calculated and the results are shown in Table 2. Two highest participation factors of each mode of each subspace are displayed in Table 2. It can be observed that, the participation factors changed with the wind power variation, but they do not change a lot. The measurement used for training the CART is selected according to the highest participation factors of four inter-area oscillations. Hence, measurement of bus frequencies of machine G9, G13, G14 and G15 is used to form the CART.

Table 2. Participation factors of different generators in different subspaces.

Sub. No.	Mode 1		Mode 2		Mode 3		Mode 4	
	Gen. No.	PF	Gen. No.	PF	Gen. No.	PF	Gen. No.	PF
Sub. 1	15	1.00	13	1.00	14	1.00	13	1.00
	14	0.37	9	0.19	16	0.49	9	0.28
Sub. 2	15	1.00	13	1.00	14	1.00	13	1.00
	14	0.38	9	0.20	16	0.68	14	0.19
Sub. 3	15	1.00	13	1.00	14	1.00	13	1.00
	14	0.38	9	0.27	16	0.82	16	0.21
Sub. 4	15	1.00	13	1.00	16	1.00	13	1.00
	14	0.39	9	0.29	14	0.97	16	0.23
Sub. 5	15	1.00	13	1.00	16	1.00	13	1.00
	14	0.39	9	0.29	14	0.78	15	0.38
Sub. 6	15	1.00	13	1.00	16	1.00	13	1.00
	14	0.39	9	0.28	14	0.67	15	0.61
Sub. 7	15	1.00	13	1.00	16	1.00	13	1.00
	14	0.39	9	0.27	14	0.61	14	0.87
Sub. 8	15	1.00	13	1.00	16	1.00	14	1.00
	14	0.39	9	0.28	14	0.58	15	0.95
Sub. 9	15	1.00	13	1.00	16	1.00	14	1.00
	14	0.39	9	0.29	14	0.59	15	1.00
Sub. 10	15	1.00	13	1.00	16	1.00	15	1.00
	14	0.40	9	0.35	14	0.62	14	0.68

The sample rate of each measurement is 30 samples per second. For each subspace of system operating point, 100 initial sampled data with 5% noise are collected. Thus 1000 trajectories for 10 subspaces are employed to form the initial learning dataset of the CART. The initial learning data is a matrix with 1000 rows and 120 columns. 1000 rows are the product of 100 initial sampled data and 10 subspaces. 120 columns are the product of 30 samples and four measurements. Therefore, a hyper-plane of classification is formed in a 120-dimensional space. There are 45 hyper-planes to classify two subspaces. As this is a case of high-dimensional data set. Formation of the CART needs to consider the compromise between the accuracy and tree complexity. A small-size CART cannot capture all the dynamic behavior and a large-size CART would bring about the incorrect identification due to the over-fitting [24]. A plot of the misclassification rate for each sub-tree is shown in Figure 9, from which it can be seen that the optimum number of the decision tree is 10 for 10 subspaces. The minimal mismatching rate of the CART is approximately 0.0433 which indicates the 95.67% probability of correct subspace matching.

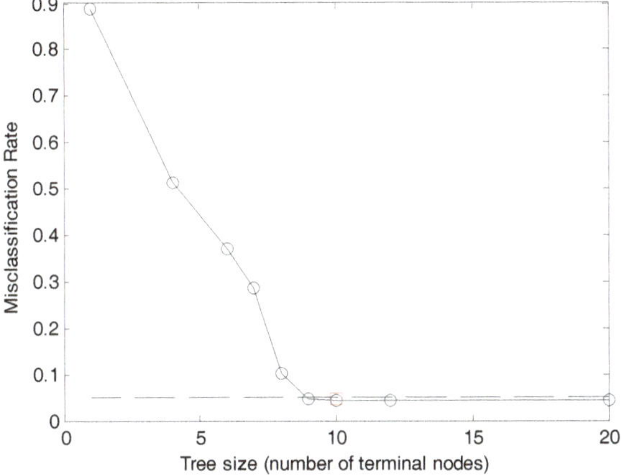

Figure 9. Cross-validation estimate of the misclassification rate.

From Figure 9, it can also be seen that 10 subspaces appeared at the positions of terminal nodes and could be classified with high accuracy, which means the number of subspaces is not too big. Since the big number of subspaces make the CART complex and difficult to classify with high accuracy. Therefore, the number of subspaces 10 is a suitable number in this paper.

After the training process, the classification tree is formed as shown in Figure 10, and the splitting rules for each node are generated at the same time. The structure of formed CART for the test system is shown in Figure 10. It has 10 terminal nodes which represent 10 operating subspaces. The rectangle blocks represent the root node and intermediate nodes. The ellipse blocks are the terminal nodes. At each separation point, there is a splitting rule. The inequations are the splitting rules of each intermediate node which are generated by training process. For example, the distance vector is dropped down to the CART. And the splitting rule of the 1st node circled by a red ellipse block is the distance to the 1st hyper-plane. At the root node, the splitting rule is $d_1 < 0.0263$ and $d_1 \geq 0.0263$. If d_1 is smaller than 0.0263 then the distance vector will be dropped down to node 2 for further splitting. If d_1 is bigger than 0.0263, then the distance vector will be dropped down to Sub. 1 which is a terminal node and the identification process is stopped. In this way, the current system operating point is identified to be in subspace 1. The distance vector from an operating point to hyper-planes is expressed as $\mathbf{d} = [d_1, d_2, \cdots, d_{45}]$, d_i represents the distance to the *i*-th hyper-plane. Because there are 10 operating subspaces, the total number of hyperplanes are 45, which could classify each two subspaces. In this way, the distance vector actually include 45 distance parameters.

After CART training, the regression test process is performed to test the performance of off line of CART. The test results are shown in Table 3. Outputs of wind power are systematically modified to generate 5000 different system operating points. From each subspace, 500 different system operating points are generated. From Table 3, it can be observed that, 4803 points out of 5000 system states are classified to the correct subspace, and 197 cases are classified to the wrong subspaces. The accuracy of regression test is about 98.9%. The misclassified cases are mostly the points around the boundaries. Also, Figure 8 shows that even the boundaries points are classified to be neighbourhood subspaces, and the neighbourhood set of PSSs can also provide satisfying damping. Therefore, good performance of CART when it is applied offline and online can be guaranteed.

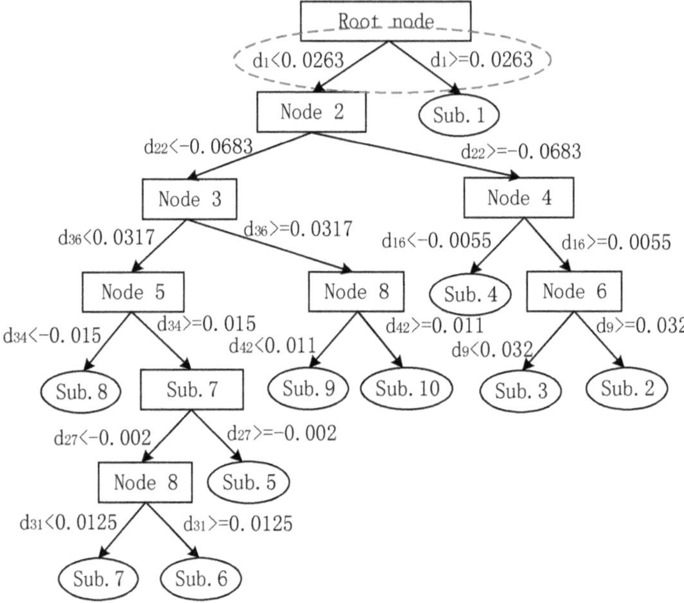

Figure 10. The CART for distinguishing different subspaces of operating point.

Table 3. Cases from subspaces test results.

Substation No.	1	2	3	4	5	6	7	8	9	10
Classified Sub.1	492	10	0	0	0	0	0	0	0	0
Classified Sub. 2	8	478	8	1	0	0	0	0	0	0
Classified Sub. 3	0	11	481	10	0	0	0	0	0	0
Classified Sub. 4	0	1	9	474	15	0	0	0	0	0
Classified Sub. 5	0	0	0	15	468	15	0	0	0	0
Classified Sub. 6	0	0	0	0	17	471	12	0	0	0
Classified Sub. 7	0	0	0	0	0	13	479	13	0	0
Classified Sub. 8	0	0	0	0	0	1	9	477	6	0
Classified Sub. 9	0	0	0	0	0	0	0	10	488	5
Classified Sub.10	0	0	0	0	0	0	0	0	5	495

3.3. Simulation Results and Discussions

The test power system operates initially in subspace 9 with the output of wind power being 3600 MW. The simulation runs for 35 s with wind power output dropped to 1350 MW at 5 s of simulation due to the disconnection of part of wind power caused by fault. At this time, the wide-area information from PMUs is sent to the CART for identification of variation of system operating point. The outputs from the CART are shown in Figure 11, from which it can be seen that, initially the CART cannot identify the correct system operating subspace. With the increase of number of sampling data, the identification result of the CART converges to the correct system operating subspace. After $t = 6$ s, the outputs of the CART indicate that the system operates in subspace 4. Then the multiple PSSs designed for subspace 4 are activated and switched on.

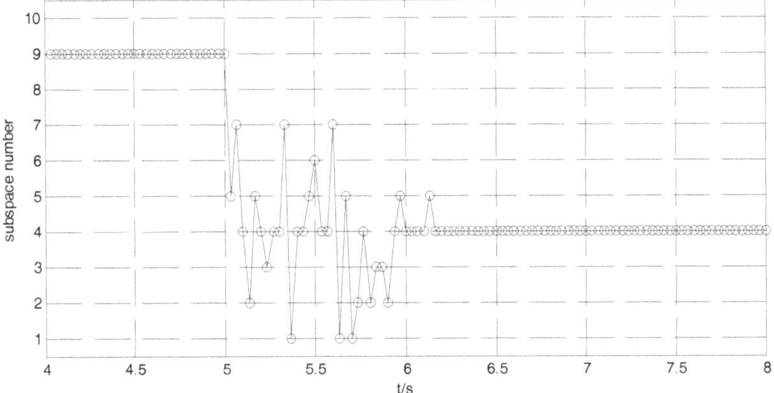

Figure 11. Output trajectory of the CART.

In the simulation part, the wind power output is changed at $t = 5$ s, and the PSSs are switched at $t = 7$ s. The time from wind power output changed to the controllers was switched on is about 2 s, which considers communication delay. From Figure 11, the identification process cost about 1 s, the left 1 s takes the communication delay into account.

Results of dynamic simulation are shown in Figure 12, where the dotted curves are the results with PSSs designed for subspace 9 unchanged (fixed-parameter PSSs). The solid curves are the results with the adaptive PSSs proposed in this paper. It can be seen that the CART based adaptive control scheme can effectively track the variation of power system operating point caused by variable grid-connected wind power, and switch on the appropriate PSSs adaptively to suppress power system inter-area power oscillation.

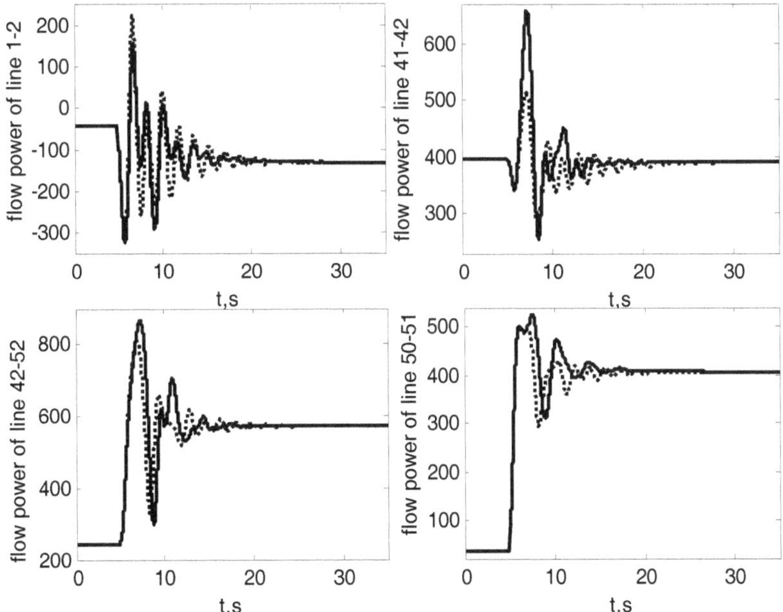

Figure 12. Dynamic responses in case of changing wind power outputs.

It is worth noting that CART plays a role of mapping the measurement space to state variable space actually. Although the wind power pattern could be different, the state variables could not drift in a dramatic manner. On one hand, from the output of CART as shown in Figure 11, it can be seen that in the initial stage, the output could not converge to a constant number, it needs a while to converge. After about 1 s, the output of CART converge to a constant number. The corresponding set of PSSs are switched on when the output of regression tree converged. On the other hand, in the process of training the CART, the initial sampled data take 5% noise into count. Therefore, the errors in adjustments could be avoided through modeling the noise to the training dataset.

To further demonstrate the damping improvement, the Prony analysis is performed to analyze the real-time responses and the results are shown in Table 4. Also, the small signal stability analysis is carried out to calculate the oscillation frequency and damping ratios. It can be observed from the Prony analysis and eigenvalue analysis that, the adaptive controllers provide better performance.

Table 4. Prony analysis and eigenanalysis results.

| Mode No. | Adaptive PSSs | | | | Fixed PSSs | | | |
| | Eigenanalysis | | Prony Analysis | | Eigenanalysis | | Prony Analysis | |
	f (Hz)	Damp (%)	f (Hz)	Damp (%)	f (Hz)	Damp (%)	f (Hz)	Damp (%)
Mode 1	0.5589	40.78	0.538	46.064	0.4978	30.46	0.479	20.145
Mode 2	0.6175	10.08	0.617	10.412	0.5536	7.24	0.565	6.88
Mode 3	0.6687	16.13	0.633	17.953	0.5986	15.20	0.614	12.989
Mode 4	0.8942	6.55	0.854	6.666	0.8725	5.64	0.887	5.994

3.4. Test System with Multiple Wind Farms

The 16-generator system with three wind farms added at different buses is used as the test system for investigating the effectiveness of the proposed adaptive control. Assume the first wind farm is added at bus 42, the 2nd wind farm is connected at bus 42, and the 3rd wind farm is connected at bus 52. The operating subspaces are divided with different wind farm outputs as shown in Table 5. The set of PSSs are predesigned for different subspaces and the new CART are also trained off-line with dataset obtained from this test system. The time domain simulations are carried out in the case of multi wind farms connected to different buses.

Table 5. The Subspaces with different WF outputs.

Sub. No.	1st WF Outputs	2nd WF Outputs	3rd WF Outputs
1	450	450	450
2	450	450	900
3	450	900	450
4	450	900	900
5	900	450	450
6	900	450	900
7	900	900	450
8	900	900	900

Assume the power system is operating at the output of three wind farms are all 450 MW, which means the power system is operating at the first space. When $t = 5$ s the output of 1st wind farm increased to 900 MW. When $t = 15$ s the output of 2nd wind farm increased to 900 MW. When $t = 25$ s the output of 3rd wind farm increased to 900 MW. The output of CART are shown in Figure 13, from which it can be seen that the initial output of the CART shows the power system is operating in subspace 1 before $t = 5$ s. After 5 s, the output shows the power system operating subspace changed to subspace 5. After 15 s the output of CART demonstrate the subspace 7 and then subspace 8 after 25 s. The output trajectory of CART shows that CART could track the operating subspace adaptive to the changing wind power outputs and give the correct switch order.

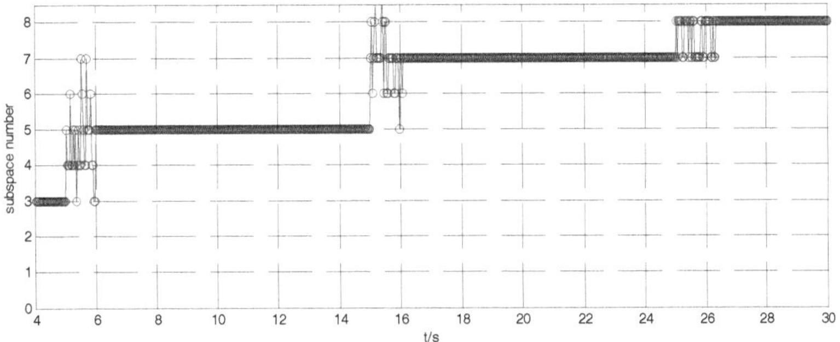

Figure 13. Output trajectory of the CART in the case of multiple wind farms.

The dynamic response of relative angular of generators 1 and 16 is displayed in Figure 14. The dashed curve is the dynamic response of relative angular with the fixed set of PSSs predesigned for subspace 3. The solid curve is the relative angular with the adaptive control method proposed in this paper. From this figure, it can be seen that the adaptive control method show good performance even if the multiple wind farms were connected to different buses.

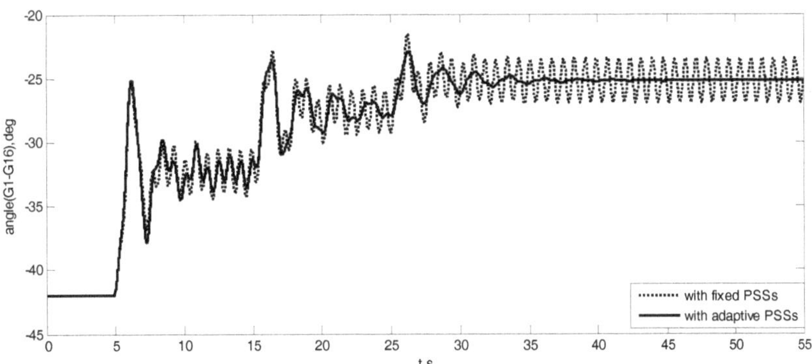

Figure 14. Dynamic responses in case of changing wind power outputs.

4. Conclusions

An adaptive damping control robust to stochastic varying operating conditions caused by integrated wind power is proposed in this paper. Firstly, the space of operating point is divided into operating subspaces by the even interval of wind power outputs. Secondly, coordinated PSSs are pre-designed for each operating subspace. Thirdly, a classification tree is formed by training the distances to the hyperplanes, and the regression tree is used for regression to identify the subspaces with the help of on-line measurement from PMUs. In this way, the dynamic behavior of a varying power system is tracked and the appropriate coordinated PSSs are switched into service adaptively. The simulation results of modified test system integrated with wind generation are presented in the paper to demonstrate the robustness of the proposed adaptive damping control based on CART to the variation of system operating point in suppressing system low-frequency power oscillations when grid-connected wind power varies. Furthermore, the proposed adaptive damping control also show the good performance in the case of multiple wind farms connected at different buses.

Author Contributions: For research articles with several authors, a short paragraph specifying their individual contributions must be provided. The following statements should be used "Conceptualization, J. D. and T.W.; Methodology, J.Y.; Software, J.Y.; Validation, F.C., and S.P.; Formal Analysis, J.D.; Investigation, J.D.; Resources, S.P.; Data Curation, S.P.; Writing-Original Draft Preparation, T.W.; Writing-Review & Editing, J.Y.; Visualization, J.S.; Supervision, F.C.; Project Administration, T.W.; Funding Acquisition, T.W.", please turn to the CRediT taxonomy for the term explanation. Authorship must be limited to those who have contributed substantially to the work reported.

Funding: This research was funded by Fundamental Research Funds for the Central Universities grant number 2018MS006.

Conflicts of Interest: The authors declare no conflict of interest.

Nomenclature

$\Delta\omega$	generator speed deviation, p.u.
ΔP	electromagnetic power deviation, p.u.
T	time constants of washout blocks, s
N	number of low-pass filters
K	the gain of PSS
α	the measurements from circles
β	the measurements from stars
μ_α	the means of measurements α
μ_β	the means of measurements β
Σ_α	the covariance of measurements of subspaces α
Σ_β	the covariance of measurements of subspaces β
$\mathbf{M}$	the classification line
$\mathbf{W}$	the normal vector
d	the distance vector
$\mathbf{N}$	the direction vector of hyper-plane
$\mathbf{C}_{mid}$	the middle point
f	frequency of oscillation mode, Hz

References

1. Kunder, P. *Power System Stability and Control*; McGraw-Hill: New York, NY, USA 1994.
2. Yu, X.; Zhang, W.; Zang, H.; Yang, H. Wind power interval forecasting based on confidence interval optimization. *Energies* **2018**, *11*, 3336. [CrossRef]
3. Bai, W.; Lee, D.; Lee, K.Y. Stochastic dynamic AC optimal power flow based on a multivariate short-term wind power scenario forecasting model. *Energies* **2018**, *12*, 2138. [CrossRef]
4. Zhao, X.; Lin, Z.; Fu, B.; He, L.; Fang, N. Research on automatic generation control with wind power participation based on predictive optimal 2-degree-of-freedom PID strategy for multi-area interconnected power system. *Energies* **2018**, *11*, 3325. [CrossRef]
5. Divshali, P.H.; Choi, B.J.; Liang, H. Multi-agent transactive energy management system considering high levels of renewable energy source and electric vehicles. *IET Gener. Transm. Distrib* **2017**, *11*, 3713–3721. [CrossRef]
6. Athari, M.H.; Wang, Z. Impacts of wind power uncertainty on grid vulnerability to cascading overload failures. *IEEE Trans. Sustain. Energy* **2018**, *9*, 128–137. [CrossRef]
7. Zhang, J.; Ju, P.; Yu, Y.; Wu, F. Responses and stability of power system under small Gauss type random excitation. *Sci. China Technol. Sci.* **2012**, *55*, 1873–1880. [CrossRef]
8. Charles, K.; Urasaki, N.; Senjyu, T.; Elsayed Lotfy, M.; Liu, L. Robust load frequency control schemes in power system using optimized PID and model predictive controllers. *Energies* **2018**, *11*, 3070. [CrossRef]
9. Mansiri, K.; Sukchai, S.; Sirisamphanwong, C. Fuzzy control for smart PV-battery system management to stabilize grid voltage of 22 kV distribution system in Thailand. *Energies* **2018**, *11*, 1730. [CrossRef]
10. Majumder, R.; Chaudhuri, B.; Pal, B.C. A probabilistic approach to model-based adaptive control for damping of interarea oscillations. *IEEE Trans. Power Syst.* **2005**, *20*, 367–374. [CrossRef]

11. Chaudhuri, B.; Majumder, R.; Pal, B.C. Application of multiple-model adaptive control strategy for robust damping of interarea oscillations in power system. *IEEE Trans. Contr. Syst. Technol.* **2004**, *12*, 727–736. [CrossRef]

12. Wang, T.; Pal, A.; Thorp, J.; Yang, Y. Use of polytopic convexity in developing an adaptive interarea oscillation damping scheme. *IEEE Trans. Power Syst.* **2017**, *32*, 2509–2520. [CrossRef]

13. Ye, H.; Liu, Y. Design of model predictive controllers for adaptive damping of inter-area oscillations. *Int. J. Electr. Power Energy Syst.* **2013**, *45*, 509–518. [CrossRef]

14. Bernabeu, E.E.; Thorp, J.S.; Centeno, V. Methodology for a security/dependability adaptive protection scheme based on data mining. *IEEE Trans. Power Syst.* **2012**, *27*, 104–111. [CrossRef]

15. Shi, X.; Lei, X.; Huang, Q.; Huang, S.; Ren, K.; Hu, Y. Hourly day-ahead wind power prediction using the hybrid model of variational model decomposition and long short-term memory. *Energies* **2018**, *11*, 3227. [CrossRef]

16. Cai, K.; Alalibo, B.P.; Cao, W.; Liu, Z.; Wang, Z.; Li, G. Hybrid approach for detecting and classifying power quality disturbances based on the variational mode decomposition and deep stochastic configuration network. *Energies* **2018**, *11*, 3040. [CrossRef]

17. Rogers, G. *Power System Oscillations*; Kluwer: Boston, MA, USA, 2000.

18. Salford Systems, CART. 2008. Available online: www.salford-systems.com (accessed on 12 May 2008).

19. Fisher, R. The use of multiple measurements in taxonomic problems. *Ann. Eugen.* **1936**, *7*, 179–188. [CrossRef]

20. He, J.; Lu, C.; Wu, X.; Li, P.; Wu, J. Design and experiment of wide area HVDC supplementary damping controller considering time delay in China southern power grid. *IET Gener. Transm. Distrib.* **2009**, *3*, 17–25. [CrossRef]

21. Wang, J.; Fu, C.; Zhang, Y. Design of WAMS-based multiple HVDC damping control system. *IEEE Trans. Smart Grid* **2011**, *2*, 363–374. [CrossRef]

22. Huo, J.; Xue, A.; Wang, Q.; Bi, T.; Li, C. Research on clustering algorithm for dynamic equivalence of doubly-fed wind farms. In Proceedings of the 2014 International Conference on Power System Technology, Chengdu, China, 20–22 October 2014.

23. Yi, D.; Liu, C.; Tian, X. The power output characteristics of Jiuquan wind power base and its reactive power compensation. In Proceedings of the 2013 IEEE PES Asia-Pacific Power and Energy Engineering Conference, Kowloon, China, 8–11 December 2013.

24. Bernabeu, E. Methodology for a Security-Dependability Adaptive Protection Scheme Based on Data Mining. Ph.D. Thesis, Department of Electrical and Computer Engineering, Virginia Polytechnic Institute and State University, Blacksburg, VA, USA, 2009.

Article

Design and Analysis of a Repetitive Current Controller for a Single-Phase Bridgeless SEPIC PFC Converter

Jinwoo Kim [1], Sanghun Han [1], Wontae Cho [1], Younghoon Cho [1,*] and Hyunsoo Koh [2]

[1]   Department of Electrical Engineering, Konkuk University, Seoul 05029, Korea; jinu@konkuk.ac.kr (J.K.); steven@konkuk.ac.kr (S.H.); free0415@konkuk.ac.kr (W.C.)
[2]   Texas Instruments, 12500 TI Boulevard, Dallas, TX 75243, USA; Hyunsoo@ti.com
*   Correspondence: yhcho98@konkuk.ac.kr; Tel.: +82-10-6207-0431

Received: 16 November 2018; Accepted: 26 December 2018; Published: 31 December 2018

Abstract: This paper studies a repetitive controller design scheme for a bridgeless single-ended primary inductor converter (SEPIC) power factor correction (PFC) converter to mitigate input current distortions. A small signal modeling of the converter is performed by a fifth-order model. Since the fifth-order model is complex to be applied in designing a current controller, the model is approximated to a third-order model. Using the third-order model, the repetitive controller is designed to reduce the input current distortion. Then, the stability of the repetitive controller is verified with an error transfer function. The proposed controller performance is validated by simulation, and the experiment results show that the input current total harmonic distortion (THD) is improved by applying the proposed controller for an 800 W bridgeless SEPIC PFC converter prototype.

Keywords: AC-DC converters; bridgeless SEPIC PFC converter; repetitive controller; current distortion; current controller design

---

1. Introduction

Until now, many power conversion systems (PCS) are connected to grid and harmonic pollution becomes an issue [1,2]. General PCS needs AC-DC conversion to operate with the grid, and the easiest way is using a diode rectifier. However, it can decrease the power quality due to harmonics of the input current of the diode rectifier [3]. Thus, to improve the quality of the input current, the grid-tied power factor correction (PFC) converter can be applied. The PFC converter can achieve a unity power factor operation with the sinusoidal input current, which has low harmonics. Generally, the boost PFC converter is used, due to the simplicity, efficiency and low cost [4–6].

There are many systems using the PFC converter, such as switching mode power supply (SMPS) and LED applications, etc. [7–9]. These applications demand a low DC output voltage while the input voltage is relatively high. However, the general boost PFC converter only generates higher output voltage than the grid peak-voltage, and it needs two stages for step-down operation. It is difficult to expect the high efficiency because the number of PCS has been increased. In this case, the step-down PFC converter which has only one stage can be used. There are step-down PFC converters such as buck, buck-boost, Cuk and single-ended primary inductor converter (SEPIC) [10–28]. The buck PFC converter is simple as boost PFC converter with high efficiency, but it only can generate the lower voltage than the input voltage and does not create the input current around in zero-crossing point. On the other hand, the buck-boost, Cuk, and SEPIC have no limit to generate the input current and can operate in both step-up and step-down modes. So, the output voltage with wide range can be generated.

Recently, there have been continuing researches of the buck-boost, Cuk, and SEPIC PFC converters. The buck-boost PFC converter needs an input filter due to pulsating input current. On the contrary, the Cuk and SEPIC converters generate the continuous input current, and do not need an additional input filter which can decrease the efficiency of the system. However, the Cuk converter's output voltage is inversed comparable table [19].

In [20–22], SEPIC and Cuk PFC converter topologies utilize the diode bridge at input side, which causes additional conduction losses. On the other hand, bridgeless SEPIC and Cuk PFC converter topologies are proposed in [23–28]. A Cuk PFC converter in [23,24] consists of 2 switches and 3 diodes. A SEPIC PFC converter in [25,26] has 2 switches and 3 diodes. The circuits using more switches have also been introduced in [27,28]. In this paper, a bridgeless SEPIC PFC converter with single switch and 5 diodes is utilized in Figure 1 [29]. Accordingly, the converter has low switching loss.

In general, the PFC converter has a unidirectional power transfer characteristic, because it is operated in unity power factor condition. Therefore, the PFC converter works in continuous conduction mode (CCM) and discontinuous conduction mode (DCM). The DCM operation is the cause of the input current distortion in low power condition. The input current distortion due to the DCM operation includes high-order harmonics, which are higher than the bandwidth of a typical current controller such as a proportional-integral (PI) controller or a proportional-resonant (PR) controller. To reduce the input current distortion, the current controller should be designed considering the plant model in DCM as well as in CCM [27,30,31]. In [27], the feed-forward is utilized in the SEPIC PFC converter for compensating the input current distortion. However, deriving an accurate DCM model is more difficult than a CCM model, which is relatively easy to interpret.

A repetitive controller can be applied to compensate the harmonic components caused by DCM operation. The repetitive controller has a high gain for the harmonics corresponding to multiples of the fundamental frequency [32–35]. Thus, if a stable repetitive controller design is guaranteed, the current distortion can be improved without the complicated analysis under DCM condition. In this paper, the design method of the current controller with repetitive controller is introduced. Also, a simplified third-order model of the SEPIC PFC converter is proposed.

This paper is organized as follows. Section 2 provides an analysis of the operating modes and the transfer function of the bridgeless SEPIC PFC converter. In Section 3, the proposed current controller design is described. The results of the simulations and the experiments are presented in Sections 4 and 5 to verify the performance of the proposed current controller.

2. Bridgeless SEPIC PFC Converter with RC Damping Circuits

2.1. Circuit Structure and Mode Analysis

Figure 1 shows the bridgeless SEPIC PFC converter dealt in this paper. A RC damping circuit is equipped to suppress the high-order resonance of the converter which will be described in a later section. Compared to a traditional SEPIC DC–DC converter, the bridgeless SEPIC PFC converter contains the blocking diodes, D_1 and D_2, and the freewheeling diodes, D_p and D_n. With this configuration, only the single switching device S_c can be utilized for both positive and negative input voltage cycles.

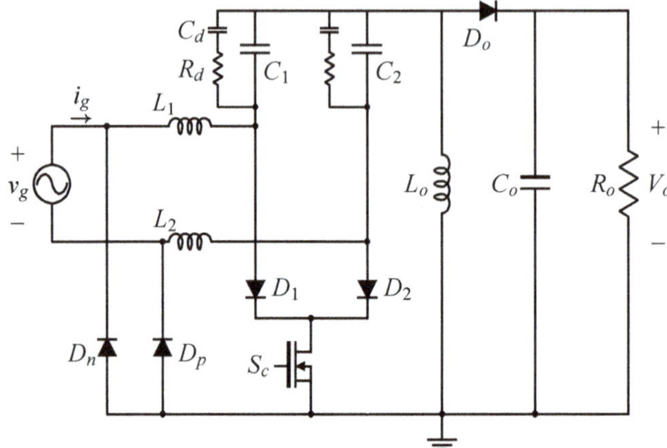

Figure 1. Topology of the bridgeless SEPIC PFC converter with the RC damping circuits. SEPIC: single-ended primary inductor converter; PFC: power factor correction.

Figure 2 represents the operation mode analysis of the bridgeless SEPIC PFC converter under a positive input voltage cycle. In Figure 2a, the current conduction paths are indicated when S_c turns on. In this case, D_n, D_2, and D_o are biased reversely. In this case, four paths are existent. Path 1 consists of the input source, L_1, D_1, S_c, and D_p. In this path, the input energy is stored in L_1. Through path 2, C_1 is charged by the energy stored in L_o. The RC damping circuit existing in this path dampens the resonant peak caused by C_1, L_1, and L_o. So, it can increase the stability of the circuit in the control viewpoint. For path 3, the energy exchange occurs between C_1, C_2, and L_2. Again, C_1 is charged while C_2 is discharged. Unlike L_o, the input source does not contribute charging of the energy for L_2. In this interval, the load R_o is supplied by C_o via path 4. When S_c is opened, the current conduction paths change as shown in Figure 2b. In this case, D_o turns on, and D_1, D_2, S_c, and D_n are blocked and the energy stored in L_1, L_2, C_1, C_2, and L_o is transferred to the load side including C_o and R_o. Similarly, the analyses for negative input voltage cycles which are not discussed here can be also performed.

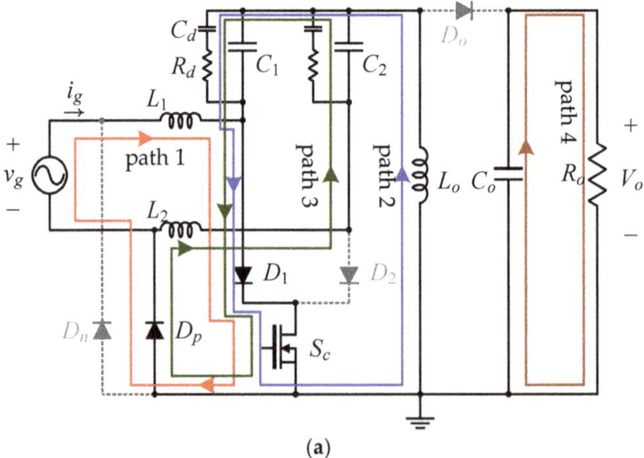

(a)

Figure 2. *Cont.*

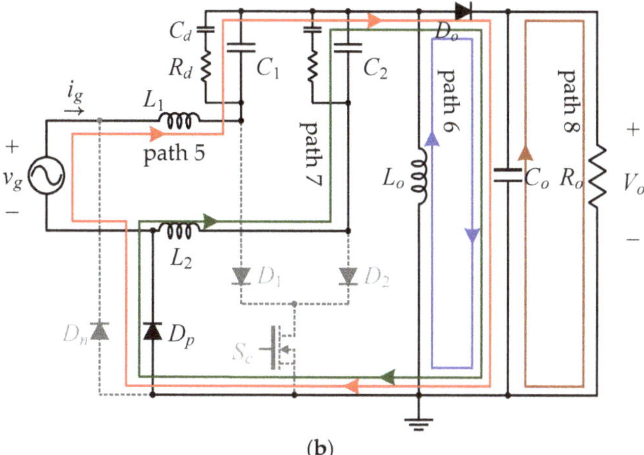

Figure 2. Operation modes of the bridgeless SEPIC PFC converter under positive voltage cycles: (**a**) when S_c is turned on; (**b**) when S_c is turned off.

2.2. Control Model Derivation of the Bridgeless SEPIC PFC Converter

Figure 3 illustrates the equivalent circuits of the bridgeless SEPIC PFC converter with the damping circuit according to the switching operations. Since paths 3 and 7 are the leakage current paths and the leakage current is very smaller than the input current, it can be ignored. So, the operation of the converter is basically identical to traditional SEPIC PFC converters except the RC damping circuit is included. In order to see the effect of the damping circuit and design the controller, the control-to-inductor current model for the SEPIC PFC converter is derived.

When S_c is turned on, the following equations can be obtained. At this moment, the voltages of L_1 and L_o are derived as follows:

$$V_{in} = L_1 \frac{di_{L1}}{dt} \tag{1}$$

$$v_{C1} = L_o \frac{di_{Lo}}{dt} \tag{2}$$

The capacitor currents i_{C1}, i_{Co} and i_{Cd} are represented as follows:

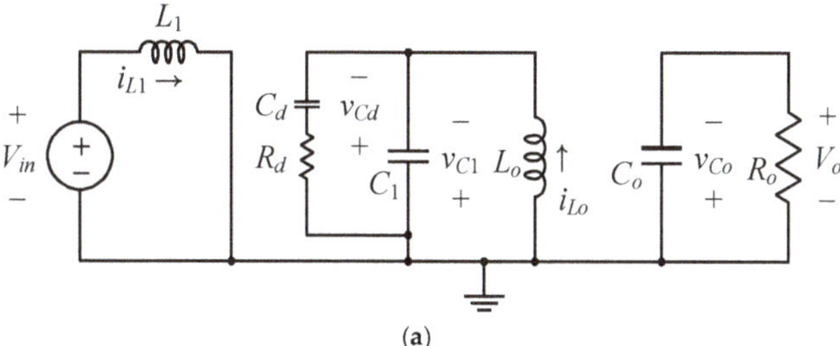

Figure 3. *Cont.*

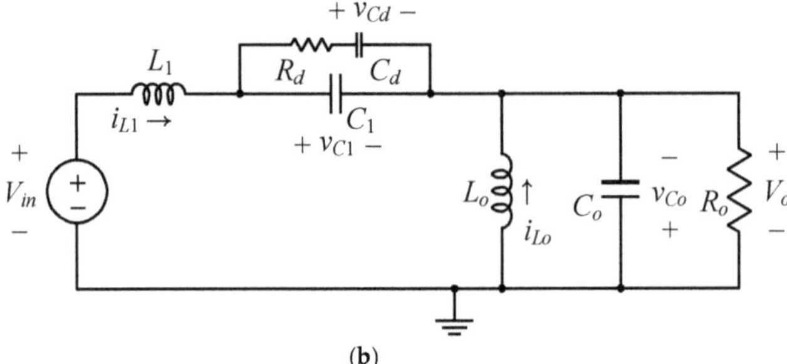

Figure 3. Equivalent circuits of the bridgeless SEPIC PFC converter with the damping circuit: (**a**) when S_c is turned on; (**b**) when S_c is turned off.

$$-i_{Lo} - \frac{v_{C1} - v_{Cd}}{R_d} = C_1 \frac{dv_{C1}}{dt} \tag{3}$$

$$\frac{v_{C1} - v_{Cd}}{R_d} = C_d \frac{dv_{Cd}}{dt} \tag{4}$$

$$-\frac{v_{Co}}{R_o} = C_o \frac{dv_{Co}}{dt} = -\frac{V_o}{R_o} \tag{5}$$

When S_c is turned off, the voltages of L_1 and L_o are expressed as below:

$$V_{in} - v_{C1} - v_{Co} = L_1 \frac{di_{L1}}{dt} \tag{6}$$

$$-v_{Co} = L_o \frac{di_{Lo}}{dt} \tag{7}$$

and the currents of C_1, C_d and C_o are written as:

$$i_{L1} - \frac{v_{C1} - v_{Cd}}{R_d} = C_1 \frac{dv_{C1}}{dt} \tag{8}$$

$$\frac{v_{C1} - v_{Cd}}{R_d} = C_d \frac{dv_{Cd}}{dt} \tag{9}$$

$$i_{L1} + i_{Lo} - \frac{V_{Co}}{R_o} = C_o \frac{dv_{Co}}{dt} \tag{10}$$

Using Equations (1)–(5), then state–space matrix for S_c on time can be rewritten as follows:

$$A_1 = \begin{bmatrix} 0 & 0 & 0 & 0 & 0 \\ 0 & 0 & \frac{1}{L_o} & 0 & 0 \\ 0 & -\frac{1}{C_o} & -\frac{1}{R_d C_1} & \frac{1}{R_d C_1} & 0 \\ 0 & 0 & \frac{1}{R_d C_d} & -\frac{1}{R_d C_d} & 0 \\ 0 & 0 & 0 & 0 & -\frac{1}{R_o C_o} \end{bmatrix} \quad B_1 = \begin{bmatrix} \frac{1}{L_1} \\ 0 \\ 0 \\ 0 \\ 0 \end{bmatrix} \tag{11}$$

and, the state–space matrix for S_c off time can be derived as below:

$$A_2 = \begin{bmatrix} 0 & 0 & -\frac{1}{L_1} & 0 & -\frac{1}{L_1} \\ 0 & 0 & 0 & 0 & -\frac{1}{L_o} \\ \frac{1}{C_1} & 0 & -\frac{1}{R_d C_1} & \frac{1}{R_d C_1} & 0 \\ 0 & 0 & \frac{1}{R_d C_d} & -\frac{1}{R_d C_d} & 0 \\ \frac{1}{C_o} & \frac{1}{C_o} & 0 & 0 & -\frac{1}{R_o C_o} \end{bmatrix} \quad B_2 = \begin{bmatrix} \frac{1}{L_1} \\ 0 \\ 0 \\ 0 \\ 0 \end{bmatrix} \tag{12}$$

$$x = \begin{bmatrix} i_{L1} \\ i_{Lo} \\ v_{C1} \\ v_{Cd} \\ v_{Co} \end{bmatrix} \quad u = [V_{in}] \quad \dot{x} = \begin{bmatrix} \frac{di_{L1}}{dt} \\ \frac{di_{Lo}}{dt} \\ \frac{dv_{C1}}{dt} \\ \frac{dv_{Cd}}{dt} \\ \frac{dv_{Co}}{dt} \end{bmatrix} \tag{13}$$

$$\frac{x(s)}{d(s)} = (sI - A)^{-1}\{(A_1 - A_2)X + (B_1 - B_2)V_{in}\} \tag{14}$$

The control-to-inductor current model can be obtained by Equations (11)–(14) as a fifth-order model. Similarly, the control-to-inductor current model without the RC damping also can be derived as a fourth-order model [29]. Figure 4 compares the control models of the bridgeless SEPIC PFC converters with the RC damping and without damping in frequency domain with parameters in Table 1. The model with the RC damping has only one resonant frequency in 800 Hz and the resonance at higher frequency is damped. On the contrary, the model without the RC damping has two resonance points in about 800 Hz and 5 kHz with high Q factor. This second resonance point with high frequency which is over than the current control bandwidth cannot be controlled and oscillates the system current unexpectedly. Therefore, the RC damping should be included to damp the second resonant peak.

Table 1. System Parameters.

Parameters	Values
Switching frequency (f_{sw})	72 kHz
Sampling frequency (f_s)	24 kHz
Input root mean square (RMS) voltage (v_g)	120 V/60 Hz
Output voltage (V_o) (buck/boost)	80 V/220 V
Input filter inductance (L_1, L_2)	1 mH
Output inductance (L_o)	1 mH
Damping resistance (R_d)	60 Ω
Energy transfer capacitance (C_1, C_2)	0.47 μF
Damping capacitance (C_d)	2.2 μF
Output capacitance (C_o)	2.6 mF

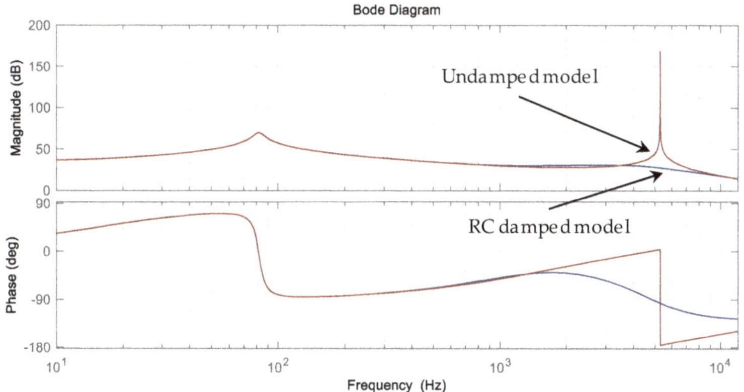

Figure 4. Frequency responses of the RC damped and the undamped models.

2.3. Control Model Approximation of the Bridgeless SEPIC PFC Converter

The control-to-inductor current model with the RC damping is the fifth-order model. Since the original undamped model is the fourth model, it became more complexed than the original undamped model. Thus, it is difficult to analyze the frequency response of the controller with the fifth model so, approximation needs to be adapted to design the controller easily. Since the capacitance of C_1 and C_d

is much lower value than the other components, the multiple of C_1 and C_d is sufficiently small to be ignored. Similarly, the multiple of the L_1 and L_o is low value as C_d or C_1, so $C_1L_1L_o$ and $C_dL_1L_o$ also can be substituted with zero. With these processes, the approximate model can be represented as follows:

$$G_{id}(s) = \frac{i_g(s)}{d(s)} = \frac{N_1 s^3 + N_2 s^2 + N_3 s + N_4}{D_1 s^3 + D_2 s^2 + D_3 s + D_4} \tag{15}$$

$$\begin{aligned}
N_1 &= C_o L_o R_o \{(C_1 + C_d)(V_{C1} + V_{Co}) + C_d R_d D'(I_{L1} + I_{Lo})\} \\
N_3 &= L_o D'(I_{L1} + I_{Lo}) + D(C_d R_d + C_o R_o)(V_{C1} + V_{Co}) + C_d R_d R_o D'(I_{L1} + I_{Lo}) \\
N_3 &= L_o D'(I_{L1} + I_{Lo}) + D(C_d R_d + C_o R_o)(V_{C1} + V_{Co}) + C_d R_d R_o D'(I_{L1} + I_{Lo}) \\
N_4 &= D(V_{C1} + V_{Co}) + R_o D'(I_{L1} + I_{Lo})
\end{aligned} \tag{16}$$

$$\begin{aligned}
D_1 &= C_d C_o R_d R_o \{L_o(1 - 2D) + D^2(L_1 + L_o)\} \\
D_2 &= R_o(L_1 D^2 + L_o D'^2)(C_1 + C_d + C_o) + L_1 R_o(1 - 2D)(C_1 + C_d) + C_d R_d \{L_o(1 - 2D) + D^2(L_1 + L_o)\} \\
D_3 &= L_o(1 - 2D) + D^2(L_1 + L_o) + C_d R_d R_o D'^2 \\
D_4 &= R_o D'^2 \\
D' &= 1 - D
\end{aligned} \tag{17}$$

The frequency responses of the original fifth-order model and the approximated third-order model in buck and boost modes are represented in Figure 5. As can be seen, the third-order model is very well matched with the fifth-order model until 600 Hz. Where the frequency is beyond 600 Hz, the fifth-order model contains the damped resonance and phase delay, which do not appear in the third-order model. However, the third-order model is enough to design the current controller, because the current control bandwidth is not very high.

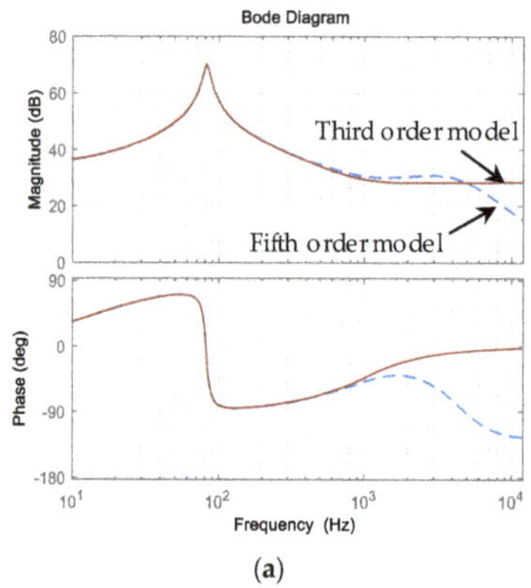

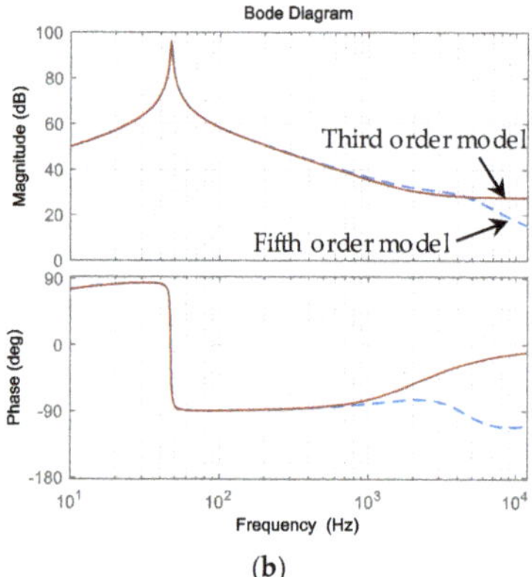

Figure 5. Frequency responses of the third- and the fifth-order RC damped models: (**a**) Buck mode; (**b**) Boost mode.

3. Proposed Current Controller

3.1. Traditional Current Controller Design

In order to control the bridgeless SEPIC PFC converter, a digital controller is implemented. Therefore, the approximated $G_{id}(s)$ is transformed on the z-domain and delays due to unit calculation and digital pulse-width modulation (PWM) update should be also considered [36]. The z-domain control-to-inductor current model $G_{id}(z)$ is derived as follows:

$$G_{id}(z) = z^{-1} G_{id}(s)\big|_{s=(z-1)/T_s} \tag{18}$$

where T_s is the sampling period. The digital delay is represented by a unit delay.

The controller structure for regulating the current is shown in Figure 6. The input current reference i_g^* and the input current i_g are input to the controller as absolute values. The duty reference d_{ref} is generated by the current controller $G_{cc}(z)$ and the feed-forward duty d_{ff}. The feed-forward duty compensates for the disturbance caused by the input voltage [37], and is calculated as below:

$$d_{ff} = \frac{V_o}{|v_g| + V_o} \tag{19}$$

where v_g is the input voltage, and V_o is the output voltage.

For the stable current controller design, the frequency response of the open-loop gain $T_i(z)$ consisting of $G_{cc}(z)$ and $G_{id}(z)$ should be analyzed. The open-loop gain $T_{i,P}(z)$ and $T_{i,PI}(z)$ of a proportional (P) controller and a proportional-integral (PI) controller are obtained as:

$$T_{i,P}(z) = K_p G_{id}(z) \tag{20}$$

$$T_{i,PI}(z) = \left(K_p + K_i \frac{T_s z}{z-1} \right) G_{id}(z) \tag{21}$$

where K_p is a proportional gain, and K_i is an integral gain. Since the approximated $G_{id}(s)$ is consistent with the original fifth-order model up to about 600 Hz, the controller is designed accordingly. Figure 7 shows the frequency response of $T_{i,P}(z)$ and $T_{i,PI}(z)$ when K_p and K_i are selected as 0.01 and 60, respectively. In the buck mode, the crossover frequency of $T_{i,P}(z)$ is 269 Hz, which is the bandwidth of the controller. The phase margin Φ_{pm} at the crossover frequency is 96.7 deg. For $T_{i,PI}(z)$, the crossover frequency is 530 Hz and the phase margin is 45.3 deg. Thus, both controllers designed are stable in the buck mode. Also, $T_{i,p}(z)$ and $T_{i,PI}(z)$ of the boost mode are stable, but the crossover frequency is wider than the buck mode. As a result, characteristic of the controller can be superior in the boost mode.

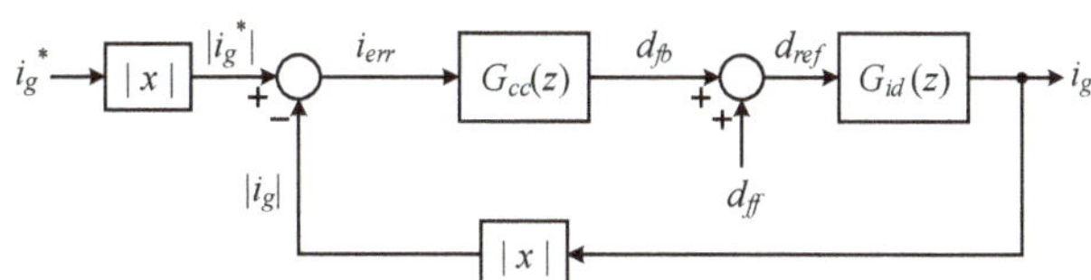

Figure 6. Current controller structure.

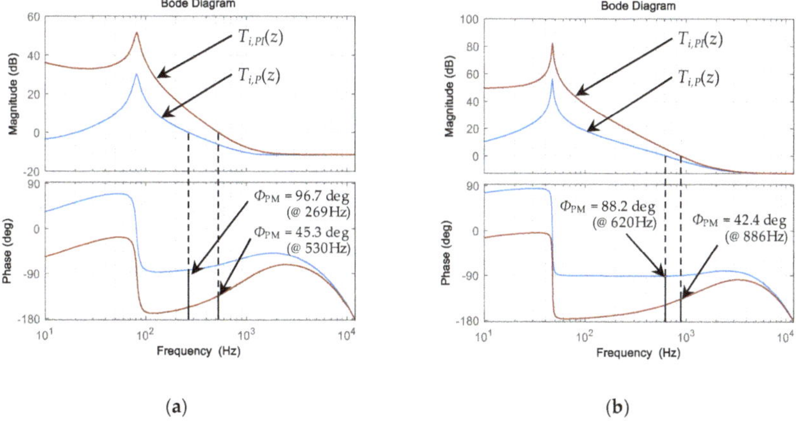

Figure 7. Frequency response of $T_i(z)$: (**a**) Buck mode; (**b**) Boost mode.

3.2. Repetitive Controller Design

The repetitive controller has excellent performance in eliminating periodic errors [32–35]. The proposed current controller is shown in Figure 8, which consists of the repetitive controller and the P controller in parallel. The repetitive controller is composed of repetitive controller gain K_{rp}, the number of samples N, the number of samples for phase leading L, and stabilization filter $q(z)$. The transfer function of the repetitive controller is derived as below:

$$G_{rp}(z) = \frac{d_{rp}}{i_{err}} = K_{rp}\frac{z^L}{z^N - q(z)} \tag{22}$$

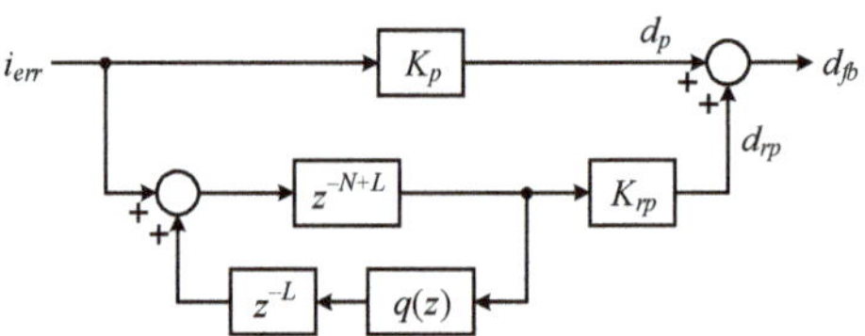

Figure 8. Proposed current controller.

The number of samples N of the repetitive controller is determined by the fundamental frequency f_r of the current reference to be controlled and the sampling frequency f_s as follows:

$$N = \frac{f_s}{f_r} \tag{23}$$

As shown in Figure 6, the current controller regulates the absolute value of the input current. Thus, compared with the frequency of the input current reference, the fundamental frequency of the repetitive controller is doubled. The number of samples for phase leading L is chosen as 2 to compensate for the digital delay of 1.5 T_s. The stabilization filter $q(z)$ is used to ensure the stability of the repetitive controller for the very high order harmonics that cannot be regulated [32]. In general, the following zero-phase delay low pass filter is selected as $q(z)$:

$$q(z) = 0.25z^{-1} + 0.5 + 0.25z \tag{24}$$

The remaining parameter of the repetitive controller is K_{rp}, which determines the stability of the repetitive controller. In order to select K_{rp}, the transfer function of the input current reference to error $G_e(z)$ should be considered, and it is obtained as:

$$G_e(z) = \frac{i_{err}}{i_g^*} = G_{ep}(z)G_{erp}(z) \tag{25}$$

where $G_{ep}(z)$ and $G_{erp}(z)$ are expressed as below:

$$G_{ep}(z) = \frac{1}{1 + T_{i,P}(z)} \tag{26}$$

$$G_{erp}(z) = \frac{z^N - q(z)}{z^N - H(z)} \tag{27}$$

where $H(z)$ is defined as follows:

$$H(z) = q(z) - K_{rp}z^L\frac{G_{id}(z)}{1 + T_{i,P}(z)} \tag{28}$$

For $G_e(z)$ to be stable, all poles must be located within the unit circle in the z-domain. If the P controller is designed to be stable, the poles of $G_{ep}(z)$ are in the unit circle. It can be ensured by selecting an appropriate K_p through the open-loop gain analysis as described above. Therefore, in order for $G_e(z)$ to be stable, the stability of $G_{erp}(z)$ must be guaranteed. According to the small gain theorem, $G_{erp}(z)$ is stable if the magnitude of $H(z)$ is less than 1 [35]. Figure 9 shows the root trajectories of $H(z)$ depending on K_{rp} up to the Nyquist frequency in buck and boost modes. When K_{rp} is 0.021, $G_{erp}(z)$ is unstable because the root trajectories of $H(z)$ deviate from the unit circle in both modes. When K_{rp} is 0.02, the magnitude of $H(z)$ is smaller than 1 in buck mode, but not in boost mode. Therefore, K_{rp} must be less than 0.02 for $G_e(z)$ to be stable in both modes. The frequency response of $G_e(z)$ is illustrated in Figure 10 when K_{rp} is 0.01. Since the fundamental frequency is 120 Hz, it can be seen that the errors of the multiples of fundamental frequency are removed.

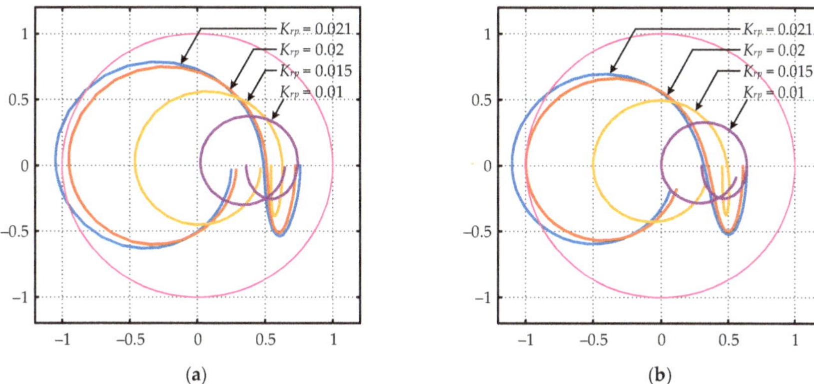

Figure 9. Root trajectories of $H(z)$: (**a**) Buck mode; (**b**) Boost mode.

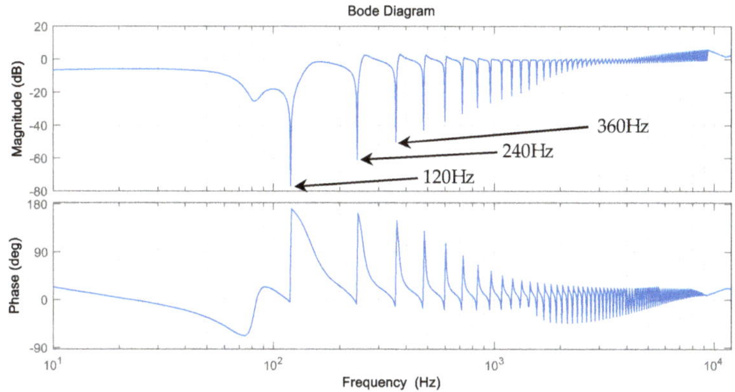

Figure 10. Frequency response of $G_e(z)$.

4. Simulation Results

In order to verify the performance of the proposed controller, the simulation studies have been performed using the simulation software package PSIM. All parameters used in the simulation are shown in Table 1.

Figures 11 and 12 show the input current and the current error when operating in buck mode and boost mode at the full load condition under 800 W and light load condition under 100 W.

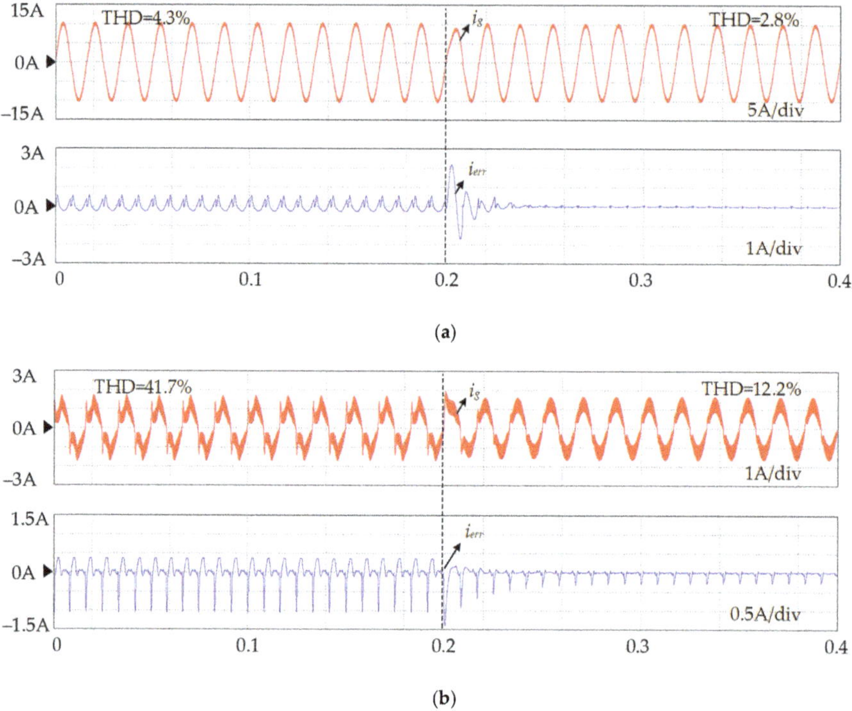

Figure 11. Simulation results applying repetitive controller at 0.2 s in buck mode: (**a**) Input current and current error at load = 800 W; (**b**) Input current and current error at load = 100 W.

Before $t = 0.2$ s, the input current is regulated with the PI controller. At $t = 0.2$ s, the PI controller is substituted with the repetitive controller in parallel with P controller. In Figure 11, before applying the repetitive controller, the magnitude of the current error is less than 1 A. However, after applying the repetitive controller, the magnitude of the current error is limited to 0.5 A. The current errors considerably decrease after $t = 0.2$ s in both load conditions. Especially, under the light load condition, the distortion of the input current is significantly reduced. Also, the total harmonic distortion (THD) of the input current is improved from 4.3% to 2.8% at the full load condition and from 41.7% to 12.2% at light load condition.

Figure 12 depicts the input current and the current error when operating in boost mode at the full load condition under 800 W and light load condition under 100 W. In boost mode, after applying the repetitive controller the magnitude of the input current error is similar with that in buck mode. The THD of the input current under heavy and light load conditions improved 4.4% to 4.3% and 45.8% to 34.8%, respectively. In boost mode, the input current ripple is larger than in buck mode, because the voltage across L_1 is higher due to output voltage according to Equation (6). So, the THD of the input current is higher than in buck mode.

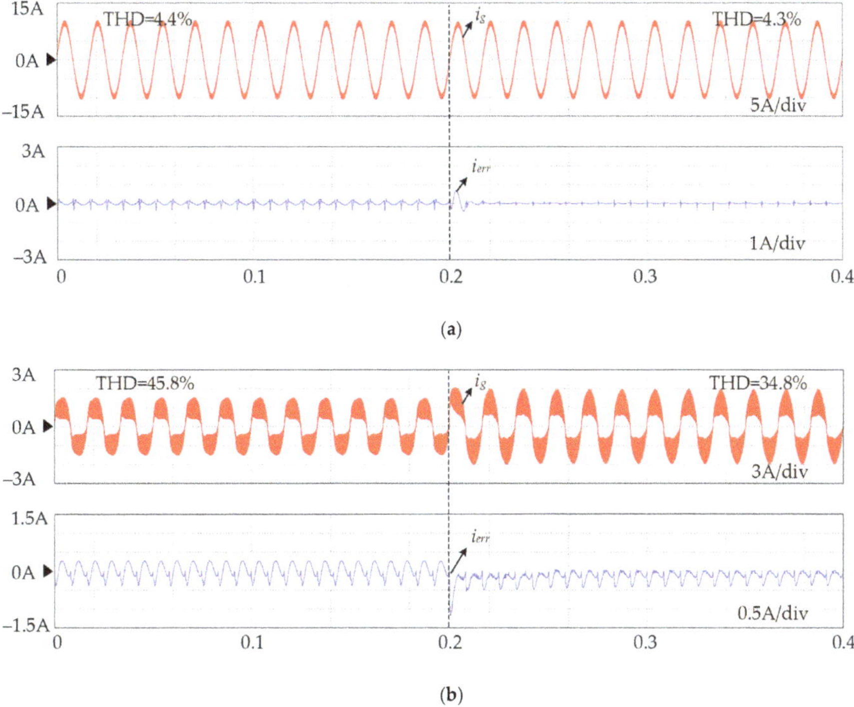

Figure 12. Simulation results applying repetitive controller at 0.2 s in boost mode: (**a**) Input current and current error at load = 800 W; (**b**) Input current and current error at load = 100 W.

5. Experimental Results

The parameter values for hardware are the same values in Figure 1, and a TMS320F28335 digital signal processor (DSP) of Texas Instruments (Dallas, TX, USA) was adopted to implement the digital controller. The bridgeless SEPIC PFC converter consists of a silicon carbide (SiC) MOSFET C2M004120D and four SiC schottky diodes C4D20120D, C3D16065A which are manufactured from Cree. The input voltage of the converter is supplied by Programmable AC power source model 61704. The SEPIC PFC prototype system has been tested from 100 W to 800 W in both buck and boost mode. The output voltages of buck and boost mode are 80 V and 220 V each.

Figures 13 and 14 illustrate the experimental results without the repetitive controller and with the repetitive controller. Figure 13 shows the input current and the current error under 100 W load condition in buck mode and in boost mode. In Figure 13a, without the repetitive controller, the input current is regulated in phase with the input voltage. However, there is the current distortion near the zero-crossing point of the input current, and the peak-to-peak value of the current error is less than 1.8 A. On the other hand, with the repetitive controller the current error is reduced to 0.39 A and the waveform of the input current is significantly improved in Figure 13b. Figure 13c,d shows the experimental results in boost mode without and with the repetitive controller, respectively. In boost mode, the magnitude of current error is 1.53 A without the repetitive controller, but only 0.47 A is measured by applying the repetitive controller.

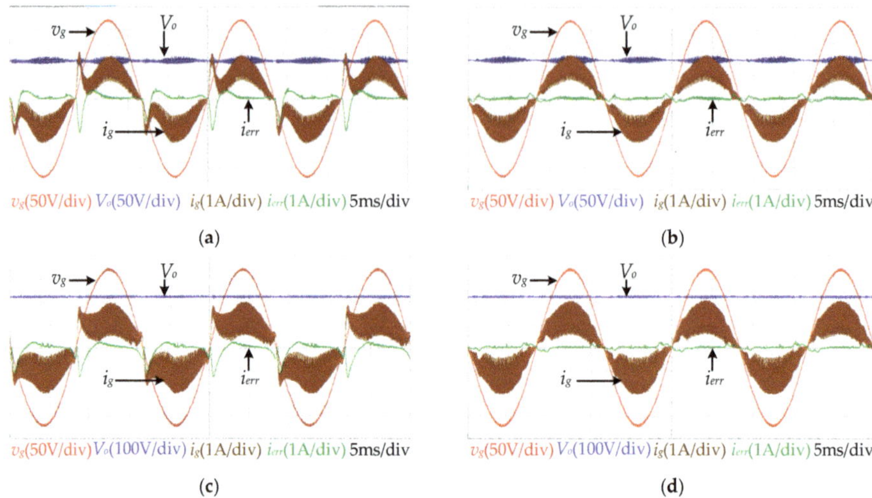

v_g(50V/div) V_o(50V/div) i_g(1A/div) i_{err}(1A/div) 5ms/div

(a)

v_g(50V/div) V_o(50V/div) i_g(1A/div) i_{err}(1A/div) 5ms/div

(b)

v_g(50V/div)V_o(100V/div) i_g(1A/div) i_{err}(1A/div) 5ms/div

(c)

v_g(50V/div)V_o(100V/div)i_g(1A/div) i_{err}(1A/div) 5ms/div

(d)

Figure 13. Experimental results under the 100 W condition in buck/boost mode under $v_g = 120$ V$_{rms}$, $V_o = 80$ V(buck)/$V_o = 220$ V(boost): (**a**) without the repetitive controller in buck mode; (**b**) with the repetitive controller in buck mode; (**c**) without the repetitive controller in boost mode; (**d**) with the repetitive controller in boost mode.

Figure 14 represents the input current and the current error under 800 W load condition in buck mode and boost mode. Figure 14a,c shows that the PI controller works well, and the addition of repetitive controller can be seen to reduce both size of current error and the input current THD as shown in Figure 14b,d. In Figure 14a that is buck mode operation, the current error is 1.35 A but after adding the repetitive controller, the current error is changed to 1.32 A as shown in Figure 14b. Similarly, Figure 14c which is the boost mode shows the current error magnitude of 1.94 A. But in Figure 14d, when the repetitive controller is applied, the current error was read as 1.12 A. Accordingly, Figures 13 and 14 show that the proposed repetitive control method improves the input current quality.

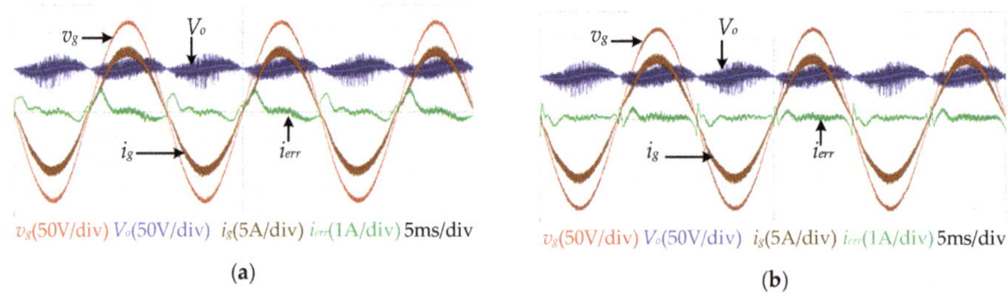

v_g(50V/div) V_o(50V/div) i_g(5A/div) i_{err}(1A/div) 5ms/div

(a)

v_g(50V/div) V_o(50V/div) i_g(5A/div) i_{err}(1A/div) 5ms/div

(b)

Figure 14. *Cont.*

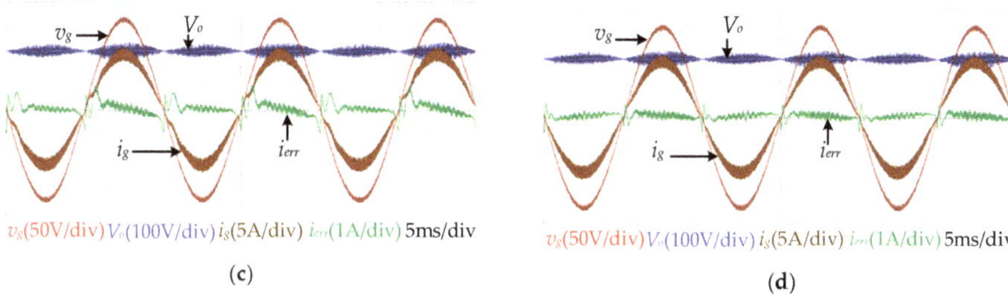

Figure 14. Experimental results under the 800 W condition in buck/boost mode under v_g = 120 V$_{rms}$, V_o = 80 V(buck)/V_o = 220 V(boost): (**a**) without the repetitive controller in buck mode; (**b**) with the repetitive controller in buck mode; (**c**) without the repetitive controller in boost mode; (**d**) with the repetitive controller in boost mode.

Under various load conditions, the input current THD comparisons between the conventional PI controller and the proposed control scheme are shown in Figure 15. Figure 15a shows the buck operation mode and Figure 15b shows the boost operation mode in the bridgeless SEPIC PFC converter. It also shows that the proposed repetitive controller has much improved THD than the conventional PI controller.

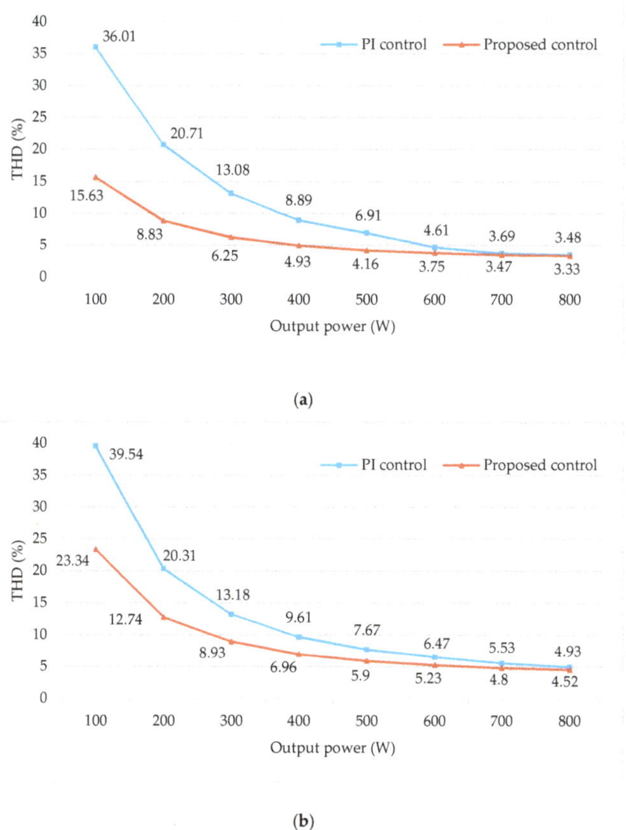

Figure 15. Input current total harmonic distortion comparison results at different load conditions: (**a**) THDs in buck mode; (**b**) THDs in boost mode.

6. Conclusions

The general PFC converters have been studied to boost the output voltage. On the other hand, it only can operate in boost mode, not in buck mode. To step down the output voltage, the general PFC converters must operate with the DC–DC converters. However, the SEPIC PFC converter can operate in buck and boost mode itself, without another system. In this paper, the bridgeless SEPIC PFC converter with RC damping topology has been discussed. The operation modes of the SEPIC converter and the control-to-inductor current model are described in detail. Also, the approximation of the current control model was proceeded, and it has been used to design and to analyze the stability of the current controller. By using this approximation model, the repetitive control scheme was evaluated with the error transfer function. The repetitive control parameters were derived by these analyses and the implementation of the digital controller is also discussed. The simulation and the experimental results verified the repetitive controller performance in 100 W to 800 W load conditions in buck mode and boost mode. As a result, the THDs of the input current are significantly decreased by the proposed repetitive controller in both buck and boost modes. Also, the experimental result shows that the controller based on simplified model is well designed.

Author Contributions: J.K., S.H., and W.C. implemented the system, performed the experiments, and wrote this paper. H.K. and Y.C. assisted with the idea development and the paper writing.

Funding: This work was supported by the "Human Resources Program in Energy Technology" of the Korea Institute of Energy Technology Evaluation and Planning (KETEP), granted financial resource from the Ministry of Trade, Industry & Energy, Republic of Korea (No. 20174030201660) and the "Human Resources Program in Energy Technology" of the Korea Institute of Energy Technology Evaluation and Planning (KETEP), granted financial resource from the Ministry of Trade, Industry & Energy, Republic of Korea (No. 20184030202270).

Conflicts of Interest: The authors declare no conflict interest.

References

1. Choi, W.; Lee, W.; Han, D.; Sarlioglu, B. New Configuration of Multifunctional Grid-Connected Inverter to Improve Both Current-Based and Voltage-Based Power Quality. *IEEE Trans. Ind. Appl.* **2018**, *54*, 6374–6382. [CrossRef]

2. Yao, Z.; Xiao, L. Control of Single-Phase Grid-Connected Inverters with Nonlinear Loads. *IEEE Trans. Ind. Electron.* **2013**, *60*, 1384–1389. [CrossRef]

3. Ismail, E.H. Bridgeless SEPIC Rectifier with Unity Power Factor and Reduced Conduction Losses. *IEEE Trans. Ind. Electron.* **2009**, *56*, 1147–1157. [CrossRef]

4. Musavi, F.; Edington, M.; Eberle, W.; Dunford, W.G. Control Loop Design for a PFC Boost Converter with Ripple Steering. *IEEE Trans. Ind. Appl.* **2013**, *49*, 118–126. [CrossRef]

5. Lee, J.; Chae, H. 6.6-kW Onboard Charger Design Using DCM PFC Converter with Harmonic Modulation Technique and Two-Stage DC/DC Converter. *IEEE Trans. Ind. Electron.* **2014**, *61*, 1243–1252. [CrossRef]

6. Xie, X.; Zhao, C.; Zheng, L.; Liu, S. An Improved Buck PFC Converter with High Power Factor. *IEEE Trans. Power Electron.* **2013**, *28*, 2277–2284. [CrossRef]

7. Tian, G.; Qi, W.; Yan, Y.; Jiang, Y.Z. High power factor LED power supply based on SEPIC converter. *Electron. Lett.* **2014**, *50*, 1866–1868. [CrossRef]

8. Wang, Y.; Huang, J.; Shi, G.; Wang, W.; Xu, D. A Single-Stage Single-Switch LED Driver Based on the Integrated SEPIC Circuit and Class-E Converter. *IEEE Trans. Power Electron.* **2016**, *31*, 5814–5824. [CrossRef]

9. Ma, H.; Lai, J.; Feng, Q.; Yu, W.; Zheng, C.; Zhao, Z. A Novel Valley-Fill SEPIC-Derived Power Supply without Electrolytic Capacitor for LED Lighting Application. *IEEE Trans. Power Electron.* **2012**, *27*, 3057–3071. [CrossRef]

10. Jang, Y.; Jovanović, M.M. Bridgeless High-Power-Factor Buck Converter. *IEEE Trans. Power Electron.* **2011**, *26*, 602–611. [CrossRef]

11. Foroozeshfar, R.; Adib, E.; Farzanehfard, H. New single-stage, single-switch, soft-switching three-phase SEPIC and Cuk-type power factor correction converters. *IET Power Electron.* **2014**, *7*, 1878–1885. [CrossRef]

12. Darwish, A.; Holliday, D.; Ahmed, S.; Massoud, A.M.; Williams, B.W. A Single-Stage Three-Phase Inverter Based on Cuk Converters for PV Applications. *IEEE J. Emerg. Sel. Top. Power Electron.* **2014**, *2*, 797–807. [CrossRef]
13. Yang, J.; Do, H. Bridgeless SEPIC Converter with a Ripple-Free Input Current. *IEEE Trans. Power Electron.* **2013**, *28*, 3388–3394. [CrossRef]
14. Diab, M.S.; Elserougi, A.; Massoud, A.M.; Abdel-Khalik, A.S.; Ahmed, S. A Four-Switch Three-Phase SEPIC-Based Inverter. *IEEE Trans. Power Electron.* **2015**, *30*, 4891–4905. [CrossRef]
15. Costa, P.J.S.; Font, C.H.I.; Lazzarin, T.B. Single-Phase Hybrid Switched-Capacitor Voltage-Doubler SEPIC PFC Rectifiers. *IEEE Trans. Power Electron.* **2018**, *33*, 5118–5130. [CrossRef]
16. Costa, P.J.S.; Font, C.H.I.; Lazzarin, T.B. A Family of Single-Phase Voltage-Doubler High-Power-Factor SEPIC Rectifiers Operating in DCM. *IEEE Trans. Power Electron.* **2017**, *32*, 4279–4290. [CrossRef]
17. Lee, S.; Do, H. Zero-Ripple Input-Current High-Step-Up Boost–SEPIC DC–DC Converter with Reduced Switch-Voltage Stress. *IEEE Trans. Power Electron.* **2017**, *32*, 6170–6177. [CrossRef]
18. Moradpour, R.; Ardi, H.; Tavakoli, A. Design and Implementation of a New SEPIC-Based High Step-Up DC/DC Converter for Renewable Energy Applications. *IEEE Trans. Ind. Electron.* **2018**, *65*, 1290–1297. [CrossRef]
19. Chiang, S.J.; Shieh, H.; Chen, M. Modeling and Control of PV Charger System with SEPIC Converter. *IEEE Trans. Ind. Electron.* **2009**, *56*, 4344–4353. [CrossRef]
20. Kamnarn, U.; Chunkag, V. Analysis and Design of a Modular Three-Phase AC-to-DC Converter Using CUK Rectifier Module with Nearly Unity Power Factor and Fast Dynamic Response. *IEEE Trans. Power Electron.* **2009**, *24*, 2000–2012. [CrossRef]
21. Kwon, J.; Choi, W.; Lee, J.; Kim, E.; Kwon, B. Continuous-conduction-mode SEPIC converter with low reverse-recovery loss for power factor correction. *IEE Proc.—Electr. Power Appl.* **2006**, *153*, 673–681. [CrossRef]
22. Chiu, H.; Lo, Y.; Chen, J.; Cheng, S.; Lin, C.; Mou, S. A High-Efficiency Dimmable LED Driver for Low-Power Lighting Applications. *IEEE Trans. Ind. Electron.* **2010**, *57*, 735–743. [CrossRef]
23. Liu, Y.; Sun, Y.; Su, M. A Control Method for Bridgeless Cuk/Sepic PFC Rectifier to Achieve Power Decoupling. *IEEE Trans. Ind. Electron.* **2017**, *64*, 7272–7276. [CrossRef]
24. Fardoun, A.A.; Ismail, E.H.; Sabzali, A.J.; Al-Saffar, M.A. New Efficient Bridgeless Cuk Rectifiers for PFC Applications. *IEEE Trans. Power Electron.* **2012**, *27*, 3292–3301. [CrossRef]
25. Mahdavi, M.; Farzanehfard, H. Bridgeless SEPIC PFC Rectifier with Reduced Components and Conduction Losses. *IEEE Trans. Ind. Electron.* **2011**, *58*, 4153–4160. [CrossRef]
26. Sabzali, A.J.; Ismail, E.H.; Al-Saffar, M.A.; Fardoun, A.A. New Bridgeless DCM Sepic and Cuk PFC Rectifiers with Low Conduction and Switching Losses. *IEEE Trans. Ind. Appl.* **2011**, *47*, 873–881. [CrossRef]
27. Liao, Y.; Jhu, J. Analysis and implementation of a bridgeless sepic AC/DC converter with power factor correction and extended gain. In Proceedings of the 2017 IEEE Applied Power Electronics Conference and Exposition (APEC), Tampa, FL, USA, 26–30 March 2017; pp. 416–423.
28. Onal, Y.; Sozer, Y. Bridgeless SEPIC PFC converter for low total harmonic distortion and high power factor. In Proceedings of the 2016 IEEE Applied Power Electronics Conference and Exposition (APEC), Long Beach, CA, USA, 20–24 March 2016; pp. 2693–2699.
29. Koh, H.; Cho, Y.; Lai, J.; Chen, R.; Zheng, C. Z-domain modeling and control design of single-switch bridgeless SEPIC PFC converter with damping circuit. In Proceedings of the 2013 Twenty-Eighth Annual IEEE Applied Power Electronics Conference and Exposition (APEC), Long Beach, CA, USA, 17–21 March 2013; pp. 2744–2748.
30. Clark, C.W.; Musavi, F.; Eberle, W. Digital DCM Detection and Mixed Conduction Mode Control for Boost PFC Converters. *IEEE Trans. Power Electron.* **2014**, *29*, 347–355. [CrossRef]
31. Youn, H.; Park, J.; Park, K.; Baek, J.; Moon, G. A Digital Predictive Peak Current Control for Power Factor Correction with Low-Input Current Distortion. *IEEE Trans. Power Electron.* **2016**, *31*, 900–912. [CrossRef]
32. Yang, S.; Wang, P.; Tang, Y.; Zagrodnik, M.; Hu, X.; Tseng, K.J. Circulating Current Suppression in Modular Multilevel Converters with Even-Harmonic Repetitive Control. *IEEE Trans. Ind. Appl.* **2018**, *54*, 298–309. [CrossRef]
33. Cho, Y.; Lai, J. Digital Plug-In Repetitive Controller for Single-Phase Bridgeless PFC Converters. *IEEE Trans. Power Electron.* **2013**, *28*, 165–175. [CrossRef]

34. Cho, Y. Dual-buck residential photovoltaic inverter with a high-accuracy repetitive current controller. *Renew. Energy* **2017**, *101*, 168–181. [CrossRef]
35. Yang, S.; Wang, P.; Tang, Y.; Zhang, L. Explicit Phase Lead Filter Design in Repetitive Control for Voltage Harmonic Mitigation of VSI-Based Islanded Microgrids. *IEEE Trans. Ind. Electron.* **2017**, *64*, 817–826. [CrossRef]
36. Kumar, M.; Gupta, R. Time-Domain Analysis of Sampling Effect in DPWM of DC–DC Converters. *IEEE Trans. Ind. Electron.* **2015**, *62*, 6915–6924. [CrossRef]
37. Shayestehfard, A.; Mekhilef, S.; Mokhlis, H. IZDPWM-Based Feedforward Controller for Grid-Connected Inverters Under Unbalanced and Distorted Conditions. *IEEE Trans. Ind. Electron.* **2017**, *64*, 14–21. [CrossRef]

Article

A Neural Network-Based Four Phases Interleaved Boost Converter for Fuel Cell System Applications

El Manaa Barhoumi [1,2,*], Ikram Ben Belgacem [3], Abla Khiareddine [4], Manaf Zghaibeh [1] and Iskander Tlili [5]

[1] Department of Electrical and Computer Engineering, College of Engineering, Dhofar University, Salalah 211, Oman; mzghaibeh@du.edu.om
[2] Laboratoire Analyse, Conception et Commande des Systèmes (LR11ES20), Ecole Nationale d'Ingénieurs de Tunis, Université de Tunis El Manar, Tunis 1002, Tunisia
[3] Laboratoire de Génie Mécanique, Ecole Nationale d'Ingénieurs de Monastir, Université de Monastir, Monastir 5019, Tunisia; ikrambenbelgacem@gmail.com
[4] Research Unit on Study of Industrial Systems and Renewable Energy (ESIER), Université de Monastir, National Engineering School of Monastir, Université de Monastir, Monastir 5019, Tunisia; khiareddine_abla@yahoo.fr
[5] Energy and Thermal Systems Laboratory, National Engineering School of Monastir, Street Ibn El Jazzar, Monastir 5019, Tunisia; Iskander.Tlili@enim.rnu.tn
* Correspondence: ebarhoumi@du.edu.om; Tel.: +968-98190380

Received: 26 October 2018; Accepted: 4 December 2018; Published: 6 December 2018

Abstract: This paper presents a simple strategy for controlling an interleaved boost converter that is used to reduce the current fluctuations in proton exchange membrane fuel cells, with high impact on the fuel cell lifetime. To keep the output voltage at the desired reference value under the strong fluctuations of the fuel flow rate, fuel supply pressure, and temperature, a neural network controller is developed and implemented using Matlab-Simulink (R2012b, MathWorks limited, London, UK). The advantage of this controller resides in its simplicity, where limited number of tests are carried out using Matlab-Simulink to construct it. To investigate the robustness of the proposed converter and the neural network controller, strong variations of the fuel flow rate, fuel supply pressure, temperature and air supply pressure are applied to both the fuel cell and the neural network controller of the converter. The simulation results show the effectiveness and the robustness of the both the proposed controller and converter to control the load voltage and minimize the current and voltage ripples. As a result of that, fuel cell current oscillations are considerably reduced on the one hand, while on the other hand, the load voltage is stabilized during transient variations of the fuel cell inputs.

Keywords: proton exchange membrane fuel cell; four phases interleaved boost converter; neural network controller

---

1. Introduction

Fuel cell technologies are becoming used in many industrial applications due to the cleanliness, high reliability and high performance of such electrical generators [1]. During the last decades, many kinds of fuel cells were developed and used. However, the proton exchange membrane fuel cell (PEMFC) has proved to have a higher efficiency when compared to other types of fuel cells [2]. In addition to its long life time, the PEMFC is characterized by its high power density at low operating temperature [3]. Moreover, the PEMFCs have good dynamic responses during instantaneous power demand variation. Nowadays, the PEMFCs are connected to hybrid renewable energy sources with energy storage systems like batteries and super capacitors [4]. Usually, such hybrid systems are used in hybrid electric vehicles to improve the performance of the global system during the peak power demand transient and instantaneous variation [5]. PEMFCs are used in electrical and hybrid cars such as Toyota, Honda, Hyundai, Nissan and Ford fuel cells cars developed their during the last years [5]. Under no load conditions, a PEMFC cell generates a low direct current voltage, approximately 1 V/cell. Typically, a sufficient number of cells are connected in series to increase the PEMFC voltage. However, the PEMFC voltage is still not enough for high load voltage demands. To increase the PEMFC voltage with the aim to meet the requirements of the load, a boost DC-DC converter is used to control the flow of the power from the PEMFC to the load and to prevent the PEMFC from overloading [6,7]. The power converters associated with fuel cells should have specific characteristics. Undeniably, the fuel cell is very sensitive to low and high frequency currents created by power electronic converters and loads [8]. Low and high frequency components of currents reduce the PEMFC output power and decrease the durability of the membranes [9]. To improve the lifespan of PEMFCs, many solutions were proposed to reduce the ripple and harmonic components of the current, as reported in [10–12]. One of the various proposed power converters is the multiphasing or interleaved DC-DC boost converter [13]. This converter consists of a parallel connection of simple boost converters, hence, it allows to minimize the ripple and harmonic contents of the voltage and current. The control of a multiphase interleaved boost DC-DC converter for classic voltage sources, i.e., batteries, usually consists of an inner current control loop and outer voltage control loop [14]. This kind of control has shown good performance in regulating the output voltage [15,16]. However, the control of the interleaved boost converter associated with PEMFC should follow different strategies due the non-linear characteristics of the PEMFC [17,18]. Indeed, any increase of the load current increases the PEMFC current. Hence, the PEMFC voltage decreases and the desired load voltage becomes unachievable [19]. The control of a superconducting magnetic energy storage (SMES) system for hybrid energy storage systems using fuel cells allows generating or absorbing load pulses to protect the fuel cell [20]. The proposed method shows good efficiency to control the variability of the load demand by using the load following control for auxiliary energy source. However, it could not be applied in a system using only a fuel cell. A boost converter working in differential and common mode is analysed in [21]. In differential mode, the converter allows one to regulate the ac output voltage. The common mode is adopted for current ripple reduction. The method is based on the use of repetitive controllers. Furthermore, a buck-boost DC–DC converter having a regulated output voltage was presented in [22]. The buck-boost was used as a second converter after a single-ended primary-inductance converter (SEPIC). The controller was designed in a way to have a regulated output voltage. In [23], a three phase interleaved boost converter was proposed and analyzed. The simulation of the average model of the converter has shown a good performance in reducing the current ripple. On the other side of the fuel cell, the temperature should be stabilized in optimal range using appropriate techniques of control. Indeed, a control of temperature allows a 10% increasing of the output power [24]. In [25], a real time optimization strategy was adopted to find the optimal value of the fuel flow rate. A maximum power point technique based on neural network was developed for the control of three phase interleaved boost converter. This method allowed to extract the maximum power from the fuel cell system under different temperature conditions. The voltage across the load was not regulated at reference value [26]. The sliding mode control of a coupled interleaved boost converter associated with a PEMFC system allows reducing the current ripples and regulating the load voltage at the presence of variable load. However, the adapted controller did not show the effect of the source conditions variation, i.e., temperature, pressure on the regulated output voltage [27].

In this paper, the objectives of the proposed converter and its controller are the mitigation of the load pulses as well as the regulation of the output voltage. In comparison with the method and results presented in [21], the proposed solution consists of the generating the required duty cycle to regulate the output voltage. The proposed PWM allows minimizing the ripple of current for all output voltage. This work adopts a non-conventional control technique based on the use of a neural network to control the interleaved boost DC-DC converter associated to the PEMFC. The remaining of the paper is organized as follows. Section 2 is dedicated to the presentation and the modelling of the PEMFC. In Section 3, the interleaved boost DC-DC converter is presented, analysed and simulated using Matlab-Simulink. In Section 4, the impact of external parameters on both load voltage and PEMFC voltage is quantified through simulations. The design of the new controller of the interleaved boost converter is presented in Section 5. Simulations results of the control of PEMFC-boost converter are presented and analysed in Section 6. Finally, conclusions and recommendations are summarized in Section 7.

2. Fuel Cell Modeling

In this paper, the considered model was developed in [1,2]. The differences in equations of the model are summarized in the following expressions. The output voltage of an elementary cell is given by [1]:

$$V_{Cell} = E_0 - \Delta V \tag{1}$$

where E_0 is the reversible voltage of the cell, called also the thermodynamic potential of the cell. This voltage is given by [2]:

$$E_0 = 1.229 - 0.85 \times 10^{-3} \times (T - 298.15) + 4.31 \times 10^{-5} \times T \times [\log(P_{H_2}) + 0.5 \log(P_{O_2})] \tag{2}$$

The voltage drop in the PEMC due to the electrical and chemical factors is the sum of three voltages: activation voltage, ohmic voltage and the concentration voltage. Hence, the cell voltage is expressed as [1,2]:

$$V_{Cell} = E_0 - V_{act} - V_{ohm} - V_{conc} \tag{3}$$

The activation losses are described by the activation overvoltage, V_{act} [1]:

$$V_{act} = -\left(\xi_1 + \xi_2 T + \xi_3 T \log(c_{O_2}) + \xi_4 T \log(i_{FC})\right) \tag{4}$$

The term i_{FC} is the fuel cell stack current. ξ_1, ξ_3 and ξ_4 are constants, given respectively by [1,2]:

$$\xi_1 = -0.948$$
$$\xi_3 = 7.6 \times 10^{-3}$$
$$\xi_4 = -1.93 \times 10^{-4}$$

The parameter ξ_2 depends to the membrane area and the concentration of hydrogen [1,2].

$$\xi_2 = 0.00286 + 0.0002 \ \log(A_{cell}) + 4.3 \ 10^{-5} \log(C_{H2}) \tag{5}$$

The oxygen concentration in the catalytic interface of the cathode is expressed by [1,2]:

$$c_{O_2} = \frac{P_{O_2}}{5.08 \ 10^6 \exp\left(-\frac{498}{T}\right)} \tag{6}$$

where P_{O_2} is the pressure of oxygen in the catalytic interface of the cathode.

The ohmic voltage is given by [2]:

$$V_{ohm} = i_{FC} R_{ohm} = i_{FC}(R_{mem} + R_e) \tag{7}$$

where R_{mem} is the equivalent resistance of the membrane expressed as in [1]:

$$R_{mem} = \frac{\rho_{mem} l}{A} \tag{8}$$

where l and A are respectively the thickness and the area of the membrane. The resistivity of the membrane, ρ_{mem}, is given by [2]:

$$\rho_{mem} = 186\left[1 + 0.03\left(\tfrac{i_{FC}}{A}\right) + 0.062\left(\tfrac{T}{303}\right)^2\left(\tfrac{i_{FC}}{A}\right)^{2.5}\right]\Big/\left[\Psi - 0.634 - 3\left(\tfrac{i_{FC}}{A}\right)\right]\cdot\exp\left[4.18\left(\tfrac{T-303}{T}\right)\right] \quad (9)$$

As a result of the concentration of reactants consumed in the reaction, a voltage called concentration overvoltage is defined and given by [2]:

$$V_{conc} = -B\log\left(1 - \frac{\left(\tfrac{i_{FC}}{A}\right)}{\left(\tfrac{i_{FC}}{A}\right)_{max}}\right) \quad (10)$$

The factor B is a parameter that depends to the fuel cell. The PEMFC is selected based on the maximum power required by the DC bus, which is about 1 kW. The required voltage on the DC bus is 60 V. Then, the PEMFC having the parameters given in Table 1 is selected.

Table 1. PEMFC Electrical Parameters.

Parameter	Value
Stack rating voltage (V)	24
Power (kW)	1.26
Stack rating Current (A)	52
Maximum Current (A)	100
Maximum voltage (V)	42
Number of Cells	42
Nominal stack Efficiency	46%
Time constant	1 ms

3. Interleaved Boost Converter

The classic boost converter used to step up the voltage is shown in Figure 1. An insulated gate bipolar transistor (IGBT), inductor, capacitor and a diode, essentially compose the boost converter. The output voltage is controlled through the pulse width modulation (PWM) system.

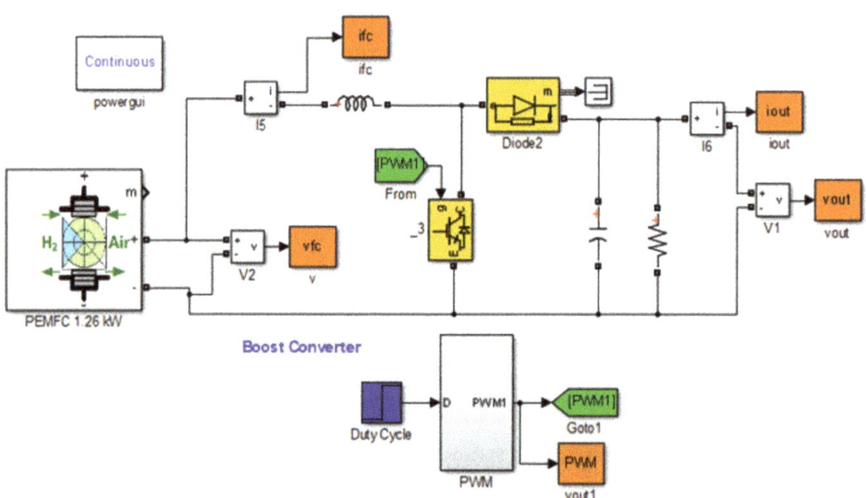

Figure 1. Fuel Cell-Boost Converter System.

A constant signal equal to the duty cycle is compared to the carrier wave to generate the PWM signal to control the IGBT. During DT, the switch is closed and the diode is reverse biased. Accordingly, the variation of the inductor current is given by [6]:

$$(\Delta i_L)_{Closed} = \frac{V_{Load}DT}{L} \tag{11}$$

When the switch is opened, the diode becomes forward biased to ensure a path for the current. Thus, the variation of the inductor current in $(1 - D)T$ is given by [6]:

$$(\Delta i_L)_{Opened} = \frac{(V_{Stack} - V_{Load})(1 - D)T}{L} \tag{12}$$

The average change current in the inductor is equal to zero. Therefore, the load voltage is expressed as follows [8]:

$$V_{Load} = \frac{1}{1 - D}V_{Stack} \tag{13}$$

The maximum voltage of the PEMFC is 42 V. A boost converter is suitable to step up the voltage to reach the load voltage of 60 V. The inductor is selected to minimize the load current ripples to less than 0.5 A. The capacitor is sized to reduce the load voltage to less than 5 V. For a switching frequency and a duty cycle of 25 kHz and 0.5 respectively, the inductor and capacitor values are calculated. The boost converter parameters are summarized in Table 2. The fuel cell boost converter system is simulated in Matlab-Simulink as shown in Figure 1. The aim of this simulation is to evaluate the ripple of the load voltage, stack voltage, load current and stack current. The simulation results are presented in Figure 2a,b.

Table 2. Boost Converter parameters.

Parameter	Symbol	Value
Inductance (mH)	L	1
Capacitance (µF)	C	50
Input Voltage (V)	V_{Stack}	—
Load resistance (Ω)	R	10
Duty Cycle	D	0.5
Frequency (kHz)	f	25

Figure 2a shows a load current ripple equals to 0.6 A. The load voltage oscillation is equal to 6 V as shown in Figure 2b. In the design of DC/DC converters, reducing the voltage and current oscillations leads to the selection of the best values of the inductor and the capacitor based on the following equations [6]:

$$L_{min} = \frac{D(1 - D)^2 R}{2f} \tag{14}$$

$$C_{min} = \frac{D}{R\left(\frac{\Delta V_{Load}}{V_{Load}}\right)f} \tag{15}$$

To minimize the inductor current ripple and the output voltage ripple to the desired values, the inductor and capacitor must be resized according to Equations (14) and (15). Normally, the major problem exists due to the size, weight and cost of high power inductor and capacitor.

Indeed, using high power inductor and capacitor will significantly increase the weight of the DC/DC converter. However, it is crucial to minimize the ripple and harmonic content of the current in the circuit in order to protect the fuel cell as well as to increase the life time of other components. Another solution was proposed to reduce the ripple current and harmonics is based on the use of what is called an interleaved power converter.

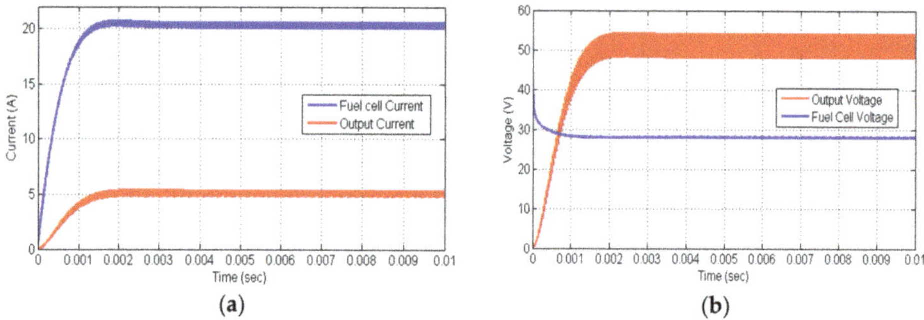

Figure 2. (**a**) Variation of currents; (**b**) Variation of voltages.

Simulation of the Four Phases Interleaved Boost Converter

Interleaved converters, also called multiphasic, are used to minimize the voltage and current ripples using the same filter components [6]. Therefore, using such interleaved converters allows reducing the size of filter components. The proposed interleaved boost converter is shown in Figure 3. The new boost converter is formed by a parallel combination of four sets of diodes, switches and inductors connected to a common capacitor and load. The PWM signals for the control of the four IGBTs is based on the PWM signal generated to control the first IGBT. Each IGBT control is shifted by a delay time equal to the fourth of the period. For a duty cycle of 0.25, the commutation sequence of the IGBTs is given as follows:

$$\begin{cases} 0 \leq t \leq \frac{T}{4} : & IGBT_1 \\ \frac{T}{4} \leq t \leq \frac{T}{2} : & IGBT_2 \\ \frac{T}{2} \leq t \leq \frac{3T}{4} : & IGBT_3 \\ \frac{3T}{4} \leq t \leq T : & IGBT_4 \end{cases} \tag{16}$$

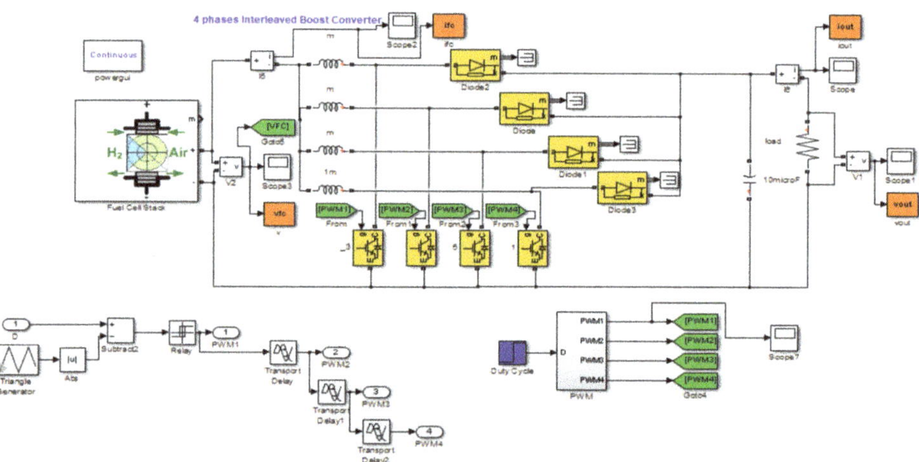

Figure 3. Interleaved Boost Converter and PWM simulation under Matlab-Simulink.

For a duty cycle of 20%, the four PWM signals are shown in Figure 4. The IGBTs are operating at 90° out of phase producing currents that are 90° out of phase. The load current is the sum of the four inductors current. Hence, the resultant load current has a smaller ripple and a frequency which is four times larger than that of the load current of the single phase boost converter. Figure 3 shows the four phases boost converter and the PWM control for the four IGBTs implemented in Matlab-Simulink.

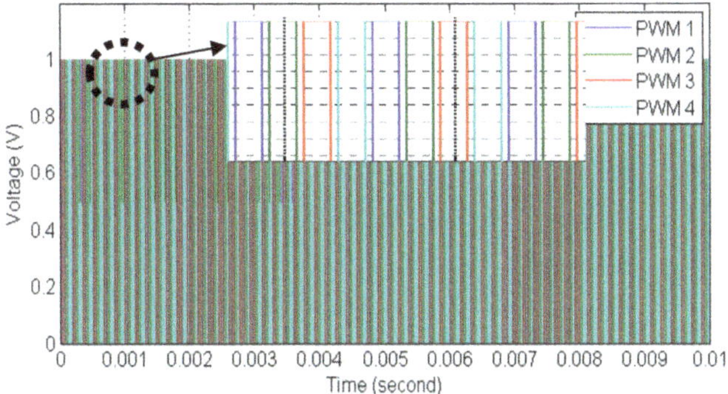

Figure 4. PWM control signals for duty cycle equal to 0.2.

Figure 5a shows the variation of the stack current and electrical current in the load. It is clear in the figure that the current oscillations are reduced to less than 0.2 A. Moreover, the variation in current generated from the PEMFC is reduced. Therefore, the stack current is perfectly smoothed. This allows to increase the life time of the PEMFC. Figure 5b shows the voltage waveforms. Using the proposed four phases interleaved boost converter allows decreasing the load voltage ripple to less than 2 V. Moreover, the PEMFC voltage ripple is decreased considerably.

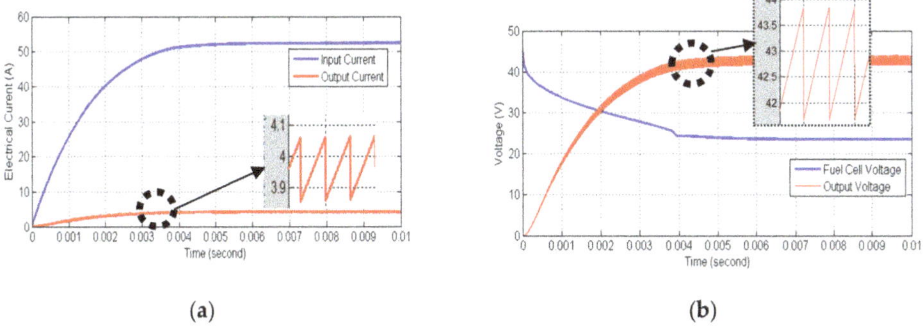

(a) (b)

Figure 5. (**a**) Currents for duty cycle 0.5, (**b**) Voltages for duty cycle 0.5.

4. Impact of External Parameters on the Output Voltage

To control the PEMFC voltage, it is required to supervise the inputs parameters of the fuel cell. Basically, the stack voltage depends to the pressure and flow rate of the fuel and the air supply. On the other hand, static characteristics show that the temperature of working environment has a strong effect on the PEMFC voltage. This section concerns the dynamic simulation of the fuel cell when varying different variables such as the fuel flow rate, the pressure, the air supply pressure, and the temperature of working environment. Hence, some simulations were carried out in the Matlab-Simulink environment to show the effect of the variation of the physical inputs on the load and PEMFC voltage. In this simulation the duty cycle is fixed at 0.5. Figure 6a shows the variation of the output voltage under temperature variation in three stages. As shown, the temperature has a strong effect on the output voltage of the fuel cell and the load voltage. Evidently, for a temperature equal to 273 K, the PEMFC voltage and the load voltage are equal to 18 V and 35 V, respectively.

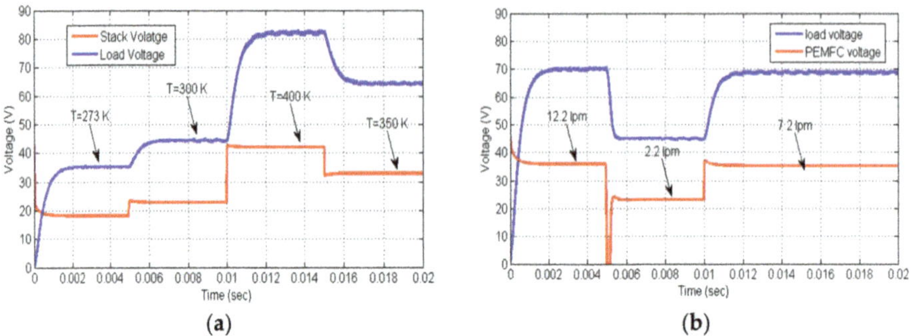

(a) (b)

Figure 6. PEMFC and load voltages for (**a**) temperature variation (**b**) variation of fuel flow rate.

The figure illustrates that an increase by 23 K increases the load voltage to 44 V. At 0.01 s a strong temperature increment of 100 K is applied. As a result of that, the load voltage has stepped up to more than 80 V. Reducing the temperature to 350 K, allows stepping down the load voltage to 65.02 V. Both PEMFC voltage and load voltage vary with the stack temperature. To maintain the load voltage to a desired value, it is required to update the duty cycle each time. The variation of the load voltage and PEMFC voltage for different fuel flow rate and constant duty cycle is presented in Figure 6b. The simulation results show the high effect of the fuel flow rate on both PEMFC voltage and load voltage. For a fuel flow rate equal to 12.2 lpm, the load voltage is about 70 V for a fixed duty cycle equal to 0.5. Due to a decline of fuel flow rate to a value of 2.2 lpm, the load voltage decreases to 45 V. In recapitulation, the load voltage and the PEMFC voltage are both dependent on the input variables of the PEMFC.

Figure 7a,b show the variation of the PEMFC and load voltages versus time for different fuel flow rate and different fuel supply pressure respectively. It is clear from these results that the PEMFC voltage is dependent on the fuel supply pressure and fuel flow rate. However, this dependence is non-linear, as shown. Undeniably, for 7 lpm as fuel flow rate, the PEMFC voltage is 15 V. An increase by 3 lpm in the fuel flow rate allows to increase the voltage to 20 V. However increasing the fuel flow rate from 17 lpm to 20 lpm increases the voltage by almost 1 V. This result confirms the nonlinearity of the variation of the voltage versus the variation of the fuel flow rate. The same comments are deduced from the results given in Figure 7b.

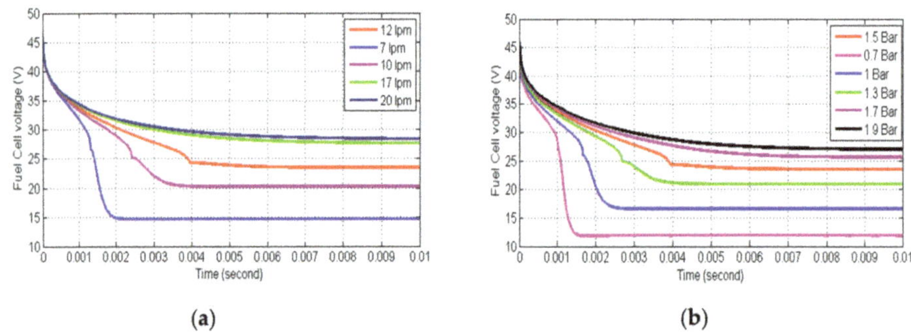

(a) (b)

Figure 7. PEMFC voltage for: (**a**) different fuel flow rate; (**b**) different fuel supply pressure.

5. Neural Network Regulation

As presented in the last section, the load voltage depends on external variables like the temperature, fuel supply pressure, fuel flow rate and other PEMFC input variables. Therefore, any variation of the input variables affects considerably the PEMFC voltage and consequently the load voltage. However, in all industrial applications, especially the hybrid or electric vehicle, it is required to regulate the continuous bus. As a solution, the duty cycle must be updated to keep the load voltage to a desired output. Usually the proportional integral derivative (PID) controller and the proportional integral (PI) controller are used to adjust the duty cycle. However, due to the multiple inputs of the PEMFC like the hydrogen supply pressure, hydrogen flow rate, temperature, air supply pressure and air flow rate, it is compulsory to use a neural network to control the duty cycle.

The artificial neural network (ANN) was developed and recognized as efficient approach for the control of nonlinear systems [28]. During the last decades, the ANN contributed to the improvement of industrial applications where the systems presented complexity in control and modeling [29]. The ANN was used to control the power converters in many cases [29,30]. Due to the multiple inputs of the fuel cell, the ANN is preferred in this case to control power converter by generating the duty cycle. Then, it seems clear that the inputs of the ANN will be the same inputs of the fuel cell in addition to the desired load voltage. The ANN controller will calculate the duty cycle according the fuel cell inputs and the desired load voltage. To achieve this objective, many tests were carried out in aim to prepare a database for the ANN.

The methodology consists on running the simulation for different values of duty cycle while keeping the same values of the inputs of the PEMFC and measuring the output voltage each time. The test is repeated for different values of the temperature, fuel supply pressure, fuel flow rate and air supply pressure. All these parameters can change at any time due to wrong manipulation or any fault in the external equipment used with the fuel cell such as the fuel compressor or the air-conditioning system. Figure 8a shows the variation of the duty cycle versus the load voltage for different values of the temperature. The data extracted from this graph formulate the first data to build the ANN. The inputs of the ANN are fuel flow rate, airflow rate, temperature, fuel supply pressure, air supply pressure and desired load voltage. The output of the ANN is the duty cycle. According to the values of the inputs, the ANN controller calculates the required duty cycle. To obtain a load voltage equal to 80 V, the duty cycle should be adjusted to 0.42 if the temperature of the PEMFC is about 400 K. To get the same load voltage for a temperature equal to 273 K, a duty cycle of 0.8 should be applied to the power converter. Therefore, it is required to adjust the duty cycle according to the desired load voltage and the temperature of the PEMFC. Figure 8b shows the variation of the duty cycle versus the load voltage for different values of the fuel flow rate. It is clear in Figure 8b below that for lower load voltage, the fuel flow rate does not have a strong effect on the duty cycle value. For the voltages less than 60 V, the graphs are superposed. However, for load voltage more than 65 V, the effect of the fuel flow rate is observed. Then, for higher fuel flow rate, the desired load voltage is obtained with smallest duty cycle. These samples of date are introduced to build the ANN controller.

Figure 8c shows the variation of the duty cycle versus the load voltage for three different values of the fuel supply pressure. The load voltage values are obtained by running the simulation for each value of the pressure and varying the duty cycle from 0.1 to 0.9. The duty cycle is dependent on the load voltage and the fuel supply pressure. Figure 8d shows the variation of the duty cycle versus the load voltage for different values of the air supply pressure. The air supply pressure can be affected by external air parameters. The variation of the air supply pressure affects the fuel cell output voltage. Hence, it is required to study the effect of this parameter on the load voltage. The results presented in Figure 8d below show a non-linearity of the variation of the duty cycle versus the load voltage for each value of the air supply pressure. These samples of results obtained from simulations will be considered in the ANN design. In addition to other PEMFC inputs, the duty cycle must be updated according to the air supply pressure.

To design the ANN controller, all the data corresponding to the variation of the duty cycle versus the load voltage for different values of PEMFC inputs are considered. The ANN is trained with the back propagation neural network (BPNN) method. The BPNN has been adopted for the control of power converters and electrical machines in [28]. The method has been proved its efficiency. To measure the error during the training of the ANN controller, the mean squared error method is adopted. As shown in Figure 9, the best training performance, achieved at epoch 2000, is 4.11×10^{-7}. After training, the ANN controller is ready to be tested. In next section, the ANN connected to the fuel cell and the power converter is simulated to test the robustness of the developed control.

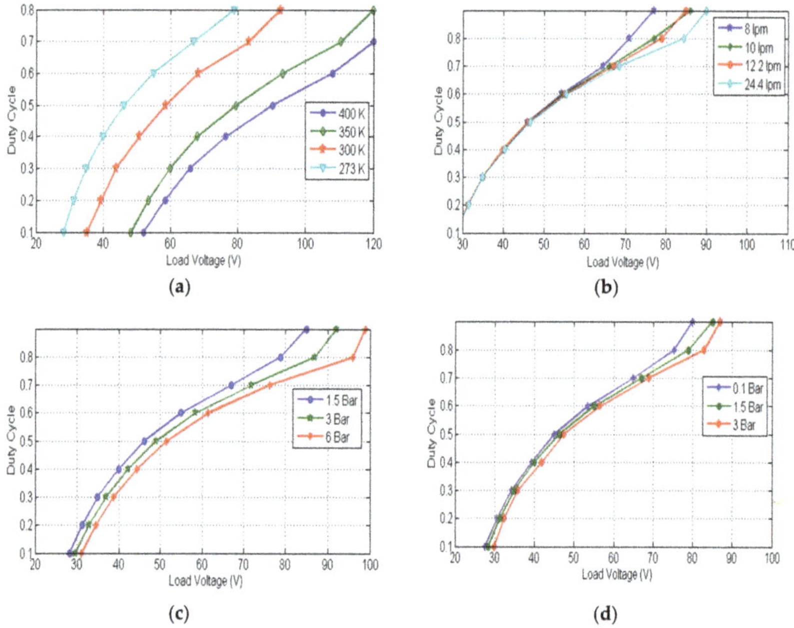

Figure 8. Duty cycle for: (**a**) different temperature values, (**b**) different values of the fuel flow rate, (**c**) different values of the fuel supply pressure, (**d**) different values of the air supply pressure.

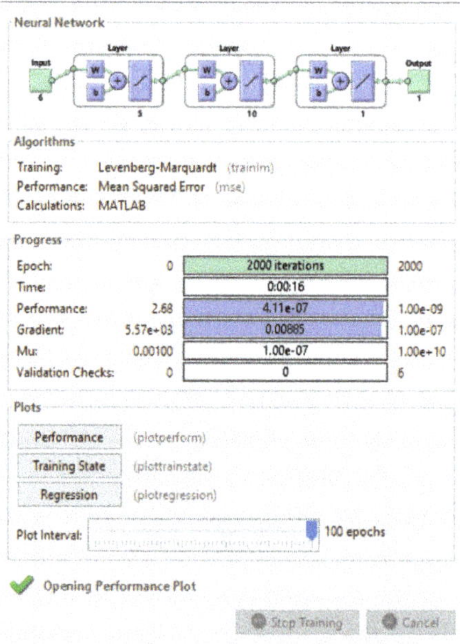

Figure 9. Implementation of the Artificial Neural Network under the Matlab-Simulink.

6. Implementation of the Neural Network Controller

6.1. Effect of the Load Impedance Variation

Figure 10 shows the ANN controller implemented in Matlab-Simulink and associated with the interleaved boost converter and the PEMFC. The inputs of the ANN controller are the same inputs of the PEMFC in addition to the desired load voltage. Thus, any variation of the PEMFC inputs will affect the output of the ANN controller. Figure 11 shows the response of the load voltage to the variation of the load impedance. For all selected values of the load resistance, we can satisfactorily obtain the desired load voltage of 60 V. In the case of decreasing the load resistance from 15 Ω to 10 Ω, an overshoot of 26% is observed. The settling time is about 0.002 s. However, decreasing the load resistance from 10 Ω to 5 Ω will increase the overshoot and the settling time to 50% and 0.0036 s respectively. Overall, the reference voltage is reached and the regulator presents good robustness to regulate the load voltage.

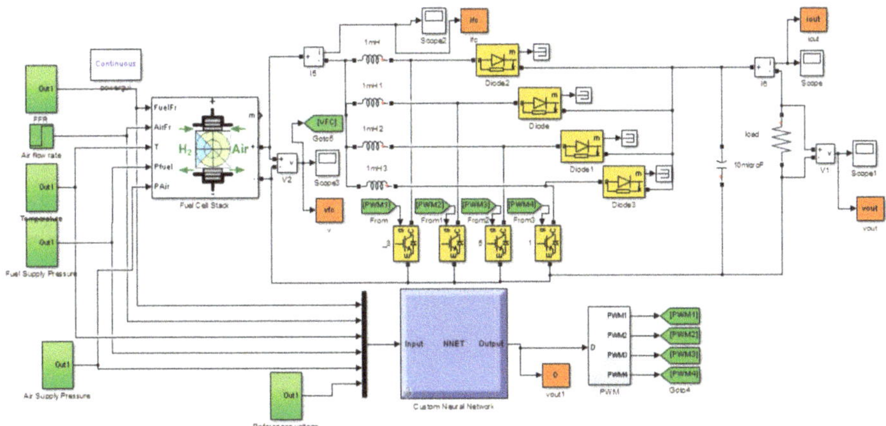

Figure 10. Fuel Cell- Boost Converter System and Neural Network Controller.

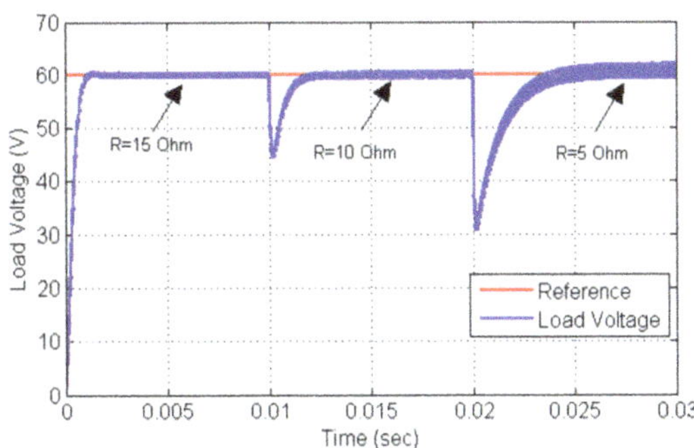

Figure 11. Output voltage response to the variation of the load resistance for desired output of 60 V.

6.2. Effect of the Temperature Variation

To study the effect of the temperature variation on the load voltage, the temperature is increased from 273 K to 300 K at 0.005 s in first time. At 0.01 s an increment of 100 K is applied to the temperature input. Finally, the temperature is reduced to 350 K at 0.015 s. the temperature variation is shown in Figure 12a. The load voltage reference is set to 60 V. Figure 12b shows the variation of the duty cycle according to the variation of the temperature. Therefore, any increase of the temperature corresponds to a decrease of the duty cycle. Consequently, the ANN controller calculated to an optimal duty cycle of 60 V as load voltage.

The dynamic responses of the load voltage and the PEMFC voltage are presented in Figure 12c. The PEMFC voltage varies according to any change of the temperature. However, the load voltage is kept at the desired voltage. This is due to the perfect variation of the duty cycle according to any variation of the PEMFC inputs. An overshoots of 25% and 41% are observed at 0.005 s and 0.01 s respectively. These overshoots are satisfactorily attenuated. The dynamic responses of the PEMFC and load currents presented in Figure 12d show the strong variation of the PEMFC current due the variation of the PEMFC voltage. The load current is kept constant. The simulation results show the efficiency of the ANN controller in regulating the load voltage under strong variation of the PEMFC temperature.

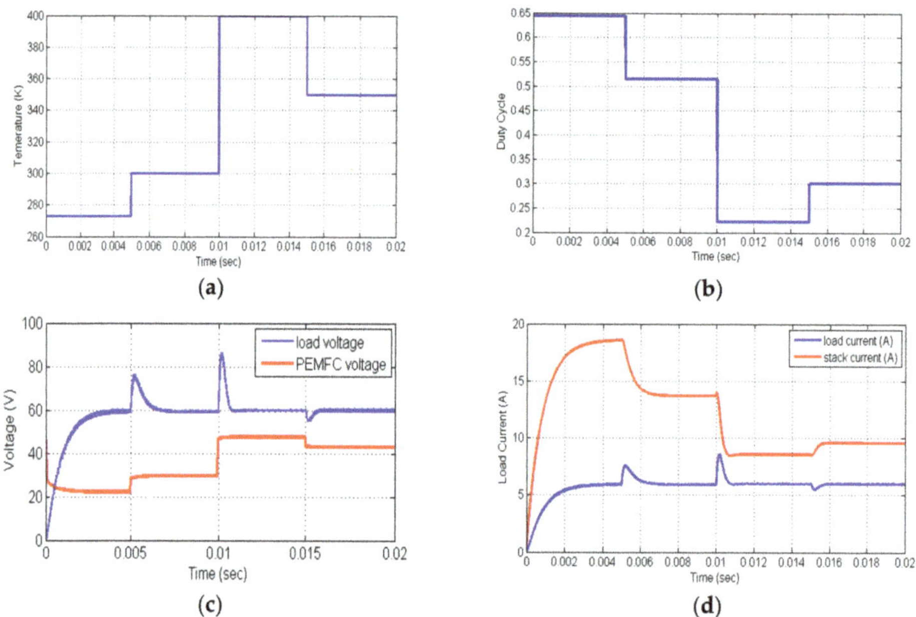

Figure 12. (**a**) Temperature variation, (**b**) Variation of the duty cycle, (**c**) voltage variation, (**d**) current variation.

6.3. Effect of the Fuel Flow Rate

The nominal value of hydrogen flow rate is about 12.2 lpm. However, this parameter can be influenced at any time due to any fault in the compressor system. To test the response of the system when varying the fuel flow rate, we apply the increments shown in Figure 13a to the PEMFC and the ANN controller. The result presented in Figure 13b shows the variation of the duty cycle according to the variation of the fuel flow rate. Here, the ANN controller proves its robustness under different fuel flow rates. Figure 13c shows that the load voltage is kept to the desired voltage.

Figure 13d shows the variation of the currents in both PEMFC and load. The PEMFC voltage has increased and decreased due to the variation of the fuel flow rate value. Even though the PEMFC output current has increased and decreased, the load current and voltage are retained at the same value. Thus, this test shows the efficiency of the ANN face to fuel flow rate variation.

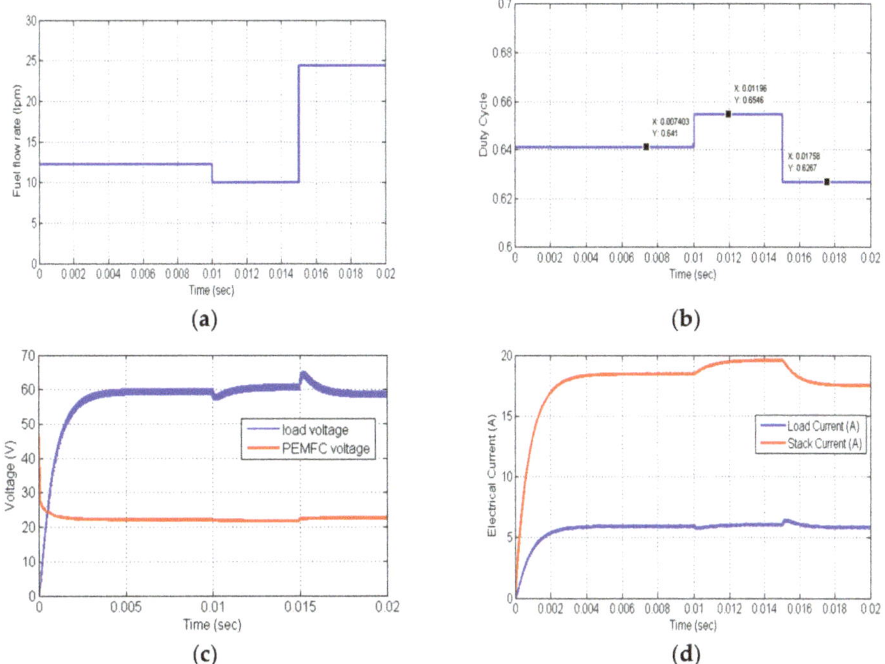

Figure 13. (**a**) Variation of the fuel flow rate, (**b**) Variation of the duty cycle, (**c**) voltage variation, (**d**) current variation.

6.4. Variation of the Fuel Supply Pressure

Fuel supply pressure drops and increases are terms used to describe the reduction and the increase in the fuel pressure due to the compressor operation. Usually, a properly designed compressor system used for hydrogen supply to the PEMFC must have a constant pressure. However, any fault due to excessive usage can happen and affect the control of the fuel supply pressure. Excessive fuel pressure variation will contribute to poor PEMFC performance and excessive energy losses.

The proposed ANN should take in account the variation of the fuel supply pressure. To test the effect of the fuel supply pressure on the load voltage after implementation of the ANN controller, the increments of Figure 14a are applied to both PEMFC and ANN controller. The simulation results presented in Figure 14b, and Figure 14c show the effectiveness of the ANN controller in updating the duty cycle to regulate the load voltage to 60 V. Figure 14d shows the strong variations of the fuel cell current due to an increase of the fuel supply pressure. The ANN controller adjusts the duty cycle each time to regulate the load voltage.

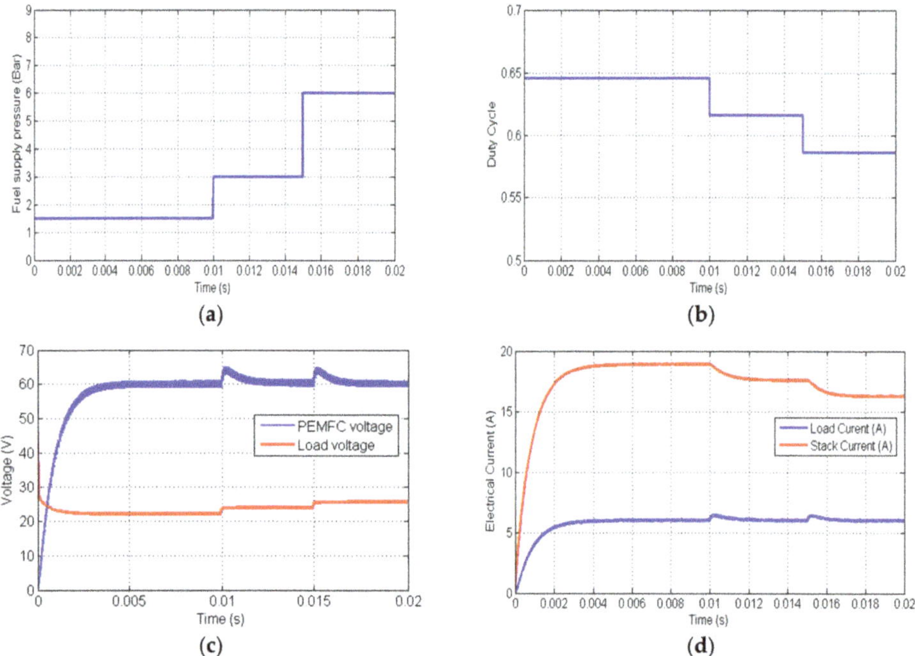

Figure 14. (**a**) Variation of the fuel supply pressure, (**b**) Variation of the duty cycle, (**c**) voltage variation, (**d**) current variation.

6.5. Variation of the Air Supply Pressure

On the other hand, the air supply pressure is also another factor affecting the performance of the PEMFC in best conditions. Any increase or decrease of the air supply pressure has an effect on the PEMFC voltage. Figure 15a shows the variation of the air supply pressure applied as input for the PEMFC and the ANN controller. Firstly, the air supply pressure is fixed to 1 bar. At the time 0.01 s, the air supply pressure is dropped to 0.1 bar. At the time 0.015 s, an increment of 1.9 bar is applied. Then the final value of the air supply pressure is 2 bar. The output of the ANN controller is presented in Figure 15b. The duty cycle is updated each time to keep the load voltage at the desired value, as shown in Figure 15c, under the variations of the PEMFC voltage resulted from the variation of the air supply pressure.

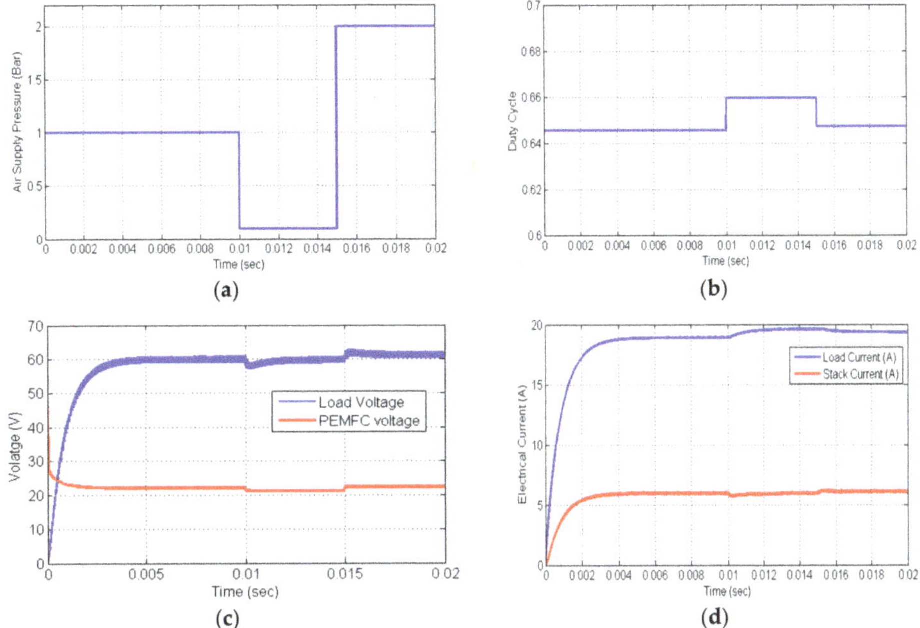

Figure 15. (**a**) Variation of the air supply pressure, (**b**) Variation of the duty cycle, (**c**) voltage variation, (**d**) current variation.

To test the efficiency of the ANN controller under the variation of the desired output voltage, strong variations of the reference voltage are applied to the controller. The obtained simulation results are shown in Figure 16. The output voltage tracks the reference for different values. Therefore, this result shows a good efficiency of the proposed controller to regulate the output voltage according to the load requirements.

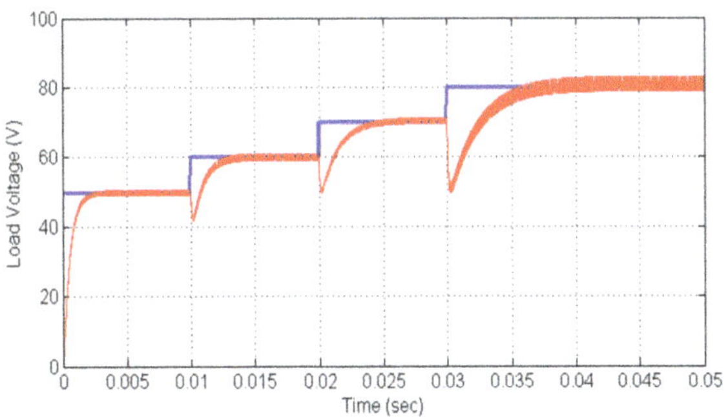

Figure 16. Response of load voltage to reference variation.

The current ripple of the proposed converter-based neural network regulator for PEMFC applications is compared with current ripples obtained with techniques presented in [9,26,27]. The comparison results are listed in Table 3. It is observed that using the proposed converter allows to obtain the lowest current ripples for the PEMFC. A comparative analysis of the proposed controller with proposed techniques in [20,23,25–27] in terms of capability of load voltage regulation in the case of load and source conditions variation is presented in Table 4. It is clear that the proposed controller in this paper has good capabilities to regulate the load voltage under the variation of the load and source conditions. However, the other proposed techniques present some limits. Indeed, the variations of source conditions were not considered in some analysis.

Table 3. Fuel cell current ripple of the of the proposed converter and techniques presented in [9,26] and [27].

Type of Control	DC Link Voltage	Fuel Cell Current Ripple
Multidevice Interleaved DC/DC Converter for Fuel Cell Hybrid Electric Vehicles [9]	400 V	0.7 A
Maximum Power Point Technique [26]	220 V	0.1 A
Sliding mode control [27]	100 V	0.3 A
Proposed ANN controller for the IBC	60 V	0.01 A

Table 4. Comparison of the proposed controller and proposed techniques in [20,23,25,26] and [27].

Type of Control	Load Voltage Regulation in the Presence of the Variation of					
	Load Impedance	Temperature	Hydrogen Pressure	Hydrogen Flow Rate	Air Pressure	Air Flow Rate
Controlling the SMES hybrid energy storage system [20]	P	NP	NP	NP	NP	NP
Proportional Integral Controller [23]	P	NP	NP	NP	NP	NP
Optimization of the PEMFC Hybrid System [25]	P	NP	NP	P	NP	P
Maximum Power Point Technique [26]	P	P	NP	NP	NP	NP
Sliding mode control [27]	P	NP	NP	NP	NP	NP
Proposed ANN controller	P	P	P	P	P	P

Notes: N.P: Not Performed; P: Performed.

7. Conclusions

To minimize the current and voltage ripples as well as regulating the load voltage, a four phases interleaved boost converter was proposed in this paper. The proposed converter has reduced the ripples of load voltage to less than 2 V. Therefore, using the interleaved boost converter minimized the PEMFC voltage and current ripples. A simple PWM control technique was developed and implemented under Matlab-Simulink for the control of the interleaved boost converter. Simulation results of the open loop control of the interleaved boost converter have shown the incapability to maintain the load voltage at desired value. Indeed, the variation of any external parameter increased or decreased the fuel cell voltage and consequently the load voltage. To overcome this issue, an artificial neural network controller was developed to regulate the load voltage at desired reference. The database used to build this ANN is composed of samples of duty cycle values, load voltage and input parameters of the PEMFC. Compared to other techniques used for the regulation of the fuel supply pressure, fuel flow rate, temperature and air supply pressure based on the use of proportional integral controller to regulate each variable, the proposed strategy allowed to regulate the load voltage with unique controller based on neural network. The simulation results of the interleaved boost converter controlled by the neural network controller show the efficiency and the robustness of the proposed converter to regulate the load voltage as well as minimizing the voltage ripples. The proposed controller showed that using limit number of tests allows to develop efficient ANN controller for the regulation of the load voltage. Therefore, the proposed method will be applied to other PEMFCs having different power range. In future work, a prototype will be developed to test the efficiency of such controllers in real conditions. The behaviour of PEMFC-supercapacitor hybrid system will be investigated to design adequate power converter suitable for such power equipment.

Author Contributions: Conceptualization and methodology, E.M.B. and I.B.B.; Software, A.K. and E.M.B.; Discussion, M.Z.; Review, I.T. and E.M.B.

Funding: This research received no external funding.

Conflicts of Interest: There is no conflict of interest.

References

1. Mann, R.F.; Amphlett, J.C.; Hooper, M.A.I.; Jensen, H.M.; Peppley, B.A.; Roberge, P.R. Development and application of a generalised steady-state electrochemical model for a PEM fuel cell. *J. Power Sources* **2000**, *86*, 173–180. [CrossRef]

2. Correa, J.M.; Farret, F.A.; Canha, L.N.; Simoes, M.G. An electrochemical-based fuel-cell model suitable for electrical engineering automation approach. *IEEE Trans. Ind. Electron.* **2004**, *51*, 1103–1112. [CrossRef]

3. Jin, K.; Ruan, X.; Yang, M.; Xu, M. A hybrid fuel cell power system. *IEEE Trans. Ind. Electron.* **2009**, *56*, 1212–1222. [CrossRef]

4. Khiareddine, A.; Salah, C.B.; Rekioua, D.; Mimouni, M.F. Sizing methodology for hybrid photovoltaic/wind/hydrogen/battery integrated to energy management strategy for pumping system. *Energy* **2018**, *153*, 743–762. [CrossRef]

5. Fuel Cell Electric Vehicles, Reports Fuel Cell Today. Available online: http://www.fuelcelltoday.com (accessed on 15 September 2018).

6. Hart, D.W. *Power Electronics*; Mc Graw Hill: New York, NY, USA, 2010; ISSN 978-0-07-338067.

7. Rabello, A.L.; Co, M.A.; Simonetti, D.S.L.; Vieira, J.L.F. An isolated DC–DC boost converter using two cascade control loops. In Proceedings of the IEEE International Symposium on Industrial Electronics, Guimaraes, Portugal, 7–11 July 1997; Volume 2, pp. 452–456.

8. Sundar, G.; Karthick, N.; Reddy, S.R. High step-up DC–DC converter for ac photovoltaic module with MPPT control. *J. Electr. Eng.* **2014**, *65*, 248–253. [CrossRef]

9. Hegazy, O.; van Mierlo, J.; Lataire, P. Analysis, Modeling, and Implementation of a Multidevice Interleaved DC/DC Converter for Fuel Cell Hybrid Electric Vehicles. *IEEE Trans. Power Electron.* **2012**, *27*, 2237–2245. [CrossRef]

10. Lee, S.; Park, J.; Choi, S. A three-phase current-fed push–pull DC–DC converter with active clamp for fuel cell applications. *IEEE Trans. Power Electron.* **2011**, *26*, 2266–2277. [CrossRef]

11. Lim, T.C.; Williams, B.W.; Finney, S.J.; Zhang, H.B.; Croser, C. Energy recovery snubber circuit for a dc-dc push–pull converter. *IET Trans. Power Electron.* **2012**, *5*, 863–872. [CrossRef]

12. Whitaker, B.; Martin, D.; Cilio, E. Extending the operational limits of the push-pull converter with SiC devices and an active energy recovery clamp circuit. In Proceedings of the 2015 IEEE Applied Power Electronics Conference and Exposition (APEC), Charlotte, NC, USA, 15–19 March 2015; pp. 2023–2038.

13. Omkun, S.; Sirisamphanwong, C.; Sukchai, S. A DSP-based interleaved boost DC-DC converter for fuel cell applications. *Int. J. Hydrog. Energy* **2015**, *40*, 6391–6404. [CrossRef]

14. Thounthong, P.; Rael, S.; Davat, B. Behaviour of a PEMFC supplying a low voltage static converter. *J. Power Sources* **2006**, *156*, 119–125.

15. Saha, S.S. Efficient soft-switched boost converter for fuel cell applications. *Int. J. Hydrog. Energy* **2011**, *36*, 1710–1719. [CrossRef]

16. Giaouris, D.; Stergiopoulos, F.; Ziogou, C.; Ipsakis, D.; Banerjee, S.; Zahawi, B.; Pickert, V.; Voutetakis, S.; Papadopoulou, S. Nonlinear stability analysis and a new design methodology for a PEM fuel cell fed DC-DC boost converter. *Int. J. Hydrog. Energy* **2012**, *37*, 18205–18215. [CrossRef]

17. Hinaje, M.; Sadli, I.; Martin, J.P.; Thounthong, P.; Rafel, S.; Davat, B. Online humidification diagnosis of a PEMFC using a static DC-DC converter. *Int. J. Hydrog. Energy* **2009**, *34*, 2718–2723. [CrossRef]

18. Kirculies, M. Fuel cell power for maritime applications. *Fuel Cells Bull.* **2005**, *2005*, 12–15.

19. Thounthong, P.; Davat, B.; Rae, S.; Sethakul, P. Fuel Cell high power applications. *IEEE Trans. Ind. Electron. Mag.* **2009**, *3*, 32–46.

20. Bizon, N. Effective mitigation of the load pulses by controlling the battery/SMES hybrid energy storage system. *Appl. Energy* **2018**, *229*, 459–473. [CrossRef]

21. Lu, N.; Yang, S.; Tang, Y. Ripple Current Reduction for Fuel-Cell-Powered Single-Phase Uninterruptible Power Supplies. *IEEE Trans. Ind. Electron.* **2017**, *64*, 6607–6617. [CrossRef]

22. Kaouane, M.; Khelifaa, A.B.; Cheritib, A. Regulated output voltage double switch Buck-Boost converter for photovoltaic energy application. *Int. J. Hydrog. Energy* **2016**, *41*, 20847–20857. [CrossRef]
23. Farhani, S.; Amari, M.; Marzougui, H.; Bacha, F. Analysis, modeling and implementation of an interleaved boost DC-DC converter for fuel cell used in electric vehicle. *Int. J. Hydrog. Energy* **2017**, *42*, 28852–28864. [CrossRef]
24. Ou, K.; Yuan, W.-W.; Choi, M.; Yang, S.; Kim, Y.-B. Performance increase for an open-cathode PEM fuel cell with humidity and temperature control. *Int. J. Hydrog. Energy* **2017**, *42*, 29852–29862. [CrossRef]
25. Bizon, N.; Mazare, A.G.; Ionescu, L.M.; Enescu, F.M. Optimization of the Proton Exchange Membrane Fuel Cell Hybrid Power System for Residential Buildings. *Energy Convers. Manag.* **2018**, *163*, 22–37. [CrossRef]
26. Reddy, K.J.; Sudhakar, N. High Voltage Gain Interleaved Boost Converter with Neural Network Based MPPT Controller for Fuel Cell Based Electric Vehicle Applications. *IEEE Access* **2018**, *6*, 3899–3908. [CrossRef]
27. Wang, W.Y.; Ding, Y.H.; Ke, X.; Ma, X. Sliding mode control of direct coupled interleaved boost converter for fuel cell. In *IOP Conference Series: Earth and Environmental Science*; Conference 1; IOP Publishing: Bristol, UK, 2017; Volume 100.
28. Barhoumi, E.M.; Ben Salah, B. Modelling and Control of a new Linear Switched Reluctance Motor for Shunting the Railways Channels. *Int. Rev. Model. Simul. (IREMOS)* **2011**, *4*, 2012–2019.
29. Barhoumi, E.M.; Ben Salah, B. New Positioning Control of Stepper Motor using BP Neural Networks. *J. Emerg. Trends Comput. Inf. Sci.* **2011**, *2*, 300–306.
30. Barhoumi, E.M.; Ben Salah, B. Design and Simulation of a new Linear Switched Reluctance Motor for Shunting the Railways Channels. *Int. Rev. Model. Simul. (IREMOS)* **2011**, *3*, 1072–1078.

Article

Study of the Intelligent Behavior of a Maximum Photovoltaic Energy Tracking Fuzzy Controller

Gul Filiz Tchoketch Kebir [1,2,*], **Cherif Larbes** [2], **Adrian Ilinca** [1], **Thameur Obeidi** [1,2] and **Selma Tchoketch Kebir** [2]

[1] Wind Energy Research Laboratory, Université du Québec à Rimouski, 300, Allée des Ursulines, Rimouski, QC G5L 3A1, Canada; adrian_ilinca@uqar.ca (A.I.); thameur.obeidi@uqar.ca (T.O.)
[2] Laboratoire des Dispositifs de Communication et de Conversion Photovoltaïque, Département d'Électronique, École Nationale Polytechnique, 10, Avenue Hassen Badi, El Harrach, Alger 16200, Algerie; cherif.larbes@gmail.com (C.L.); selma.tchoketch_kebir@g.enp.edu.dz (S.T.K.)
* Correspondence: GulFiliz.TchoketchKebir@uqar.ca; Tel.: +1-418-721-7783

Received: 16 October 2018; Accepted: 17 November 2018; Published: 23 November 2018

Abstract: The Maximum Power Point Tracking (MPPT) strategy is commonly used to maximize the produced power from photovoltaic generators. In this paper, we proposed a control method with a fuzzy logic approach that offers significantly high performance to get a maximum power output tracking, which entails a maximum speed of power achievement, a good stability, and a high robustness. We use a fuzzy controller, which is based on a special choice of a combination of inputs and outputs. The choice of inputs and outputs, as well as fuzzy rules, was based on the principles of mathematical analysis of the derived functions (slope) for the purpose of finding the optimum. Also, we have proved that we can achieve the best results and answers from the system photovoltaic (PV) with the simplest fuzzy model possible by using only 3 sets of linguistic variables to decompose the membership functions of the inputs and outputs of the fuzzy controller. We compare this powerful controller with conventional perturb and observe (P&O) controllers. Then, we make use of a Matlab-Simulink® model to simulate the behavior of the PV generator and power converter, voltage, and current, using both the P&O and our fuzzy logic-based controller. Relative performances are analyzed and compared under different scenarios for fixed or varied climatic conditions.

Keywords: fuzzy logic controller; MPPT: maximum power point tracking; photovoltaic system; step-up boost converter

---

1. Introduction

Solar energy conversion using photovoltaic (PV) generators has lately been in accelerated development, both for small and large installations. This clean, quiet and low-maintenance energy source has seen the largest growth rate with a renewable and continuous price reduction. Its further development requires improvement of conversion efficiency and component cost reduction. The electrical energy extracted by the photovoltaic generators depends on a complex equation relating the solar radiation, the temperature, and the total resistance of the circuit, which results in a nonlinear variation of the output power P as a function of the circuit voltage V in the form $P = f(V)$ [1,2]. There is a unique point, under given irradiation and temperature conditions, where the generator produces maximum power, named the MPP (maximum power point). This MPP is reached when the power's rate of change as a function of voltage is equal to zero. The nonlinear relationship of the power output from the PV generator with respect to environmental conditions renders the conversion efficiency of solar generators relatively low, so power extraction optimization becomes a key issue in solar energy conversion [3,4]. This paper focuses on the development of a coupled fuzzy logic–mathematical analysis method as a maximum power point tracking (MPPT) technique to increase the power extracted by the PV generator.

2. Motivation

In the case of considering photovoltaic power output with respect to voltage for a particular solar generator under varying irradiation and temperature levels, we note that there is a unique point where maximum power can be harvested (Figure 1) [2,4].

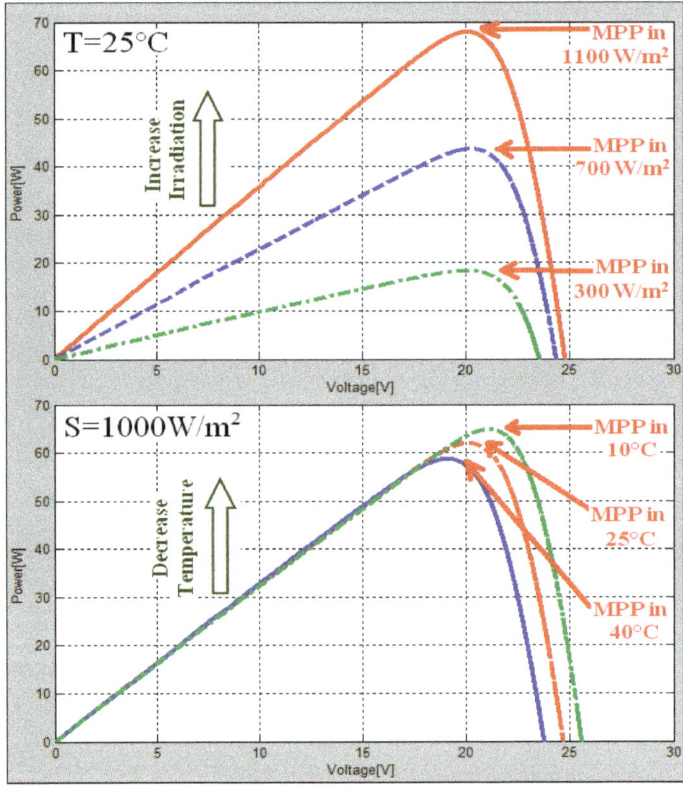

Figure 1. Variation of the maximum power point (MPP) with variations of irradiation and temperature.

A similar MPP tracking analysis can be performed by considering an *I-V* curve, as shown in Figure 2 below. If we consider irradiation S, a temperature T, and a varying resistive load R_i, then the solar cell provides a short-circuit current I_{SC} and an open circuit voltage V_{OC}. We note that there exists an MPP that can be identified from the *I-V* curve. Whatever the approach, *P-V* or *I-V*, the tracking of gradient variation of *I* or *V* enables us to identify the maximum power point from a PV generator [1,3]. In the literature [2], there are a number of MPPT (Maximum Power Point Tracking) techniques used to optimize the efficiency of photovoltaic systems.

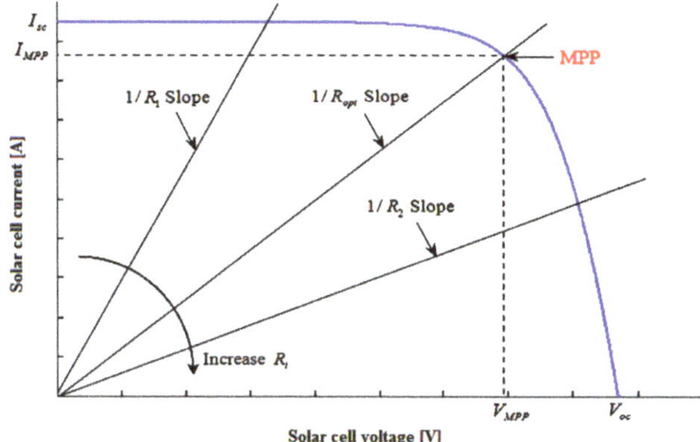

Figure 2. Load effect on *I-V* photovoltaic characteristics.

Photovoltaic systems are generally connected to static converters (DC-DC) driven by programmed controllers that continuously analyze the power output from the solar generator. MPPT controllers adjust the parameters to extract maximum energy, whatever the load and atmospheric conditions are [5]. The MPPT methods portrayed in the different studies use different techniques and algorithms which widely differ in performance, such as convergence speed, implementation complexity, accuracy, and most importantly, the cost of implementation of the whole setup [6]. In the following paragraphs, we briefly recall the principles of some of the most popular MPPT tracking algorithms.

The "Hill Climbing/P&O Method" [7–10]: The principle of this algorithm is to calculate the power provided by the PV panel at time k, following a disturbance effect on the voltage of the PV panel while acting on the duty cycle, D. This is compared to the previous measurement at the moment $k - 1$. If the power increases, we approach the MPP, and the variation of the duty cycle is maintained in the same direction. On the contrary, if the power decreases, we move away from the MPP. So, we have to reverse the direction of the change in the duty cycle.

The "Incremental Conductance Method" [11,12]: The principle of this algorithm is based on the knowledge of the value of the conductance $G = I/V$ on the increment of the conductance dG to deduce the position of the operating point relative to the MPP. If dG is greater than the opposite of the conductance $-G$, the duty cycle is decreased. On the other hand, if dG is lower than $-G$, the duty cycle is increased. This process is repeated until reaching the MPP.

The "Fractional Open-Circuit Voltage Method" [2,4]: This method is based on the relation $V_{MPP} = \alpha \times V_{OC}$, where α is a voltage factor depending on the characteristics of the PV cell. To deduce the optimal voltage, the V_{OC} voltage must be measured. As a result, the operating point of the panel is kept close to the MPP by adjusting the panel voltage to the calculated optimal voltage. This is achieved by cyclically acting on the duty cycle to reach the optimum voltage.

The "Fractional Short-Circuit Current Method" [6,13]: This technique is based on the relation $I_{MPP} = \alpha \times I_{SC}$, where α is a current factor depending on the characteristics of the PV cell. The optimum operating point is obtained by bringing the current of the panel to the optimum current, changing the duty cycle until the panel reaches the optimum value.

Algorithms based on fuzzy logic [3,14–16]: MPPT control techniques based on fuzzy logic have recently been introduced because they offer the advantage of robust control and do not require exact knowledge of the mathematical model of the system. In addition, they improve performances (convergence speed, accuracy, ease of implementation, and low cost).

Other MPPT techniques include the "Artificial Intelligence Algorithms" [10,17]. These new technology MPPT algorithms are inspired by nature and biological structures. Among them we can mention the "particle-swarm-optimisation-based MPPT" [5,18], "genetic algorithms" [19], "neural networks" [12,20], and the "hybrid methods" [5,21].

According to the literature [4,22–24], we used a comparative study in Table 1 between the most used methods in terms of technical knowledge of PV panel parameters, complexity, speed, and accuracy.

Table 1. A comparative table of MPPT's characteristics.

MPPT Algorithms	Perturb & Observe	Incremental Conductance	Fractional Open-Circuit Voltage	Fractional Short-Circuit Current	Fuzzy Logic Control	Neural Network	Particle Swarm Optimization
Convergence speed	Varies	Varies	Medium	Medium	Fast	Fast	Fast
Implementation complexity	Low	Medium	Low	Medium	High	High	Medium
True MPPT	Yes	Yes	No	No	Yes	Yes	Yes
Sensed parameters	Voltage Current	Voltage Current	Voltage	Current	Varies	Varies	Varies
Efficiency (%)	Medium	Medium	Low	Low	Very high	Very high	high
Cost	Moderate	Moderate	Cheap	Cheap	Expensive	Expensive	Expensive
Stability	Not stable	Stable	Not stable	Not stable	Very stable	Very stable	Very stable

This paper is organized as follows: Section 3 is reserved for the study of the photovoltaic system, starting with a presentation of the photovoltaic panel. Next we explain all the parts constituting the architecture and the functioning of a PV-MPPT system. To improve the MPPT techniques' relative performances (convergence speed, accuracy, ease of implementation, and low cost), we have developed a control method using fuzzy logic that has been applied to a step-up boost MPPT for PV generators in Section 4. In Section 5, we talk about the most popular conventional MPPT controller based on the P&O algorithm. These techniques are studied and analyzed both theoretically and by simulation using Matlab-Simulink® (R2018a, Mathworks, Natick, MA, USA) in Section 6. Then, a comparison is presented of the performance of both methods.

3. Challenges in Exploiting the Maximum Energy from the Photovoltaic System

Our analysis is performed on the most sophisticated and widespread real photovoltaic cell model, consisting of two-diodes [1,25] as illustrated in Figure 3:

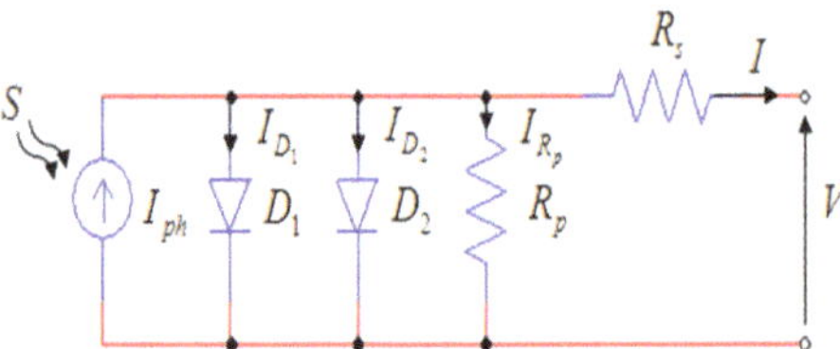

Figure 3. The two-diode circuit model of a photovoltaic cell.

Equation (1) expresses the mathematical relationship of the circuit output current in terms of the circuit parameters [25]:

$$I = S \cdot I_{ph}(T) - I_{S_1}\left[e^{\frac{q(V+IR_S)}{n_1 kT}} - 1\right] - I_{S_2}\left[e^{\frac{q(V+IR_S)}{n_2 kT}} - 1\right] - \frac{V + IR_S}{R_P} \tag{1}$$

where:

$$I_{ph}(T) = I_{ph}\Big|_{(T=298K)}\left[1 + (T - 298) \cdot (5 \cdot 10^{-4})\right] \tag{2}$$

$$I_{S_1} = K_1 T^3 e^{-\frac{Eg}{kT}} \tag{3}$$

$$I_{S_2} = K_2 T^{\frac{5}{2}} e^{-\frac{Eg}{kT}} \tag{4}$$

I and V are the output current and output voltage of the photovoltaic cell, S is the irradiance, T is the absolute temperature in Kelvin, $I_{ph}(T)$ is the generated photo-current, I_{s1} and I_{s2} are the diode saturation currents and the reverse diode saturation currents, n_1 and n_2 are the diode ideality factors,

R_s the series resistance, and R_p the parallel resistance. E_g is the band-gap energy of the semiconductor, q is the elementary charge constant (1.602×10^{-19} C) and k is the Boltzmann constant (1.38×10^{-23} J/K), $K_1 = 1.2$ A/cm^{2}K^3 and $K_2 = 2.9 \times 10^5$ A/cm^{2}K$^{5/2}$.

Equation (1) leads to a generalized equation of the entire photovoltaic panel with z photovoltaic cells, connected in series [1,25]:

$$I = S \cdot I_{ph}(T) - I_{s1}\left[e^{\frac{q(V+IzR_s)}{zn_1kT}} - 1\right] - I_{s2}\left[e^{\frac{q(V+IzR_s)}{zn_2kT}} - 1\right] - \frac{V + IzR_s}{zR_p} \tag{5}$$

From Equation (5), we note that the output current of a photovoltaic panel connected to a load R_i is highly dependent on the *I-V* variation of this load (Figure 2). Furthermore, Equation (5) illustrates that the *I-V* and *P-V* characteristics of the PV module vary not only with the connected load, but also with temperature and solar irradiance. Therefore, for each temperature and irradiance condition, it is necessary to track the corresponding MPP. Figure 1 illustrates the existence of an MPP on the *P-V* characteristic of PV generator, with variable irradiance and temperature according to Equation (5).

To force the PV system to operate in its MPP region according to incident irradiation and temperature, it is necessary to include a maximum power point tracking (MPPT) device between the PV module and the load (Figure 4). The MPPT device consists of a DC-DC converter which can be buck-type, boost-type, or buck-boost type [23,26]. The step-up boost converter has been chosen in this work.

The transducer captures the instantaneous values of current I and voltage V from the PV array, Which are used by the computing circuit inputs for the calculation of the inputs of the fuzzy logic controller. The control output must be injected into another circuit of calculation to determine the duty ratio D, which will be used at the end of the process by the drive gate to control directly the Mosfet of the boost converter (Figure 4).

The DC-DC converter is included between the array of photovoltaic cells and the energy storage unit (load) to match the voltage of the solar array with the battery voltage. If the duty ratio D of the converter is varied by a control circuit to constantly adjust the operating voltage of the solar panel to its point of maximum power V_{MPP}, that means it is operated as a maximum power point tracker MPPT (Figure 5).

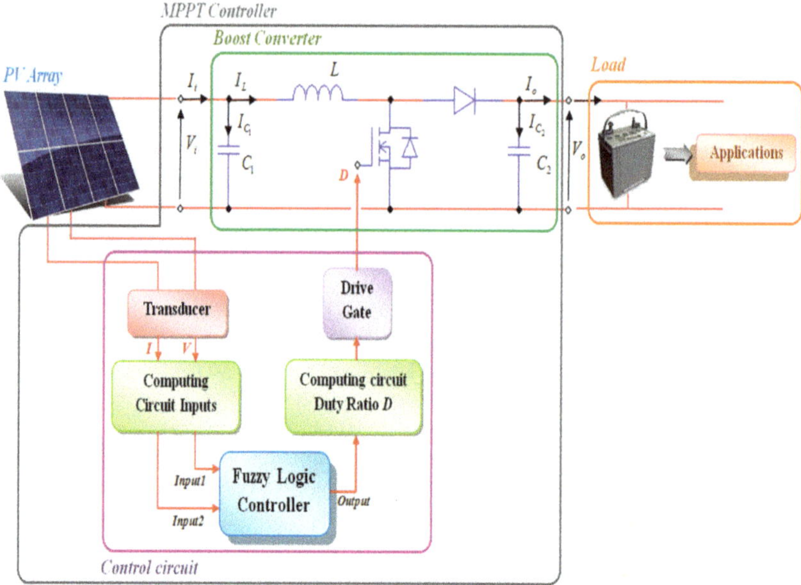

Figure 4. Photovoltaic maximum power point tracker (MPPT) architecture.

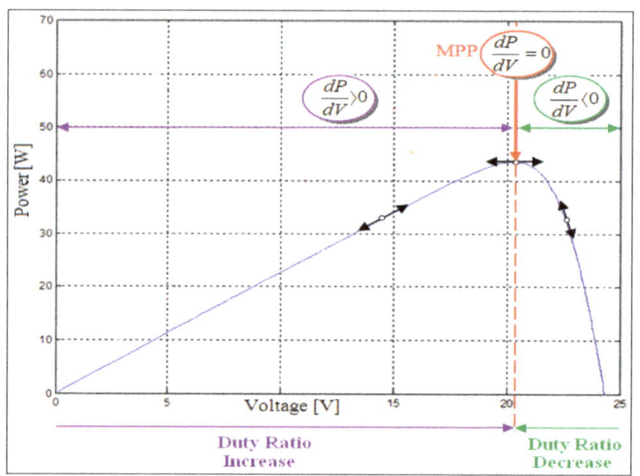

Figure 5. The direction change of the duty ratio D for tracking the MPP.

The DC-DC switching converter consists of capacitors, inductors, and switches. Ideally, the power consumption of all these devices is very low, which is the reason for the efficiency of DC-DC switching converters [25,27]. A metal oxide semiconductor field effect transistor (MOSFET) is used as a switching semiconductor device since it is easy to control using a pulse width modulation (PWM) signal generated by the controller. During the operation of the converter, the switch will be geared at a constant frequency f with an on-time value DT and an off-time value $(1-D)T$, where T is the switching period and D is the duty ratio of the switch ($D \in {]}0,1{[}$) (Figure 6).

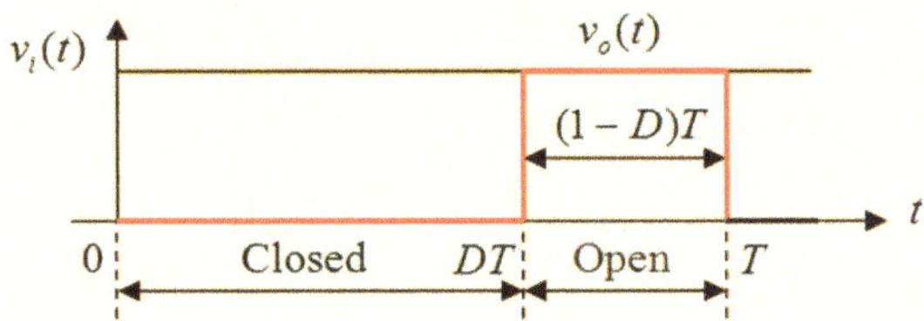

Figure 6. Output voltage $v_o(t)$ of the ideal DC-DC switching converter.

Figure 7 illustrates the step-up boost converter circuit used in the MPPT technique.

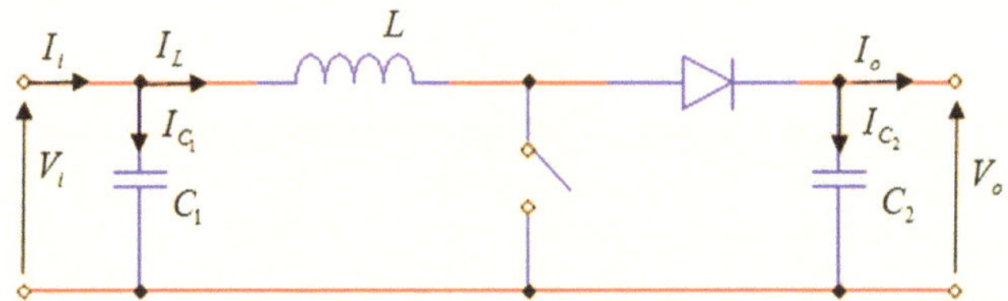

Figure 7. The ideal boost converter circuit.

Mathematical equations of the step-up Boost converter used in the Matlab-Simulink model are as follows:

$$\frac{V_o}{V_i} = \frac{1}{(1-D)} \tag{6}$$

$$I_L = I_i - C_1 \frac{dV_i}{dt} \tag{7}$$

$$I_o = (1-D)I_L - C_2 \frac{dV_o}{dt} \tag{8}$$

$$V_i = (1-D)V_o + L\frac{dI_L}{dt} \tag{9}$$

It is understood from Equation (6) that an increase in duty ratio results in an increase of the output voltage of the boost converter, and vice-versa. Hence, MPPT device instantly controls the decrease and increase of the duty ratio D in order to push the operating point to the MPP (Figure 5).

4. The Fuzzy Logic MPPT-Based Controller

4.1. Methodology

The mathematical study of the *P-V* characteristic illustrated in Figure 5 leads us to the choice of the following MPPT algorithm:

(1)   The analysis of the slope $m(p_i)$ at the p_i point on the *P-V* characteristic (Figure 5) is used to locate the actual operation point p_i. According to this data, the controller will decide whether to increase or decrease the voltage by varying the duty ratio D.

(2)   Analysis of the rate of change of the slope at the p_i point $\Delta m(p_i) = s(p_i)$ expresses the rate of the approach and distancing of the MPP. This parameter is also included in the controller for faster MPP searching.

4.2. The Configuration of the Fuzzy Controller

Fuzzy systems are good models for nonlinear systems. Fuzzy models are based on fuzzy rules. These fuzzy rules provide information about uncertain nonlinear systems [28]. A fuzzy logic controller consists of three main operations: "Fuzzification", "inferencing", and "defuzzification" [29,30]. The input data are fed into a fuzzy logic-based system where physical quantities are represented as linguistic variables with appropriate membership functions. These linguistic variables are then used in the antecedents (If-Part) of a set of fuzzy "If-Then" rules within an inference engine to result in a new set of fuzzy linguistic variables, or a consequent (Then-Part) [31]. Figure 8 illustrates the schematic representation of the fuzzy controller:

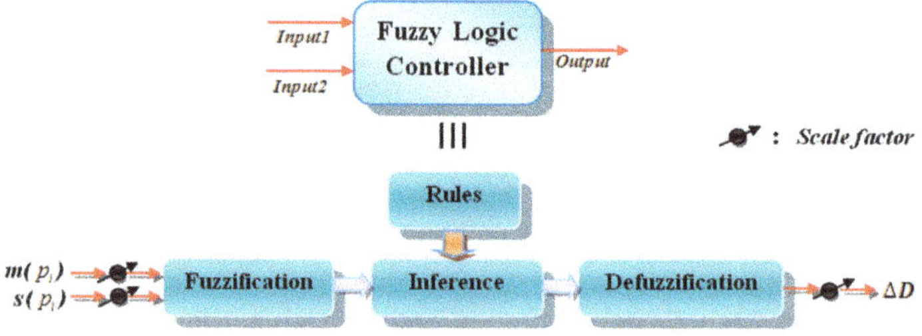

Figure 8. Fuzzy Controller configuration.

4.2.1. Fuzzification

The control circuit instantaneously measures the voltage $V(i)$ and current $I(i)$ of the photovoltaic generator and calculates power as $P(i) = I(i) \times V(i)$. As explained in Section 4.1, the controller analyses $input_1(i)$ that represents the slope of the current operating point on the *P-V* curve ($m(p_i)$) and $input_2(i)$ that represents the rate of approaching or distancing of the point p_i toward MPP. The fuzzy controller takes instantaneous measurements of these two input values and then decides and calculates the output, $\Delta D(i)$ which is actually the change of the duty ratio of the MOSFET. The input and output variables of the fuzzy controller are expressed in terms of membership functions. Determination of the range of fuzzy linguistic variables that composes the membership functions of the input and output variables of the fuzzy controller is based on the experiences and observations of automation specialists who works with the PV system [31,32], as well as on the right choice of the rules of inference.

In other words, our observations suggest that the value of the slope of a point p_i on the curve in Figure 5 (which represents $input_1$) will be positive, negative, or zero (zero is the MPP). The value of change of the slope of two points p_i and p_{i-1} on the same curve (and which represents $input_2$) will be either positive, negative or zero. The fuzzy controller will decide to increase, decrease, or stabilize the output of the command, which is ΔD.

Therefore, in order to achieve the best possible results from our simulations experiments, and after several calculations and tests on our PV system, we decided to choose the decomposition of the following membership functions shown in Figure 9.

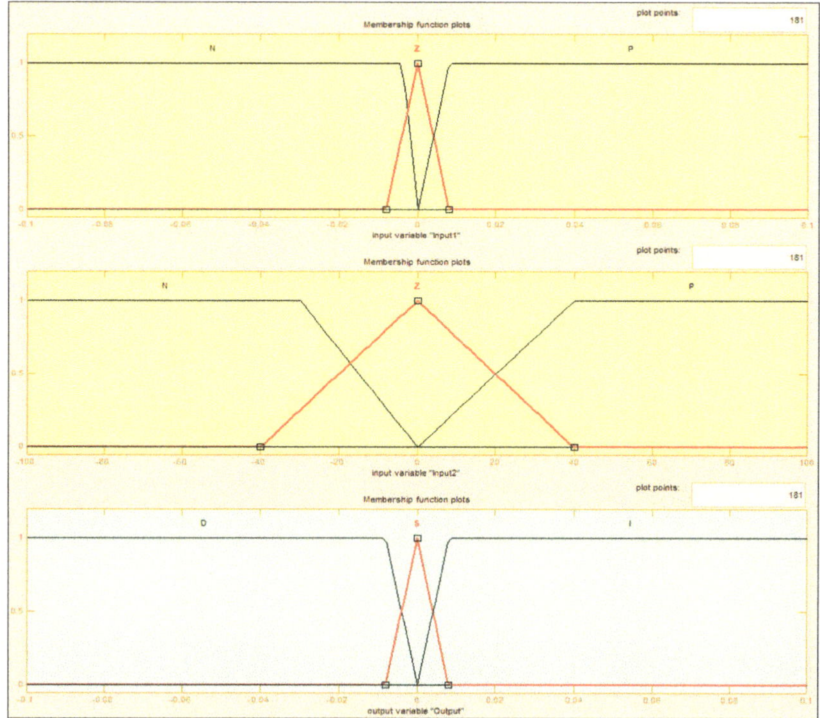

Figure 9. Membership functions of the two entries: Input$_1$, input$_2$, and the output, with three sets of linguistic variables.

We propose to define the membership functions of the inputs and the output in terms of a set of linguistic variables:

(1) Input$_1$: N: Negative, Z: Zero, P: Positive,
(2) Input$_2$: N: Negative, Z: Zero, P: Positive,
(3) Output: D: Decrease, S: Stabilize, I: Increase.

The real values of input$_1$, input$_2$, and the output are normalized by an input scaling factor [32,33]. In this system, the input scaling factor has been designed as follows:

- Input$_1$ values are between -0.1 and 0.1;
- Input$_2$ values are between -100 and 100;
- Output values are between -0.1 and 0.1.

In the literature [31], different forms of membership functions may exist: Trapezoidal, triangular, rectangular, bell-shaped, concave shapes, etc. Triangular or trapezoidal shapes are generally used in this work as membership functions. The choice of the functions is also based on users' experience. Membership functions need to overlap to enable partial inclusion of the same linguistic variable at the same time in two different fuzzy sets [1,19,31].

4.2.2. The Inference Method

The inference method works in such a way that a change in the duty ratio of the boost chopper leads to the voltage V_{MPP} corresponding to the MPP. Following the analysis of an exhaustive number of combinations of input variables and an analysis of the corresponding outputs, we propose decision inference rules, illustrated in Figure 10:

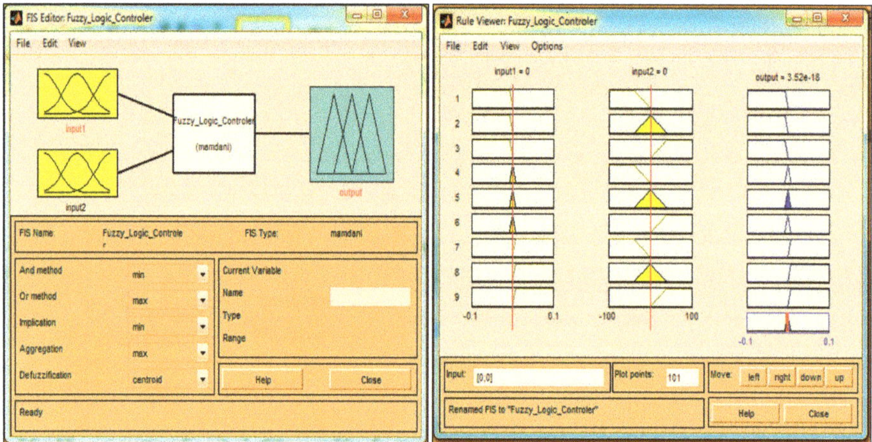

Figure 10. Proposed fuzzy rules decisions.

In this work, we have used the Mamdani method [31] for fuzzy inference with the max-min operation fuzzy composition law, as illustrated in Figure 11.

Figure 11. Max-min composition for the calculation of ΔD output.

4.2.3. Defuzzification

Following the inferencing operation, the controller output is expressed as a linguistic variable curve. Defuzzification methods are then used to calculate and decode the linguistic variable to a numerical value. In this work, we use a centroid method [31], which determines the crisp controller output as the value of the center of gravity of the final combined fuzzy set.

5. Extract of the MPP Using the Perturb and Observe (P&O) and Fuzzy Methods

Since the 1970's, the P&O (perturb and observe) method has been the most widely used approach in MPPT [5,12]. There are several variants of the P&O method, including the one described in Figure 12 below, whose results are compared with our fuzzy logic-based MPPT model.

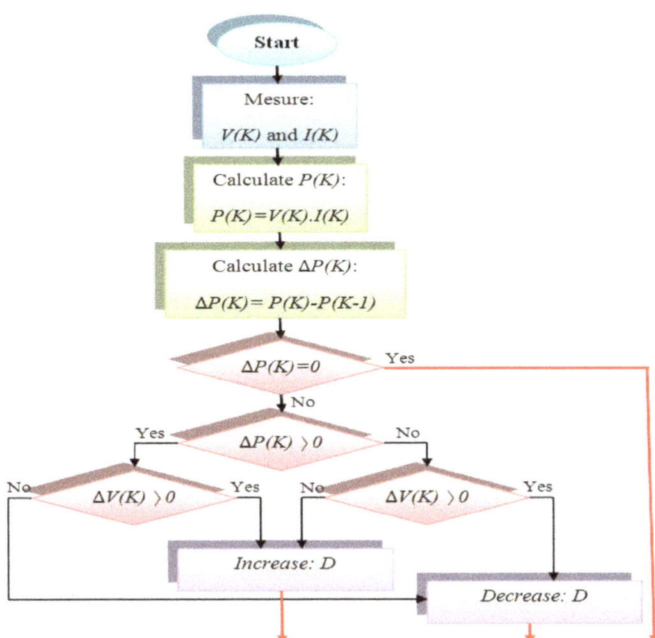

Figure 12. Flowchart of the perturb and observe (P&O) MPPT algorithm used.

The P&O method uses an algorithm that infers based on the duty ratio (increases or decreases) until the MPP is reached. As illustrated in Figure 12, $V(K)$ and $I(K)$ are continuously monitored, and the array output power $P(K)$ is calculated. This instantaneous value $P(K)$ is compared with the previously measured value of $P(K-1)$. If the two measured values are identical, this means that the maximum power point has been reached and no change is applied to the duty ratio. In the case where the output power and the voltage $V(K)$ have changed between two successive measurements and in the same direction, the duty ratio is increased. If $\Delta P(K)$ increases while $V(K)$ decreases and vice-versa, then the duty ratio is decreased [1,25].

In this paper, we compare the MPPT performance of the traditional P&O method with our fuzzy logic-based method. We illustrate in Figure 13 the fuzzy-based MPPT method and in Figure 14 the P&O MPPT method, as implemented on Simulink-Matlab®.

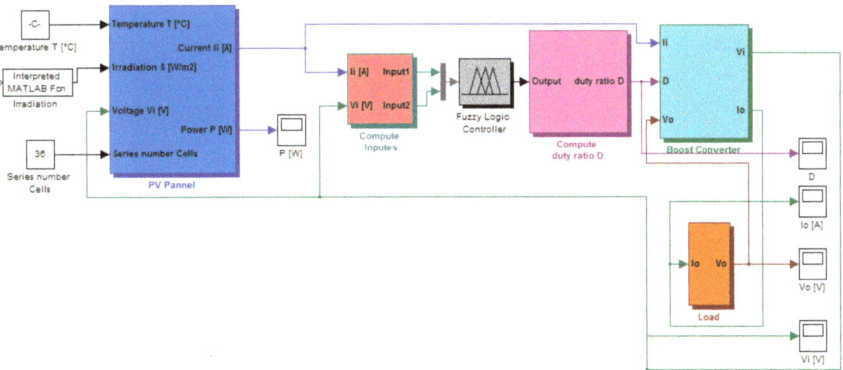

Figure 13. PV total system Simulink representation with a fuzzy logic controller.

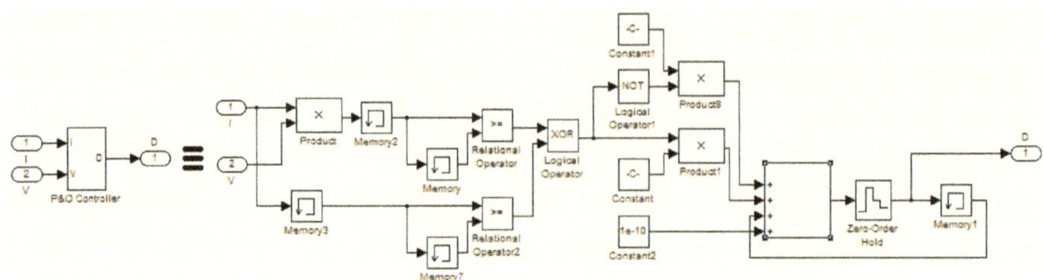

Figure 14. Details of the MPPT subsystems of the P&O Controller.

6. Simulations & Discussions

The fuzzy logic-based MPPT model has been built to increase efficiency for variable climatic conditions. Hence, the ambient temperature and incident irradiation on the PV panel is defined as an array of instantaneous input values. The mathematical representation of the PV system is defined in Equations (2)–(5), implemented together with the following parameters:

(1)   The number of PV modules connected in series is 14;
(2)   the number of photovoltaic cells in each PV module, connected in series $z = 36$;
(3) $R_p = 30\ \Omega$, $R_s = 15 \times 10^{-3}\ \Omega$, $E_g = 1.1\ \mathrm{eV}$, $n_1 = 1$, $n_2 = 2$, $k = 1.380 \times 10^{-23}\ \mathrm{J/K}$;
(4) $q = 1.602 \times 10^{-19}\ \mathrm{C}$;
(5) $I_{ph}\Big|_{(T=298\cdot\mathrm{K})} = 3.25\ \mathrm{A}$
(6)   the initial value of duty ratio was 0.1.

For the PV boost converter, Equations (6)–(9) were implemented together with the following numerical values [1,19]: $C_1 = 5.6\ \mathrm{mF}$, $C_2 = C_1$, $L = 3.5\ \mathrm{mH}$. For the PV load block, Equation (10) has been implemented together with the parametric definitions from Equation (11):

$$Z(s) = \frac{a_2 s^2 + a_1 s + a_0}{b_2 s^2 + b_1 s + b_0} \tag{10}$$

$$
\begin{aligned}
a_2 &= R_{bs} R_{b1} R_{bp} C_{b1} C_{bp}, \\
a_1 &= R_{bs} R_{b1} C_{b1} + R_{bs} R_{bp} C_{bp} + R_{b1} R_{bp} C_{bp} + R_{bp} R_{b1} C_{b1}, \\
a_0 &= R_{bs} + R_{b1} + R_{bp}, \\
b_2 &= R_{b1} R_{bp} C_{b1} C_{bp}, \\
b_1 &= R_{b1} C_{b1} + R_{bp} C_{bp}, \\
b_0 &= 1.
\end{aligned}
\tag{11}
$$

Equation (10) and the parametric definitions (Equation (11)) were used in previous works [1,19,25]. To model lead-acid batteries, the following numerical values were used to complete the model: $R_{bs} = 0.0013\ \Omega$, $R_{b1} = 2.84\ \Omega$, $R_{bp} = 10 \times 10^3\ \Omega$, $C_{b1} = 2.5\ \mathrm{mF}$, $C_{bp} = 2 \times 45 \times 9 \times 12 \times 36{,}000/(125^2 - 90^2) = 4.650\ \mathrm{KF}$.

6.1. Simulation Results for Fixed Climatic Conditions

To evaluate the fuzzy logic-based MPPT system, we analyzed its power extraction capabilities and stability versus the traditional P&O controller. In this particular simulation, the PV model described previously has been simulated with fuzzy logic and P&O controllers for fixed climatic conditions, i.e., an irradiance of 1000 W/m^2 and temperature of 25 °C. The results are illustrated in Figure 15. For PV power output, the fuzzy logic-based MPPT method achieves maximum power output faster than the P&O controller (2.4 s compared to 12.3 s). Moreover, the fuzzy logic-based MPPT controller shows better performance not only in set point achievement, but also in stability and robustness (mitigation of power fluctuation). The PV generator reaches its maximum stable power output just after a minor overshoot at t = 2.1 s, and the output remains stable within a 0.0001 W range. In the meantime, the P&O controller is slower to reach its set point and is subject to significant oscillations prior to stability achievement. Moreover, a steady regime is subject to a 0.0002 W continuous oscillation. Similar behavior is observed with the PV voltage output, while the P&O controller achieves its maximum set point after 15 s, compared to a rapid 2.4 s with a fuzzy logic controller. Furthermore, the P&O controller is subjected to an important overshoot prior to stabilization with a continuous 0.02 V oscillation, compared to virtually no oscillation in the case of our fuzzy controller. The same trend is noticed with the converter output voltage, PV module current, and converter current, while the fuzzy logic-based controller shows amazingly better performance than the P&O controller in speed for maximum power achievement, stability, and robustness.

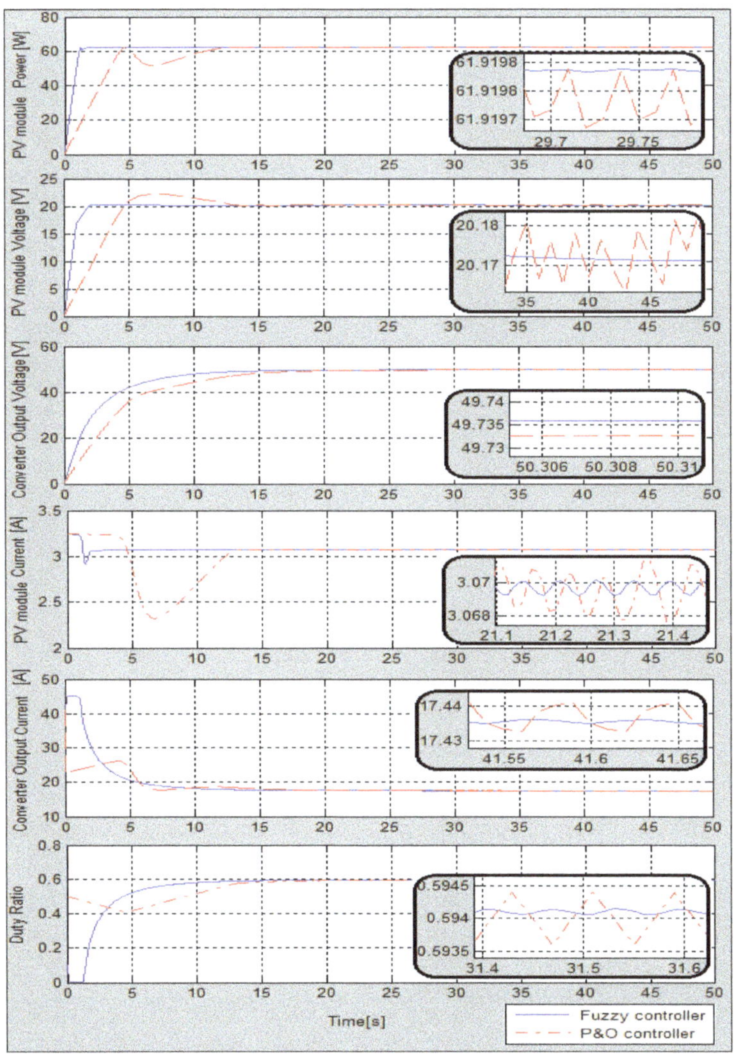

Figure 15. Simulation results for fixed climatic conditions: Irradiance $S = 1000 \text{ W/m}^2$ and temperature $T = 25\ ^\circ\text{C}$.

Performance improvement is the result of a faster and more appropriate variation of the duty ratio in the case of the fuzzy logic controller.

6.2. Simulation Results for the Changing Climatic Conditions

6.2.1. Simulation Results for a Fixed Temperature at 25 °C and Fast Increase of Irradiance from 500 Wm^{-2} to 1100 Wm^{-2}

In this case, the irradiance was quickly increased from $S = 500$ Wm^{-2} to $S = 1100$ Wm^{-2} via a step function at $t = 30$ s. As illustrated in Figure 16, the fuzzy logic-based MPPT method shows much better performance than the P&O controller. The fuzzy controller responds to the irradiance change virtually instantaneously and regains stability with amazing robustness for PV module power and voltage output (reduced power oscillation). The P&O controller takes longer to achieve stability, which occurs after signal oscillations in the case of the PV power output and after overshoot in the case of the PV voltage output. We note that the duty ratio variation by the fuzzy logic controller is much more rapid than that of the P&O controller when they detect the irradiance change. The duty ratio gradient decreases in the case of the fuzzy controller as compared to a constant gradient in the case of the P&O controller. This probably helps with both the speed of maximum power achievement and oscillation and overshoot mitigation.

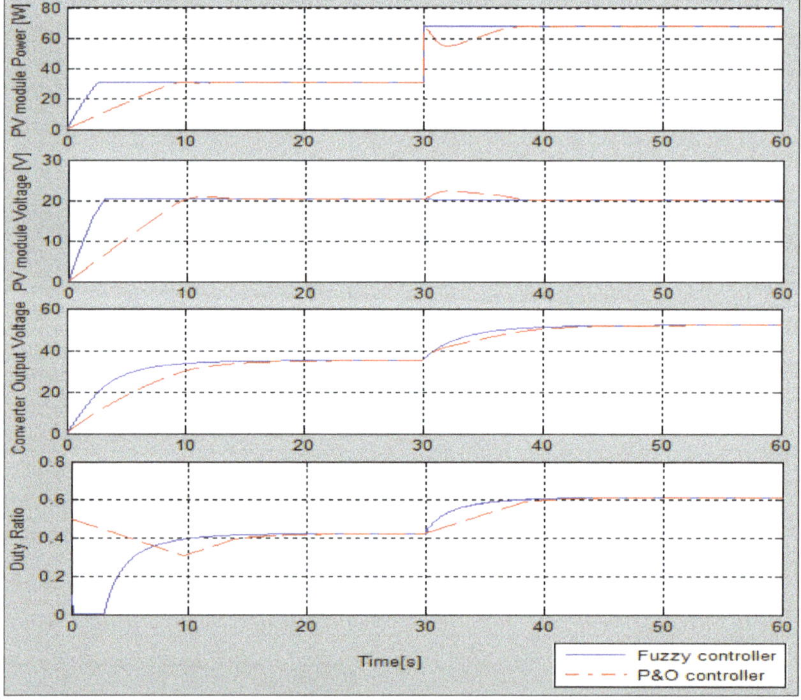

Figure 16. Simulation results with a fast increase in irradiance at $t = 30$ s from $S = 500$ W/m^2 to $S = 1100$ W/m^2 at constant temperature $T = 25$ °C.

6.2.2. Simulation Results for a Fixed Temperature at 25 °C and Slow Increase of Irradiance from 500 Wm^{-2} to 650 Wm^{-2}

In this case, we evaluate the relative performance of the P&O and a fuzzy logic-based controller for fixed temperature and slow irradiance increase from 500 Wm^{-2} to 650 Wm^{-2}. As illustrated in

Figure 17, the irradiance is slowly and continuously increased from $S = 500$ Wm^{-2} at $t = 20$ s up to $S = 650$ Wm^{-2} at $t = 80$ s. In this case, we see that both controllers show almost similar performance.

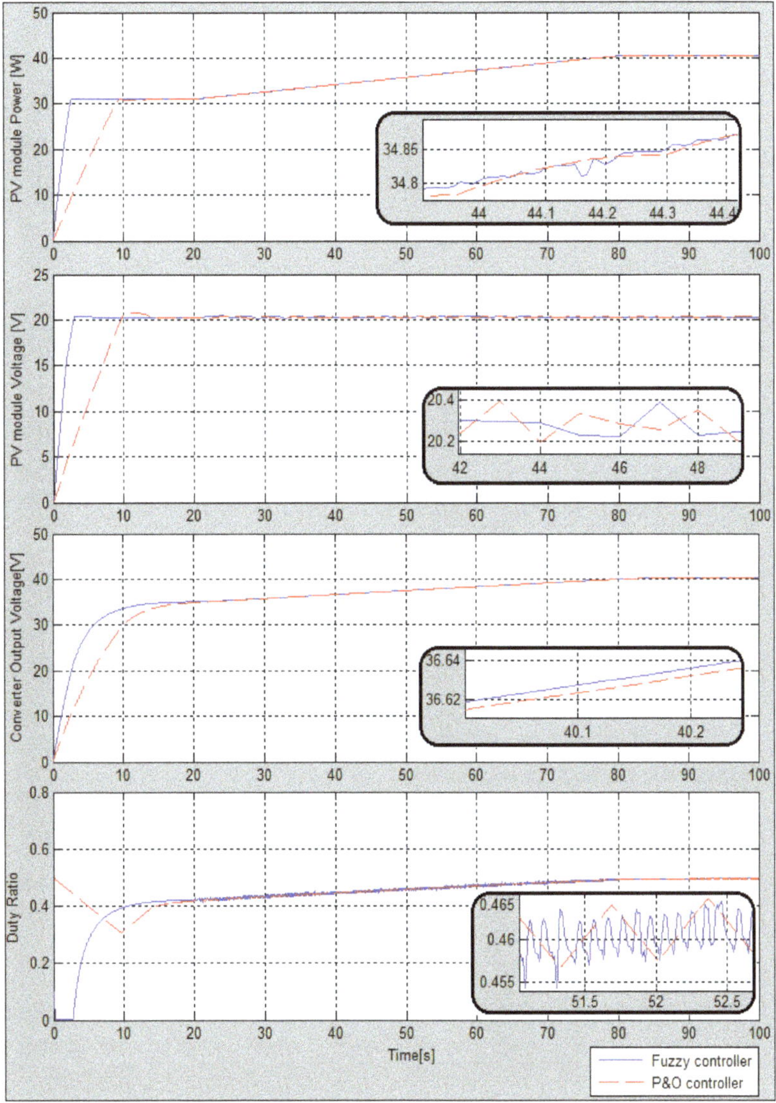

Figure 17. Simulation results with slow increase from $S = 500$ W/m^2 ($t = 20$ s) to $S = 650$ W/m^2 ($t = 80$ s) at constant temperature $T = 25$ °C.

6.2.3. Simulation Results for Fixed Irradiance at 1000 Wm^{-2} and Fast Temperature Decrease from 40 °C to 10 °C.

In this case, the temperature is decreased quickly via a step function at $t = 30$ s while keeping irradiance fixed at 1000 Wm^{-2}. We note similar observations for the case with quick irradiance increase with fixed temperature. The fuzzy-based MPPT controller reacts quickly to the change via a much more aggressive duty ratio change Figure 18. This leads to a faster maximum power output achievement with comparable stability with the P&O controller.

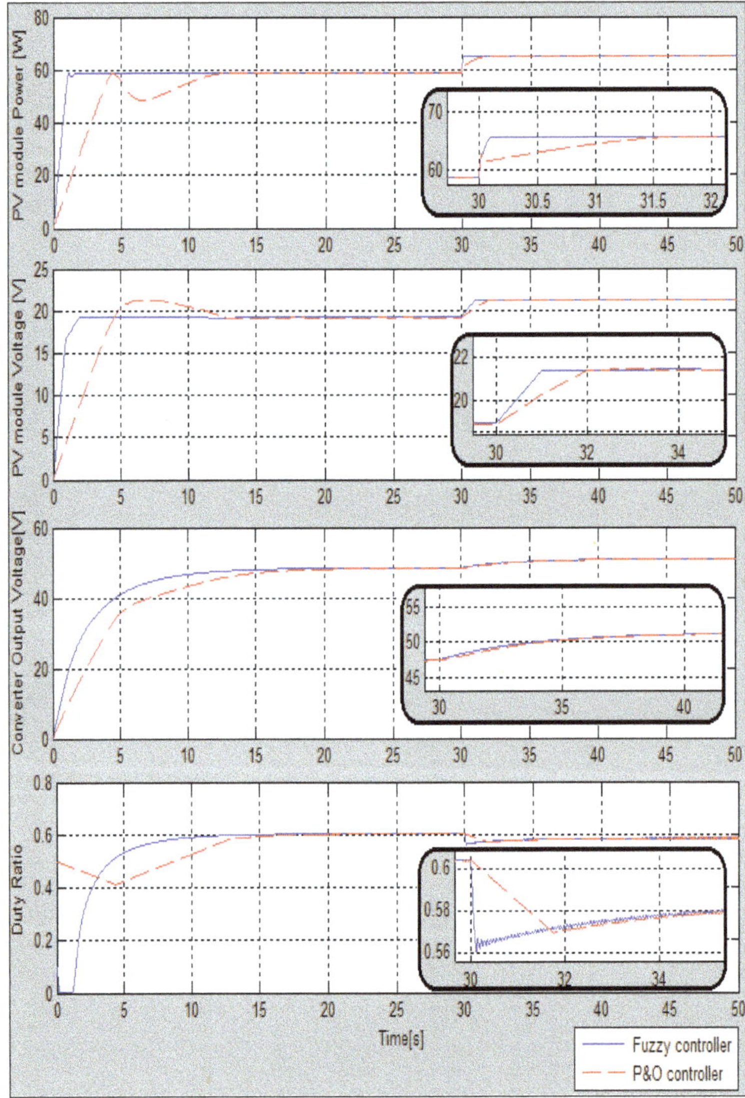

Figure 18. Simulation results with a fast decrease of temperature at $t = 30$ s from $T = 40\,^{\circ}\text{C}$ to $T = 10\,^{\circ}\text{C}$, with a constant irradiance of $S = 1000\ \text{W}/\text{m}^2$.

6.2.4. Simulation Results for Fixed Irradiance at 1000 Wm^{-2} and Slow Temperature Increase from 25 $^{\circ}$C to 30 $^{\circ}$C

In this case, as seen in Figure 19, both P&O and fuzzy logic controllers show comparable performance in PV power output achievement and stabilization. However, a notable difference appears in the case of the PV voltage output. The fuzzy controller shows no significant overshoot compared to P&O controller. Moreover, better performance is shown by the fuzzy logic controller when it comes to voltage output in the converter.

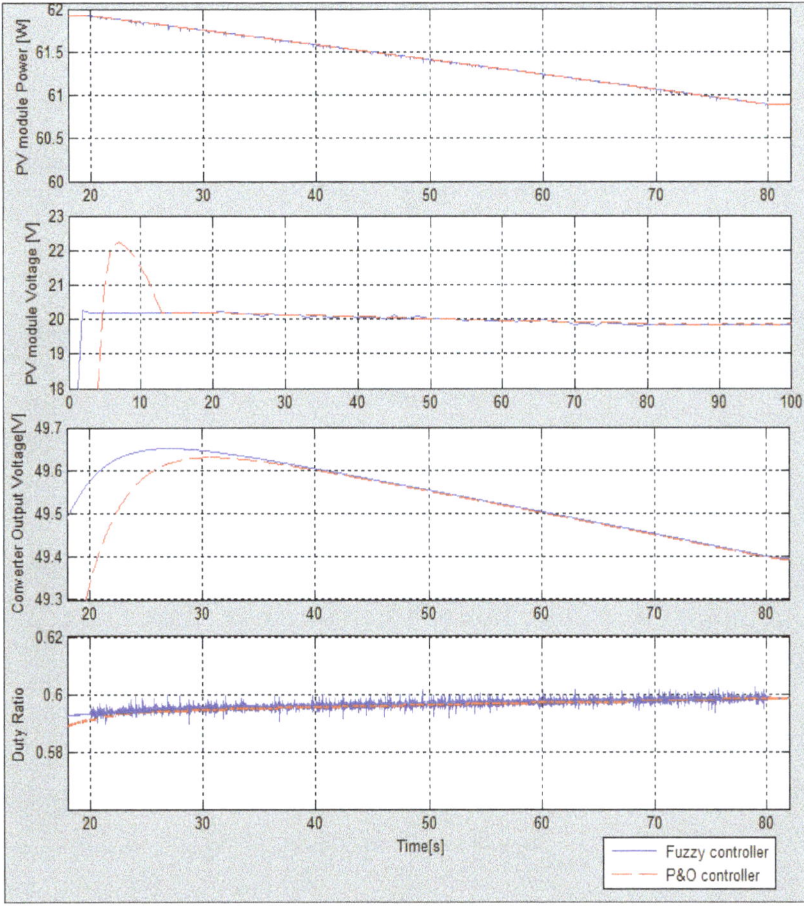

Figure 19. Simulation results with a slow increase in temperature from $T = 25\,^\circ$C ($t = 20$ s) to $T = 30\,^\circ$C ($t = 80$ s), with fixed irradiance of $S = 1000$ W/m^2.

7. Conclusions

The cost of solar energy is a major issue when it comes to its potential for greater development. Maximum power extraction is, therefore, an important parameter that influences the total production of PV systems and enables better payback of PV projects. In this paper, we have presented a fuzzy logic method which achieves faster and more stable power output at MPP from PV modules. In order to illustrate the performance of this controller, a Matlab-Simulink® model was built, and simulations were done for different operation scenarios. The results were compared with commonly used P&O controllers. Simulation results proved higher efficiency in maximum power tracking for the fuzzy logic controller. The simulations showed that most significant performance differences were achieved with rapidly varying parameters that influence power output (temperature, irradiance). Moreover, the fuzzy logic-based controller, as compared to the P&O controller, shows better performance in maximum power tracking time delay, stability, and robustness in all cases. Better stability and robustness performances from the fuzzy logic-based controller offer major advantages in mitigation of power fluctuation. The fuzzy logic algorithm is a robust and efficient algorithm. Indeed, this algorithm works at the optimal point without oscillations. In addition, it is characterized by good transient behavior. However, the implementation of this type algorithm is easier than conventional algorithms.

We can conclude that the use of fuzzy logic for the control MPPT presents a very interesting advantage, because there are always amazing results for the acceleration of the speed of MPP pursuit, the stability, achieved through the elimination of oscillations in steady state, and robustness. These amazing results are obtained after multiple tests by the engineer user's experience, who decide the designs of the fuzzy regulator, but the disadvantage is that with each model of the photovoltaic system, we must study and specify the engineer's own parameters and membership functions and the rules of his own fuzzy controller that help to achieve the MPP. For perspective, we propose a generalized study which can contain a global and generalized fuzzy model for any model of a photovoltaic system, if possible. The analysis in this paper should be useful for MPPT users, designers, and commercial PV manufacturers.

Author Contributions: Supervision, C.L. and A.I.; Visualization, T.O. and S.T.K.; Writing—original draft, G.F.T.K.

Funding: This research received no external funding.

Conflicts of Interest: The authors declare no conflict of interest.

Abbreviations

List of abreviation and symbols:

PV	Photovoltaic
MPPT	Maximum power point tracking
MPP	Maximum Power Point
P&O	Perturb and observe
MOSFET	Metal Oxide Semiconductor Field Effect Transistor frequency
D	Duty cycle
I_{SC}	Short circuit current
V_{OC}	Open circuit voltage
q	Electron charge
k	Boltzman's constant
Eg	band-gap energy of the semiconductor
V	Input voltage
I	Output current
I_{RS}	Reverse saturation current
I_S	Saturation current
PWM	Pulse width modulation
S	Irradiation
T	Temperature
P-V	photovoltaic characteristic $P(V)$
I-V	photovoltaic characteristic $I(V)$
z	number of photovoltaic cells connected in series
Rs	the series resistance of photovoltaic cells
Rp	parallel resistance of photovoltaic cells

References

1. Cheikh, M.A.; Larbes, C.; Kebir, G.T.; Zerguerras, A. Maximum power point tracking using a fuzzy logic control scheme. *Rev. Des Energ. Renouv.* **2007**, *10*, 387–395.
2. Khaligh, A.; Onar, O.C. *Energy Harvesting: Solar, Wind, and Ocean Energy Conversion Systems*; CRC Press: Boca Raton, FL, USA, 2017.
3. Takun, P.; Kaitwanidvilai, S.; Jettanasen, C. Maximum Power Point Tracking Using Fuzzy Logic Control for Photovoltaic Systems. In Proceedings of the International Multiconference of Engineers and Computer Scientists, Hong Kong, China, 16–18 March 2011.

4. Pandey, S.; Jena, P.K. A Review on Maximum Power Point Tracking Techniques for Photovoltaic Systems. *Int. Res. J. Eng. Technol.* **2017**, *4*, 1715–1720.

5. Rai, A.; Awasthi, B.; Singh, S.; Dwivedi, C.K. A Review of Maximum Power Point Tracking Techniques for Photovoltaic System. *Int. J. Eng. Res.* **2016**, *5*, 539–545.

6. Esram, T.; Chapman, P.L. Comparison of Photovoltaic Array Maximum Power Point Tracking Techniques. *IEEE Trans. Energy Convers.* **2007**, *22*, 439–449. [CrossRef]

7. Hua, C.; Shen, C. Comparative study of peak power tracking techniques for solar storage system. In Proceedings of the APEC'98 Thirteenth Annual Applied Power Electronics Conference and Exposition, Anaheim, CA, USA, 15–19 February 1998.

8. Subudhi, B.; Pradhan, R. A comparative study on maximum power point tracking techniques for photovoltaic power systems. *IEEE Trans. Sustain. energy* **2013**, *4*, 89–98. [CrossRef]

9. Onat, N. Recent Developments in Maximum Power Point Tracking Technologies for Photovoltaic Systems. *Int. J. Photoenergy* **2010**, *2010*, 245316. [CrossRef]

10. Oulcaid, M.; El Fadil, H.; Yahya, A.; Giri, F. Maximum Power Point Tracking Algorithm for Photovoltaic Systems under Partial Shaded Conditions. *IFAC-PapersOnLine* **2016**, *49*, 217–222. [CrossRef]

11. Khalifa, M.A.; Saied, K.M.; Bitro, S.S.; Anwar, M.; Nizam, M. PV Power System Using Maximum Power Point Tracking (Increment Conductance Algorithm). *Int. J. Innov. Res. Sci. Eng. Technol.* **2014**, *3*, 12267–12275.

12. Kamarzaman, N.A.; Tan, C.W. A comprehensive review of maximum power point tracking algorithms for photovoltaic systems. *Renew. Sustain. Energy Rev.* **2014**, *37*, 585–598. [CrossRef]

13. Garg, V.K. A review paper on various types of MPPT techniques for PV system. *Int. J. Innov. Eng. Res. Technol.* **2014**, *4*, 320–330.

14. Allataifeh, A.A.; Bataineh, K.; Al-Khedher, M. Maximum Power Point Tracking Using Fuzzy Logic Controller under Partial Conditions. *Smart Grid Renew. Energy* **2015**, *6*, 1–13. [CrossRef]

15. Gheibi, A.; Mohammadi, S.M.A. Maximum Power Point Tracking of Photovoltaic Generation Based on the Type 2 Fuzzy Logic Control Method. *Energy Procedia* **2011**, *12*, 538–546. [CrossRef]

16. El Khateb, A.H.; Rahim, N.A.; Selvaraj, J. Type-2 Fuzzy Logic Approach of a Maximum Power Point Tracking Employing SEPIC Converter for Photovoltaic System. *J. Clean Energy Technol.* **2013**, *1*. [CrossRef]

17. Luo, F.L.; Hong, Y. *Renewable energy systems: advanced conversion technologies and applications*; CRC Press: Boca Raton, FL, USA, 2016.

18. Cheng, P.C.; Peng, B.R.; Liu, Y.H.; Cheng, Y.S.; Huang, J.W. Optimization of a Fuzzy-Logic-Control-Based MPPT Algorithm Using the Particle Swarm Optimization Technique. *Energies* **2015**, *8*, 5338–5360. [CrossRef]

19. Larbes, C.; Cheikh, S.M.A.; Obeidi, T.; Zerguerras, A. Genetic algorithms optimized fuzzy logic control for the maximum power point tracking in photovoltaic system. *Renew. Energy* **2009**, *34*, 2093–2100. [CrossRef]

20. Mahdavi, M.; Li, L.; Zhu, J.; Mekhilef, S. An Adaptive Neuro-Fuzzy Controller for Maximum Power Point Tracking of Photovoltaic Systems. In Proceedings of the TENCON 2015—2015 IEEE Region 10 Conference, Macao, China, 1–4 November 2015.

21. Reisia, A.R.; Moradi, M.H.; Jamasb, S. Classification and comparison of maximum power point tracking techniques for photovoltaic system: A review. *Renew. Sustain. Energy Rev.* **2013**, *19*, 433–443. [CrossRef]

22. Kumar, J.; Bahrani, P. Comprehensive Review on maximum power point tracking methods for SPV system. *Int. Res. J. Eng. Technol.* **2017**, *4*, 1634–1639.

23. Abbes, H.; Abid, H.; Loukil, K.; Toumi, A.; Abid, M. Etude comparative de cinq algorithmes de commande MPPT pour un système photovoltaïque. *Revue des Energ. Renouv.* **2014**, *17*, 435–445.

24. Ezinwanne, O.; Fu, Z.; Li, Z. Energy performance and cost comparison of MPPT techniques for photovoltaics and other applications. *Energy Procedia* **2017**, *107*, 297–303. [CrossRef]

25. Knopf, H. Analysis, Simulation, and Evaluation of Maximum Power Point Tracking (MPPT) Methods for a Solar Powered Vehicle. Computer-Produced Typeface. Master's Thesis, Portland State University, Portland, OR, USA, 1999.

26. Farahat, M.A.; Metwally, H.M.B.; Mohamed, A.A.E. Optimal choice and design of different topologies of DC-DC converter used in PV systems, at different climatic conditions in Egypt. *Renew. Energy* **2012**, *43*, 393–402. [CrossRef]

27. Allouache, H.; Zegaoui, A.; Boutoubat, M.; Bokhtache, A.A.; Kessaissia, F.Z.; Charles, J.P.; Aillerie, M. Distributed photovoltaic architecture powering a DC bus: Impact of duty cycle and load variations on the efficiency of the generator. *AIP Conf. Proc.* **2018**, *1968*. [CrossRef]

28. Jafari, R.; Yu, W. Fuzzy Modeling for Uncertainty Nonlinear Systems with Fuzzy Equations. *Math. Probl. Eng.* **2017**, *2017*, 8594738. [CrossRef]
29. Jose, P.; Jose, P.R. Grid Connected Photovoltaic System with Fuzzy Logic Control Based MPPT. *Int. J. Eng. Innov. Technol.* **2014**, *3*, 142–148.
30. Algazar, M.M.; El-Halim, H.A.; Salem, M.E.E.K. Maximum power point tracking using fuzzy logic control. *Int. J. Electr. Power Energy Syst.* **2012**, *39*, 21–28. [CrossRef]
31. Bühler, H. *Réglage par Logique Floue*; Presses Polytechniques et Universitaires Romandes: Lausanne, Switzerland, 1994.
32. Patcharaprakiti, N.; Premrudeepreechacharn, S.; Sriuthaisiriwong, Y. Maximum power point tracking using adaptive fuzzy logic control for grid-connected photovoltaic system. *Renew. Energy* **2005**, *30*, 1771–1788. [CrossRef]
33. Won, C.Y.; Kim, D.H.; Kim, S.C.; Kim, W.S.; Kim, H.S. A new maximum power point tracker of photovoltaic arrays using fuzzy controller. In Proceedings of the 1994 Power Electronics Specialist Conference, Taipei, Taiwan, 20–25 June 1994.

Article

A Load-Balance System Design of Microgrid Cluster Based on Hierarchical Petri Nets

Jose R. Sicchar [1,*], **Carlos T. Da Costa Jr.** [2], **Jose R. Silva** [3], **Raimundo C. Oliveira** [4] **and Werbeston D. Oliveira** [5]

[1] Control and Automation Engineering Department, High School Technology,
     University of the Amazon State, Manaus 69050-025, Brazil
[2] Electrical Engineering Faculty, Institute of Technology, Federal University of Pará,
     Belém 66075-110, Brazil; cartav@ufpa.br
[3] Mechatronic Department, Polytechnic High School, University of São Paulo,
     São Paulo 05508-900, Brazil; reinaldo@usp.br
[4] Computation Engineering Department, University of the Amazon State, Manaus 69050-025, Brazil;
     rcoliveira@uea.edu.br
[5] Electric Engineering Department, Federal University of Amapá,
     Macapá, Amapá 68903-419, Brazil; wdoliveira@unifap.br
*   Correspondence: jvilchez@uea.edu.br; Tel.: +55-92-99221-2832

Received: 17 October 2018; Accepted: 19 November 2018; Published: 22 November 2018

Abstract: In the new paradigm of urban microgrids, load-balancing control becomes essential to ensure the balance and quality of energy consumption. Thus, phase-load balance method becomes an alternative solution in the absence of distributed generation sources. Development of efficient and robust load-balancing control algorithms becomes useful for guaranteeing the load balance between phases and consumers, as well as to establish an automatic integration between the secondary grid and the supervisory center. This article presents a new phase-balancing control model based on hierarchical Petri nets (PNs) to encapsulate procedures and subroutines, and to verify the properties of a combined algorithm system, identifying the load imbalance in phases and improving the selection process of single-phase consumer units for switching, which is based on load-imbalance level and its future state of load consumption. A reliable flow of automated procedures is obtained, which effectively guarantees the load equalization in the low-voltage grid.

Keywords: hierarchical Petri nets; urban microgrids; phase-load balancing

---

1. Introduction

Electric energy distribution in low voltage (LV) can be enhanced by a distributed architecture based on *urban microgrids*(UMG) [1–4]. A microgrid is essentially a cluster of residential consumers where at least some consumers possess local energy sources and a storage system. Energy supply in this system is a balance between electric power provided by a power line and that obtained from domestic loads generated by user sources [5–7]. Supervision and control of energy flow is managed by a **Microgrid Central Control**(MGCC) [8–10], which manages the balance between energy consumption, the main supply, and energy from microgrid components [11–14].

Nevertheless, in the legacy LV system [15] the phase-load imbalance is a drawback, especially because domestic loads generated by single-phase consumers affect grid phase stability, and the energy quality supplied [16,17]. Thus, some methods of solving this problem are highlighted in the the electrical current injection from distributed generation microgrids [18–20], the coordinated load balance [16,21], the integrated multimicrogrid control [12,22–24] and the load phase balance [25–27].

In the case of urban microgrids with distributed generation, the load-balancing method is based on the "electric current injection" in consumer unit phases, as well as in the phases of the LV grid, compensating for the imbalance of load and voltage. However, it is necessary to use a complex AC/DC–DC/AC signal converter control architecture called Microgrid Central Control (MGCC) [28], frequency inverters [29] and, in particular, supervision and control algorithms that optimize power and electric current flow [20]. The MGCC usually manages this automated solution flow, which does not always guarantee the efficient control of the phase shift effects between the main electrical current and the injected electrical current [30].

The load-balance procedure based on the "coordinated load balance" offers a wide range of control features for current injection, working synchronously with the grid transformer [16], with frequency compensation between the grid phases and consumer units, along with phase compensation between the grids' electrical current and the electric current injected [31]. Ensuring robustness and load balancing, however, requires a complex central control and supervision structure with local (distributed) controllers with high-reliability algorithms [32] that ensure automated operational integration at all control and supervisory levels.

Another method of load balance based on "integrated multimicrogrids control" is being widely used because of the large mix of micro-sources of energy to be applied for load-balancing [22,29,33], along with frequency and phase compensation in the grid and consumer units [34], also requiring a complex architecture with control and supervision algorithms that efficiently coordinate current injection and frequency and phase compensation in the LV grid [9,11,35], as well such as a large number of distributed generation units [36], which in fact means a great limitation for a large-scale implementation in developing countries [7,37].

An alternative to implementing the above-mentioned techniques is phase-load balancing, which consists of switching single-phase consumer units to the phases of the LV grid that are balanced. The procedure is based on the use of identification algorithms and load transfer management, aiming at minimizing current and load consumption [38] or voltage and load [27]. In both cases, the voltage and load equilibrium state in the grid phases is guaranteed; however, the switching choice is based only on current load consumption of consumers' units, disregarding the imbalance level and the future states of load consumption, which could contribute to the robustness of the system to eventual consumption peaks and to the durability of the load stability over time.

By contrast, it has been observed that the use of Petri nets (PNs) in complex systems is quite broad [39], due to its formal modeling, simulation and property verification capabilities [40–42], which allows development and verification of intelligent algorithms for control and supervision of application in smart grids [43,44]. The formal verification of routine flow allows evaluation of incidences, conflicts, deadlocks, loops, and reachability [45] of all stages and subroutines, as well as evaluation of inviolable flows and cycles of the algorithm in all its hierarchical levels [46], and also the automatic integration workflow with the control and supervision systems of an urban microgrid [47].

Thus, the use of PNs can contribute to the solution of the lack of automation in the operational procedures of load balancing in urban microgrids and especially in the LV grid [48], such as in the case of the legacy Brazilian LV distribution grid [15], with partially automated flows and manual methods without automatic full flow with the central supervisory system. Therefore, the existence of an intelligent algorithm that allows automation of the load-balancing procedures in the LV grid, as well as automatic integration at all levels of the grid control and supervision, would generate a great improvement in the legacy methods of load balancing, with correct, reliable and efficient processes, guaranteeing the load stability, as well as the streamlining of operational procedures in case of possible problems of load and voltage imbalance in the grid, and even emergency situations such as the burning of the transformer, among others.

In this article, we present a new system design, based on hierarchical PNs, of an intelligent algorithm to automate the load-balancing process, in order to provide reliable and effective procedures and to integrate efficiently the automation workflow in the legacy Brazilian LV grid.

The authors believe that the main contribution of this paper lies in providing a formal process-automation model that optimizes and integrates the workflow of a load-balancing control system in the legacy LV grid. The proposed control system is based on combined algorithms to minimize load consumption in the grid phases (feeders), through following programmable procedures: "load transfer in the grid feeders", which is based on a fuzzy inference to identify and perform the load transfer or between feeders; "imbalanced consumer unit identification", which is based on a fuzzy inference system to detect load-imbalance level in consumer units; "load forecast in consumers units", which is based on a Markov chain algorithm that forecasts monthly levels of discrete states on load consumption; and "switch selection" which is based on an optimal choice algorithm of imbalanced consumer unit with high load consumption.

The main contribution of this paper can be summarized as follows:

- A novel system design of a load-balance system integrated with the legacy LV system and urban microgrids is proposed. This is validated in Petri nets, emphasizing the novel form of encapsulating combined algorithms, evidenced by hierarchy levels of integration [43];
- The reachability graph and place-invariant analysis for property verification and the experimental assessment of robustness and efficiency of the load-balance algorithm is used. In addition, simulation dynamic tests are applied in a real case study of a LV grid of a city in the north of Brazil, for performance analyses of the proposed algorithm. Stored data about user consumption and grid feeders were used for simulation and analysis.
- A new method of choosing single-phase consumer units for the load-balancing process based on the imbalance levels and future states of load consumption, resulting in the efficient attenuation of the load average imbalance between LV feeders is applied, in comparison to the legacy system method and the bibliographic revision, which consider in both cases only the current load consumption.

The proposed system was validated efficiently through obtained results, providing an efficient and an automated reliable workflow for the load-balancing process in the legacy LV grid, which may also became an alternative load-balancing control procedure at the MGCC, in the urban microgrid context to operate as a coordinated control system with the current injection system of microgrids.

The remainder of this article is organized as follows: Section 2 explains the related background. In Section 3 the load-balancing control architecture is presented. In Section 4 the experimental system design validation and simulation dynamic results are presented and discussed. Finally, the conclusions and suggestions of this study are presented.

2. Background

In this section, we address some related issues that support the proposal. First, we present the state of art regarding urban microgrids. Next we address the load-imbalance problem in the LV grid. Finally, we address some definitions about hierarchical Petri nets for use in this research.

2.1. Urban Microgrid in the Smart-Grid Context

The urban microgrid (UMG) is a special instance of the smart-grid concept, derived from the special architecture of the LV grid inherited from a legacy system existing in several countries [15] and practically in all BRIC (Brazil, Russia, India, China) countries. It imposes that modern forms of power generation in urban unities be integrated with this legacy system to provide a hybrid LV system.

Figure 1 shows a schematic arrangement of the urban microgrid, which is controlled by the MGCC [49]. The UMG derives from a "point of in-common coupling" of the primary grid [18,28,37]. A distributed algorithm is executed by Local Controllers (LC) with a bi-directional communication network [36]. The main goal is to control the energy consumed by domestic loads and integrate the energy flow from distributed energy resources with power converters, and surplus energy into storage systems. This overall integrated control is managed by a Local Controller Supervisor (LCS), which works as an interface with smart meters [13,19].

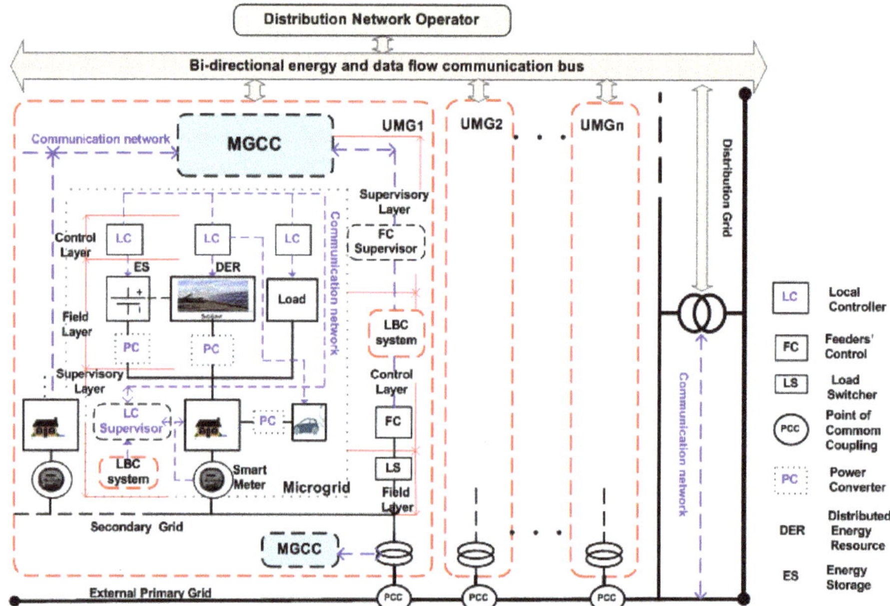

Figure 1. General Architecture of Urban Microgrid.

The proposed system is adapted in a global architecture for UMG. Its main component is the LCS, which supervises energy consumption in residential units to identify imbalanced load feeders assuming that a selection can be made for switching in the LV grid. Simultaneously, the Feeder Control Supervisor (FCS) identifies the grid feeder load imbalance, and coordinates the load transfer to reestablish the steady state [38].

2.2. Load Imbalance in Low-Voltage Grid Feeders

In the legacy low-voltage grid the "feeder load imbalance" constitutes a power consumption flow problem, as shown in Figure 2.

Generally, it is caused by growing disorder and by unplanned consumption of domestic loads in residences [17]. In extreme situations, this can affect the power supply, especially in the equilibrium between grid feeders. The transformer can be burned if this problem is not solved in good time [50].

The *Phase-Load Balancing* technique based on automatic load switching is an interesting approach for addressing this problem [16], and is an alternative technique to the legacy method used in the most part in LV Brazilian grids [15]. This implies that overloaded single-phase consumer units are switched to a feeder with a lower load level using some electronic switching device, as shown in Figure 2. This uses a control algorithm to automate the load and electrical current minimization [38] or voltage and load [27].

In this paper, instead of load balancing introduced by distributed-resource power injection [10], an automated approach of *phase-load balancing* method will use a control system based on a combined algorithm, addressed in detail in Section 3.3 [51]. In this specific case, we address the control system design using a hierarchical Petri net to achieve an automated and efficient flow for phase-load balancing in the LV grid.

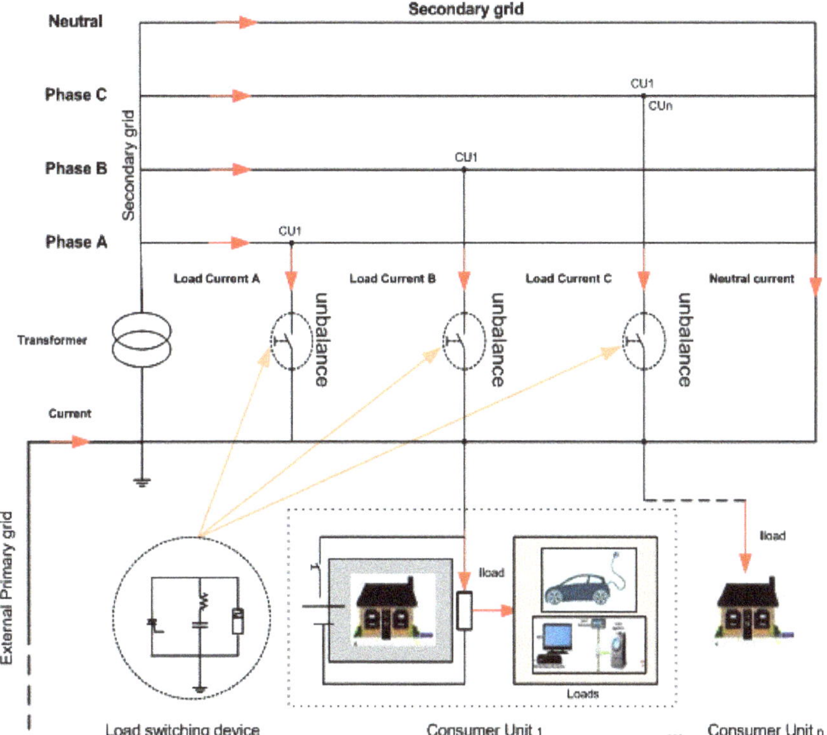

Figure 2. Load imbalance in secondary grid.

2.3. Hierarchical Petri Nets

Formal models of complex systems—such as urban microgrid control—are validated from static and dynamic points of view by Petri net property analysis and workflow. The use of Petri net extensions could facilitate this process, introducing hierarchical abstraction or time. Among all extensions, we would detach hierarchy and the introduction of time as key issues to support UMG design. The former would be advisable to treat large systems and the latter to open space to optimization in the service provided to user unities (specifically concerning the LBS). Thus, the use of PNs is very suitable since it is a formal method widely adapted to requirements of engineering, due to its wide range of environments for the modeling of dynamics [40,45].

The hierarchical approach would also fit the architecture imposed on the retrofitted (and automated) legacy system [52] and to the identification of points to couple the LBS service. In this article, we will consider a simple case of PNs with four hierarchical levels, which are integrated—the legacy LV grid, the MGCC system, the proposed algorithm system as part of the MGCC, and a fuzzy inference (as part of the proposed system)—for load-balance procedures in LV grid feeders.

2.3.1. Hierarchical Petri Net Definition (HPN)

HPN can be defined as follows.

- *Definition 2.3.1.1. Hierarchical Petri Net (HPN).* A HPN is a 6-tuple, according to expression Equation (1):

$$N = (P, T, A, w, M_0, F) \tag{1}$$

Such that

1. The 5-tuple

$$B = (P, T, A, w, M_0) \tag{2}$$

 is a marked Petri net, where:

 - P is a finite set of places, $P \neq \varnothing$;
 - T is a finite set of transitions, $T \neq \varnothing$;
 - $A \subseteq (PxT) \cup TxP)$ is the set of arcs from places to transitions and from transitions to places;
 - $w : A \rightarrow \{1, 2, 3, \cdots\}$ is a weight function on the arcs, and
 - M_0 is the initial marking of the PN [42]

2. F is a function *Place-Bounded Substitution* that ensures that a subnet Y limited by transitions can be replaced by a place s generating another net: $N' = \{P', T', F'\}$, where:

 - $P' = P \backslash S_y \cup \{s\}$, where S_y is the set of places in Y;
 - $T' = T \cup T_Y$, where T_Y are the transitions in Y;
 - $F' = F \backslash Int(Y)$, where $Int(Y)$ is the inner arcs set of Y [46]

In this paper, the use of an HPN is justified by the automated load-balanced flow-integrated system design, where the proposed algorithm system is a subnet of the UMG architecture, which in turn is part also of the LV legacy system. Thus, assessment validation and property verification can be through hierarchy propagation of lower subnets from macro-places using the PBS method.

Thus, through the structure defined in Equation (1), it is possible to model the states and intervals of operations and routines of the workflow, in the form of "P" places, "T" transitions, along with the start and end relationship between each of them in "A" arcs, the sequence order of the workflow in "M_0" marking, involving the flows of each level of the integrated distribution system: LV network, microgrid, Load-Balance Control (LBC) system and subsystem of load transfer in "F" hierarchical subnetworks.

In this case, the system design will begin in the legacy LV grid structure, as the first hierarchical level of integrated system, considering the supervision center as the system beginning, i.e., the place and initial marking of the network. The second hierarchical level will be started from the transformers of the LV grid, i.e., the MGCC subnet, in which all the physical structure of automation and control of the load-balancing system will be represented. The third hierarchical level will start from the MGCC control device, i.e., the subnet of the proposed balancing system. In this third subnet, all the programmable steps of the proposed combined algorithm will be represented. Finally, the embedded algorithm subnet used to identify and transfer loads between grid feeders represents the fourth hierarchical level as the formal system design of one of these steps.

3. The Load-Balance Control System (LBC)

The proposed system is called load-balancing control (LBC), and is based on a combined algorithm with four stages according to Figure 3, which aims to automate the procedures related to load-imbalance identification in the grid feeders and consumer units, as well as the consumer unit arrangement for the switching process, which is based on load-imbalance level identification and load forecast in the single-phase consumption units [51]. Thus, the system design will be based on this architecture and the LBC system flowchart, as shown in Figure 4.

3.1. LBC Architecture

Figure 3 shows the LBC system architecture.

The LBC system interacts with the concessionaire measurement interface, and is composed of:

- Feeder Control Supervisor (FCS). This manages the procedure that identifies the load imbalances in grid feeders, once given from the central control of the legacy LV system. This is based on a fuzzy inference system [38]. Processed load data are collected offline from the MGCC information system. In cases of load imbalance in feeders, it will activate the Local Control Supervisor.

- Local Control Supervisor (LCS). This is activated in cases of imbalance in grid feeders, and performs the load-imbalance identification (based on a fuzzy inference) and load forecasting (based on Markov chains) in single-phase consumption units [51], delivering it as a result in the LC. Data processed as energy, energy variation, and load variation are collected offline from the MGCC information system. Temperature variation and energy price variation are collected offline externally from the meteorology and rnergy market information centers, respectively.

- Local Controller (LC). This receives from the LCS the future states of load consumption and the load-imbalance levels in the single-phase consumers, to chose a switching arrangement. The choice criterion implies selecting consumers that present the highest level of load imbalance and also the highest future state of load consumption in each consumption unit. The choice is checked with the load transfer levels indicated by the FCS in each phase of the grid. The final result obtained the switching arrangement of the consumer units, returning the load stability to the MGCC information system and to the legacy system.

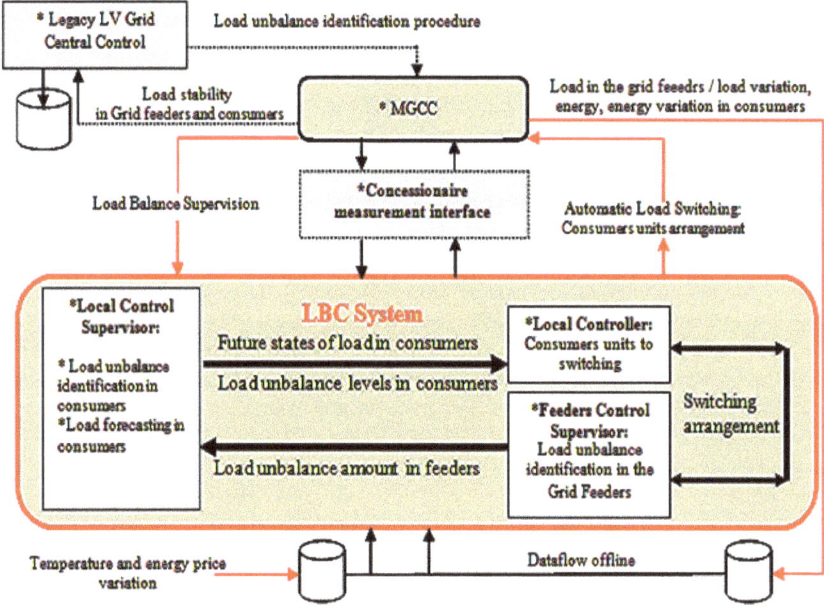

Figure 3. LBC Architecture.

3.2. High-Level Flowchart of the LBC System

Figure 4 shows in detail the high-level flowchart of the LBC system, as an alternative control to the load-balancing process for the UMG; thus, the LBC system can also be inserted as an interface in the legacy LV grid.

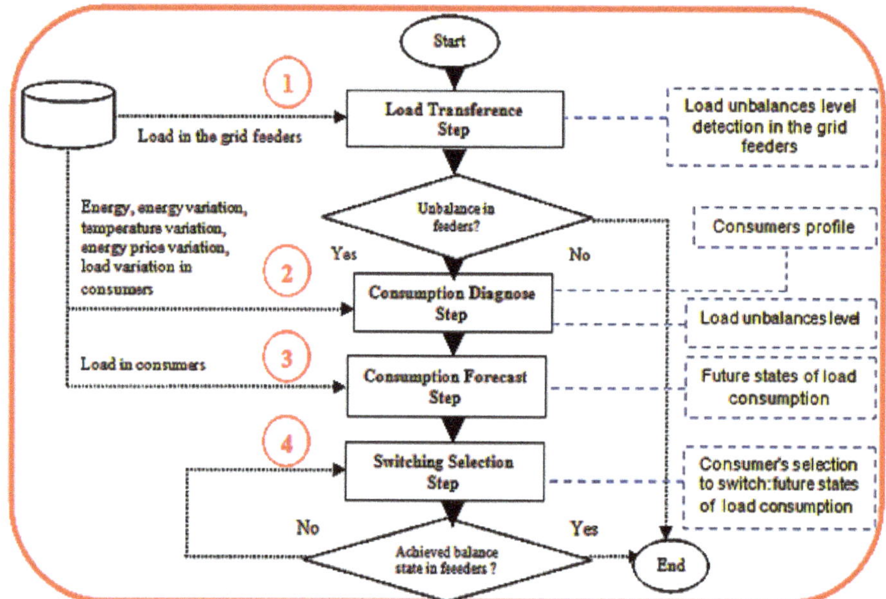

Figure 4. LBC system high-level flowchart.

The LBC high-level flowchart is explained as follows:

Step 1. Load Transference. This flow is started when the "Load" level consumptions in each grid feeder (from the database) are processed in the LBC Fuzzy inference (explained in detail in the following subsection) in order to detect load imbalance. As a result, it is informed whether feeders are balanced or not. Thus, both situations are informed by the FCS. In cases of load imbalance in some grid feeders, the second modular step will be started. Otherwise, the process will be ended.

Step 2. Consumption Diagnosis. This module is activated when one of the grid feeders is imbalanced. It is processed in the load-imbalance inference (LUI), also explained in the following subsection, to identify the consumer profile and the load-imbalance level in single-phase consumer units. This result will be used to improve the consumer unit arrangement choice, for the switching process on the grid feeders.

Step 3. Consumption Forecast. This step detects the future load consumption in the single-phase consumption units with load-imbalance levels detected in the previous step. The load future consumption results, along with the load-imbalance level, are used for the consumer unit switching selection on the grid feeders.

Step 4. Switch Selection. This last module assists in obtaining a reliable combination for switching selection of consumer units. It is based on the load future consumption, in each single-phase consumer unit with load-imbalanced level detected. In the case of not finding a good arrangement, a new one will be found, as indicated in the following section. Otherwise, the process will be ended.

3.3. Combined Algorithms Flowchart of the LBC System

Figure 5 shows in detail the integration flowchart of the combined algorithms that combines the four programmable steps of the LBC system.

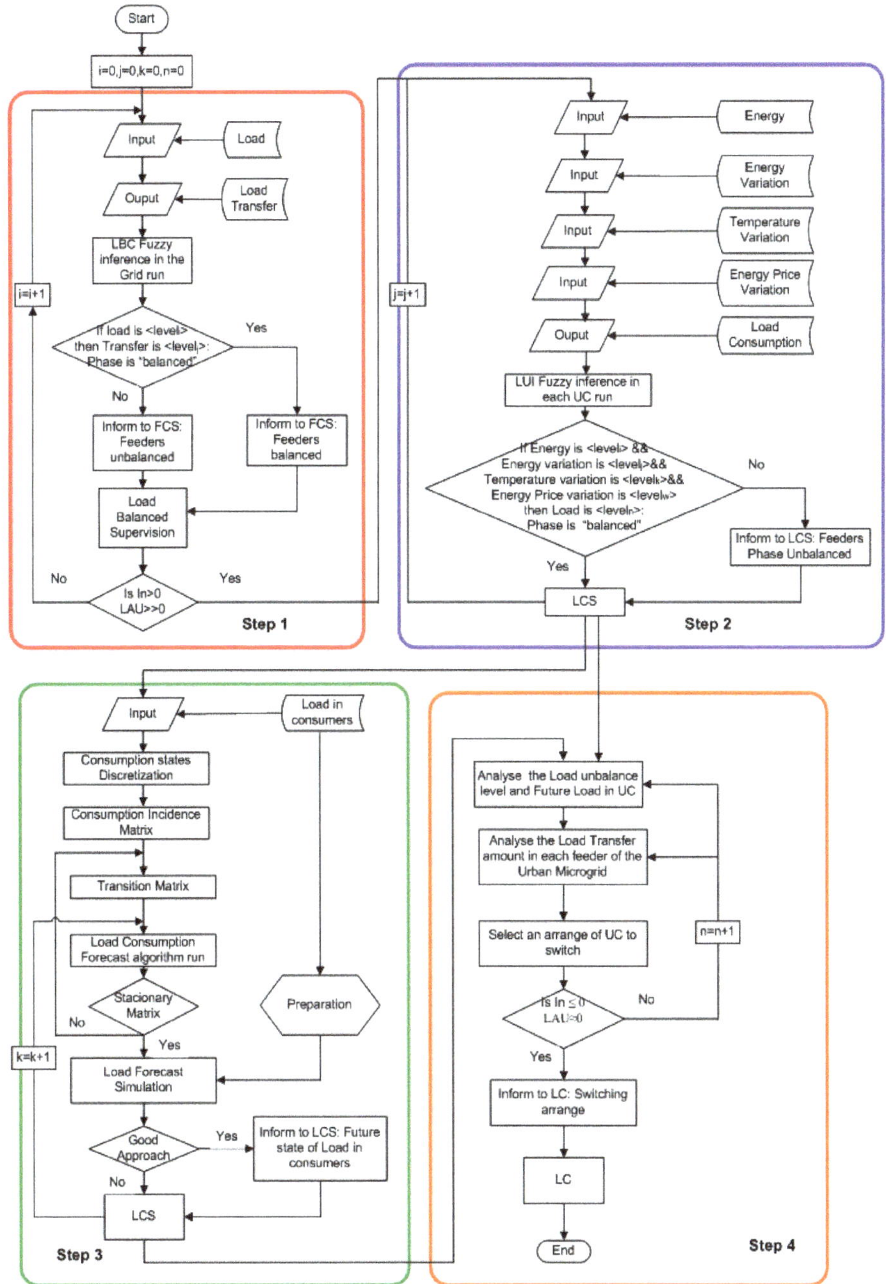

Figure 5. Combined algorithms flowchart of the LBC system.

Step 1. **Load Transference inference**. This first algorithm is highlighted in the red rectangle in Figure 5. This is based on a Mamdani's fuzzy inference with only an input called "Load" and an output called "Load Transfer" [51]. The input variable has eight S_{1_i} membership sets, which represent the possible load level consumptions x_i in each grid feeder with their respective μ_i membership degree. This is defined according to Equation (3).

$$S_{1i} = \{(x_i, \mu_i(x_i)|x_i \in \text{``Load''}\} \tag{3}$$

where: $i = 1 \ldots 8$.

The output variable has also eight S_{1j} membership subsets, which represent the possible load transference levels y_j to each grid feeder. This is defined according Equation (4).

$$S_{1j} = \{(y_j, \mu_j(y_j)|y_j \in \text{``Load Transfer''}\} \tag{4}$$

where: $j = 1 \ldots 8$.

Thus, both variables are inferred according to Equation (5).

$$\textbf{If ``}Load\text{''} is \text{``}x_i''\textbf{ then ``}Load\ Transfer\text{''} is \text{``}y_j'' \tag{5}$$

After this process, the FCS is informed that feeders are balanced or lso that feeders are imbalanced. Thus, both situations are informed as to the Load-Balanced Supervision. In cases of load imbalance, the third step will be started. Otherwise, the process will be started again to a new load-imbalance identification procedure.

Step 2. **Load-Imbalance inference**. This second module is highlighted in blue in the Figure 5 and is activated when one of the grid feeders is imbalanced. This is applied only in single-phase consumption units, and this is also based on a Mamdani's fuzzy inference with four inputs called "Energy", "Energy variation", "Temperature variation", and "Energy price variation", and one output called "Load variation" [53]. The input variable definitions are as follows.

- "Energy". This first input variable has three S_{2ai} membership sets, which represent the possible "energy" level consumption x_{ai} in each grid feeder with their respective μ_{ai} membership degree, according to Equation (6).

$$S_{2ai} = \{(x_{ai}, \mu_{ai}(x_{ai})|x_{ai} \in \text{``Energy''}\} \tag{6}$$

where: $i = 1 \ldots 3$.

- "Energy variation". This second input variable also has three S_{2ai} membership sets, which represent the possible "energy variation" levels x_{bi} in each grid feeder with their respective μ_{bi} membership degree, according to Equation (7).

$$S_{2bi} = \{(x_{bi}, \mu_{bi}(x_{bi})|x_{bi} \in \text{``Energy variation''}\} \tag{7}$$

where: $i = 1 \ldots 3$.

- "Temperature variation". This third input variable has also three S_{2ci} membership sets, which represent the possible "temperature variation" x_{ci} which affect the consumer units with their respective μ_{ci} membership degree, according to Equation (8).

$$S_{2ci} = \{(x_{ci}, \mu_{ci}(x_{ci})|x_{ci} \in \text{``Temperature variation''}\} \tag{8}$$

where: $i = 1 \ldots 3$.

- "Energy Price variation". This fourth input variable has also three $S_{2_{di}}$ membership sets, which represent the possible "energy price variation" x_{di} which affect the consumer units with their respective μ_{di} membership degree, according to Equation (9).

$$S_{2_{di}} = \{(x_{di}, \mu_{di}(x_{di})|x_{di} \in \text{"Energy price variation"}\} \tag{9}$$

where: $i = 1 \ldots 3$.

The output variable has three S_{2j} membership sets, which represent the possible "load variation" y_j in each consumer units with their respective μ_j membership degree, according to Equation (10).

$$S_{2j} = \{(y_j; \mu_j(y_j)|y_j \in \text{"Current variation"}\} \tag{10}$$

where: $j = 1 \ldots 8$.

Thus, these variables are inferred according to Equation (11).

$$\textbf{If } \text{"\textit{Energy}" is "}x_{ai}'' \textbf{ and } \text{"\textit{Energy variation}" is "}x_{bi}'' \textbf{ and } \text{"\textit{Temperature variation}" is "}x_{ci}'' \textbf{ and}$$
$$\text{"\textit{Energy price variation}" is "}x_{di}'' \textbf{ then } \text{"\textit{Load variation}" is "}y_{2j}'' \tag{11}$$

In the case of the single-phase consumption units being balanced, the LCS is informed and the process will be started again. Otherwise, the LC is informed of this diagnosis and the process will be started from the third module.

Step 3. **Consumption Forecast**. This third algorithm is highlighted in green in the Figure 5. This step detects the future states of load consumption, for the best choice of the single-phase consumption units to the switching procedure.

This is based on Markov chains performing the load consumption forecast in each "F_{ij}" consumer feeder. According to Equation (12), the load data-flow is prepared and is inserted in the input to the Consumption States Discretization. Thus, based on the "π_{ij}" incidence jump probabilities, the Consumption Incidence Matrix is formed to achieve each "$X(k + n)$" future state (low, medium, and high) from the previous state "$X(k)$". Then, as a result, the Transition Matrix is obtained which starts the Load Consumption Forecast algorithm.

$$C_{F_{\pi_{ij}}}(n) = P\{X(k + n) = j | X(k) = i \tag{12}$$

where: $C_{F_{\pi_{ij}}}(n) \geq 0$

In cases of not obtaining the Stationary Matrix, the flow will be started again from the Transition Matrix step. Otherwise, the Load Forecast Simulation will be started. In cases of obtaining a good approach, the LCS will be informed of the Future State of Load (FLS) for each consumption unit. Otherwise, the algorithm will run again. The load temporal series validation along a specific period is performed beforehand, training a dataset to establish a reliable forecast model to forecast the FSL. A 48-month data history of load consumption, to forecast 12 months of future consumption, will be used in each consumer in this specific case.

Step 4. **Switch Selection**. This last module is highlighted in orange in the Figure 5. This assists in obtaining an optimal combination to selection of "i" single-phase consumer units to switching process, according to Equation (13).

$$L_i = \alpha.min(L_i) + \beta.min(FL_i) \tag{13}$$

Analyzed in Equation (13) is the load variation level detected "L_i" and the FLS "FL_i" (low, medium, and high), in each consumption unit of the imbalanced feeder, choosing the "i" consumer unit that indicates the highest level of "L_i" and "FL_i". Then, observed in Equation (14) is a restriction of equality, such that the "L_i" total load amount of the chosen consumers should not be greater than the load transfer level "P_j" indicated at each "j" phase. In cases of not obtaining a good arrangement, the process will be started. Otherwise, the process will be ended.

$$\sum_{i=1}^{n} L_i \leq P_j \tag{14}$$

4. LBC System Implementation Results in Hierarchical Petri Nets

4.1. System Design Method

For design purposes, Figure 6 shows the flowchart describing the system design applied method.

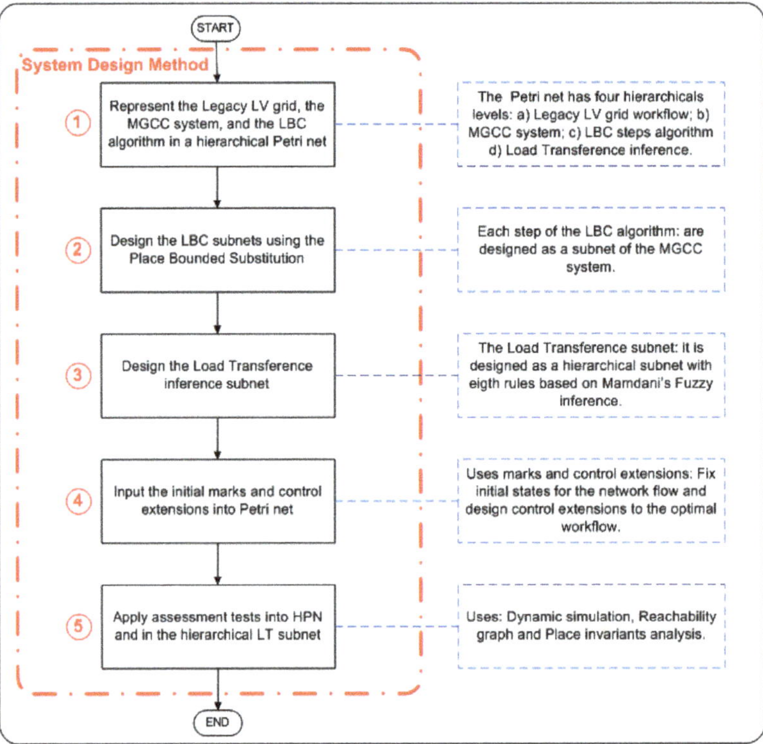

Figure 6. Flowchart describing the system design method.

First, system flow integration is represented: the legacy LV grid flow, the MGCC system, the LBC algorithm, and the LT subnet. Each of them composes a level of HPN. In addition, the Load Transference Inference of the LBC system represents a fourth hierarchical subnet, which highlights the formal system design of inference rules for load-balancing procedure in the LV grid. Second, the LBC system design as a subsystem of the MGCC system is performed. Their subnets are represented using thePBS method. Third, the Load Transference Inference system design as a hierarchical LBC subnet is performed, with eight rules (based on Mamdanis' fuzzy inference) to identify load-imbalance level in the grid feeders. Fourth, marks (tokens) on the initial states on network are placed, as well as the control extensions for dynamic simulation. Finally, assessment tests into HPN and the LT subnet are applied. In this case, in both evaluated PNs, the dynamic system simulation, the reachability states analysis to evaluate tangibility of states (places) over HPN and on its subnets, and the place-invariant analysis to verify compliance of automation routines workflow in the whole HPN and also into its hierarchical subnets will be applied.

4.2. Dynamic System Design

Figure 7 shows the LBC system design modeling in an HPN.

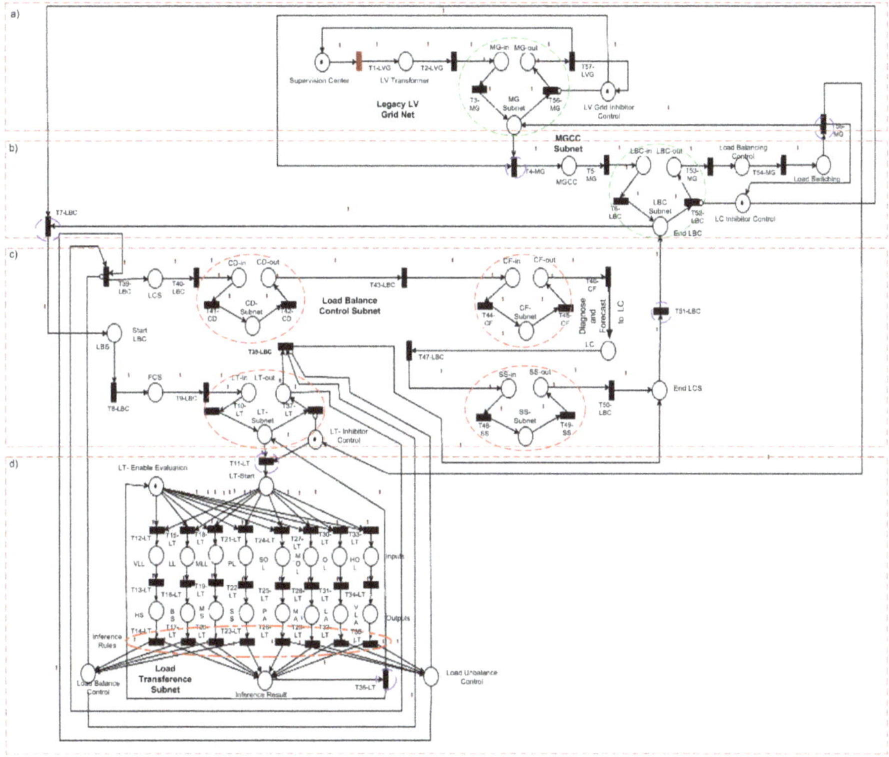

Figure 7. HPN System design: (**a**) Legacy LV Grid net; (**b**) MGCC subnet; (**c**) LBC subnet; (**d**) LT subnet.

This model describes four levels of hierarchy:

(a) **Legacy LV net**. This is showen in Figure 7a on the HPN first level. It represents the currently electrical LV grid, with following operating flow:

- Supervision Center place. This represents the substation supervision center. This starts the whole HPN, and indicates the initial point of load-balancing verification process.
- LV Transformer place. This represents each LV transformer. This has an interface with the load consumption supervision in the secondary grid inner-installed [15]. This also starts control and supervision integration between the new UMG architecture and the MGCC.
- MGCC macro-place. This represents the start of the second HPN level, which is represented by the Place-Bounded Substitution method, highlighted in green circles, with an input place, "MGU_{in}" and an output place, "MG_{out}" and an "MG Subnet" place. This is bounded by borders formed by the "$T4 - MG$" and "$T55 - MG$" transitions highlighted in blue.
- LV Inhibitor Control. This extension control activates the "$T4 - MG$" transition and inhibits the "$T56 - MG$" transition, ensuring workflow from the first hierarchical level to the second

subnet, avoiding the return to the first subnet without running the full load-balancing flow. It allows only operation flow completion once it is performed, through the "$T55_{MG}$" transition, which, in addition, returns tokens to the LV inhibit control, thus forming an automatic retention property of tokens for this subnet.

(b)   **MGCC subnet**. Figure 7b shows this second hierarchical level of HPN, and represents the MGCC architecture addressed by the load-balance control, which follows the operating flow below:

- MGCC place. This represents the MGCC information system [28]. This started the load-balancing procedure in LV grid feeders.
- LBC macro-place. This represents the start of the third HPN level. This is also based on the Place-Bounded Substitution method, highlighted in green circles, with an input place, "LBC_{in}" and an output place, "LBC_{out}" and the "LBC Subnet" place. This is bounded by borders formed by the "$T7 - LBC$" and "$T51 - LBC$" transitions, highlighted in blue.
- LC Inhibitor Control. This extension control activates the "$T7 - LBC$" transition and inhibits the "$T52 - MG$" transition, ensuring workflow from the second hierarchical level to the third subnet, avoiding return to the second subnet without running full LBC system flow, allowing only flow completion once it is performed, through the "$T51 - LBC$" transition.
- Load-Switching Control place. This represents the load-switching control of consumer units to some LV grid feeders, according to the final result of the LBC system.
- Load-Switching place. This represents the load switching in each LV grid feeder. The final result of the load-balancing process is transferred to the MG subnet through the "$T55_{MG}$" transition. In addition, this transition returns a token to the LC inhibit control, thus also forming an automatic retention property of tokens for this subnet.

(c)   **LBC subnet**. Figure 7c shows this third subnet, highlighting in red circles the four LBC flowcharts addressed in Section 3.2 as some specific subnets, derived from a macro-place based on the PBS method. This follows operating flow below:

- LBS place. This represents the Load-Balance Supervision of the LBC system, and the initial state of the third subnet workflow.
- FCS place. This represents the Feeder Control Supervision (FCS) and from it starts the load-balance procedure in each grid feeder.
- LT macro-place. This represents the start of the fourth HPN level. This is also represented by the PBS method, highlighted in red circles, with an input place, "LT_{in}" and an output place, "LT_{out}" and the "LT Subnet" place. This is bounded by borders formed by input transitions "$T11 - LT$" and by output transition "$T36 - LT$".
- LCS place. This transmits the final detection result of load imbalances from the LT subnet and activates the following subnet.
- Consumption diagnosis (CD) macro-place. This represents the start of "step 2", called the Consumption Diagnosis subnet. This is also represented by the PBS method, highlighted in red circles, with an input place, "CD_{in}" and an output place, "CD_{out}" and the CD Subnet place.
- Consumption forecast (CF) macro-place. This starts the "step 3", called the Consumption Forecast subnet. This is also represented by the PBS method, highlighted in red circles, with an input place, "CF_{in}" and an output place, "CF_{out}" and the CF Subnet place.
- LC place. This sends the procedure results from CD and CF subnet to the SS subnet.
- SS macro-place. This represents the start of "step 4", called the Switch Selection subnet. This is also represented by the PBS method, highlighted in red circles, with an input place, "SS_{in}" and an output place, "SS_{out}" and the SS Subnet place.

- End LCS place. This represents the final workflow of the LT subnet and transmits as a return to the MGCC subnet.

(d) **LT subnet**. Figure 7d shows the fourth subnet. It represents the load transference based on a Mamdanis' fuzzy machine, which is composed of eight inputs which represent different load levels in each grid feeder. This is shown also in detail in Table 1:

- VLL place. This represents the heavily less-loaded level in each grid feeder.
- LL place. This represents the less-loaded level in each grid feeder.
- MLL place. This represents the medium less-loaded level in each grid feeder.
- PL place. This represents the perfectly loaded level in each grid feeder.
- SOL place. This represents the slightly overloaded level in each grid feeder.
- MOL place. This represents the medium overloaded in each grid feeder.
- OL place. This represents the overloaded level in each grid feeder.
- HL place. This represents the heavily overloaded level in each grid feeder.

and by eight outputs that represents different load transfer amounts to each grid feeder. This is shown also in detail in Table 2:

- HS place. This represents the high subtraction of load level in each grid feeder.
- BS place. This represents the big subtraction of load level in each grid feeder.
- MS place. This represents the medium subtraction of load level in each grid feeder.
- SS place. This represents the slight subtraction of load level in each grid feeder.
- PA place. This represents the perfect addition of load level in each grid feeder.
- MA place. This represents the medium addition of load level in each grid feeder.
- LA place. This represents the large addition of load level in each grid feeder.
- VLA place. This represents the very large addition of load level in each grid feeder.

Thus, finally, eight associated inference rules are obtained, showed also in detail in Table 3. Each rule is activated one at a time by LT enable evaluation extension control.

- "$T14 - LT$" transition. This represents the first rule and implies that if load level is "VLL" then "VLA" of load in some grid feeder will be transferred.
- "$T17 - LT$" transition. This represents the second rule and implies that if load level is "LL" then "LA" of load in some grid feeder will be transferred.
- "$T20 - LT$" transition. This represents the third rule and implies that if load level is "MLL" then "MA" of load in some grid feeder will be transferred.
- "$T23 - LT$" transition. This represents the fourth rule and implies that if load level is "PL" then "PA" of load in some grid feeder will be transferred.
- "$T26 - LT$" transition. This represents the fifth rule and implies that if load level is "SOL" then "SS" of load in some grid feeder will be transferred.
- "$T29 - LT$" transition. This represents the sixth rule and implies that if load level is "MOL" then "MS" of load in some grid feeder.
- "$T32 - LT$" transition. This represents the seventh rule and implies that if load level is "OL" then will be transferred "S" of load in some grid feeder.
- "$T35 - LT$" transition. This represents the eighth rule and implies that if load level is "HOL" then will be transferred "HS" of load in some grid feeder will be transferred.

Thus, as a result of these rules, two possible workflows are addressed:

- In cases of load imbalance one of last four rules ("$T26 - LT$", "$T29 - LT$", "$T32 - LT$", "$T35 - LT$") will be activated by Load-Imbalance Control following transition "$T39 - LBC$", activating remaining LBC subnets in sequence until the ending process in the "end LCS" place.

- Otherwise, in cases of load balance, one of first four rules ("$T14 - LT$", "$T17 - LT$", "$T20 - LT$", "$T23 - LT$") will be activated by load-balance control, following transition "$T38 - LBC$" and ending the workflow in the "end LCS" place.

4.3. System Design Validation

This section addresses the discussion of dynamic performance of the proposed system implementation. Dynamic simulation assessment, reachability and coverability graph was applied in this case, and the place-invariant analysis was used to verify some properties. In addition, these will also be used to analyze the load transference subnet. A free version of Pipe 4.3.0 was applied a simple case of an extended (hierarchical) play-transition or HPN.

4.3.1. HPN Implementation Analyze

- Dynamic simulation. Figure 7 shows the HPN simulation workflow. Thus, the integrated automation flow between the legacy LV system, the MGCC architecture, the LBC system, and the load transference inference was validated. Thus, it was verified as an extended simple PN with a four-level hierarchy, where each subnet is started from a special macro-place based on the Place-Bounded Substitution method. Through dynamic simulation, all lower hierarchical flows were verified in each subnet and their integration with all upper levels of HPN, complying efficiently the integral workflow addressed in Section 4.2. In addition, several simulations with 10,000 firings were carried out with 50 ms time delay between each firing, and have not been registered as "no stop being" and deadlocks.
- Reachability Graph. Figure 8 shows the reachability graph of the HPN system design.

This represents the PN reachable state diagram obtained from its initial state "S_0", indicated by the red arrow. Through this diagram, it was verified that all 52 network states and 57 transitions were reached and covered, without deadlock and conflicts. Thus, Figure 8 shows also two load-imbalance workflow verifications, as a result of the "LT" subnet from the "33" place (highlighted in red circles), called "LT_{out}". In this case, both are highlighted in green circles by the "$T38 - LBC$" transition (no load imbalances) and by the "$T39 - LBC$" transition (with load imbalances). In addition, Figure 8 also shows the reachability and coverability of all 19 states of the LT subnet, demarcated from the $T11 - LT$" transition until the $T36 - LT$" transition.

- Place-invariant analysis. Place (P) invariant analysis was performed to verify bounded and liveliness properties of HPN, and especially some automation workflow, which is a set of places marked with the same constant token consumption, ensuring the net completion cycle. In this case, two place-invariant equations were obtained.

Equation (15) shows the first P-invariant that verifies the first automation workflow related to the LT subnet flow: the LT Enable Evaluation starts the LT inference as a control extension, activating one of the possible eight load diagnostic rules to obtain the inference result. This P-invariant shows a lower flow for the load-balancing procedure, and completes marking condition for this cycle equal to "1", while performing an LBC order in the LV grid. Figure 9 shows this workflow highlighted with a red line.

$$
\begin{aligned}
&M(LT - Enable\ Evaluation) + M(VLL) + M(HS) + \\
&M(LL) + M(BS) + M(MLL) + M(MS) + \\
&M(PL) + M(SS) + M(SOL) + M(PA) + \\
&M(MOL) + M(MA) + M(OL) + M(LA) + \\
&M(HOL) + M(VLA) + M(Inference\ Result) = 1
\end{aligned}
\tag{15}
$$

Equation (16) shows the second place-invariant, which verifies the whole flow integration of all hierarchical levels: from the Supervision Center place, the LV transformer place, the MGCC Subnet, and the LBC Subnet, to the LT Subnet to perform the load transference procedure.

$$
\begin{aligned}
&M(Supervision\ Center) + M(LV\ Transformer) + \\
&M(MG_{in}) + M(MG\ Subnet) + M(MGCC) + \\
&M(LBC_{in}) + M(LBC\ Subnet) + M(LBS) + \\
&M(FCS) + M(LT_{in}) + M(LT\ Subnet) + \\
&M(LT - Start) + M(VLL) + M(HS) + \\
&M(LL) + M(BS) + M(MLL) + M(MS) + \\
&M(PL) + M(SS) + M(SOL) + M(PA) + \\
&M(MOL) + M(MA) + M(OL) + M(LA) + \\
&M(HOL) + M(VLA) + M(Inference\ Result) + \\
&M(LT_{out}) + M(LCS) + M(CD_{in}) + \\
&M(CD\ Subnet) + M(CD_{out}) + M(CF_{in}) + \\
&M(CF\ Subnet) + M(CF_{out}) + M(LC) + \\
&M(SS_{in}) + M(SS\ Subnet) + M(SS_{out}) + \\
&M(EndLBS) + M(LBC_{out}) + M(Load\ Balancing\ Control) + \\
&M(Load\ Switching) + M(MG_{out}) = 1
\end{aligned}
\tag{16}
$$

In cases of load imbalances, the CD Subnet, CF Subnet and SS Subnet are activated. The final result is sent for implementation to the load-balance control and the load-switching places. The final report is sent to the MGCC. Thus, a balanced marking for the cycle is equal to "1", and shows that the balanced cycle is a sequential system as already expected. In this case, we also conclude that the combined net including LBC and the coupling with the legacy system is also sequential. Figure 9 shows the balanced workflow highlighted with a blue line.

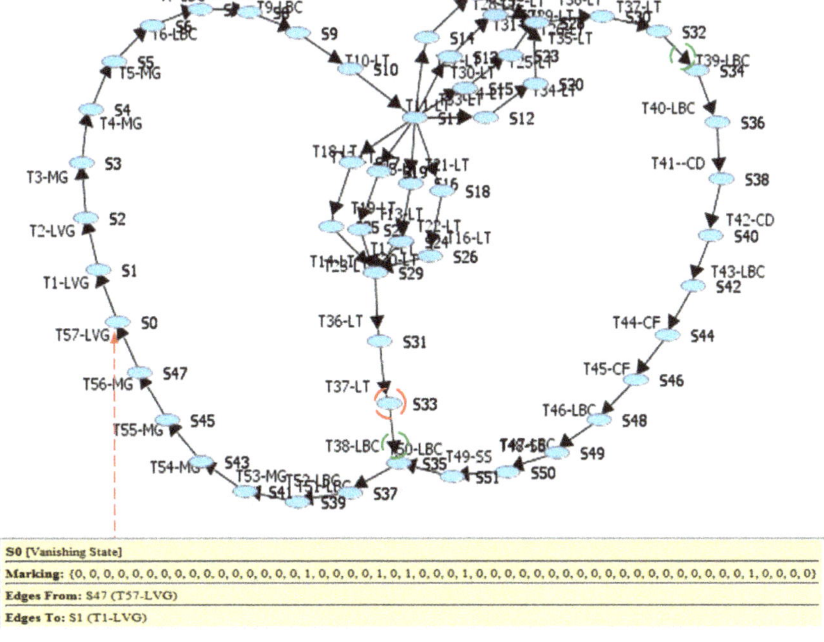

Figure 8. HPN Reachability graph.

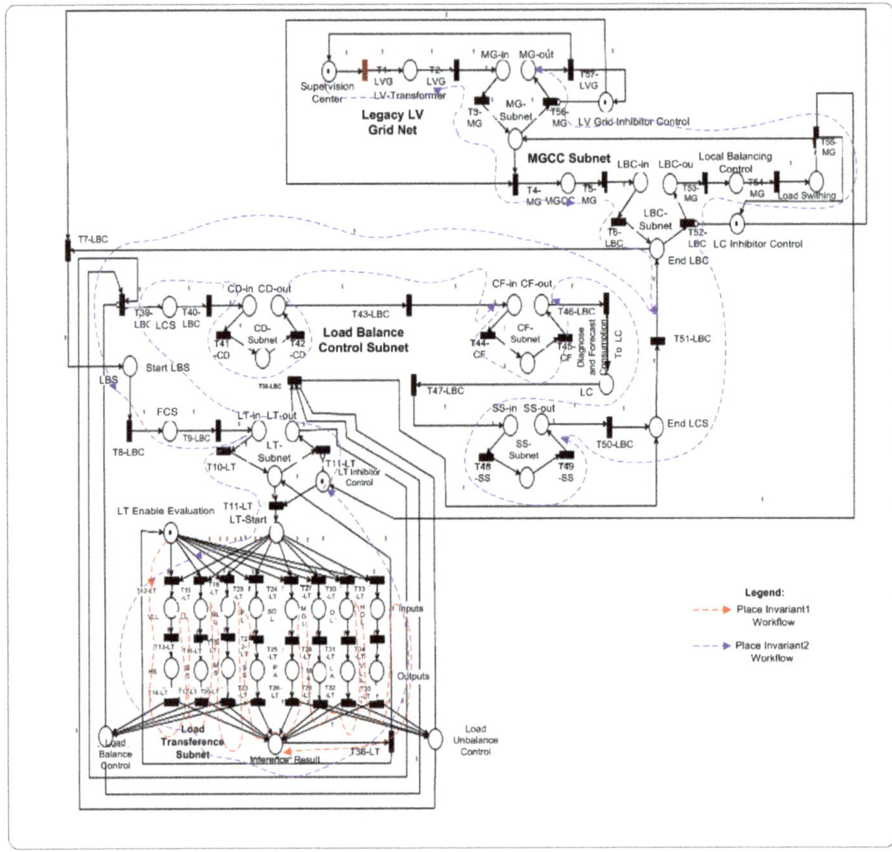

Figure 9. HPN Place-invariant workflow representation.

4.3.2. Load Transfer Implementation Analysis

- Dynamic simulation. In this section, the hierarchy is applied in subnets separately to verify the validation and verification into HPN subnets, and to assess the internal automation flow verification with whole network workflows. In this case, Figure 10a shows the LT network system design in detail, with the eight inputs and outputs respectively, as well as, the eight rules inferred highlighted in red.

The pertinence functions of input and output variables are allocated in eight triangular sets, to obtain a homogeneous distribution of the load-imbalance levels in feeders, in the case of the input variable, as well as the transfer levels to load addition or subtraction in feeders, to the output variable. This distribution is reported in Siti [38], where load balancing is applied in a LV circuit, which results in a homogeneous load balancing between feeders, with the lowest load average imbalance level.

Figure 10b shows the membership functions of the input variable parameter "load". Thus, its distribution ranges values are divided into eight sets, and 39.9 Kilo-Watts (KW) was determined as the maximum amount of load allowed in feeders based on the technical data of a 110 kVA transformer with 60 KW of active power [51]. Considering the origin of the first triangular set at 0 KW, the load concentration division was developed manually on the fuzzy toolbox in the MATLAB environment, for each set, and the best obtained distribution is shown in Table 1.

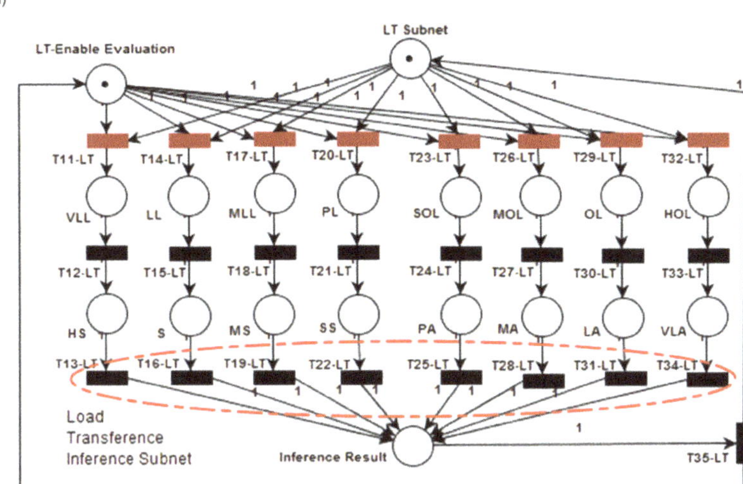

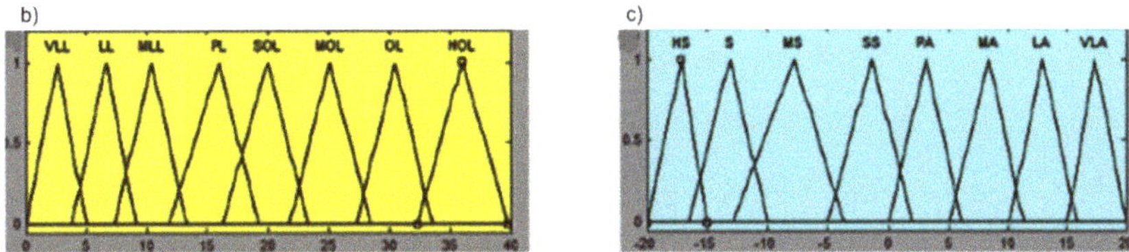

Figure 10. Load Transference subnet: (**a**) LT PN; (**b**) Membership function for input parameters; (**c**) Membership function for output parameters.

Figure 10c shows the membership functions of the output variable parameter "load transfer". Similarly, its distribution range values are also divided into eight sets, and "load addition" was considered to the load transfer, in cases of placing additional load in a balance phase, as was "load subtraction" in cases of withdrawing load at an imbalanced phase. -20 KW was determined as the maximum amount of load subtraction, based on the technical data of a LV grid with 110 Kilo-Volts-Amperes (KVA) transformer, and as the maximum amount of load addition, 20 KW [51], as shown in Table 2.

Table 1. Input fuzzy nomenclature.

Inp	Desc	Fuzzy Nom	Kw Range
1	Heavily Less-Loaded	VLL	0–5
2	Less-Loaded	LL	3.8–9.0
3	Medium Less-Loaded	MLL	7.3–13.3
4	Perfectly Loaded	PL	11.8–19.3
5	Slightly Overloaded	SOL	16.3–23.3
6	Medium Overloaded	MOL	21.7–28.4
7	Overloaded	OL	21.2–33.4
8	Heavily Overloaded	HL	32.3–39.8

Table 3 shows the *fuzzy* rules for the LT system. Thus, it was verified as an simple PN subnet with 19 places and 25 transitions. This is started in the LT Subnet place as an extended hierarchical level of the LT macro-place of HPN. Through dynamic simulation, all inference evaluation flows were verified, and the inference results were also verified as transitions highlighted in red circles. In addition, several simulations with 10,000 firings were carried out with 50 ms delay between each firing, and have not been registered as "no stop being" and deadlocks.

Table 2. Output fuzzy nomenclature.

Out	Desc	Fuzzy Nom	Kw Range
1	High subtraction	HS	-20 to -15.3
2	Big subtraction	BS	-16.5 to -10
3	Medium subtraction	MS	-12.9 to -3.6
4	Slight subtraction	SS	-4.9 to -2
5	Perfect Addition	PA	0–6
6	Medium Addition	MA	5.0–11.2
7	Large Addition	LA	10.1–15.7
8	Very large addition	VLA	15–20

Table 3. Fuzzy rules.

Rule	If Input	Is	Then Output	Is
1	"Load"	VLL	"Transfer"	VLA
2	"Load"	LL	"Transfer"	LA
3	"Load"	MLL	"Transfer"	MA
4	"Load"	PL	"Transfer"	PA
5	"Load"	SOL	"Transfer"	SS
6	"Load"	MOL	"Transfer"	MS
7	"Load"	OL	"Transfer"	BS
8	"Load"	HOL	"Transfer"	HS

- Reachability graph. Figure 11 shows the reachability graph of the LT subnet. It represents the PN reachable diagram obtained from its initial state "S_0" highlighted by a red circle, which also represents the initial marking of this PN. Through Figure 11 it is verified, the reachability and coverability of all 19 states and 25 transitions of the LT subnet were reached and covered without deadlock and conflicts. Thus, being verified also each possible rule inferred in order to each specific level of load concentration (input variables) and load transference amount (output variables) of the "Fuzzy" system design, addressed in Section 4.2.

On the other hand, it is observed that this result coincides with the reachability and coverability states diagram found in the HPN net to the LT subnet, shown in detail in Figure 8, due to the hierarchy propagation to the lower subnets, thus verifying the tangibility of the evaluation of inputs, outputs, and rules of the load-balancing inference system to grid feeders.

- Place-invariant analysis. Equation (17) shows the first P-invariant that verifies the first automation workflow of the LT Subnet.

$$\begin{aligned}
&M(LT - Enable\ Evaluation) + M(VLL) + M(HS)+ \\
&M(LL) + M(BS) + M(MLL) + M(MS)+ \\
&M(PL) + M(SS) + M(SOL) + M(PA)+ \\
&M(MOL) + M(MA) + M(OL) + M(LA)+ \\
&M(HOL) + M(VLA) + M(Inference\ Result) = 1
\end{aligned} \tag{17}$$

In this case, evaluation of each inference rule based on the inputs and outputs variables following the stream is verified: "LT Enable evaluation", which evaluates "VLL place" and "HS place" to perform the first inference rule, "LL place" and "BS place" to perform the second inference rule, "MLL place" and "MS place" to perform the third inference rule, "PL place" and "SS place" to perform the fourth inference rule, "SOL place" and "PA place" to perform the fifth inference rule, "MOL place" and "MA place" to perform the sixth inference rule, "OL place" and "LA place" to perform the seventh inference rule and "HOL place" and "VLA" to perform the eighth inference rule.

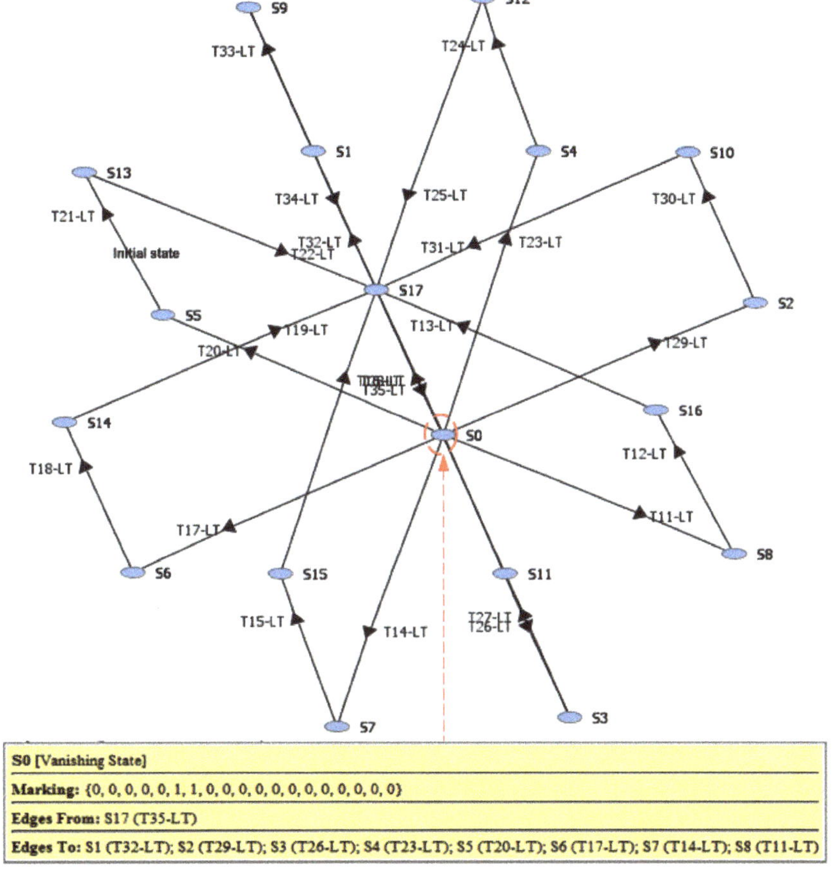

Figure 11. Reachability graph of the Load Transference subnet.

The final result is sent to the inference result. Thus, the complete marking condition of this sequence of places is be equal to "1", while performing this evaluation process.

In addition, the internal flow of load-imbalance identification, through propagation of hierarchy of the "LT macro-place" to the Load Transference Subnet is also verified in this subnet. Figure 12 shows this workflow highlighted with a red line.

By contrast, Equation (18) shows the second P-invariant, which verifies the workflow of the Load Transference Subnet: from input variables are started from the "LT Subnet place": "VLL place", "LL place", "MLL place", "PL place", "SOL place", "MOL place", "OL place", "HOL place", and started also the outputs variables: "HS place", "BS place", "MS place", "SS place", "PA place", "MA place", "LA place", and "VLA place".

$$
\begin{aligned}
&M(LT\ Subnet) + M(VLL) + M(HS) + \\
&M(LL) + M(BS) + M(MLL) + M(MS) + \\
&M(PL) + M(SS) + M(SOL) + M(PA) + \\
&M(MOL) + M(MA) + M(OL) + M(LA) + \\
&M(HOL) + M(VLA) + M(Inference\ Result) = 1
\end{aligned}
\tag{18}
$$

Thus, these are associated with the inference rules that are transferred as a result of the "inference result place". This procedure is performed through the internal control "LT Enable Evaluation" place, which enables only a rule selection after the evaluation of each one, thus, emulating a fuzzy Mamdani inference, evaluating each condition (set) of the membership function of the input variable "Load", and each condition (set) of the relevance function of the output variable, "Load Transfer".

This guarantees the best choice evaluation for load transfer, according to the situation identified in each phase, regardless of load subtraction, when the phase is imbalanced, or load addition when there is a balanced phase.

Thus, this validates the cycle and completes marking condition of the sequence be equal to "1", the final result back to the "LT Subnet place" from where it will be propagated to the upper-hierarchical levels of HPN by the "LT macro-place". Figure 12 shows this place-invariant flow, highlighted with a blue line.

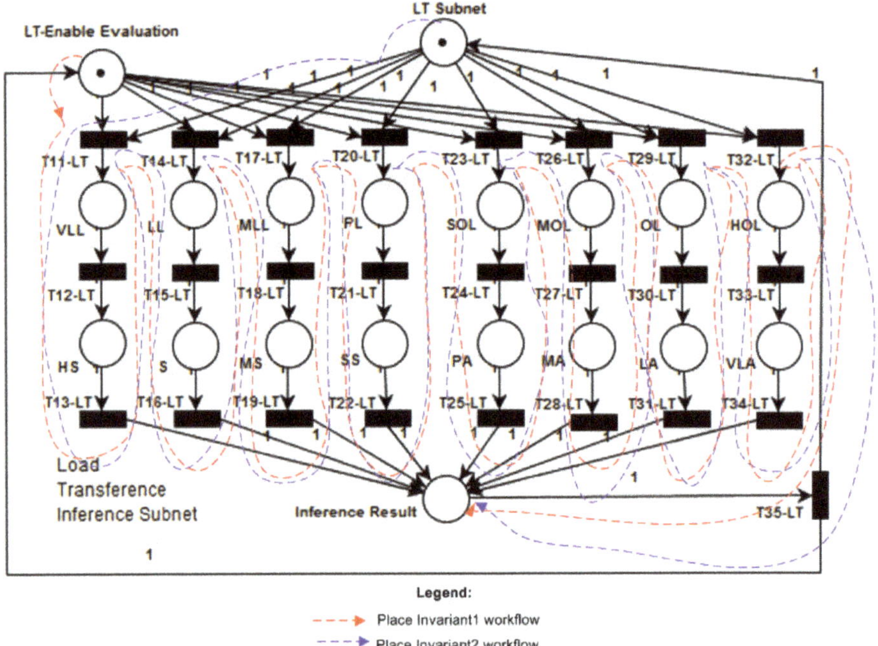

Figure 12. LT subnet- Place-Invariant workflow representation.

Thus, the automation flow verification of the integrated system (HPN) and the load transfer algorithm (LT subnet) are validated by the evaluation of the two place invariants obtained in each case, respectively. The existence of these place invariants demonstrate, first, that the workflow of the load transfer system works inviolably, reliably and efficiently, without risks of stoppages or infinite cycles, and that the workflow of the integrated system also works inviolably, reliably and efficiently without risk of stoppages or infinite cycles.

4.4. LBC HPN Performance Results Evaluation

Based on the results obtained through the dynamic simulation, the reachability and coverability graph and the place-invariant analyses, it is possible to perform the following dynamic and performance evaluation of the integrated network (HPN) and the LT subnet.

Through dynamic simulation, it was possible to observe that the integrated workflow of the main integrated network (HPN) and the load transference algorithm (LT subnet) reached all its states and transitions, without identification of conflicts, bottlenecks, siphons, and temporary shutdowns or deadlock. This guarantees an efficient and faultless automatic flow of the algorithm proposed in all its internal stages, and especially in relation to its integrated automation flow with the upper-hierarchical levels—in this case, the MGCC system and the legacy LV grid.

The obtained results show that the attainability and coverage graph was generated, which ensures that the HPN and the LT subnet growth are limited, respectively, without the incidence of infinite cycles of routines (which may be caused by trapped siphons and deadlock), therefore making both networks limited. On the other hand, the results also ensure that all states and transitions are reached, in the integrated workflow and in each upper-hierarchical level, thus verifying the hierarchy propagation in both networks.

As a result of the place-invariant analysis, it is first verified that the internal workflow of the LT subnet is guaranteed inviolably without stops, conflicts, or deadlock, due to the constant marking consumption of the places set which constitute it, as indicated in Equation (15) and (17), therefore ensuring, in each case, the only admissible flow for the load transfer process at the grid feeders.

It is also verified that the integrated workflow for the HPN was guaranteed in an inviolable way without stops, conflicts, or deadlock, with the incidence of constant consumption of mark in the places that constitute it, as indicated in Equations (16) and (18). Through this the invariant workflow between the proposed balancing algorithm and the integrated system was checked, as well as the hierarchy propagation in each layer of the network and of integral flow, thereby ensuring the only admissible flow for the integrated automation process between the LBC system and the upper-hierarchical supervision and control levels.

Therefore, based on the above-mentioned analyses, a load-balancing algorithm in the LV grid was obtained, with reliable, efficient, and secure flow, without conflicts, stops, or deadlock, acting in an efficient and secure manner with the internal steps (algorithms and subroutines) and with the upper control and supervision systems of the MGCC and the legacy BT network.

4.5. LBC Simulation Results and Test Performance Evaluation

To validate the performing of the obtained system design, the LBC system was submitted to a simulation study with real data (referring to load consumption in September of 2015), in a LV circuit of Manaus city (a north Brazilian city) with load consumption data of 51 consumers, a transformer of 110 KVA, with almost 67 KW of active power.

The load distribution in each grid feeder is broken down as shown in Table 4, and it is verified that there is a phase-load imbalance, as indicated in Equation (19) that shows an initial absolute balance (IAB) level per phase of 13.33 KW.

$$\frac{IAB}{phase} = \frac{(|F_A - F_B| + |F_B - F_C| + |F_C - F_A|)}{3} = 13.33 \ (\text{KW}) \tag{19}$$

The neutral current is I_N, determined by the currents in each phase, according to Equation (20).

$$I_N = I_{F_A} + I_{F_B} + I_{F_C} = 38.28(A) \tag{20}$$

As a next step, we will show the results obtained from the application of the LBC algorithm applied in the load-balancing process of the circuit shown in Table 4, identifying the load amount to be subtracted in the imbalanced phases and added in the balanced phases. The concentration levels and the future states of the load consumption states are then considered for the choice of single-phase consumers for switching process in the grid feeders, according to the workflow of the LBC algorithm validated in Section 4.3.

Thus, the results of each step of the LBC algorithm are shown, considering for this purpose its application in the LV grid feeder under study, and the performance evaluation of the CD step and the CF step of one of the single-phase consumer units (CU) of the phase A, with 0.5 KW (360 KWh) highlighted in bold in Table 4.

Table 5 shows the load transfer for each grid feeder: 20 KW to phase A, 21 KW to phase B and 20 KW to phase C, in comparison with the original imbalanced. Subtract 12 KW from phase A, and add 4 KW in phase B and 8 KW in phase C, respectively.

Table 4. Load Consumption Data in a LV grid.

CU-P_A	KW	CU-P_B	KW	CU-P_C	KW
1	2.0	21	0.6	21	0.5
2	2.3	22	0.1	37	0.1
3	1.6	23	0.6	38	1.3
4	1.2	24	1.0	24	0.8
5	1.0	5	0.6	39	0.2
6	1.8	25	0.1	40	0.6
7	1.8	26	0.1	41	0.1
8	1.5	27	1.5	42	0.1
9	0.7	9	0.5	9	0.2
10	2.5	28	1.7	43	1.8
11	2.0	29	1.0	44	0.1
12	0.2	30	1.2	45	0.6
13	1.8	31	1.5	46	0.1
14	2.5	32	0.1	47	1.6
15	2.4	33	0.1	48	0.5
16	2.7	34	0.1	49	1.4
17	1.0	17	1.0	17	0.5
18	**0.5**	35	1.7	50	0.2
19	1.5	36	2.5	51	1.0
20	1.,0	20	1.0	20	0.3
P_A	32	P_B	17	P_C	12

Table 5. LT-step results.

Scenary	I_N (A)	Phase A (KW)	Phase B (KW)	Phase C (KW)	LAU (KW)
Imbalanced	38.28	32	17	12	13.3
LBC	0	20	21	20	0.6
Load Transfer		12	4	8	

Thus, the procedure for the efficient load transfer between phases implied in verifying the diagnosis and prediction of the future states of load consumption, in the single-phase consuming units of the phases where the loads were subtracted, in this case phase A. Table 5 shows the load transfer for each grid feeder: 20 KW to phase A, 21 KW to phase B and 20 KW to phase C, in comparison with the original imbalance. Subtract 12 KW from phase A and add 4 KW in phase B and 8 KW in phase C, respectively, eliminating the neutral current and significantly attenuating the average imbalanced load, around 0.6 KW.

The results of the CD step are shown in Table 6, and indicate the load limits allowed in each phase for three discrete levels of consumption (low, medium, and high), depending on the energy variation (EV), the temperature variation (TV), and energy price variation (EP) as addressed in Section 3.3. On the other hand, the CF step results indicate the monthly consumption forecast with twelve steps forward, i.e., the future value of load (FL) for three states of consumption (low, medium, and high). Based on these two results, a future consumption matrix for 12 months of 2015 was implemented and is shown in Table 7, where the first column indicates the discrete load consumption states projected for each month. The asterisk values (of the consumption states) are the values, where the forecast was not correct.

In other columns are the EV, TV, EP variation (PV) and load variation (LV).

As indicated in the last column, the diagnosis for switching selection can be "To switch" (S) in case of load variation indicating a value greater than 0.3. Otherwise, "Do not switch" (NS).

Table 6. Load level limits in single-phase.

Load Level Consumption	Load Variation (%)
Low	>0.2
Medium	>0.3
High	>0.4

Table 7. Future Consumption Matrix of single-phase CU.

Month	CF Step	EV	TV	PV	LV	Diag
J	Low	0	0, 1	0.2	<0.3	NS
F	Medium	0	0.1	0.1	<0.3	NS
M	Medium	0.3	0.1	0.2	<0.3	NS
A	Medium	0.3	0.2	0.1	<0.3	NS
M	Medium	0.5	0.2	0.1	<0.3	NS
J	Medium	0.1	0.3	0.1	<0.3	NS
J	Medium	0.2	0.3	0.2	<0.3	NS
A	Medium	0.35	0.4	0.3	>0.3	S
S	High	0.36	0.4	0.3	>0.3	S
O	High	0.37	0.4	0.3	>0.3	S
N	Medium	0.2	0.3	0.3	<0.3	NS
D	Medium	0.3	0.2	0.3	<0.3	NS

Table 7 shows the results applied in the single-phase CU "18" of phase A with 0.5 KW, with 360 KWh of energy consumption for the month of September. From a history of consumption of 48 months, the discrete consumption states of low consumption (100 KWh), medium consumption (165 KWh), and high consumption (240 KWh) are distributed, obtaining the future consumption projections for each month of 2015 according to the second column of Table 7, through the algorithm indicated in Equation (12).

It shows the future consumption matrix for this consumer unit, specifying in the month of September (study analysis period) to switch (S) because the load variation in the phase is greater than 0.3, due to "High" value of FL in this month, applying the same procedure to the other single-phase CU of this phase, as shown in Table 8.

Table 8. Diagnosis Matrix for Load Transfer of Phase A.

CU	Diag	Load (KW)	CU	Diag	Load (KW)
1	S	2.0	11	S	2.0
2	S	2.3	12	S	0.2
3	NS	1.6	13	S	1.8
4	S	1.2	14	S	2.5
6	NS	1.8	15	NS	2.4
7	NS	1.8	16	S	2.7
8	S	1.5	18	S	0.5
10	S	2.5	19	S	1.5

In this case, the single-phase CU, 2, 10, 14, 16, 18, and 19 were subtracted from phase A, totaling 12 KW. In phase B, the CU 10 and 19 were added, totaling 4 KW, and in phase C, the CU 2, 14, 16 and 18 were added, totaling 8 KW. This results in a final load-balance state as shown in Table 5.

To compare the results, we consider the legacy load-balancing method, a method based on a fuzzy balancing algorithm [38], a LBC algorithm approach (LBC1), and a LBC2 approach, considering an optimal solution to load imbalance. All performances of each applied method were developed using the MATLAB environment. These results are showed in Table 9.

Table 9. Load-Balance Performance.

Param	Imbalance	Legacy	Fuzzy	LBC1	LBC2
L_{P_A} KW	32	25	22	20	19.8
L_{P_B} KW	17	12	19	21	20
L_{P_C} KW	12	24	20	20	20
I_N A	38.3	0	0	0	0
IAB KW	13.3	–	–	–	–
LAU KW	–	8.7	2.0	0.6	0.1

Figure 13a shows the load transfer in each phase, according to each of the methods applied, the load distribution represented in green, obtained by the proposed system (LBC1).

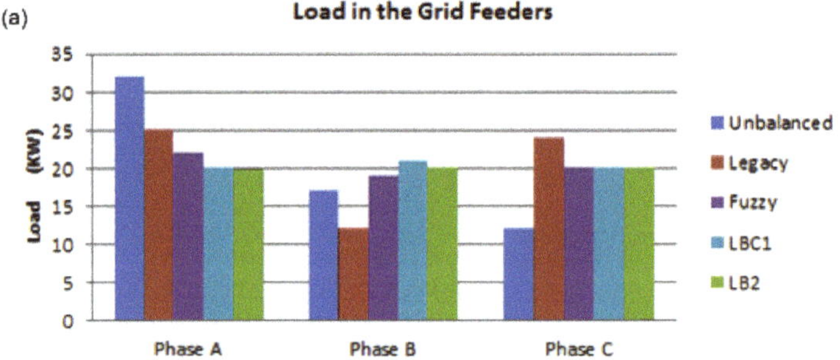

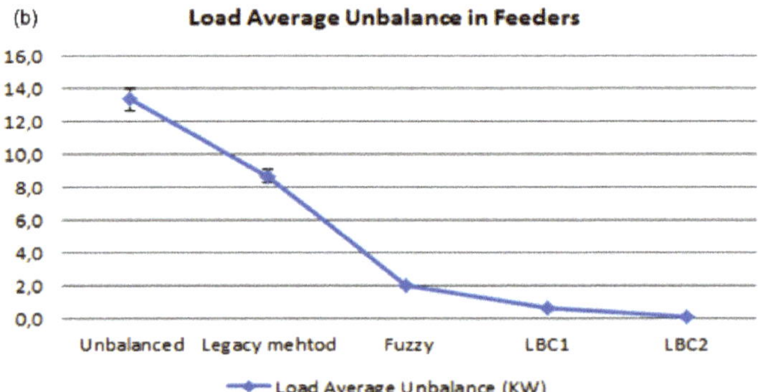

Figure 13. LBC system validation: (**a**) Load in the Grid Feeders; (**b**) Load absolute imbalance.

In addition, the LBC system (LBC1) reached the lowest mean load-imbalance value, around 0.6, compared to the legacy system method results of around 8.7 KW , and a fuzzy control algorithm, with around 2 KW, therefore proving its efficient validation which is showed also in Figure 13b. Finally, through a second system application (LBC2) a lowest load average imbalance (LAU) value around 0.1 KW was obtained, indicating a load division of 19.8 KW in phase A and 20 KW in phases B and C respectively. However, this solution is applicable when switching from the consumer unit "4" with 1.2 KW to another LV circuit.

Based on results obtained, the efficiency of the proposed method is demonstrated, in relation to the intelligent identification of load transfer in each phase of the LV circuits, and the choice of the CU for the switching process, according to their level of concentration and future states of the load consumption, thus ensuring an effective and reliable balancing process, as well as the load balancing in feeders.

4.6. Future Practical Implementation in the Control System of UMGs

The validated model becomes an alternative control proposal for load balancing in the legacy LV grid where the implementation of microgrids and distributed sources of power generation is not currently implemented. Thus, it can act as an alternative control resource in synchronization with the current injection of microgrids, the integrated coordination of control and the integrated coordination of multimicrogrids, as discussed in Section 1 and the general architecture of urban microgrids, as also discussed in Section 2.1 and according to the bibliographic review [21,31,32]. Where it would constitute a distributed control system connected with the CU, the LV transformer, the MGCC system, and the supervision center of the whole electrical system, in order to guarantee the efficient acquisition of consumption data, the load-balancing application and the switching selection according to the load consumption matrix of each single-phase consumer unit.

By contrast, the implementation of the LBC system as a combined system of integrated algorithms will mainly obey the workflows validated in this work, through specific semantics of structural language translation of embedded systems still in development by the authors. Each step of the combined algorithm will have operational modularity synchronized with all stages of the system. Its experimental validation will be carried out first in circuits of the LV legacy grid, and later in urban microgrids, in order to validate its effectiveness as an alternative control for the balancing of LV grids.

5. Conclusions

In this paper, a new system design of a distributed control system for load-balance procedure in the LV grid has been presented. This is composed of combined algorithms, called LBC system, contributing to the load amount identification to transfer between feeders, and, with the single-phase consumer unit selection, to the switch operation of load-balance procedure. In this case, a hierarchical PN approach was used first to represent and to validate the workflow of each inner algorithm of the control system. In this case, the inner algorithm of load transfer identification (LT subnet) was developed to highlight the fuzzy inference employed in the intelligent identification of the amount of load to be withdrawn or added in LV grid feeders. In addition, we represent and also validate the integrated workflow of the proposed system with the upper-hierarchical levels, as the MGCC system, and the legacy LV grid. The PBS method was used to represent the hierarchical-level connection of network. This was developed using macro-places formed by an input and output, as well as the initial location of the lower subnet.

Both networks were tested through dynamic simulation, the application of the reachability and coverability graph, and the place-invariant analysis. Verifying reliable and reliable dynamic performance in both, free of conflicts, stops and deadlock, the attainability of all its states and transitions was also verified, identifying that both are limited and safe networks. Finally, two inviolable workflows were identified in both networks, which guarantee the efficient execution of the load transfer algorithm and its evaluation of each fuzzy inference rule used to identify load transfer, respectively, as well as the integrated workflow between the LBC system with upper-hierarchical control and supervision levels in the MGCC and the LV grid. This provided an efficient and reliable load-balancing algorithm that ensures a single and admissible load-balancing (automation cycle) solution to the integrated control workflow, as well as a unique and admissible inference rule to the load transfer.

The combined algorithm of the LBC system was also tested by dynamic simulation above the historical load of a LV circuit with a transformer of 110 KVA, which presented load imbalance between its phases.

The results showed the identification of the load transfer amount in each phase, as well as the limits of variation of load in relation to the discrete states of consumption in each phase, the future consumption matrix that indicates the switching diagnosis of each single-phase consumer unit in relation to their limits of load variation, and the future load consumption states. The consumer unit selection was based on the diagnosis of this matrix for the month of September 2015. The performance of the LBC system (LBC1) was compared along with the legacy load-balancing method, a fuzzy controller, in relation to the load transfer in each phase, and the load average imbalance (LAU) value. The LBC system presented the lowest LAU, around 0.6 KW, compared to the other applied methods. A second application of the LBC system (LBC2) was also tested, presenting the lowest mean load-imbalance value, around 0.1 KW, demonstrating the efficiency of the proposed system.

For future work, the authors propose the development of a coordinated control system to represent the electrical current injection from microgrids and the LBC system as a simultaneous, integrated and automated operation flowchart, in order to efficiently ensure the load management consumption and greater load stability in UMG. This new model will be developed using timed transitions with fixed intervals of operation, to emulate the workflow temporal integration along the integrated UMG and each lower hierarchical layer.

Author Contributions: Conceptualization, J.R.S. (Jose R. Sicchar); Data curation, J.R.S. (Jose R. Sicchar); Formal analysis, J.R.S. (Jose R. Sicchar), J.R.S. (Jose R. Silva) and C.T.D.C.J.; Investigation, J.R.S. (Jose R. Sicchar); Methodology, J.R.S. (Jose R. Sicchar), J.R.S. (Jose R. Silva), R.C.O. and W.D.O.; Supervision, C.T.D.C.J. and J.R.S. (Jose R. Silva); Validation, J.R.S. (Jose R. Sicchar); Data Analyzed, J.R.S. (Jose R. Sicchar), J.R.S. (Jose R. Silva) and C.T.D.C.J.; Visualization, J.R.S. (Jose R. Silva), R.C.O. and W.D.O.; Writing—original draft, J.R.S. (Jose R. Sicchar), J.R.S. (Jose R. Silva), C.T.D.C.J., R.C.O. and W.D.O.

Acknowledgments: The authors thank to UEA, UFPA, USP and FAPEAM, for allowing scientific achievement of this proposal.

Conflicts of Interest: The authors declare no conflict of interest.

References

1. Bracco, S.; Brignone, M.; Delfino, F.; Procopio, R. An energy management system for the savona campus smart polygeneration microgrid. *IEEE Syst. J.* **2017**, *11*, 1799–1809. [CrossRef]

2. Jones, K.; Bartell, S.; Nugent, D.; Hart, J.; Shrestha, A. The urban microgrid: Smart legal and regulatory policies to support electric grid resiliency and climate mitigation. *Fordham Urb LJ* **2013**, *41*, 1695–1670.

3. Ma, T.; Wu, J.; Niu, X. Reliability assessment indices and method for urban microgrid. *CIRED-Open Access Proc. J.* **2017**, *2017*, 837–840. [CrossRef]

4. Siirto, O.; Hyvärinen, M.; Loukkalahti, M.; Hämäläinen, A.; Lehtonen, M. Improving reliability in an urban network. *Electr. Power Syst. Res.* **2015**, *120*, 47–55. [CrossRef]

5. Chen, Y.; Hu, M. Balancing collective and individual interests in transactive energy management of interconnected micro-grid clusters. *Energy* **2016**, *109*, 1075–1085. [CrossRef]

6. Kantamneni, A.; Brown, L.; Parker, G.; Weaver, W. Survey of multi-agent systems for microgrid control. *Eng. Appl. Artif. Intell.* **2015**, *45*, 192–203. [CrossRef]

7. Khan, A.; Naeem, M.; Iqbal, M.; Qaisar, S.; Anpalagan, A. A compendium of optimization objectives, constraints, tools and algorithms for energy management in microgrids. *Renew. Sustain. Energy Rev.* **2016**, *58*, 1664–1683. [CrossRef]

8. Kaur, A.; Kaushal, J.; Basak, P. A review on microgrid central controller. *Renew. Sustain. Energy Rev.* **2016**, *55*, 338–345. [CrossRef]

9. Mariam, L.; Basu, M.; Conlon, M. Microgrid: Architecture, policy and future trends. *Renew. Sustain. Energy Rev.* **2016**, *64*, 477–489. [CrossRef]

10. Wu, P.; Huang, W.; Tai, N.; Liang, S. A novel design of architecture and control for multiple microgrids with hybrid AC/DC connection. *Appl. Energy* **2018**, *210*, 1002–1016. [CrossRef]

11. Almada, J.; Leão, R.; Sampaio, R.; Barroso, G. A centralized and heuristic approach for energy management of an AC microgrid. *Renew. Sustain. Energy Rev.* **2016**, *60*, 1396–1404. [CrossRef]

12. Khan, M.R.; Jidin, R.; Pasupuleti, J. Multi-agent based distributed control architecture for microgrid energy management and optimization. *Energy Convers. Manag.* **2016**, *112*, 288–307. [CrossRef]

13. Mahmoud, M.; Hussain, S.; Abido, M. Modeling and control of microgrid: An overview. *J. Frankl. Inst.* **2014**, *351*, 2822–2859. [CrossRef]

14. Martin-Martínez, F.; Sánchez-Miralles, A.; Rivier, M. A literature review of Microgrids: A functional layer based classification. *Renew. Sustain. Energy Rev.* **2016**, *62*, 1133–1153. [CrossRef]

15. Gomes, R.; Silva, J.; da Costa, C.; da Silva, P. Automation Meta-System Applied to Smart Grid Convergence of Low Voltage Distribution Legacy Grids. In Proceedings of the 2017 5th IEEE International Conference on Smart Energy Grid Engineering, Oshawa, ON, Canada, 14–17 August 2017; pp. 400–413.

16. Safitri, N.; Shahnia, F.; Masoum, M. Coordination of single-phase rooftop PVs in unbalanced three-phase residential feeders for voltage profiles improvement. *Aust. J. Electr. Electron. Eng.* **2016**, *13*, 77–90. [CrossRef]

17. Sharma, I.; Canizares, C.; Bhattacharya, K. Smart charging of PEVs penetrating into residential distribution systems. *IEEE Trans. Smart Grid* **2014**, *5*, 1196–1209. [CrossRef]

18. Patrao, I.; Figueres, E.; Garcerá, G.; González-Medina, R. Microgrid architectures for low voltage distributed generation. *Renew. Sustain. Energy Rev.* **2015**, *43*, 415–424. [CrossRef]

19. Quesada, J.; Sebastián, R.; Castro, M.; Sainz, J. Control of inverters in a low voltage microgrid with distributed battery energy storage. Part I: Primary control. *Electr. Power Syst. Res.* **2014**, *114*, 126–135. [CrossRef]

20. Strasser, T.; Andrén, F.; Kathan, J.; Cecati, C.; Buccella, C.; Siano, P.; Leitao, P.; Zhabelova, G.; Vyatkin, V.; Vrba, P.; et al. A review of architectures and concepts for intelligence in future electric energy systems. *IEEE Trans. Ind. Electron.* **2015**, *62*, 2424–2438. [CrossRef]

21. Lyu, Z.; Wei, Q.; Zhang, Y.; Zhao, J.; Manla, E. Adaptive Virtual Impedance Droop Control Based on Consensus Control of Reactive Current. *Energies* **2018**, *11*, 1801. [CrossRef]

22. Xu, Z.; Zhang, Y.; Liang, Y.; Zeng, Z.; Yang, P.; Peng, J.; He, T.; Chen, J. Multi-timescale coordinated optimization of hybrid three-phase/single-phase multimicrogrids. *Int. Trans. Electr. Energy Syst.* **2018**, *28*, e2499. [CrossRef]

23. Hosseinzadeh, M.; Salmasi, F. Fault-tolerant supervisory controller for a hybrid AC/DC micro-grid. *IEEE Trans. Smart Grid* **2018**, *9*, 2809–2823. [CrossRef]

24. Hosseinzadeh, M.; Salmasi, F. Robust optimal power management system for a hybrid AC/DC micro-grid. *IEEE Trans. Sustain. Energy* **2015**, *6*, 675–687. [CrossRef]

25. Siti, W.; Jimoh, A.; Nicolae, D. Phase load balancing in the secondary distribution network using a fuzzy logic and a combinatorial optimization based on the newton raphson. In Proceedings of the International Conference on Intelligent Data Engineering and Automated Learning, Burgos, Spain, 23–26 September 2009; pp. 92–100.

26. Siti, W.; Nicolae, D.V.; Jimoh, A.; Ukil, A. Reconfiguration and load balancing in the LV and MV distribution networks for optimal performance. *IEEE Trans. Power Deliv.* **2007**, *22*, 2534–2540. [CrossRef]

27. Shahnia, F.; Ghosh, A.; Ledwich, G.; Zare, F. Voltage unbalance improvement in low voltage residential feeders with rooftop PVs using custom power devices. *Int. J. Electr. Power Energy Syst.* **2014**, *55*, 362–377. [CrossRef]

28. Meng, L.; Sanseverino, E.; Luna, A.; Dragicevic, T.; Vasquez, J.; Guerrero, J. Microgrid supervisory controllers and energy management systems: A literature review. *Renew. Sustain. Energy Rev.* **2016**, *60*, 1263–1273. [CrossRef]

29. Evangelopoulos, V.; Georgilakis, P.; Hatziargyriou, N.D. Optimal operation of smart distribution networks: A review of models, methods and future research. *Electr. Power Syst. Res.* **2016**, *140*, 95–106. [CrossRef]

30. Xiao, F.; Ai, Q. New modeling framework considering economy, uncertainty, and security for estimating the dynamic interchange capability of multi-microgrids. *Electr. Power Syst. Res.* **2017**, *152*, 237–248. [CrossRef]

31. Dong, H.; Yuan, S.; Han, Z.; Cai, Z.; Jia, G.; Ge, Y. A Comprehensive Strategy for Accurate Reactive Power Distribution, Stability Improvement, and Harmonic Suppression of Multi-Inverter-Based Micro-Grid. *Energies* **2018**, *11*, 745. [CrossRef]

32. El-Hendawi, M.; Gabbar, A.H.; El-Saady, G.; Ibrahim, E. Control and EMS of a Grid-Connected Microgrid with Economical Analysis. *Energies* **2018**, *11*, 129. [CrossRef]

33. Tungadio, D.H.; Bansal, R.C.; Siti, M.W. Optimal control of active power of two micro-grids interconnected with two AC tie-lines. *Electr. Power Compon. Syst.* **2017**, *45*, 2188–2199. [CrossRef]

34. Hosseinimehr, T.; Ghosh, A.; Shahnia, F. Cooperative control of battery energy storage systems in microgrids. *Int. J. Electr. Power Energy Syst.* **2017**, *87*, 109–120. [CrossRef]
35. Goyal, M.; Ghosh, A. Microgrids interconnection to support mutually during any contingency. *Sustain. Energy Grids Netw.* **2016**, *6*, 100–108. [CrossRef]
36. Reddy, K.; Kumar, M.; Mallick, T.; Sharon, H.; Lokeswaran, S. A review of Integration, Control, Communication and Metering (ICCM) of renewable energy based smart grid. *Renew. Sustain. Energy Rev.* **2014**, *38*, 180–192. [CrossRef]
37. Nunna, H.K.; Saklani, A.M.; Sesetti, A.; Battula, S.; Doolla, S.; Srinivasan, D. Multi-agent based Demand Response management system for combined operation of smart microgrids. *Sustain. Energy Grids Netw.* **2016**, *6*, 25–34. [CrossRef]
38. Siti, W.; Jimoh, A.; Nicolae, D. Distribution network phase load balancing as a combinatorial optimization problem using fuzzy logic and Newton–Raphson. *Electr. Power Syst. Res.* **2011**, *81*, 1079–1087. [CrossRef]
39. Li, Z.; Zhou, M.C.; Wu, N.; Huang, Y.S. Special Issue on Modeling, Simulation, Operation and Control of Discrete Event Systems. *Appl. Sci.* **2018**, *8*, 202. [CrossRef]
40. Ázquez, C.R.; Ramírez-Treviño, A.; Silva, M. Controllability of timed continuous Petri nets with uncontrollable transitions. *Int. J. Control* **2014**, *87*, 537–552. [CrossRef]
41. Popova-Zeugmann, L. Time petri nets. In *Time and Petri Nets*; Springer: Berlin, Germany, 2013; pp. 31–137.
42. Murata, T. Petri nets: Properties, analysis and applications. *Proc. IEEE* **1989**, *77*, 541–580. [CrossRef]
43. Wang, L.; Chen, Q.; Gao, Z.; Niu, L.; Zhao, Y.; Ma, Z.; Wu, D. Knowledge representation and general Petri net models for power grid fault diagnosis. *IET Gener. Transm. Distrib.* **2015**, *9*, 866–873. [CrossRef]
44. Kyriakarakos, G.; Dounisios, A.; Konstantinos, K.A.; Papadakis, G. A fuzzy cognitive maps–petri nets energy management system for autonomous polygeneration microgrids. *Appl. Soft Comput.* **2012**, *12*, 3785–3797. [CrossRef]
45. Zhang, X.; Zheng, H.; Liu, Y. A petri-net based context-aware workflow system for smart home. In Proceedings of the 2012 IEEE 26th International Parallel and Distributed Processing Symposium Workshops and & PhD Forum (IPDPSW), Shanghai, China, 21–25 May 2012; pp. 2336–2342.
46. Heiner, M.; Deussen, P.; Spranger, J. A case study in design and verification of manufacturing system control software with hierarchical Petri nets. *Int. J. Adv. Manuf. Technol.* **1999**, *15*, 139–152. [CrossRef]
47. Mladjao, M.; Abdel-Moumen, D. New Robust Energy Management Model for Interconnected Power Networks Using Petri Nets Approach. *Smart Grid Renew. Energy* **2016**, *7*, 46. [CrossRef]
48. Negrete-Pincetic, M.; Nayyar, A.; Poolla, K.; Salah, F.; Varaiya, P. Rate-constrained energy services in electricity. *IEEE Trans. Smart Grid* **2016**, *9*, 2894–2907. [CrossRef]
49. Kanchev, H.; Colas, F.; Lazarov, V.; Francois, B. Emission reduction and economical optimization of an urban microgrid operation including dispatched PV-based active generators. *IEEE Trans. Sustain. Energy* **2014**, *5*, 1397–1405. [CrossRef]
50. Huang, Y.; Mao, S.; Nelms, R. Smooth scheduling for electricity distribution in the smart grid. *IEEE Syst. J.* **2015**, *9*, 966–977. [CrossRef]
51. Sicchar, J.; da Costa, C.; Silva, J. Controlador local de balanceamento de carga para µGRIDS. In Proceedings of the XIII Simpósio Brasileiro de Automação Inteligente (SBAI), Porto Alegre, Brazil, 1–4 October 2017; pp. 1–6.
52. Gomes, L.; Barros, J. On structuring mechanisms for Petri nets based system design. In Proceedings of the 2003 IEEE Conference on Emerging Technologies and Factory Automation, Lisbon, Portugal, 16–19 September 2003; Volume 2, pp. 431–438.
53. Sicchar, J.; da Costa, C.; Silva, J.; Freitas, R.D. Gerenciamento de Consumo de Energia em Residências com Frame GCR. *Proc. XII SBAI* **2015**, *1*, 1–6.

Article

A High-Efficiency Bidirectional Active Balance for Electric Vehicle Battery Packs Based on Model Predictive Control

Shixin Song [1], Feng Xiao [2,*], Silun Peng [2], Chuanxue Song [2] and Yulong Shao [3]

[1] School of Mechanical Science and Engineering, Jilin University, Changchun 130022, China; songshx202@126.com
[2] State Key Laboratory of Automotive Simulation and Control, Jilin University, Changchun 130022, China; pengsilun@jlu.edu.cn (S.P.); songchx@126.com (C.S.)
[3] Zhengzhou Yutong Bus Co., Ltd., Zhengzhou 450016, China; shaoyl13@mails.jlu.edu.cn
* Correspondence: xiaofengjl@jlu.edu.cn; Tel.: +86-137-5647-3659

Received: 28 October 2018; Accepted: 15 November 2018; Published: 20 November 2018

Abstract: This study designs an active equilibrium control strategy based on model predictive control (MPC) for series battery packs. To shorten equalisation time and reduce unnecessary energy consumption, bidirectional active equalisation is modelled and analysed, and the model predictive control algorithm is then applied to the established state space equation. The optimisation problem that minimises the equilibrium time is transformed to a linear programming form in each cycle. By solving the linear programming problem online, a group of control optimal solutions is found and the series equalisation problem is decoupled. The equalisation time is shortened by dynamically adjusting the equalisation current. Simulation results show that the MPC algorithm can avoid unnecessary energy transfer and shorten equalisation time. The bench experimental result shows that the equilibrium time is reduced by 31%, verifying the rationality of the MPC strategy.

Keywords: electric vehicle; battery packs; active balance; model predictive control

---

1. Introduction

With the development of the automobile industry, energy conservation and environmental protection technologies have become widespread concerns. Traditional internal combustion engine cars can no longer adapt to the development of the future automobile industry because of their low energy efficiency and exhaust emission pollution, while batteries and electrochemical capacitors have low-cost and environmentally friendly performance features [1,2]. In this situation, new energy vehicles have become the focus of global automotive companies and research institutions [3]. As the main form of new energy vehicles, electric vehicles have received widespread attention. As a means of transport, electric vehicles use electric power instead of fossil fuel to work and generate truly zero emissions and pollution [4,5]. At the same time, lithium cells have become major power cells because of their high energy density and current charge and discharge characteristics [6–8]. In addition, the self-discharge rate of these batteries is small, and they have a long service life. To meet the power demand, the voltage of battery pack is approximately several hundred volts [9–12]. For a long driving range, the total capacity is approximately 100 Kwh. A single battery cell would be insufficient, so hundreds/thousands of cells must be combined as a battery pack for use. The problem is that battery cells can vary, despite the use of advanced production processes. That is, each battery is not guaranteed to be exactly the same as another. Filtering the cells before use is possible to ensure that all the cells of the entire battery pack are identical. However, distinct working conditions of the battery cells, such as temperature and current, vary due to the different layouts of battery packs. Variability of cells in a battery packs is inevitable because the internal resistance of each cell in a series battery pack is different [13]. The energy flowing through each cell is distinct, and these discrepancies will increase with use, which will inevitably further increase the inconsistency of the battery cells within the battery pack. If the inconsistency of the battery cells is handled improperly, then certain cells will be fully charged before the others. In such a case, charging should be stopped to protect the strong cells from being overcharged, which will result in other cells being partially charged. Similarly, during discharge, certain cells will run out of energy before others. The process should be stopped to protect the poor cells from over-discharge, which will leave the energy of other cells unused. All in all, this variability of cells will lead to a substantial decrease in the total capacity of the battery pack. Therefore, the battery balance problem is an important task of battery management system, which protects the battery pack and extends battery life [14]. The energy equalisation technology in a Battery Management System (BMS) ensures that the energy in each cell remains identical by discharging the high-energy cells or charging the low-energy cells, reducing voltage or battery difference between the battery cells and keeping it within reasonable range. In this way, the battery pack can be fully charged and discharged, utilising the capacity of the battery pack to the maximum extent without adversely affecting battery life.

Through energy transfer, overcharging or over-discharging of individual cells is avoided, overall performance of a battery pack is improved, and service life of the battery pack is fully enhanced, thereby boosting the performance and cruising range of electric vehicles.

This study designs an active equilibrium control strategy based on model prediction for series battery packs. Section 2 introduces the current equilibrium algorithm and presents is advantages and disadvantages. Section 3 designs a set of equalisation boards using the principle of distributed equalisation. Section 4 introduces the MPC algorithm and applies it on the DC–DC control method. Section 5 verifies the proposed method, and Section 6 summarises the study.

2. Comparison of Series Balance Schemes

Depending on how to deal with the superfluous energy of battery cells, the balance scheme is divided into passive and active [15,16].

2.1. Passive Balance

This type of balance is called passive because the superfluous energy of battery cells is consumed through a bypass resistance. That is, superfluous energy is consumed by generating heat through a shunt resistance to achieve consistency in the cells' voltage. This method is effective when a battery pack contains only one cell with superfluous power. However, if that cell has low power, then it will cause the rest of the cells to discharge, which will generate substantial heat [17]. This method will critically reduce energy utilisation efficiency. In addition, the heating phenomenon will become severe, which will further cause an imbalance in temperature distribution. Thus, heat management is necessary for a passive balance.

2.2. Active Balance

Figure 1 shows that the capacitor and each cell are connected in parallel through the control of switch circuit. The advantages of this scheme are the number of capacitor is few, the principle is simple and energy transfer efficiency is high. The setup also has disadvantages, such as several control switches are used, heavy noise is generated during operation and the switches have relatively short life. In addition, when the battery scale is large, no batteries can work simultaneously, and the work efficiency of balance is low. Figure 2 presents the second capacitor-type balance scheme. Each pair of adjacent battery cells can transfer energy through the capacitor between them. This scheme uses as many capacitors as the number of cells in series. By controlling a single-pole double-throw switch, parallel switching between capacitors and battery cells can be achieved. However, when the first and last battery monomers need to be balanced, the transfer energy will pass through all capacitors and battery cells, which reduces the balance efficiency [18].

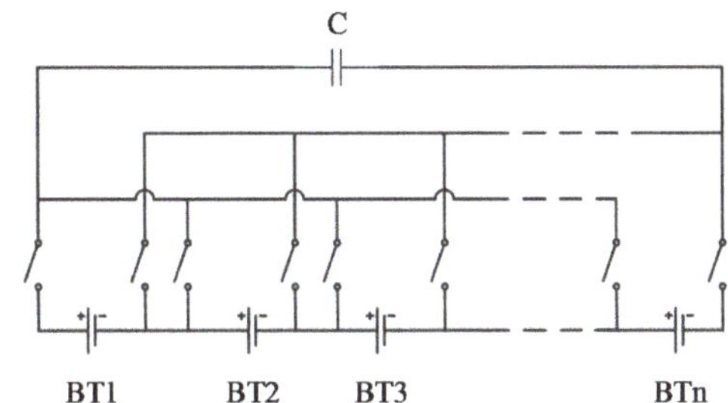

Figure 1. Principle of flying capacitor balance.

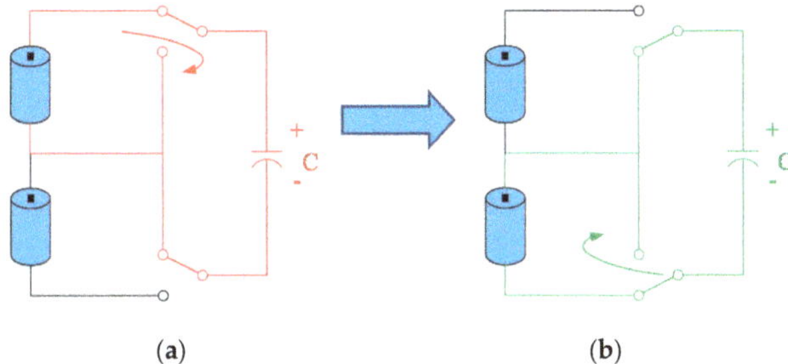

Figure 2. Principle of switch capacitor balance. (**a**) Energy transfer from cell to capacitor; (**b**) Energy transfer from capacitor to cell.

According to the number of inductors, the inductor-type active balance is classified into two types, namely, inductors in the adjacent cells (Figure 3) and an inductor in the battery stack (Figure 4). Figure 3 displays the balance principle of inductors in the adjacent cells. The advantage of this balance method is that it is easy to expand, and the conductor has a high transfer efficiency. However, the electricity transfer between non-adjacent cells will pass through all cells and inductors, which will reduce the efficiency and complicate the control strategy.

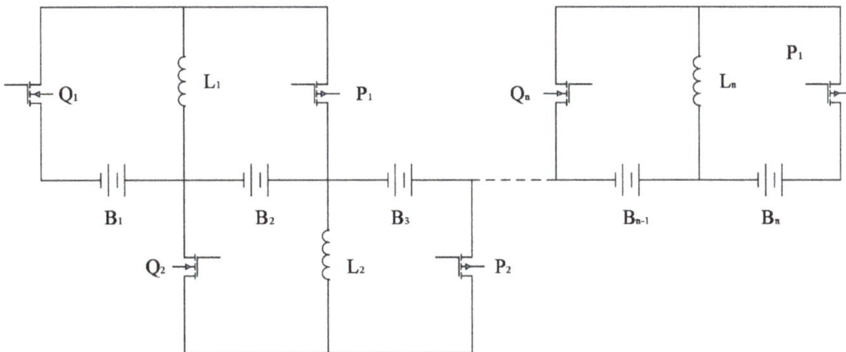

Figure 3. Balance principle of inductors in adjacent cells.

Figure 4 illustrates the balance principle of an inductor in a battery stack. The advantage of this balance method is the high-energy transfer efficiency [19]. However, only one balance circuit works at a time, and it is unsuitable for large battery stacks [20].

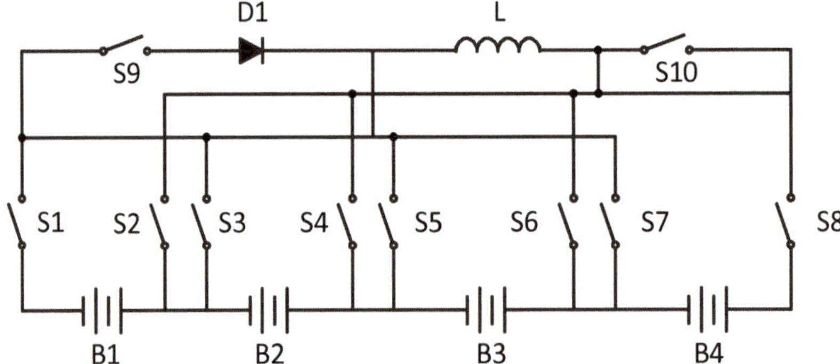

Figure 4. Balance principle of one inductor in a battery stack.

Figure 5 shows the principle of multi-winding transformer balance [21]. The advantages are its simple control principle and high efficiency, whereas the drawbacks are the high manufacturing requirement of the transformer and unsuitability for expanding the battery pack.

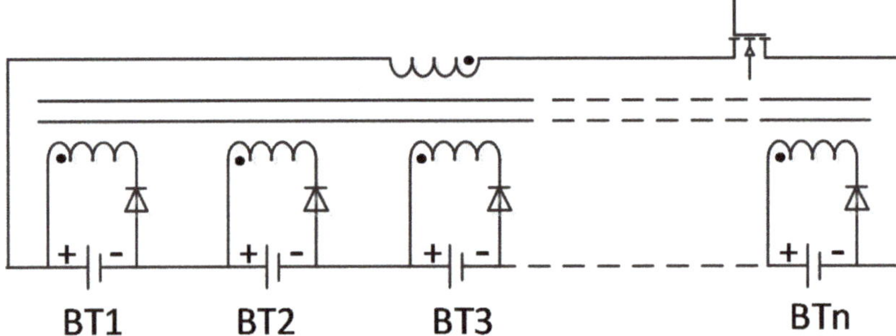

Figure 5. Principle of multi-winding transformer balance.

Figure 6 provides the principle of the multi-transformer balance. The expanding the battery pack will be easy [22]. However, the costs for the transformers are high.

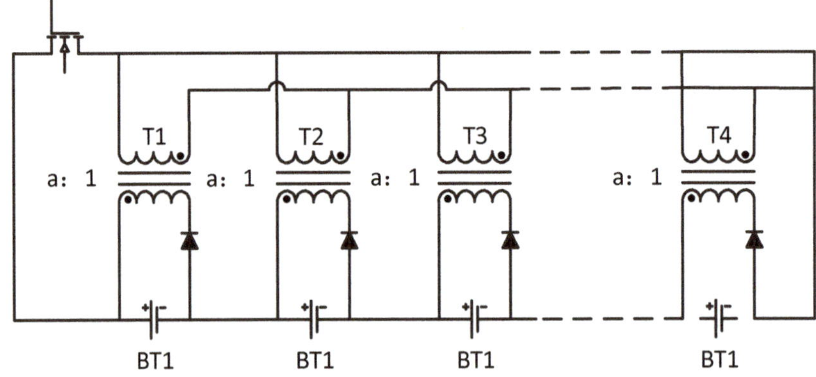

Figure 6. Principle of multi-transformer balance.

Figure 7 demonstrates the principle of the switching transformer balance. The advantage of this balance type is that the energy transfer efficiency is high, but only one balance circuit works at a time and expanding the battery pack is difficult.

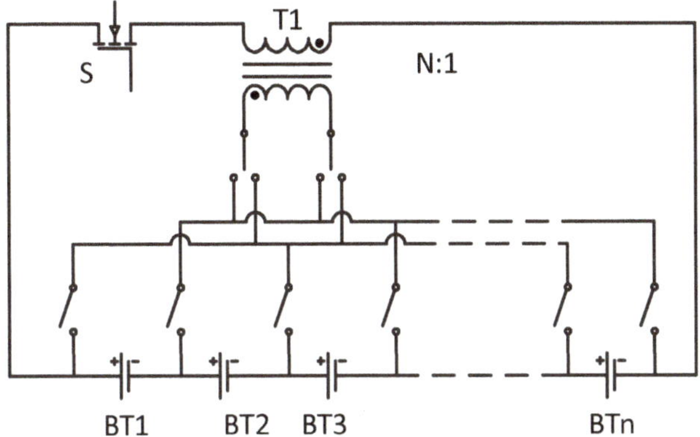

Figure 7. Principle of switching transformer balance.

Figure 8 shows the principle of the distributed DC–DC active balance. The electrical energy in this scheme can only be transferred between adjacent cells [23,24]. A large balance current can be achieved due to the high power of the DC–DC module, so this scheme is especially suitable for the internal balance of large capacity cells. However, this scheme also has an obvious disadvantage: the electric energy needs to pass through all battery cells as it is transferred from the top to bottom of the battery stack. This mechanism will reduce balance efficiency.

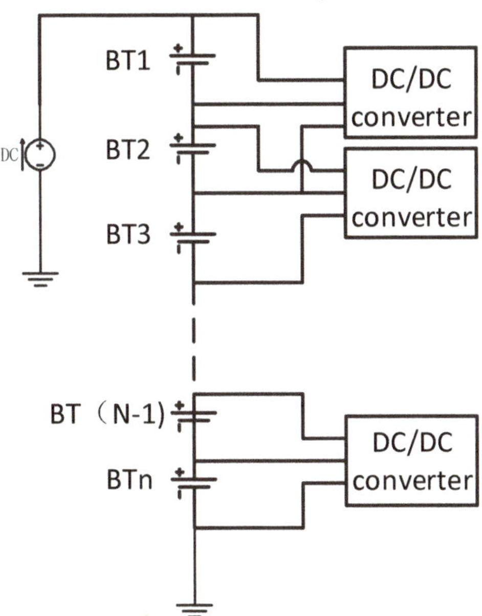

Figure 8. Principle of distributed DC–DC active balance.

Figure 9 provides the principle of the centralised DC–DC active balance [25]. The advantages are that the control strategy is flexible and changeable and the power of DC–DC is high. The drawbacks are that the cost of DC–DC is high, and it is unsuitable for large-scale battery stacks.

All in all, passive energy balance is unsuitable for large capacity cells because of its low energy transfer efficiency and long balance time.

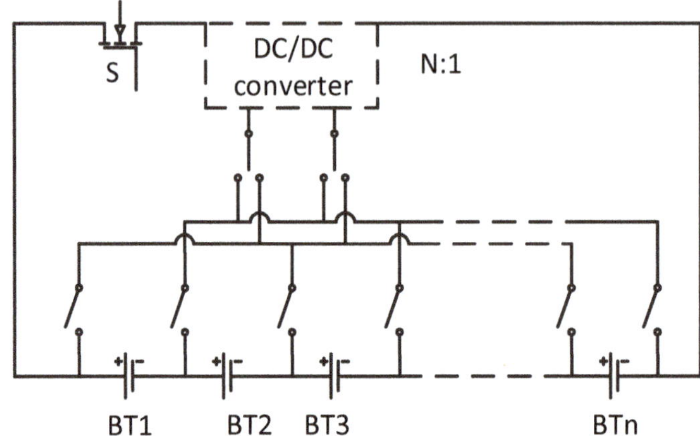

Figure 9. Principle of centralised DC–DC active balance.

In 2011, Kim [20] designed an automatic charge equalization circuit based on regulated voltage source for series connected lithium-ion batteries. In 2015, Shang [26] designed a cell-to-cell battery equalizer with zero-current switching and zero-voltage gap based on quasi-resonant LC converter and boost converter; and Wang [25] designed a novel active equalization method for lithium-ion batteries in electric vehicles. In 2017, Mccurlie [27] studied the fast model predictive control for redistributive lithium-ion battery balancing. For active energy balance, extra energy is difficult to be transferred quickly in the existing technology, and the minimum energy transfer times is not guaranteed [23]. This study designs an active equilibrium control strategy based on model prediction for series battery packs. We focus on the analysis of the energy balance between tandem cells, which rely on fly-back DC–DC to achieve high-efficiency bidirectional active equalisation. Furthermore, we develop a set of BMS, which is suitable for 132 series cells. To shorten equalisation time and reduce unnecessary energy consumption, bidirectional active equalisation is modelled and analysed, and the process is described using state-space equations. Then, the MPC algorithm is applied to the established state-space equations according to the MPC principle. A closed-loop control of the equalisation is established, and a fast MPC algorithm is adopted to equalise the SOC of the battery pack. The optimisation problem that minimised the equilibrium time is transformed into linear programming in each cycle process. By solving the linear programming problem online, a group of control optimal solutions is explored, and the series equalisation problem is decoupled. The first element of the local optimal solutions is applied to the controlled equalisation circuit, and the equalisation time is shortened by dynamically adjusting the equalisation current.

3. Implementation of Bidirectional Active Equilibrium

This section aims to achieve bidirectional active equalisation. Bidirectional active equalisation is realised using a BMS slave controller. However, we need the BMS master controller to send out control commands. For example, the main controller sends out instructions to collect data and analyses the data returned by the slave controller. The slave controller then controls bidirectional active balancing. The BMS master controller is essential to the normal work of the slave controller for bidirectional active equilibrium.

3.1. Function and Principle of BMS Controllers

This study designs an active equalisation circuit that can achieve bidirectional balancing of battery cells and be used in large-scale battery packs. The main functions are:

(1)   acquisition of voltage, current and temperature;
(2)   protection function;
(3)   SOC and SOH estimation function;
(4)   equilibrium function.

The scheme adopts an improved distributed DC–DC equalisation scheme. Each battery cell is assigned to a DC–DC equaliser, and each equaliser can operate independently. Figure 10 shows the principle underlying the charge transfer between individual cells and a large neighbouring battery stack. The operating principle of the cell discharge is as follows. Taking cell 1 as an example, the switch G1P controls the battery coil to close when cell 1 needs be discharged. Then, cell 1 starts to charge the DC–DC coil (T1), and the magnetic field energy stored in the inductor reaches the maximum value when the charging current reaches the set peak value. During this time, switch G1P is turned off, and the switch G1S is turned on. The energy in the inductor primary winding is transferred to the secondary coil on the side of the battery pack, and the battery module (monitor 1 to battery 12) is charged through switch G1S. When the charging current drops to 0, the G1S is turned off and G1P is turned on synchronously, and cell 1 starts charging the DC–DC again. This process is repeated until the voltage or power of cell 1 recovers to the set level.

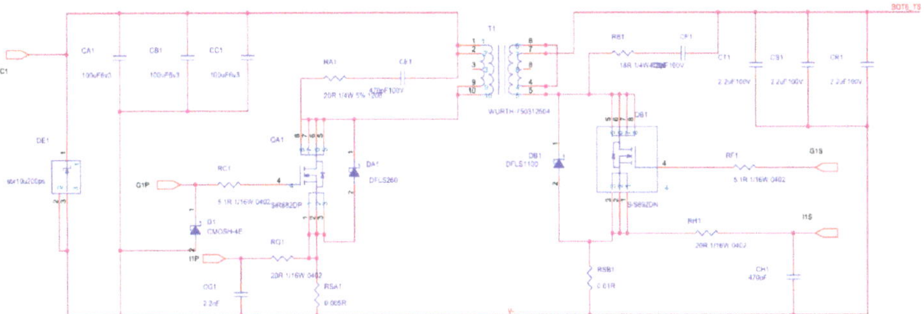

Figure 10. Bidirectional equilibrium schematic.

The principle of charging the cells is similar to the aforementioned process. Still taking cell 1 as an example, the controller controls the G1S to close when cell 1 needs to be charged. The DC–DC module (T1) firstly takes power from the battery stack (cells 1–12). The energy in the DC–DC module reaches the maximum value when the charging current reaches the maximum value. During this time, G1S is turned off and G1P is closed synchronously. Then, the energy in the DC–DC is converted into electrical energy and begins to charge cell 1. When the charging current drops to 0, G1P is closed again; G1S is turned on synchronously and obtains energy from the battery stack (cells 1–12) again. This process is repeated until the voltage or power of cell 1 recovers to the set level.

3.2. Hardware Design of Bidirectional Active Balance BMS

The main control unit obtains the voltages, current and ambient temperature of the battery pack; and protects the battery pack from being overused, estimates the battery status (SOC, SOH); and communicates with other electronic control units. In this paper, a high-efficiency, low-power 32-bit STM32F446RET is used as the BMS main control unit microprocessor due to the heavy load. Its CPU speed reaches 84 MHz, flash memory reaches 256 bytes, RAM capacity is 64 bytes. It has three USART interfaces, one SDIO interface, three IIC serial bus interfaces, four SPI serial peripheral communication interfaces and one 16-channel 12-bit 2.4MSPS A/D converter. To save the last SOC and SOH values of the battery at the time of power-off, a peripheral 256-byte EEPROM chip is used. Figure 11 presents the main chip and peripheral circuit structure.

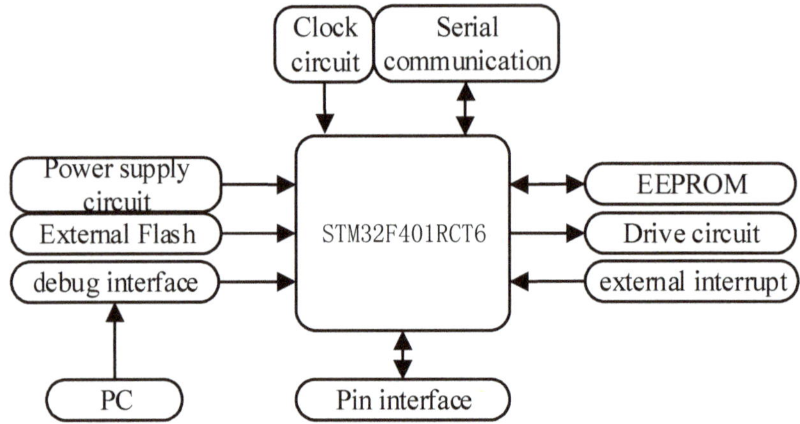

Figure 11. Block diagram of main control unit.

This study adopts Linear Technology's LTC6804 chip to collect voltage. It is one of the few monitoring chips that can simultaneously collect the voltage of 12 series cells. The measurement error is less than 1.2 mV, and the measurement time of all voltages is less than 290 μs. Therefore, we propose to use this chip to measure the voltage of each battery module with 12 cells. The current of the battery pack is not only an important parameter monitored by the battery protection system but also important information for estimating battery capacity. This study uses a Hall-type current sensor DHAB S/24, which has a dual-range to achieve high accuracy of current acquisition: range 1 is ±75 A, absolute error is ±1.5 A (25 °C). Range 2 is ±500 A with an absolute error of ±5 A (25 °C). This study uses the Bq76PL536 chip to collect the temperature of the main control board; the temperature of cells is collected by the input port of the LTC6804 chip. The balance scheme uses a modular design, in which every 12 cells as a module, and a slave controller to manage the module. One LTC6804 chip is used to read the voltage of 12 cells. Two LTC3300s are used to control the operation of 12 DC–DC to achieve equalisation of individual cells. The protection function is implemented with relays, which employs the EV150-AAD. Figure 12 shows the design of the slave controller, which can achieve a maximum of 4.2 A equalisation current.

Figure 12. Slave controller.

The LTC3300's current acquisition and DC–DC loop traces greatly affect current equalisation. If the current is inaccurate, then it will directly affect the control of the equalisation DC–DC. If the impedance of the DC–DC loop is large, then the equalisation current will be reduced. Therefore, attention should be paid to the route of these lines to maximise the function of the DC–DC module.

Figure 13 shows the design for the BMS master controller, which communicates with multiple slave controllers through daisy-chain SPI communication technology, controls slave controllers for voltage and temperature acquisition and sends equalisation commands to related slave controllers.

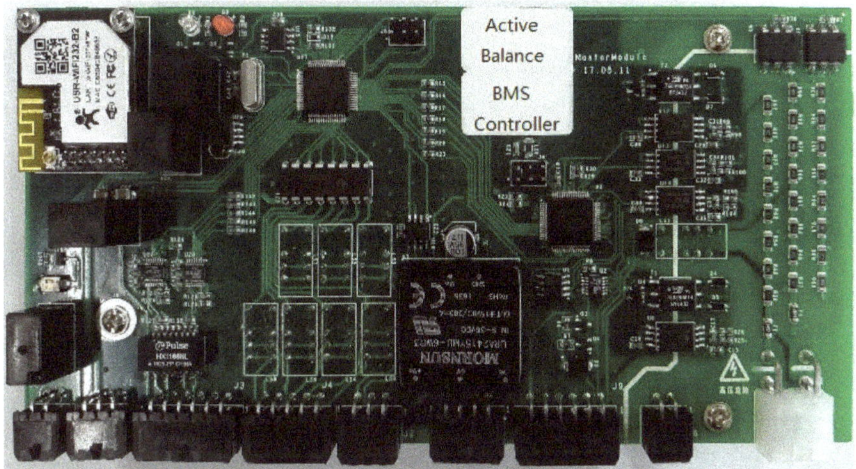

Figure 13. Master controller.

The BMS master controller obtains the cells' voltages, current and temperature from the slave controllers. Using the information, the master controller protects the cells from overuse, estimates the SOC of the battery and balances the cells if necessary.

3.3. Experiment Results and Analysis

To verify the equalisation effect, we designed an experimental control process, as shown in Figure 14. The monitoring interface communicates with the master controller using the CAN bus, and the master controller communicates with the slave controller using CAT-5 twisted-pair wires.

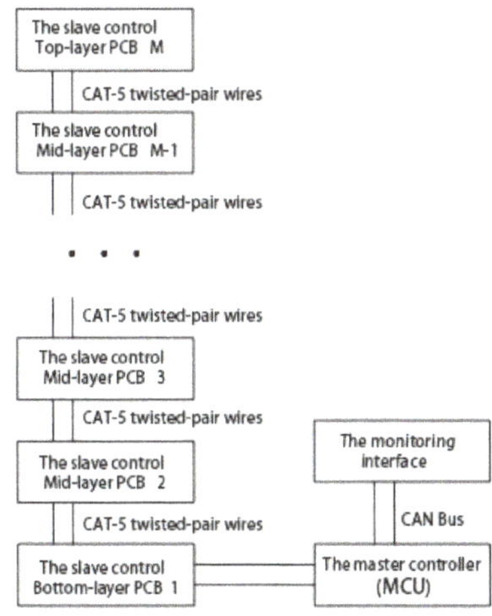

Figure 14. Experimental communication control process.

Figure 15 displays the test bench and the management of a battery pack (containing 132 lithium titanate cells). The process comprises 11 slave controllers, and each one controls 12 cells. The daisy-chain communication is adopted between the slave controllers, and the master controller manages all cells through the lowest-end slave controller.

Figure 15. Slave controllers and battery pack.

At the beginning of the experiment, the voltage of the cells in the battery pack is inconsistent. The BMS balances the 132 series cells and collects the voltage of the battery pack. Figure 16 provides the voltages of cells from 1 to 12. The diagram shows that the voltage of cell 1 is gradually pulled up during the equilibrium process. Figure 17 shows the beginning of the equilibrium process, in which the voltage difference among the 12 cells reaches 130 mV. At the end of the process, the maximum voltage difference is controlled within 20 mV, as shown in Figure 18.

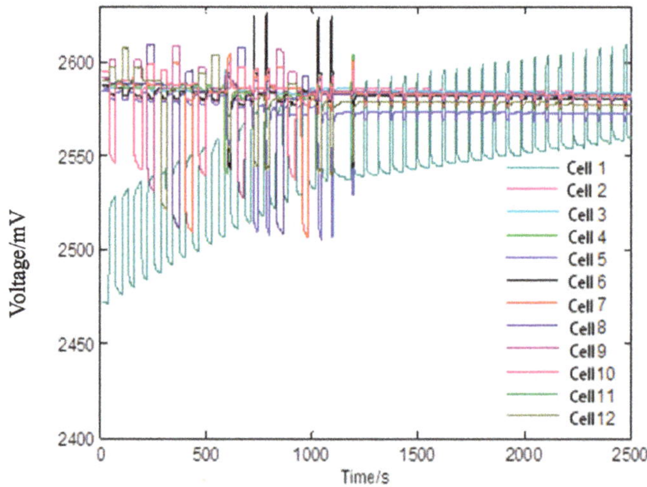

Figure 16. Voltage equalisation curves of the first battery module.

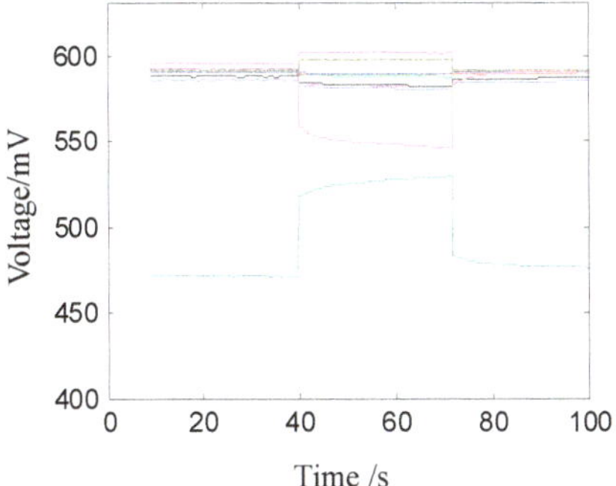

Figure 17. Equilibrium at initial stage.

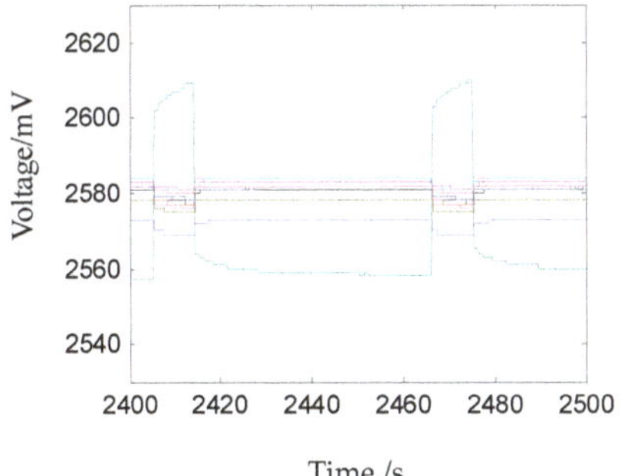

Figure 18. Equilibrium at final stage.

4. Model Prediction Control Algorithm

4.1. Basic Principle of Model Prediction Control

Model predictive control (MPC) was developed in the 1970s, mainly due to the requirements of social industrial control and progress in technology, although it is not for control theory. The application for MPC is extensive since its early development. People have turned their attention to MPC methods because they are incapable of building accurate models in complex systems. The MPC algorithm adopts the rolling optimisation concept. In this manner, the requirements of the model accuracy are low, and the amount of real-time calculations is considerably low. Thus, its workload of calculation is less than the traditional optimisation algorithm, and its control effect is moderately well.

Finding a global optimal control variable for a system is difficult for MPC. Fortunately, industrial control does not need such a global optimal control variable as long as a control variable that satisfies the constraint is available in a finitely predicted time domain and makes the cost function locally optimal.

The MPC process can be divided into three steps. Firstly, based on the prediction model and current initial conditions, MPC forecasts the future output within a limited time domain. Secondly, it works out a set of local optimal control variables that can satisfy the constraints of the system and minimise the cost function according to the future model. Thirdly, the first set of control variables in the local control variables obtained during the second step is applied to the controlled system. At the next sampling time, the predictive model is modified based on the actual output value of the system. Then, the process is repeated over and over.

In Figure 19, x_0 represents the state of the system in the k moment, u is the system input, y is the system output, H_p is prediction in the time domain and H_c is the control output for the optimal predictive in time domain.

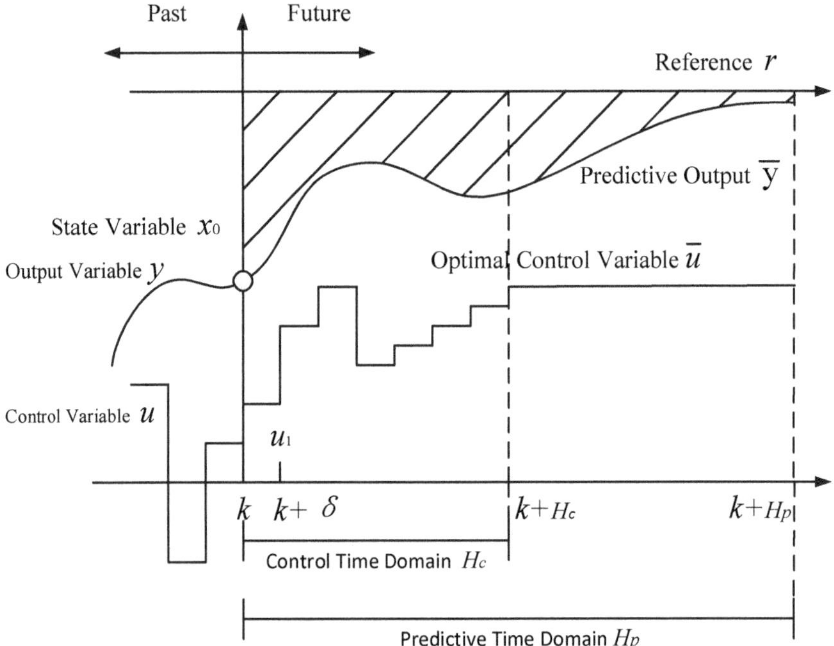

Figure 19. Schematic of model predictive control.

If the model of the controlled system can be expressed by a standard state space model, according to measurement information $y(k)$ and status information $x(k)$, then the predictive model predicts system output in p cycles and solves the optimisation problem, such that the predicted output $\overline{y}$ and set target r in the prediction period have the least errors in the time domain Hp. If the optimal solution to the optimisation problem is control sequence $[u(k)\ u(k+1)\ldots u(k+c)]$, then the first element $u(k)$ of the control sequence acts on the controlled object as the current control input and discards other elements.

4.2. Nonlinear Model Predictive Control

Assuming that all state variables of the system are measurable, the problem of nonlinear MPC is described as follows:

$$\begin{cases} x(k+1) = f(x(k), u(k)), k \geq 0 \\ y_c(k) = g_c(x(k), u(k)) \\ y_b(k) = g_b(x(k), u(k)) \end{cases} \tag{1}$$

where $x(k) \in R^n$ is the state variable of the predictive model, $y_c(k) \in R^c$ is the system output of the predictive model under control input and $y_b(k) \in R^b$ is the system output of the predictive model under constraint input.

Actuators have maximum values of output and increments, and their upper and lower limits are constant values. We assume that the controller's control value, control increments, and output constraints are as follows:

$$\begin{cases} u_{min} \leq u(k) \leq u_{max}, \forall k \geq 0 \\ \Delta u_{min} \leq \Delta u(k) \leq \Delta u_{max}, \forall k \geq 0 \\ y_{min} \leq y_b(k) \leq y_{max}(k), \forall k \geq 0 \end{cases} \tag{2}$$

We also assume that the state of the nonlinear system at the current k time can be observed and is $x(k)$; thus, the discrete model-based optimisation problem can be expressed by the following equation:

$$\min_{U_k} J(x(k), U_k) \tag{3}$$

If the control and prediction times are h_c and c_p, respectively, then the constraint in time domain is:

$$\begin{cases} u_{min} \le \overline{u}(k+i) \le u_{max}, 0 \le i \le h_c \\ \Delta u_{min} \le \Delta \overline{u}(k+i) \le u_{max}, \\ \Delta \overline{u}(k+i) = \overline{u}(k+i) - \overline{u}(k+i-1), \\ y_{min}(k+i) \le \overline{y}_b(k+i) \le y_{max}(k+i), 0 \le i \le h_p \\ \Delta \overline{u}(k+i) = 0, h_c \le i \le h_p \end{cases} \tag{4}$$

The cost function is:

$$J(x(k), U_k) = \sum_{i=1}^{h_p} \|\overline{y}_c(k+i) - r(k+i)\|_Q^2 + \sum_{i=1}^{h_c-1} \left(\|\overline{u}(k+i) - u_r(k+i)\|_R^2 + \|\Delta \overline{u}(k+i)\|_2^S \right) \tag{5}$$

where Q, R and S are the weighting matrix of the system cost function; $r(k+i)$ is the expected reference output; and $u_r(k+i)$ is the reference input corresponding to the expected reference output. $\overline{y}_c(k+i)$ and $\overline{y}_b(k+i)$ are the predicted control and constraint output, respectively, which can be updated using the following equations:

$$\begin{cases} \overline{x}(i+1) = f(\overline{x}(i), \overline{u}(i)), k \le i \le k+h_p, \overline{x}(k) = x(k) \\ \overline{y}_c(i) = g_c(\overline{x}(i), \overline{u}(i)) \\ \overline{y}_b(i) = g_b(\overline{x}(i), \overline{u}(i)) \end{cases} \tag{6}$$

where $x(k)$ is the state variable of the system at the current time k, which can be used as the initial condition of the prediction model and starting point for predicting the future output, and $\overline{u}(k+i)$ is the predictive control input, which takes on the following form:

$$\overline{u}(k+i) = \overline{u}_i, i = 0, 1, \cdots, h_c - 1 \tag{7}$$

where $\overline{u}_0$ and $\overline{u}_1 \cdots \overline{u}_{h_c-1}$ are the independent variables of the cost function. These variables compose vector U_k. The optimal solution for the cost function is U_k^*:

$$U_k = \begin{bmatrix} \overline{u}_0 \\ \overline{u}_1 \\ \vdots \\ \overline{u}_{h_c-1} \end{bmatrix}, U_k^* = \begin{bmatrix} \overline{u}_0^* \\ \overline{u}_1^* \\ \vdots \\ \overline{u}_{h_c-1}^* \end{bmatrix} \tag{8}$$

According to the MPC algorithm, the first component of solution U_k^* is applied to the system. That is, the current control quantity is

$$u(k) = \overline{u}_0^* \tag{9}$$

In practical engineering problems, the system's cost functions and models are often nonlinear. Therefore, the designed controller is typically nonlinear. Solving the optimal solution of the cost function by means of a numerical calculation method is difficult. Furthermore, solving it online is challenging. Scholars use various methods to linearise the problems involved and then employ linear programming to find the optimal solution.

4.3. Optimal Solution for Linear Programming

The linear programming problem is an important part of operations research. Under certain constraint conditions, linear programming guides scholars to find a decision method that has the least cost through mathematical calculation, which highlights the important role of limited resources.

Therefore, linear programming problems have widespread applications in various fields, such as path selection, industrial production allocation, business management and so on.

Linear programming problems are divided into two forms. The first form is that of the target function, which has a large output with limited resources. The other one is concerned with achieving the goal with minimum consumption. The former is a maximum value problem, whereas the latter is a minimum value problem. Equations (10) and (11) display the standard form of the linear programming problem model as follows:

$$min\ z = \sum_{j=1}^{n} c_j x_j \tag{10}$$

$$s.t. \begin{cases} \sum_{j=1}^{n} a_{ij} x_j \leq b_i (i = 1, 2, \cdots, m) \\ x_j \geq 0 (j = 1, 2, \cdots, n) \end{cases} \tag{11}$$

The matrix form is shown in the following equations:

$$Min\ Z = CX \tag{12}$$

$$\begin{cases} \sum_{j=0}^{n} p_j x_j \leq b \\ X \geq 0 \end{cases} \tag{13}$$

where:

$$C = (c_1, c_2, L, c_n)$$

$$X = \begin{bmatrix} x_1 \\ x_2 \\ \vdots \\ x_n \end{bmatrix}, p_j = \begin{bmatrix} a_{1j} \\ a_{2j} \\ \vdots \\ a_{mj} \end{bmatrix} (j = 1, 2, \cdots, n), b = \begin{bmatrix} b_1 \\ b_2 \\ \vdots \\ b_m \end{bmatrix}$$

In this study, we solve the linear programming problem using MATLAB functions. The steps of the solution are as follows. Firstly, we find a feasible solution using the iterative method, then judge whether it is the optimal solution. If not, then we continue to iterate until an optimal solution is found or determined unsolved.

The following equations show the standard form of linear programming in MATLAB:

$$min\ z = f^T x \tag{14}$$

$$s.t \begin{cases} Ax \leq b \\ A_{eq} x = b_{eq} \\ lb \leq x \leq ub \end{cases} \tag{15}$$

The following equations show correlation function, that is:

$$[x, fval, exitflag] = linprog(f, A, b, Aeq, beq, lb, ub)$$

where *lb* and *ub* are the constraint lower and upper limits of the variable *x*. Conversely, *x* is the optimal solution of the objective function, *Fval* is the minimum value of the objective function and *exitflag* is the state of the solution.

In the MPC for equilibrium, we intend to minimise the equilibrium time. Hence, our cost function can be expressed as:

$$Min(|u_1| + |u_2| + \ldots + |u_n|) \tag{16}$$

To convert this problem into a standard linear programming problem, w_i and v_i can be expressed as in the following equations:

$$W_i = \frac{u_i + |u_i|}{2} \text{ and } V_i = \frac{|u_i| - u_i}{2}$$

then:

$$u_i = W_i - V_i \text{ and } |u_i| = W_i + V_i$$

where $W_i > 0$ and $V_i > 0$.

If:

$$\begin{cases} w = [w_1, w_2, \ldots w_n]^T \\ v = [v_1, v_2, \ldots v_n]^T \end{cases} \tag{17}$$

then, the aforementioned problem can be converted into:

$$Min \sum_{i=1}^{n} (w_i + v_i) \tag{18}$$

$$s.t \begin{cases} A(w - v) \le b \\ w, v \ge 0 \end{cases} \tag{19}$$

The constraint function can be further expressed as:

$$s.t \begin{cases} [A \ -A] \begin{bmatrix} w \\ v \end{bmatrix} \le b \\ w, v \ge 0 \end{cases} \tag{20}$$

In this manner, optimisation can be performed using MATLAB's *linprog()* function.

5. Model Predictive Control for Active Equilibrium

Section 3 demonstrates the discharging of the highest voltage battery and charging of the lowest voltage battery. However, if multiple cells need to be balanced simultaneously, then achieving consistency among series cells at a minimum cost (time or energy consumption) is a new problem. We aim to balance all cells in the shortest possible time or with minimal energy loss. Therefore, the goal of battery management system equalisation in general is to select the minimum equilibrium time or energy consumption in equilibrium or a combination of both. To quickly achieve the consistency of the battery and reduce the amount of calculation, this paper chooses the minimum equilibrium time as the objective function of the series equilibrium control. The general control method uses SOC as the equilibrium control condition. That is, the SOC of each battery is compared with the average value. When the SOC of the battery cell is higher than the average value, the battery is discharged, whereas when the SOC of the battery is lower than the average value, the battery needs to be charged. If the equalisation control period is set to a fixed value ΔT, based on the general control method, when a battery needs to be balanced, then it may be charged or discharged for the whole period of ΔT. This leads to the repeated charge and discharge of the battery cell during the entire equilibrium process, which not only reduces the efficiency of the balance but also affects the life of the battery. Based on the MPC method, the MPC controller calculates the equalisation current in each period that is required to complete the equalisation in each cycle based on the difference between the single SOC and average values. The general control method can only calculate whether it needs equalisation but cannot control the magnitude of the equilibrium current. The control of the equalisation current can be realised through PWM. That is, the dynamic adjustment of the equalisation current is achieved by controlling the duty cycle of the equalisation switch in one cycle [28].

5.1. Battery Pack Equalisation Modelling

Suppose the bidirectional DC–DC equalisation system has n series battery in series and m group equalisation channels. The capacity of cell 1 to cell n can be represented by a diagonal matrix Q_x based on the metrics outlined in [27,29]. The SOC of each battery is defined as $x(t)$.

The amount of electricity flowing through each string of cells can be expressed as:

$$Q_x x(t) \in R^n, \tag{21}$$

where:

$$Q_x = \begin{bmatrix} \hat{C}_1 & 0 & \cdots & 0 \\ 0 & \hat{C}_2 & \cdots & 0 \\ \vdots & \vdots & \ddots & \vdots \\ 0 & 0 & \cdots & \hat{C}_n \end{bmatrix} \in R^{n \times n}, x(t) = \begin{bmatrix} x_1 & x_2 & \cdots & x_n \end{bmatrix}^T$$

The SOC of the battery is a number between 0 and 1, with 0 indicating that the battery power is exhausted and 1 indicating that the battery is fully charged.

If the difference in SOC among cells is large, then energy transfer is needed, and charge is transferred between m channels.

If a diagonal matrix Q_u is used to represent the maximum equalisation current of channels 1 to m, the equalisation current of each channel after normalisation is expressed by $u(t)$.

The actual equalisation current can be expressed as:

$$Q_x * x(t) \in R^n \tag{22}$$

where:

$$Q_n = \begin{bmatrix} \hat{L}_{l1} & 0 & \cdots & 0 \\ 0 & \hat{l}_{l2} & \cdots & 0 \\ \vdots & \vdots & \ddots & \vdots \\ 0 & 0 & \cdots & \hat{L}_{l3} \end{bmatrix} \in R^{m \times m} \tag{23}$$

According to the principle of distributed bidirectional equalisation, which is introduced in the previous section, the electricity discharged from the cell with the highest SOC is transferred to the entire battery pack. If the total transferred electricity is 1, then each cell (including the discharged cell) in the battery pack obtains $1/n$ of power. Hence $1-1/n$ electricity is transferred from the discharged cell, and other cells receive $1/n$.

Similarly, the battery with the lowest SOC obtains energy from the entire battery pack. If the total transferred electricity is 1, then each cell (including the charged cell) in the battery pack loses $1/n$ of the energy, such that the charged cell receives $1-1/n$ of power, and other cells $-1/n$.

The matrix $T \in R^{n \times m}$ is used to describe the balanced energy transferred between the cells and indicates the connection between n series of cells and m groups of channels:

$$T = \begin{bmatrix} \frac{1}{n} - 1 & \frac{1}{n} & \cdots & \frac{1}{n} \\ \frac{1}{n} & \frac{1}{n} - 1 & \cdots & \frac{1}{n} \\ \vdots & \vdots & \frac{1}{n} - 1 & \vdots \\ \frac{1}{n} & \frac{1}{n} & \cdots & \frac{1}{n} - 1 \end{bmatrix} \in R^{n \times m} \tag{24}$$

The amount of electricity transferred in unit time Δt can be expressed as $E = TQ_u u(t)\Delta t$, where $u(t) < 0$ means the battery is being charged, where as $u(t) < 0$ means the battery is being discharged.

The goal of equalisation is to realise that the difference between the average value of the SOC and SOC of the cells is less than the threshold value within a short time. That is:

$$y(t) = \begin{bmatrix} \frac{x_1+x_2+\cdots+x_n}{n} - x_1 \\ \frac{x_1+x_2+\cdots+x_n}{n} - x_2 \\ \cdots \\ \frac{x_1+x_2+\cdots+x_n}{n} - x_n \end{bmatrix} = \begin{bmatrix} \frac{1}{n}-1 & \frac{1}{n} & \cdots & \frac{1}{n} \\ \frac{1}{n} & \frac{1}{n}-1 & L & \frac{1}{n} \\ \vdots & \vdots & \frac{1}{n}-1 & \vdots \\ \frac{1}{n} & \frac{1}{n} & \cdots & \frac{1}{n}-1 \end{bmatrix}. \tag{25}$$

which is close to the target value 0.

If battery capacity $x(t)$ is selected as the state variable, control current $u(t)$ is used as the input control variable and $y(t)$ is the system output variable, then the state control equation of the equalisation can be expressed as

$$\begin{cases} x(t+1) = Ax(t) + Bu(t) \\ \quad\quad y(t) = Cx(t) \end{cases} \tag{26}$$

ere $A = E_n$, $B = Q_x^T E = Q_x^T T Q_u u(t) \Delta t$ and $C = T$.

The system constraint is: $u(t) \in \{u \in R^m | -1 \leq u \leq 1\}$, that is, a limit is set for the equalisation current of each channel.

5.2. Simulation Experiment Verification

In the previous two sections, the principle of distributed bidirectional DC–DC equilibrium process and model prediction are analysed. In this section, we utilise the battery model in Figure 20 and use Simulink to model the two parts. For simplicity, we select a six-series battery pack for analysis, as shown in Figure 20.

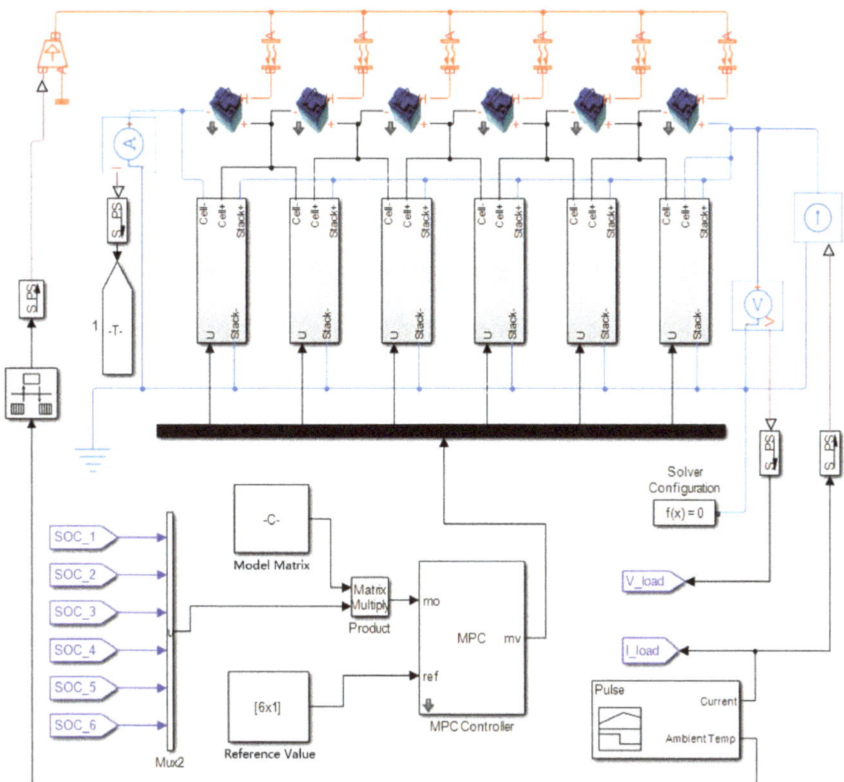

Figure 20. Distributed DC–DC equalisation model.

Figure 20 shows that the signal generator in the lower right corner generates the battery pack's operating current and ambient temperature. The six cells work in series. Each battery has a bidirectional DC–DC equaliser connected to both ends of the cell. The equaliser evens out the cell according to the balanced signal outputted by the MPC controller.

When $u(t) > 0$, the battery is charged, and the charge time in each controller cycle is $|u(t)| * \Delta T$. Furthermore, when $u(t) < 0$, the battery is discharged, and the discharge time in each controller cycle is $|u(t)| * \Delta T$. Table 1 shows the initial state of the cells.

Table 1. Initial state of cells.

Index	Cell 1	Cell 2	Cell 3	Cell 4	Cell 5	Cell 6
SOC	0.956	0.869	0.725	0.678	0.627	0.428
Voltage (V)	2.687	2.634	2.528	2.501	2.255	2.164

Equilibrium aims to maintain the SOC difference among all cells at less than 3%.

Figure 20 further shows that the bidirectional DC–DC is simulated by the model, where ports 1 and 2 are connected to the negative and positive electrodes of the cell, respectively, whereas ports 3 and 4 connect the positive and negative poles of the battery pack ports, respectively. When the battery is charged, the maximum current of the bidirectional DC–DC primary coil (battery side) is 4 A. According to the voltage and current of the battery side, the output power of DC–DC (battery side) can be calculated. Then, according to the conversion efficiency, we can calculate the input power of DC–DC. The output current of the battery pack can be calculated based on the voltage of the battery pack.

Similarly, when the battery is discharged, the DC–DC (battery pack side) input power can be calculated based on the maximum current of the designed DC–DC circuit (battery pack side). Then, through conversion efficiency, the output power of the DC–DC (battery side) can be obtained, then the output current of the battery can be calculated. We set the equalisation control period $\Delta T = 45$ s.

The contrast test adopts SOC as the equilibrium control condition. Based on the general control method, if a battery requires balance, then it may be charged or discharged all the time.

Based on the MPC method, the MPC controller can calculate the equilibrium current $u(t) \in \{u \in R^m | -1 \leq u \leq 1\}$ in each calculation period based on the difference between single-cell SOC and average value.

The general control method can only calculate the need for equalisation and cannot control the amplitude of the balanced current.

Controlling the equalisation current can be realised by PWM. That is, the dynamic adjustment of the equalisation current is achieved by controlling the duty cycle of the equalisation switch in one cycle. The DC–DC simulation model is shown in Figure 21.

Figures 22 and 23 show the equalisation effect of two equalisation strategies. The two pictures show that these equilibrium strategies can achieve the goal of equilibrium. Compared to the two algorithms, the equalisation circuit is either idle or equalising with the largest balance ability for the common control algorithm in a control cycle. In addition, when a cell is charged, it will discharge the battery that does not need equalisation because electricity is removed from the entire battery pack. As a result of energy loss, they need to be rebalanced, and vice versa. In this manner, battery cells are often charged and discharged repeatedly, which consumes energy, impairs the life of the battery and prolongs equilibration time. By contrast, the MPC algorithm decouples the equalisation and calculates the equalisation current required for each cell to reach a balanced state in every cycle. The required equalising current is then achieved by controlling the duty cycle of the equalising circuit in one control period using the PWM wave. This process avoids repeated charge and discharge of a battery so that the battery pack reaches equilibrium as quickly as possible.

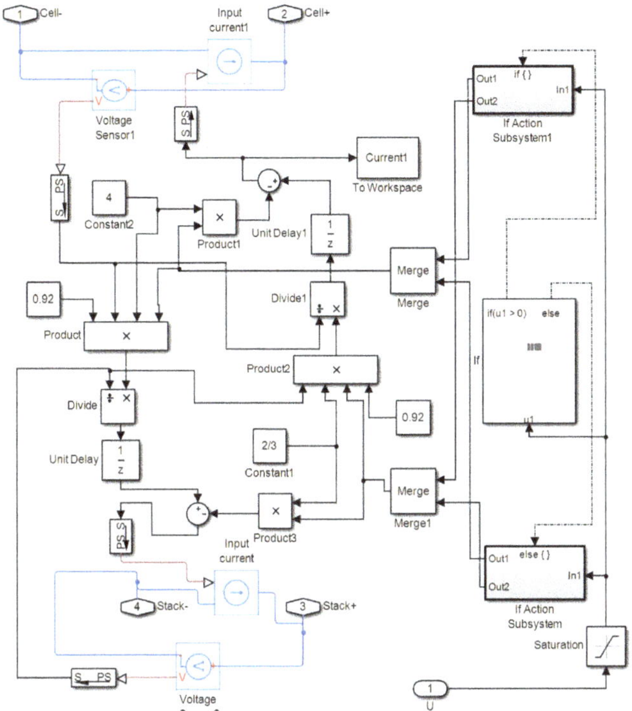

Figure 21. DC–DC simulation model.

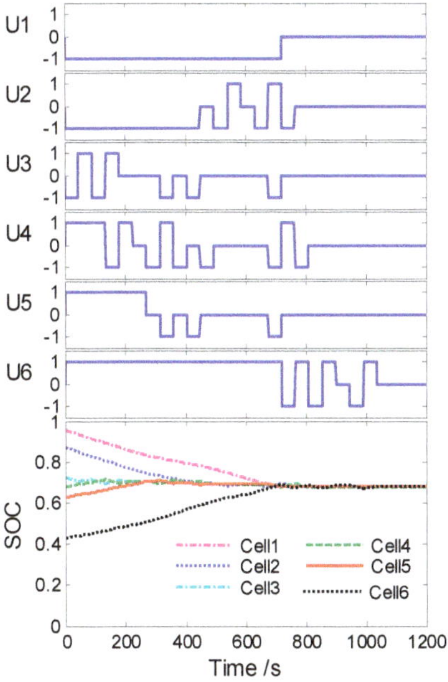

Figure 22. Based on the general control.

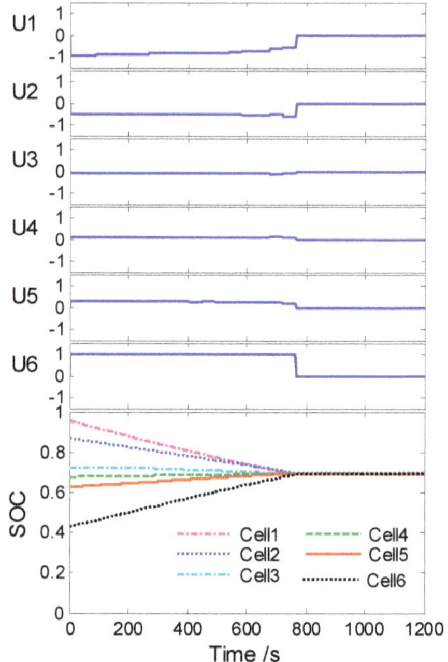

Figure 23. Based on the MPC.

5.3. Bench Test and Result Analysis

To prove the effectiveness of the proposed control algorithm, this section builds a small battery experimental bench to evaluate the effect of balanced control based on MPC by comparing the collected equalisation current and SOC variation. To measure the equilibrium current of each path, a self-built multi-channel isolated current sensor is used to measure the balanced current, and the NI data acquisition system is used to collect the current sensor output sampling voltage value and convert it into corresponding current value. Battery voltage equalisation and SOC are obtained through the BMS board. Figure 24 presents the 14-channel Hall-type current sensors, whereas Figure 25 displays the NI data acquisition system.

Figure 24. 14-channel Hall-type current sensors.

Figure 26 shows that the battery pack contains 24 series cells (lithium titanate), and the capacity of each battery is 2.9 Ah. The maximum balanced capacity of the bidirectional synchronous fly-back balanced circuit is ±4 A. The cell and neighbouring module can be treated as a C2S topology. The fast

MPC algorithm, based on the SOC of each battery, determines the balance current needed for each cell. Then, adjusting the balance current is achieved by controlling the time of equilibrium circuit that is working in a balanced cycle, and the battery cell is balanced according to the local optimal equilibrium current.

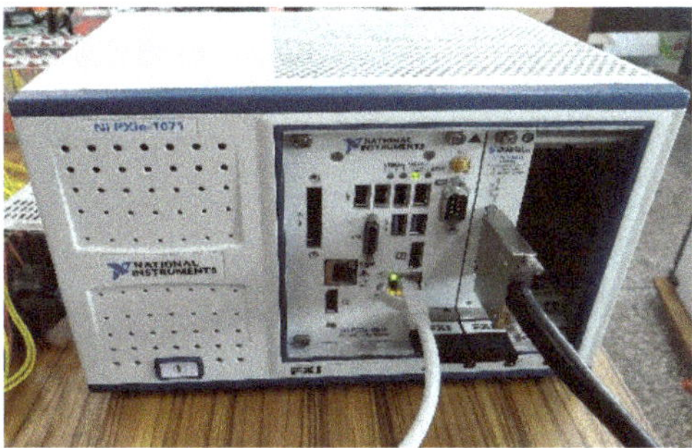

Figure 25. NI data acquisition system.

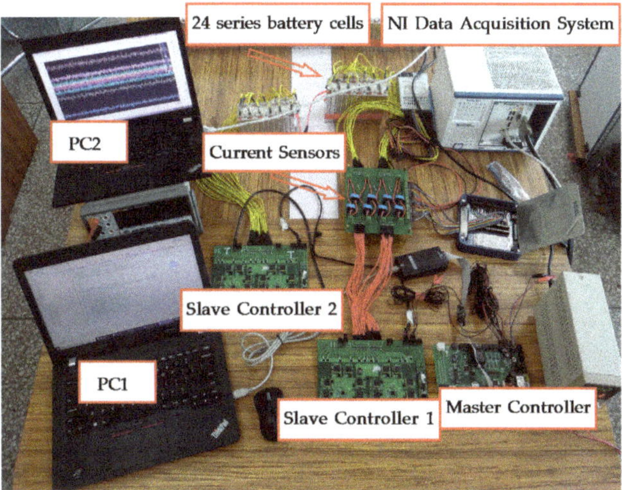

Figure 26. Active balance test bench.

The specific implementation is to use the MPC algorithm and LP solution function, which is written by the S function and translated into C code using the MATLAB automatic code generation tool. Finally, the C code was downloaded to the BMS main control circuit board through the Keil compiler.

Figure 26 illustrates the experimental bench.

The battery pack consists of two 12 series modules. Twenty-four strings of cells and two slave controllers are used, and a 14-channel Hall-type current sensor is utilised to measure the magnitude of the equalisation current in controller 1. PC1 is used to debug the master controller, which controls the equalisation function of controllers 1 and 2, whereas PC2 is used to record the collected current value.

The voltage values of the first six cell cells are set according to Table 1, and the voltage difference of the cells is approximately 0.5 V. Figure 25 demonstrates the balanced current collected by the NI data acquisition system. This figure shows that, based on the common equalisation control rules, the five other channel currents frequently change direction during equalisation except for the equalisation current direction of cell 1. By contrast, the equalisation currents for all cells do not change direction based on the equalisation process of the MPC rules. The current in the equilibrium period in a graph

(a) is either a positive or negative maximum; whereas in graph (b), the equilibrium current of each circuit in a control period is controlled by the duty cycle in a control period. The duty ratio is adjusted to 1 when the maximum balance ability is required. Figure 27b shows the equilibrium current of cell 6. In Table 2, by comparison, under the same initial conditions, the equilibrium time in graph (a) is 1030 s, whereas that in graph (b) is 710 s. The equilibrium time is reduced by 31%.

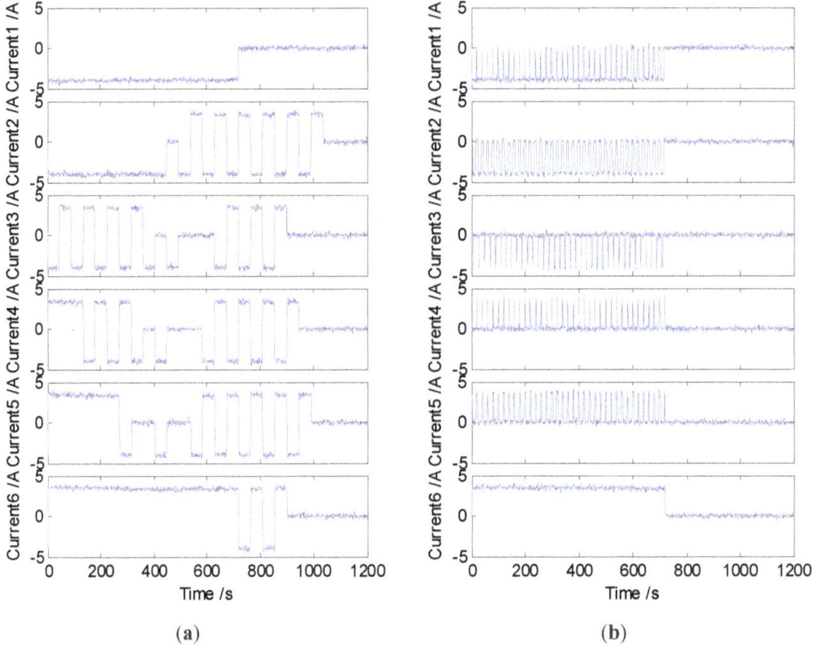

(a)   (b)

Figure 27. Comparison of actual current measurement. (**a**) Based on the general control method; (**b**) Based on the MPC method.

Table 2. Equilibrium time comparison.

Method	General Control	MPC	Time Reduced
Time (s)	1030	710	31%

If we use common control rules to balance the current between cells in the battery pack without decoupling the equilibrium process, then the direction of the balanced current will change frequently. This case makes the battery cells charge and discharge repeatedly, damages the life of the battery and increases the balance time. Hence, this type of control is not advisable despite its simplicity. Using equalisation based on the MPC rules, the overall analysis of the battery cell's equalisation is performed. The mutual influence between the balanced cells is considered and the decoupling of the equilibrium process is achieved. Therefore, the equalisation current, which is calculated by the optimal algorithm, is used. Repeated changes do not appear in the balance current direction. In addition, the equalisation current is controlled by the duty cycle, and an adjustable equalisation current is output on the fixed balance capability of the hardware, which is an important innovation of this study.

6. Summary

This study introduces the method of using MPC algorithm on distributed DC–DC equalisation circuit and compares the effect of using the algorithm. The optimisation results show that the

MPC algorithm can avoid unnecessary energy transfer and shorten equalisation time. The MPC equilibrium scheme is compared with the common equilibrium mode through bench experiments. The experimental result indicates that the equilibrium time is reduced by 31%, which verifies the rationality of the use of MPC algorithm. The algorithm presented in this study can be applied to the active equalisation of a series of battery packs and realise energy transfer with minimum time and high efficiency to make the battery reach equilibrium. This paper improve the active equivalent efficiency through MPC, which can be used as new engineering technology. The key of this method is that the balance current is adjustable. However, the computation process of local optimal solutions is a time consuming process; In the future, other optimization algorithm should be tried to reduce the computation time, which will shorten equilibrium period and increase efficiency further.

Author Contributions: S.S.: testing and writing manuscript; F.X.: designing the control strategies and writing manuscript; S.P.: developing the control system and data analysis; C.S.: organizing the research work and designing the control strategies; Y.S.: developing the control system, designing and building the test bench.

Funding: This research is funded by the Key Tackling Item in Science and Technology Department of Jilin Province, China Grant number [20150204017GX], Jilin Provincial Natural Science Foundation, China Grant number [20150101037JC].

Conflicts of Interest: The authors declare no conflict of interest.

References

1. Repp, S.; Harputlu, E.; Gurgen, S.; Castellano, M.; Kremer, N.; Pompe, N.; Woerner, J.; Hoffmann, A.; Thomann, R.; Emen, F.M.; et al. Synergetic effects of Fe^{3+} doped spinel $Li_4Ti_5O_{12}$ nanoparticles on reduced graphene oxide for high surface electrode hybrid supercapacitors. *Nanoscale* **2018**, *10*, 1877–1884. [CrossRef] [PubMed]
2. Wang, X.; Cheng, K.W.E.; Fong, Y.C. Non-Equal Voltage Cell Balancing for Battery and Super-Capacitor Source Package Management System Using Tapped Inductor Techniques. *Energies* **2018**, *11*, 1037. [CrossRef]
3. Genc, R.; Alas, M.O.; Harputlu, E.; Repp, S.; Kremer, N.; Castellano, M.; Colak, S.G.; Ocakoglu, K.; Erdem, E. High-Capacitance Hybrid Supercapacitor Based on Multi-Colored Fluorescent Carbon-Dots. *Sci. Rep.* **2017**, *7*, 11222. [CrossRef] [PubMed]
4. Hoque, M.M.; Hannan, M.A.; Mohamed, A.; Ayob, A. Battery charge equalization controller in electric vehicle applications: A review. *Renew. Sustain. Energ. Rev.* **2017**, *75*, 1363–1385. [CrossRef]
5. Omar, N.; Daowd, M.; Hegazy, O.; Mulder, G.; Timmermans, J.M.; Coosemans, T.; Van den Bossche, P.; Van Mierlo, J. Standardization Work for BEV and HEV Applications: Critical Appraisal of Recent Traction Battery Documents. *Energies* **2012**, *5*, 138–156. [CrossRef]
6. Diao, W.P.; Xue, N.; Bhattacharjee, V.; Jiang, J.C.; Karabasoglu, O.; Pecht, M. Active battery cell equalization based on residual available energy maximization. *Appl. Energy* **2018**, *210*, 690–698. [CrossRef]
7. Wang, Y.J.; Zhang, C.B.; Chen, Z.H. An adaptive remaining energy prediction approach for lithium-ion batteries in electric vehicles. *J. Power Sources* **2016**, *305*, 80–88. [CrossRef]
8. Wang, S.L.; Shang, L.P.; Li, Z.F.; Deng, H.; Li, J.C. Online dynamic equalization adjustment of high-power lithium-ion battery packs based on the state of balance estimation. *Appl. Energy* **2016**, *166*, 44–58. [CrossRef]
9. Moo, C.S.; Hsieh, Y.C.; Tsai, I.S.; Cheng, J.C. Dynamic charge equalisation for series-connected batteries. *IEE P-Elect. Power Appl.* **2003**, *150*, 501–505. [CrossRef]
10. Kutkut, N.H.; Wiegman, H.L.N.; Divan, D.M.; Novotny, D.W. Design considerations for charge equalization of an electric vehicle battery system. *IEEE Trans. Ind. Appl.* **1999**, *35*, 28–35. [CrossRef]
11. Wu, Z.; Ling, R.; Tang, R.L. Dynamic battery equalization with energy and time efficiency for electric vehicles. *Energy* **2017**, *141*, 937–948. [CrossRef]
12. Fong, Y.C.C.; Cheng, K.W.E.; Raman, S.R.; Wang, X. Multi-Port Zero-Current Switching Switched-Capacitor Converters for Battery Management Applications. *Energies* **2018**, *11*, 1934. [CrossRef]
13. Moo, C.S.; Hsieh, Y.C.; Tsai, I.S. Charge equalization for series-connected batteries. *IEEE Trans. Aerospace Electron. Syst.* **2003**, *39*, 704–710. [CrossRef]
14. Ma, Y.; Duan, P.; Sun, Y.S.; Chen, H. Equalization of Lithium-Ion Battery Pack Based on Fuzzy Logic Control in Electric Vehicle. *IEEE Trans. Ind. Electron.* **2018**, *65*, 6762–6771. [CrossRef]

15. Fan, F.L.; Tai, N.L.; Zheng, X.D.; Huang, W.T.; Shi, J.X. Equalization Strategy for Multi-Battery Energy Storage Systems Using Maximum Consistency Tracking Algorithm of the Conditional Depreciation. *IEEE Trans. Energy Convers.* **2018**, *33*, 1242–1254. [CrossRef]

16. Chen, Y.; Liu, X.F.; Fathy, H.K.; Zou, J.M.; Yang, S.Y. A graph-theoretic framework for analyzing the speeds and efficiencies of battery pack equalization circuits. *Int. J. Electron. Power Energy Syst.* **2018**, *98*, 85–99. [CrossRef]

17. Liu, X.T.; Wan, Z.H.; He, Y.; Zheng, X.X.; Zeng, G.J.; Zhang, J.F. A Unified Control Strategy for Inductor-Based Active Battery Equalisation Schemes. *Energies* **2018**, *11*, 405. [CrossRef]

18. Kobzev, G.A. Switched-capacitor systems for battery equalization. In Proceedings of the 6th International Scientific and Practical Conference of Students, Post-graduates and Young Scientists. Modern Techniques and Technology. MTT'2000 (Cat. No.00EX369), Tomsk, Russia, 3 March 2000; pp. 57–59.

19. Baronti, F.; Roncella, R.; Saletti, R. Performance comparison of active balancing techniques for lithium-ion batteries. *J. Power Sources* **2014**, *267*, 603–609. [CrossRef]

20. Kim, M.Y.; Kim, J.W.; Kim, C.H.; Cho, S.Y.; Moon, G.W. Automatic Charge Equalization Circuit Based on Regulated Voltage Source for Series Connected Lithium-Ion Batteries. In Proceedings of the IEEE International Conference on Power Electronics and Ecce Asia, Jeju, Korea, 30 May–3 June 2011; pp. 2248–2255.

21. Kutkut, N.H. A Modular Nondissipative Current Diverter for EV Battery Charge Equalization. In Proceedings of the APEC 1998 Thirteenth Annual Applied Power Electronics Conference and Exposition, Anaheim, CA, USA, 15–19 February 1998; Volume 2, pp. 686–690.

22. Kutkut, N.H.; Divan, D.M. Dynamic equalization techniques for series battery stacks. In Proceedings of the 1996 International Telecommunications Energy Conference, Boston, MA, USA, 6–10 October 1996; 1996; pp. 514–521.

23. Zhang, Z.L.; Gui, H.D.; Gu, D.J.; Yang, Y.; Ren, X.Y. A Hierarchical Active Balancing Architecture for Lithium-Ion Batteries. *IEEE Trans. Power Electron.* **2017**, *32*, 2757–2768. [CrossRef]

24. Narayanaswamy, S.; Kauer, M.; Steinhorst, S.; Lukasiewycz, M.; Chakraborty, S. Modular Active Charge Balancing for Scalable Battery Packs. *IEEE Trans. Large Scale Integr. Syst.* **2017**, *25*, 974–987. [CrossRef]

25. Wang, Y.J.; Zhang, C.B.; Chen, Z.H.; Xie, J.; Zhang, X. A novel active equalization method for lithium-ion batteries in electric vehicles. *Appl Energy* **2015**, *145*, 36–42. [CrossRef]

26. Shang, Y.L.; Zhang, C.H.; Cui, N.X.; Guerrero, J.M. A Cell-to-Cell Battery Equalizer With Zero-Current Switching and Zero-Voltage Gap Based on Quasi-Resonant LC Converter and Boost Converter. *IEEE Trans. Power Electron.* **2015**, *30*, 3731–3747. [CrossRef]

27. McCurlie, L.; Preindl, M.; Emadi, A. Fast Model Predictive Control for Redistributive Lithium-Ion Battery Balancing. *IEEE Trans. Ind. Electron* **2017**, *64*, 1350–1357. [CrossRef]

28. Caspar, M.; Hohmann, S. Optimal Cell Balancing with Model-based Cascade Control by Duty Cycle Adaption. *IFAC Proc. Vol.* **2014**, *47*, 10311–10318. [CrossRef]

29. Preindl, M.; Danielson, C.; Borrelli, F. Performance Evaluation of Battery Balancing Hardware. In Proceedings of the 2013 European Control Conference (ECC), Zurich, Switzerland, 17–19 July 2013; pp. 4065–4070.

 *energies*

Article

Total Suspended Particle Emissions Modelling in an Industrial Boiler

Guillermo Ronquillo-Lomeli [1,*], Gilberto Herrera-Ruiz [2], José Gabriel Ríos-Moreno [2], Irving Alfredo Alejandro Ramirez-Maya [1] and Mario Trejo-Perea [2]

[1] Department of Energy, Center for Engineering and Industrial Development, Santiago de Querétaro 76125, México; irving.ramirez@cidesi.edu.mx

[2] Faculty of Engineering, Autonomous University of Queretaro, Santiago de Querétaro 76010, México; gherrera@uaq.mx (G.H.-R.); riosg@uaq.mx (J.G.R.-M.); mtp@uaq.mx (M.T.-P.)

* Correspondence: gronquillo@cidesi.edu.mx; Tel.: +52-442-211-9800

Received: 3 October 2018; Accepted: 4 November 2018; Published: 9 November 2018

Abstract: Particulate matter emission into the atmosphere is a massive-scale problem. Fossil fuel combustion is an important source of this kind of pollution. The knowledge of total suspended particle (TSP) emissions is the first step for TSP control. The formation of TSP emissions is poorly understood; therefore new approaches for TSP emissions source modelling are required. TSP modelling is a multi-variable non-linear problem that would only require basic information on boiler operation. This work reports the development of a non-linear model for TSP emissions estimation from an industrial boiler based on a one-layer neural network. Expansion polynomial basic functions combined with an orthogonal least-square and model structure selection approach were used for modelling. The model required five independent boiler variables for TSP emissions estimation. Data from the data acquisition system of a 350 MW industrial boiler were used for model development and validation. The results show that polynomial expansion basic functions are an excellent approach to solve modelling problems related to complex non-linear systems in the industry.

Keywords: system identification; parameter estimation; system modelling; model reduction; polynomial expansion; orthogonal least square; industrial process

---

1. Introduction

Fossil fuels burning in boilers results in the release of combustion gases into the atmosphere containing a gaseous phase and suspended solids or liquid particles pollutants. Environmental regulations have been defined to limit ambient pollutions that can be emitted into the environment around the world.

In order to control pollution in industrial boilers, the first step is to employ an adequate measurement system. A typical system for polluting emissions measurements in industrial boilers consists of Continuous Emissions Monitoring Systems (CEMS). CEMS are nevertheless operationally expensive, and their readings reliability is sensitive to environmental conditions, with difficulties related to their installation and maintenance. Predictive Emissions Monitoring Systems (PEMS) are a recent approach solution planned to replace online analyzers like CEMS by estimating emissions concentrations from process data. PEMS provide a relationship between the process and the emissions through nonlinear modelling data.

Methods based on the use of measurements data from a process have been used for process modelling. Modelling based on linear systems are popular for their simplicity; however, in complex problems with strong non-linearity, the results are limited. It is necessary, then, to develop methodologies that allow non-linear modeling, seeking to preserve the advantages of linear models.

There are different tools to solve the modelling identification problem, depending on the system complexity and knowledge. If the identification is based only on input and output measured data, considering there is no knowledge about the system physics, the identification process is classically called black box modelling. When detail system physics is known, purely mathematical models can be calculated, and the identification process is called white box modelling.

In the process of modelling pollutant emissions in the combustion process, it is practically impossible to have the knowledge of all the thermal, physical, and chemical processes that could exist, and, although this knowledge is available, it is always dispersed. This complicates the development of a complete description of the system in terms of a continuous-time mathematical model. Transferring the knowledge required for a description in discrete time is often difficult, since it is often lost in the process of discretization. Getting a black box model in complex systems is not easy to achieve, even if one has experience and knowledge of the system. Although the basic understanding of the system dynamics sometimes is not available, it is always useful and helps to establish the model structure.

Air quality is a widespread concern, which is generated mainly by population centralization in large cities, which increases traffic, industrialization, and the indiscriminate use of conventional energy [1,2]. Technological advances and the development of clean energy are outweighed by the indiscriminate use of fossil fuels. As a result, the emission of dangerous particles into the atmosphere is increasing [3], having catastrophic effects on all scales and bringing unimaginable damage to public health [4]. Environmental pollution is regulated by global organizations, but most countries do not satisfy the requirements [1,5], and, in addition, in some cases there are records of measurements that widely exceed the limits, thereby causing millions of deaths [4,6,7]. One of the most dangerous environmental pollutants with indiscriminate emissions is particulate matter (PM), which includes large particles (PM_{10}), small particles ($PM_{2.5}$), and fine particles (<100 nm) that are not regulated [8], which makes the problem larger [9]. A World Health Organization report on environmental atmosphere contamination stated that the average density of PM_{10} was augmented about 5% between 2008 and 2013 in 720 cities around the world [10]. It has been reported that a diminution in the content of PM_{10} by 5 μg per cubic meter in Europe would avoid between 3000 and 8000 annual premature deaths [11]. Alike estimates for $PM_{2.5}$ advise a decrease of 7 to 8 months in the life likelihood [12]. In equivalent studies of fine particulate matter, it is estimated that they also increase the damage to health together with large particles, whose health effect is known, and, in consequence, are expected to increase the costs to maintain public health and avoid premature deaths [6].

Total suspended particles (TSP) emissions formation is widely reported but poorly understood. In order to reduce TSP emissions, theoretical TSP generation models have been studied [13] and mainly discuss theories for linking discrete to continuum modelling but do not propose a general theory at the micro- and macroscale levels. Regression modelling of the spatial particulate matter has been developed [14] for modelling the concentrations of suspended particles in the time and space domains in specific localities with a high population concentration. The studies on modelling TSP emissions generally focus on analyzing the effects of suspended particles on the population health [15–20]. Modelling of TSP emission sources has been studied [21], conducting research on source apportionment of ambient particulate matter in Europe using receptor models such as principal component analysis, enrichment factors, classical factor analysis, and positive matrix factorization, [22] and building and applying a numerical air quality model that relies on scientific first principles to predict the concentration particles. A study [23] reported a computer-controlled ambient-simulation method to determine the source characteristic profiles of emissions from an oil-fired boiler through isokinetic withdrawal.

In relation to non-linear modelling, artificial neural networks (ANN) are widely applied in engineering processes, in particular for TSP modelling, and neural networks with retro-propagation, such as multi-layer perceptron (MLP), are the most used. However, there is a great diversity of types of neural networks for non-linear modelling based on data. For instance, Ye [24] presented Bayesian–Gaussian neural networks (BGNN), a new methodology for the application of neural networks improved in training time, minimum location, and auto-tuning for online applications. A redefinition of the BGNN algorithm was presented by Liu [25] using genetic algorithms (GA) for the offline adjustment of the threshold matrix and a sliding data window for online applications suitable for non-linear systems that change over time, a feature that makes this algorithm very attractive for online dynamic systems modelling.

The radial basic function (RBF) network offers an alternative in signal processing applications using two-layer neural networks. Common learning algorithms for the RBF network are based on the selection of random functions centers, which has some drawbacks. Reference [26] presents a method using orthogonal least squares, selecting the basic functions centers radially one by one, until an adequate network is built. Kassam [27] presents a training algorithm for the RBF network based on the stochastic gradient (SG) error and shows its versatility in non-linear signal processing applications. Because of its numerical features, the stochastic gradient algorithm is a training algorithm widely used in networks with adaptive radial basis functions, but it presents a compromise between convergence speed and precision due to fixed values in the steps size. Zeng [28] solved this problem by presenting the convex combination of multiple RBF (MCRBF) network algorithm, applying the SG learning algorithm and varying the configuration parameters.

The Volterra polynomials have also been used for system identification [29]. Here, the orthogonal least-squares method is used for offline model framework determination.

Considering that the combustion process is complex, multivariable, and non-linear, which hinders the application of white-box modelling techniques, fitting models for PEMS development is a challenge. The goal of this work was to develop a nonlinear model for TSP emissions in a 350 MW conventional boiler that burns heavy fuel oil in order to provide a good TSP prediction that can be used in PEMS applications using operational boiler data. The non-linear model proposed is based on the polynomial expansion of a multiple-variable function in the neural network framework, considering that there is a finite number of basic functions for an ANN to estimate a non-linear function. The orthogonal least-squares algorithm was used as a parameter estimator because it allows to easily determine a reduced subset of basic functions that best represent the TSP emission dynamics. A classical MLP three-layer ANN was implemented and compared with the polynomial expansion network proposed.

2. Methodology

Regardless of the fact that all systems are non-linear, most of the literature on systems identification refers to the identification of linear systems. The main reason for this is that the assumptions may become very restrictive because of the process complexity, which forces the designer to use strong simplifications or fix the model components. Also, process innovation for improvement together with diverse local environments often results in significant differences between two apparently similar plants.

Power plant equipment and installation are usually fitted to suit the local conditions of a specific place. The construction depends on factors such as fuel availability, innovations and local ambient conditions towards better thermal efficiency and emission control, etc. To make the existing models adequate for different constructions, redesigning and tuning are required. Model equations solving might also add problems to highly detailed first-principle models. Mathematical knowledge is required to develop the model, and time-consuming interactive computations need to be performed.

Identification is the experimental approach to process modelling [30,31]; this approach includes the following steps:

- Experimentation
- Structure model selection
- Parameter estimation
- Model validation

An experimental design must be carried out in order to obtain data that represent the behavior of the process in the whole operating field of the dynamic system. In other words, it is necessary to establish physically support values throughout the input range and measure the effect on the output. The corresponding input and output data sets are finally used to infer a system model. Some parameters to be defined in the experimental phase are: tests preparation, sampling time choice, suitable experiments design, and data pre-processing. Pre-processing data includes, for example, testing the response time, removal of irregularities, control of noise and another unexpected behaviors of the data.

Defining the model framework is called structure model selection. That is a framework that must be explored to get a good model where the model input–output signals and the internal interaction of the model are determined. The model structure is derived using prior knowledge. In general, structure model selection implies the selection of the model approach (multilayer perceptron networks, radial basis function, etc.) considered appropriate to describe the system, and the selection of a subset of this structure model defining the appropriate number of parameters for a specific problem.

The values of the unknown parameters of a parametrized model structure are estimated. Normally, the model that best performs according to the design specifications is chosen. The design specifications can be expressed in many different forms; preferably, they should be formulated with the final model application in mind. In general, the model is selected on the basis of the best possible model predictions following some criterion of error magnitude measurement between the observed output and the model estimation. The method for calculating the model parameters of the selected structure is developed according to the statistical theory and is called estimation. The equivalent process in the ANN theory is often denominated training or learning.

Once the model has been determined, it must be tested to verify whether it meets the design specifications, by evaluating precision, robustness, convergence, and good generalization abilities (interpolation). Validation is closely related to the final model application. The validation criteria of the tracking error compensation algorithm will be defined on the basis of the characteristics of the current compensation algorithms of relevance.

2.1. Polynomial Basic Functions

The model structure is a candidate model set, i.e., a set within which a model must be sought. In general, the problem with the model selection implies a model family selection. For TSP modelling, a polynomial expansions network was used as a model family because of its ability to model non-linear dynamics.

Consider a non-linear dynamic multiple-inputs–simple-output (MISO) system, which is represented by

$$y_t = f(u_1(1), u_1(2), \ldots, u_1(t - n_1), \cdots, u_m(1), u_m(2), \ldots, u_m(t - n_m)), \tag{1}$$

where $f(\cdot)$ is a non-linear function, u is the input vector, n_p is the delay samples number in the p-th input, and m is the input variable number.

Define u_t as

$$u_t = [u_1(1), u_1(2), \ldots, u_1(t - n_1), \cdots, u_m(1), u_m(2), \ldots, u_m(t - n_m)], \tag{2}$$

re-index u_t as

$$u_t = [u_1, u_2, \cdots, u_n], \tag{3}$$

where $n = n_1 + n_2 + \cdots + n_m$, and the Equation (1) is

$$y_t = f(u_t). \tag{4}$$

Polynomial multivariable expansions have been suggested as candidate base functions [32,33] and they are usually applied in function structure, mainly in one-input variable functions [34]. Recently, the polynomial basis function (PBF), in the multivariable functions context, has been launched within the neural network model structure. Its functional representation is described by

$$f(u_t) = \hat{f}(u_t, \Theta) + e\left(u_t^k\right),\tag{5}$$

$$\hat{f}(u_t, \Theta) = \theta_0 + \sum_{i=1}^{n} \theta_i u_i + \sum_{i_1=1}^{n} \sum_{i_2=i_1}^{n} \theta_{i_1 i_2} u_{i_1} u_{i_2} + \cdots + \sum_{i_1=1}^{n} \sum_{i_2=i_1}^{n} \cdots \sum_{i_k=i_{k-1}}^{n} \theta_{i_1 i_2 \cdots i_k} u_{i_1} u_{i_2} \cdots u_{i_k} = \sum_{j=1}^{N} \theta_j \varphi_j(u_t),\tag{6}$$

where $\Theta = \{\theta_j\}$ is the concatenated parameters, $\{\varphi_j\}$ is the set of basic functions formed from the polynomial input terms, $N = \frac{(n+k)!}{n!k!}$ is the polynomial basic functions number, k is the polynomial expansion order, and $e(u_t^k)$ indicates the approximation error generated by the order k from the input vector. The basic functions are polynomials of some specific order of the input vector $u_t \in \mathcal{R}^n$. This process can be viewed as the transformation of the multivariable input vector to a space of higher dimensions.

There are a certain number of basic functions to approximate a non-linear function with accuracy [35]. However, a practical method is necessary to determine these basic functions. The necessary approximation precision can be achieved by an acceptable number of linearly independent non-linear basic functions.

Polynomial expansion base functions offer a good approximation of non-linear functions. The structure of non-linear identification is shown in Figure 1. It is assumed that the non-linear function $\hat{f}(\cdot)$ in the polynomial expansion basic functions is estimated by a single-layer ANN, which is a linear combination of non-linear polynomials.

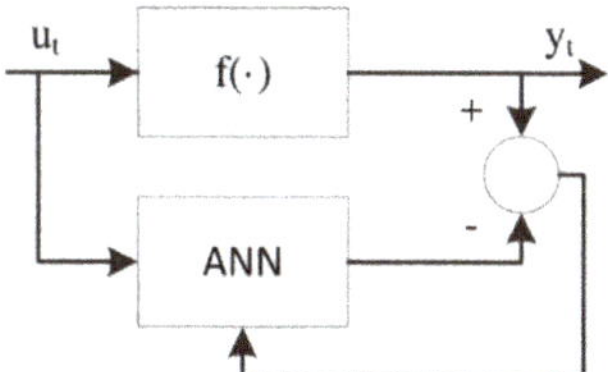

Figure 1. Identification based on neural networks.

With the increase in order k, the basic functions number N gets bigger and bigger. So, the problem is function $\hat{f}(u_t)$ estimation using a suitable ANN, dimensioned such that the estimate precision is according to the specified requirements. The framework model selection and parameters estimation of one-layer ANN are detailed here.

2.2. Model Structure Selection and Parameter Estimation

Obtaining the model that represents the TSP emissions dynamic of a boiler with good precision is a problem that requires looking for the best model structure, which means defining the number of parameters (basic functions) appropriate for the model and estimating the parameters to obtain exact values for the model.

There are many ways to the selection of basic functions. In this case, the structure selection was executed offline using the orthogonal least-squares algorithm [36] to determine the more meaningful basic functions number for TSP modelling.

This assumes that the data $(y_t, u_t, t = 1, 2, \cdots, M)$ from the input and output systems are known. On the basis of Equation (6), the estimation function can be formulated in vector form through:

$$Y = \Phi(u)\Theta + E\left(u^k\right), \tag{7}$$

where the input vector $Y \mathcal{R}^{M \times 1}$, the parameter vector $\Theta \mathcal{R}^{N \times 1}$, the error vector $E(u^k) \mathcal{R}^{M \times 1}$, and the basic functions matrix $\Phi \mathcal{R}^{M \times N}$ are

$$Y = [y_1\, y_2\, \cdots\, y_M]^T, \tag{8}$$

$$\Theta = [\theta_1\, \theta_2\, \cdots\, \theta_N]^T, \tag{9}$$

$$E\left(u^k\right) = \left[e\left(u_1^k\right) e\left(u_2^k\right) \cdots e\left(u_M^k\right)\right]^T, \tag{10}$$

$$\Phi(x) = \begin{bmatrix} \varphi_1(u_1) & \varphi_2(u_1) & \cdots & \varphi_N(u_1) \\ \varphi_1(u_2) & \varphi_2(u_2) & \cdots & \varphi_N(u_2) \\ \vdots & \vdots & \ddots & \vdots \\ \varphi_1(u_M) & \varphi_2(u_M) & \cdots & \varphi_N(u_M) \end{bmatrix}. \tag{11}$$

The parameter vector Θ is generally found by optimizing the error vector norm, that is,

$$\hat{\Theta} = \mathrm{argmin}_W \|Y - \Phi(u)W\|_2, \tag{12}$$

obtaining the least-squares solution.

The vector of $\Phi_i = [\varphi_i(u_1), \varphi_i(u_2), \cdots, \varphi_i(u_M)]^T$, for $i = 1, 2, \cdots, N$, forms a basic vector set, and the orthogonal least-squares solution $\hat{\Theta}$ satisfies the condition that $\Phi(u)\hat{\Theta}$ will be the projection of Y on the space generated by the basic function vectors $\{\Phi_i\}$. The orthogonal least-squares method implies the basic vector set transformation $\{\Phi_i\}$ into an orthogonal basic vector set, and, therefore, makes it possible to compute the individual impact on the output, for all base vectors. An orthogonal factorization of $\Phi(u)$ can be obtained by means of a construction known as Gram–Schmidt orthogonalization process.

First, set $P_1 = \Phi_1$. The following vectors are then given inductively in the following way: suppose that $P_1, \cdots, P_m$ $(1 \leq m \leq n)$ have been chosen, so that for each k $\{P_1, \cdots, P_m\}$, $(1 \leq k \leq m)$ is an orthogonal base for the vector subspace that is generated by that $\Phi_1, \cdots, \Phi_k$. For building the next vector P_{m+1}, set

$$P_{m+1} = \Phi_{m+1} - \sum_{k=1}^{m} \frac{\Phi_{m+1}, P_k}{\|P_k\|^2} P_k, \tag{13}$$

So $P_{m+1} \neq 0$, since, otherwise, P_{m+1} is a linear combination of $P_1, \cdots, P_m$. Furthermore, if $1 \leq j \leq m$, then

$$\langle P_{m+1}, P_j \rangle = \langle \Phi_{m+1}, P_j \rangle - \sum_{k=1}^{m} \frac{\langle \Phi_{m+1}, P_k \rangle}{\|P_k\|^2} \langle P_k, P_j \rangle = \langle \Phi_{m-1}, P_j \rangle - \langle \Phi_{m-1}, P_j \rangle = 0. \tag{14}$$

Therefore, $\{P_1, \cdots, P_{m+1}\}$ is an orthogonal set consisting of $m + 1$ nonzero vectors in the subspace generated by $\Phi_1, \cdots, \Phi_{m+1}$. Because orthogonal nonzero vectors are linearly independent, this is a base for this subspace. So, the vectors $P_1, \cdots, P_n$ can be constructed one after the other according to the Equation (14). In general, we have, for $n = N$:

$$P_1 = \Phi_1, \tag{15}$$

$$P_2 = \Phi_2 - \frac{\langle \Phi_2, P_1 \rangle}{\|P_1\|^2} P_1, \tag{16}$$

$$P_3 = \Phi_3 - \frac{\langle \Phi_3, P_1 \rangle}{||P_1||^2} P_1 - \frac{\langle \Phi_3, P_2 \rangle}{||P_2||^2} P_2, \tag{17}$$

$$P_N = \Phi_N - \sum_{k=1}^{N-1} \frac{\langle \Phi_N, P_k \rangle}{||P_k||^2} P_k. \tag{18}$$

The matrix $\Phi(u)$ can be written as

$$\Phi(u) = PQ, \tag{19}$$

where the matrix $P = [P_1, P_2, \cdots, P_N]$ has the size $M \times N$ with orthogonal columns, and Q is a unitary $N \times N$ upper triangular matrix with 1 on the main diagonal and 0 below the main diagonal.

$$Q = \begin{bmatrix} 1 & \frac{\langle \Phi_2, P_1 \rangle}{||P_1||^2} P_1 & \frac{\langle \Phi_3, P_1 \rangle}{||P_1||^2} P_1 & & \frac{\langle \Phi_N, P_1 \rangle}{||P_1||^2} P_1 \\ 0 & 1 & \frac{\langle \Phi_3, P_2 \rangle}{||P_2||^2} P_1 & \cdots & \frac{\langle \Phi_N, P_2 \rangle}{||P_2||^2} P_2 \\ 0 & 0 & 1 & & \frac{\langle \Phi_N, P_{N-1} \rangle}{||P_{N-1}||^2} P_{N-1} \\ & & \vdots & \ddots & \vdots \\ 0 & 0 & 0 & \cdots & 1 \end{bmatrix}. \tag{20}$$

The properties of orthogonality of P are advantageous from the logarithmic point of view, in such a way that Equation (7) can be represented by

$$Y = PQ\Theta + E\left(u^k\right), \tag{21}$$

and, defining $W = Q\Theta$, the Equation (21) can be represented by

$$Y = PW + E\left(u^k\right), \tag{22}$$

and

$$\Theta = Q^{-1}W, \tag{23}$$

where $W = [w_1, w_2, \cdots, w_N]^T R^{N \times 1}$. The vector W can be deduced as the optimal estimate $\hat{W} = [\hat{w}_1, \hat{w}_2, \cdots, \hat{w}_N]^T$ as follows:

From Equation (22), can be represented by

$$\hat{Y} = P\hat{W}, \tag{24}$$

pre-multiplying Equation (24) by P^T

$$P^T \hat{Y} = P^T P \hat{W}, \tag{25}$$

in vector form

$$\begin{bmatrix} P_1^T \hat{Y} \\ \vdots \\ P_N^T \hat{Y} \end{bmatrix} = \begin{bmatrix} P_1^T P_1 \hat{w}_1 \\ \vdots \\ P_N^T P_N \hat{w}_N \end{bmatrix}, \tag{26}$$

$$\hat{w}_i = \frac{P_i^T \hat{Y}}{P_i^T P_i}, \text{ for } i = 1, 2, \cdots, N. \tag{27}$$

So, $||Y - P\hat{W}||_2$ is minimal. The equivalent optimal parameters vector is

$$\hat{\Theta} = Q^{-1}\hat{W}. \tag{28}$$

The orthogonalization Gram–Schmidt algorithm can be applied to calculate Equation (28), therefore, to solve the least squares algorithm and to estimate $\hat{\Theta}$ and evaluate the variance for each basic function φ_i.

The variance of the output can be written as

$$\frac{1}{M}Y^{T}Y = \frac{1}{M}\sum_{i=1}^{N} w_i^2 P_i^T P_i + \frac{1}{M}\left(E\left(u^k\right)\right)^{T} E\left(u^k\right).$$ (29)

Note that $\sum_{i=1}^{N} w_i^2 P_i^T P_i / M$ is the part of the variance of the looked-for output which can be represented by the basic functions, and $E^T E / M$ is the variance not represented by y_t. Thus, $w_i^2 P_i^T P_i$ is the increment of the variance of the desired output represented by P_i, and the reduction ratio of the error due to P_i can be defined by:

$$r_i = \frac{\hat{w}_i^2 P_i^T P_i}{Y^T Y},$$ (30)

This ratio allows a simple and efficient mechanism to search for a significant basic functions subset. The implementation is based on the classical Gram–Schmidt method [36,37], see Appendix A.

PBFs order changing will result in an error reduction ratio change, r_i. For PBFs, there are $N!$ combinations; $r_i^{(j)}$ denotes the error reduction r_i corresponding to the j-th PBFs arrangement. The conventional Gram–Schmidt algorithm can be applied to find the actual arrangement of the basic functions $\varphi_1(u_t), \varphi_2(u_t), \cdots, \varphi_N(u_t)$, which represents the best arrangement, in such a way that

$$\sum_{i=1}^{k} r_i \geq \sum_{i=1}^{k} r_i^{(j)} \text{ for } j \neq 0, \ j = 1, 2, \cdots N!, \ k = 1, 2, \cdots, N.$$ (31)

So, the priority of the basic functions is determined. Therefore, the best PBFs arrangement is denoted by $\varphi_1(u_t), \varphi_2(u_t), \cdots, \varphi_N(u_t)$, and the corresponding parameter vector is Θ.

3. TSP Modelling

3.1. TSP Emission Measurement Instrument

A PM instrument was used to measure TSP emissions from 350 MW Unit. The instrument installed is a Neo Laser Dust Long-Path monitor, based on a diode laser technology. The instrument was installed on the appropriate stack elevation to prevent interference with the probe due to problems of flow turbulences, cyclonic flow, and fluctuating PM stratification, see Figure 2. The measuring principle of this instrument is based on the particles property to absorb and scatter transmitted light. A red laser light is transmitted through the process flue gas, and two separate detectors detect the forward scattered light and direct light transmitted. Both signals are functions of the number of dust particles contained in the flue gas. Because of the low range of concentration levels used in the scattering mode, the PM probe was setup to operate on the direct light signal mode. This mode works for an active range from 0 to 10,000 mg/Nm^3. Given the known PM emissions range at Unit, the instrument output current loop (from 4 to 20 mA) was adjusted from 0 to 800 mg/Nm^3, to increase sensitivity. The response time of this instrument is 1–2 s. The detection limit is 0.5 mg/Nm^3, and the accuracy is ± 0.5 mg/Nm^3 or ± 5 percent of the reading, whichever is higher. The PM Neo instrument consists of a transmitter and a receiver unit. Both the transmitter and the receiver are installed onto stack flanges. The optical path length for this probe can be adjusted from 3 to 6 m. For this particular application, it was setup at 5.5 m. The instrument has also an air purging system, preventing dust from fouling the optical windows and interfering with the measurements.

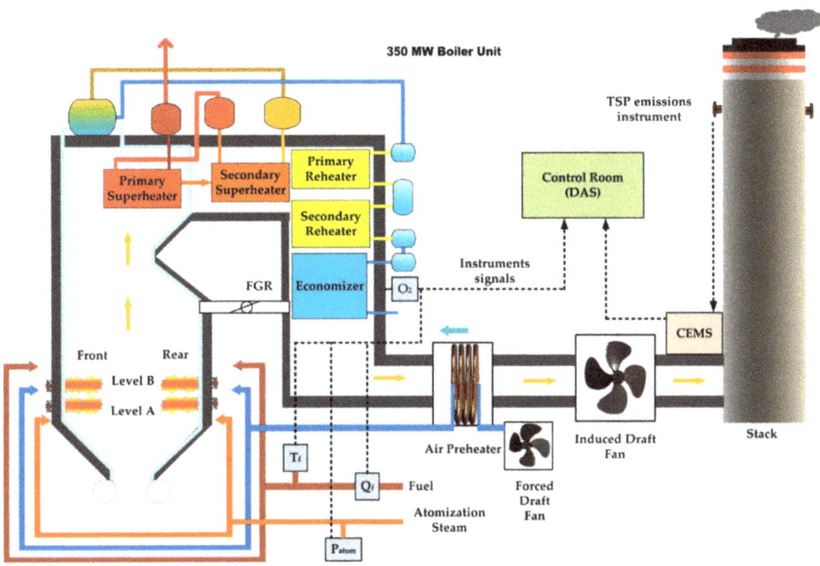

Figure 2. Boiler unit configuration. TSP: total suspended particles, CEMS: continuous emissions monitoring systems, T_f: fuel temperature, Q_f: fuel mass flow, P_{atom}: atomization pressure, O_2: oxygen excess, FGR: flue gas recirculation, DAS: data acquisition system.

3.2. Experimental Test

The boiler was an opposed wall-fired 350 MW unit, designed and manufactured by Babcock Hitachi. The unit had a balanced draft furnace. The firing system was composed of 12 conventional oil-fired combustion cells (two burners per cell, six cells per wall), with one common secondary air register per burner pairs. Secondary air was introduced at the adjacent side of the combustion walls; however, the wind box was of a wrap-around design to minimize wind-box air stratification. The boiler was a single reheat unit, with hot reheat steam temperatures of 540 °C. The boiler was equipped with flue gas recirculation (FGR) for steam temperature control, as well with attemperating sprays and sootblowers. There was neither particulate nor SO_2 control on this unit. This unit fired Mexican and imported high-sulfur (~3–4 percent sulfur) heavy fuel oil (Bunker C). For testing, TSP data were acquired over a range that included low, mid, and high TSP concentrations. To have good variability in the data, a range of unit operating conditions were used, including economizer oxygen excess, steam atomization pressure, fuel temperature, etc.; more boiler details can be consulted in reference [38]. Figure 2 shows a simplified boiler scheme including the used instruments.

The excess air could be manipulated within the capabilities of the O_2 trim control. Fuel combustion parameters, such as the fuel temperature and the steam atomization pressure variables, could also be manipulated to the maximum extent allowed by the firing system (physical limitations, design parameters, etc.). The FGR damper and secondary air register biasing could be performed within the available capability of the FGR damper and secondary air register control. Non-controllable variables (unit load, fuel flow, etc.) were monitored in order to characterize the effect on TSP emissions. The minimum and maximum levels were determined so not to break any operational constraints imposed by the boiler's limits (i.e., minimum wind-box pressure, maximum steam temperatures, maximum furnace exit gas temperature, etc.).

Forty-eight parametric tests were performed at full load with all cells in service. For all runs, the main variables data were collected for at least a 15 min, once the steady-state unit condition was achieved. A combination of data were acquired automatically by the data acquisition system (DAS) and the continuous emissions monitoring (CEM) system, including TSP emissions, Unit load (Load), oxygen excess (O_2), atomization pressure (P_{atom}), fuel mass flow (Q_f), fuel temperature (T_f). The tests at 100 percent load included combinations of oxygen excess, fuel temperature, and atomization pressure. The series of tests conducted at full load to evaluate the effect of atomization pressure and fuel temperature on boiler emissions and performance was accomplished at different levels of excess O_2, in a range from 0.6 to 1.8 percent. Tests with the FGR registers were performed at full load, only to observe the effect on TSP emissions and Unit performance. In the TSP emissions model, the FGR variable was not included, since, at this load, the flue gas recirculation was not manipulated. The influence of the FGR gate opening and levels of excess O_2 in TSP emissions is shown in Figure 3. The average difference in TSP emissions between the open and closed FGR damper conditions was approximately 40 mg/Nm^3, significant enough to consider a modified operation with the OFF/ON damper closed. Obviously, the manipulation of this damper modified other important parameters, such as NOx emissions. TSP emissions increased as the FGR gate opened and the main steam temperature control required a higher steam attemperation flow. On the other hand, a greater impact on TSP emissions was observed as the O_2 excess decreased, with a clearly non-linear behavior. Manipulation of the fuel temperature was accomplished at different levels of excess O_2, in a range from 0.6 to 1.7 percent, while the atomization pressure was kept at 15.1 bar and the OFF/ON FGR discharge damper was shut. The effect of the fuel temperature on TSP resulted in an average reduction of approximately 0.8 mg/Nm^3 per °C increase in fuel temperature, or a decrease of approximately 20 mg/Nm^3 in TSP for the tested range of fuel temperature from 115 to 142 °C.

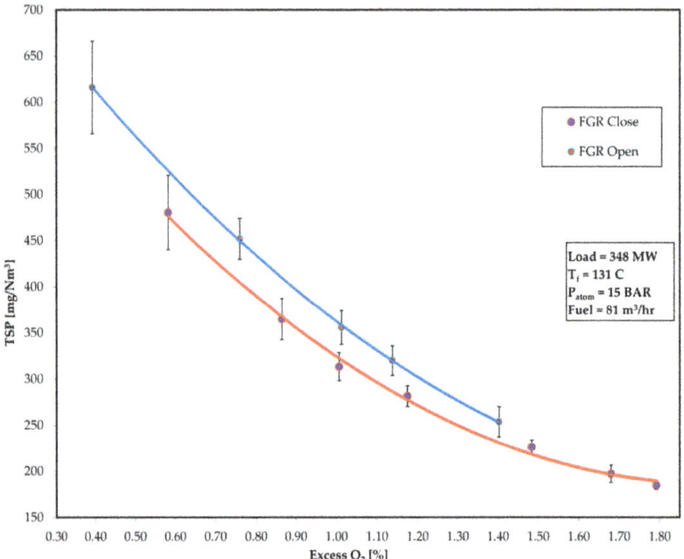

Figure 3. Effect of O_2 excess and FGR gate opening level on TSP emissions.

A series of tests were conducted at full load to evaluate the effect of atomization pressure and fuel temperature on boiler emissions and performance. The manipulation of the atomization pressure was accomplished at different levels of excess O_2 in a range from 0.6 to 1.8 percent, while the fuel temperature was kept at 130.8 °C and the OFF/ON FGR discharge damper was shut. The results of the atomization pressure tests on TSP are shown in Figure 4. The effect of the steam atomization pressure on TSP was of reducing TSP emissions as much as 15 mg/Nm^3 per bar increase in the atomization pressure at the low O_2 levels. The effect of steam atomization on nitrogen oxides (NOx) emissions was less significant than for TSP, representing approximately 1% increase in NOx emissions per bar increase in atomization pressure, independent of the excess O_2 level.

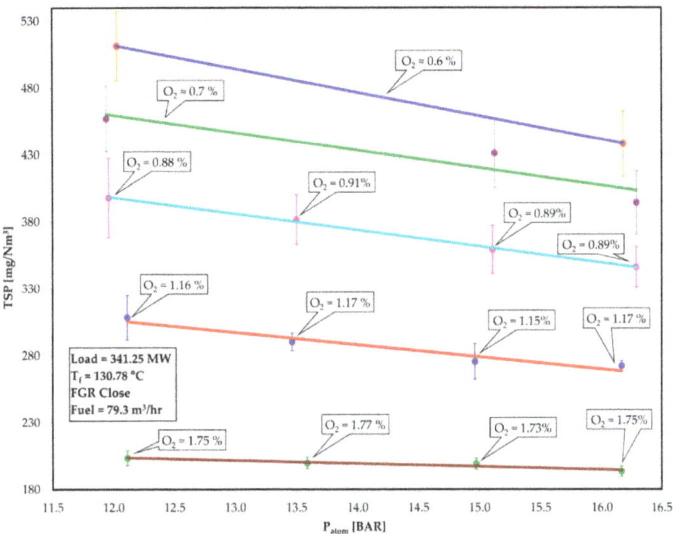

Figure 4. Steam atomization pressure effects on TSP emissions.

3.3. TSP Emissions Modelling and Validation

From the parametric tests results, the variables that had independent effects on TSP emissions in the combustion process of the unit were: O_2 excess, P_{atom}, T_f, Load, and Q_f. Considering these variables as input, the proposed PBF–ANN was developed. From Equation (2), m = 5, $n_1 = n_2 = n_3 = n_4 = n_5 = 1$, n = 5, and $u_t = [u_1, u_2, u_3, u_4, u_5]^T$, where $u_1 = $ Load, $u_2 = Q_f$, $u_3 = P_{atom}$, $u_4 = T_f$, $u_5 = O_2$. Figure 5 shows an outline of the structure of the ANN for the implementation of the model defined in the Equation (6).

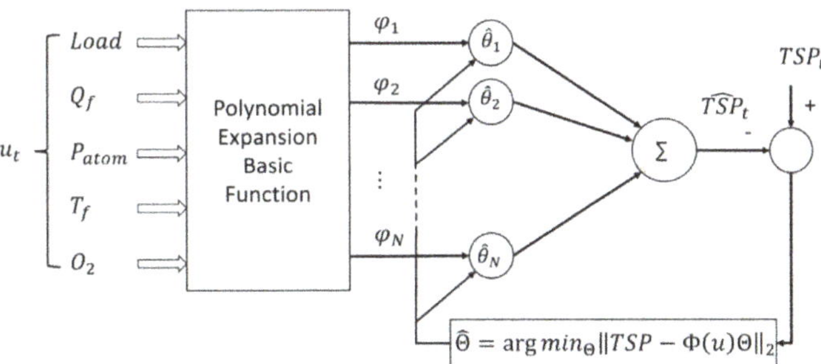

Figure 5. Polynomial basis function (PBF)–artificial neural networks (ANN) model structure for TSP modelling.

The basic functions number of the polynomial expansion for order k = 3 is N = $\frac{(5+3)!}{5!3!}$ = 56.

The classical Gram–Schmidt method for Equations (A1)–(A11) was implemented in MatLab® software in order to get the highest priority basic functions subset and parameter estimation Θ.

The classical MLP–ANN was implemented with input, hidden, and output layers. The input layer had six neurons (one for each model input variable, O_2 excess, P_{atom}, T_f, Load, and Q_f). The hidden layer was defined with N neurons, and the output layer was with one neuron; the TSP prediction was defined as $T\hat{S}P_t$. The input to the output relationship is characterized by $\hat{PST}_t = \frac{1}{1-e^{-f_s}}$, $f_s = \sum_{i=1}^{N} s_i \hat{v}_i + \hat{v}_{N+1}$, $s_i = \frac{1}{1-e^{-r_j}}$, $r_j = \sum_{i=1}^{5} x_i w_{ij} + w_{6,j}$. Figure 6 shows a graphical representation of the MLP–ANN structure.

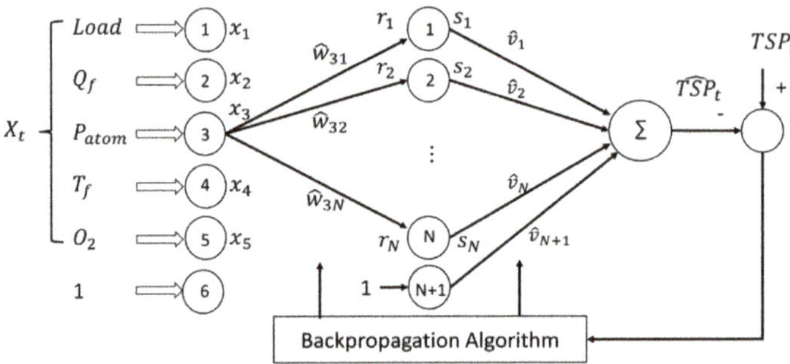

Figure 6. MLP–ANN model structure for TSP modelling.

The MLP model training process was made with a backpropagation algorithm implemented in MatLab® software (The MathWorks, Inc., Natick, MA, USA) in order to build the model and to evaluate the best hidden layer neural N value for modelling TSP emissions.

In order to model the validation, the set of parametric tests was divided into two subsets: the first subset for the training model (80% of parametric tests) and the second subset for the validation model (20% of parametric tests).

For evaluating the predictive accuracy between the PBF and the MLP models, two statistical metrics were selected: the root-mean-square error (RMSE) and the correlation coefficients (R^2). R^2 indicates the correlation between observed and predicted data, and RMSE measures error-based model accuracy. $R^2 = 1$ gives the results of the predicted value equal to the measured value.

R^2 is defined as

$$R^2 = 1 - \frac{\sqrt{\sum_{k=1}^{M}(y_k - \hat{y}_k)^2}}{\sqrt{\sum_{k=1}^{M}(y_k - \bar{y}_k)^2}}, \tag{32}$$

and RMSE is

$$RMSE = \sqrt{\frac{\sum_{k=1}^{M}(\hat{y}_k - y_k)^2}{M}} \tag{33}$$

where M is the number of dataset elements for the estimation, y_k is the actual TSP value (observed output), $\hat{y}_k$ is the model-predicted TSP value (estimate output), and $\bar{y}_k$ is the mean of the observed data.

Several PBF and MPL–ANN were simulated with different model sizes, varying radically the first most significant base functions number L in the PBF model and the hidden layers neuron number N in the MLP model for computing statistical metrics. Table 1 shows the model performance results. The PBF models with L = 10 to 25 presented a limited improvement of only $R^2 = 0.0077$ and RMSE of 1.51 mg/Nm^3, and the MLP models were similar with a minor variation of $R^2 = 0.0082$, from N = 2 to N = 6.

Table 1. PBF and MLP–ANNs models performance results. RMSE: root-mean-square error, R^2: correlation coefficient

Model	Model Size	R^2	RMSE
	L = 5	0.9161	30.59
	L = 10	0.9236	29.19
PBF	L = 15	0.9274	28.47
	L = 20	0.9297	28.01
	L = 25	0.9313	27.68
	N = 1	0.8628	39.08
	N = 2	0.8642	38.88
MLP	N = 3	0.8622	39.17
	N = 4	0.8607	39.38
	N = 5	0.8581	39.75
	N = 6	0.8560	40.04

TSP emissions model predictions with PBF and MLP–ANNs using validation data were computed. The first most significant base functions L = 25 and hidden layers neuron N = 2 specific numbers were used for PBF and MLP model simulation, respectively. Figure 7 shows a comparative simulation from the PBF and MLP models employing the validation data. The validation data matched the different boiler operating conditions where the economizer oxygen excess changed from 0.4 to 1.2, and the steam atomization pressure changed from 11.9 to 16.4 bar, leading to TSP emission extremes.

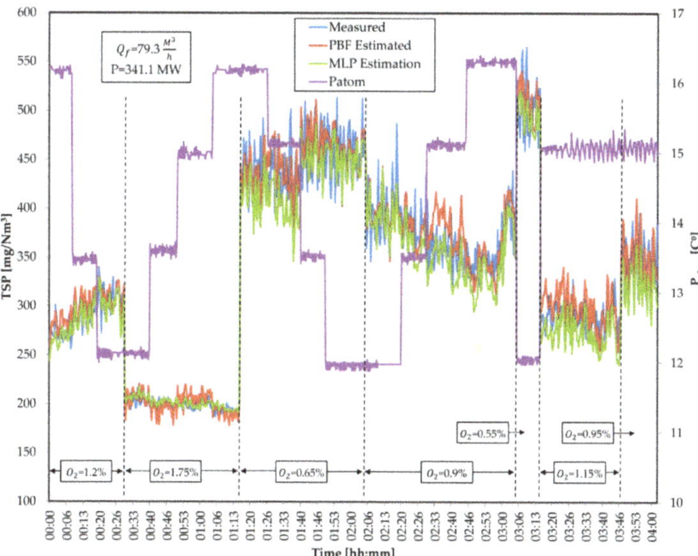

Figure 7. Measured (validation subset) and estimated TSP with the PBF and MLP models.

The correlation between TSP emission prediction and TSP measured values for both PBF and MLP models is shown in Figure 8; for the best PBF and MLP structure models, i.e., L = 25 and N = 2, respectively, the correlation linear trend was drawn in order to have a framework for the comparison. Here, we could appreciate that the correlation coefficient for the PBF model was R^2 = 0.9313, and that for the PBF model was R^2 = 0.8642. In high TSP emissions boiler operating conditions, a larger error was observed.

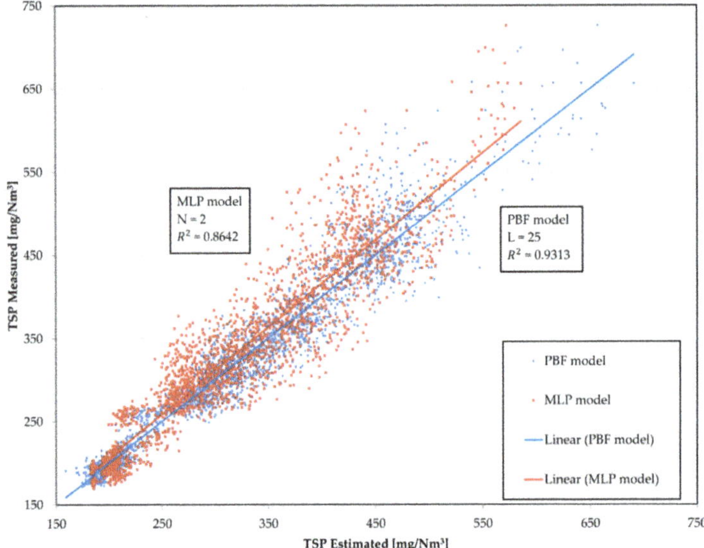

Figure 8. TSP emission models performance.

In the MLP model, the fifty-six basic functions generated form the polynomial expansion basic function were not necessary for TSP prediction. The approach proposed allowed reducing the model size. The TSP model structure could be simplified, taking only the most relevant basic functions and discarding the basic functions that contributed little or nothing to the system dynamics.

Using only the 10 most significant basic functions was enough to have a good model for TSP emission predictions. The model parameters were estimated applying the orthogonal Gram–Schmidt method-adapted algorithm. The model structure order selected and the corresponding parameters are shown in Table 2. The model is a three-order function.

Table 2. TSP emissions model with the 10 most significant PBF.

Priority Order i	φ_i	$\hat{\theta}_i$
1	1	0.5365
2	u_4	−1.5361
3	u_4^2	0.4770
4	$u_2^2 u_3$	−0.8133
5	$u_1^2 u_2$	1.0052
6	$u_4 u_5^2$	−0.0265
7	$u_3 u_4 u_5$	0.1460
8	$u_1 u_4$	0.0741
9	$u_1^2 u_4$	−0.3164
10	$u_1^2 u_5$	−0.2501

4. Conclusions

Orthogonal least-square algorithms are a great tool that provide extra information about internal model behaviors. Finite expansion polynomial basic functions can be implemented with one-layer ANN and agree with the universal approximation theorem. The user can decide the model complexity accuracy for model selection, selecting the polynomial order and most significant basic function number.

PBF networks provide a viable alternative for the complex multivariable non-linear systems modelling in the industry, where there is little or no process knowledge. The model structure developed allows TSP emissions estimation due to fossil fuel combustion. Non-linear combinations of oxygen excess, atomization pressure, fuel temperature, load, and flow fuel values are sufficiently informative to predict TSP emissions with excellent precision. The TSP emissions estimation can be improved by increasing the training set with experimental tests. The estimation algorithm requires few resources for its implementation, providing a viable alternative for estimating pollutant emissions into the atmosphere during fossil fuel combustion processes.

This methodology can be replicated for other pollutants estimation emitted into the atmosphere in combustion processes, such as, for example, NOx and CO emissions, etc., and has a great application potential, regarding process design, process control combustion optimization, emission control, fault detection, etc., that would bring great benefits in the prevention of diseases, climate change, etc.

Author Contributions: Conceptualization, G.R.-L. and G.H.-R.; methodology, G.R.-L. and G.H.-R.; software, I.A.A.R.-M.; validation, J.G.R.-M. and M.T.-P.; formal analysis, G.R.-L. and G.H.-R.; investigation, G.R.-L.; resources, M.T.-P.; data curation, G.R.-L. and I.A.A.R.-M.; writing—original draft preparation, G.R.-L.; writing—review and editing, G.H.-R. and J.G.R.-M.; visualization, J.G.R.-M., I.A.A.R.-M. and M.T.-P.; supervision, G.R.-L.

Funding: This work was supported by the CONACYT-CFE sectorial foundation of México (Grant: FSIDTE-CFE 2009-C08-120854).

Conflicts of Interest: The authors declare no conflict of interest.

Appendix A

Error reduction ratio implementation was based on the classical Gram–Schmidt method:
In the first step, for $i = 1, 2, \cdots, N$, calculate

$$P_1^{(i)} = \Phi_i, \tag{A1}$$

$$W_1^{(i)} = \frac{Y^T P_1^{(i)}}{\left(P_1^{(i)}\right)^T P_1^{(i)}}, \tag{A2}$$

$$r_1^{(i)} = \frac{\left(w_1^{(i)}\right)^2 \left(P_1^{(i)}\right)^T P_1^{(i)}}{Y^T T}, \tag{A3}$$

find

$$s_1 = \text{argmáx}\left\{ r_1^{(i)} i = 1, 2, \ldots, N \right\}. \tag{A4}$$

In the k-th step, where $k \geq 2$ for $i = 1, 2, \cdots, N$, $i \neq s_1, \cdots, i \neq s_{k-1}$, compute

$$\alpha_{jk}^{(i)} = \frac{\Phi_i^T P_j}{\left(P_j\right)^T P_j}, \quad j = 1, 2, \cdots, k, \tag{A5}$$

$$P_k^{(i)} = \Phi_i - \sum_{j=1}^{k-1} \alpha_{jk}^{(i)} P_j, \tag{A6}$$

$$W_k^{(i)} = \frac{Y^T P_k^{(i)}}{\left(P_k^{(i)}\right)^T P_k^{(i)}}, \tag{A7}$$

$$r_k^{(i)} = \frac{\left(w_k^{(i)}\right)^2 \left(P_k^{(i)}\right)^T P_k^{(i)}}{Y^T T}, \tag{A8}$$

find

$$s_k = \text{argmax}\left\{ r_k^{(i)}, \; i = 1, 2, \ldots, N, \; i \neq s_1, \ldots, \; i \neq s_{k-1} \right\}, \tag{A9}$$

and select

$$P_k = P_k^{(s_k)} = \Phi_{s_k} - \sum_{j=1}^{k-1} \alpha_{jk}^{(s_k)} P_j. \tag{A10}$$

The algorithm ends at the L-th step in L limit or when at minimal error e is achieved,

$$1 - \sum_{j=1}^{L} r_j < e. \tag{A11}$$

References

1. Kumar, P.; Jain, S.; Gurjar, B.R.; Sharma, P.; Khare, M.; Morawska, L.; Britter, R. New Directions: Can a "blue sky" return to Indian megacities? *Atmos. Environ.* **2013**, *71*, 198–201. [CrossRef]
2. Molina, L.T.; Molina, M.J.; Slott, R.S.; Kolb, C.E.; Gbor, P.K.; Meng, F.; Singh, R.B.; Galvez, O.; Sloan, J.J.; Anderson, W.; et al. Air Quality in Selected Megacities. *J. Air Waste Manag. Assoc.* **2004**, *54*, 1–73. [CrossRef]
3. Akimoto, H. Global Air Quality and Pollution. *Science* **2003**, *302*, 1716–1719. [CrossRef] [PubMed]
4. Lim, S.S.; Vos, T.; Flaxman, A.D.; Danaei, G.; Shibuya, K.; Adair-Rohani, H.; Amann, M.; Anderson, H.R.; Andrews, K.G.; Aryee, M.; et al. A comparative risk assessment of burden of disease and injury attributable to 67 risk factors and risk factor clusters in 21 regions, 1990–2010: A systematic analysis for the Global Burden of Disease Study 2010. *Lancet* **2012**, *380*, 2224–2260. [CrossRef]
5. Sharma, P.; Sharma, P.; Jain, S.; Kumar, P. An integrated statistical approach for evaluating the exceedence of criteria pollutants in the ambient air of megacity Delhi. *Atmos. Environ.* **2013**, *70*, 7–17. [CrossRef]
6. Kumar, P.; Morawska, L.; Birmili, W.; Paasonen, P.; Hu, M.; Kulmala, M.; Harrison, R.M.; Norford, L.; Britter, R. Ultrafine particles in cities. *Environ. Int.* **2014**, *66*, 1–10. [CrossRef] [PubMed]
7. White, R.M.; Paprotny, I.; Doering, F.; Cascio, W.E.; Solomon, P.A.; Gundel, L.A. Sensors and "apps" for community-based: Atmospheric monitoring. *EM Air Waste Manag. Assoc. Mag. Environ. Manag.* **2012**, *5*, 36–40.
8. Kittelson, D.B.; Watts, W.F.; Johnson, J.P. Nanoparticle emissions on Minnesota highways. *Atmos. Environ.* **2004**, *38*, 9–19. [CrossRef]
9. Heal, M.R.; Kumar, P.; Harrison, R.M. Particles, air quality, policy and health. *Chem. Soc. Rev.* **2012**, *41*, 6606–6630. [CrossRef] [PubMed]
10. WHO. Ambient (Outdoor) Air Pollution Database by Country and City. United Nations 2014. Available online: http://www.who.int/phe/health_topics/outdoorair/databases/cities-2014/en/ (accessed on 14 June 2018).
11. Medina, S.; Plasencia, A.; Ballester, F.; Mücke, H.G.; Schwartz, J. Apheis: Public health impact of PM10 in 19 European cities. *J. Epidemiol. Community Health* **2004**, *58*, 831–836. [CrossRef] [PubMed]
12. DEFRA. Automatic Urban and Rural Network (AURN) 2018. Available online: https://uk-air.defra.gov.uk/networks/network-info?view=aurn (accessed on 14 June 2018).
13. Zhu, H.P.; Zhou, Z.Y.; Yang, R.Y.; Yu, A.B. Discrete particle simulation of particulate systems: Theoretical developments. *Chem. Eng. Sci.* **2007**, *62*, 3378–3396. [CrossRef]
14. Merbitz, H.; Fritz, S.; Schneider, C. Mobile measurements and regression modeling of the spatial particulate matter variability in an urban area. *Sci. Total Environ.* **2012**, *438*, 389–403. [CrossRef] [PubMed]
15. Aragón, F.M.; Miranda, J.J.; Oliva, P. Particulate matter and labor supply: The role of caregiving and non-linearities. *J. Environ. Econ. Manag.* **2017**, *86*, 295–309. [CrossRef]
16. Drewnick, F.; Schwab, J.J.; Jayne, J.T.; Canagaratna, M.; Worsnop, D.R.; Demerjian, K.L. Measurement of Ambient Aerosol Composition During the PMTACS-NY 2001 Using an Aerosol Mass Spectrometer. Part I: Mass Concentrations Special Issue of Aerosol Science and Technology on Findings from the Fine Particulate Matter Supersites Program. *Aerosol Sci. Technol.* **2004**, *38*, 92–103. [CrossRef]
17. Gozzi, F.; Della Ventura, G.; Marcelli, A. Mobile monitoring of particulate matter: State of art and perspectives. *Atmos. Pollut. Res.* **2016**, *7*, 228–234. [CrossRef]

18. Honarvar, A.R.; Sami, A. Towards Sustainable Smart City by Particulate Matter Prediction Using Urban Big Data, Excluding Expensive Air Pollution Infrastructures. *Big Data Res.* **2018**, *1*, 1–10. [CrossRef]

19. Soni, M.; Payra, S.; Verma, S. Particulate matter estimation over a semi arid region Jaipur, India using satellite AOD and meteorological parameters. *Atmos. Pollut. Res.* **2018**, *9*, 949–958. [CrossRef]

20. Vakili, M.; Sabbagh-Yazdi, S.R.; Khosrojerdi, S.; Kalhor, K. Evaluating the effect of particulate matter pollution on estimation of daily global solar radiation using artificial neural network modeling based on meteorological data. *J. Clean. Prod.* **2017**, *141*, 1275–1285. [CrossRef]

21. Belis, C.A.; Karagulian, F.; Larsen, B.R.; Hopke, P.K. Critical review and meta-analysis of ambient particulate matter source apportionment using receptor models in Europe. *Atmos. Environ.* **2013**, *69*, 94–108. [CrossRef]

22. Jia, J.; Cheng, S.; Yao, S.; Xu, T.; Zhang, T.; Ma, Y.; Wang, H.; Duan, W. Emission characteristics and chemical components of size-segregated particulate matter in iron and steel industry. *Atmos. Environ.* **2018**, *182*, 115–127. [CrossRef]

23. Lee, S.W.; Pomalis, R.; Kan, B. A new methodology for source characterization of oil combustion particulate matter. *Fuel Process. Technol.* **2000**, *65–66*, 189–202. [CrossRef]

24. Ye, H.; Nicolai, R.; Reh, L. A Bayesian–Gaussian neural network and its applications in process engineering. *Chem. Eng. Process.* **1998**, *37*, 439–449. [CrossRef]

25. Liu, Y.; Peng, C. Time-Variation Nonlinear System Identification Based on Bayesian-Gaussian Neural Network. In Proceedings of the 2009 Fifth International Conference on Natural Computation, Tianjin, China, 14–16 August 2009; pp. 353–357. [CrossRef]

26. Chen, S.; Cowan, C.N.; Grant, P.M. Orthogonal least squares learning algorithm for radial basis function networks. *IEEE Trans. Neural Netw.* **1991**, *2*, 302–309. [CrossRef] [PubMed]

27. Kassam, S.A.; Cha, I. Radial basis function networks in nonlinear signal processing applications. *Signals Syst. Comput.* **1993**, *2*, 1021–1025.

28. Zeng, X.; Zhao, H.; Jin, W. Nonlinear Plant Identifier Using the Multiple Adaptive RBF Network Convex Combinations. In *Advances in Computer Science and Information Engineering*; Springer: Berlin/Heidelberg, Germany, 2012; Volume 2, pp. 185–191.

29. Liu, G.P. *Nonlinear Identification and Control: A Nueral Network Approach*; Springer: London, UK; New York, NY, USA, 2001.

30. Astrom, K.J.; Wittenmark, B. *Computer-Controlled Systems: Theory and Design*, 2nd ed.; Prentice Hall International: Upper Saddle River, NJ, USA, 1990.

31. Ljung, L. *System Identification*; Prentice Hall International: Upper Saddle River, NJ, USA, 1998.

32. Duda, P.E.; Hart, P.E.; Stork, D.G. Multilayer Neural Networks. In *Pattern Classification*; Wiley & Sons: Hoboken, NJ, USA, 2000.

33. Kohonen, K. *Self-Organiation and Asociative Memory*, 2nd ed.; Springer: New York, NJ, USA, 1984.

34. Powell, M.J.D. *Approximation Theory and Methods*; Cambridge University Press: Cambridge, UK, 1981.

35. Haykin, S. *Neural Networks: A Comprehensive Foundation*, 2nd ed.; Prentice Hall International: New York, NJ, USA, 1999; Volume 13.

36. Billings, S.A.; Chen, S.; Korenberg, M.J. Identification of MIMO non-linear systems using a forward-regression orthogonal estimator. *Int. J. Control* **1989**, *49*, 2157–2189. [CrossRef]

37. Billings, S.A.; Korenberg, M.J.; Chen, S. Identification of non-linear output-affine systems using an orthogonal least-squares algorithm. *Int. J. Syst. Sci.* **1988**, *19*, 1559–1568. [CrossRef]

38. Ronquillo-Lomeli, G.; Romero, C.E.; Yao, Z.; Si, F.; Coria-Silva, R.; Hernandez-Rosales, F.; José Sanchez-Gaytana, L.; Trejo-Moralesa, A. On-line flame signal time series analysis for oil-fired burner optimization. *Fuel* **2015**, *158*, 416–423. [CrossRef]

Article

Multi-Step Short-Term Power Consumption Forecasting with a Hybrid Deep Learning Strategy

Ke Yan [1], Xudong Wang [1], Yang Du [2], Ning Jin [1,*], Haichao Huang [3] and Hangxia Zhou [1]

[1]   Key Laboratory of Electromagnetic Wave Information Technology and Metrology of Zhejiang Province, College of Information Engineering, China Jiliang University, Hangzhou 310018, China; yanke@cjlu.edu.cn (K.Y.); wxd.1992825@163.com (X.W.); zhx@cjlu.edu.cn (H.Z.)

[2]   Department of Electrical and Electronic Engineering, Xi'an Jiaotong-Liverpool University, Suzhou 215123, China; Yang.Du@xjtlu.edu.cn

[3]   State Grid Zhejiang Electric Power Co., Ltd, Hangzhou 310007, China; jdhhc@163.com

*   Correspondence: jinning1117@cjlu.edu.cn; Tel.: +86-571-8691-4503

Received: 30 September 2018; Accepted: 7 November 2018; Published: 8 November 2018

Abstract: Electric power consumption short-term forecasting for individual households is an important and challenging topic in the fields of AI-enhanced energy saving, smart grid planning, sustainable energy usage and electricity market bidding system design. Due to the variability of each household's personalized activity, difficulties exist for traditional methods, such as auto-regressive moving average models, machine learning methods and non-deep neural networks, to provide accurate prediction for single household electric power consumption. Recent works show that the long short term memory (LSTM) neural network outperforms most of those traditional methods for power consumption forecasting problems. Nevertheless, two research gaps remain as unsolved problems in the literature. First, the prediction accuracy is still not reaching the practical level for real-world industrial applications. Second, most existing works only work on the one-step forecasting problem; the forecasting time is too short for practical usage. In this study, a hybrid deep learning neural network framework that combines convolutional neural network (CNN) with LSTM is proposed to further improve the prediction accuracy. The original short-term forecasting strategy is extended to a multi-step forecasting strategy to introduce more response time for electricity market bidding. Five real-world household power consumption datasets are studied, the proposed hybrid deep learning neural network outperforms most of the existing approaches, including auto-regressive integrated moving average (ARIMA) model, persistent model, support vector regression (SVR) and LSTM alone. In addition, we show a *k*-step power consumption forecasting strategy to promote the proposed framework for real-world application usage.

Keywords: electric power consumption; multi-step forecasting; long short term memory; convolutional neural network

---

1. Introduction

Artificial intelligence (AI) enhanced electric power consumption short-term forecasting is an important technique for smart grid planning, sustainable energy usage and electricity market bidding system design. Existing work shows that 20% extra energy output is required to overcome a 5% integrated residential electric power consumption peak increment without effective power consumption forecasting [1]. The advanced metering infrastructure (AMI) introduces the possibility to learn power consumption pattern for each residential house from its historical data. The resulting power consumption prediction provides an important hint for both the power suppliers and consumers to maintain a sustainable environment for energy saving, management and scheduling [2,3].

Efficient and precise power consumption forecasting is always demanded in dynamic electricity market bidding system design [4–6]. However, both manual or automated power bidding requires response time for computerized calculation. Most existing works only perform one-step power load forecasting, which enquire immediate response from the participants. A multi-step forecasting strategy can be more preferred under this situation. In summary, for individual household electric power consumption prediction, two main challenges exist in the literature:

1. High prediction accuracy. The volatility level of single household power consumption is high due to the irregular human behaviours. Moreover, the source data is usually univariate, consisting only power consumption records in kilowatts (kws), which increases the difficulty for accurate power consumption forecasting.
2. Multi-step forecasting. Most existing load forecasting works focus on one-step forecasting solutions. A longer time forecasting solution is required to facilitate real-world application usage, such as the dynamic electricity market bidding system design.

Traditional electric power forecasting methods overcome the uncertainty by integrating the overall power consumption of a large group of households or clustering similar pattern customers into sub-groups to reduce the irregularity. However, during the development process of smart grid, the accurate prediction of a household electric power consumption is highly demanded, which may come out with a customized electricity price plan for that particular household. Moreover, univariate data forecasting remains as one of the most challenging problems in the field of machine learning, since most of the dependent variables are unknown, such as the electric current, voltage, weather conditions, etc. [7]. Classic univariate forecasting methods are usually applied to cases that either the rest of the features are too difficult to be measured or there are too many variables to be measured, e.g., the stock market indices forecasting problems [8]. Flexibility of those univariate forecasting methods is introduced while no extra information is required. The proposed approach can be plugged into management system for other households power consumption forecasting as long as the historical data is available in the system.

In recent years, deep learning neural networks (DLNNs) became increasingly attractive throughout the world and were extensively employed in a large number of application fields, including natural language processing (NLP) [9], image object detection [10], time series analysis [11], etc. For individual household power consumption forecasting problems, recent works reported that the long short term memory (LSTM) neural network provides extremely high accuracy on prediction [2,12,13]. Experimental results show that, by using the conventional LSTM neural network alone, the prediction accuracy outperforms most of the traditional statistical and machine learning methods, including auto-regressive integrated moving average (ARIMA) model [14], support vector machine (SVM) [15], non-deep artificial neural networks (ANNs) [16] and their combinations [17], because of the extra neighboring time frame states dependencies introduced by memory gates in recurrent neural network (RNN). However, even recent works, such as [2,12,13] focus on short-term forecasting strategy, which forecast power load only one step further. For particular applications, such as electricity market bidding system design, a longer time forecasting strategy can be more preferred.

Moreover, LSTM neural network is a special form of RNN [18]; and there exist other types of DLNNs, such as convolution neural networks (CNNs) [19] and deep belief nets (DBNs) [20]. The temporal CNN, which consists of a special 1-D convolution operation, is also reported to be potentially useful for time series prediction problems [21]. In the field of NLP, there are suggestions to combine the temporal CNN with RNN to obtain more precise classification results [22].

1.1. Related Works

Electric power consumption forecasting is useful in many application areas. Besides electricity market bidding, it can also be applied to demand side management for transcative grid [3] and power ramp rate control [23]. Conventional forecasting methods include support vector regression (SVR),

ANNs, fuzzy logic methods [24] and time series analysis methods, such as autoregressive integrated moving average (ARIMA) [25], autoregressive method with exogenous variables [26,27] and grey models (GMs) [28]. As early as 2007, Ediger and Akar [29] started to use ARIMA and seasonal ARIMA methods to forecast the energy consumption by fuel until the year 2020 in Turkey. Yuan et al. [14] compared the results of China's primary energy consumption forecasting using ARIMA and GM (1,1). Both methods work well; and a hybrid method combining the two methods was also proposed to show the best mean absolute percent error (MAPE) value they could achieve. Oğcu et al. [30] compared ANN and support vector regression (SVR) models in forecasting electricity consumption of Turkey. For performance measurement, the mean absolute percentage error (MAPE) rates are used; and the SVR model showed a 0.6% better performance than ANN. Rodrigues et al. [31] designed an ANN energy consumption model consisting of a single hidden layer with 20 neurons to forecast 93 households energy consumptions in Portugal. Experimental results showed an averaged MAPE value of 4.2% for daily energy consumption forecasting in between of the 93 households. Deb et al. [32] compared ANN and an adaptive neuro-fuzzy interface system for energy consumption forecasting of three institution buildings in Singapore, and showed high forecasting accuracy. Wang and Hu [33] proposed hybrid forecasting method combining ARIMA model, extreme learning machine (ELM), SVRs and Gaussian process regression model for short-term wind speed forecasting problem. All individual base forecasting models are integrated in a non-linear way, where the experimental results showed the forecasting accuracy and reliability of the proposed hybrid method.

Deep learning neural networks are modern popular machine learning techniques dealing with big data with high classification and prediction accuracy, which has been widely applied in many fields, such as stock indices forecasting [34,35], wind speed prediction [36,37], solar irradiance forecasting [38,39], etc. In recent years, with the fast development of smart grid technology, DLNNs are widely employed to solve power consumption forecasting problems, both for industrial and residential buildings; and because of the significantly more internal hidden layers and computations compared to classic ANNs, DLNN is applied to more challenging problems, such as power consumption forecasting for individual households [21]. Ryu et al. [40] trained DLNN with single household electricity consumption data in 2016 and showed that the DLNN can produce better prediction accuracy compared with shallow neural network (SNN), double seasonal Holt–Winters (DSHW) model and the autoregressive integrated moving average (ARIMA). Shi et al. [12] proposed a pooling-based deep recurrent neural network to capture the uncertainty of single household load forecasting problem and applied the proposed method on 920 Ireland customers' smart meter data. Experimental results show that the proposed deep learning neural network outperforms most classic data-driven forecasting methods, including ARIMA, SVR and RNN. Kong et al. [13] straightly applied a two-hidden-layer LSTM to single household power consumption forecasting problems; and compared their results with back-propagation neural network (BPNN), k-nearest neighbor regression (KNN) and extreme learning machine (ELM) to show the large forecasting accuracy improvement by using LSTM.

1.2. Contributions

In this study, a hybrid deep learning neural network framework combining LSTM neural network with CNN is designed to deal with the single household power consumption forecasting problem. The conventional LSTM neural network is extended by adding a pre-processing phase using CNN. The pre-processing phase extracts useful features from the original data and more importantly, converts the univariate data into multi-dimensional by 1-D convolution, which potentially enhances the prediction capability of the LSTM neural network. To evaluate the performance of the proposed framework, a series of experiments were performed based on five real-world households electric power consumption data collected by the UK-DALE dataset [41]. The experimental results show that the proposed hybrid DLNN framework outperforms most of the existing approaches in the literature, including auto-regressive integrated moving average (ARIMA) model, support vector regression (SVR) and LSTM alone with three measurement metrics, including root-mean-square error (RMSE), mean absolute error (MAE) and mean absolute percentage error (MAPE). The scientific impacts of this work to the literature involve:

- A 1-D convolutional neural network is introduced to pre-process the univariate dataset and convert the original data into multi-dimensional features after two layers of temporal convolution operations.
- A hybrid deep neural network is designed to forecasting power consumption for individual household. Experimental results show that the proposed framework outperforms most of the existing approaches including ARIMA, SVR and LSTM.
- A k-step forecasting strategy is designed to introduce k forecasting points/values simultaneously. The value of k is determined to be less than or equal to the number of cores/threads to maintain the efficiency. The actual forecasting period/response time depends on the power consumption recording interval and the value of k. Compared with traditional one-step forecasting strategies, the k-step forecasting solution provides more response time for dynamic electricity market bidding.

Five individual households located in UK are studied to show the effectiveness and robustness of the proposed hybrid DLNN structure design. The study of multi-step electric power consumption forecasting strategy can be useful in customizing the smart grid planning and electricity market bidding system design.

2. Materials and Methods

Long short term memory (LSTM) and convolutional neural network (CNN) are two hot branches of deep learning neural network and they have attracted wide attention across the world in recent years. In this study, aiming at solving the high volatility and uncertainty of single household power consumption forecasting problem, we combine LSTM and CNN to form a hybrid deep learning approach that is able to provide more accurate and robust forecasting result compared with traditional approaches.

With five real-world household power consumption data, the proposed framework pre-processes the raw data with CNN and uses the output of CNN to train the LSTM model.

2.1. Data Description

The power consumption data collected from five households located in London, UK was original published by Kelly and Knottenbelt [41]. In the original dataset, smart meters are used to collect power consumption data from each individual electric power device, such as television, air-con, fridge and so on. We utilize the aggregate power consumption data for the five households only. The original collection frequency is 6 s. We merge the data to convert it to time series datasets with time intervals at 5 min. Since the data lengths vary from different households, we select a continuous time period consisting of 12,000 data samples for each household. Out of 12,000 data samples, 10,800 data samples are used to train the proposed DLNN framework; and the remaining 1,200 data samples are retained for testing and verification purposes for each household.

2.2. Long Short Term Memory based Recurrent Neural Network

Long short term memory (LSTM) model is a special form of the recurrent neural network (RNN) that provides feedback at each neuron. The output of RNN is not only dependent on the current neuron input and weight but also dependent on previous neuron inputs. Therefore, theoretically speaking, the RNN structure is typically suitable for processing time series data. However, when dealing with a long and correlated series of data samples, exploding and vanishing gradients problems appear [42], which later becomes the cutting point for LSTM model to be introduced [43].

To overcome the vanishing gradients problem of RNN model, LSTM contains internal loops that maintain useful information and abandon garbages. There are four important elements in the flowchart of LSTM model: cell status, input gate, forget gate and output gate (Figure 1). The input, forget and output gates are used to control the update, maintenance and deletion of information contained in cell status. The forward computation process can be denoted as:

$$f_t = \sigma(W_f \cdot [h_{t-1}, x_t] + b_f), \tag{1}$$

$$i_t = \sigma(W_i \cdot [h_{t-1}, x_t] + b_i), \tag{2}$$

$$\tilde{C}_t = tanh(W_C \cdot [h_{t-1}, x_t] + b_C), \tag{3}$$

$$C_t = f_t \cdot C_{t-1} + i_t \cdot \tilde{C}_t, \tag{4}$$

$$o_t = \sigma(W_o \cdot [h_{t-1}, x_t] + b_o), \tag{5}$$

$$h_t = tanh(C_t), \tag{6}$$

where C_t, C_{t-1} and $\tilde{C}_t$ represent current cell status value, last time frame cell status value and the update for the current cell status value, respectively. The notations f_t, i_t and o_t represent forget gate, input gate and output gate, respectively. With proper parameter settings, the output value h_t is calculated based on $\tilde{C}_t$ and C_{t-1} values according to Equations (4) and (6). All weights, including: W_f, W_i, W_C and W_o, are updated based on the difference between the output value and the actual value following back-propagation through time (BPTT) algorithm [44].

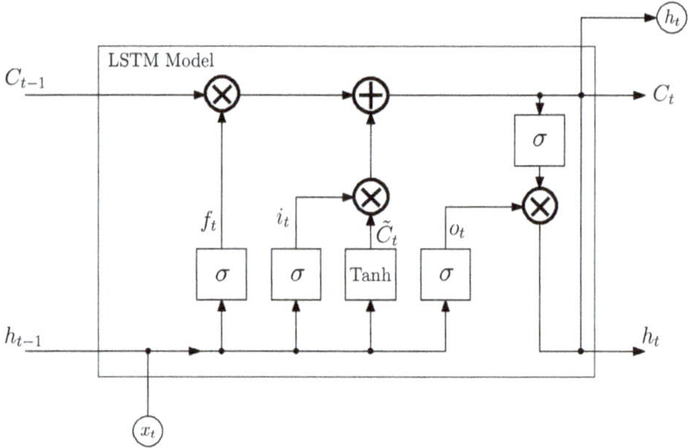

Figure 1. The internal structure of LSTM model.

2.3. Temporal Convolutional Neural Network

Convolutional neural network (CNN) is probably the most commonly used deep learning neural network which is currently mainly applied to image recognition/classification topics in the field of computer vision. With a large quantity of input raw data samples, CNN is usually capable to extract useful subsets of the input data efficiently. Generally speaking, CNN is still a feed-forward neural network, which is extended from multi-layer neural network (MLNN). The main difference between CNN and the traditional MLNN is that CNN has the properties of sparse interaction and parameter sharing [45].

Traditional MLNN uses full connection strategy to build the neural network between input layer and output layer, which means that each output neuron has the chance to interact with each input neuron. Suppose that there are m inputs and n outputs, the weight matrix has $m \times n$ entries. CNN reduces the weight matrix size from $m \times n$ to $k \times n$ by setting up a convolutional kernel with size $k \times k$. Moreover, the convolutional kernel is shared by all inputs, which means that there is only one weight matrix with size $k \times n$ to be learned from the training process. The two properties of CNN increases the training efficiency for parameter optimization; under the same computational complexity, the CNN is able to train a neural network with more hidden layers, or, in other words, a deeper neural network.

Temporal convolutional neural network introduces a special 1-D convolution, which is suitable for processing univariate time series data. Instead of using a $k \times k$ convolutional kernel as in the traditional CNN, the temporal CNN uses a kernel size of $k \times 1$. Suppose that the input data fits function $g(x) \in [l, 1] \to \mathbb{R}$; the convolutional kernel function is $f(x) \in [k, 1] \to \mathbb{R}$. The 1-D convolution mapping between the input and kernel $h(x) \in [(l - k)/d + 1, 1] \to R$ with step size d can be written as:

$$h(y) = \Sigma_{x=1}^{k} f(x) \cdot g(y \cdot d - x + k - d + 1).$$

After the temporal convolutional operation, the original univariate dataset can be expanded to a m-dimensional feature dataset. In this way, the temporal CNN applies 1-D convolution to time series data and expand the univariate dataset to multi-dimensional extracted features (first phase in Figure 2); and the expanded features are found to be more suitable for prediction using LSTM.

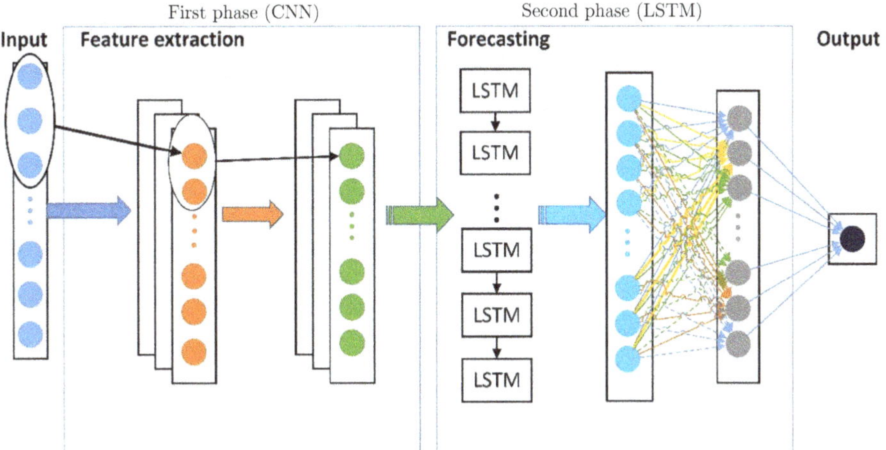

Figure 2. The proposed hybrid DNN power consumption forecasting framework.

2.4. CNN-LSTM Forecasting Framework

To attack the two challenges (volatility and univariate data) that we mentioned in Section 1, a hybrid deep neural network (DNN) combining CNN with LSTM is proposed. The structure of the hybrid DNN framework is depicted in Figure 2. In the pre-processing phase, CNN extract important information from the input data and most importantly, re-organize the univariate input data to multi-dimensional batches using convolution (Figure 2). In the second phase, the re-organized batches are input into LSTM units to perform forecasting.

From Figure 2, a two-hidden-layer temporal CNN is used to pre-process the input dataset. It is noted that the traditional temporal CNN usually includes pooling operations to prevent over-fitting when the number of hidden layer is greater than five. In this study, we omit the pooling operation to maximally retain the extracted features.

After pre-processing the input data, a LSTM neural network is designed to train and forecast the power consumption for individual household. The training process of LSTM structure is shown in Figure 3, where the extracted features from the first phase are treated as inputs to train the LSTM model. A dropout layer is added to the LSTM neural network to prevent overfitting. The loss value, which is the difference between the predicted output y_p and the expected output y_e, is computed to optimize the weights of all LSTM units. The optimization process follows the gradient descent optimization algorithm named RMSprop, which is commonly used for weight optimization of deep neural networks [46].

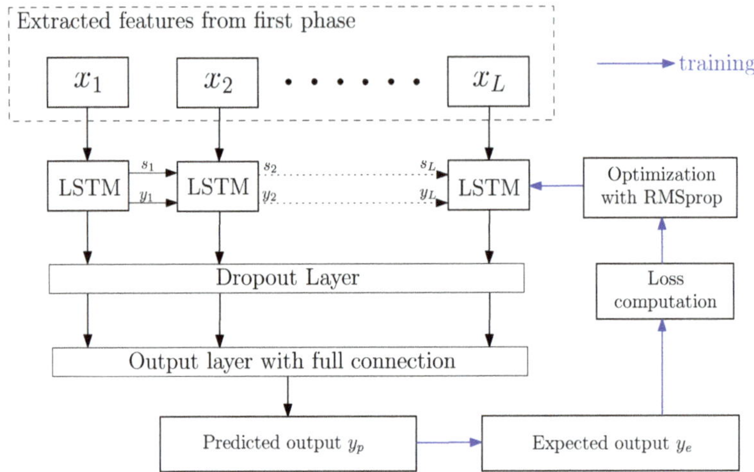

Figure 3. The training process of the LSTM model.

2.5. A k-Step Power Consumption Forecasting Strategy

Traditional power consumption forecasting approaches focus on one-step forecasting solutions [2,12,13]. For very short step size, such as 5 min, the response time can be too short for manual/automated electricity market bidding. In this study, we design a k-step power consumption forecasting strategy, which predicts k future data points simultaneously. Preassumption is made that the historical data is long enough to perform the data re-organization step.

Recall that the original power consumption data collected by UK-DALE has the step size at 6 s. The original data can be re-organized into different datasets with step size at n min, $2n$ min, ... kn min. In this study, we focus on $n = 5$. For each dataset, a core or a thread can be assigned to perform CNN-LSTM power consumption forecasting. The combinational result of all calculations from k cores provides a k-step power consumption forecasting solution, i.e., forecasting power consumption data points at 5 min, 10 min, ... until $5k$ min in the future. Detailed algorithm of the proposed k-step power consumption forecasting strategy is shown in Algorithm 1.

Algorithm 1 A k-step power consumption forecasting strategy

Input: The UK-DALE dataset.
Output: Data points at 5 min, 10 min, ...$5k$ min.
Initialization: re-organize the original data into k different datasets according to specified step sizes.
While There are unassigned datasets and there are free threads/cores
    Assign any unassigned dataset to a free thread/core.
    Apply the proposed CNN-LSTM framework to the specific dataset and obtain one-step forecasting result.
end-While
Combine all one-step forecasting results to obtain a k-step power consumption forecasting result.

Using the concurrent programming, we claim that the efficiency of the proposed k-step forecasting algorithm is competitive to the traditional one-step forecasting algorithms, given that the value of k is less than or equivalent to the number of threads/cores.

3. Results

The proposed hybrid DNN framework is implemented using Python 3.5.2 (64-bit) with PyCharm Community Edition 2016.3.2. The hardware configuration includes an Intel Core i7-7700 CPU @2.80GHz, 8G RAM and a NVIDIA GeForce GTX1050 graphics card. The proposed hybrid DNN framework is built based on the open source deep learning tool Tensorflow, proposed by Google [47] with Keras [48] version 2.0.8 as the front-end interface.

The prediction results of the proposed CNN-LSTM are compared with modern existing methods, including ARIMA model, SVR and LSTM. The prediction performances are evaluated using error metrics [49]. Three error metrics are calculated, including root-mean-square error (RMSE), mean absolute error (MAE) and mean absolute percentage error (MAPE). Generally speaking, smaller values of the error metrics present higher prediction accuracy. The formulations of the above three metrics are listed in Equations (7)–(9):

$$RMSE = \sqrt{\frac{\Sigma_{i=1}^{N}(\hat{y}_i - y_i)^2}{N}}, \tag{7}$$

$$MAE = \frac{\Sigma_{i=1}^{N}|\hat{y}_i - y_i|}{N}, \tag{8}$$

$$MAPE = \frac{\Sigma_{i=1}^{N}\left|\frac{\hat{y}_i - y_i}{y_i}\right| \times 100}{N}, \tag{9}$$

where y_i is an actual testing sample value; $\hat{y}_i$ is the prediction result of y_i; and N is the total number of testing samples.

All error metrics values of the five compared methods with all five households data described in Section 2.1 are listed in Table 1. The averaged computational time for each prediction point is recorded in Table 2 for all compared methods except the persistence model, since the persistence simply takes the previous time stamp's data as the prediction result [50]. On average, and most of the cases in Table 1, the proposed CNN-LSTM framework outperforms all other compared forecasting methods with reasonable computational time (around 0.06 s for each prediction). Compared with SVR, the proposed framework has slightly higher MAE and RMSE values for households 2 and 4. From the data description of UK-DALE project, the power consumption curves of households 2 and 4 are less volatile; and the power consumption curves of households 1, 3 and 5 are relatively more active. The prediction results suggest that the deep learning methods are more suitable for volatile data description. Moreover, for MAPE, which measures the relative errors of the prediction results, the proposed CNN-LSTM framework shows lower error rates compared with all other methods for all five households.

Figures 4–6 show the detailed prediction results for households 1, 3 and 5. The actual power consumption curves are shown in black color; and the CNN-LSTM prediction results are shown in red. In general, from Figures 4–6, the proposed CNN-LSTM method shows lower prediction errors and consequently higher prediction accuracy compared with ARIMA model, SVR and LSTM for all fives houses power consumption data collected by the UK-DALE dataset, which suggests that the proposed method is more robust than other methods for short-term power consumption forecasting.

In addition, we show the k-step power consumption forecasting results for k value up to 6. Table 3 shows RMSE and MAPE values for each house, while the value of k increases from 2 to 6. In Figure 7, twenty groups of 6-step power consumption forecasting results are depicted with training data omitted. It can be easily observed that the k-step forecasting algorithm produces more steps of forecasting results with acceptable compared to traditional one-step forecasting approaches.

Table 1. Prediction results comparison between ARIMA, SVR, LSTM, persistent model (persis.) and the proposed method (propos.), using RMSE, MAE and MAPE.

Dataset	RMSE					MAE					MAPE				
	ARIMA	SVR	LSTM	Persis.	Propos.	ARIMA	SVR	LSTM	Persis.	Propos.	ARIMA	SVR	LSTM	Persis.	Propos.
Hse 1	0.0305	0.034	0.0299	0.0335	0.0304	0.0151	0.0156	0.0149	0.0153	0.0140	20.8196	18.3178	20.8594	21.3442	18.1268
Hse 2	0.0027	0.0014	0.0023	0.0038	0.0016	0.0024	0.0011	0.0018	0.0018	0.0010	12.5666	5.2907	7.3485	7.7049	3.7647
Hse 3	0.0122	0.0144	0.0168	0.0356	0.0117	0.0072	0.0078	0.0090	0.0168	0.0069	8.7881	8.6152	8.7619	16.9004	8.5512
Hse 4	0.0075	0.0058	0.0070	0.0072	0.0064	0.0054	0.0036	0.0050	0.0046	0.0042	27.1986	14.2533	22.9484	17.5189	15.3256
Hse 5	0.0070	0.0073	0.0069	0.0071	0.0060	0.0043	0.0037	0.0032	0.0028	0.0028	8.4620	6.7788	5.8760	5.0290	5.0226
Average	0.0120	0.0127	0.0126	0.0175	0.0112	0.0069	0.0064	0.0068	0.0083	0.0058	15.5670	10.6512	13.1589	13.6995	10.1582

Table 2. Averaged computational time (in seconds) taken by CNN-LSTM, LSTM, SVR and ARIMA models for each predicted data point.

Approach	House 1	House 2	House 3	House 4	House 5	Average
CNN-LSTM	0.0652	0.0631	0.0591	0.0631	0.0656	0.0632
LSTM	0.0059	0.0036	0.0044	0.0044	0.0038	0.0021
SVR	0.0075	0.0065	0.0045	0.0095	0.005	0.0066
ARIMA	0.7493	0.6898	0.6918	0.7109	0.8783	0.7440

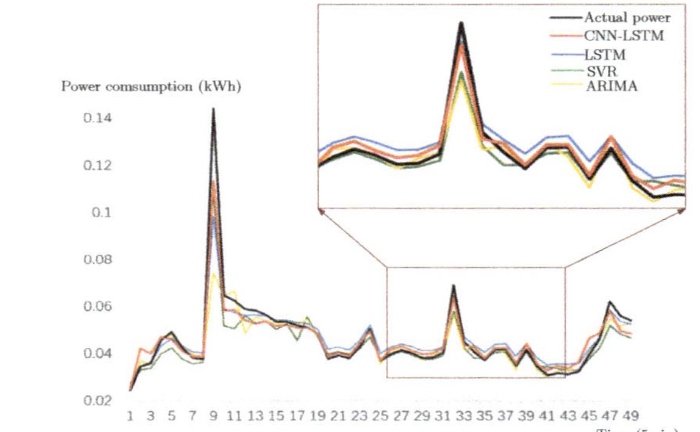

Figure 4. The prediction results for household 1 power consumption data using various methods. The dark red box shows a zoom-in region of the prediction results.

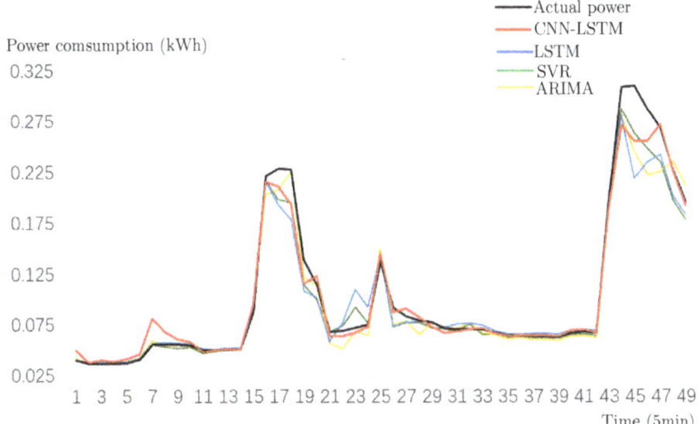

Figure 5. The prediction results for household 3 power consumption data.

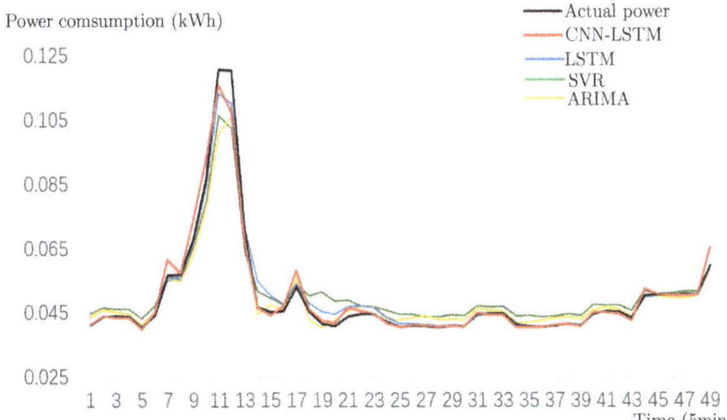

Figure 6. The prediction results for household 5 power consumption data.

Table 3. Averaged RMSE and MAPE values for each household data. The value of k increases from 2 to 6.

Metric	RMSE					MAPE				
Dataset	$k = 2$	$k = 3$	$k = 4$	$k = 5$	$k = 6$	$k = 2$	$k = 3$	$k = 4$	$k = 5$	$k = 6$
House 1	0.0341	0.0339	0.0478	0.0508	0.0577	18.42	18.98	19.66	19.89	20.15
House 2	0.0017	0.0021	0.0024	0.0026	0.0025	3.91	4.33	4.64	4.90	4.85
House 3	0.0120	0.0274	0.0284	0.0236	0.0256	9.24	10.31	10.50	11.98	10.98
House 4	0.0068	0.0069	0.0072	0.0070	0.0071	15.41	15.25	16.38	15.79	16.08
House 5	0.0067	0.0068	0.0079	0.0879	0.0100	5.53	5.78	6.03	6.41	6.64

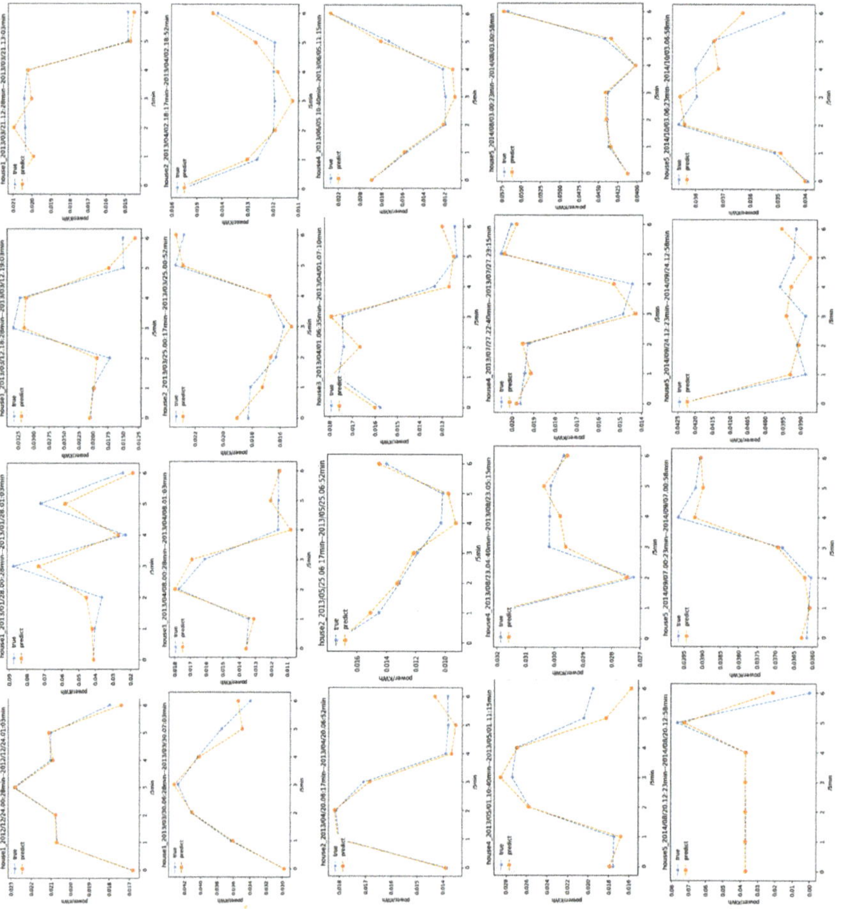

Figure 7. Twenty groups of *k*-step (*k* = 6) power consumption forecasting results for different household datasets, showing the performance and robustness of the proposed CNN-LSTM framework.

Based on Algorithm 1, the k-step forecasting method repetitively runs the proposed CNN-LSTM framework. The average error and the average running time for the k-step algorithm will be very close to the original one-step CNN-LSTM framework, given that the value of k is less than or equivalent to the number of cores/threads.

Considering a very small power consumption interval at 5 min, the proposed method demonstrates a 30 min response time forecasting for dynamic electricity market bidding, which can be potentially useful in real-world applications [6]. The 30 min forecasting period is the necessary response time that we considered in this experimental section. Nevertheless, the 30 min response time can be further extended in two ways:

- First, the $5 \times 6 = 30$ min can be extended with larger k value. In order to keep our computation in real-time, we force the value of k to be less than or equivalent to the number of cores/threads. The response time can be extended with more powerful CPU.
- Second, the $5 \times 6 = 30$ min can also be extend using a coarser time interval, e.g., 15 min resolution instead of 5 min. For $k = 6$, the proposed k-step forecasting algorithm provides a one-and-a-half-hour response time for market bidding.

The project page and source code of the proposed CNN-LSTM framework is freely available online at: http://www.keddiyan.com/files/PowerForecast.html.

4. Conclusions and Future Work

This study proposed a novel hybrid deep learning neural network framework combining convolutional neural network (CNN) and long short term memory (LSTM) neural work to deal with univariate and volatile residential power consumption forecasting. Recent works already show that by LSTM neural network alone, high prediction accuracy for power consumption forecasting can be achieved [2,12,13]. We further demonstrate that the hybrid framework that was proposed in this study outperforms the conventional LSTM neural network. The CNN extracts the most useful information from the original raw data and converts the univariate single household power consumption dataset into multi-dimensional data, which potentially facilitates the prediction performance of LSTM.

Figure 8 shows the prediction accuracy improvement from conventional LSTM to CNN-LSTM using MAPE as a measurement metric. The results were obtained based on five real-world households power consumption data collected by the UK-DALE project. The proposed CNN-LSTM framework is 13.1%, 48.8%, 2.4%, 33.2% and 14.5% lower than LSTM, respectively, for the five tested households, using MAPE as the error metric, which demonstrates the usefulness of the proposed method for maintaining a sustainable balance between energy consumption and savings.

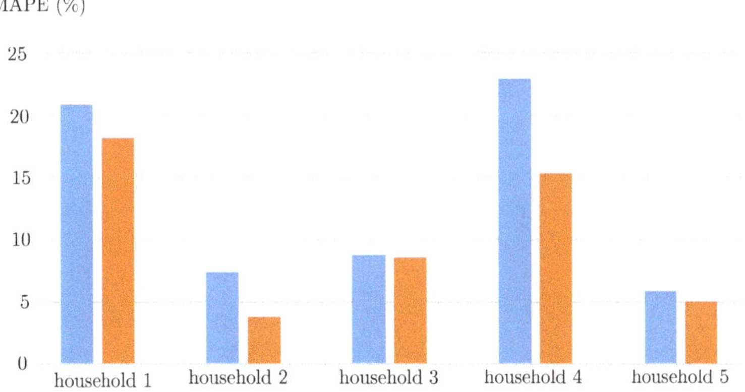

Figure 8. Experimental result comparison between LSTM and CNN-LSTM using MAPE as an error metric.

Instead of adopting the traditional one-step forecasting approaches, our method proposes a k-step forecasting algorithm for small step sizes, e.g., 6-step or 30 min forecasting period compared with the original 5 min short-term forecasting, to provide more response time for real-world applications, such as the electricity market bidding system design. The experimental results show the effectiveness for the proposed k-step forecasting algorithm for longer time period power consumption forecasting. The proposed approach can be further extended to deal with longer response time by varying different time interval resolutions and k values.

For the future work of this study, we intend to apply the proposed CNN-LSTM framework to more sophisticated real-world load datasets to verify the robustness of the proposed framework.

Author Contributions: Conceived and designed the algorithms: K.Y., X.W. and Y.D.; Performed the simulations: K.Y., X.W., N.J. and H.H.; Processed and Analyzed the data: X.W. and H.Z. Wrote the paper: K.Y., X.W.and Y.D.; Provide ideas to improve the computational approach: K.Y. and Y.D.

Funding: This work was supported by the National Natural Science Foundation of China (No. 61850410531, No. 61803315 and No. 61602431), research development fund of XJTLU (RDF-17-01-28), Key Program Special Fund in XJTLU (KSF-A-11), Jiangsu Science and Technology Program (SBK2018042034) and Zhejiang Provincial Basic Public Welfare Research Project (from Natural Science Foundation of Zhejiang Province) (No. LGF18F020017).

Conflicts of Interest: The authors declare no conflict of interest.

References

1. Gyamfi, S.; Krumdieck, S. Scenario analysis of residential demand response at network peak periods. *Electr. Power Syst. Res.* **2012**, *93*, 32–38. [CrossRef]
2. Kong, W.; Dong, Z.Y.; Hill, D.J.; Luo, F.; Xu, Y. Short-term residential load forecasting based on resident behaviour learning. *IEEE Trans. Power Syst.* **2018**, *33*, 1087–1088. [CrossRef]
3. Chen, S.; Liu, C. From demand response to transactive energy: State of the art. *J. Mod. Power Syst. Clean Energy* **2017**, *5*, 10–19. [CrossRef]
4. Gountis, V.P.; Bakirtzis, A.G. Bidding strategies for electricity producers in a competitive electricity marketplace. *IEEE Trans. Power Syst.* **2004**, *19*, 356–365. [CrossRef]
5. Kian, A.R.; Cruz, J.B., Jr. Bidding strategies in dynamic electricity markets. *Decis. Support Syst.* **2005**, *40*, 543–551. [CrossRef]
6. Siano, P. Demand response and smart grids—A survey. *Renew. Sustain. Energy Rev.* **2014**, *30*, 461–478. [CrossRef]
7. Hu, M.; Ji, Z.; Yan, K.; Guo, Y.; Feng, X.; Gong, J.; Zhao, X.; Dong, L. Detecting anomalies in time deries fata via a meta-feature based approach. *IEEE Access* **2018**, *6*, 27760–27776. [CrossRef]
8. Hsieh, T.J.; Hsiao, H.F.; Yeh, W.C. Forecasting stock markets using wavelet transforms and recurrent neural networks: An integrated system based on artificial bee colony algorithm. *Appl. Soft Comput.* **2011**, *11*, 2510–2525. [CrossRef]
9. Socher, R.; Lin, C.C.; Manning, C.; Ng, A.Y. Parsing natural scenes and natural language with recursive neural networks. In Proceedings of the 28th International Conference on Machine Learning (ICML-11), Bellevue, WA, USA, 28 June–2 July 2011; pp. 129–136.
10. Shao, L.; Cai, Z.; Liu, L.; Lu, K. Performance evaluation of deep feature learning for RGB-D image/video classification. *Inform. Sci.* **2017**, *385*, 266–283. [CrossRef]
11. Kuremoto, T.; Kimura, S.; Kobayashi, K.; Obayashi, M. Time series forecasting using a deep belief network with restricted Boltzmann machines. *Neurocomputing* **2014**, *137*, 47–56. [CrossRef]
12. Shi, H.; Xu, M.; Li, R. Deep learning for household load forecasting—A novel pooling deep RNN. *IEEE Trans. Smart Grid* **2018**, *9*, 5271–5280. [CrossRef]
13. Kong, W.; Dong, Z.Y.; Jia, Y.; Hill, D.J.; Xu, Y.; Zhang, Y. Short-term residential load forecasting based on LSTM recurrent neural network. *IEEE Trans. Smart Grid* **2017**. [CrossRef]
14. Yuan, C.; Liu, S.; Fang, Z. Comparison of China's primary energy consumption forecasting by using ARIMA (the autoregressive integrated moving average) model and GM (1, 1) model. *Energy* **2016**, *100*, 384–390. [CrossRef]
15. Yan, K.; Du, Y.; Ren, Z. MPPT perturbation optimization of photovoltaic power systems based on solar irradiance data classification. *IEEE Trans. Sustain. Energy* **2018**. [CrossRef]

16. Du, Y.; Yan, K.; Ren, Z.; Xiao, W. Designing localized MPPT for PV systems using fuzzy-weighted extreme learning machine. *Energies* **2018**, *11*, 2615. [CrossRef]

17. Shen, W.; Babushkin, V.; Aung, Z.; Woon, W.L. An ensemble model for day-ahead electricity demand time series forecasting. In Proceedings of the Fourth International Conference on Future Energy Systems, Berkeley, CA, USA, 21–24 May 2013; pp. 51–62.

18. Funahashi, K.I.; Nakamura, Y. Approximation of dynamical systems by continuous time recurrent neural networks. *Neural Netw.* **1993**, *6*, 801–806. [CrossRef]

19. Krizhevsky, A.; Sutskever, I.; Hinton, G.E. Imagenet classification with deep convolutional neural networks. In Proceedings of the 25th International Conference on Neural Information Processing Systems, Lake Tahoe, NV, USA, 3–6 December 2012; pp. 1097–1105.

20. Hinton, G.E.; Osindero, S.; Teh, Y.W. A fast learning algorithm for deep belief nets. *Neural Comput.* **2006**, *18*, 1527–1554. [CrossRef] [PubMed]

21. Almalaq, A.; Edwards, G. A Review of deep learning methods applied on load forecasting. In Proceedings of the 2017 16th IEEE International Conference on Machine Learning and Applications (ICMLA), Cancun, Mexico, 18–21 December 2017; pp. 511–516.

22. Wang, J.; Yu, L.C.; Lai, K.R.; Zhang, X. Dimensional sentiment analysis using a regional CNN-LSTM model. In Proceedings of the 54th Annual Meeting of the Association for Computational Linguistics, Berlin, Germany, 7–12 August 2016; pp. 225–230.

23. Chen, X.; Du, Y.; Wen, H.; Jiang, L.; Xiao, W. Forecasting based power ramp-rate control strategies for utility-scale PV systems. *IEEE Trans. Ind. Electron.* **2018**, *3*, 1862–1871 [CrossRef]

24. Karnik, N.N.; Mendel, J.M. Applications of type-2 fuzzy logic systems to forecasting of time-series. *Inform. Sci.* **1999**, *120*, 89–111. [CrossRef]

25. Conejo, A.J.; Plazas, M.A.; Espinola, R.; Molina, A.B. Day-ahead electricity price forecasting using the wavelet transform and ARIMA models. *IEEE Trans. Power Syst.* **2005**, *20*, 1035–1042. [CrossRef]

26. Yan, K.; Ji, Z.; Shen, W. Online fault detection methods for chillers combining extended kalman filter and recursive one-class SVM. *Neurocomputing* **2017**, *228*, 205–212. [CrossRef]

27. Yan, K.; Shen, W.; Mulumba, T.; Afshari, A. ARX model based fault detection and diagnosis for chillers using support vector machines. *Energy Build.* **2014**, *81*, 287–295. [CrossRef]

28. Kumar, U.; Jain, V. Time series models (Grey-Markov, Grey Model with rolling mechanism and singular spectrum analysis) to forecast energy consumption in India. *Energy* **2010**, *35*, 1709–1716. [CrossRef]

29. Ediger, V.Ş.; Akar, S. ARIMA forecasting of primary energy demand by fuel in Turkey. *Energy Policy* **2007**, *35*, 1701–1708. [CrossRef]

30. Oğcu, G.; Demirel, O.F.; Zaim, S. Forecasting electricity consumption with neural networks and support vector regression. *Procedia-Soc. Behav. Sci.* **2012**, *58*, 1576–1585. [CrossRef]

31. Rodrigues, F.; Cardeira, C.; Calado, J.M.F. The daily and hourly energy consumption and load forecasting using artificial neural network method: A case study using a set of 93 households in Portugal. *Energy Procedia* **2014**, *62*, 220–229. [CrossRef]

32. Deb, C.; Eang, L.S.; Yang, J.; Santamouris, M. Forecasting energy consumption of institutional buildings in Singapore. *Procedia Eng.* **2015**, *121*, 1734–1740. [CrossRef]

33. Wang, J.; Hu, J. A robust combination approach for short-term wind speed forecasting and analysis—Combination of the ARIMA (Autoregressive Integrated Moving Average), ELM (Extreme Learning Machine), SVM (Support Vector Machine) and LSSVM (Least Square SVM) forecasts using a GPR (Gaussian Process Regression) model. *Energy* **2015**, *93*, 41–56.

34. Armano, G.; Marchesi, M.; Murru, A. A hybrid genetic-neural architecture for stock indexes forecasting. *Inform. Sci.* **2005**, *170*, 3–33. [CrossRef]

35. Rather, A.M.; Agarwal, A.; Sastry, V. Recurrent neural network and a hybrid model for prediction of stock returns. *Expert Syst. Appl.* **2015**, *42*, 3234–3241. [CrossRef]

36. Wang, H.; Wang, G.; Li, G.; Peng, J.; Liu, Y. Deep belief network based deterministic and probabilistic wind speed forecasting approach. *Appl. Energy* **2016**, *182*, 80–93. [CrossRef]

37. Khodayar, M.; Kaynak, O.; Khodayar, M.E. Rough deep neural architecture for short-term wind speed forecasting. *IEEE Trans. Ind. Inf.* **2017**, *13*, 2770–2779. [CrossRef]

38. Voyant, C.; Notton, G.; Kalogirou, S.; Nivet, M.L.; Paoli, C.; Motte, F.; Fouilloy, A. Machine learning methods for solar radiation forecasting: A review. *Renew. Energy* **2017**, *105*, 569–582. [CrossRef]

39. Alzahrani, A.; Shamsi, P.; Dagli, C.; Ferdowsi, M. Solar irradiance forecasting using deep neural networks. *Procedia Comput. Sci.* **2017**, *114*, 304–313. [CrossRef]

40. Ryu, S.; Noh, J.; Kim, H. Deep neural network based demand side short term load forecasting. *Energies* **2016**, *10*, 3. [CrossRef]

41. Kelly, J.; Knottenbelt, W. The UK-DALE dataset, domestic appliance-level electricity demand and whole-house demand from five UK homes. *Sci. Data* **2015**. [CrossRef] [PubMed]

42. Jozefowicz, R.; Zaremba, W.; Sutskever, I. An empirical exploration of recurrent network architectures. In Proceedings of the 32nd International Conference on International Conference on Machine Learning, Lille, France, 6–11 July 2015; pp. 2342–2350.

43. Hochreiter, S.; Schmidhuber, J. Long short-term memory. *Neural Comput.* **1997**, *9*, 1735–1780. [CrossRef] [PubMed]

44. Werbos, P.J. Backpropagation through time: What it does and how to do it. *Proc. IEEE* **1990**, *78*, 1550–1560. [CrossRef]

45. Ketkar, N. Convolutional Neural Networks. In *Deep Learning with Python*; Springer, Berlin, Germany, 2017; pp. 63–78.

46. Goodfellow, I.; Bengio, Y.; Courville, A.; Bengio, Y. *Deep Learning*; MIT Press, Cambridge, UK, 2016; Volume 1.

47. Abadi, M.; Barham, P.; Chen, J.; Chen, Z.; Davis, A.; Dean, J.; Devin, M.; Ghemawat, S.; Irving, G.; Isard, M., et al. TensorFlow: A System for Large-Scale Machine Learning. In Proceedings of the 12th USENIX Conference on Operating Systems Design and Implementation, Savannah, GA, USA, 2–4 November 2016; pp. 265–283.

48. Keras: The Python Deep Learning library. Available online: http://keras.io/ (accessed on 1 November 2018).

49. Brailsford, T.J.; Faff, R.W. An evaluation of volatility forecasting techniques. *J. Bank. Financ.* **1996**, *20*, 419–438. [CrossRef]

50. Kavasseri, R.G.; Seetharaman, K. Day-ahead wind speed forecasting using f-ARIMA models. *Renew. Energy* **2009**, *34*, 1388–1393. [CrossRef]

Article

Integrated Energy Micro-Grid Planning Using Electricity, Heating and Cooling Demands

He Huang [1,*], DaPeng Liang [1] and Zhen Tong [2]

[1] School of Management, Harbin Institute of Technology, Harbin 150001, China; ldp0920@hit.edu.cn
[2] Electric Power Development Research Institute, China Electricity Council, Beijing 100761, China; tongzhen@cec.org.cn
* Correspondence: superrio@yeah.net; Tel.: +86-138-1041-3900

Received: 12 September 2018; Accepted: 17 October 2018; Published: 18 October 2018

Abstract: Many research works have demonstrated that taking the combined cooling, heating and power system (CCHP) as the core equipment, an integrated energy system (IES), which provides multiple energy flows by a combination of different energy production equipment can bring obvious benefit to energy efficiency, CO_2 emission reduction and operational economy in urban areas. Compared with isolated IES, an integrated energy micro-grid (IEMG) which is formed by connecting multiple regions' IES together, through a distribution and thermal network, can further improve the reliability, flexibility, cleanliness and the economy of a regional energy supply. Based on the existing IES model, this paper describes the basic structure of IEMG and built an IEMG planning model. The planning was based on the mixed integer linear programming. Economically, construction planning configuration are calculated by using known electricity, heating and cooling loads information and the given multiple equipment selection schemes. Finally, the model is validated by a case study, which includes heating, cooling, transitional and extreme load scenarios, proved the feasibility of planning model. The results show that the application of IEMG can effectively improve the economy of a regional energy supply.

Keywords: energy internet; multi-energy complementary; integrated energy systems; distribution network planning

---

1. Introduction

Increasing pressure on energy resources endowment and environmental problems resulting from the use of the energy internet (EI) are a major focus of energy researchers and practitioners [1]. Combined cooling, heating, and power unit (CCHP) technology integrates production of power from electrical and thermal systems, and solves problems caused by their separate decision-making frameworks. With support of a CCHP, an integrated energy systems (IES) can provide multiple energy flows (electricity, heating, steam, cooling, and desalination) by combining different energy production equipment (natural gas, solar, wind, etc.). It has become widely accepted as one of the most efficient examples of integration of multiple energy sources [2]. The biggest benefit of IES is that different kinds of energy production systems are no longer planned separately or operated independently. It thus takes the overall process of energy production—from generation and transmission to consumption—into full consideration during the stages of planning, construction, and operation [3,4]. Numerous cases testify to the strength of IES (with a CCHP at its core) in urban areas where it improves primary energy use efficiency, CO_2 emission reduction, and the operational economy [5,6]. Its position in the energy network has become increasingly important.

IES theories can be summarized as electricity–heat IES and electricity–gas–heat IES [7]. The author of [8] expounded the importance of electricity and heating networks combination analysis. The physical interaction between electricity and heating networks are discussed by Pan et al. in [9]. Zeng et al. [10]

designed an analysis framework considering bi-directional energy conversion, which unified the power- and gas-flow of an electricity-gas IES. The author of [11] explored the impacts of gas composition on both electricity and gas networks by a decomposed method of power- and gas-flow analysis. References [12,13] proposed a steady-state energy-flow analysis of IES containing electricity, gas and heat based on the Newton–Raphson method. Some other studies have focused on expanding the components of IES. An IES model based on a micro-grid—including the combination of the CCHP and renewable energies—was established [14,15], and the benefits of the model for increasing the utilization rate of renewable energy and reducing the energy consumption of the CCHP were proven. The authors in [16] describes an IES operation optimization model including photovoltaic and battery energy storage, and illustrated that battery lifetime loss is an inevitable factor in the optimization model.

On the basis of previous studies of IES analysis and modelling, planning is also one of the steady-state research directions. An electricity distribution lines and elements expansion model, which provides optimal reconfiguration in electricity and natural gas distribution systems within energy hubs, is proposed in [17]. Multi-area and multistage gas and electricity infrastructures integrate the expansion planning model from the central decision maker perspective and is formulated in [18]. Based on the planned power generating units, Ref. [19] built a centralized expansion planning model that integrates gas and electricity distribution networks. A joint expansion planning model of combined gas and electricity networks with the objective function of maximizing the social welfare is presented in [20]. A multi-period integrated framework incorporates a three-level procedure to solve the generation, transmission and natural gas grid expansion planning for large-scale systems is introduced in [21]. In the content of uncertainties from increasing utilization of natural gas in an electric power system, a novel expansion co-planning framework is proposed in [22] to address the integrated gas–electricity expansion planning that considers maximization of the cost/benefit ratio with a market price of gas and electricity as several scenarios. An integrated electricity and natural gas transportation system planning algorithm that is based on a two-stage robust optimization problem is provided in [23] for enhancing the power grid resilience in extreme conditions. In this model, a variable uncertainty set is involved to describe the interactions among power grid expansion states and extreme events and its case study result shows the benefit of integrated planning on improving power grid resilience. The centralized expansion planning model upon a two-stage stochastic optimization framework is established in [24], and the model provides the tradeoff of building natural gas facilities versus electric facilities under the uncertainty of demand growth. A long-term planning model of gas distribution pipelines, gas-fired power generators, and capacitor banks is presented in [25], which is solved by a sequential planning approach, and its result proves a relationship between the expansion plans and the reliability policies of a distribution utility. A co-expansion planning of gas and electricity systems based on a multi-attribute decision-making method (MADM) is introduced in [26] by the analytical hierarchy process of the central entity and privacy of gas and electricity energy parties, and the model demonstrated the effectiveness of the proposed MADM method.

Previous research on the planning and operation optimization of IES have laid a foundation. Thus, based on the theory of electricity–heat IES, this paper proposes an integrated energy micro-grid model containing distributed energy resources, and taking into account various load conditions. It can be used in IEMG planning and operation optimization. To prove the validity of the model, an IEMG planning framework by using electricity, heating and cooling demands are introduced. Through a mixed integer linear programming method (MILP), the economy of construction and operation of IEMG is optimized and analyzed. The main characteristics are as follows. Firstly, from the perspective of regional integrated energy suppliers, comprehensive planning was carried out to fulfill the power electricity, heating, and cooling loads in multiple regions, including economic analysis of the construction and operation costs of IEMG. Secondly, more equipment options are available for the planning selection, including the addition of solar power, substation expansion, and additional CCHPs, gas boilers, an absorption chiller, or air conditioning. Thirdly, in terms of the operational strategy, the model divides the energy cycle into winter, summer, and an interim period, according to changes in load demands. An extreme load scenario was added to further guarantee the accuracy and reliability of the planning results. Finally, this paper verifies the IEMG planning model with practical examples to prove its significance in guiding the construction and operation of IEMG for regional integrated energy suppliers.

2. The Integrated Energy Micro-Grid

2.1. Structure of the Integrated Energy System (IES) and the Integrated Energy Micro-Grid (IEMG)

IES usually consists of a CCHP, distributed power sources (adjustable and/or non-adjustable), the electrical load, heating load, cooling load, thermal network, and the electrical network. It can also be connected to the external power grid by a transformer substation. The Busbar structure for IES is shown in Figure 1 [27].

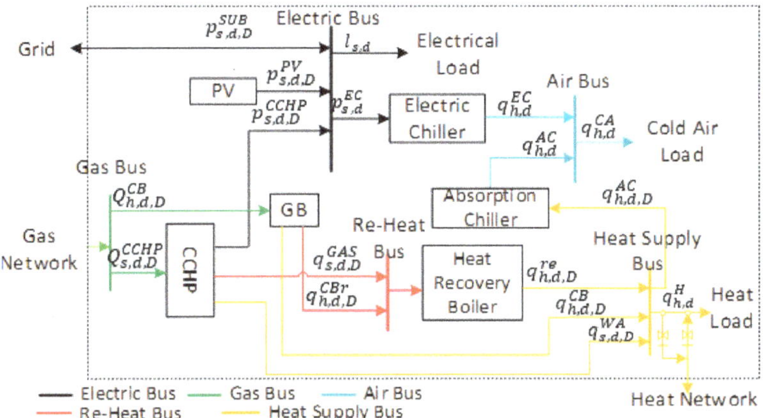

Figure 1. The Busbar structure for the integrated energy systems (IES).

In IES, the demands of electricity, heating, and cooling loads can be satisfied simultaneously. The electricity demand is generated by the CCHP, distributed power resources (photovoltaic (PV) cells, for example), and an external grid (when the load demand exceeds the total electricity capacity), and so on. The heating demands are fulfilled by the CCHP and the gas boiler, and the cooling demands are met by both an electric chiller and an absorption chiller. The heat recovery boiler (HRB) acts as a waste heat recovery facility, and can collect the waste heat generated by both the CCHP and the gas boiler, which can significantly improve the amount of heat used in the system. The power grid, the thermal equipment linking to the link energy production equipment, and the different loads operate together to achieve energy circulation through the whole system. It is necessary to emphasize that the interaction between the CCHP and the thermal network is bidirectional, so there is a switch apparatus between them to achieve directional selectivity. The heat networks of different regions can also transfer heat through a switch apparatus between the heat exchanger and the heat load.

A schematic diagram of the IEMG is shown in Figure 2. The IEGM connects several regions' IES (here called the subarea of the IEMG) by a micro-grid, a heat network, and a natural gas pipeline network, to make a scheduling balance within the whole region possible. Thus, the IEMG regards the multiple IES as a controllable whole—they can be safely connected to the low-voltage distribution grid and operate in a flexible manner. Meanwhile, through the coordinated control of the equipment in these regions, the IEMG can provide a more economical, efficient, and reliable supply of energy for different kinds of loads. Furthermore, it can connect to the external power grid and the thermal network, and can thus purchase electricity and heat energy when the overall output in the region is insufficient, or sell surplus energy to external buyers. It can, therefore, significantly improve the economic efficiency for integrated energy suppliers. To summarize, due to the multi-region and multi-energy complementation, the electricity reliability, economical efficiency, and comprehensive utilization rate of energy within the region of the IEMG is effectively improved.

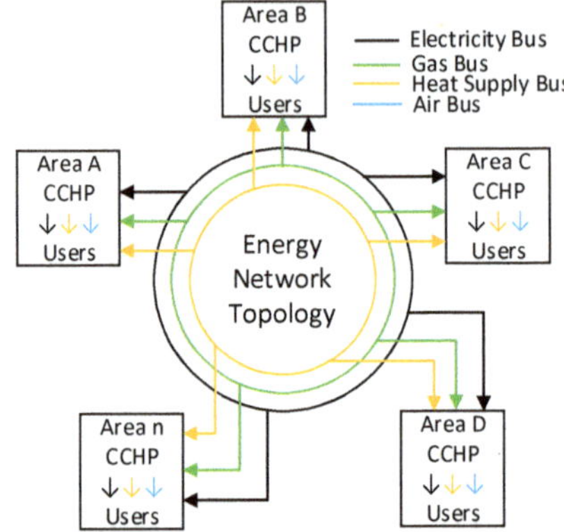

Figure 2. Diagram of the integrated energy micro-grid (IEMG).

2.2. The Principle Elements of the IEMG

The IEMG model in this paper includes three main parts: energy generation, the transformation network, and loads. Generation refers to the electricity, heat, and other energy supply equipment, including the gas combustion engine, CCHP, photovoltaics, substation, and boiler. The network denotes the electrical power grid and the thermal network. The loads are the electrical, heating, and cooling loads. This section models these devices mathematically.

2.2.1. The Generation Equipment

(1)   Combined Cooling, Heating and Power System (CCHP)

In this paper, the CCHP is built using a constant efficiency model, and the relationship between its thermoelectric power and fuel consumption is established through approximation of a linear function [28,29].

The fuel combustion of the CCHP can be calculated by the following:

$$Q_{s,d,D}^{CCHP} = V_{s,d,D}^{CCHP} \theta_{NG} / 3.6$$

Accordingly, its electrical output is:

$$p_{s,d,D}^{CCHP} = \alpha_{d,D}^{GE} Q_{s,d,D}^{CCHP} + \beta_{d,D}^{GE}$$

The net calorific value of the waste heat is:

$$q_{s,d,D}^{GAS} = \alpha_{d,D}^{GAS} Q_{s,d,D}^{CCHP} + \beta_{d,D}^{GAS}$$

The net calorific value of the jacket-cooling water is:

$$q_{s,d,D}^{WA} = \alpha_{d,D}^{WA} Q_{s,d,D}^{CCHP} + \beta_{d,D}^{WA}$$

The expression of fuel combustion describes the total calorific value $Q_{s,d,D}^{CCHP}$ converted from the combustion inflow $V_{s,d,D}^{CCHP}$ (unit: m^3/h) per unit time, where θ_{NG} is the average calorific value (which is a constant, 32.967 MJ/m^3 for natural gas). The other three expressions describe the electrical power, and the available calorific value in the waste heat generated by the CCHP. In these equations, $p_{s,d,D}^{CCHP}$ is the power output of the CCHP; $q_{s,d,D}^{GAS}$ and $q_{s,d,D}^{WA}$ are the net calorific values of the waste heat and the jacket-cooling water, respectively, in kW. The parameters α and β are two known coefficients that are used to fit the performance of the CCHP. In practice, the output of the CCHP is subject to other technical constraints, including the service life of the unit, the maximum and minimum output limitation, the ramp rate of the unit, the minimum continuous running time, and the minimum continuous downtime. These constraints and parameters are selected differently during the different optimizing purposes. The main purpose of this paper is the planning and optimization of the IEMG. Therefore, the specific constraints and parameter selection will be detailed later.

(2)   Distributed Generation

The distributed generation in this paper is the PV power generation system, and the mathematical model applied is as follows. The output of a PV system is affected by weather, temperature, and solar illumination. If the PV output is $p_{s,d,D}^{PV}$, it can be modeled as:

$$p_{s,d,D}^{PV} = \frac{\zeta_t}{\zeta_{t,s}} A \cdot f_{pv} \eta \left[1 + \alpha_p (T_s - T_{stc}) \right]$$

where ζ_t is the actual illumination intensity during the tth hour (kW); $\zeta_{t,s}$ is the illumination intensity under standard conditions; A is the total area of the PV panels, which is: $A = \sum_{m=1}^{M} A_m$, in which, A_m is the area of a single panel; f_{pv} is the power derating factor of the PV system, denoting the ratio of the actual output power to the rated output power, which is used to represent the power loss caused by dirt, rainwater, or snow on the PV panels, and by the aging of the panels (its value here is taken to be 0.9); η is the overall conversion efficiency of the PV panels: $\eta = \frac{1}{A} \sum_{m=1}^{M} A_m \eta_m$, in which η_m is the conversion efficiency of a single panel (kW); α_p is the power temperature coefficient (%/°C) (which is generally -0.47); T_{stc} is the reference temperature of the PV generation system measured under standard conditions (25 °C here); T_s can be calculated by:

$$T_s = T_a + 0.0138(1 + 0.031 T_a)(1 - 0.042 v_{pv}) \cdot A$$

where T_a is the ambient temperature (T_a) and v_{pv} is the wind velocity (m/s).

After building the model of the PV system, it is necessary to describe the relationship between solar radiation and the output of the PV system. The beta-distribution probability function can be used to express the output variation of the PV generation system. According to [30]:

$$f(P_M) = \frac{\Gamma(\alpha + \beta)}{\Gamma(\alpha)\Gamma(\beta)} \left(\frac{P_M}{R_M} \right)^{\alpha - 1} \left(1 - \frac{P_M}{R_M} \right)^{\beta - 1}$$

where α and β are the shape parameters of the beta-distribution; P_M is the total power of the PV array, and R_M is the maximum power that the PV array can output.

(3)   Gas Boiler (GB)

The heat required in the IEMG is mainly produced by two devices, a CCHP and a gas boiler. During production, the thermal energy is mainly provided by the CCHP. Once it cannot meet the heating load, the gas boiler (GB) can convert the chemical energy of the fuel into thermal energy with a high conversion efficiency, to achieve the thermal balance of the system. Assuming that the gas boiler converts the energy of the natural gas into heat at a constant conversion efficiency, the thermal power of the natural gas that is consumed by the gas boiler is:

$$Q_{h,d,D}^{CB} = V_{h,d,D}^{CB} \theta_{NG} / 3.6$$

and the heat supply efficiency is:

$$q_{h,d,D}^{CB} = \eta_{d,D}^{GB} Q_{h,d,D}^{CB}$$

where $Q_{h,d,D}^{CB}$ is the thermal power of the fuel consumed by the boiler (kW); $q_{h,d,D}^{CB}$ is the boiler's heat output (kW); and $\eta_{d,D}^{GB}$ is the heating efficiency coefficient of the boiler.

Similar to the CCHP, some waste heat of the gas boiler can be reused. The formula for the available waste heat efficiency is:

$$q_{h,d,D}^{CBr} = \eta_{d,D}^{GBr} Q_{h,d,D}^{CB}$$

where $q_{h,d,D}^{CBr}$ is the total available waste heat of the boiler, and $\eta_{d,D}^{GBr}$ is the waste heat energy efficiency coefficient of the boiler.

(4) Heat Recovery Boiler

The recycling of waste heat is an important means for improving energy efficiency. In IES and IEMGs, a heat recovery boiler (HRB) is used to collect waste heat from the system. The model of the heat recovery boiler is:

$$q_{h,d}^{re} = \eta_{re}\left(q_{s,d,D}^{GAS} + q_{h,d,D}^{CBr}\right)$$

where $q_{h,d}^{re}$ is the output of the heat recovery boiler (kW), $q_{s,d,D}^{GAS}$ is the waste heat from CCHP (kW); $q_{h,d,D}^{CBr}$ is the waste heat from gas boiler (kW); η_{re} is the thermal efficiency of the equipment.

(5) Chiller

There are two kinds of chillers commonly used in IES and IEMGs. These are the absorption chiller (AC) and the electric chiller (EC). ACs are driven by a thermal medium, such as lithium bromide or ammonia solution; during operation, the working medium vaporization absorbs a lot of heat from the refrigerant water, so as to achieve cooling. The refrigeration principle of the EC involves, first, compressing the gas refrigerant by electricity, then discharging the refrigerant into a condenser. Under set pressure and temperature conditions, the low temperature and low pressure refrigerant cools the air or the condensed water in the condenser to achieve a cooling effect.

The models of the two chillers are as follows. For an AC:

$$q_{h,d}^{AC} = q_{h,d,D}^{AC,in} \cdot \eta_{d,D}^{AC}$$

where $q_{s,d}^{AC}$ is the cooling output (kW); $q_{h,d,D}^{AC,in}$ is the heat input (kW); and $\eta_{d,D}^{AC}$ is the refrigeration coefficient, which is the ratio of the heat input to the cooling output, and it is usually used to measure the performance of an AC. For an EC:

$$q_{s,d}^{EC} = p_{h,d,D}^{EC,in} \cdot \eta_{d,D}^{COP,EC}$$

where $q_{s,d}^{EC}$ is the cooling output (kW); $p_{h,d,D}^{EC,in}$ is the electric power input (kW); and $\eta_{d,D}^{COP,EC}$ is the refrigeration coefficient of the EC.

2.2.2. Energy Network Model

The mathematical expression for the energy network, which connects different devices and different regions in the IEMG, can be represented by network topology. Its connection mode is described by the incidence matrix of the topology structure. It consists of the combination of several IES, as shown in Figure 1, with the framework of Figure 2 as an example. As in Figure 1, suppose that the equipment and energy network connection (both from the power grid and the thermal network) in the IEMG are nodes. Each pipe or power line serves as a branch, taking the flow direction of the working medium as the branch direction. A basic model of the energy network expressed by the incidence matrix is by the following. Assume that V is the node set of the network. and E is the set of the power lines or pipes (referred to in the following as the set of edges). The energy network can be represented by the v by e incidence matrix $A = \left[A_{ij}\right]$. Thus, each node of the network is a row of the incidence matrix, and each edge is a column [31]. The relationship between the nodes and edges can be indicated by the sign of A_{ij}. When $A_{ij} = 1$, the node v_i is linked with edge e_j and the direction points away from v_i. When $A_{ij} = -1$, v_i is also linked with e_j and the direction points towards v_i. When $A_{ij} = 0$, v_i and e_j are not linked.

Through this approach, the incidence matrix can represent any connection modes of the network system, as with the energy network. However, in order to describe the energy network more accurately, the incidence matrix needs to be further expanded. Based on the form of the incidence matrix A, it is split into the start incidence matrix A_1 and the end point incidence matrix A_2, to represent the node set of the starting or ending points of the power and pipe lines, respectively. Therefore, A_1 and A_2 are defined as follows:

$$A_1 = (a_{zk})_{n \times m} \in \{0, 1\}^{n \times m}$$

$$a_{zk} = \begin{cases} 1, & (z, c) = E_k \\ 0, & \text{else} \end{cases}$$

and

$$A_2 = (a_{zk})_{n \times m} \in \{0, -1\}^{n \times m}$$

$$a_{zk} = \begin{cases} -1, & (c, z) = E_k \\ 0, & \text{else} \end{cases}$$

Hence, suppose that the basic loop set of $G(V, E)$ is L containing p elements, and its basic loop matrix is $B = [B_{hk}]$. Thus, in the matrix B, each element B_{hk} describes the relationship of the loop $L_h (L_h \in L, h = 1, 2, \ldots, p)$ with edge k (a branch of the grid or pipeline in the thermal network). When $B_{hk} = 1$, the loop L_h is in the same direction as edge k; when $B_{hk} = -1$, the loop L_h is opposite to the edge k. If $B_{hk} = 0$, the edge k is not in the loop L_h. This method can be used to describe most of the network system. The energy network based on the incidence matrix is summarized by [32], and the matrix can express the energy network as:

$$\begin{cases} AH = 0 \\ B\Delta X = 0 \end{cases}$$

where A is the incidence matrix of the energy network, B is the basic loop matrix, H is the energy extensive flow matrix, and ΔX is the energy-intensive difference matrix. The equivalent transfer characteristics for incompressible fluids in the energy network is:

$$H = H^* = \frac{X_{A1} - X_{A2}}{R},$$

$$R = \frac{L}{KS}$$

where H is the flow of energy transferred in the network, H^* is the equivalent extensive energy flow in the transfer process, R is the transmission resistance, L and S are the length and cross-sectional area of the transmission line, and K is the transfer coefficient of extensibility.

By combining these two equations, the energy transfer characteristic equation set for the energy network can be established in order to describe the energy transfer state at each node. The advantage of the incidence matrix is that it turns the topology relationship, and the structure of the nodes and edges in the network, into variables in a matrix which is convenient for calculations. It is also helpful for making real-time variations of the connection mode during network analysis and, therefore, it simplifies and expands the analysis of the energy network system. In addition, the incidence matrix can be used to calculate the power flow at any position of the energy network, which can improve the accuracy of energy network planning and operation optimization. On the basis of the incidence matrix expression for the energy network, models of the electrical network and the thermal network, and their constraints can be determined, as follows.

(1) Electrical Network Model

Based on the above, by taking into account Kirchhoff's current law (KCL), Kirchhoff's voltage law (KVL), and electricity flow constraints for network systems, the power grid model of the IEMG containing the distributed generation sources can be derived.

a. The Kirchhoff's current constraint (KCL):

$$\sum S_t^{EA} f_t^{EA} + \sum G_t^E + r_t = d_t^E$$

This equation reflects the equilibrium relationship between the inlet current and the outlet current at any node in scenario t. In the equation, S_t^{EA} is the node-branch incidence matrix of the power grid in scenario t, f_t^{EA} is the branch current, G_t^E is the input power of the node from the power generation equipment, r_t is the lost electrical load, and d_t^E is the electrical load at the node.

b. Kirchhoff's voltage constraint (KVL) and the voltage magnitude constraint are:

$$Z_{j,t}^{EA} f_{j,t}^{EA} + \left[S_t^{EA} \right]_{rowj}^T V_{j,t} = 0$$

$$V_{min} \leq V_j \leq V_{max}$$

Here, row j is the j-th column, T refers to the transpose, V is the column vector of the node voltage, and Z is the line impedance. The inequality defines the magnitudes of the node voltage in which V_{min} and V_{max} are the maximum and the minimum voltage magnitudes, respectively.

(2) Thermal Network Model

Similar to the above, using the energy network incidence matrix and Kirchhoff's laws, a model describing the working principles of the thermal network can be established. Three functions are used to define the transmission flow of the working medium, the relationship of heat and flow, and the change of the transmission pressure. In addition, apart from the relationship between the heat transfer and the mass flow, it is necessary to consider the corresponding heat loss [33]. The thermal network model is as follows.

a. The transmission flow constraint:

$$\sum S_t^{HA} q_t^{HA} + \sum G_t^H = d_t^H$$

where S_t^{HA} is the node–branch incidence matrix of the thermal network in scenario t, q_t^{HA} is the energy flow in the branch pipe line in scenario t, G_t^E is the input power of the node from heat generation equipment, and d_t^H is the thermal load at the node.

b. The heat-flow constraint:

According to the equivalent energy transfer characteristic equation for the thermal network, the relationship between the available heat and the flow is:

$$q_t^{HA} = \frac{P_t}{k(T_{A1} - T_{A2})}$$

where P_t is the energy (heat) intensity in the pipeline section, k is the specific heat capacity of the working medium, T_{A1} and T_{A2} are the feed-water temperature and return-water temperature at the node, respectively.

c. The transmission pressure constraint:

After the relationship between the heat and the flow in the thermal network has been determined, the heat intensity at the node conforms to the following heat balance constraint:

$$Z_{j,t}^{HA} q_{j,t}^{HA} + \left[S_t^{HA}\right]_{rowj}^{T} P_{j,t} = 0$$

where $Z_{j,t}^{HA}$ is the demand for the thermal intensity at node j during period t; $q_{j,t}^{HA}$ is the energy flow at node j; $P_{j,t}$ is the heat column vector in the pipeline section connected to node j.

2.3. Energy Balance of the IEMG

From the configuration and structure of the IEMG, as shown in Figure 3, the electrical/cooling/heating loads should be balanced in any district, or balanced over the whole region, and in all of the different scenarios this principle is basically the same. The energy balance equations in the IEMG are introduced below.

a. Balance of the electrical load (in all scenarios):

$$p_{s,d}^{SUB} + p_{s,d}^{PV} + p_{s,d}^{CCHP} = l_{s,d} + p_{s,d}^{EC}$$

The electrical load balance shall be satisfied in any scenario s ($s = c, h, t, e$). p_s^{SUB} is the power from the external grid, $p_{s,d}^{PV}$ is the PV generation power, $p_{s,d}^{CCHP}$ is the power of the CCHP (kW), $l_{s,d}$ is the pure electrical load, and $p_{s,d}^{EC}$ is the power of the electric chiller (kW).

b. Balance of the cooling load (in scenarios of a cooling supply period):

The demand for cooling is satisfied by two devices: the electric chiller and the heat adsorption chiller:

$$q_{h,d}^{CA} = q_{h,d}^{EC} + q_{h,d}^{AC}$$

where $q_{h,d}^{CA}$ is the total load of cooling, $q_{h,d}^{EC}$ is the input power of the electric chiller, and $q_{h,d}^{AC}$ is the input power of the adsorption chiller.

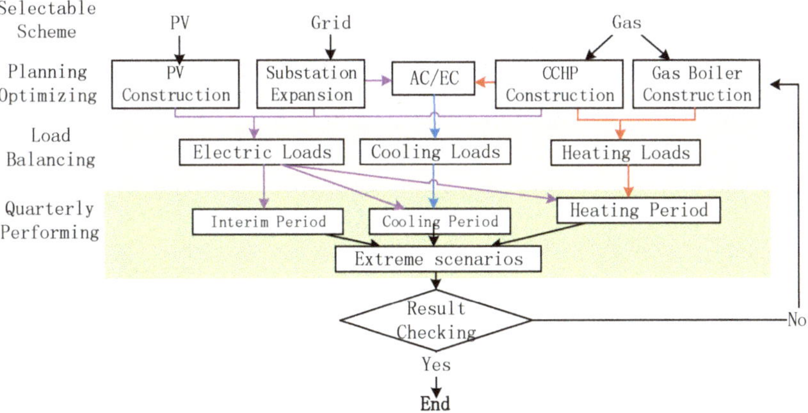

Figure 3. Optimization strategy of IEMG planning.

c. The balance of heating load (in the scenario of a heating supply period):

The heating load is satisfied by the CCHP and the gas boiler:

$$q_{h,d}^{H} = q_{h,d}^{CCHP} + q_{h,d}^{GB}$$

where $q_{h,d}^{H}$ is the total heating load, $q_{h,d}^{CCHP}$ is the heat supplied by the CCHP, and $q_{h,d}^{GB}$ is the heat from the gas boiler.

3. The IEMG Planning Optimization Model

3.1. Planning Process and Framework

In this study, the procedure for IEMG planning is summarized as follows:

(1) Extract the regional division, loads, and other planning-related data and information, and carry out the overall regional energy supply equipment configuration (for electricity, heating, and cooling).

(2) Obtain the overall configuration capacity of the energy supply equipment from step 1, combined with the load characteristics of each region, and do the regional equipment type selection and capacity optimization.

(3) According to the equipment selection and capacity optimizing results, deduce the electricity, heating and the cooling load balance operation simulation of each region, and output the results.

(4) Deduce the load balance operation simulation on the basis of quarterly and extreme scenarios, and output the results.

(5) Test and determine whether the regional and quarterly simulation results conform to the energy flow and all other constraints.

(6) If the result does not satisfy all the constraints, adjust the selection and capacity results until all constraints are met, then output the corresponding configuration.

In the steps above, Step (1) is preparatory work. Its main function is to determine the macroscopic capacity of the whole region on the basis of the known information, in order to narrow the scope of subsequent optimization. Step (2) determines the equipment selection and the installed capacity of each region on the basis of the macro-planning results. Steps (3) and (4) formulate the operation strategy and calculate the system operation cost. This is done through regional and situational operation simulation, using the planning scheme and the scenarios determined in the previous steps. Steps (5) and (6) ensure that the results meet the requirements of the constraints, and improve the accuracy of the optimization. The process is also shown in Figure 3.

Assume that an IEMG satisfies its electrical/heating/cooling loads through PV panels, natural gas (for the CCHP and gas boiler), as well as by purchasing electricity from the external grid. It is then necessary to consider the IEMG plan from the aspect of expanding its original capacity or building a new transformer substation and PV system to supply electricity. A new CCHP construction can meet the electrical/cooling/heating energy demands. Increasing the number of gas boilers compensates for heating between the CCHP heating output and the heating demand. Adding more chilling equipment can satisfy the cooling load. Accordingly, the decision process of the IEMG planning model is given in Figure 4.

Thus, the decision variables in the model can be classified into two types: construction and operation.

The constructional decision variables are mainly binary, where '0' and '1' mean to undo/do a decision, respectively. To be specific, in the type selection option d, $x_{d,D}^{CCHP}$ is the decision variable for whether to invest in the new CCHP in district D or not; similarly, $x_{d,D}^{GE}$ is the decision variable of any other power generator in district D. $x_{d,D}^{GH}$ is the decision variable of the heating generator, $x_{d,D}^{GC}$ is the decision variable of the cooling generator, and x_j^{SUB} is the decision variable of a new or expanded transformer substation.

The operation decision variables are continuous and include: the electrical generation output of the CCHP, g_s^{CCHP}; the heating generation output of the CCHP, q_s^{CCHP}; the heating power of the gas boiler, q_s^{GB}; the power of the electricity purchased from the external gird to the substation, g_s^{SUB}; and the power of the cooling generation equipment, q_s^C.

In addition, there are four typical load periods mentioned during the optimization, which measure the economics of the operational strategies. These are the transitional period ($s = t$), the cooling supply period ($s = c$), the heating supply period ($s = h$), and the extreme period ($s = e$). The extreme period indicates the unusual and sudden situation in which high cooling supplementation is required in summer, and makes sure that the results of the planning are reliable under extreme conditions.

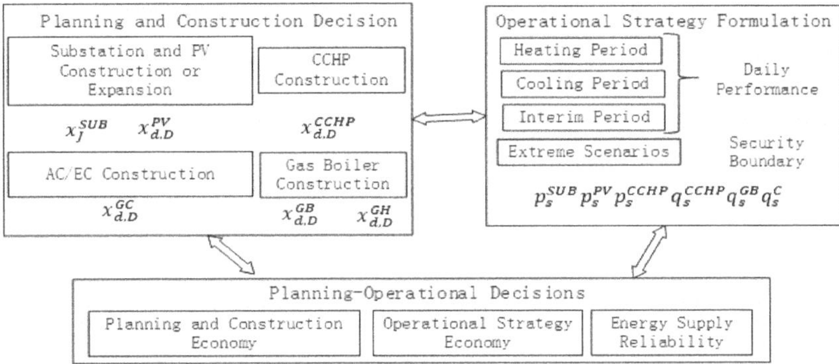

Figure 4. The main variables and decisions of the IEMG planning model.

3.2. The Objective Function

The overall objective of the planning model is to meet the maximum energy needs of the whole region while minimizing the sum of the construction cost and the operation cost. Hence, the objective function in this model consists of three parts: the planned construction cost, the planned operation cost, and the value of lost loads,

$$\min C_{INV} + C_{OPE} + C_{VOLL}$$

where C_{INV} is the planned construction cost, C_{OPE} is the planned operation cost, and C_{VOLL} is the value of lost load. The calculation of each part is as follows.

(1)   The planned construction cost:

The function of the planning model is to select the economic optimal among several construction schemes, then it should be noticed that there are many different supplement construction portfolios. The equation below gives the cost which is determined for construction. Now, assume that plan D is known to be executed, thus, the planned construction cost C_{INV} of plan D includes the construction cost of the CCHP, the construction cost of the electrical/cooling/heating generator, and the expansion cost of the transformer substation:

$$C_{INV} = \sum_{d \in \Omega} \sum_{D \in \phi_d^{CCHP}} M_{d,D}^{PV} x_{d,D}^{PV} + \sum_{J \in \phi^{SUB}} M_J^{SUB} x_J^{SUB} + \sum_{d \in \Omega} \sum_{D \in \phi_d^{CCHP}} M_{d,D}^{CCHP} x_{d,D}^{CCHP}$$
$$+ \sum_{d \in \Omega} \sum_{D \in \phi_d^{CCHP}} M_{d,D}^{GB} x_{d,D}^{GB} + \sum_{d \in \Omega} \sum_{D \in \phi_d^{CCHP}} M_{d,D}^{GH} x_{d,D}^{GH} + \sum_{d \in \Omega} \sum_{D \in \phi_d^{CCHP}} M_{d,D}^{GC} x_{d,D}^{GC}$$

where $M_{d,D}^{PV}$, $M_{d,D}^{CCHP}$, $M_{d,D}^{GB}$, $M_{d,D}^{GH}$, and $M_{d,D}^{GC}$ are the construction costs of PV panels, the CCHP, the gas boiler, the heating equipment, and the cooling equipment, respectively, in region d. M_J^{SUB} is the cost of expanding the transformer substation (J represents plan J); and x is a binary decision variable, where the cost is taken into account when the value is 1.

(2)   The planned operation cost:

Here, equipment maintenance and depreciation costs are put aside. The planned operation cost C_{OPE} includes the cost of the fuel for the CCHP and the gas boiler, and the cost of purchased electricity:

$$C_{OPE} = \sum_r \frac{r}{(1+i)^r} \sum_{d \in \Omega} \sum_{s=c,h,t} \varepsilon_s \left(Pr^{GAS} F_{s,d}^{fuel} + Pr^{SUB} p_{s,d}^{SUB} \right)$$

where r represents the system run cycle, $\sum_r \frac{r}{(1+i)^r}$ is the total net cost of the annual operation, i is the discount rate, and ε_s is the proportional contribution of scenario s to the entire planning period. For instance, when the planning period is one year (12 months), if the heating supply period contains four months from 15 November to 15 March, the proportion is $4/12 = 0.333$; if the cooling supply period contains three months from 15 June to 15 September, the proportion is $3/12 = 0.25$, then the transitional period contains the other five months and the proportion is $5/12 = 0.417$. Pr^{GAS} and Pr^{SUB} are the prices of natural gas and external electricity, respectively. $F_{s,d}^{fuel}$ is the fuel consumption per unit time in district d, which consists of the fuel consumed by the CCHP and the gas boiler:

$$F_{s,d}^{fuel} = \sum_{D \in \phi_d^{CCHP}} F_{s,d,D}^{CCHP} + \sum_{D \in \phi_d^{GB}} F_{s,d,D}^{GB}.$$

$g_{s,d}^{SUB}$ is the quantity of the electricity purchased from the external grid by a substation in district d.

(3)   The value of the lost load:

In this part, C_{VOLL} indicates the compensation cost for unsatisfied electrical/heating/cooling loads, which are not supplied during scenario s in district d.

$$C_{VOLL} = P^{VOLL} \sum_{d \in \Omega} \sum_s R_{d,s}, \quad s = c, h, t, e$$

Here, $R_{d,s}$ is the capacity of the lost loads and P^{VOLL} is the cost coefficient of the lost loads. It should be pointed out that P^{VOLL} is set to a relatively high value, in order to avoid load loss during operation.

3.3. Constraint Conditions

In planning the IEMG, the variation of generation equipment parameters in the model should be within a certain range. Their constraint conditions are given in the following.

a. The permeability constraint on the distributed generation (DG):

The proportion of the DG to the total installed capacity should be within a certain range:

$$p_{d,D}^{PV,min} \leq p_{d,D}^{PV} \leq p_{d,D}^{pv,max}$$

$$v_{d,D}^{PV,min} \leq v_{d,D}^{PV} \leq v_{d,D}^{PV,max}$$

$$N_{d,D}^{PV,min} \leq NG_{d,D}^{PV} \leq N_{d,D}^{PV,max}$$

where in the plan D for region d, $p_{d,D}^{PV,min}$ and $p_{d,D}^{pv,max}$ are the lower and upper limits of the active power of the DG (kW). $v_{d,D}^{PV,min}$ and $v_{d,D}^{PV,max}$ are the lower and upper limits of the reactive power (kW) of DG, respectively. $p_{d,D}^{PV}$ and $v_{d,D}^{PV}$ are the actual active power and reactive power of DG in district d (kW). $N_{d,D}^{PV,min}$ and $N_{d,D}^{PV,max}$ are the lower and upper limits of the number of the DG in the system. $NG_{d,D}^{PV}$ is the number of DGs in district d.

b. Operational constraints for the CCHP:

The operational constraints for the CCHP include the maximum and the minimum outputs, the ramp rate, and the maximum and the minimum continuous running times. They are given by:

$$\sum_{D\in\phi_d^{CCHP}} x_{d,D}^{CCHP} p_{min,d,D}^{CCHP} \leq p_{d,D}^{CCHP} \leq \sum_{D\in\phi_d^{CCHP}} x_{d,D}^{CCHP} p_{max,d,D}^{CCHP}$$

where $p_{min,d,D}^{CCHP}$ and $p_{max,d,D}^{CCHP}$ are the lower and the upper limits of the CCHP's output (kW).

The number of generation units in each district needs to be limited, due to geographical factors. In this model, we only limit the total number of CCHP units, and allow one per district:

$$\sum_{D\in\phi_d^{CCHP}} X_{d,D}^{CCHP} \leq 1$$

where $X_{d,D}^{CCHP}$ is the total number of CCHP generation units in district d. This constraint can be adjusted according to planning requirements.

c. Operation constraints on the gas boiler:

The output heating power of the gas boiler during operation should be no larger than its rated power:

$$0 \leq q_{h,D}^{GB} \leq \sum_{D\in\phi_d^{GB}} x_{d,D}^{GB} q_{max,d,D}^{GB}$$

where $q_{max,d,D}^{GB}$ is the rated power.

Similarly, the construction constraint for the gas boiler is:

$$\sum_{D\in\phi_d^{GB}} X_{d,D}^{GB} \leq 1.$$

This ensures that the boilers in district d are of the same capacity.

d. Constraints on the chillers:

The power of the adsorption chiller and the electrical chiller during operation should be no larger than their rated power:

$$0 \leq q_{s,d}^{AC/EC} \leq q_{max,s,d}^{AC/EC}$$

where $q_{max,s,d}^{AC/EC}$ is the rated power of the chiller (kW).

e. The power flow constraint:

The power flow of the system should be limited according to the magnitude of the current in the electrical network:

$$\left| f^{EA} \right| \leq y^{EA} f_{max}^{EA}$$

where f_{max}^{EA} is the upper limit of the current magnitude and y^{EA} is the conductance value at the corresponding node.

f. The balance constraint for heat loss in the thermal network:

If there is too much heat loss in the pipelines of the thermal network, the temperature of the working medium in the pipelines will become lower than the temperature of the working medium in the return-water system. As a result, the thermal network will be ineffective. To ensure the efficiency of the thermal network, we therefore need to ensure that the power (temperature) of the usable heat in the pipelines is higher than a critical value and lower than the maximum power that can be transferred in the pipelines:

$$P_t^{min} \leq P_{i,t}^* \leq P_t^{max}$$

where $P_{i,t}^*$ is the power of the usable heat in the working medium at node j, P_t^{min} is the lower critical value of the power of the usable heat, and P_t^{max} is the maximum power of the usable heat. If the working medium flows away from node i then the value of $P_{i,t}^*$ is positive, otherwise it is negative.

g. The supply capacity constraint for the transformer substation (in the external grid):

In the transformer substation, the total power supply capacity should not be greater than the product of the load and the capacity–load ratio, expressed as:

$$\gamma_{min} p_s^{SUB} \leq p_0^{SUB} + \sum_{J \in \varnothing^{SUB}} x_J^{SUB} p_J^{SUB} \leq \gamma_{max} p_s^{SUB}$$

$$\sum_{J \in \varnothing^{SUB}} x_J^{SUB} \leq 1$$

The first inequality describes the relationship between the original power supply capacity p_0^{SUB} and the expanded capacity $\sum_{J \in \varnothing^{SUB}} x_J^{SUB} p_J^{SUB}$. The sum is the expanded total power supply capacity. To ensure accuracy and effectiveness in planning, the product of the total power supply capacity and capacity–load ratio γ should be valid for the extreme load scenario ($s = e$). The range of the capacity–load ratio γ in this model is 1.8~2.1. The second inequality means that only one transformer substation expansion plan in the set $\varnothing^{SUB}$ will be carried out.

3.4. Calculation Method

Generally, dynamic programming (DP) algorithms could be implemented to energy planning optimization [34]. Here, we used the mixed integer linear programming (MILP) method to solve the IEMG planning model. The model involves the following decision variables: the output of the PV power system, the input and outputs of the CCHP, the electricity purchased from or sold to the external power grid, the input of the conversion equipment, and the input and output of the gas boiler. The model can be solved using a mature algorithm, or directly by commercial software, such as CPLEX, GUROBI, and LINGO [35]. In this study, the model was built by the software MatLab and Yalmip, and solved by the optimization software GUROBI. The process of the optimization algorithm is as in Figure 5.

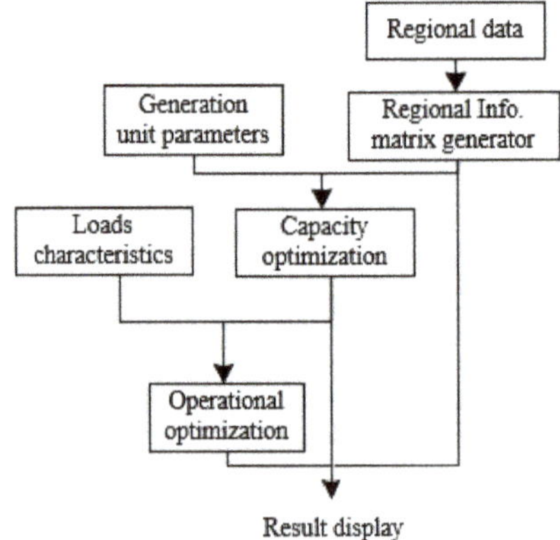

Figure 5. The optimization algorithm process of IEMG planning.

4. Case Study

4.1. Case Description

In this work, a new development area of a municipality was taken as a case study. The planning data and the predicted annual saturated electrical/cooling/heating load data were already known, as shown in Figure 6.

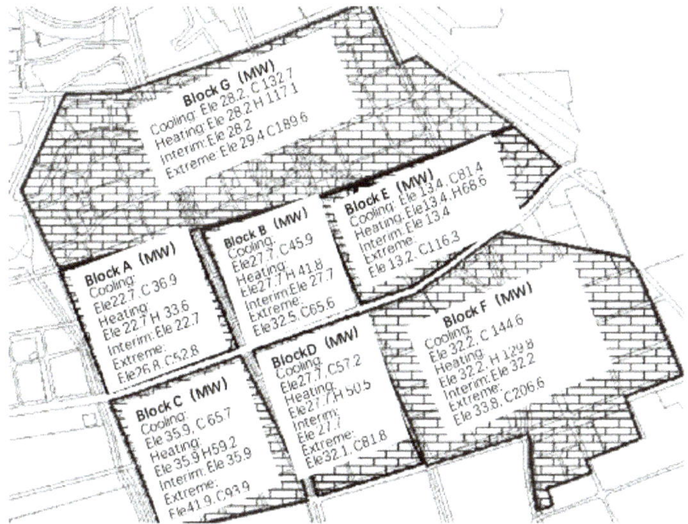

Figure 6. Map showing the energy supply planning information of the case study area.

From Figure 5, the data of the loads were classified under four scenarios: cooling period, heating period, transitional period, and extreme cooling period. It should be noted that during the calculation the load on the air conditioner should be subtracted during the cooling period.

Alternative planning options for the energy supply in the region are listed in Table 1. The parameters of the different CCHP units are listed in Table 2. The parameters used for the 9.5 MW CCHP unit are the same as those for the 5 MW unit.

Table 1. Alternative planning options for the energy supply equipment. Combined cooling, heating, and power unit (CCHP), photovoltaic (PV), absorption chiller (AC), electric chiller (EC), heat recovery boiler (HRB); USD: US dollars.

Substation Expansion			CCHP		
Scheme No.	Capacity (MVA)	Construction Cost	Scheme No.	Capacity (MW)	Construction Cost
1	1 × 50	125	1	1	185
2	2 × 50	250	2	2	375
3	3 × 50	375	3	3	550
4	4 × 50	500	4	5	850
5	5 × 50	625	5	9.5	1600
6	6 × 50	750			
Gas Boiler			**Others**		
Scheme No.	Alternative Options (MW)	Construction Cost	Equipment	Construction Cost	Unit
1	50	440	PV	130	10^4 USD/MW
2	100	900	AC	10	10^4 USD/MW
3	150	1350	EC	15	10^4 USD/MW
4	200	1750	HRB	2.5	10^4 USD/MW

(unit: 10^4 USD).

Table 2. Performance parameter of the CCHP.

Full Capacity (MW)	Characteristic Function Coefficient					
	α^{GE}	β^{GE}	α^{GAS}	β^{GE}	α^{WA}	β^{WA}
1	0.421	-222.411	0.211	3.624	0.149	81.731
2	0.466	-657.422	0.219	13.644	0.151	90.737
3	0.479	-758.940	0.208	94.383	0.153	173.562
5	0.472	-896.264	0.207	125.832	0.149	204.828
9.5	0.470	-915.262	0.204	130.261	0.146	220.152

Other operational parameters included the calorific value and price of gas: 32.967 MJ/m^3 and 0.5 USD/m^3, respectively; the external electricity price in this area was a commercial price, which was 0.16 USD/kWh (the purchase price from the weighted average of the peak–valley electricity prices). The total planning period was 10 years and the annual discount rate i was 5%. To minimize the lost load, the value of P^{VOLL} was set to $150{,}000 \times 10^4$ USD/MW.

Four cases were set to make the calculation of the planning more accurate:

CASE 1 Electricity is only purchased from the external power grid, without consideration of PV and CCHP;

CASE 2 A 4 MW PV system at least is built in each district and electricity can be purchased from the external power grid, without consideration of the CCHP;

CASE 3 CCHP construction is considered and electricity can be purchased from the external power grid when the output of the CCHP is insufficient, without consideration of PV;

CASE 4 A 4 MW PV generation system at least is built in each district, a CCHP is considered, and electricity can be purchased from the external power grid when the outputs of the CCHP and the PV are insufficient.

4.2. Results and Analysis

The IEMG planning result for each region as in Figure 7, and the overall economic results are shown in Table 3.

	Sub	PV	CCHP	GB	AC	EC			Sub	PV	CCHP	GB	AC	EC
A	50	0	0	100	53	0		**E**	0	0	0	150	115	0
	50	4	0	100	65	0			0	8	0	150	115	0
	0	0	9.5	50	55	0			0	0	9.5	100	100	0
	0	4	9.5	50	55	0			0	4	9.5	100	100	0
B	50	0	0	100	65	0		**F**	50	0	0	200	205	0
	50	4	0	100	65	0			50	8	0	200	200	5
	0	0	9.5	50	50	10			50	0	9.5	150	150	60
	0	8	9.5	50	50	10			50	8	9.5	150	150	60
C	50	0	0	150	95	0		**G**	50	0	0	200	188	0
	50	8	0	150	88	5			0	4	0	200	188	0
	50	0	9.5	50	60	40			0	0	9.5	100	100	80
	50	8	9.5	50	60	40			0	4	9.5	100	100	80
D	50	0	0	150	85	0			Case1					
	50	4	0	150	85	0			Case2					
	50	0	9.5	100	80	5			Case3					
	0	8	9.5	100	80	5			Case4					

Figure 7. The IEMG planning result of each region in different cases (Unit: MW).

Table 3. Cost comparison of the different case study scenarios.

Comparative Case		CASE 1	CASE 2	CASE 3	CASE 4
Selected Scheme		No PV No CCHP	With PV No CCHP	No PV With CCHP	PV And CCHP
Construction Cost	PV	0	3718.75	0	3718.75
	CCHP	0	0	11,221.875	11,221.875
	GB&HRB	11,161.72	11,161.72	6696.40625	6696.40625
	Substation	750	625	375	250
	AC/EC	7496.41	7496.41	6591.40625	6591.40625
	In Total	19,408.13	23,001.88	24,884.6875	28,478.4375
Operational Cost	PV	0	42,276.42	0	42,276.42
	CCHP	0	0	74,857.50	65,116.65
	Purchased electricity	322,498.77	257,164.66	206,985.31	160,298.32
	GB&HRB	53,500.78	53,500.78	44,561.415	44,561.41
	In Total	375,999.53	352,941.86	351,288.91	340,731.23
Total Cost		395,407.66	395,407.66	375,943.73	376,173.59

(unit: 10^4 USD).

From Figure 7 and Table 3, the results showed that: in CASE 1, the capacity of the transformer substation is 6 × 50 MVA, the construction cost is 0.194 billion USD, and the total planning cost, including the operation cost, is 3.954 billion USD. In CASE 2, a 4 MW PV generation source is constructed in each of the seven districts, providing 28 MW in total, and the capacity of the substation is 5 × 50 MVA, the construction cost is 0.230 billion USD, and the total planning cost is 3.954 billion USD. In CASE 3, a 9.5 MW CCHP is constructed in each district (66.5 MW in total), the capacity of the substation is 3 × 50 MVA, the construction cost is 0.249 billion USD, and the total planning cost is 3.759 billion USD. In CASE 4, a 4 MW PV source and a 9.5 MW CCHP are constructed in each district, the capacity of the substation is 2 × 50 MVA, the construction cost is 0.285 billion USD, and the total planning cost is 3.762 billion USD.

In CASE 1, the demands for electricity are satisfied by purchasing electricity from the external power grid and heating is from the gas boiler. This scheme involved the least equipment. The system's structure was relatively simple, and the construction cost was, therefore, the lowest, but the energy supply form was simple and the operational cost was relatively high: 3.225 billion and 0.535 billion USD for electricity and heat, respectively, and 3.760 billion USD in total (which is the highest of the four CASES). In CASE 2, due to PV sources in each district (28 MW in total), the construction cost of the whole system was increased by 18.52% compared to CASE 1. However, the operation cost of the external electricity was lower, and the thermal generation's cost remained unchanged, leading to CASE 2 having a 5.31% reduction in operation cost and a 4.13% decrease in total planning costs. In CASE 3, a CCHP system was added to supply electricity and thermal energy. Compared to CASE 1, the construction cost was 28.22% higher, but due to the application of the CCHP the cost of purchasing electricity was 31.71% lower. The cost of the heating supply was 16.70% lower, and thus the total cost decreased by 4.86%. In CASE 4, both PV and CCHP are constructed, which meant that the scheme combines the characteristics of CASES 2 and 3. Therefore, compared to CASE 1, the construction cost was 46.73%, higher but the operation cost decreased by 9.37%. Lower operating costs offset the higher construction costs, making CASE 4 the lowest costing among the four cases. Therefore, in CASES 2, 3, and 4, the total planning–operation costs were decreased by 3.74%, 4.86%, and 6.63% compared to CASE 1, respectively. To sum up, the application of PV and the CCHP in the IEMG is efficient in reducing the total planning–operation cost as a whole.

5. Conclusions

This paper presents an IEMG planning model with distributed photovoltaic by MILP. First of all, the model determines the capacity construction configuration of the energy production equipment by known electricity, heating and cooling loads. Second, to further improve the feasibility of the planning results, the calculated capacity allocation is put into operational cost analysis of heating, cooling, transitional and extreme load scenarios. The model takes district energy suppliers as the main investors; the optimized capacity configuration can meet the overall energy demand of the region in different scenarios and, at the same time, give the construction and operation cost of different sub-regions. A case study is given to prove the validity of the model. The case study is in a seven sub-district development zone, and four comparison schemes are given: CASE 1 (electricity supplied by an external power grid), CASE 2 (supplied by an external grid and PV), CASE 3 (supplied by an external grid and CCHP), and CASE 4 (supplied by all above equipment). The calculations show that the ranking of the total costs is CASE 1 > CASE 2 > CASE 3 > CASE 4. Compared to CASE 1, the total planning–operation costs in the other three cases are decreased by 3.74%, 4.86%, and 6.63%, respectively, which reflects the fact that the construction of the distributed PV and CCHP generation sources are beneficial for reducing the total planning–operation costs. From the results of the model calculation, the model we have proposed can be seen as a theoretical reference for the planning of multi-district IES (an IEMG in this paper).

The study could be further improved in the following aspects: the electrical network model and the thermal network model are relatively simple, and only the constraints of the power flow in the power grid and the thermal network are discussed, without consideration of the variation of energy quantity flow rate, the variations of temperature and pressure in the thermal network, or time delay in the thermal network's heat transmission. In addition, the planning and construction of the framework of the energy network is not fully investigated.

Author Contributions: Methodology, H.H. and Z.T.; Writing-Original Draft Preparation, H.H.; Writing-Review and Editing, all the authors.

Funding: This research was funded by National Natural Science Foundation of China under Grant No. 71774039.

Conflicts of Interest: The authors declare no conflict of interest.

References

1. Mei, S.; Li, R.; Xue, X.; Chen, Y.; Lu, Q.; Chen, X.; Carsten, D.A.; Li, R.; Chen, L. Paving the Way to Smart Micro Energy Grid: Concepts, Design Principles, and Engineering Practices. *CSEE J. Power Energy Syst.* **2017**, *3*, 440–449. [CrossRef]

2. Lund, H.; Münster, E. Integrated energy systems and local energy markets. *Energy Policy* **2006**, *34*, 1152–1160. [CrossRef]

3. Wang, Y.; Huang, Y.; Wang, Y.; Yu, H.; Li, R.; Song, S. Energy Management for Smart Multi-Energy Complementary Micro-Grid in the Presence of Demand Response. *Energies* **2018**, *11*, 974. [CrossRef]

4. Ming, N.; Wei, H.; Guo, J.; Su, L. Research on Economic Operation of Grid-Connected Micro-grid. *Power Syst. Technol.* **2010**, *34*, 38–42.

5. Han, L.; Wang, F.; Tian, C. Economic Evaluation of Actively Consuming Wind Power for an Integrated Energy System Based on Game Theory. *Energies* **2018**, *11*, 1476. [CrossRef]

6. Chicco, G.; Mancarella, P. Distributed Multi-generation: A Comprehensive View. *Renew. Sustain. Energy Rev.* **2009**, *13*, 535–551. [CrossRef]

7. Liu, X.; Wang, S.; Sun, J. Energy Management for Community Energy Network with CHP Based on Cooperative Game. *Energies* **2018**, *11*, 1066. [CrossRef]

8. Liu, X.Z.; Wu, J.Z.; Jenkins, N.; Bagdanavicius, A. Combined Analysis of Electricity and Heat Networks. *Appl. Energy* **2016**, *162*, 1238–1250. [CrossRef]

9. Pan, Z.G.; Guo, Q.L.; Sun, H.B. Interactions of District Electricity and Heating Systems Considering Time-scale Characteristics Based on Quasi-Steady Multi-Energy Flow. *Appl. Energy* **2016**, *167*, 230–243. [CrossRef]

10. Zeng, Q.; Fang, J.K.; Li, J.H.; Chen, Z. Steady-state Analysis of the Integrated Natural Gas and Electric Power System with Bi-directional Energy Conversion. *Appl. Energy* **2016**, *184*, 1483–1492. [CrossRef]

11. Wang, W.L.; Wang, D.; Jia, H.J.; Chen, Z.Y.; Guo, B.Q.; Zhou, H.M.; Fang, M.W. Steady State Analysis of Electricity-gas Regional Integrated Energy System with Consideration of NGS Networks Status. *Proc. CSEE* **2017**, *37*, 1293–1304.

12. Shabanpour-Haghighi, A.; Seifi, A.R. An Integrated Steady-state Operation Assessment of Electrical, Natural Gas and District Heating Networks. *IEEE Trans. Power Syst.* **2016**, *31*, 3636–3647. [CrossRef]

13. Wang, Y.R.; Zeng, B.; Guo, J.; Shi, J.Q.; Zhang, J.H. Multi-energy Flow Calculation Method for Integrated Energy System Containing Electricity, Heat and Gas. *Power Syst. Technol.* **2016**, *40*, 2942–2950.

14. Wang, R.; Gu, W.; Wu, Z. Economic and Optimal Operation of Combined Heat and Power Microgrid with Renewable Energy Resources. *Autom. Electr. Power Syst.* **2011**, *35*, 22–27.

15. Liu, X.; Wu, H. A Control Strategy Operation Optimization of Combine Cooling Heating and Power System Considering Solar Comprehensive Utilization. *Autom. Electr. Power Syst.* **2015**, *39*, 1–6. [CrossRef]

16. Wang, Y.; Yu, H.; Yong, M.; Huang, Y.; Zhang, F.; Wang, X. Optimal Scheduling of Integrated Energy Systems with Combined Heat and Power Generation, Photovoltaic and Energy Storage Considering Battery Lifetime Loss. *Energies* **2018**, *11*, 1676. [CrossRef]

17. Zhou, X.; Guo, C.; Wang, Y.; Li, W. Optimal Expansion Co-planning of Reconfigurable Electricity and Natural Gas Distribution Systems Incorporating Energy Hubs. *Energies* **2017**, *10*, 124. [CrossRef]

18. Unsihuay-Vila, C.; Marangon-Lima, J.W.; de Souza, A.Z.; Perez-Arriaga, I.J.; Balestrassi, P.P. A Model to Long-term, Multiarea, Multistage, and Integrated Expansion Planning of Electricity and Natural Gas Systems. *IEEE Trans. Power Syst.* **2010**, *25*, 1154–1168. [CrossRef]

19. Saldarriaga, C.A.; Hincapié, R.A.; Salazar, H. A Holistic Approach for Planning Natural Gas and Electricity Distribution Networks. *IEEE Trans. Power Syst.* **2013**, *28*, 4052–4063. [CrossRef]

20. Qiu, J.; Dong, Z.Y.; Zhao, J.H.; Xu, Y.; Zheng, Y.; Li, C.; Wong, K.P. Multi-stage Flexible Expansion Co-planning Under Uncertainties in a Combined Electricity and Gas Market. *IEEE Trans. Power Syst.* **2015**, *30*, 2119–2129. [CrossRef]

21. Barati, F.; Seifi, H.; Sepasian, M.S.; Nateghi, A.; Shafie-khah, M.; Catalao, J.P.S. Multi-period Integrated Framework of Generation, Transmission, and Natural Gas Grid Expansion Planning for Large-scale Systems. *IEEE Trans. Power Syst.* **2015**, *30*, 2527–2537. [CrossRef]

22. Qiu, J.; Dong, Z.Y.; Zhao, J.H.; Meng, K.; Zheng, Y.; Hill, D.J. Low Carbon Oriented Expansion Planning of Integrated Gas and Power Systems. *IEEE Trans. Power Syst.* **2015**, *30*, 1035–1046. [CrossRef]

23. Shao, C.; Shahidehpour, M.; Wang, X.; Wang, X.; Wang, B. Integrated Planning of Electricity and Natural Gas Transportation Systems for Enhancing the Power Grid Resilience. *IEEE Trans. Power Syst.* **2017**, *32*, 4418–4429. [CrossRef]

24. Zhao, B.; Conejo, A.J.; Sioshansi, R. Coordinated Expansion Planning of Natural Gas and Electric Power Systems. *IEEE Trans. Power Syst.* **2017**, *33*, 3064–3075. [CrossRef]

25. Odetayo, B.; MacCormack, J.; Rosehart, W.D.; Zareipour, H. A Sequential Planning Approach for Distributed Generation and Natural Gas Networks. *Energy* **2017**, *127*, 428–437. [CrossRef]

26. Khaligh, V.; Buygi, M.; Anvari-Moghaddam, A.; Guerrero, M. A Multi-Attribute Expansion Planning Model for Integrated Gas–Electricity System. *Energies* **2018**, *11*, 2573. [CrossRef]

27. Li, G.; Wang, R.; Zhang, T.; Ming, M.; Sciubba, E. Multi-Objective Optimal Design of Renewable Energy Integrated CCHP System Using PICEA-g. *Energies* **2018**, *11*, 743. [CrossRef]

28. Wu, D.; Wang, R. Combined Cooling, Heating and Power: A Review. *Prog. Energy Combust. Sci.* **2006**, *32*, 459–495. [CrossRef]

29. Jradi, M.; Riffat, S. Tri-generation systems: Energy policies, prime movers, cooling technologies, configurations and operation strategies. *Renew. Sustain. Energy Rev.* **2014**, *32*, 396–415. [CrossRef]

30. Bao, Z.; Zhou, Q.; Yang, Z.; Yang, Q.; Xu, L.; Wu, T. A Multi Time-scale and Multi Energy-type Coordinated Microgrid Scheduling Solution-Part I: Model and Methodology. *IEEE Trans. Power Syst.* **2015**, *30*, 2257–2266. [CrossRef]

31. Li, Z.; Jenkins, N.; Wu, W.; Shahidehpour, M.; Wang, J.; Zhang, B. Combined Heat and Power Dispatch Considering Pipeline Energy Storage of District Heating Network. *IEEE Trans. Sustain. Energy* **2016**, *71*, 12–22. [CrossRef]

32. Li, J.; Fang, J.; Zeng, Q.; Chen, Z. Optimal Operation of the Integrated Electrical and Heating Systems to Accommodate the Intermittent Renewable Sources. *Appl. Energy* **2016**, *167*, 244–254. [CrossRef]

33. Bloess, A.; Schill, W.; Zerrahn, A. Power-to-heat for renewable energy integration: A review of technologies, modeling approaches, and flexibility potentials. *Appl. Energy* **2017**, *212*, 1611–1626. [CrossRef]

34. Dolara, A.; Grimaccia, F.; Magistrati, G.; Marchegiani, G. Optimization Models for Islanded Micro-Grids: A Comparative Analysis between Linear Programming and Mixed Integer Programming. *Energies* **2017**, *10*, 241. [CrossRef]

35. Gu, W.; Wang, Z.; Wu, Z.; Luo, Z.; Tang, Y.; Wang, J. An online optimal dispatch schedule for CCHP microgrids based on model predictive control. *IEEE Trans. Smart Grid* **2017**, *8*, 2332–2342. [CrossRef]

Article

Improved Adaptive Backstepping Sliding Mode Control of Static Var Compensator

Qingyu Su, Fei Dong and Xueqiang Shen *

School of Automation Engineering, Northeast Electric Power University, Jilin 132012, China;
suqingyu@neepu.edu.cn (Q.S.); dff0728@163.com (F.D.)
* Correspondence: shenxueqiang@neepu.edu.cn; Tel.: +86-138-4322-5717

Received: 27 August 2018; Accepted: 12 October 2018; Published: 14 October 2018

Abstract: The stability of a single machine infinite bus system with a static var compensator is proposed by an improved adaptive backstepping algorithm, which includes error compensation, sliding mode control and a κ-class function. First, storage functions of the control system are constructed based on modified adaptive backstepping sliding mode control and Lyapunov methods. Then, adaptive backstepping method is used to obtain nonlinear controller and parameter adaptation rate for static var compensator system. The results of simulation show that the improved adaptive backstepping sliding mode variable control based on error compensation is effective. Finally, we get a conclusion that the improved method differs from the traditional adaptive backstepping method. The improved adaptive backstepping sliding mode variable control based on error compensation method preserves effective non-linearities and real-time estimation of parameters, and this method provides effective stability and convergence.

Keywords: adaptive backstepping; nonlinear power systems; sliding mode control; error compensation; κ-class function

---

1. Introduction

With the development of economy and the various fields of modern life, especially industry, the requirement of electric power is more and more important. The modern life has higher requirements for the stability of power system [1]. Static var compensator (SVC) is one of the important members in flexible transmission system and it is a device which can control reactive power in power grid [2]. SVC is according to the reactive power to compensate automatically and it is from the grid to absorb reactive power to maintain voltage stability of instruction. Moreover, it is good for power grid reactive power balance [3]. When the system fails, svc can stabilize the system by adjusting reactive power. Therefore, studying SVC's control law has important significance in improve the power system's stability.

At present, scholars have studied many control methods in SVC [4]. For example, the traditional proportion, integral and differential (PID) is designed by the local linearization of the model, it cannot adapt to changing the power system operating point. The passage in [5] designs SVC's control law with exact linearization method. The passage in [6] puts the direct feedback linearization theory into the design of the SVC controller and gets a good effect. But they are designed for the linear systems. We know some nonlinear characteristics of the nonlinear systems are good for the design of the control law. The passage in [7] designs the robust controller and the parameters controller based on the uncertain equivalence principle of adaptive mechanism and gets a good effect. The passage in [8] studies the nonlinear control with a single machine infinite bus based on the theory of generalized Hamilton system method. The passage uses the neural network algorithm to estimate the transmission capacity margin of the power system with SVC in [9].

In engineering practice, many control systems have nonlinear [10–12] characteristics. Since the transmission power of the power system is directly proportional to the sinusoidal value of the power system, if we want to study the large range of motion in power system, we must consider the influence of nonlinear characteristics. Nonlinear phenomenon is widespread in nature and nonlinear system is the most general but linear system is just one special case. Nonlinear adaptive method is applied in many fields, for example, gain-scheduled control [13–15], feedback linearization control [16], sliding mode control [17–19], nonlinear adaptive control [20–23].

The method of backstepping [24,25] is decomposing a complex nonlinear system into no more than n subsystems, and then designing for each subsystem function and intermediate virtual variables, finally backing to the entire system and integrating them together to complete the whole design of the control. As for a single machine infinite bus system, the adaptive backstepping method [26,27] can reduce the design burden on the stability control and parameter estimation. However, it has defects in the stability of a system. Thus this paper puts forward a new adaptive backstepping sliding mode control for nonlinear systems [28,29], sliding mode variable control is highly robust and has high control accuracy for internal parameters perturbations and external disturbances.

First, this paper introduces an improved adaptive backstepping method based on error compensation (ABEC) and this method considers the damping coefficients. Then this paper introduces the improved adaptive backstepping sliding mode variable control based on error compensation method (ABSMVCEC). This method can get the system stable more quickly. In addition, the κ-class function is added to the selection of the intermediate virtual variable function, which speeds up the convergence of the system. Finally, the conclusion is obtained by simulation results.

In this paper, Section 2 introduces the model of a single machine infinite bus (SMIB) system with SVC suitable for controller design. Section 3 presents a new method's design details, ABEC. Section 4 proposes another new method, adaptive backstepping sliding mode control based on error compensation. It designs nonlinear controller and stability proof using the Lyapunov stability criterion. The effective of the controller is verified by simulation in Section 5, and Section 6 is the conclusions.

2. Model of MSIB System with SVC and Control Objective

2.1. Model of SMIB System with SVC

Considering a SMIB system, in the middle of the transmission lines connected to TCR-FC (thyristor control fixed reactor in parallel capacitor group) type of compensation device, the principle diagram and equivalent circuit as shown in Figure 1. Hypothesis generator transient electric potential E_q' and mechanical power P_m are constant, the single machine infinite system with SVC dynamic equation can be represented as:

$$
\begin{aligned}
\dot{\delta} &= \omega - \omega_0, \\
\dot{\omega} &= -\frac{D}{H}(\omega - \omega_0) + \frac{w_0}{H}P_m - \frac{w_0}{H}E_q'V_s y_{svc}sin_\delta, \\
\dot{y_{svc}} &= \frac{1}{T_{svc}}(-y_{svc} + y_{svc0} + u),
\end{aligned}
\tag{1}
$$

where w and D are the speed and the damping coefficient for SVC, δ is the angle, H is the rotational inertia of the rotor, P_m is the output mechanical power of the prime mover, V_s and T_{svc} are the infinite bus voltage and the inertial time constant, E_q' is the inner voltage of the generator shaft, B_L and B_C are the inductance susceptance and the susceptance of the capacitors, $y_{svc} = \frac{1}{X_1 + X_2 + X_1 X_2 (B_L + B_C)}$ is the susceptance of the whole system, $X_1 = X_d' + X_T + X_L, X_2 = X_L, B_L + B_C$ is the equivalent reactance in SVC, and u is the equivalent amount of control. Considering the damping coefficient D is not easy to measure, thus we let the $\theta = -\frac{D}{H}$ for uncertain constant parameters. Selecting the state variables $x_1 = \delta - \delta_0, x_2 = \omega - \omega_0, x_3 = y_{svc} - y_{svc0}, (\delta_0, \omega_0, y_{svc0})$ is the operating point of the system. We custom $a_0 = \frac{w_0}{H}P_m, k = \frac{w_0 E_q'V_s}{H}$. Then, the system can be transformed as

$$
\begin{aligned}
\dot{x_1} &= x_2 \\
\dot{x_2} &= \theta x_2 + a_0 - k(x_3 + y_{svc0})sin(x_1 + \delta_0) \\
\dot{x_3} &= -\frac{1}{T_{svc}}x_3 + \frac{1}{T_{svc}}u
\end{aligned}
\tag{2}
$$

Remark 1. *The model of the power system* (1) *is a simpilied three-order model. And it has been quoted several times in the journal literature, such as the references [17,19,21]. Among them, the model of reference [19] is the same as the reference [17], and the model* (1) *in this paper is the same as [17,19]. Different from [21], the reference [17] assumes the transient potential* E'_q *and the mechanical power* P_m *of the generator are constant. The principle diagram and equivalent circuit of the system* (1) *is shown in Figure 1. In order to study the performance of the system* (1) *conveniently, we translated the three-order physical* (1) *into the three-order mathematical as shown in* (2).

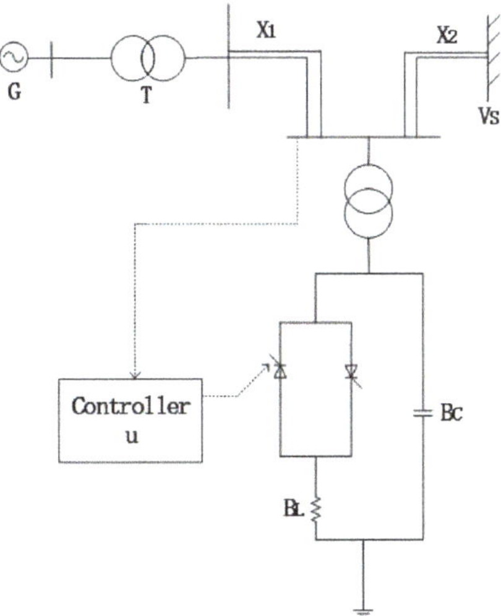

Figure 1. Single machine infinite bus system with SVC.

2.2. The Statement of Problem and Control Objective

Our control object is designing an adaptive backstepping controller u to make the system stable. This paper researches the nonlinear controller and a parameter updating law for the single machine bus system with SVC. Using the improved ABEC method improves the external disturbances. To get a higher control accuracy for internal parameters perturbations, we put sliding mode control based on error compensation into the traditional adaptive backstepping method. Finally, the three methods are compared by simulation. The structure diagram of SVC system is shown in Figure 2.

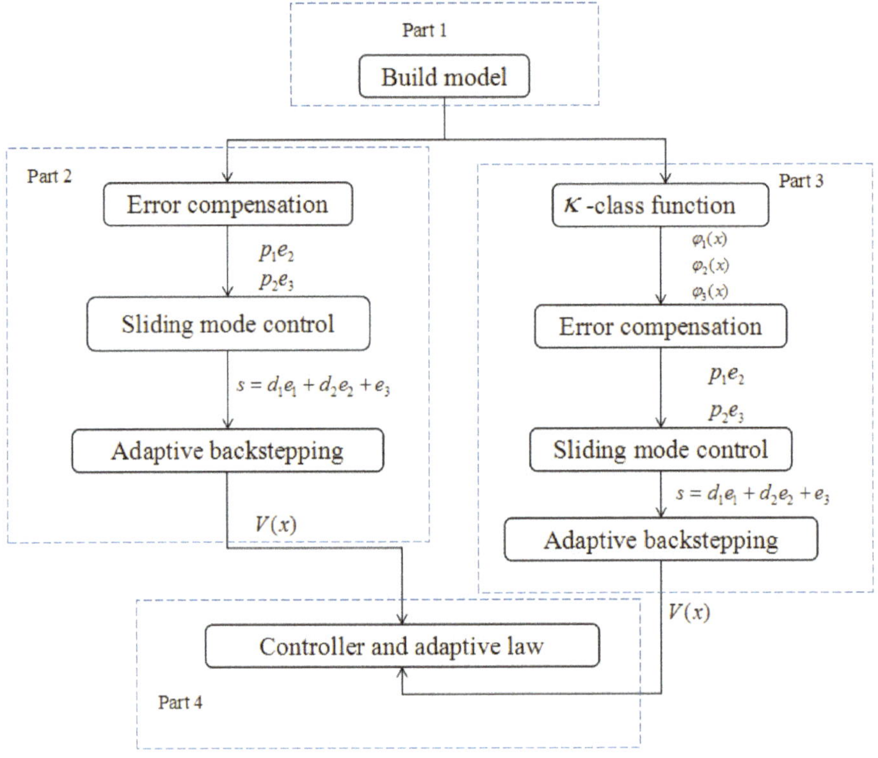

Figure 2. The structure diagram of SVC system.

3. Design of Adaptive Backstepping Controller Based on Error Compensation

The basic principle of backstepping is to decompose the system into subsystems of n which do not exceed the order of the system. First, we design a part of the Lyapunov function and virtual variables for each subsystem. Then, we can back to the whole system while each subsystem is stable. Finally a nonlinear controller is designed to stabilize the system. Because there is an error between theory and real engineering, this section will consider the influence of error. We will put error compensation into the traditional adaptive backstepping method. The following four steps are designed for backstepping controller.

Definition 1. *For system* (2), *we select* x_2^* *and* x_3^* *as the virtual stabilization functions. Then, we can obtain error variables* e_i, $i = 1,2,3$, *as shown by the following equations:*

$$
\begin{aligned}
e_1 &= x_1 \\
e_2 &= x_2 - x_2^* \\
e_3 &= x_3 - x_3^*
\end{aligned}
\tag{3}
$$

Step 1: Firstly, choose the virtual control of x_2 as x_2^*,

$$
x_2^* = -(\varphi_1(|e_1|) + c_1)e_1 - p_1 e_2
\tag{4}
$$

where c_1 is a positive constant, $\varphi_1(\cdot)$ is a class-κ function to be designed, and $p_1 e_2$ is an error compensation to eliminate system shake problem. Then, based on Definition 1, getting

$$\dot{e}_1 = -(\varphi_1(|e_1|) + c_1)e_1 + (1 - p_1)e_2 \tag{5}$$

Choose the first storage Lyapunov function

$$V_1 = \tfrac{1}{2}e_1^2 \tag{6}$$

thus, the derivative of V_1 is

$$\dot{V}_1 = -(\varphi_1(|e_1|) + c_1)e_1^2 + (1 - p_1)e_1 e_2 \tag{7}$$

Select $\varphi_1(\cdot)$ as $\varphi_1(|e_1|) = \tfrac{1}{3}\varepsilon_1 e_1^2$, where ε_1 is a positive constant. Absolutely, when $e_2 = 0$, there is $\dot{V}_1 < 0$, satisfying the stability condition.

Step 2: Choose a Lyaponov function

$$V_2 = V_1 + \tfrac{1}{2}e_2^2 \tag{8}$$

we know,

$$\begin{aligned}
\dot{e}_2 &= \dot{x}_2 - \dot{x}_2^* \\
&= \theta x_2 + a_0 - k(x_3 + y)sin(x_1 + \delta_0) + \varepsilon_1 e_1^2 x_2 + c_1 x_2 + p_1 \dot{e}_2
\end{aligned} \tag{9}$$

then result,

$$\begin{aligned}
\dot{e}_2 &= \dot{x}_2 - \dot{x}_2^* \\
&= \tfrac{1}{1-p_1}[\theta x_2 + a_0 - k(x_3 + y)sin(x_1 + \delta_0) + c_1 x_2 + \varepsilon_1 e_1^2 x_2]
\end{aligned} \tag{10}$$

The derivative of V_2 is

$$\begin{aligned}
\dot{V}_2 &= \dot{V}_1 + e_1 e_2 \\
&= -(\varphi_1(|e_1|) + c_1)e_1^2 + \tfrac{1}{1-p_1}[(1 - p_1)^2 e_1 + \theta x_2 + a_0 - k(x_3 + y)sin(x_1 + \delta_0) + c_1 x_2 + \varepsilon_1 e_1^2 x_2]
\end{aligned} \tag{11}$$

Moreover, by Definition 1: $e_3 = x_3 - x_3^*$, choose virtual stabilization function:

$$x_3^* = \tfrac{1}{ksin(x_1+\delta_0)}[(1 - p_1)^2 e_1 + \hat{\theta} x_2 + a_0 + c_1 x_2 + (\varphi_2(|e_2|) + c_2)e_2 + p_2 e_3 + \varepsilon_1 e_1^2 x_2] - y \tag{12}$$

among them c_2, p_2 are constants, and $\varphi_2(\cdot)$ is a class-κ function to be designed. Then selecting $\varphi_2(\cdot)$ as $\varphi_2(|e_2|) = \tfrac{1}{3}\varepsilon_2 e_2^2$, where ε_2 is a positive constant, $\hat{\theta}$ is the estimate of θ, $\tilde{\theta}$ is estimated error and $\tilde{\theta} = \theta - \hat{\theta}$, thus,

$$\dot{V}_2 = -(\varphi_1(|e_1|) + c_1)e_1^2 + \tfrac{e_2}{1-p_1}[\tilde{\theta}x_2 - (\varphi_2(|e_2|) + c_2)e_2 - p_2 e_3 - ksin(x_1 + \delta_0)e_3] \tag{13}$$

Step 3: For the whole system (2), select the new Lyapunov function

$$V = V_2 + \tfrac{1}{2}e_3^2 + \tfrac{1}{2\rho}\tilde{\theta}^2 \tag{14}$$

where $\rho > 0$ is given adaptive gain parameter, while

$$\begin{aligned}
\dot{e}_3 &= \dot{x}_3 - \dot{x}_3^* \\
&= \tfrac{ksin(x_1+\delta_0)}{ksin(x_1+\delta_0)+p_2}\{-\tfrac{1}{T}x_3 + \tfrac{1}{T}u - \tfrac{1}{ksin(x_1+\delta_0)}[(1 - p_1)^2 x_2 + \dot{\hat{\theta}} x_2 + 2\varepsilon_1 e_1 x_2^2 \\
&\quad + \tfrac{1}{1-p_1}(\varepsilon_2 e_2^2 + c_2)(\varepsilon_1 e_1^2 x_2 + c_1 x_2) + (\hat{\theta} + c_1 + \tfrac{c_2}{1-p_1} + \varepsilon_1 e_1^2 + \tfrac{\varepsilon_2 e_2^2}{1-p_1})(\theta x_2 \\
&\quad + a_0 - k(x_3 + y)sin(x_1 + \delta_0))] + \tfrac{cos(x_1+\delta_0)x_2}{ksin^2(x_1+\delta_0)}[(1 - p_1)^2 e_1 + \hat{\theta} x_2 \\
&\quad + a_0 + \varepsilon_1 e_1^2 x_2 + c_1 x_2 + (\varphi_2(|e_2|) + c_2)e_2 + p_2 e_3]\}
\end{aligned} \tag{15}$$

where p_2e_3 is error compensation term, it can compensate the influence of unknown error in the process of dynamic stability. Since $\theta = \hat{\theta} + \tilde{\theta}$, the derivative of V is:

$$
\begin{aligned}
\dot{V} = {} & -(\varphi_1(|e_1|) + c_1)e_1^2 - \tfrac{1}{1-p_1}(\varphi_2(|e_2|) + c_2)e_2^2 + \tfrac{e_2}{1-p_1}\tilde{\theta}x_2 - \tfrac{1}{\rho}\tilde{\theta}\dot{\hat{\theta}} - \tfrac{e_3\tilde{\theta}x_2}{ksin(x_1+\delta_0)+p_2}(\hat{\theta} + \varepsilon_1 e_1^2 \\
& + c_1 + \tfrac{\varepsilon_2 e_2^2}{1-p_1} + \tfrac{c_2}{1-p_1}) + \tfrac{ksin(x_1+\delta_0)e_3}{ksin(x_1+\delta_0)+p_2}\{-\tfrac{ksin(x_1+\delta_0)+p_2}{1-p_1}e_2 + \tfrac{1}{T}u - \tfrac{1}{T}x_3 - \tfrac{1}{ksin(x_1+\delta_0)}[(1-p_1)^2 x_2 \\
& + \hat{\theta}x_2 + 2\varepsilon_1 e_1 x_2^2 + \tfrac{1}{1-p_1}(\varepsilon_2 e_2^2 + c_2)(\varepsilon_1 e_1^2 x_2 + c_1 x_2) + (\hat{\theta} + c_1 + \tfrac{c_2}{1-p_1} + \varepsilon_1 e_1^2 + \tfrac{\varepsilon_2 e_2^2}{1-p_1})(\hat{\theta}x_2 \\
& + a_0 - k(x_3+y)sin(x_1+\delta_0))] + \tfrac{cos(x_1+\delta_0)x_2}{ksin^2(x_1+\delta_0)}[(1-p_1)^2 e_1 + \hat{\theta}x_2 + a_0 + \varepsilon_1 e_1^2 x_2 \\
& + c_1 x_2 + (\varphi_2(|e_2|) + c_2)e_2 + p_2 e_3]\}
\end{aligned}
\tag{16}
$$

Considering (17), get the parameters replacement law

$$
\dot{\hat{\theta}} = \rho[\tfrac{e_2}{1-p_2} - \tfrac{e_3}{ksin(x_1+\delta_0)+p_2}(\hat{\theta} + \varepsilon_1 e_1^2 + c_1 + \tfrac{\varepsilon_2 e_2^2}{1-p_1} + \tfrac{c_2}{1-p_1})]x_2
\tag{17}
$$

Then

$$
\dot{V} = -(\varphi_1(|e_1|) + c_1)e_1^2 - \tfrac{1}{1-p_1}(\varphi_2(|e_2|) + c_2)e_2^2 - (\varphi_3(|e_3|) + c_3)e_3^2
\tag{18}
$$

where $\varphi_3(\cdot)$ is a class-κ function to be designed. Then we can select $\varphi_3(|e_3|) = \tfrac{1}{3}\varepsilon_3 e_3^2$, and ε_3 is a positive constant. If there exist new control parameters p_i and constant $c_i, 0 < p_i, c_i < 1$, it is clearly, getting $\dot{V} < 0$ by (19).

Step 4: Finally, considering the control problem based on Section 2, an adaptive backstepping controller u can be got, as the following shown. The control law u and parameters replacement law $\hat{\theta}$ based on adaptive backstepping controller of SVC system are follows:

$$
\begin{aligned}
u = {} & T\{\tfrac{ksin(x_1+\delta_0)+p_2}{1-p_1}e_2 + \tfrac{1}{T}x_3 + \tfrac{1}{ksin(x_1+\delta_0)}[(1-p_1)^2 x_2 + \hat{\theta}x_2 + 2\varepsilon_1 e_1 x_2^2 + \tfrac{1}{1-p_1}(\varepsilon_2 e_2^2 + c_2)(\varepsilon_1 e_1^2 x_2 \\
& + c_1 x_2) + (\hat{\theta} + c_1 + \tfrac{c_2}{1-p_1} + \varepsilon_1 e_1^2 + \tfrac{\varepsilon_2 e_2^2}{1-p_1})(\hat{\theta}x_2 + a_0 - k(x_3+y)sin(x_1+\delta_0))] \\
& - \tfrac{cos(x_1+\delta_0)x_2}{ksin^2(x_1+\delta_0)}[(1-p_1)^2 e_1 + \hat{\theta}x_2 + a_0 + \varepsilon_1 e_1^2 x_2 + c_1 x_2 + (\varphi_2(|e_2|) + c_2)e_2 + p_2 e_3] \\
& - \tfrac{ksin(x_1+\delta_0)+p_2}{ksin(x_1+\delta_0)}(\varphi_3(|e_3|) + c_3)e_3\}
\end{aligned}
\tag{19}
$$

The closed-loop system under the new coordinates (e_1, e_2, e_3) is as follows:

$$
\begin{aligned}
\dot{e}_1 &= -(\varphi_1(|e_1|) + c_1)e_1 + (1-p_1)e_1 e_2 \\
\dot{e}_2 &= \tfrac{1}{1-p_1}[\tilde{\theta}x_2 - (1-p_1)^2 e_1 - (\varphi_2(|e_2|) + c_2)e_2 - p_2 e_3 - ksin(x_1+\delta_0)e_3] \\
\dot{e}_3 &= \tfrac{ksin(x_1+\delta_0)}{1-p_1}e_2 - \tfrac{1}{ksin(x_1+\delta_0)+p_2}(\hat{\theta} + \varepsilon_1 e_1^2 + c_1 + \tfrac{\varepsilon_2 e_2^2}{1-p_1} + \tfrac{c_2}{1-p_1})\tilde{\theta}x_2 - (\varphi_3(|e_3|) + c_3)e_3
\end{aligned}
\tag{20}
$$

Theorem 1. *When we consider the model of SVC system* (1) *under the influence of controller u* (19) *and parameters $\hat{\theta}$* (17), *the closed-loop system* (20) *is asymptotically stable nearby the origin.*

Proof. By (25), we know $V(t) \leq V(0)$. Namely, e_1, e_2, x_1, x_2 are bounded. Then we define $\Gamma = -\dot{V}$, $\int_0^t \Gamma(\tau)d\tau = V(0) - V(t)$. Because V(0) is bounded, V(t) is also decreasing and bounded, we know that $\lim\limits_{t\to\infty} \int_0^t \Gamma(\tau)d\tau < \infty$. In addition, $\dot{\Gamma}$ is bounded, thus, $\lim\limits_{t\to\infty} \Gamma = 0$ is proved by Barbalat lemma. When $t \to \infty$, there are $e_1 \to 0, e_2 \to 0, x_1 \to 0, x_2 \to 0$. Based on the definitions of x_1, x_2, x_3 and x_2^*, x_3^*, we know that $e_3 \to 0$, x_3 is also bounded. $\square$

Remark 2. *If $\delta = k\pi, k = 0, 1, 2, 3...$, then $sin(x_1 + \delta_0) = 0$, the power system will be lost stability and there is no longer normal operation. Fortunately, in support of the normal conditions in the system $0 < \delta < \pi$, $sin(x_1 + \delta_0) \neq 0$ can be guaranteed.*

Remark 3. *When the coefficient of error compensation $p_1, p_2 = 0$, the ABSMVCEC method changes into the traditional backstepping control method. In other words, the traditional backstepping method is a special case of Theorem 1.*

Remark 4. *We take a class-κ function $\varphi_i(\cdot)$, where $i = 1, 2, ...n$, during the process of recursive design of update law. By the function of $\varphi_i(\cdot)$ we can speed up the convergence rate in error. But error is decrease with the increase of time, $\dot{V}$ becomes the traditional adaptive backstepping gradually.*

Remark 5. *In the improved adaptive backstepping method above, if $\varphi_i(\cdot) \equiv 0$, it equivalents to the traditional methods.*

4. Design of Adaptive Backstepping Sliding Mode Variable Structure Controller Based on Error Compensation

To get a higher control accuracy for internal parameters perturbations, we put sliding mode variable structure control based on error compensation into the traditional adaptive backstepping method.

The ABSMVCEC method is the same as the chapters mentioned before, thus we start from step 3 to introduce the improved method.

Step 3: Choose the sliding mode surface $s = d_1 e_1 + d_2 e_2 + e_3 = 0$, d_1 and d_2 are constants respectively. The whole system Lyapunove function is given by

$$V = V_2 + \tfrac{1}{2}s^2 + \tfrac{1}{2\rho}\tilde{\theta}^2 \tag{21}$$

while noting that $\dot{s} = d_1\dot{e}_1 + d_2\dot{e}_2 + \dot{e}_3$. The derivative of V is

$$\dot{V} = \dot{V}_2 + s\dot{s} - \tfrac{1}{\rho}\tilde{\theta}\dot{\hat{\theta}} \tag{22}$$

since $\theta = \hat{\theta} + \tilde{\theta}$ and $e_3 = s - d_1 e_1 - d_2 e_2$, therefore,

$$
\begin{aligned}
\dot{V} =\ & -(-(\varphi_1(|e_1|)+c_1) - \tfrac{ksin(x_1+\delta_0)+p_2}{2(1-p_1)}d_1)e_1^2 - [(\tfrac{1}{1-p_1}(\varphi_2(|e_2|)+c_2) - \tfrac{ksin(x_1+\delta_0)+p_2}{2(1-p_1)}d_1 \\
& - \tfrac{ksin(x_1+\delta_0)+p_2}{1-p_1}d_2]e_2^2 - \tfrac{ksin(x_1+\delta_0)+p_2}{2(1-p_1)}d_1(e_1-e_2)^2 + \tfrac{e_2}{1-p_1}\tilde{\theta}x_2 - \tfrac{1}{\rho}\tilde{\theta}\dot{\hat{\theta}} + \tfrac{d_2 s}{1-p_1}\tilde{\theta}x_2 \\
& - \tfrac{s}{ksin(x_1+\delta_0)+p_2}(\hat{\theta}+\varepsilon_1 e_1^2 + c_1 + \tfrac{\varepsilon_2 e_2^2}{1-p_1} + \tfrac{c_2}{1-p_1})\tilde{\theta}x_2 + s\{-\tfrac{e_2}{1-p_1}p_2 - \tfrac{ksin(x_1+\delta_0)}{1-p_1}e_2 \\
& + d_1 x_2 + \tfrac{d_2}{1-p_1}[\hat{\theta}x_2 + a_0 - k(x_3+y)sin(x_1+\delta_0) + c_1 x_2 + \varepsilon_1 e_1^2 x_2] \\
& + \tfrac{ksin(x_1+\delta_0)}{ksin(x_1+\delta_0)+p_2}\{-\tfrac{1}{T}x_3 + \tfrac{1}{T}u + \tfrac{cos(x_1+\delta_0)x_2}{ksin^2(x_1+\delta_0)}[(1-p_1)^2 e_1 + \hat{\theta}x_2 + a_0 \\
& + \varepsilon_1 e_1^2 x_2 + c_1 x_2 + (\varphi_2(|e_2|)+c_2)e_2 + p_2 e_3] - \tfrac{1}{ksin(x_1+\delta_0)}[(1-p_1)^2 x_2 \\
& + \hat{\theta}x_2 + 2\varepsilon_1 e_1 x_2^2 + \tfrac{1}{1-p_1}(\varepsilon_2 e_2^2 + c_2)(\varepsilon_1 e_1^2 x_2 + c_1 x_2) + (\hat{\theta} + c_1 + \tfrac{c_2}{1-p_1} + \varepsilon_1 e_1^2 \\
& + \tfrac{\varepsilon_2 e_2^2}{1-p_1})(\hat{\theta}x_2 + a_0 - k(x_3+y)sin(x_1+\delta_0))]\}\}
\end{aligned}
\tag{23}
$$

The parameter replacement law is that

$$
\dot{\hat{\theta}} = \rho[\tfrac{e_2}{1-p_1} + \tfrac{d_2 s}{1-p_1} - \tfrac{s}{ksin(x_1+\delta_0)+p_2}(\hat{\theta}+\varepsilon_1 e_1^2 + c_1 + \tfrac{\varepsilon_2 e_2^2}{1-p_1} + \tfrac{c_2}{1-p_1})]x_2 \tag{24}
$$

Then

$$
\begin{aligned}
\dot{V} =\ & -(-(\varphi_1(|e_1|)+c_1) - \tfrac{ksin(x_1+\delta_0)+p_2}{2(1-p_1)}d_1)e_1^2 - [(\tfrac{1}{1-p_1}(\varphi_2(|e_2|)+c_2) - \tfrac{ksin(x_1+\delta_0)+p_2}{2(1-p_1)}d_1 \\
& - \tfrac{ksin(x_1+\delta_0)+p_2}{1-p_1}d_2]e_2^2 - \tfrac{ksin(x_1+\delta_0)+p_2}{2(1-p_1)}d_1(e_1-e_2)^2 - (\varphi_3(|e_3|)+\beta)s^2
\end{aligned}
\tag{25}
$$

Select the same class-κ function $\varphi_1(\cdot)$, $\varphi_2(\cdot)$ as $\varphi_1(|e_1|) = \tfrac{1}{3}\varepsilon_1 e_1^2$, $\varphi_2(|e_2|) = \tfrac{1}{3}\varepsilon_2 e_2^2$, select $\varphi_3(\cdot)$ as $\varphi_3(|e_3|) = \tfrac{1}{3}\varepsilon_3 e_3^2$, where $\varepsilon_1, \varepsilon_2, \varepsilon_3 > 0$.

Step 4: Obviously, $\dot{V} < 0$ by (25), then an adaptive backstepping controller u can be got according to Section 2. We can choose the proper parameters $c_i, d_i, p_i, i = 1, 2$, which meet conditions

(1) $\quad (\varphi_1(|e_1|)+c_1) - \tfrac{ksin(x_1+\delta_0)+p_2}{2(1-p_1)}d_1 \geq 0$

(2) $\quad \frac{1}{1-p_1}(\varphi_2(|e_2|)+c_2) - \frac{ksin(x_1+\delta_0)+p_2}{2(1-p_1)}d_1 - \frac{ksin(x_1+\delta_0)+p_2}{1-p_1}d_2 \geq 0$

(3) $\quad \frac{ksin(x_1+\delta_0)+p_2}{2(1-p_1)}d_1 \geq 0$

We know, $\beta > 0$ is a positive parameter of the sliding mode, and the other parameter variables should be given according to control requirements.

The control law u based on ABSMVCEC method is as follows:

$$
\begin{aligned}
u &= T\{[\frac{ksin(x_1+\delta_0)+p_2}{ksin(x_1+\delta_0)}[\frac{ksin(x_1+\delta_0)+p_2}{1-p_1}e_2 - d_1x_2 - \frac{d_2}{1-p_1}[\hat{\theta}x_2 + a_0 - k(x_3+y)sin(x_1+\delta_0) + \varepsilon_1 e_1^2 x_2 \\
&+ c_1x_2]] + \frac{1}{T}x_3 - \frac{cos(x_1+\delta_0)x_2}{ksin^2(x_1+\delta_0)}[(1-p_1)^2e_1 + \hat{\theta}x_2 + a_0 + \varepsilon_1 e_1^2 x_2 + c_1x_2 + (\varphi_2(|e_2|)+c_2)e_2 + p_2e_3] \\
&+ \frac{1}{ksin(x_1+\delta_0)}[(1-p_1)^2x_2 + \hat{\theta}x_2 + 2\varepsilon_1 e_1x_2^2 + \frac{1}{1-p_1}(\varepsilon_2 e_2^2 + c_2)(\varepsilon_1 e_1^2 x_2 + c_1x_2) + (\hat{\theta} + c_1 + \frac{c_2}{1-p_1} \\
&+ \varepsilon_1 e_1^2 + \frac{\varepsilon_2 e_2^2}{1-p_1})(\hat{\theta}x_2 + a_0 - k(x_3+y)sin(x_1+\delta_0))] - \frac{ksin(x_1+\delta_0)+p_2}{ksin(x_1+\delta_0)}(\varphi_3(|e_3|)+\beta)s\}
\end{aligned}
\tag{26}
$$

The closed-loop system under the new coordinates (e_1, e_2, e_3) are

$$
\begin{aligned}
\dot{e}_1 &= -(\varphi_1(|e_1|)+c_1)e_1 + (1-p_1)e_1e_2 \\
\dot{e}_2 &= \frac{1}{1-p_1}[\tilde{\theta}x_2 - (1-p_1)^2e_1 - (\varphi_2(|e_2|)+c_2)e_2 - p_2e_3 - ksin(x_1+\delta_0)e_3] \\
\dot{e}_3 &= \frac{ksin(x_1+\delta_0)+p_2}{1-p_1}e_2 - d_1x_2 - \frac{d_2}{1-p_1}[\tilde{\theta}x_2 + a_0 - k(x_3+y)sin(x_1+\delta_0) + \varepsilon_1 e_1^2 x_2 + c_1x_2] \\
&\quad - \frac{1}{ksin(x_1+\delta_0)+p_2}(\hat{\theta} + \varepsilon_1 e_1^2 + c_1 + \frac{\varepsilon_2 e_2^2}{1-p_1} + \frac{c_2}{1-p_1})\tilde{\theta}x_2 - (\varphi_3(|e_3|)+\beta)s
\end{aligned}
\tag{27}
$$

Theorem 2. *When we consider mode for the valve control system* (1) *with its equivalent from* (2)*, the closed-loop error system* (27) *is asymptotically stable nearby the origin under the influence of the control law u* (26)*.*

Proof. By (25), we know $V(t) \leq V(0)$. Namely, e_1, e_2, x_1, x_2 are bounded. Then we define $\Gamma = -\dot{V}$, $\int_0^t \Gamma(\tau)d\tau = V(0) - V(t)$. Because V(0) is bounded, V(t) is also decreasing and bounded, we know that $\lim\limits_{t\to\infty} \int_0^t \Gamma(\tau)d\tau < \infty$. In addition, $\dot{\Gamma}$ is bounded, thus, $\lim\limits_{t\to\infty} \Gamma = 0$ is proved by Barbalat lemma. When $t \to \infty$, there are $e_1 \to 0, e_2 \to 0, x_1 \to 0, x_2 \to 0$. Based on the definitions of x_1, x_2, x_3 and x_2^*, x_3^*, we know that $e_3 \to 0$, x_3 is also bounded. $\square$

Remark 6. *We add the parameter $\varepsilon_i, i = 1, 2, 3$ in the design process. We could select the smaller parameter ε_i when the error is larger, thus it makes the controller gain does not increase a lot. In order to improve the transient performance advantage, we could select the bigger ε_i when the error is smaller. That is to say, the corresponding speed of the system can be increased or decreased by adjusting the gain parameter ε_i.*

Remark 7. *When the parameters $p_1, p_2 = 0, d_1, d_2 = 0, \varepsilon_1 = 0, \varepsilon_2 = 0, \varepsilon_3 = 0$, the ABSMVCEC method becomes the traditional backstepping control. In other words, the traditional backstepping sliding mode method is a special case of Theorem* 2*.*

5. Simulation

In order to verify the effectiveness of the improved method, we have carried on the simulation to the system. These are system parameters which are chosen from [17]: $c_i(i = 1, 2, 3)$ is positive constant, taken $c_1 = 2.5; c_2 = 2; c_3 = 2; p_i(i = 1, 2)$ is a positive error constant, taken $p_1 = 0.85; p_2 = 0.45; \rho$ is adaptive gain parameter, taken $\rho = 1; T$ is the inertial time constant, taken $T = 0.02; V$ is infinite bus voltage, taken $V = 1; E$ is transient electric potential, $E = 0.05; H$ is moment of inertia, $H = 20; w$ is speed of the generator rotor, y_{svc} is the susceptance of the overall system, δ is the angle of the generator rotor, and w_0, y_0, θ_0 is the work point when the system is stable, taken $w_0 = 314; y_0 = 0.814; \theta_0 = 57$.

As shown in Figures 3 and 4, the black line represents the traditional adaptive backstepping method, the green line represents the ABEC method, and the red line represents the ABSMVCEC method.

As shown in Figure 3a, setting the initial value of the power angle to 57.3, we can see the power angle δ_0 is 57.3 when the system is stable from Figure 3a. That is to say x_1 gradually tends to 0, thus the error gradually tends to 0, and it makes the system stable. We can see the improved ABSMVCEC method makes transient power angle performance better compared with the method of adaptive backstepping in [17] and the improved ABEC method from Figure 3a. In addition, the improved method of ABSMVCEC has faster convergence rate and smaller amplitude of vibration compared with the other two methods. In addition, we can see from Figure 4a that the effect of the power angle δ is also the same under large perturbations. When there is a large perturbation, the traditional method fluctuates greatly, but the improved method can resist the perturbation very well.

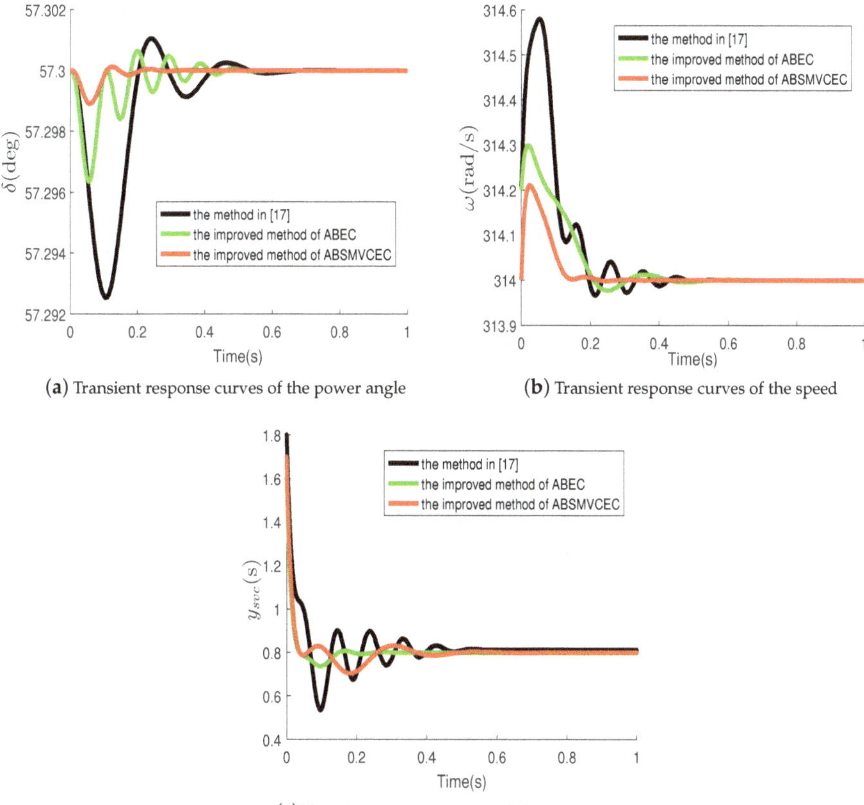

(**a**) Transient response curves of the power angle

(**b**) Transient response curves of the speed

(**c**) Transient response curves of the susceptance

Figure 3. (**a–c**) are simulation results for the system (1) in small perturbations.

As shown in Figure 3b, we set the initial value of the speed of the generator rotor to 314, and we can see the operating point of the power system (w_0) is gradually tends to 314, that is to say x_2 gradually tends to 0, thus error gradually tends to 0, the system gradually stable. It is clearly to see that three controllers settle down to their steady state values. However, the third method can make the system stable more quickly, meanwhile, the transient stability is smaller and smoother than above two methods. In a word, the third method shows a faster convergence speed and a better transient performance. In addition, we can see from Figure 4b that the effect of the speed ω is also the same under large perturbations. When there is a large perturbation, the traditional method fluctuates greatly, but the improved method can resist the perturbation very well.

As shown in Figure 3c, the susceptance y_{svc} tends to 0.8, stabile basically. In addition, we can see the three methods all can get the system stable. It is obviously that the ABSMVCEC method has a faster convergence speed and brings a better steady state. Also, we can see from Figure 4c that the effect of the susceptance y_{svc} is also the same under large perturbations. When there is a large perturbation, the traditional method fluctuates greatly, but the improved method can resist the perturbation very well.

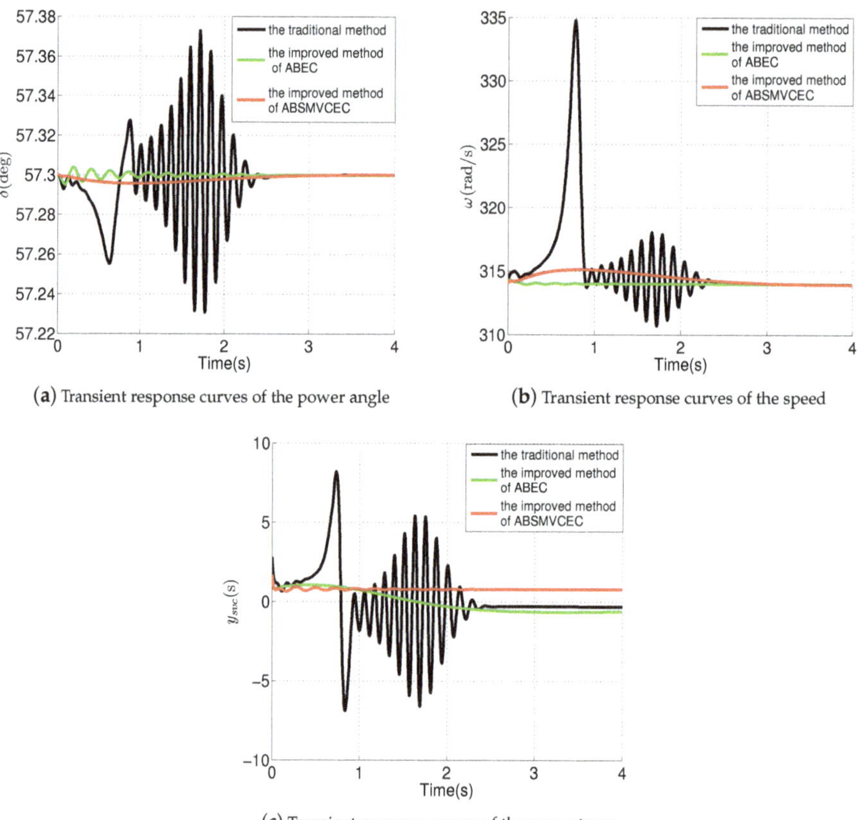

(**a**) Transient response curves of the power angle   (**b**) Transient response curves of the speed

(**c**) Transient response curves of the susceptance

Figure 4. (a–c) are simulation results for the system (1) in large perturbations.

6. Conclusions

This paper investigates the nonlinear controller for SVC system with the ABSMVCEC method. From the result of simulations, we can see that the two new methods all have better influences on the nonlinear power system; in particular, the ABSMVCEC method is more effective in improving the transient stability. In order to avoid one-sided pursuit of transient response speed and make the controller gain too heavy for implementation, this paper designs parameters ε_i, $i = 1, 2, 3$, to adjust the contradiction.

Author Contributions: Q.S., F.D. and X.S. have proposed and validated the main idea. F.D. have developed the simulation and written the remaining manuscript. All authors together organized and refined the manuscript in the present form.

Funding: This work was supported by the National Natural Science Foundation of China (61503071,61873057), the National Key Research and Development Program under grant (2016YFB0900104), Natural Science Foundation of Jilin Province (20180520211JH) and Science Research of Education Department of Jilin Province (201693, JJKH20170106KJ), Jilin City Science and Technology Bureau (201831727, 201831731).

Acknowledgments: The authors would like to thank the anonymous reviewers and the associate editors for their helpful remarks that improved the presentation of the paper.

Conflicts of Interest: The authors declare no conflict of interest.

Abbreviations

The following abbreviations are used in this manuscript:

SVC	Static Var Compensator
ABSMVCEC	Adaptive Backstepping Sliding Mode Variable Control Based On Error Compensation
PID	Proportion Integral And Differential
ABEC	Adaptive Backstepping Method Based On Error Compensation
SMIB	Single Machine Infinite Bus

References

1. Benidris, M.; Elsaiah, S.; Sulaeman, S.; Mitra, J. Transient Stability of Distributed Generators in the Presence of Energy Storage Devices. In Proceedings of the 44th IEEE North American Power Symposium, Champaign, IL, USA, 9–11 September 2012.
2. Caciotta, M.; Leccese, F.; Trifiro, A. From power quality to perceived power quality. In Proceedings of the IASTED International Conference on Energy and Power Systems, Chiang Mai, Thailand, 29–31 March 2006; pp. 94–102.
3. De Araujo, V.V.; Simas Filho, E.F.; Oliveira, A.; de Oliveira, W.L.A.; Esteves, L.T.C. Integrated circuit for real-time poly-phase power quality monitoring. *Microelectron. J.* **2018**, *79*, 57–63. [CrossRef]
4. Kundur, P. *Power System Stability and Control*; Et Power Electronics; McGraw-Hill Education: New York, NY, USA, 2007; Volume 7, pp. 100–103.
5. Su, J.; Chen, C. Study on SVC control for power system with nonlinear loads. *Autom. Electri. Power Syst.* **2002**, *26*, 12–15.
6. Li, S.; Su, Z. Static state feedback control for saddle-node bifurcation in a simple power system. In Proceedings of the International Conference on Power Electronics and Intelligent Transportation System, Shenzhen, China, 19–20 December 2010; pp. 158–160.
7. Hwang, Y.H.; Park, K.K.; Yang, H.W. Robust adaptive backstepping control for efficiency optimization of induction motors with uncertainties. In Proceedings of the ieee International Symposium on Industrial Electronics, Cambridge, UK, 30 June–2 July 2008; pp. 878–883.
8. Okano, K.; Hagino, K. Adaptive control design approximating solution of Hamilton-Jacobi-Bellman equation for nonlinear strict-feedback system with uncertainties. In Proceedings of the Annual Conference on Sice Conference, Tokyo, Japan, 20–22 August 2008; pp. 204–208.
9. Modi, P.K.; Singh, S.P.; Sharma, J.D. Fuzzy neural network based voltage stability evaluation of power systems with SVC. *Appl. Soft Comput.* **2008**, *8*, 657–665. [CrossRef]
10. Dimirovski, G.; Yuan, W.; Wen, L. Adaptive Back-Stepping Design of TCSC Robust Nonlinear Control for Power Systems. *Intell. Autom. Soft Comput.* **2006**, *12*, 75–87. [CrossRef]
11. Shang, A.L.; Wang, Z. Adaptive backstepping second order sliding mode control of non-linear systems. *Int. J. Model. Identif. Control* **2013**, *19*, 195–201. [CrossRef]
12. Shi, C.X.; Yang, G.H.; Li, X.J. Robust adaptive backstepping control for hierarchical multi-agent systems with signed weights and system uncertainties. *IET Control Theory Appl.* **2017**, *11*, 2743–2752. [CrossRef]
13. Mitra, A.; Mukherjee, M.; Naik, K. Enhancement of power system transient stability using a novel adaptive backstepping control law. In Proceedings of the Third International Conference on Computer, Communication, Control and Information Technology, West-Bengal, India, 7–8 February 2015; pp.1–5.
14. Ma osa, V.; Ikhouane, F.; Rodellar, J. Control of uncertain non-linear systems via adaptive backstepping. *J. Sound Vib.* **2005**, *280*, 657–680.
15. Wang, H.M.; Huang, T.L.; Tsai, C.M. Power system stabilizer design using adaptive backstepping controller. In Proceedings of the Fifth International Conference on Power Electronics and Drive Systems, Singapore, 17–20 November 2003.

16. Benayache, R.; Chrifialaoui, L.; Bussy, P. Adaptive backstepping sliding mode control for hydraulic system without overparametrisation. *Int. J. Model. Identif. Control* **2012**, *16*, 60–69. [CrossRef]

17. Sun, L.-Y. *Based on the Backstepping Method of Power System Nonlinear Robust Adaptive Control Design*; Northeastern University: Boston, MA, USA, 2009; pp. 50–56.

18. Zhang, T.; Dong, C. Compound control system design based on adaptive backstepping theory. *J. Beijing Univ. Aeronaut. Astronaut.* **2013**, *39*, 902–906.

19. Mao, J.; Sun, Y.-K.; Wu, G.-Q.; Liu, X.-F. A variable structure robust control method in SVC application for performance improvement. In Proceedings of the 2007 Chinese Control Conference, Harbin, China, 7–11 August 2006; pp. 67–70.

20. Hagras, A.; Zaid, S.; Ekousy, A.A. Performance comparison of shunt active power filter for interval type-2 fuzzy and adaptive backstepping controllers. *Int. J. Model. Identif. Control* **2014**, *21*, 270–287. [CrossRef]

21. Salma, K.; Nouha, B.; Souhir, S.; Larbi, C.-A.; MBA, K. Transient stability enhancement and voltage regulation in SMIB power system using SVC with PI controller. In Proceedings of the International Conference on Systems and Control, Wuhan, China, 7–9 May 2017; pp. 115–120.

22. Yimin, L.; Yang, Y.; Li, L. Adaptive Backstepping Fuzzy Control Based on Type-2 Fuzzy System. *J. Appl. Math.* **2012**, *2012*, 295–305.

23. Lin, F.J.; Shen, P.H.; Hsu, S.P. Adaptive backstepping sliding mode control for linear induction motor drive. *Proc. Electr. Power Appl.* **2002**, *149*, 184–194. [CrossRef]

24. Yu, H.; Wang, J.; Deng, B. Adaptive backstepping sliding mode control for chaos synchronization of two coupled neurons in the external electrical stimulation. *Commun. Nonlinear Sci. Numer. Simul.* **2012**, *17*, 1344–1354. [CrossRef]

25. Song, Z.; Sun, K. Adaptive backstepping sliding mode control with fuzzy monitoring strategy for a kind of mechanical system. *ISA Trans.* **2014**, *53*, 125–133. [CrossRef] [PubMed]

26. Zhao, D.; Zou, T.; Li, S. Adaptive backstepping sliding mode control for leader-follower multi-agent systems. *Control Theory Appl. IET* **2012**, *6*, 1109–1117. [CrossRef]

27. Gong, X.; Hou, Z.C.; Zhao, C.J. Adaptive Backstepping Sliding Mode Trajectory Tracking Control for a Quad-rotor. *Int. J. Autom. Comput.* **2012**, *9*, 555–560. [CrossRef]

28. Dong, L.; Tang, W.C. Adaptive backstepping sliding mode control of flexible ball screw drives with time-varying parametric uncertainties and disturbances. *ISA Trans.* **2014**, *53*, 110–116. [CrossRef] [PubMed]

29. Ma, L.; Schilling, K.; Schmid, C. Adaptive Backstepping Sliding Mode Control with Gaussian Networks for a Class of Nonlinear Systems with Mismatched Uncertainties. In Proceedings of the IEEE 2005 and 2005 European Control Conference on Decision and Control, Seville, Spain, 12–15 December 2005; pp. 5504–5509.

Article

Medium-Voltage AC Static Switch Solution to Feed Neutral Section in a High-Speed Railway System

Jose Maria Canales *, Iosu Aizpuru [†], Unai Iraola [†], Jon Andoni Barrena [†] and Manex Barrenetxea [†]

Mondragon Goi Eskola Politeknikoa, Mondragon University, 20500 Mondragon, Spain; iaizpuru@mondragon.edu (I.A.); uiraola@mondragon.edu (U.I.); jabarrena@mondragon.edu (J.A.B.); mbarrenetxeai@mondragon.edu (M.B.)
* Correspondence: jmcanales@mondragon.edu; Tel.: +34-943-794-700
† These authors contributed equally to this work.

Received: 12 September 2018; Accepted: 9 October 2018; Published: 12 October 2018

Abstract: A high-speed train (HST) is a single-phase load supplied by a three-phase AC grid. The HST produces unbalanced three-line currents affecting the power quality of the grid. To balance the asymmetries on average, railway feeding sections are supplied that rotate the three phases of the grid. An electric isolation segment, called the neutral section (NS), between different sections is necessary. The HST must pass through this 1.6 km NS without power supply. In this paper, a medium-voltage AC static switch solution to feed the high-speed train in the NS is proposed. Thyristor technology is selected to design the 25 KVAC static switch. A medium-voltage power electronics procedure design is proposed to ensure proper operation in the final application. An NS operation is analyzed to identify impacts within the electric system and solution requirements are developed. Then, a low-scale prototype is used to experimentally validate the solution based on thyristor technology and the medium-voltage AC static switch is designed. Limitations on power and voltage at the Mondragon University Medium-Voltage Laboratory do not allow testing of the AC static switch at nominal conditions. A partial test procedure to test sections of the AC static switch is proposed and applied to validate the solution. Finally, experimental results for the Cordoba–Malaga (Spain) high-speed railway in real conditions with an HST crossing the NS at 300 km/h are shown.

Keywords: railway; high-speed railway; neutral section; medium voltage; thyristor; AC static switch

---

1. Introduction

In 2003, more than 100,000 km of railway operated at 25 KVAC 50 Hz worldwide. This supposes 44.8% of the total line length of the electric railway system [1]. Using a 50 Hz public grid to feed a 25 KVAC railway eliminates the considerable cost of independent power generation, for example, 15 KVAC 16²/3 Hz and DC railway systems [2]. The 25 KVAC railway system is a single-phase load supplied by a medium-voltage three-phase AC grid, and it is produced to unbalance the three-line currents. The grid current asymmetry produces negative effects in transformers and rotary electric machines, reducing the power quality of the public grid [3]. A solution is to connect 25 KVAC 50 Hz electric railway systems to national main grids with very high power capacity. It is necessary to balance the energy consumption asymmetry on average from the main three-phase grid, and the solution consists in supplying each railway section from different phases of the main grid to reduce the imbalance. The voltages between sections have the same value, 25 KVAC, with different electric phases, and it is required for an electrical isolation segment between sections, called neutral sections (NSs). Trains must cross each NS without voltage supply for approximately 1.6 km [4]. Figure 1 shows the typical electric supply topology from the main grid to feed one 25 KVAC 50 Hz high-speed railway

(HSR). The length of each section is about 35 km, and it is supplied by rotating the phases of the main grid through the single-phase transformers included in the traction substations.

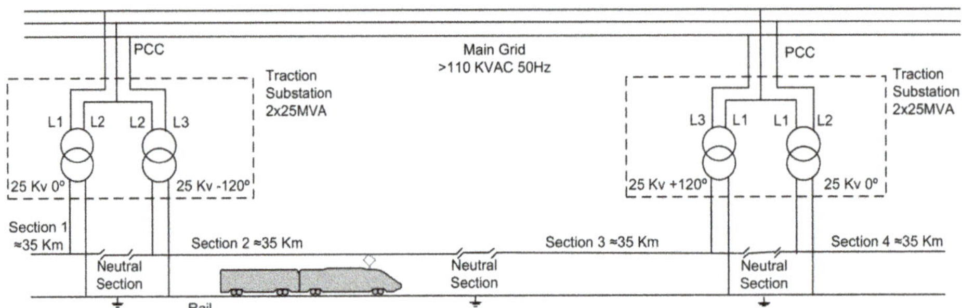

Figure 1. 1 × 25 KVAC 50 Hz high-speed railway electrical topology supply from the main grid.

When the train is near the NS, its main breaker is opened and the high-speed train (HST) is disconnected from the power supply. In this condition, the HST works in regenerative mode to keep its auxiliary power supply fed, and a little loss of speed, around 9 km/h [4], is produced. Once the train goes out of the NS, its main breaker is closed, the power supply is reestablished in the HST, and traction power can be applied again.

The operation through the NS causes several problems for the HST:

- When the HST main breaker is not open in the NS and the train continues feeding, an arc flash in the catenary is produced, and in some case can be broken [5–7].
- At slow speeds, the train can be stopped in the NS. The train needs an operation of 10 min to go out from the NS [8].
- The pantograph contact transition between the section and the NS catenary produces a voltage feeding of the NS. This results in electromagnetic interference (EMI) and affects the signaling and security of the railway system [5,9].
- Every 35 km, the HST crosses a neutral section and its main breaker is operated with one connection and one disconnect. The main breaker operates approximately 10,000 times/year, the mechanical lifetime is reduced, and the maintenance operation costs are increased [6].
- The closing instant of the main breaker produces inrush current in the HST traction transformer. It can be an inrush current about 10 to 25 times greater than the nominal current of the train. These high values of current cause high mechanical efforts by the transformer conductors, accelerating the aging of the transformer and main breaker. Occasionally, it can activate the main breaker current protection and the train is stopped [6,10,11].
- The opening instant of the main breaker demagnetizes the traction transformer, which can result in overvoltage. An electric arc is formed in main breaker contacts and accelerates its aging [12].

According to the problems identified above, a compensation strategies review and some power converter solutions were shown in [9], which can reduce the impact on the main grid and eliminate the neutral section. These solutions [9,13–18], based on very high-power converters, come with considerable cost. Engineers of the Japanese National Railways Tokaido Shinkansen HSR installed two switch breakers to feed NSs from the adjacent catenaries. For commutation assurance between catenary voltages and to avoid short circuit, 0.3 s without voltage in an NS are required [11,19]. The circuit breakers must work for each train crossing the NS. Therefore, on the Tokaido Shinkansen, they must be replaced every five years. The Central Japan Railway Company [10] proposed the use of static switches based on thyristors in NSs to replace the circuit breakers and reduce the inrush current on the transformer into the train. The study includes a demonstration of suppressing the inrush current on a transformer with a thyristor prototype at low voltage. Korea Railroad Research Institute [7] proposed the use of static switches with thyristors and a sequence to trigger when the HST is running through the NS, but the on and off transition between static switches was not clear.

This paper describes the design and validation of a medium-voltage AC static switch solution to feed a neutral section in an HSR system. Very little work on this topic is found in the literature; in [6] a prototype design with some experimental results are presented, but it is a low-voltage prototype, and they do not mention any of the problems that arise at medium-voltage levels, such as isolation issues, or how to test and validate the prototype in the laboratory. In [10] a theoretical concept of the solution is presented, but they do not mention any of the issues concerning real implementation, such as the uncertainty of the commutation of the switches.

Therefore, several challenges must be solved and are presented in this paper:

- Designing thyristor drivers with high-voltage isolation to achieve triggering of the thyristor valves at any load current, even at zero load current.
- Designing a test methodology for validation of the AC static switch at nominal conditions in the laboratory, reducing the risks and validation times.
- Validating the proposed solution in real application tests on the Cordoba–Malaga High-Speed Railway (Spain).

2. Solution Concept

The solution is based on two static switches, S1 and S2, connected to the neutral section, and each one to a different section; the topology is shown in Figure 2. Four position detectors, D1 to D4, are necessary to identify the position and direction of the train running on the NS.

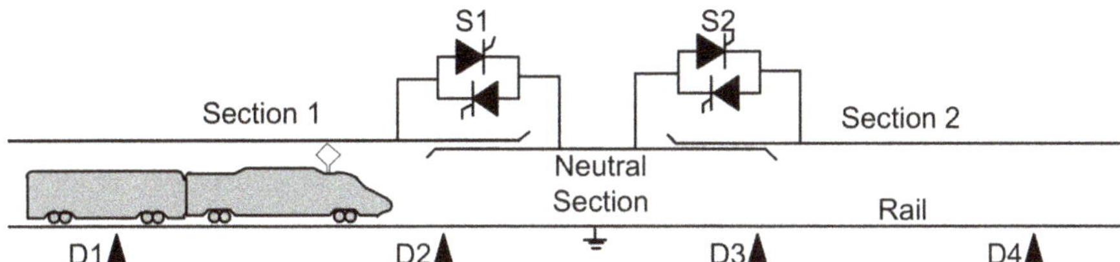

Figure 2. Medium-voltage AC static switch solution to feed the neutral section (NS) in a high-speed railway (HSR) system.

2.1. Operation Mode

In the case of train runs from section 1 to section 2, the presence of the train near the NS is detected by D1. Then, D3 detects when the entire train is running in the NS. Finally, the train leaving the NS is detected by D4. In the reverse direction, the train's arrival to the NS is detected by D4. D2 identifies when the entire train is in the NS. Finally, the departure from the NS is detected by D1.

Table 1 shows the different states of S1 and S2 static switches depending on the train position in the neutral section and the direction of the train running from section 1 to section 2.

Table 1. Transition states when a train runs through the neutral section from section 1 to section 2.

State	Train Position	S1	S2	V_{NS}	I_{NS}
1	Train is not present	Off	Off	$(V_{S1}-V_{S2})/2$	0
2	D1 detects train arriving to NS	On	Off	V_{S1}	0
3	Train runs from section 1 to NS	On	Off	V_{S1}	I_{train}
4	D3 detects entire train in NS and 60 millisecond timing	Off	Off	$(V_{S1}-V_{S2})/2$	0
5	Trains runs from NS to section 2	Off	On	V_{S2}	I_{train}
6	Train leaves from NS to section 2	Off	On	V_{S2}	0
7	D4 detects train is not present	Off	Off	$(V_{S1}-V_{S2})/2$	0

In states 1, 4, and 7, the two switches are in the open position and an AC voltage, V_{NS}, appears in the NS because static switches in the open state have high impedance. In this condition, the current in the NS, I_{NS}, is null. State 4 has a 60 ms timing period to detect zero current in the NS. This means that the S1 switch is off and S2 can be switched on, avoiding a short-circuit risk.

In states 2 and 6, one static switch, either S1 or S2, is in close position and the NS is fed from one section. In this period, the pantograph of the train is always in contact with one section catenary and can be in contact simultaneously with the NS catenary, then the current I_{NS} is null because the current can travel more easily through the section catenary than through the NS and static switch.

In states 3 and 5, the NS feeds from one section through one static switch, S1 or S2. The pantograph of the train is only in contact with the NS catenary and the train current is established through the NS and one static switch. Figure 3 shows the current flow in the different states of Table 1.

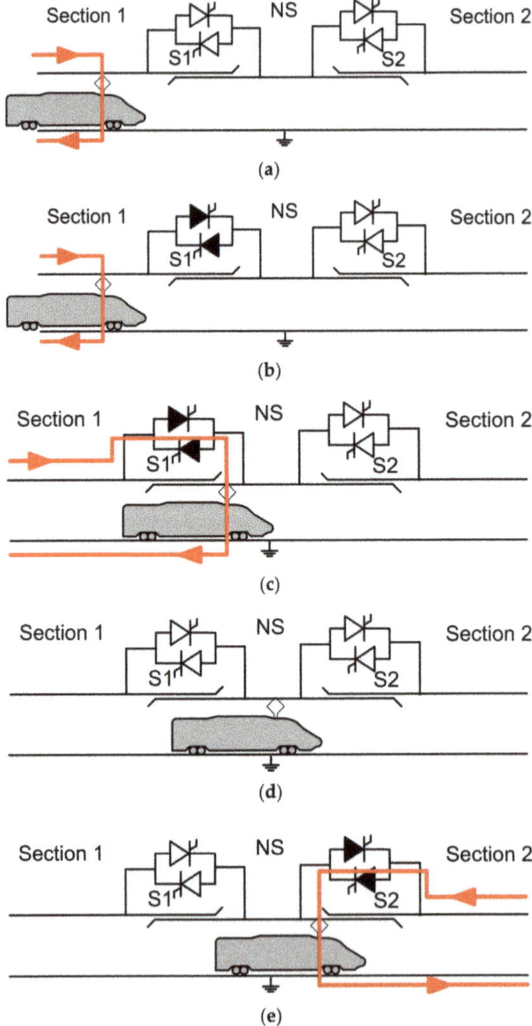

Figure 3. *Cont.*

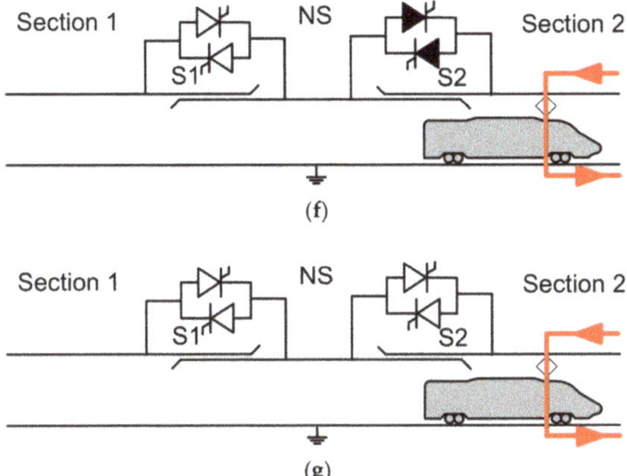

Figure 3. Current flow during transition states when a train runs through the neutral section from section 1 to section 2 in correspondence with Table 1. (**a**) State 1; (**b**) State 2; (**c**) State 3; (**d**) State 4; (**e**) State 5; (**f**) State 6; (**g**) State 7.

2.2. AC Static Switch Technology

In the 1970s, the advancement of thyristor technology enabled the development of static var compensator (SVC) as a solution to control the voltage stability in AC electrical transmissions lines. An SVC is based on thyristor-controlled reactors (TCRs), thyristor switched capacitors (TSCs), and/or fixed capacitors (FCs) tuned to harmonic filters. The TCR regulates the reactive power controlling the firing angle of the thyristor, and the TSC connects its capacitor the instant the voltage transient is at its minimum. On the TCR and TSC, the thyristors are used like AC static switches. The typical voltage of the SVC is 30 KVAC, but there are some installations where the voltage is up to 69 KVAC [15].

The thyristor technology presents some interesting properties for AC static switch applications. The thyristor can support AC voltage, due to its ability to block reverse and forward voltage. The switching on of the thyristor is controllable with the firing command, and it is possible to synchronize with the AC voltage to limit the transformer inrush current. The switching off is produced when the thyristor current is null, and this is a great advantage because it guarantees that the traction transformer will be switched off without the voltage being transient. Finally, the blocking voltage of commercial thyristors is up to 9500 V with a relatively low economical cost in the 6500 V range.

To achieve a medium-voltage AC switch, single-phase assemblies connected in series are used. Each assembly is composed of two antiparallel thyristors and one RC snubber for dv/dt limitation and to guarantee the voltage balance of assemblies.

3. AC Switch for Neutral Section in High-Speed Railway System Requirements

The Spanish high-speed railway system is considered for design of the AC static switch for neutral sections. The railway system voltage topology is 2 × 25 KVAC 50 Hz and the high-speed train maximum power is 12 MVA. The neutral section characteristics are shown in Table 2.

Table 2. Neutral section characteristics.

Characteristic	Data
Maximum train speed	300 km/h
Neutral section length	405 m
Number of pantographs connected	2
Distance between pantographs	395 m

The maximum switch voltage requirement is based on the 29 KVAC maximum voltage in the 2 × 25 KVAC railway systems defined in European Standard EN 50160. When one static switch is closed, the second static switch must support the 50 KVAC voltage between section 1 and section 2. The number of neutral section operations is according to the train timetable. The time to switch between S1 and S2 is up to 120 milliseconds and depends on the train speed and the travelled distance where the two pantographs are connected simultaneously to the NS catenary. Table 3 shows the requirements obtained from the above characteristics.

Table 3. AC static switch for neutral section requirements.

Requirement	Value
Number of AC static switches	2
Maximum RMS switching voltage	50 KVAC
Maximum RMS switching current	685 A
Frequency	50 Hz
Ambient temperature	0 to 45 °C
Number of operations	Every 5 min
Switching period	<120 ms

Furthermore, the AC static switch must limit the transient electricity in the connection and disconnection of the train in the NS. In transition state 4, the switching off must commutate at zero current to limit the transient voltage in the traction transformer of the train. In transition state 5, the switching on must be synchronized with the maximum amplitude of voltage to avoid the inrush current of the traction transformer.

The last requirement is to integrate the complete solution of AC static switches for the NS in one standard container for easy transportation and installation in the application location.

4. AC Static Switch Design

The design process is divided into three sections: power electronics, control system, and final assembly.

4.1. Power Electronics

The AC static switch is made of assemblies of two antiparallel thyristors and one RC snubber. The number of assemblies connected in series is calculated by Equation (1):

$$n = \frac{V_{\text{switch rms}} \cdot \sqrt{2} \cdot k}{V_{\text{DRM,RRM}}} + 1 \tag{1}$$

where a voltage safety factor k = 1.8 is applied to switch 50 KVAC with commercial thyristors of 6500 volts. It results in 21 assemblies connected in series, meaning that each thyristor in nominal conditions supports a maximum voltage of 3.367 V, which represents a final safety factor of 1.93, enough to support voltage spikes. Considering an AC current of 685 A, thyristor model 5STP03D6500 from ABB in press-pack format with diameter 34 mm is selected.

Two criteria for the selection of the RC snubber are considered. For calculating the capacity value, the maximum dispersion of the thyristor recovery charges, Q_{rec}, is taken into account. Considering a maximum thyristor voltage of 5 KV, V_{max}, and 10% of capacitor tolerance with Equation (2) 1 μF of snubber capacitor value is obtained:

$$C_{\text{snubber}} \geq \frac{(n-1) \cdot Q_{\text{rec}}}{(1-b) \cdot \left(n \cdot V_{\text{thyristor}} - n \cdot V_{\text{max}}\right)} \tag{2}$$

The value of the snubber resistance is calculated by Equation (3) obtained from the mathematical expression of electric current transition during the switching off of the thyristor.

$$R_{snubber} \geq 2\zeta \sqrt{L/(n \cdot C_{snubber})} \tag{3}$$

Considering a damping factor, ζ, between 0.5 and 1 and the magnetizing inductor of the traction transformer, the snubber resistance result is 150 ohms.

The mode of operation of the AC switch generates little power losses in each thyristor and has a system of cooling by natural convection. Figure 4a shows the basic module of the AC switch; it is performed by three aluminum heat sinks, two thyristors, two drivers, and one RC snubber. Each thyristor is mounted between two heat sinks by a mechanical clamp. Figure 4c shows the complete solution of 50 KVAC static switch; it is performed by three stacks connected in series. Each stack is composed of seven basic modules connected in series and is supported by post insulators to get the electric isolation from the earth.

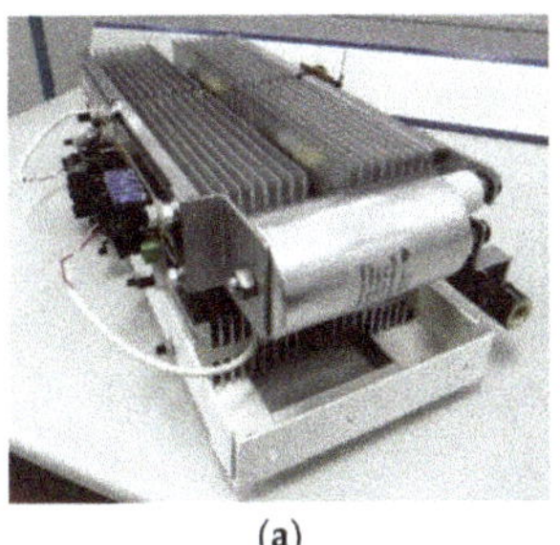
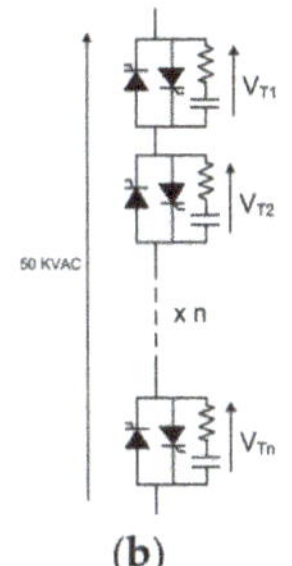

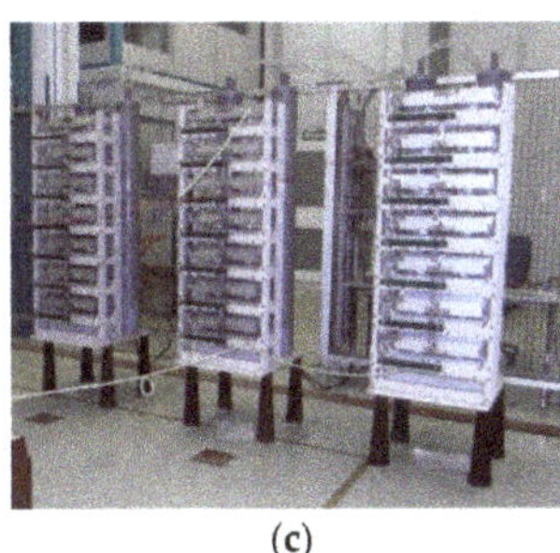

(a) (b) (c)

Figure 4. (**a**) Basic module of AC static switch; (**b**) electric scheme of 50 KVAC static switch with n = 21; (**c**) 50 KVAC static switch composed of three stacks of seven basic modules.

4.2. Control System

The control system is divided into two blocks: the thyristor driver and the controller. The complexity of the driver thyristor is the high-voltage isolation requirement. The controller is based on the classical field-programmable gate array (FPGA) with digital signal processor (DSP) control topology with the advantage of algorithm execution in a very short time.

4.2.1. Thyristor Driver

The thyristor driver must transform the control command into electrical pulses to trigger the thyristor. The main requirements of the thyristor driver for the AC static switch in the NS are as follows:

- It must guarantee isolation to 52 KVAC, related to the maximum AC voltage between different sections.
- High-frequency pulses must trigger, to guarantee thyristor conduction with any amplitude and phase of the main current, including zero current. The frequency pulse triggering must be above 5 KHz.
- In serial connected thyristors, manufacturers recommend applying a peak gate current about 5 A and 2 A/μs minimum slope.
- The maximum dispersion time between the firing pulses of serial connected thyristors is 1 μs.

The driver consists of two parts: the power supply and the trigger circuit [20].

The power supply is based on current transformers at high frequency. An AC current source feeds a high-voltage conductor, which crosses the toroidal magnetic core, where a current is induced into secondary winding to feed each thyristor driver. Figure 5 shows the topology to feed seven thyristor drivers, half of one stack of seven basic modules. The AC HF current source includes an LLC resonant tank working at 25 KHz, which feeds the high-voltage conductor. This topology is suitable for working in an open loop and with variable loads such as thyristor drivers [21]. Seven toroidal transformers are crossed by the high-voltage conductor and immersed in mineral oil to guarantee the isolation requirement.

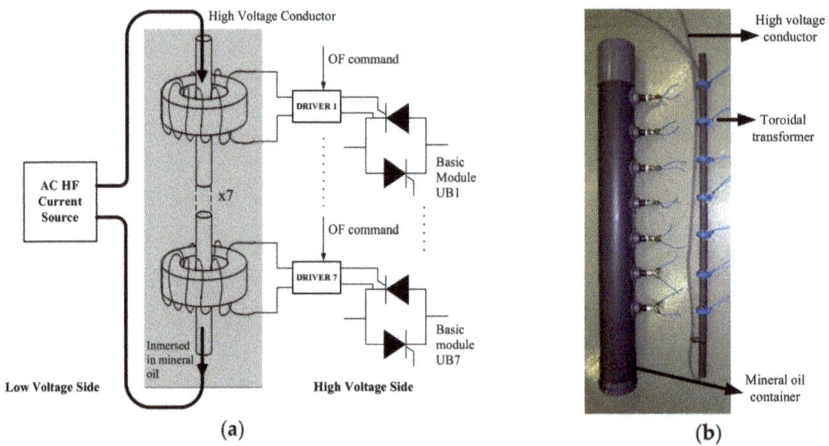

Figure 5. (**a**) AC current transformer topology with high-voltage isolation to feed thyristor drivers; (**b**) toroidal transformer assembly with high-voltage conductor. The assembly is included in the mineral oil container.

The trigger circuit transforms the optical fiber command in 5 KHz electrical gate impulses with the current amplitude and slope specified. The 42 drivers of one AC static switch receive the same optical fiber command with time dispersion less than 1 μs.

4.2.2. Controller

The controller is based on a hardware platform that includes a personal computer (PC), DSP, and FPGA. The PC is the user interface for monitoring and parameterizing the full system. The DSP operates the information between the PC and FPGA and manages the external alarms. The FPGA executes the state machine according to travel direction and train position and controls the on/off positions of switches S1 and S2. Other auxiliary functions executed in the FPGA are switching the thyristors on and off in synchronization with voltage phase, generating optical fiber signals for thyristor firing, calculating current RMS value, and treating alarms and data management with DSP.

The state machine executes the sequence listed in Table 1 when the train travels from section 1 to section 2 and executes the reverse sequence when the train travels from section 2 to section 1.

The synchronization is based on the phase locked loop (PLL) of monophase voltages of section 1 and section 2. In state 5 in Table 1, the switching on is synchronized to avoid the inrush current of traction transformer. In state 7, the switching off of S2 is at low current. Before state 7, switch S1 is blocked and switch S2 is closed and the train is out of the NS; in this condition, the current path is through switch S2 and the RC snubber of switch S1. The RC snubber limits to a low AC current value near the holding current characteristic of the thyristors connected in series. When the thyristor firing is removed, the anode–cathode current must be lower than the holding current characteristic of the thyristor to get the blocking. Dispersion of the holding current characteristics can cause blocking of several thyristors and conduction of the rest. To get simultaneous blocking of all thyristors connected in series, the thyristor firing is removed when the anode–cathode current is near zero, synchronized with the maximum voltage amplitude blocked by switch S1.

4.3. Final Assembly

The solution to feed the neutral section was integrated into a 20-foot standard container. The container included two medium-voltage AC static switches, AC voltage and current measurements, and a full control system. The heat produced by RC snubbers, thyristors, and the control system was removed from the container with a fan. Fiber optic Ethernet was used to communicate externally with the control system.

Out of the container, medium-voltage aerial protections such as switch breakers for two poles and electrical disconnect switches of three poles were connected. The switch breaker connected the container to the voltages of sections 1 and 2 and protected within short circuits of the container. The electrical disconnect switch connected the container to the voltages of section 1, section 2, and the neutral section, which was opened when the switch connected the power terminals of the container to earth protection.

5. Testing

The process of experimental validation was divided into two parts: laboratory tests and final test in real conditions with train and railway infrastructure.

The laboratory tests were based on partial testing processes due to the limited power of the facility. The sequence of the proposed partial tests guaranteed success in the final application. The power and voltage levels of the solution are very high and the laboratory requirements for validation are complex and expensive. A methodology was proposed in [22] based on synthetic tests for thyristor valves used in high-voltage direct current (HVDC) systems, and [23] applied the methodology to modular multi-level converters (MMC) also in HVDC application. The synthetic tests validated the ability of the power semiconductors to support the same electrical stress of the real application. In this way, the synthetic tests were applied at partial sections of thyristor valves, testing each section and its thyristors at working voltage and current levels. This work proposes to extend the synthetic tests by adding the application control to reproduce and validate the real operation in the laboratory with minimal infrastructure. The sequence of proposed partial tests guarantees success in the final application.

The test in real conditions with train and railway infrastructure demonstrates the technical feasibility of the AC static switch and the proposed validation methodology.

5.1. Laboratory Tests

In general, the electric power laboratories present capacity limitations for testing high-power equipment. In this paper, is proposed a process to testing partially the AC static switches in the Medium-Voltage Laboratory of Mondragon University. The tests are realized in laboratory-controlled conditions. Two partial testing processes were designed: functionality tests of power electronics and application conditions tests.

5.1.1. Functionality Tests of Power Electronics

Based on [22,23], the purpose of the test is to validate the electric isolation, blocking voltage, and current conduction capabilities of the AC static switch. The process starts with the test of thyristor basic modules, then thyristor stacks, and finally AC static switch, and the test sequence is as follows: isolation, voltage blocking, and current conduction. Figure 6 shows the functionality tests.

To pass the functionality test, the AC static switch must be complete successfully with the procedure in Figure 6. The proposed test procedure achieves a size reduction of the laboratory equipment used. In the voltage blocking test, the thyristors are open and only must feed the leakage current of the RC snubber and thyristor. A low-power, high–AC voltage source is necessary, 52 KV and 30 KVA. In the current conduction test, the thyristors are closed and the voltage across each thyristor is lower than 2 volts, and a low-power AC voltage source is necessary, about 50 V and 40 KVA.

DUT / Test	Isolation	Voltage Blocking	Current Conduction
	Megaohmeter	2500 VAC 50 Hz Current monitoring	685 A 50 Hz Temperature monitoring
	95 KVAC 50 Hz 1 min Current monitoring	17500 VAC 50 Hz Current and voltage balanced monitoring	685 A 50 Hz Temperature monitoring
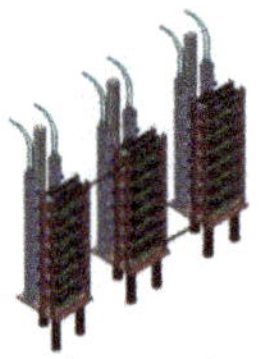	95 KVAC 50 Hz 1 min Current monitoring	52 KVAC 50 Hz Current and voltage balanced monitoring	685 A 50 Hz Temperature monitoring
DUT / Test topology	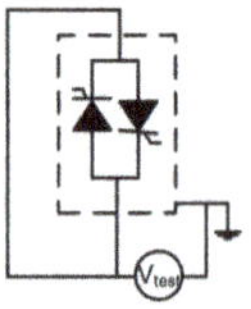	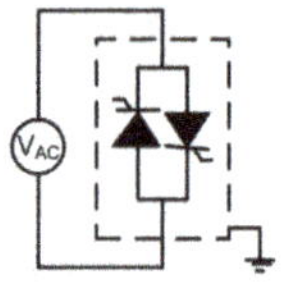	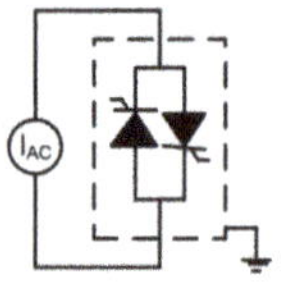

Figure 6. Procedure for the functional testing of the AC static switch. DUT, device under test.

5.1.2. Application Condition Tests

The purpose of the test is to validate the integration between the control system and power electronics, including the emulation of the train travelling through the NS. The procedure proposes a sequential test from the basic modules to AC static switch, similar to a functionality test.

The procedure sequence, Figure 7, starts with the test of the basic module; two phases of low-voltage transformer are used like voltages of sections 1 and 2, and the third phase is the rail connected to the ground. A second one-phase transformer is connected to the basic modules and ground, and this configuration emulates the train travelling into the NS. The one-phase transformer has a resistor load equivalent to the nominal current absorbed by the train. In these conditions, the PLL and state machine of the control system are validated and the commutation transitions at nominal current through the thyristors are verified. Also validated is the synchronization of the on- and off-switching of the thyristor with the voltage phase to reduce the transformer inrush current and guarantee the thyristor blocking at zero current.

The next step of the procedure uses two stacks of seven thyristors connected in series and a three-phase transformer at 17.5 KVAC. The worst case of voltage balancing occurs with switching of the thyristors connected in series at zero current and the test topology does not include the one-phase transformer with the resistor load. The test validates the commutation transitions at nominal voltage and the voltage balancing between the basic modules of thyristors.

The final procedure tests the two AC static switches at medium voltage, 30 KV, with commutation transitions at zero current. The test verifies the voltage balancing between the three stacks in the worst case of commutation. Figure 8 shows the transition from S2 to S1. S2 is switched on, the NS is fed from the section 2 voltage, and S1 blocks 30 KV. S2 is switched off without load in the maximum voltage amplitude of section 2, near the zero-crossing current. The transient response in the RC snubber results in a voltage average value in switches and NS but within the working voltage values. Then, S1 is switched on in the maximum voltage amplitude of section 1 to reduce the transformer inrush current, the NS is fed from section 1, and S2 blocks 30 KV.

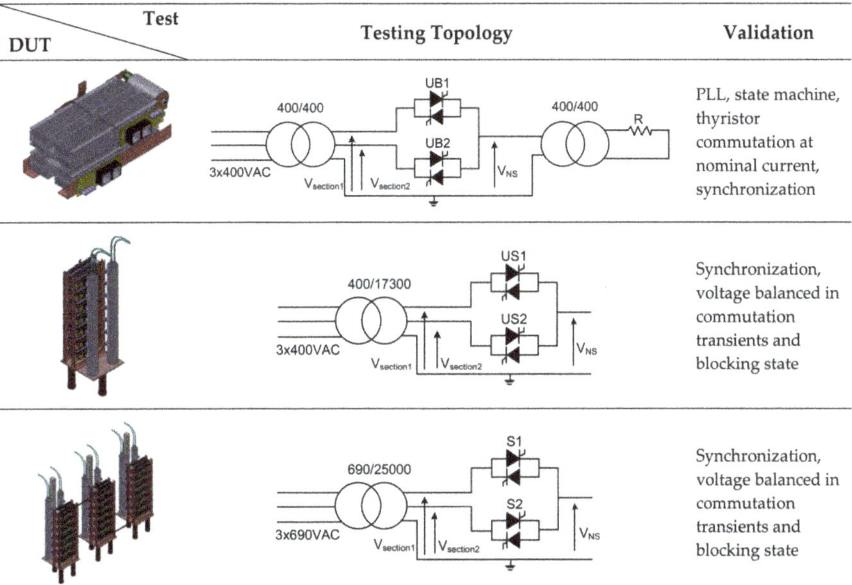

Figure 7. Application conditions for the test procedure of AC static switch. PLL, phased locked loop. DUT, device under test.

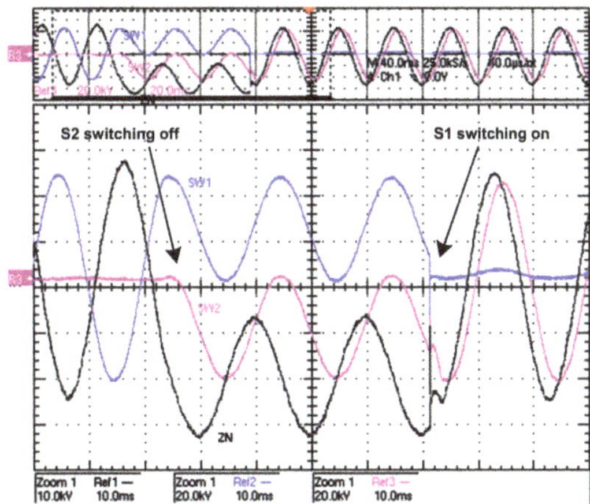

Figure 8. Switching transient between AC static switches S1 and S2 at 30 KV without load obtained in laboratory tests. SW1, S1 voltage; SW2, S2 voltage; ZN, NS voltage.

5.2. Final Application Tests

To demonstrate the technical feasibility of the AC static switch to feed the NS, real tests in the railway system were done. The container with the two AC static switches was connected to a neutral section of the Cordoba–Malaga High-Speed Railway, in the traction substation of La Roda de Andalucía. More information available in Supplementary Materials which contain Adif's press realease. The traction substation feeds two sections of the HSR and is in front of the NS where the container is located. Two one-phase transformers feed the sections at 27.5 KV and 800 m of neutral section separates the two sections. Figure 9 shows the container situation and connection to the NS. The test procedure followed three steps: control verification, commutation of NS with the same voltage in sections, and commutation of NS with different voltage phases in sections. The commutation test included the following conditions in both directions of circulation:

- Train crossing NS at 100 km/h in traction mode
- Train crossing NS at 100 km/h in regenerative mode
- Train crossing NS at 300 km/h in traction mode
- Train crossing NS at 300 km/h in regenerative mode
- Train stopping in NS
- Train stopping 10 min into NS
- Train starting from stationary position at NS

The test procedure was realized with the laboratory high-speed train from Adif, called Seneca, out of the business exploitation timetable. Seneca HST is based on Talgo 350, 8 Mw of traction power, 200 m in length, and 330 Km/h of maximum speed. The time to complete final tests was 5 days. The most critical test was Seneca crossing the NS at 300 km/h in traction mode, where the available time to commutation is minimal at nominal current level.

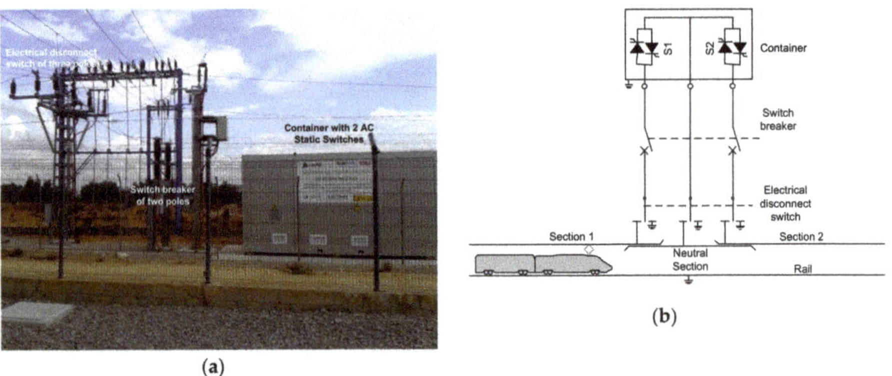

Figure 9. (**a**) Container with two AC static switches and connection to the HSR; (**b**) electrical scheme with protections and connection to HSR.

5.2.1. Control Verification

The test consisted of verifying the detection of train position and correctly executing the triggering sequence of the two AC static switches, which were electrically disconnected. Control was verified with the Seneca HST crossing the neutral section at different speeds up to 300 km/h in both directions.

5.2.2. Commutation of Neutral Section with Equal Voltages in Sections

The traction substation was configured to feed section 1 and section 2 with equal voltage, 27.5 KV. The container was energized and the voltage through the two AC switches, S1 and S2, was zero. This test condition avoids short-circuit risks between sections in case the commutation sequence fails. Three steps were proposed to complete the test:

- First the electrical isolation between active parts to ground into the container for 24 h was tested and the electrical isolation in nominal conditions of exploitation was verified.

- The next step consisted of the commutation of S1 and S2 without load when the Seneca HST was crossing the NS with its main breaker disconnected. This test was realized at different speeds in both directions and verified the correct sequence of commutation with the system energized.
- In the last step, the HST travelled through the NS electrically connected at different speeds in both directions. Further testing was done when the HST travelled in regenerative breaking mode, in which the current was in opposite phase with voltage. Also tested was the stop and boot of the HST into the neutral section. The test verified the correct sequence of commutation at nominal voltage and different currents.

Figure 10 shows the complete electric sequence of the train crossing the NS with the states from 1 to 7 described in Table 1. The HST was near the NS and caused the transition from state 1 to state 2, where switch S2 was activated to feed the NS. The instant the HST absorbed current from the NS through S1 was the transition from state 2 to state 3 and the HST was powered into the NS. Figure 10 shows the sinusoidal current absorbed by the train in phase with the voltage. The change to state 4 was caused by the detection of the complete train in the NS, and the control sent the switching-off command to S2. In the transition from state 4 to 5, the switching off at zero current crossing can be observed and the result is a null electrical transient. Once the zero current was checked in the NS for 60 milliseconds, state 5 transitioned to state 6 and S1 was triggered on the maximum voltage amplitude of section 2. The train was fed from section 1 and the inrush current of the traction transformer was avoided. In three electrical cycles, the nominal current was reestablished from the NS through S1 to the train. During state 6, the train went out from the NS and the current was reduced to zero, caused by the progressive loss of contact with the NS catenary. Finally, the detection of the complete train out of the NS caused the transition from state 6 to 7 and S1 was switched off at zero current.

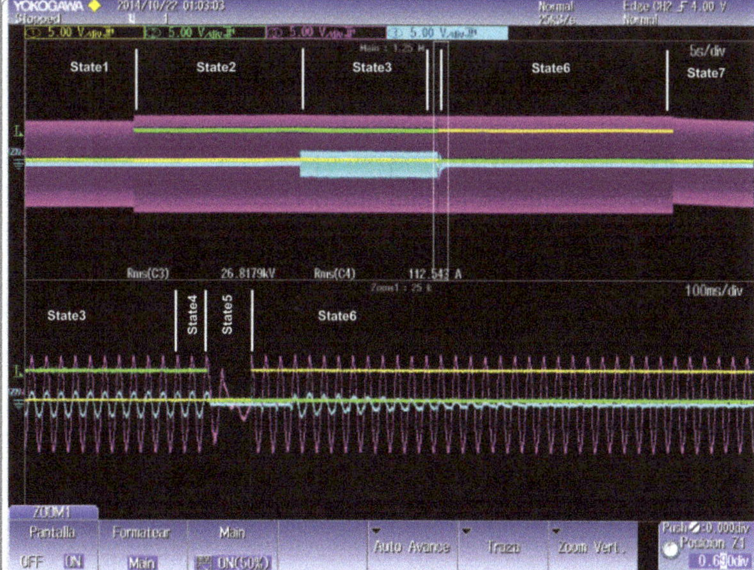

Figure 10. HST travelling at 300 km/h from section 2 to section 1. Ch1, S1 command; Ch2, S2 command; Ch3, NS voltage; Ch4, NS current.

The test with equal voltage in sections 1 and 2 verified the correct execution of the AC static switch solution with a real train with low risk of failure.

5.2.3. Commutation of Neutral Section with Section Voltages with 120° Lag

The traction substation was configured to feed section 1 and section 2 with two 27.5 KV with a lag of 120°. The Seneca HST crossed the NS electrically connected at different speeds in both directions.

Figure 11 shows the transition of the train from section 2 to section 1. The switching off of the current in state 5 was at zero current crossing and the inrush current of transformer in state 6 was null. The change from state 6 to 7 shows the voltage transient in the NS similar to the functionality test in laboratory, which is shown in Figure 9. In state 6, the current absorbed from the NS was low because the input power converter of the train reacted slowly to the change of the 120° phase lag. This effect was not present when the voltages in sections 1 and 2 were equal.

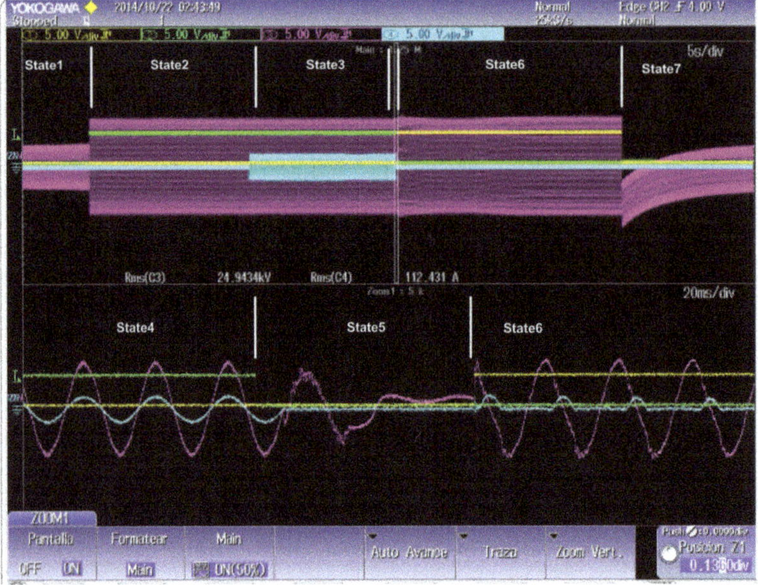

Figure 11. HST travelling at 280 km/h from section 2 to section 1. Ch1, S1 command; Ch2, S2 command; Ch3, NS voltage; Ch4, NS current.

The tests with 120° lag in voltages of sections 1 and 2 were realized successfully and show the viability of the AC static switch solution based on thyristor technology.

6. Conclusions

The solution to feed the NS in the HSR system is based on thyristors working at a medium voltage level. The concept test [10] at low voltage shows the technical viability of the solution, but the design process identifies the complexity to achieve the electrical isolation and the difficulties in testing at nominal conditions in the laboratory.

Current transformer topology with the same primary conductor, with high isolation and immersed in mineral oil, is the solution proposed to achieve electrical isolation for 52 KVAC.

To test the AC static switch, a test methodology is proposed with the aims to reduce:

- expensive equipment used in the medium-voltage laboratory,
- risk of failure, and
- validation time in the final application.

The validation procedure is based on a progressive test where the power electronic components are verified separately at nominal conditions of voltage and current:

- Functionality tests of power electronics (basic module, stack, and AC static switch) to test the ability to support nominal voltage and current.
- Application conditions tests (basic module, stack, and AC static switch) to test the control and power electronics operation.

The procedure and laboratory tests have been demonstrated and guaranteed the success of the final test in the real application. Five days were required to complete the tests on the Cordoba–Malaga High-Speed Railway.

From the application point of view, the thyristor achieves the right solution to feed high-speed trains in neutral sections, avoiding the electrical transient of connection and disconnection of traction transformers and failure of the train's main breaker. Also, the AC static switch solution is suitable for conventional railways, where the speed can be very slow and there is a risk of stopping in a neutral section without electrical power.

Supplementary Materials: Adif's press release on the If zone project are available online at http://prensa. adif.es/ade/u08/gap/prensa.nsf/Vo000A/CD63F8963B956BA7C1257AAA00434DF5?Opendocument, http:// prensa.adif.es/ade/u08/GAP/Prensa.nsf/Vo000A/7448B72B54087C76C1257DD600363D68?Opendocument, http://www.fomento.es/AZ.BBMF.Web/documentacion/pdf/A27860.pdf.

Author Contributions: J.M.C. conceived the solution for neutral sections based on solid state switches and wrote the paper. I.A. and U.I. designed and realized the experimental validation in the laboratory and in real conditions of the Cordoba–Malaga High-Speed Railway. J.A.B. contributed to the control design of the global solution, and M.B. contributed to the power electronics design including the electrical isolation.

Funding: This work was supported in part by the Spanish Ministerio de Fomento through project IFZONE (FOM 39/08).

Conflicts of Interest: The authors declare no conflict of interest.

Abbreviations

The following abbreviations are used in this manuscript:

ABB	ASEA Brown Boveri
DSP	Digital signal processor
DUT	Device under test
EMI	Electromagnetic interference
FC	Fixed capacitor
FPGA	Field-programmable gate array
HF	High Frequency
HSR	High-speed railway
HST	High-speed train
HVDC	High-voltage direct current
IFZONE	Investigation of Advanced Techniques for Railway Operation in the Neutral Sections of the high-speed railway
KVAC	Kilovolt altern current
LLC	Inductor inductor capacitor
MMC	Modular multi-level converter
NS	Neutral section
PC	Personal computer
PLL	Phase locked loop
RC	Resistor capacitor
RMS	Root mean square
SVC	Static var compensator
TCR	Thyristor-controlled reactor
TSC	Thyristor switched capacitor

References

1. Steimel, A. Under Europe's incompatible catenary voltages a review of multi-system traction technology. In Proceedings of the Electrical Systems for Aircraft, Railway and Ship Propulsion (ESARS), Bologna, Italy, 16–18 October 2012.
2. Steimel, A. Power-electronic grid supply of AC railway systems. In Proceedings of the Optimization of Electrical and Electronic Equipment (OPTIM), Brasov, Romania, 25–27 May 2012.
3. Di Manno, M.; Varilone, P.; Verde, P.; De Santis, M.; Di Perna, C.; Salemme, M. User friendly smart distributed measurement system for monitoring and assessing the electrical power quality. In Proceedings of the AEIT International Annual Conference (AEIT), Naples, Italy, 14–16 October 2015.
4. Conrado, J.; Tobajas, C.; Almenara, L.F.; Iglesias, J.; Cabello, J.; Berrios, A.; Nieva, T.; Aiarzaguena, M. "IFZONE" Project: Improving circulation in neutral sections. In Proceedings of the 9th World Congress Railway Research, Lille, France, 22–26 May 2011.
5. Chen, S.; Wen, Y. Discussions on the generation mechanism of spark phenomenon when locomotive running through an electricity neutral section. In Proceedings of the International Conference on Electromagnetics in Advanced Applications, Sydney, Australia, 20–24 September 2010; pp. 749–752.
6. Shin, H.; Cho, S.; Huh, J.; Kim, J.; Kweon, D. Application on of SFCL in Automatic Power Changeover Switch System of Electric Railways. *IEEE Trans. Appl. Supercond.* **2012**, *22*, 5600704. [CrossRef]
7. Donguk, J.; Moonseob, H. Study of composition draft on automatic changeover system in neutral section of electric railway catenary system for highspeed train line. In Proceedings of the 9th World Congress on Railway Research, Lille, France, 22–26 May 2011.
8. Zhou, F.; Li, Q.; Qiu, D. Co-Phased Traction Power System Based on Balanced Transformer and Hybrid Compensation. In Proceedings of the Asia-Pacific Power and Energy Engineering Conference, Chengdu, China, 28–31 March 2010; pp. 1–4.
9. Mousavi, S.M.; Tabakhpour, A.; Fuchs, E.F.; Al-Haddad, K. Power Quality Issues in Railway Electrification: A Comprehensive Perspective. *IEEE Trans. Ind. Electron.* **2015**, *62*, 3081–3090. [CrossRef]
10. Ajiki, K.; Mochinaga, Y.; Uzuka, T.; Saitoh, T.; Yoshiki, Y. Investigation of Shinkansen Static Changeover Switch Capable of Suppression of Exciting Inrush Current into Transformer on Train. *IEEJ Trans. Ind. Appl.* **2001**, *121*, 340–346. [CrossRef]
11. Hayashiya, H.; Ueda, Y.; Ajiki, K.; Watanabe, H.; Ando, M.; Nakajima, M. Investigation of closing surge in Shinkansen Power System and proposal of a novel power electronic electronics application for changeover section. *IEEJ Trans. Ind. Appl.* **2006**, *126*, 857–864. [CrossRef]
12. Tian, X.; Jiang, Q.; Wei, Y.; Zhang, J.; Wei, Y. Novel high speed railway uninterruptible flexible connector based on modular multilevel converter structure. In Proceedings of the International Conference on Power Electronics and ECCE Asia, Seoul, South Korea, 1–5 June 2015; pp. 952–958.
13. Wu, C.; Luo, A.; Shen, J.; Ma, F.J.; Peng, S. A Negative Sequence Compensation Method Based on a Two-Phase Three-Wire Converter for a High-Speed Railway Traction Power Supply System. *IEEE Trans. Power Electron.* **2012**, *27*, 706–717. [CrossRef]
14. Grunbaum, R. FACTS to enhance availability and stability of AC Power Transmission. In Proceedings of the Bucharest PowerTech Conference, Bucharest, Romania, 28 June–2 July 2009; pp. 1–8.
15. He, X.; Guo, A.; Peng, X.; Zhou, Y.; Shi, Z.; Shu, Z. A Traction Three-Phase to Single-Phase Cascade Converter Substation in an Advanced Traction Power Supply System. *Energies* **2015**, *8*, 9915–9929. [CrossRef]
16. He, X.; Shu, Z. Advanced Cophase Traction Power Supply System Based on Three-Phase to Single-Phase Converter. *IEEE Trans. Power Electron.* **2014**, *29*, 5323–5333. [CrossRef]
17. Ladoux, P.; Fabre, J.; Caron, H. Power Quality Improvement in ac Railway Substations: The concept of chopper-controlled impedance. *IEEE Electrif. Mag.* **2014**, *2*, 6–15. [CrossRef]
18. Han, Z.; Liu, S.; Gao, S. An Automatic System for China High-speed Multiple Unit Train Running Through Neutral Section with Electric Load. In Proceedings of the Asia-Pacific Power and Energy Engineering Conference, Chengdu, China, 28–31 March 2010; pp. 1–3.
19. Uzuka, T. Faster than a Speeding Bullet: An Overview of Japanese High-Speed Rail Technology and Electrification. *IEEE Electrif. Mag.* **2013**, *1*, 11–20. [CrossRef]
20. Wahl, F.P. Firing Series SCRs at Medium Voltage: Understanding the Topologies Ensures the Optimum Gate Drive Selection. In Proceedings of the Power Systems World Conference, Chicago, IL, USA, 31 October 2002.

21. Aizpuru, I.; Canales, J.M.; Fernandez, J. Scalable High Insulation Power Supply for Medium Voltage Power Converters. In Proceedings of the Integrated Power Electronics Systems (CIPS), Nuremberg, Germany, 6–8 March 2012; pp. 1–6.
22. Liu, L.; Yue, K.; Pang, L.; Zhang, X.; Li, Y.; Zhang, Q. A novel synthetic test system for thyristor level in the converter valve of HVDC power transmission. In Proceedings of the MATEC Web of Conferences ICMMR, Chongqing, China, 15–17 June 2016.
23. Tang, G.; He, Z.; Zha, K.; Luo, X.; Wu, Y.; Gao, C.; Wang, H. Type tests on MMC-based HVDC valve section with synthetic test circuits. *Int. Trans. Electr. Energy Syst.* **2016**, *26*, 1983–1998. [CrossRef]

Article

Data-Driven Risk Analysis for Probabilistic Three-Phase Grid-Supportive Demand Side Management

Niels Blaauwbroek *, Phuong Nguyen and Han Slootweg

Electrical Energy Systems Group, Eindhoven University of Technology, 5600 MB Eindhoven, The Netherlands; p.nguyen.hong@tue.nl (P.N.); j.g.slootweg@tue.nl (H.S.)
* Correspondence: n.blaauwbroek@tue.nl

Received: 25 August 2018; Accepted: 17 September 2018; Published: 21 September 2018

Abstract: Along with the emerging development of demand side management applications, it is still a challenge to exploit flexibility realistically to resolve or prevent specific geographical network issues due to limited situational awareness of the (unbalanced low-voltage) network as well as complex time dependent constraints. To overcome these problems, this paper presents a time-horizon three-phase grid-supportive demand side management methodology for low voltage networks by using a universal interface that is established between the demand side management application and the monitoring and network analysis tools of the network operator. Using time-horizon predictions of the system states that the probability of operational limit violations is identified. Since this analysis is computationally intensive, a data driven approach is adopted by using machine learning. Time-horizon flexibility is procured, which effectively prevents operation limit violation from occurring independent of the objective that the demand side management application has. A practical example featuring fair power sharing demonstrates the effectiveness of the presented method for resolving over-voltages and under-voltages. This is followed by conclusions and recommendations for future work.

Keywords: demand side management; operation limit violations; probabilistic power flow; network sensitivity; neural networks

---

1. Introduction

The fast growing share of distributed renewable energy sources (DRES) like solar photovoltaic (PV) and highly energy-intensive appliances and distributed energy resources (DER) such as heat pumps and electric vehicles (EVs) results in increasing uncertainties in the power flows in distribution networks, which challenges the distribution system operator (DSO) to keep the network operated within safe and secure operation limits [1–4]. Due to the high uncertainties, a continuation of the current paradigm of infrastructure over dimensioning is expected to result in high future investment costs. To cope with this, DSOs can deploy control algorithms to resolve operational limit violations, e.g., using on-load tap-changers (OLTC) [5] or reactive power control of inverters [6,7]. However, the application of such direct control actions can be highly expensive or otherwise ineffective. The consequence is that specific operation limit violations stay unresolved if the local controllers cannot effectively resolve the operation limit violation, i.e., there is a remaining operation limit violation (OLV). Alternatively, the DSO can invoke flexibility in the demand and supply of DER owners instead by using demand side management (DSM) applications [8]. Due to high R/X ratio's typical in distribution networks, active power from DSM will have a more significant influence on the system states than a reactive power control [9]. This flexibility can result from DER capable of advancing or deferring their starting time or changing the active power consumption or production during some time interval [10]. Examples include various types of controllable appliances like time shiftable appliances (e.g., freezers or washing machines) and buffering appliances involving some form of storage such as batteries of EV and heat pumps [11].

DSM applications are expected to be operated by a market actor, i.e., aggregators or energy suppliers with limited to no insight in the network operation state. As such, they tend to be unaware about specific grid related issues and physical and geographic aspects of the network during operation. Although some are designed to resolve operation limit violations [12,13], their optimization objective is often not focussed on resolving specific operation limit violations occurring at a certain geographical location. This is especially true for low-voltage (LV) networks where uncertainty is even higher due to the non-aggregated load profiles [14].

These shortcomings can be solved by using a universal interface between DSOs and DSM applications, facilitating the exchange of information on specific (predicted) network issues, and ending user flexibility. In this case, state prediction of the network system states will be of high value to account for the probabilistic system states that will occur in the near future. For coping with the high stochasticity of DER and end user behaviour, a probabilistic approach is required to determine potential network risks [15–17]. On the down side, probabilistic approaches often require some form of probabilistic power flow (PPF) calculations to evaluate the probability of having operation limit violations at certain locations in the network. Despite abundant attempts (e.g., [18–22]) to lower the computational complexity for PPF calculations for a large three-phase network, they still require a significant amount of time to complete. Performing PPF calculations for each 15-min time interval within a 24-h day-ahead period for all the networks it operates forms an enormous computational burden. Quite often, however, we are not interested in the full probability density functions produced by the PPF, but solely in the probability that a certain operation limit will be violated. Rephrasing along with control and management decisions are made based only on specific information of the operational limit violations (e.g., the probability of it exceeding a certain predefined threshold). As such, only a limited amount of information is required, which creates possibilities to speed up the computation time.

To this extent, in this work, we use machine learning to come up with predictions on whether the probability for operation limit violations exceeds a certain threshold and how DSM can bring the probability back to acceptable levels. Various machine learning techniques such as neural networks (NN) have attracted greater attention for applications in DSM. Often, their application can be found in load forecasting [23,24]. Furthermore, various types of home energy management systems implement machine learning for decision making. For example, in Reference [25], a NN is applied for the scheduling of PV panels and a storage system. This work presents the application of NN to specifically deal with geographically dependent operational limit violations in distribution networks, which results in a probabilistic approach for time-horizon DSM using a universal applicable interface between DSO and DSM applications. Specifically, the probabilistic NN based analysis will indicate whether OLV are expected to occur at certain geographical locations in the network and with what probability. If this probability exceeds a certain threshold, the DSO will request the DSM application for flexibility to reduce the probability of the violation of operation limits at acceptable levels. Based on the probability of the system states of the three-phase nodes over a day-ahead (DA) or intra-day (ID) time interval, the three-phase sensitivity of the operational limit violation faced by the DSO with respect to the active power injection by customers is derived. The DSO will use this information to specify the expected operation limit violation and how its probability can be reduced towards the DSM application by using the universal interface. The main contributions of the paper are:

- Time-horizon analysis of operation limit violations using a probabilistic NN based approach: if operational limit violations are expected with a certain probability, grid-supportive DSM is triggered by using a universal interface to reduce the probability back to acceptable levels;
- Specification of a three-phase unbalanced network operation limit violations occurring with a certain probability over time and their sensitivity with respect to active power, according to the network model in Reference [26]. This allows DSM applications to resolve network

issues by optimizing time dependent flexibility, which is independent of the objective of the DSM application;

- A demonstration will be given of a time-horizon DSM optimization resolving under and over voltages featuring fair power sharing among the different prosumers.

The remainder of the paper is organized as follows. In Section 2, the universal interface between the DSO and DSM applications is described, which allows for a generalized procurement of flexibility. Section 3 follows up with how the required amount of flexibility is triggered based on predicted system states and the network sensitivity, which is optimized according to the objective of the DSM application. Section 4 presents an alternative method based on a NN. Section 5 will present the overall DSM optimization while Section 6 gives a proof of concept using numerical simulations, which shows the applicability of the proposed approach on the thee-phase IEEE low voltage EU network. Lastly, Section 7 will elaborate further recommendations and conclusions.

2. Universal Procurement of Flexibility

This section briefly introduces the universal interface adopted in this work for establishing DSM using a probabilistic, time-horizon, grid-supportive methodology. A more in-depth discussion of this framework can be found in Reference [27]. One of the main features of this interface is that it should work efficiently with all possible DSM applications in the field. In this case, the starting point of the interface is that the exchanged information should be non-iterative and integrable in the widest selection of optimization algorithms. Therefore, for each time interval, the DSO can specify what flexibility it requires depending on the geographical locations of the flexibility within the network. After receiving this information, the DSM application can allocate the required flexibility depending on its own optimization strategy and send the results back to the DSO. The specified information by the DSO on the required flexibility consists of lower limits for certain linear combinations of changes in active power injection at the geographical locations of the customer connection points and their corresponding linear gains. From this, linear constraints can be constructed, which can be considered by virtually any DSM application. These gains are formed by the sensitivity matrix of the network, which specifies the linearized sensitivity of the system states with respect to changes in active power, as detailed in Section 3.2. As such, independent of what objective a DSM application pursues, it will be able to resolve the OLV by including constraints on the before mentioned sensitivity of active power injection towards the OLV the DSO faces. This way, the DSM application is enabled to consider the operation limits at each geographical location in the distribution network in its optimization. Depending on the DSM application in place, additional information can be included such as for bidding or pricing information. In the proposed framework of Reference [27], the procurement of flexibility takes place in two stages: (1) time-horizon flexibility based on both probabilistic power flow and machine learning during the DA/ID preventive planning phase, which is the main topic of this paper and (2) real-time flexibility based on state estimation during the corrective execution phase, as described in Reference [27].

Preventive time-horizon flexibility will be procured in on a DA/ID basis. Preventive DSM based on predictive load forecasting [28–30] will be of high importance to account for the system states and possible operational limit violations that might occur in the near future. One of the motivations for this is that prosumer flexibility in energy production or consumption is expected to involve complex time-dependent interdependencies. This means that flexibility provided within some time interval might need to be compensated by flexibility in another time interval. Furthermore, to prevent excessive life time degradation, DSOs will prefer to not switch certain controllers very frequently. This can, for example, hold true for OLTC tap position adjustments. As such, an optimization over a certain time horizon of the network system states and the available flexibility is of high importance to resolve OLVs that are expected over time. For this, probability density functions (PDF) for each of the system states can be obtained for the time interval of the DA/ID optimisation period from where the probabilistic system states can be obtained by using probabilistic power flow calculations. DSOs can make important decisions with this information for adjustments of set points of the previously mentioned local control capabilities. Whenever there are any OLV after adjustments of the local controller setpoints, the DSO can trigger time-horizon flexibility from DSM applications by specifying the specific OLV at hand along with the sensitivity regarding active power injection at each geographical network location.

Within the corrective execution phase, real-time DSM is based on actual state estimation (SE) of the network. It is expected that SE capabilities must be expanded from transmission to distribution networks to gain insight in the network system states and allow for the control applications, which is mentioned earlier [31]. Within the real-time execution phase, real-time insight in the actual system states will be crucial to trigger these control applications as well as real-time flexibility from DSM applications, which is described in Reference [27].

3. Probabilistic Grid-Supportive Flexibility

As stated, the DSO can invoke flexibility in active power injection from DSM applications to address OLVs. Based on the real-time and predictive monitoring capabilities, the interface between the DSO and the DSM application will facilitate a generalized information exchange on the OLV the DSO is facing with the sensitivity for changes in active power at specific geographical locations. As discussed, a newly proposed benchmarking method is proposed based on PPF calculations as well as a novel method using a NN. A flow diagram of the full functioning of the proposed methods covered in Sections 3–5 is shown in Figure 1. This figure illustrates both the benchmark approach and the NN approach.

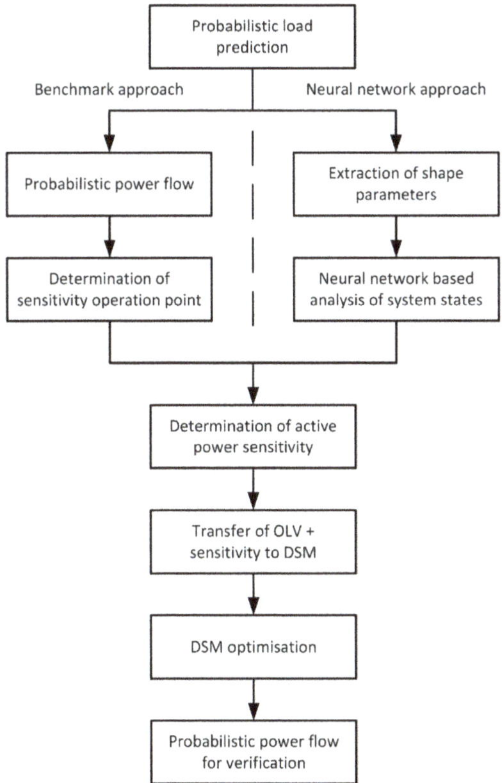

Figure 1. Comparison of the proposed grid-supportive DSM using PPF and machine learning.

3.1. Probabilistic Prediction of Operational Limit Violations

In this paper, we assume an unbalanced radial network [26] consisting of N nodes including the slack node where there are M households tapping of the feeders offering flexibility in their active power consumption or production. The slack node is located at the root of the radial feeder and is assumed to be constant. To determine the required amount of flexibility in active power consumption or production at each node in the network within the preventive planning phase, the DSO performs a probabilistic analysis of the network system states for each time interval t of the DA/ID optimization. This can be either based on the before mentioned PPF or the NN approach presented in this paper. For benchmarking purposes, in this section, first the PPF approach will be discussed, which is followed by the NN based approach in Section 4.

As an input, the PPF calculation takes the power injection PDF of each household, which is based on predictions. The PPF calculation results in a PDF for the all the system states of the network. For example, the probability density of the voltage magnitude $V_{n,p}^t$ occurring at node n, phase p, and time interval t is given by $P\left(V_{n,p}^t\right)$. Now suppose that the network has a lower limit and upper limit V^l and V^u for the voltage magnitude, then the probability of an undervoltage occurring at node n and phase p during time interval t can be calculated, according to the cumulative distribution function (CDF) $F_{V_{n,p}^t}$. This is shown below.

$$\Pr\left[V_{n,p}^t \leq V^l\right] = F_{V_{n,p}^t}\left(V^l\right) = \int_0^{V^l} P\left(V_{n,p}^t\right) dV_{n,p}^t \tag{1}$$

Similarly, for the probability of over voltages, we can state that:

$$\Pr\left[V_{n,p}^t \geq V^u\right] = 1 - F_{V_{n,p}^t}(V^u) = \int_{V^u}^{\infty} P\left(V_{n,p}^t\right) dV_{n,p}^t \tag{2}$$

Similar expressions can obviously be made for overloading or violation of voltage imbalances throughout the phases. However, in this paper, we focus only on under/over-voltages. Now, if the probability of having an under/overvoltage as expressed in Equations (1) and (2) exceeds a certain threshold value f. The DSO will opt to procure flexibility from the DSM application in order to shift the PDF of the voltage magnitudes such that the probability of having an under/overvoltage is brought down to acceptable levels. To formulate this as an easier condition to deal with, we first introduce the variables $V_{n,p}^{t,l}$ and $V_{n,p}^{t,u}$. In this case, $V_{n,p}^{t,l}$ is the voltage magnitude for which the CDF $F_{V_{n,p}^t}$ equals the acceptable probability threshold f.

$$F_{V_{n,p}^t}\left(V_{n,p}^{t,l}\right) = f \tag{3}$$

In other words, there is a probability of exactly f that the voltage at node n and phase p at time interval t will be lower or equal to $V_{n,p}^{t,l}$. Similarly, $V_{n,p}^{t,u}$, can be expressed as:

$$F_{V_{n,p}^t}\left(V_{n,p}^{t,u}\right) = 1 - f \tag{4}$$

Rephrasing in other words, there is a probability of exactly f that the voltage at node n and phase p at time interval t will be higher or equal to $V_{n,p}^{t,u}$. In case of a discontinuous CDF, $V_{n,p}^{t,l}$ and $V_{n,p}^{t,u}$ take the next available value closest to the nominal voltage. We define the vectors $V_p^{t,l}$ and $V_p^{t,u}$ of all voltages $V_{n,p}^{t,l}$ and $V_{n,p}^{t,u}$ of all nodes n (apart from the slack node) in phase p and time interval t, according to the equations below.

$$V_p^{t,l} = \left[V_{2,p}^{t,l}, V_{3,p}^{t,l}, \ldots, V_{N-1,p}^{t,l}, V_{N,p}^{t,l}\right]^T \tag{5}$$

respectively,

$$V_p^{t,u} = \left[V_{2,p}^{t,u}, V_{3,p}^{t,u}, \ldots, V_{N-1,p}^{t,u}, V_{N,p}^{t,u}\right]^T \tag{6}$$

Now, the DSO will procure flexibility for under-voltages if any of the voltage magnitudes $V_{n,p}^{t,l}$ are lower than the lower limit V^l and flexibility for overvoltages if any of the voltages $V_{n,p}^{t,u}$ are higher than the upper limit V^u. After all, in both cases, the probability of having a voltage limit violation is higher than the probability threshold f. This can be expressed as the conditions below.

$$V_{n,p}^{t,l} \leq V^u \tag{7}$$

and

$$V_{n,p}^{t,u} \geq V^u \tag{8}$$

If one of these conditions holds, the DSO needs flexibility to reduce the probability at least to the acceptable probability threshold f, which is illustrated in Figure 2 and detailed in the next sections.

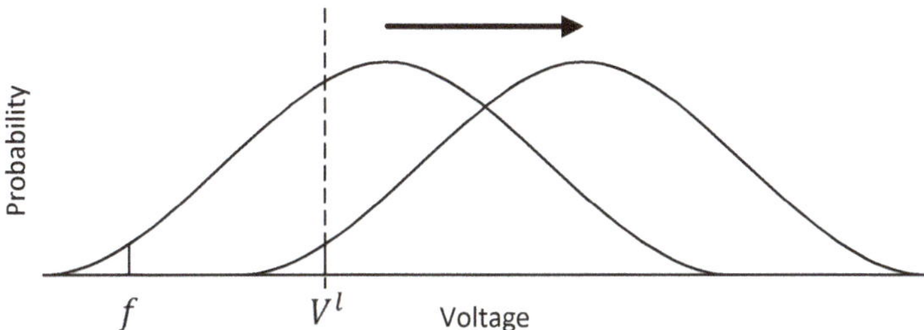

Figure 2. Illustration reducing probability of under voltages to acceptable levels.

3.2. Network Sensitivity Operation Point

To procure the right amount of flexibility from the DSM application, the DSO needs to know how much flexibility will be required to solve the OLV depending on the geographical location at which the flexibility is delivered. As stated in the introduction section, this work uses the linearized sensitivity of the network system states, which has been used effectively in other studies [27,32]. The sensitivity of the probabilistic OLV informs the DSM application what linear combinations in active power flexibility can address the OLV. To obtain the linearized sensitivity of OLVs with respect to changes in active power injection $\Delta P_{n,p}^t$ at node n, phase p, and time t, the Jacobian matrix of the partial derivatives of the power injections $P_{n,p}^t$ with respect to the violated system states are derived. As an example, concerning violation of the voltage magnitude, the partial derivatives can be expressed by using well-known power flow equations as well as expressing the nodal power injection in terms of the nodal voltage magnitudes and network parameters (which are considered static and, therefore, left out of the notation).

$$J_V^t = \begin{bmatrix} \dfrac{\partial P_{2,a}^t}{\partial V_{2,a}^t} & \dfrac{\partial P_{2,a}^t}{\partial V_{2,b}^t} & \cdots & \dfrac{\partial P_{2,a}^t}{\partial V_{N,c}^t} \\[2mm] \dfrac{\partial P_{2,b}^t}{\partial V_{2,a}^t} & \dfrac{\partial P_{2,b}^t}{\partial V_{2,b}^t} & \cdots & \dfrac{\partial P_{2,b}^t}{\partial V_{N,c}^t} \\[2mm] \vdots & \vdots & \ddots & \vdots \\[2mm] \dfrac{\partial P_{N,c}^t}{\partial V_{2,a}^t} & \dfrac{\partial P_{N,c}^t}{\partial V_{2,b}^t} & \cdots & \dfrac{\partial P_{N,c}^t}{\partial V_{N,c}^t} \end{bmatrix} \tag{9}$$

It should be noted that if $M < N$, the nodes at which no active power flexibility is available can be left out of the Jacobian.

From this point, the operation point of the partial derivatives is based on the outcome of the PPF calculation. The PPF calculation is in practice often and is completed by using Monte Carlo simulations, which results in discrete samples of the PDFs of the system states. In this paper, a pragmatic approach is taken to obtain an effective operation point from the discrete samples of the PDFs of the system states. Suppose that the Monte Carlo simulations for each time interval t result in a set $\mathcal{K}_p^t$ of discrete samples of the PDFs of the system states. For each sample $k \in \mathcal{K}_p^t$ at time interval t, we can compose the vector of all system states for each node n in phase p.

$$V_p^{t,k} = \left[V_{2,p}^{t,k}, V_{3,p}^{t,k}, \ldots, V_{N-1,p}^{t,k}, V_{N,p}^{t,k} \right]^T \tag{10}$$

We define the subset $\mathcal{K}_p^{t,l} \subset \mathcal{K}_p^t$ to be the set of samples k, which yields that the minimum value of $V_p^{t,k}$ has a probability lower or equal to f, i.e., the minimum value of $V_p^{t,k}$ is smaller or equal than the minimum value of $V_p^{t,l}$.

$$\min V_p^{t,k} \leq \min V_p^{t,l} \tag{11}$$

Similarly, the subset $\mathcal{K}_p^{t,u} \subset \mathcal{K}_p^t$ is the set of samples k, which yields that the maximum value of $V_p^{t,k}$ is larger or equal than the maximum value of $V_p^{t,u}$:

$$\max V_p^{t,k} \geq \max V_p^{t,u} \tag{12}$$

Lastly, we define $\bar{\delta}_p^{t,l}$ and $\bar{\delta}_p^{t,u}$ as the average voltage angle vector, which is averaged elementwise over all complex voltages corresponding to all k being an element of the set $\mathcal{K}^{t,l}$, respectively, $\mathcal{K}^{t,u}$.

Now, the operating point of the network for the Jacobian is chosen to be $V_p^{t,l} \angle \bar{\delta}_p^{t,l}$ or $V_p^{t,u} \angle \bar{\delta}_p^{t,u}$, i.e., the voltage vectors for which the magnitude corresponds with the probability f and the angle is averaged over the phasors of the sets $\mathcal{K}^{t,l}$ end $\mathcal{K}^{t,u}$. Note that the sensitivity will be different for under and over voltages, which are denoted as $S_{V,l}^t$ and $S_{V,u}^t$. Averaging of the angles $\bar{\delta}_p^{t,l}$ and $\bar{\delta}_p^{t,u}$ is required since there is not a single angle that corresponds with the system states for the probability f. For the voltage magnitudes, there exists a unique relation, which is expressed in Equations (3) and (4). For the angles, such a unique relation does not exist. Therefore, we average over all the values in the sets $\mathcal{K}_p^{t,u}$ and $\mathcal{K}_p^{t,l}$. Note that this is only used for determining a suitable operation point for the Jacobian and in no way aims to change the power factor of any appliance.

In relation to this, an important note should be made concerning the Jacobian. The partial derivatives forming the entries of the Jacobian differ depending on the relation between the reactive and the active power at each node and phase. This relation is determined by the appliance associated to the power injections. As an example, the power factor can remain constant for any change in the active power injection (therefore, changing the reactive power injection) or the reactive power can remain constant independent of the active power injection. This results in different partial derivatives where highly non-linear or discontinuous relations between active and reactive power will occur. This might complicate the process of deriving a suitable Jacobian. In the simulation results of this work, we assume the reactive power to remain constant. Lastly, the sensitivity with respect to the OLV is obtained by inverting the Jacobian, according to the equation below.

$$S = J^{-1} \tag{13}$$

3.3. Constraints for Demand Side Management

To reduce the probability of a specific operation limit violation to acceptable levels, the DSO will specify the required minimum required change in any network system state. As flexibility for resolving operation limit violations most likely will result in shifting power consumption to another point in time, we also need to specify the available 'capacity' in time intervals where no operation limit violations are expected. If the OLV concerns nodal voltage magnitude violations, the DSO will specify the vectors ΔV_l^t and ΔV_u^t, which indicates the voltage with which the system states are exceeded and what capacity is available. We rearrange the elements of ΔV_l^t and ΔV_u^t, in general denoted as ΔV^t, such that the vectors contain the elements for all nodes n (excluding the slack node) and all phases $p \in \{a,b,c\}$.

$$\Delta V^t = \left[\Delta V^t_{2,a}, \Delta V^t_{2,b}, \Delta V^t_{2,c}, \ldots, \Delta V^t_{N,a}, \Delta V^t_{N,b}, \Delta V^t_{N,c} \right]^T \tag{14}$$

The individual elements of these vectors are given by the equation below.

$$\Delta V^t_{l,\,n,p} = V_{\min} - V^t_{n,p} \tag{15}$$

and

$$\Delta V^t_{u,\,n,p} = V_{\max} - V^t_{n,p} \tag{16}$$

In this case, $V_{\min}$ and $V_{\max}$ are the minimum and maximum voltage limits while $V^t_{n,p}$ is the voltage before DSM at node n and phase p.

As a final step, the change in active power at time t for a certain node n, phase p, and time t is represented by $\Delta P^t_{n,p}$ where the elements for all nodes and phases together compose the vector ΔP^t.

$$\Delta P^t = \left[\Delta P^t_{2,a}, \Delta P^t_{2,b}, \Delta P^t_{2,c}, \ldots, \Delta P^t_{n,a}, \Delta P^t_{n,b}, \Delta P^t_{n,c} \right]^T \tag{17}$$

Currently, the sensitivity with respect to the OLV is obtained by inverting the Jacobian according to $S = J^{-1}$. herein this case, the linearized change in the network system states is calculated by multiplying the change in active power $\Delta P^t_{n,p}$ injection with the sensitivity matrix. For an OLV of the nodal voltage magnitude, the resulting change in the system states $\Delta V^t_{n,p}$ yields the following equation.

$$\Delta V^t = S^t_V \Delta P^t \tag{18}$$

4. Neural Network-Based Grid-Supportive Demand Side Management

As discussed, performing the PPF is computationally costly despite the many studies in literature improving its efficiency. By carrying out DA PPF calculations for all time intervals in the DA period and all the networks it is managing, this might require significant computational power. Therefore, this work introduces an NN based approach to accurately approximate the findings derived in the previous section and, therefore, drastically speeding up the required computation times. After all, the interest in this case is not on the probabilistic system states but rather the need for flexibility. For this purpose, a multi-layer NN (NN) is introduced that only needs to be trained once for a particular distribution network. After training, evaluating the NN is considerably faster than performing the PPF, which results in a significantly reduced computational effort. The next subsections will respectively discuss the NN architecture and the training of the NN.

4.1. Neural Network Architecture

The NN is designed to prevent the costly PPF within the risk analysis of the DSO. To this extent, it needs to approximate the information on the operational limit violation at hand and its sensitivity with respect to changes in active power, which is detailed in Section 3. For this purpose, in this work, a regressive NN is applied to provide the DSM application with the required information by replacing the costly PPF. The overall architecture of the network is displayed in Figure 3.

For the PPF method described in Section 3, the inputs are the PDFs of each of the households while the outputs are formed by the required change in systems states ΔV^t and the sensitivity matrix S^t_V. However, the full PDFs of the households are not very suitable to be used as an input for a NN since it normally expects a numeric input value rather than a continuous function specification. On a similar note, the output sensitivity matrix S^t_V has a number of entries equal to the square of the number of nodes, which makes it too large to be effectively approximated by a regression based NN. As a final point of concern, the required change in system states ΔV^t is a non-linear function with a discontinuous derivative and, therefore, is also not very suitable for regression analysis.

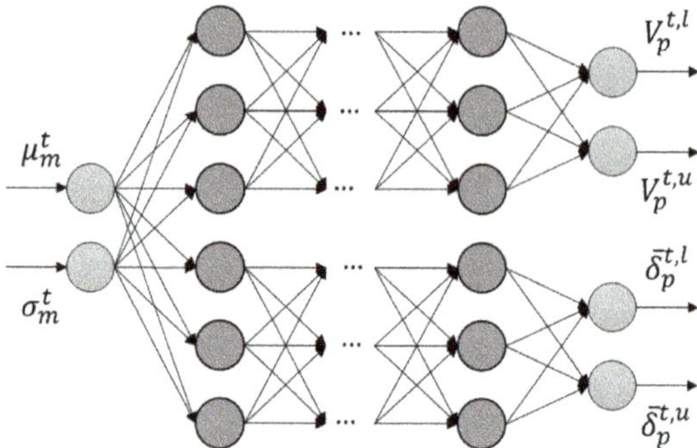

Figure 3. Schematic overview of the separated NN for the nodal voltage magnitudes and angles.

To overcome these design challenges for the inputs of the NN, the PDF of the loading of each household can be fitted on a suitable well-known default PDF like a Gaussian, beta, or Weibull distribution. Specific features or shape parameters can be extracted like the mode, mean, median, and variance or the α and β or k and λ parameters for the beta and Weibull distributions. These shape parameters can be used as inputs for the NN. For the sensitivity matrix, although large matrix inversion is involved, determining the sensitivity based on a set of operation points of the network is a simple task and relatively computationally efficient. In addition, determining the required change in the system states ΔV^t from the operation points is highly straightforward. Therefore, the NN is trained to not output the required change in the system states and corresponding sensitivity but rather approximate the operation point used for calculating the network sensitivity.

Mathematically expressed, these are $V_p^{t,l} \angle \bar{\delta}_p^{t,l}$ or $V_p^{t,u} \angle \bar{\delta}_p^{t,u}$ as introduced in Section 3, i.e., the voltage vectors for which the magnitude corresponds with the probability f and the angle is averaged over the phasors of the sets $\mathcal{K}^{t,l}$ end $\mathcal{K}^{t,u}$. The network sensitivity S_V^t can be calculated based on Equations (9) and (13). Similarly, ΔV^t is easily derived using Equations (15) and (16).

The simulation results presented in Section 6 of this work use data for the household loading PDFs that is suitable to be fitted on a Gaussian distribution. The mean μ_m^t and variance σ_m^t of the PDFs for each household m at time t are defined as the input variables of the NN. This means that the number of inputs for the NN will be equal to $2M$. Similarly, the output variables together $V_p^{t,l} \angle \bar{\delta}_p^{t,l}$ and $V_p^{t,u} \angle \bar{\delta}_p^{t,u}$ will be of size $12(N-1)$, (three phase nodes and four variables per node excluding the slack node). However, since the numerical values of the voltage magnitudes and angles are considerably different, it is better to split the NN in two separate networks of size $6(N-1)$ since, this way, performance indicators such as the mean squared error (MSE) are used more meaningful. Lastly, in the average distribution network, we do not expect differences in voltage angles up to π radians. Therefore, it is advisable to shift the voltage angles with π radians to eliminate the transition between 0 and 2π. As an alternative, one could consider working in Cartesian coordinates rather than polar coordinates. As a final step, experimental simulations will be required to determine a suitable number of hidden layers and neurons. For the experimental results presented in Section 6, it turns out that two or up to three hidden layers strike a reasonable balance between training time and accuracy by estimating the non-linear relation between the input and output variables accurately, which can be seen from the presented results in Section 6. The number of hidden neurons is highly dependent on the expected correlation of the input data and, therefore, should be determined experimentally for each network.

4.2. Neural Network Training

To train the NN with the architecture as described above, a large amount of historical data is required, which comprises the PDF of household loadings for many different situations. Based on this, the mean μ_m^t and variance σ_m^t can be derived together with the corresponding $V_p^{t,l} \angle \bar{\delta}_p^{t,l}$ or $V_p^{t,u} \angle \bar{\delta}_p^{t,u}$, as described in Section 3, which forms the training set for the NN. Clearly, it is important that the training set contains a sufficiently diverse number of situations that might occur in the network to make the system robust for unexpected events. Since the proposed NN architecture will have a significant size for larger distribution networks, it is advisable to perform the training of the NN on a graphics card to exploit the possibilities of parallelism. When doing so, from experimental results, backpropagation training using gradient derivatives and the steepest descent turns out to strike a good balance between performance and training time for the NN architecture proposed in this study. However, the convergence is very sensitive to the learning rate and, therefore, in this study, the gradient descent with an adaptive learning rate backpropagation is used, which is implemented in MATLAB (MathWorks Inc., 2018b (prerelease), Natick, MA, USA) [33].

5. Overall Demand Side Management Optimization

After the reception of the specification on the OLV and corresponding sensitivity from the DSO (either using the PPF or NN approach), the DSM application deploys an overall optimization of the available flexibility in active power injection depending on its optimization objectives and is constrained by the specified OLVs. As stated in the introduction section, the objectives can take many forms like local supply and demand matching [11], fair power sharing [34], or a market mechanism where global welfare is optimized [35]. We define the set $\mathcal{F}$ of size M of all pairs of nodes and phases $\{m,q\}$ at which flexibility is offered by prosumers. Independently of the optimization objective the DSM pursues, the general optimization function in Equation (19) can be defined alongside the corresponding constraints for voltage OLVs Equation (20).

$$\min_{\Delta P^t_{m,q}} g\left(\Delta P^t\right) \ \forall \{m,q\} \in \mathcal{F} \tag{19}$$

subjected to:

$$\Delta V^t_l \leq S^t_{V,l} \Delta P^t$$
$$\Delta V^t_u \geq S^t_{V,u} \Delta P^t \tag{20}$$
$$\text{other constraints of flexible appliances}$$

Examples of other constraints of flexible appliances can be found in Reference [36]. Simultaneous over voltages and under voltages at different nodes of the network are (though unlikely because of the huge voltage difference) possible to address if it does not render the optimization infeasible. These constraints will need to be satisfied for each time interval t within the optimization horizon in case a time-horizon optimization is concerned based on probabilistic predictions. Besides the constraints on the system states, further time-dependent appliance specific constraints for $\Delta P^t_{m,q}$ can be included. For example, this has been done in Reference [11]. The general optimization function above can be changed to specific optimization objectives depending on the goal and purposes of the DSM application. We will illustrate this by using two examples where each example can be solved using quadratic programming.

Possible objective 1: For the optimization objective of local supply and demand matching over the time-horizon from $t = vF$ to $t = vT$ within each of the households connected to the distribution network while spreading out the remaining supply and demand over time (profile flattening), one can minimize the sum of squares of the base load power vector P^t calculated with the flexibility vector ΔP^t of each of the prosumers. This is shown in the formula below.

$$\min_{\Delta P^t} \sum_{t=vF}^{vT} \left(P^t + \Delta P^t\right)\left(P^t + \Delta P^t\right)^T \tag{21}$$

If one alternatively wishes to balance the supply and demand of the distribution network as a whole, one can minimize the square of the sums of the base load power injection P^t calculated with the flexibility ΔP^t (i.e., not the sum of squares but the square of sums).

Possible objective 2: In a similar fashion, with the objective of fair power sharing, the flexibility provided by different prosumers is aimed to be mostly equal among those who can reasonably contribute to the OLV at hand, which prevents some prosumers from being asked to provide flexibility more often than others depending on their physical point of connection. Since this effect can obviously not completely rule out the physical aspects of the network, some users will be able to make a larger contribution than others (depending on the phase and location of their connection). Therefore, the least sum of squares of $\Delta P_{n,p}^t$ is taken as a balanced approximation to realize fair power sharing by automatically selecting those prosumers that can reasonably contribute to resolving the OLV at hand.

$$\min_{\Delta P^t} \sum_{t=vF}^{vT} \left(\Delta P^t \right) \left(\Delta P^t \right)^T \tag{22}$$

6. Simulation Results

This section aims to demonstrate the effectiveness of the probabilistic time-horizon DSM methodology featuring fair power sharing, which is seen in Equation (22) as the optimization objective. The DSM is applied to a three-phase implementation of the IEEE European LV test network as displayed in Figure 4. This network has a radial architecture consisting of 117 nodes and 55 prosumer households with single phase connections.

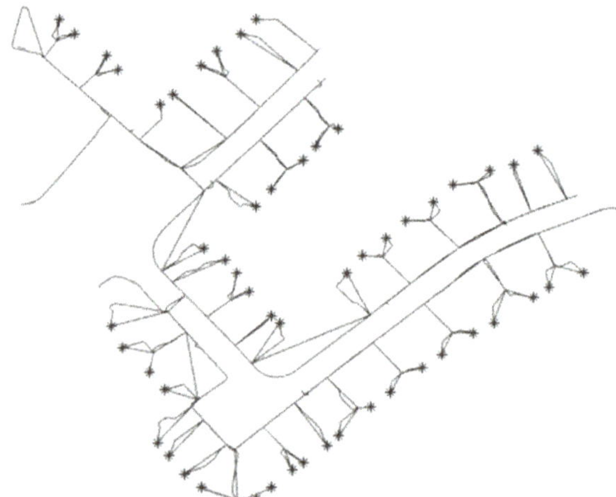

Figure 4. IEEE European LV test network.

6.1. Overall Simulations

Within the simulation setup, the slack node voltage is assumed to remain constant at 230 V and the DSO is assumed to have set voltage limits of 0, 9, and 1 p.u. (i.e., $+/-23$ V) for the whole feeder.

The acceptable probability limit f for operational limit violations of the voltage magnitudes is set to 0.1 for each time interval (i.e., 90% certainty of having no operation limit violation). Additional constraints on operational limits can be considered such as constraints for the branch current magnitude or the unbalance factor (VUF) for the voltage between the three phases. However, in this work, we focus on violations of the voltage magnitudes in either of the three phases. The assumption is that flexibility is available at each household with an upper bound of 2 kW where future work will be done to make this flexibility more realistic, which is discussed in Reference [37]. It should be noted that requests for that much flexibility will only be met in the case of very high loadings of the network such as in cases of multiple charging electrical vehicles. In those cases, 2 kW is less than a quarter of the total household loading and, therefore, considered realistic. For now, the goal is to model flexibility as realistic as possible but to show the effectiveness of the proposed probabilistic sensitivity method. The overall simulation consists of the following steps, which corresponds with the flow diagram in Figure 1.

Step 1: As a first step, the DSO performs the probabilistic load prediction to form the input distributions for the grid supportive demand side management. In the simulations performed in this work, the PDFs are normally distributed [38] where the mean and variance is different for each household and time interval and fitted from historical data obtained from the Pecan Street project [39]. The mean values of the household consumption PDFs range from -5.3 to 6.8 kW depending on the household and time of the day while the variance goes up to 5×10^6 W^2 with an average of 8×10^5 W^2. Besides the ordinary base load, a significant amount of PV generation is present in the load profiles, which is a high peak load in the evening due to electrification of cooking and heating installations.

Step 2a: For benchmarking purposes, the DSO also performs a Monte Carlo based PPF network for the DA 24-h period by using three phase unbalanced power flow calculations [26]. The Monte Carlo simulations for each time interval take 10,000 samples from the PDF of the power consumption of the connected households. From the resulting PDFs of the system states (being nodal voltages and branch currents), the DSO will determine whether the probability of operational limit violations occurring in the network is acceptable and, if not, determines the sensitivity operation point for calculation of the Jacobian.

Step 2b: In the machine learning based approach, the DSO will extract the distribution shape parameters (e.g., mean and standard deviation) and perform the NN based analysis of the IEEE European LV test network for the active power flexibility described in Section 4 based on the PDFs for each of the households. From the resulting outputs of the NN, the DSO will determine for each time interval whether the probability of any operation limit violation occurring in the network is acceptable or not.

Step 3: If the probability for operation limit violations is too high based on the results of step 2a and 2b, the DSO will determine the active power sensitivity of the OLVs at hand.

Step 4: The results of the analysis for OLVs and their active power sensitivities are sent to the DSM application.

Step 5: The DSM application optimizes the flexibility provided by the prosumers in accordance with its own optimization objective. As stated, the optimization objective featured in this work is fair power sharing among the different prosumers that can reasonably contribute in resolving the OLV at hand, which is specified in Equation (22). This way, the optimization will resolve the OLV expected in the network while dividing the burden for doing so over the different participating prosumers. For each appliance in the optimization, additional constraints can be set such as described in Reference [11].

Step 6: As a final step, the effectiveness of the allocated flexibility is assessed. The allocated flexible power comes on top of the original base load, which was represented by the PDFs of the power consumption of the households. Since this base load will still have the associated uncertainty after allocation of the flexibility, the resulting ΔP^t of the optimization is added to the original power values of the PDFs of the household consumptions. After this, there is a final verification in the simulation for both the PPF approach as well as the NN approach. It should be noted that the final verification carried out in this work is merely to verify the performance of the proposed approach and will not be carried out during a practical application. During the final verification, for both the PPF approach as well as the NN approach, a Monte Carlo simulation is carried out with the only difference that the input samples from the PDFs are now shifted over ΔP^t where ΔP^t is either the result of the PPF approach or the NN approach. The overall results of the simulation are discussed in the next section by starting off with the benchmarking results.

6.2. Benchmarking Results Using Probabilistic Power Flow

For each of the Monte Carlo simulations within the benchmarking PPF approach, the minimum respectively maximum values out of the vectors $[V_a^{t,l}; V_b^{t,l}; V_c^{t,l}]$ and $[V_a^{t,u}; V_b^{t,u}; V_c^{t,u}]$ are determined for each time interval t and their probabilities and are displayed in Figures 5 and 6, respectively. In other words, Figure 5 displays the probability of the lowest voltage, according to any node or phase within the network. In addition, shown is the probability of having a lower or equal voltage of exactly f. That means that there is a 90% chance that there will be no lower voltage than the displayed value anywhere in the network. If the displayed value is lower than the voltage limit, there is a higher than 10% chance for under voltages.

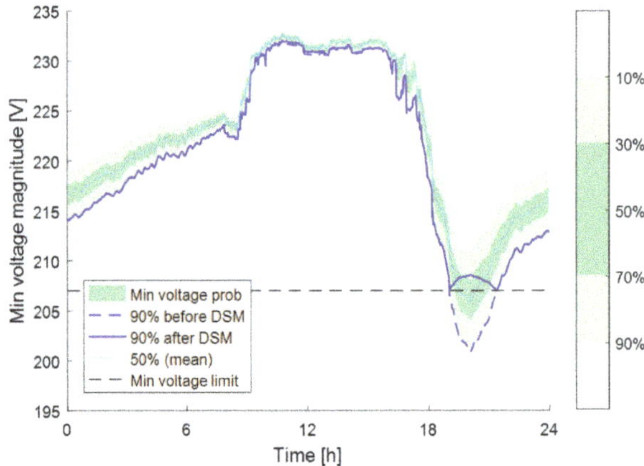

Figure 5. 90% probability minimum voltage magnitudes over time.

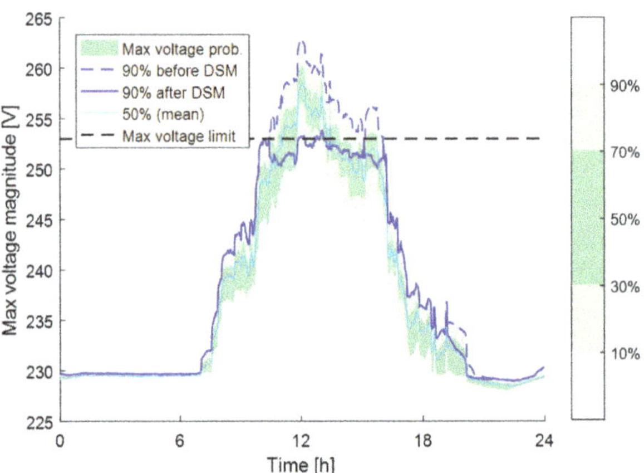

Figure 6. 90% probability maximum voltage magnitudes over time.

Similarly, Figure 6 displays the probability of the highest voltage of any node or phase including the probability of having a higher or equal voltage of exactly f (i.e., there is a 90% chance that there will be no higher voltage and a higher than 10% chance for overvoltages if the displayed value is higher than the voltage limit). The dashed blue lines represent the voltage magnitudes over time as they would be when no DSM is applied while the solid blue lines represent the voltage magnitudes after application of the fair power sharing DSM optimization. It should be noted that the displayed values can relate to any node and phase in the network. This can concern a different node and phase at each time interval and even a different node and phase before and after DSM. It cannot be seen from the graphs which node is concerned since this is changing over time depending on the loading of the network and at which node and phase the most extreme voltages (i.e., lowest or highest voltage) occur.

From the figures, we can derive that the network is expected not to be capable of facilitating the large PV infeed on a sunny day as installed for this configuration nor the energy intensive appliances that consume power in the evening since there is a higher than 10% probability of having over voltages during the middle of the day and under voltages during peak hours in the evening. During the remainder of the day, there is a less than 10% probability of having operation limit violations, which is where the displayed voltages are below the limit. After application of the DSM optimization, all OLVs have been (nearly) resolved by reducing the probability of operational limit violations for the voltage magnitude below 10% at nearly all times. Some deviations and small OLVs do remain, which can mainly be attributed to the linearization process and the probabilistic uncertainty in the PPF.

Deviations from the linearization process occur especially when the network becomes highly unbalanced, which is the case during the PV infeed. Nevertheless, the DSM application is slightly conservative most of the time, which overcompensates more when there is a more severe operational limit violation. This can be considered as favourable, as with this, the system operates on the safe side. In the rare event that a small operation limit violation will remain after the time-horizon DSM, additional corrective real-time DSM can be triggered based on state estimation.

6.3. Results Using the Neural Network

In the previous subsection, from the benchmarking results, the PPF based approach has been shown to be effective for reducing the probability of OLV to acceptable levels on a DA basis. In this case, the performance of the NN based approach is discussed where the results are presented in Figure 7 for under voltages and in Figure 8 for over voltages. This time, only the voltage corresponding with a probability exactly f is shown. Similar to Figures 5 and 6, as soon as this voltage crosses the indicated voltage limits, the probability of having a OLV higher than f and flexibility needs to be procured. In the figures, the dashed blue lines represent the voltages before DSM while the solid blue lines represent the voltages after DSM using the PPF approach. Lastly, the red lines are for the voltages after DSM using the NN based approach. The insets give a detailed comparison between both approaches during the hours at which operation limit violations take place.

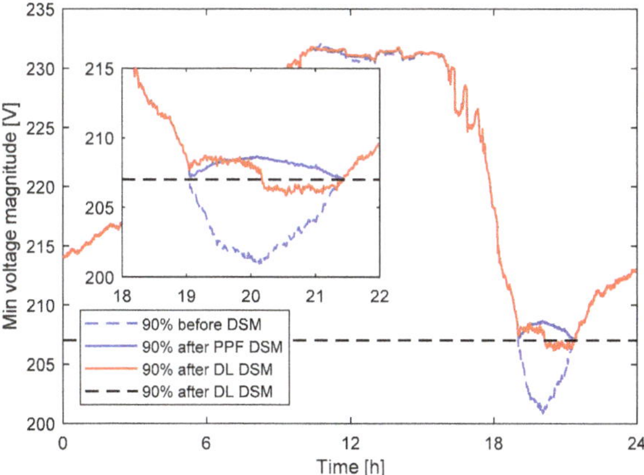

Figure 7. 90% probability minimum voltage magnitudes over time.

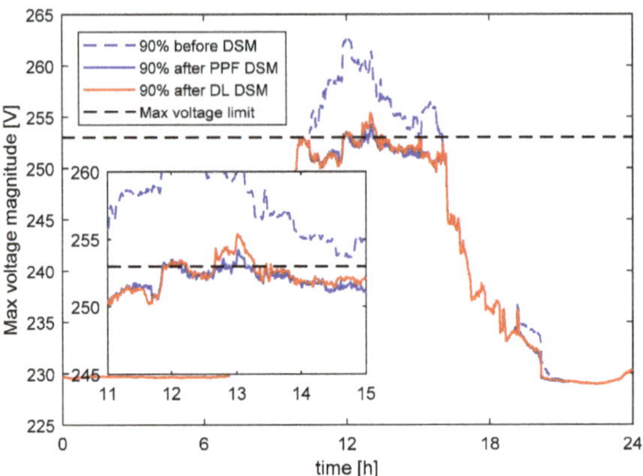

Figure 8. 90% probability maximum voltage magnitudes over time.

From the figures, it can be seen that the NN-based approach resembles the performance of the PPF approach in a good way and, in this simulation, it is especially accurate for the over voltages. During the evening hours where under voltages occur, in the second half of the under voltage period, a situation occurs in which the deviation from the PPF approach is higher (although more accurately to the operation limit). In this case, the occurring loading configuration was insufficiently present in the training data for the NN, which results in the lower accuracy.

6.4. Execution and Training Time

As has been discussed, the PPF approach is computationally intensive despite the many works in literature improving its efficiency. Carrying out DA PPF calculations for its complete distribution network, this might require significant computational power. Notwithstanding the large computation time for the PPF approach, one may argue that, since the time-horizon preventive DSM proposed in this work runs on a DA/ID basis and not in real-time, computation time is not a top priority. However, as predictions on the load probability density functions tend to be more accurate closer to the time of delivery/consumption, deferring the procurement of flexibility by the DSO as much as possible is important. The DSM application itself should be allowed time to perform its optimization. For the numerical simulations performed in this work, the PPF and DSM optimization for the 1440 time intervals takes over two hours on an Intel Core i7-4790 CPU excluding the verification step in step 4. The computation time is strongly dependent on the number of time intervals for which an OLV is expected. For the NN-based approach, the computation time reduces strongly to only under 15 min because of the fast evaluation of the NN. Besides the execution time, the NN also requires training time. By training the NN on a Nvidia GTX 1080 Ti graphics card, within one hour, training the mean squared error of the voltage magnitudes drops below 0.12 V^2 by using the mentioned gradient descent algorithm. It should be mentioned, however, that the choice for the number of layers and neurons for a particular network configuration might require repeated retraining of networks. Lastly, preparing the training data set might require way more time than the training itself. Still, each of these tasks are in principle a one-time exercise that will save considerably on the computation time during operation of the presented grid-supportive DSM.

7. Recommendations and Conclusions

This paper contributes with a probabilistic approach for grid supportive demand side management by presenting a Monte Carlo as well as a Neural Network based approach to reduce the probability of geographical dependent operation limit violations to acceptable levels. A practical case study is presented using the IEEE EU LV test feeder for which numerical simulations show that the proposed approach forms an effective method for DSOs to invoke time-horizon flexibility to address (too high probabilities of) expected operational limit violations in the network. In this case, the NN-based approach offers a significant benefit over the PPF-based approach in terms of computational complexity. Nevertheless, from these findings, this research can be extended in several directions. First, the proposed approach will need to be extended with models of actual flexible appliances to allow for more realistic modelling of the available appliance flexibility. Furthermore, the currently used constraints to prevent violation of the nodal voltage magnitudes can be extended to various other power quality related limits such as for the voltage unbalance factor between the three phases. Lastly, a more in-depth quantization of the severity of operation limit violations that can be expected in the future is required together with an analysis of whether the expected available flexibility will be enough to reduce the probability of these operational limit violations sufficiently. The proposed approach needs to be extended to deal with a shortage of flexibility by rendering the optimization infeasible. As a future solution, the optimization problem could be reformulated on the fly such that it will minimize the probability of operational limit violations.

Author Contributions: Conceptualization, N.B., P.N., and H.S. Data curation, N.B. Formal analysis, N.B. Investigation, N.B. Methodology, N.B. Project administration, P.N. Resources, N.B. Software, N.B. Supervision, P.N. and H.S. Validation, N.B. Visualization, N.B. Writing–original draft, N.B. Writing–review & editing, P.N. and H.S.

Funding: This research has been funded by The Netherlands Organization for Scientific Research (NWO) under the DISPATCH project, number 408-13-056.

Conflicts of Interest: The authors declare no conflict of interest.

References

1. Clement-Nyns, K.; Haesen, E.; Driesen, J. The impact of Charging plug-in hybrid electric vehicles on a residential distribution grid. *IEEE Trans. Power Syst.* **2010**, *25*, 371–380. [CrossRef]
2. Veldman, E.; Gibescu, M.; Slootweg, H.J.G.; Kling, W.L. Scenario-based modelling of future residential electricity demands and assessing their impact on distribution grids. *Energy Policy* **2013**, *56*, 233–247. [CrossRef]
3. Qian, K.; Zhou, C.; Allan, M.; Yuan, Y. Modeling of load demand due to EV battery charging in distribution systems. *IEEE Trans. Power Syst.* **2011**, *26*, 802–810. [CrossRef]
4. Eftekharnejad, S.; Vittal, V.; Heydt, G.T.; Keel, B.; Loehr, J. Impact of increased penetration of photovoltaic generation on power systems. *IEEE Trans. Power Syst.* **2013**, *28*, 893–901. [CrossRef]
5. Roupioz, G. Assessment of the MV/LV on-load tap changer technology as a way to increase LV hosting capacity for photovoltaic power generators. *IET Conf. Proc.* **2016**, *44*, 1–4. [CrossRef]
6. Xin, H.; Qu, Z.; Seuss, J.; Maknouninejad, A. A self-organizing strategy for power flow control of photovoltaic generators in a distribution network. *IEEE Trans. Power Syst.* **2011**, *26*, 1462–1473. [CrossRef]
7. Kobayashi, H.; Hatta, H. Reactive power control method between DG using ICT for proper voltage control of utility distribution system. In Proceedings of the 2011 IEEE Power and Energy Society General Meeting, Detroit, MI, USA, 24–29 July 2011; pp. 1–6.
8. Shariatzadeh, F.; Mandal, P.; Srivastava, A.K. Demand response for sustainable energy systems: A review, application and implementation strategy. *Renew. Sustain. Energy Rev.* **2015**, *45*, 343–350. [CrossRef]
9. Venkatesan, N.; Solanki, J.; Solanki, S.K. Residential Demand Response model and impact on voltage profile and losses of an electric distribution network. *Appl. Energy* **2012**, *96*, 84–91. [CrossRef]
10. Gelazanskas, L.; Gamage, K.A.A. Demand side management in smart grid: A review and proposals for future direction. *Sustain. Cities Soc.* **2014**, *11*, 22–30. [CrossRef]
11. Blaauwbroek, N.; Nguyen, P.H.; Konsman, M.J.; Shi, H.; Kamphuis, R.I.G.; Kling, W.L. Decentralized Resource Allocation and Load Scheduling for Multicommodity Smart Energy Systems. *IEEE Trans. Sustain. Energy* **2015**, *6*, 1506–1514. [CrossRef]
12. Zakariazadeh, A.; Homaee, O.; Jadid, S.; Siano, P. A new approach for real time voltage control using demand response in an automated distribution system. *Appl. Energy* **2014**, *117*, 157–166. [CrossRef]

13. Haque, A.N.M.M.; Nguyen, P.H.; Vo, T.H.; Bliek, F.W. Agent-based unified approach for thermal and voltage constraint management in LV distribution network. *Electr. Power Syst. Res.* **2017**, *143*, 462–473. [CrossRef]

14. Hupez, M.; De Greve, Z.; Vallee, F. A qualitative comparison of non-sequential and sequential approaches for low voltage network stochastic analyses. In Proceedings of the 2017 52nd International Universities Power Engineering Conference (UPEC), Heraklion, Greece, 28–31 August 2017; pp. 1–6.

15. Nijhuis, M.; Gibescu, M.; Cobben, S. Gaussian Mixture Based Probabilistic Load Flow for LV-Network Planning. *IEEE Trans. Power Syst.* **2017**, *32*, 2878–2886. [CrossRef]

16. Mokryani, G.; Majumdar, A.; Pal, B.C. Probabilistic method for the operation of three-phase unbalanced active distribution networks. *IET Renew. Power Gener.* **2016**, *10*, 944–954. [CrossRef]

17. Nguyen, P.H.; Blaauwbroek, N.; Nguyen, C.; Zhang, X.; Flueck, A.; Wang, X. Interfacing applications for uncertainty reduction in smart energy systems utilizing distributed intelligence. *Renew. Sustain. Energy Rev.* **2017**, *80*, 1312–1320. [CrossRef]

18. Zhou, G.; Bo, R.; Chien, L.; Zhang, X.; Yang, S.; Su, D. GPU-Accelerated Algorithm for On-line Probabilistic Power Flow. *IEEE Trans. Power Syst.* **2017**, *33*, 1132–1135. [CrossRef]

19. Hagh, M.T.; Amiyan, P.; Galvani, S.; Valizadeh, N. Probabilistic load flow using the particle swarm optimisation clustering method. *IET Gener. Transm. Distrib.* **2018**, *12*, 780–789. [CrossRef]

20. Ni, F.; Nguyen, P.; Cobben, J.F.G. Basis-Adaptive Sparse Polynomial Chaos Expansion for Probabilistic Power Flow. *IEEE Trans. Power Syst.* **2017**, *32*, 694–704. [CrossRef]

21. Fan, M.; Vittal, V.; Heydt, G.T.; Ayyanar, R. Probabilistic power flow studies for transmission systems with photovoltaic generation using cumulants. *IEEE Trans. Power Syst.* **2012**, *27*, 2251–2261. [CrossRef]

22. Huang, Y.; Xu, Q.; Yang, Y.; Jiang, X. Numerical method for probabilistic load flow computation with multiple correlated random variables. *IET Renew. Power Gener.* **2018**, *12*, 1295–1303. [CrossRef]

23. Mocanu, E.; Nguyen, P.H.; Gibescu, M.; Kling, W.L. Deep learning for estimating building energy consumption. *Sustain. Energy, Grids Netw.* **2016**, *6*, 91–99. [CrossRef]

24. Baliyan, A.; Gaurav, K.; Kumar Mishra, S. A review of short term load forecasting using artificial neural network models. *Procedia Comput. Sci.* **2015**, *48*, 121–125. [CrossRef]

25. Matallanas, E.; Castillo-Cagigal, M.; Gutiérrez, A.; Monasterio-Huelin, F.; Caamaño-Martín, E.; Masa, D.; Jiménez-Leube, J. Neural network controller for Active Demand-Side Management with PV energy in the residential sector. *Appl. Energy* **2012**, *91*, 90–97. [CrossRef]

26. Ciric, R.M.; Feltrin, A.P.; Ochoa, L.F. Power flow in four-wire distribution networks-general approach. *IEEE Trans. Power Syst.* **2003**, *18*, 1283–1290. [CrossRef]

27. Blaauwbroek, N.; Nguyen, P.; Slootweg, H. Applying demand side management using a generalised three phase grid supportive approach. In Proceedings of the 2017 IEEE International Conference on Environment and Electrical Engineering (EEEIC), Milan, Italy, 6–9 June 2017; pp. 1–6.

28. Hong, T.; Fan, S. Probabilistic electric load forecasting: A tutorial review. *Int. J. Forecast.* **2016**, *32*, 914–938. [CrossRef]

29. Wang, Y.; Zhang, N.; Chen, Q.; Kirschen, D.S.; Li, P.; Xia, Q. Data-Driven Probabilistic Net Load Forecasting With High Penetration of Behind-the-Meter PV. *IEEE Trans. Power Syst.* **2018**, *33*, 3255–3264. [CrossRef]

30. Bo, R.; Li, F. Probabilistic LMP Forecasting Considering Load Uncertainty. *IEEE Trans. Power Syst.* **2009**, *24*, 1279–1289. [CrossRef]

31. Primadianto, A.; Lu, C.-N. A Review on Distribution System State Estimation. *IEEE Trans. Power Syst.* **2017**, *32*, 3875–3883. [CrossRef]

32. Borghetti, A.; Bosetti, M.; Grillo, S.; Massucco, S.; Nucci, C.A.; Paolone, M.; Silvestro, F. Short-term scheduling and control of active distribution systems with high penetration of renewable resources. *IEEE Syst. J.* **2010**, *4*, 313–322. [CrossRef]

33. Hagan, M.T.; Demuth, H.B.; Beale, M.H. *Neural Network Design*; PWS Pub: Boston, MA, USA, 1996; ISBN 0534943322.

34. Viyathukattuva Mohamed Ali, M.M.; Nguyen, P.H.; Kling, W.L.; Chrysochos, A.I.; Papadopoulos, T.A.; Papagiannis, G.K. Fair power curtailment of distributed renewable energy sources to mitigate overvoltages in low-voltage networks. In Proceedings of the 2015 IEEE Eindhoven PowerTech, Eindhoven, The Netherlands, 29 June–2 July 2015; pp. 1–5.

35. Torbaghan, S.S.; Blaauwbroek, N.; Nguyen, P.; Gibescu, M. Local market framework for exploiting flexibility from the end users. In Proceedings of the International Conference on the European Energy Market (EEM), Porto, Portugal, 6–9 June 2016; pp. 1–6.
36. Blaauwbroek, N.; Nguyen, P.H. Optimal resource allocation and load scheduling for a multi-commodity smart energy system. In Proceedings of the 2015 IEEE Eindhoven PowerTech, Eindhoven, The Netherlands, 29 June–2 July 2015; pp. 1–6. [CrossRef]
37. Blaauwbroek, N.; Bosch, R.; Nguyen, P.; Slootweg, H. Three-phase grid supportive demand side management with appliance flexibility modelling. In Proceedings of the 2018 IEEE International Conference on Environment and Electrical Engineering (EEEIC), Palermo, Italy, 12–15 June 2018.
38. Xie, J.; Hong, T.; Laing, T.; Kang, C. On Normality Assumption in Residual Simulation for Probabilistic Load Forecasting. *IEEE Trans. Smart Grid* **2017**, *8*, 1046–1053. [CrossRef]
39. Pecan Street Inc. Energy Research. Available online: http://www.pecanstreet.org/ (accessed on 11 August 2017).

Article

New Monitoring System for Photovoltaic Power Plants' Management

Václav Beránek [1], Tomáš Olšan [2], Martin Libra [2,*], Vladislav Poulek [2], Jan Sedláček [2], Minh-Quan Dang [2] and Igor I. Tyukhov [3]

[1] Solarmonitoring, Ltd., 14700 Prague, Czech Republic; beranek@solarmon.eu
[2] Department of Physics, Czech University of Life Sciences Prague, Kamycka 129, 16500 Prague, Czech Republic; tolsan@email.cz (T.O.); poulek@tf.czu.cz (V.P.); sedlacek@tf.czu.cz (J.S.); dang@tf.czu.cz (M.-Q.D.)
[3] Department of Mechanical Engineering, San Jose State University, One Washington Square, San Jose, CA 95192-0087, USA; ityukhov@yahoo.com
* Correspondence: libra@tf.czu.cz; Tel.: +420-224383284

Received: 4 July 2018; Accepted: 17 September 2018; Published: 20 September 2018

Abstract: An innovative solar monitoring system has been developed. The system aimed at measuring the main parameters and characteristics of solar plants; collecting, diagnosing and processing data. The system communicates with the inverters, electrometers, metrological equipment and additional components of the photovoltaic arrays. The developed and constructed long working system is built on special data collecting technologies. At the generating plants, a special data logger BBbox is installed. The new monitoring system has been used to follow 65 solar plants in the Czech Republic and elsewhere for 175 MWp. As an example, we have selected 13 PV plants in this paper that are at least seven years old. The monitoring system contributes to quality management of plants, and it also provides data for scientific purposes. Production of electricity in the built PV plants reflects the expected values according to internationally used software PVGIS (version 5) during the previous seven years of operation. A comparison of important system parameters clearly shows the new solutions and benefits of the new Solarmon-2.0 monitoring system. Secured communications will increase data protection. A higher frequency of data saving allows higher accuracy of the mathematical models.

Keywords: solar monitoring system; photovoltaic array; energy management

---

1. Introduction

Growing interest and investments in photovoltaic (PV) technology as a consequence of increased lifespan and efficiency, the decrease of PV modules' price and the reduction of the environmental impact of solar systems in comparison to traditional fossil fuel technologies have led to many PV systems and big solar plants being installed.

Monitoring of PV system plants is an urgent and imperative activity for practical implementation of new ecologically clean solar plants due to the information that allows owners to maintain, operate and control these systems, to reduce maintenance costs and to avoid unwanted electric power disruptions. Different monitoring systems with various requirements have been reviewed and described in [1].

The functioning of solar photovoltaic arrays obviously requires having high quality real-time measurements (monitoring) and a data storing system for several essential parameters; see also, for example, [2–5]. A wide variety of the control and monitoring systems for photovoltaic arrays was discussed in [6–14].

The works [15,16] described Lab-VIEW software, maximum power point tracking algorithms, a monitoring system with the implementation of metrological data and solar irradiation with the help of satellites, which were employed for displaying, storing and processing the obtained monitoring data, which enabled forecasting of sunny and cloudy hours. In the future, more and more PV systems will be integrating with other renewable systems and GIS systems designed for optimal collecting and distributing of electrical power nets according to the consumer demand.

The work [17] discussed monitoring of the damaged PV solar modules by a drone. The main reasons for the failures of PV solar modules were summarized, for instance, in the report [18].

One more important example of a real-time monitoring system is rMeter™, developed by Loren Kallevig and Ron Swenson [19]. This site presents useful information on PV systems for Northern California on the web. All data are collected and stored on a server with backup. Data are collected by the data loggers and periodically pushed to the central server, where they may be accessible to the public or password protected with restricted access. These are some of the components that go into a typical data monitoring installation. Optional items such as wireless communication and a display kiosk are also included. Some owners and clients choose to display their rMeter™ data on-site to inform their employees about their assurance of energy independence. This is the paradigm of a self-contained display station.

Solar radiation monitoring is an important part for any monitoring system. The University of Oregon Solar Radiation Monitoring Laboratory is a solar radiation data center, whose goal is to provide important solar resource data for planning, designing, deployment and operation of solar electric facilities in the Pacific Northwest of the USA [20]. Creating the short- and long-term solar radiation database necessary to achieve this goal requires persistence, the maintenance of high standards and an effort to inform and educate people about the importance of a solar radiation database and how to process the versatile data. Both Internet resources [19,20] are very useful for educational activities, particularly for students of different universities in CA, like San Jose State University.

The paper [21] discussed photovoltaic module performance measurements' traceability and mentioned that improved uncertainty is a key factor for the fast-growing PV market and has a huge impact on the economy and environment.

In solar power generation systems, the existence of faults in any PV module of the array may lead to reductions in overall power generation [22]. This paper considered testing results used to ascertain the feasibility of the photovoltaic generation fault diagnosis meter. However, it is very suitable for development only into a portable solar power generation system fault and flaw diagnosis meter.

The goal of the current paper is to describe experience in creating a new expensive monitoring system. We start from the work described in [23]. In recent years, numerous photovoltaic arrays had been installed in the Czech Republic. At the beginning of 2009, there were approximately 50 MWp of PV photovoltaic arrays and systems, and only after one year in early 2010, there were 460 MWp installed. Furthermore, in early 2011, this reached 2000 MWp. The generated energy from photovoltaic arrays has subsided since 2006. However, until the period between 2007 and 2008, the dramatic decrease in the cost of PV solar modules led to the reduction of photovoltaic arrays and boosted the investment return. The boom lasted for three years from 2008 till the end of 2010, when the drop of government subsidies put an end to this boom. A similar circumstance has happened in several other European countries since the beginning of this century.

This article focuses on designing and testing a unique monitoring system with third generation software for photovoltaic arrays with the cooperation of Solarmonitoring ltd. and the Faculty of Engineering of the Czech University of Life Sciences Prague.

Instead of the usual programmable logic controller (PLC) hardware system, an innovative system using dynamic-link library (DLL) is created. Therefore, the monitoring system is not fixed on a single specific type of hardware. The DLL-based monitoring system can be used on different types of computers with different operating systems (Microsoft, Unix, etc.).

In the paper, data from 13 photovoltaic arrays during seven-year periods are presented. All photovoltaic arrays are equipped with PV solar modules placed on the ground-mounting system and face south (see below). Photovoltaic arrays with the sun-tracking mounting system [24,25] and systems with the concentration of solar radiation were also installed [26,27]. On those photovoltaic arrays, a monitoring system was also installed. However, data used for the comparison were taken from the PV system with a similar construction only. Data were compared with the predicted amount of generated energy for specific locations and photovoltaic array constructions.

The described monitoring systems were installed at 65 photovoltaic plants in total in Czechia and elsewhere (Slovakia, Romani, Chile); see, for example [28].

2. Explanation of the Monitoring System Solarmon-2.0

The new monitoring system named Solarmon-2.0 is intended for collecting, diagnosing and processing data. The system communicates with the inverters, electrometers, metrological equipment and external components of the photovoltaic arrays. The whole system is developed and built on special data collecting technology. At the photovoltaic array, a special data logger BBbox is installed. This device has similar functions as the devices used by inverter manufactures DELTA, SMA, Vacon, ABB and DANFFOS (Solar-Log, etc.). On the big photovoltaic arrays, long-distance communication technology has to be used. The data logger is programmed for standard protocols RS485, MODbus and SCADA. This is the most ideal and cost-effective solution for data collection. BBbox is able to communicate with many devices and sensors in the photovoltaic plant (Figure 1) with a responding time from 1–3 s and for distances up to 1200 m. Data signals that are connected to BBbox include outputs from the electric meters on the low voltage (LV) and high voltage (HV) side of the transformer and outputs from inverters of any kind (SMA, Fronius, Kaco, Delta, etc.), as well as a security system such as fire sensor, moving sensors, metrological sensors, thermometers, thermocouples, sensors measuring wind direction and speed and meters measuring irradiance.

Data downloaded to BBox from inverters, substations and other devices are generally processed, stored and properly labeled. Data are sent over a secure SSL channel to the database server Solarmon-2.0. The database can receive and process data from different types of data loggers (see Figure 1). BBbox can exchange information with another BBbox. This opportunity is mainly used in photovoltaic arrays with nominal power above 5 MW_p and with a large working area. Communication and transmission of data are mostly done using optical fiber.

Solarmon-2.0 receives raw data from inverters or substations, and according to their priority, the data are categorized and assigned to the appropriate sub-database. The triggers are software algorithms performing regular operations related to the defined time. The operations are for example comparison of the energy and power at inverters and strings. Individual triggers, respectively complex and sophisticated patterns, place the data for individual diagnostic tests. The developed system uses tests for calculating faults and errors, for comparison of inverters and for control of voltage on each branch to evaluate efficiency, energy losses on each individual device and to predict the power from the system. The processed output data are displayed on two types of interface: service and user.

Service interface: This is designed in more detail for service technicians, who require more comprehensive supervision over the photovoltaic array. The service mode identifies and defines the causes of the defects of described errors and the events that have occurred. The service interface allows specifying the parameters of the solar plant, which are an essential part of the algorithm (see Figure 2). The system could define the responsibility for the serviceman of a specific photovoltaic array and set up emails and phone numbers for sending the notice about errors, reports and interruptions. Furthermore, technical specifications such as the type of each panel, inverter and substation could be specified. The limit value of temperature, temperature coefficients and other input data are used to predict the performance of PV solar modules. The chart of the predicted performance is derived from the actual values recorded from the metrological station (temperature + irradiance measuring device of the panel), user-defined values of the active solar modules' area (area of silicon wafers), the performance of the solar modules, the tolerance of the solar modules, the performance characteristics of the given type of solar modules, the temperature coefficient, the number of solar modules in the string, the number of strings on the inverter and the number of inverters.

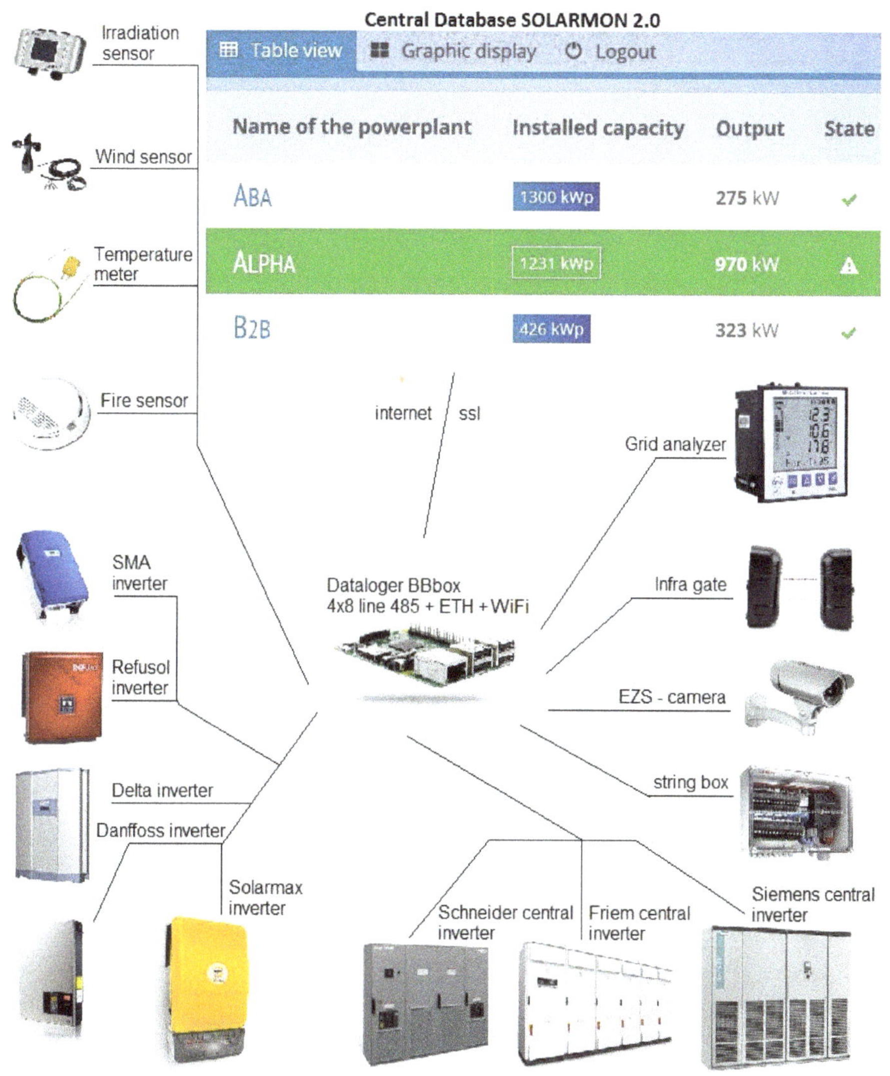

Figure 1. Scheme of the monitoring system Solarmon-2.0.

The interface for users was developed particularly for investors, bankers, insurance companies and manufacturers of solar modules. The detailed performance is shown for the owners of the photovoltaic arrays. Specifically, the current performance of inverters is shown. It also demonstrates the daily supply and energy consumption, which is an important basis for billing; there are also a monthly invoice template, reports of individual inverters, the calculation of earnings and emission allowances. There is a window called "Failure", which indicates and sends reports of serious faults and errors in the solar plant.

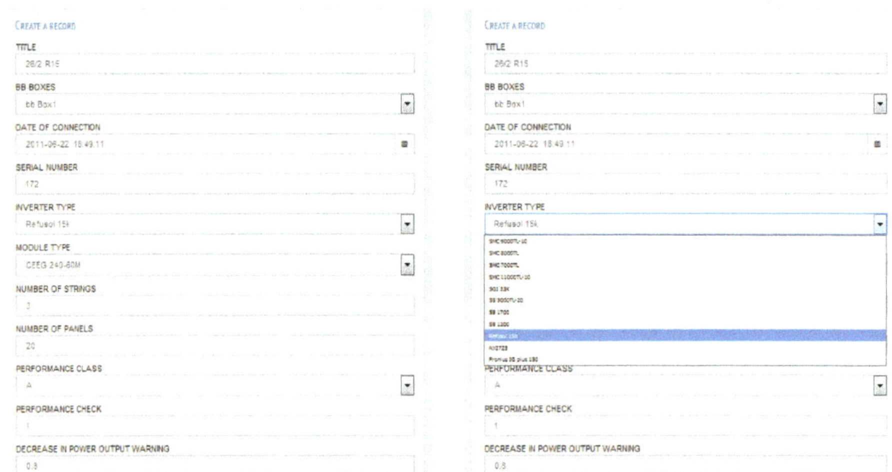

Figure 2. Service interface of the monitoring system Solarmon-2.0.

The database system Solarmon-2.0 allows one to track the current status of the individual inverters, including the possibility of a description of the fault or error related to a project number. Weather equipment monitoring provides information on wind and temperature, which is an essential basis for insurance companies (for instance, if the wind speed exceeds the hurricane level, it means the investors will lose the guarantee on a certain type of construction). Most of the solar plant mechanical structures are designed for wind speeds up to 190 km/h. According to international standard ENV 1991-2-4 "Basis of design and actions on structures-Part 2–4: Actions on structures-Wind actions" in the territory of Central Europe, construction must deal with the wind at ground level of a max speed equal to 160 km/h. A similar issue is related to the temperature; e.g., PV solar modules from CEEG Company (as well as the majority of PV solar modules worldwide) have operating temperatures from -40–$+90\,^{\circ}$C. The operating temperature range is determined mostly for encapsulation of PV cells in standard PV solar modules using ethylene vinyl acetate (EVA). A larger operating temperature range can be achieved by using a special technology for encapsulation of PV cells with polysiloxane gel. This technology is described for example in [29] for the PV solar modules and in [30,31] for Photovoltaic/Photothermal solar modules. The report [32] is related to monitoring of PV panel temperature. Whenever the PV panel temperature exceeds the operating limit, it may cause permanent damage, thereby automatically voiding the warranty. Solarmon-2.0 has the function of controlling defective invertors. This is by the comparison of the average values of actual powers (see Figure 3). Additionally, it measures and monitors voltages in each string and compares them with the limit values. The resulting records are reported and also trackable. Reports of significant failures are recorded and sent via SMS and e-mail to the responsible person or investor.

TABULAR DATA DISPLAY SETTINGS

DATE	TYPE	VALUE	
2017-03-16	From the mean value	1	Set

TABLE VIEW OF FAULTS

Inverter's name	Type of inverter	Module type	Daily energy	Efficiency	
22/205 S7 C	SMC 7000TL	CEEG 230 Wp	24.6	0.7 (67%)	error
14/2 S1,1	SB 1200	CEEG 175 Wp	4.7	0.8 (83%)	error
15/3 S1,1	SB 1200	CEEG 175 Wp	4.8	0.9 (85%)	error
12/37 F15 C	Fronius IG plus 150	CEEG 180 Wp	60.0	0.9 (86%)	error
24/12 S1,1	SB 1200	CEEG 175 Wp	4.9	0.9 (87%)	error
12/1 S1,1	SB 1200	CEEG 175 Wp	5.0	0.9 (90%)	
20/178 S7 C	SMC 7000TL	CEEG 175 Wp	35.8	0.9 (91%)	
24/3 S1,7 CANADIAN	SB 1700	C56P-230	9.1	0.9 (92%)	

Figure 3. Data output of the monitoring system Solarmon-2.0.

3. Determination of Power Losses

Losses of power can be defined in several ways. The determination of losses on the DC side of the inverter is the most important and most complicated. This is a specially developed algorithm that defines the current performance of the solar modules. Input parameters are the temperature of PV solar modules and the immediate value of the irradiance falling on the plane parallel with the solar modules. Information about the type of PV solar modules used is important to improve the accuracy of prediction. The power temperature coefficient predicts the panel behavior under certain temperature conditions. This value is based on the detection conditions of the manufacturer (most often indicating 25 °C, irradiance 1000 W/m^2, $t = 1$ s, but in some cases, the temperature is set at 300 K, which is 27 °C). The estimation of the active area of the panel is another factor. The next important parameters are the number of PV cells per the PV panel (usually 36 or 72), the number of parallel branches (usually 3 on 12, 3 on 24), the size and arrangement of the PV cells (pseudosquare) and output terminals. Finally, it is important to minimize the power variation of the solar modules (the manufacturer proclaims variation of the PV solar modules at a range of mostly around 3%). Variation can be minimized by sorting solar modules into groups according to the maximum power point current Impp (aiming to achieve maximum power in serial parallel connection). Modules are classified into four groups A–D (A = highest Impp). The sorting of PV solar modules is related to the implementation project. Then, the setting of the Solarmon-2.0 individual drives is done while putting the photovoltaic array into operation. Therefore, there is a map of each group of solar modules connected with the project. Defining the operation point MPP (maximum power point) of PV solar modules in the string is the most complicated part. V-I (Volt-Ampere) characteristics of individual PV solar modules are a little different. This is the place for further improving and subsequent calculation of prediction performance (depending on the temperature and incident energy) of one concrete type of PV solar module.

To determine the losses in DC cables, the scheme shown in Figure 4 should be used. For instance, the inverter SB 1200 has one string with 13 modules of Type CEEG 175. The measurement in front of inverters is performed at Point B. By comparing the actual performance of concrete power A and concrete drive B (B/A), this is counted as lost on the DC line. The measurement behind inverters is performed at the point C (Pac (W) from the inverter). It is therefore an alternative current output. When comparing the actual inverter output power to the input, we get the efficiency drive. From the data, we can declare that 90% load of the inverter efficiency corresponds to an accuracy of ±3%. There is a rule saying that the reduction of load inverters increases the inaccuracy of measurement. Measurement of input in the transformer with the calibrated meter point is given by D. By comparing the points D and C, the sum indicates loss on the AC side. Comparing output energy via an opto-element E, comparing the values of E/D, we get the efficiency of the transformer. Electricity connection is important also in terms of billing. Finally, the overall efficiency of the photovoltaic array can be determined by comparing the sum of E/A. The overall efficiency of the solar plant is the most important basis for the banks as lenders.

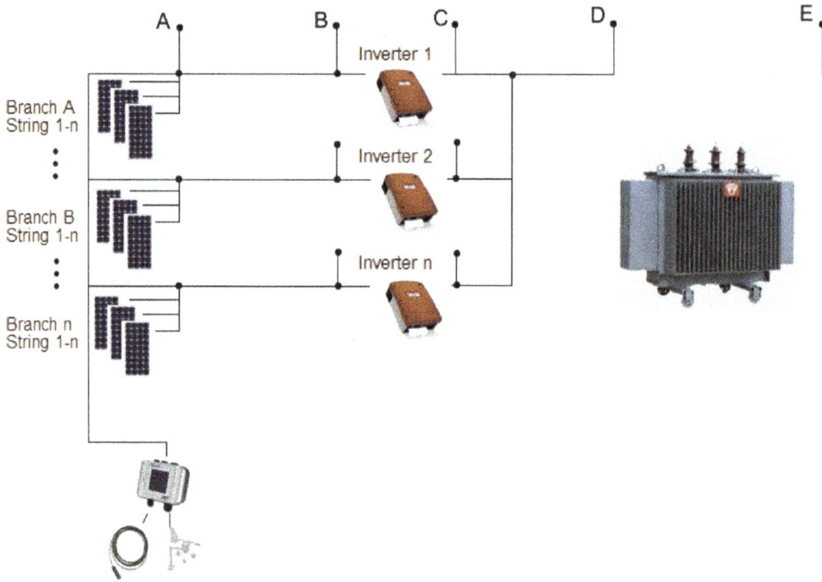

Figure 4. Scheme of the monitoring points.

4. Saving of Data and Archive

An integral part of the whole system is the back up and reviewing of the data. Solarmon-2.0 can support the database and create a preview of these files. It is a statement of values under the specified time. The data archive has four subsections: inverters, Meteo station, substation, electronic security system. Considering the data are divided into three types: daily, monthly and yearly classification, it is possible to extract data on specific drivers and summarize the overall data. Data from individual drives are very much needed to allocate the existing errors and faults. All categories of time (daily, monthly, annual) are further divided into seconds and hours and summarized. The files are saved as XLS files; therefore, it is also possible to download them.

5. Results and Discussion

The Solarmon-2.0 monitoring system as designed and described above had been installed and tested in 65 photovoltaic arrays. The testing helped to identify and eliminate installation errors and contributed to the failure-free operation and exploitation of the photovoltaic arrays. As an example, we present and discuss data from 13 selected photovoltaic arrays in this paper. These photovoltaic arrays were equipped with the new monitoring system. The data have been recorded and evaluated for the seven-year period. All these photovoltaic arrays had PV solar modules based on crystalline silicon and installed on a fixed mounting system with an angle of 35 degrees to the south. Such a fixed inclination is approximately 50° north latitude optimal in terms of the maximum annual yield of electricity. Figure 5 shows the locations where photovoltaic arrays were placed. Figure 6 shows the example of the photovoltaic array at Měnín.

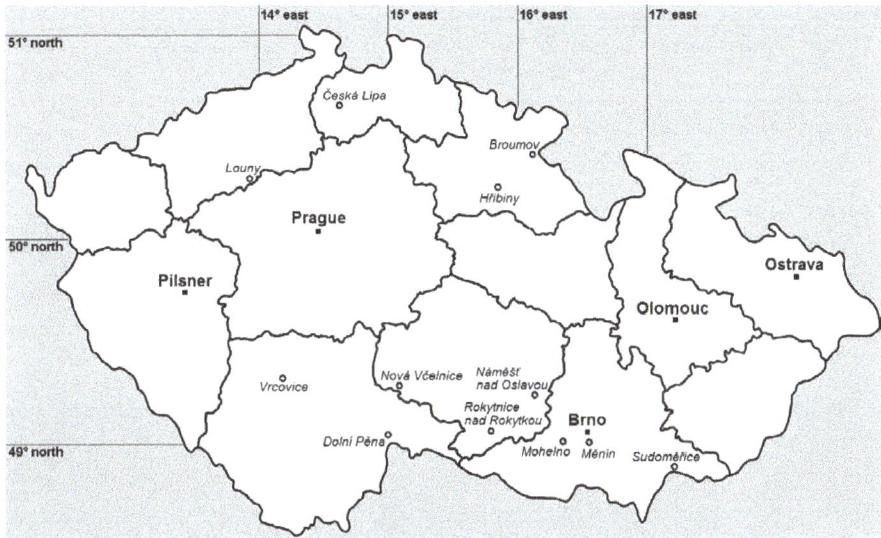

Figure 5. Position of the selected photovoltaic arrays.

Figure 6. Photovoltaic array at Měnín.

An example of the data evaluation is given in Figure 7. The figure shows the graph of generated electrical energy for each month during two selected years. The summary annual value of produced electricity is also included. The values were recalculated for 1 kW of installed nominal output photovoltaic arrays to be able to objectively compare the values. It is seen that during the observation period, the operation does not show a decrease in production of electric energy due to the aging of the PV solar modules. However, the variation of weather conditions had a strong impact on energy production. The observation period of seven years is still relatively short compared with an expected lifetime of the PV solar modules of 25 years. It is likely that the effect of the aging of the PV solar modules will eventually occur, and a decrease in energy conversion efficiency would be significantly reflected in the second half-life of the PV modules (see [33]).

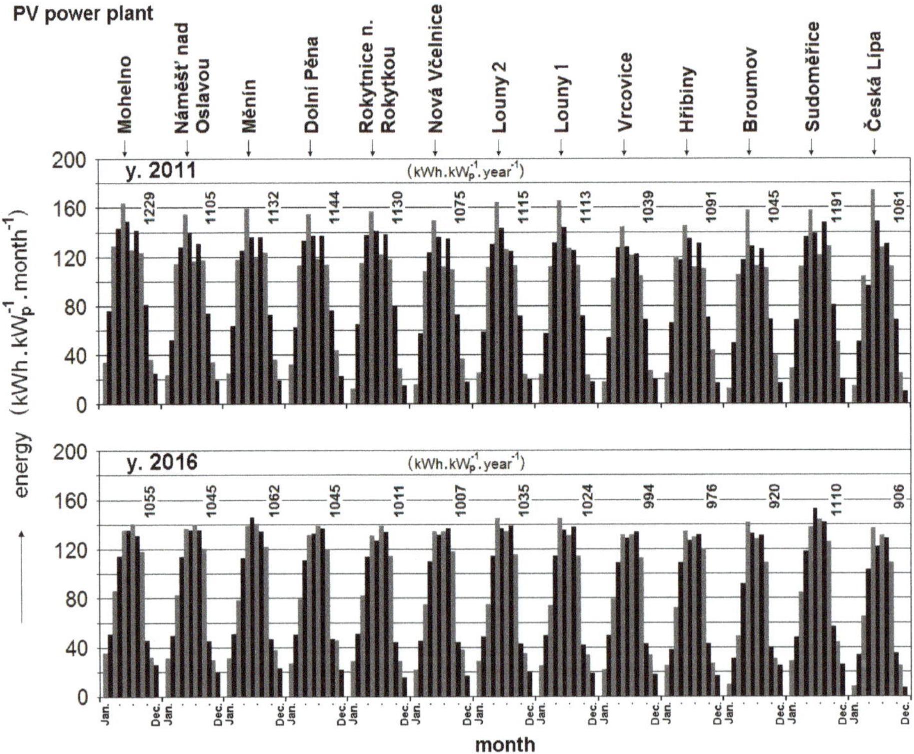

Figure 7. An example of the data evaluation from the selected PV power plants during the years 2011 and 2016; the graph of generated electrical energy for each month.

Table 1 shows the comparison between predicted and actual values of generated electrical energy from the selected photovoltaic arrays. The expected amount was determined in accordance with the internationally used software PVGIS [34] and the actual energy is calculated as the average of seven years of results (actual values for each year are shown in Table 2). It is seen that in all the above PV, the power amount of produced electric energy corresponded to the prediction; in all cases, the actual values were higher by 10–20%. It can be assumed that due to the aging of PV solar modules, the amount of electricity produced would decrease by 10–20% within 20 years of operation, which is common; see, e.g., [35,36].

Table 1. Comparison between predicted and actual values of generated electric energy from selected photovoltaic arrays in the Czech Republic.

Photovoltaic Array (Fixed Support Structures, Tilt Angle 35° Southwards)	Installed Peak Power (kW$_p$)	Technology	Prediction of Annual el. Energy Production According to PVGIS (kWh·kW$_p$$^{-1}$·year^{-1})	Average Annual el. Energy Production during Years 2011–2017 (kWh·kW$_p$$^{-1}$·year^{-1})	Difference (%)
Mohelno	1520	SMA	934	1109	+18.7
Náměšť nad Oslavou	513	SMA	926	1069	+15.4
Měnín	3274	SMA	932	1074	+15.2
Dolní Pěna	1000	KACO	921	1068	+16.0
Rokytnice n. Rokytkou	1000	SMA	935	1058	+13.2
Nová Včelnice	2400	VACON	917	1017	+10.9
Louny 1	120	SMA	875	1051	+20.1
Louny 2	500	SMA	872	1032	+18.3
Vrcovice	1750	VACON	891	1006	+12.9
Hřibiny	699	SMA	900	1010	+12.2
Broumov	426	VACON	899	981	+9.1
Sudoměřice	3568	SMA	959	1097	+14.4
Česká Lípa	1300	VACON	877	964	+9.9

Table 2. Annual electric energy production from selected photovoltaic arrays in the Czech Republic (an example of the evaluation of monitored data).

Photovoltaic Array (Fixed Support Structures, Tilt Angle 35° Southwards)	Annual el. Energy Production (kWh·kW$_p$$^{-1}$·year^{-1})						
	2011	2012	2013	2014	2015	2016	2017
Mohelno	1229	1204	1060	1005	1101	1055	1106
Náměšť nad Oslavou	1105	1151	1016	992	1081	1045	1094
Měnín	1132	1133	1002	1008	1084	1062	1093
Dolní Pěna	1144	1111	997	1020	1071	1045	1088
Rokytnice n. Rokytkou	1130	1138	1006	988	1070	1011	1064
Nová Včelnice	1075	1011	963	980	1037	1007	1046
Louny 1	1115	1081	959	985	1100	1035	1084
Louny 2	1113	1069	924	972	1092	1024	1028
Vrcovice	1039	1035	899	980	1043	994	1050
Hřibiny	1091	1031	902	1018	1078	976	975
Broumov	1045	1044	897	953	1020	920	990
Sudoměřice	1191	1012	1025	1091	1123	1110	1126
Česká Lípa	1061	1050	884	919	988	906	941

Table 3 shows an example of the comparison of parameters between the widely-used monitoring system Solar-Log 1200 and the newly-developed system Solarmon-2.0. The system Solar-Log 1200 was selected for comparison because it is widely used at present.

Table 3. Comparison of parameters between the widely-used monitoring system Solar-Log 1200 and the newly-developed system Solarmon-2.0.

	Solarmon BB Box	Solar-Log 1200
Hardware comparison	4 × 8 communication lines price ca 500 Euro communication protocols MODBUS, SCADA, RS485 secured communication	2 communication lines price ca 1000 Euro communication protocol RS485 no secured communication
	Solarmon-2.0	**Solar-Log Server**
Software comparison	responsive web design inverters' data measurement Meteo data measurement stringbox data measurement grid analyzer data measurement electric meter data measurement battery data measurement data saving each 3 s (better accuracy of the mathematical model) service log is integrated inside the monitoring system, automatic error reports, error location and online connection between Solarmon-2.0 and service engineer (advantage for large photovoltaic arrays)	special application for web, Android, Apple inverters' data measurement Meteo data measurement stringbox data measurement no grid analyzer data measurement electric meter data measurement battery data measurement data saving each 60 s (lower accuracy of the mathematical model) service log is not integrated inside the monitoring system

6. Conclusions

A new monitoring system was successfully designed and tested on 65 photovoltaic arrays in the Czech Republic and several other countries. The monitoring system contributes to quality management of plants and also provides data for scientific purposes.

The production of electricity in the photovoltaic array built by our projects reflects the expected values according to the internationally used software PVGIS during the previous six years of operation. Data evaluation and counting of the predicted decline in energy conversion efficiency of PV solar modules during the aging process make the assumption that electricity production will comply with expected values for the lifetime of all these photovoltaic arrays.

The monitoring system helped to identify and eliminate installation errors and contributed to the continuous operation of the photovoltaic arrays. If reduced output power were indicated by the monitoring system, the servicemen and engineers would check the PV plant on the spot and fix the errors. Experience with the operation will be applied to the further improvement on the new version of the monitoring system Solarmon-2.0.

The measurements show that even with degradation of photovoltaic modules by 1% per year, there is a slight increase in the energy produced from photovoltaic sources in Central Europe. The increase is due to a higher number of sun hours and little precipitation.

A comparison of important system parameters clearly shows the new solutions and benefits of the new Solarmon-2.0 monitoring system. The increased number of input lines, especially the service log, automatic error reports and their exact localization, will be appreciated by operators of large photovoltaic arrays. Secured communications will increase data protection. A higher frequency of data saving allows a higher accuracy of the mathematical models. It is also advantageous to collect additional data from other power plant components (grid analyzer and invoicing measurement).

Solarmon-2.0 uses so-called responsive web design: the system adapts to each display. Solarmon-2.0 software is available at: app.solarmon.eu (user name: demo, password: demo1234).

Author Contributions: V.B. (15%), design of solar monitoring system, Software, Visualization; T.O. (15%), data processing and analysis, on site measurements, Formal Analysis; M.L. (14%), data processing and analysis, article editing, Methodology, Data Curation, Writing-Review & Editing, Supervision, Funding Acquisition; V.P. (14%), data processing and analysis, article editing, Conceptualization, Writing-Original Draft Preparation, Project Administration; J.S. (14%), data processing and analysis, Investigation; M.-Q.D. (14%), data processing and analysis, on site measurements, Resources; I.I.T. (14%), data processing and analysis, Validation.

Funding: This research was funded by Czech University of Life Sciences Prague, grant number IGA 2018:31120/1312/3118.

Conflicts of Interest: The authors declare no conflicts of interest.

References

1. Rahman, M.M.; Selvaraja, J.; Rahima, N.A.; Hasanuzzamana, M. Global modern monitoring systems for PV based power generation: A review. *Renew. Sustain. Energy Rev.* **2018**, *82*, 4142–4158. [CrossRef]
2. Olalla, C.; Maksimovic, D.; Deline, C.H.; Martinez-Salamero, L. Impact of distributed power electronics on the lifetime and reliability of PV systems. *Prog. Photovolt. Res. Appl.* **2017**, *25*, 821–835. [CrossRef]
3. Otani, K.; Kato, K.; Takashima, T.; Yamaguchi, T.; Sakuta, K. Field Experience with Large-scale Implementation of Domestic PV Systems and with Large PV Systems on Buildings in Japan. *Prog. Photovolt. Res. Appl.* **2004**, *12*, 449–459. [CrossRef]
4. Reinders, A.H.M.E.; van Dijk, V.A.P.; Wiemken, E.; Turkenburg, W.C. Technical and Economic Analysis of Grid-connected PV Systems by Means of Simulation. *Prog. Photovolt. Res. Appl.* **1999**, *7*, 71–82. [CrossRef]
5. Guerriero, P.; Di Napoli, F.; Vallone, G. Monitoring and Diagnostics of PV Plants by a Wireless Self-Powered Sensor for Individual Solar modules. *IEEE J. Photovolt.* **2016**, *6*, 286–294. [CrossRef]
6. Ayompe, L.M.; Duffy, A.; McCormack, S.J.; Conlon, M. Measured performance of a 1.72 kW rooftop grid connected photovoltaic system in Ireland. *Energy Convers. Manag.* **2011**, *52*, 816–825. [CrossRef]
7. Boonmee, C.H.; Plangklang, B.; Watjanatepin, N. System performance of a three-phase PV-grid-connected system installed in Thailand: Data monitored analysis. *Renew. Energy* **2009**, *34*, 384–389. [CrossRef]

8. Madeti, S.R.; Singh, S.N. Monitoring system for photovoltaic plants: A review. *Renew. Sustain. Energy Rev.* **2017**, *67*, 1180–1207. [CrossRef]

9. Ventura, C.; Tina, G.M. Utility scale photovoltaic plant indices and models for on-line monitoring and fault detection purposes. *Electr. Power Syst. Res.* **2016**, *136*, 43–56. [CrossRef]

10. Yahyaoui, I.; Segatto, M.E.V. A practical technique for on-line monitoring of a photovoltaic plant connected to a single-phase grid. *Energy Convers. Manag.* **2017**, *132*, 198–206. [CrossRef]

11. Ma, S.; Chen, M.; Wu, J.; Huo, W.; Huang, L. Augmented Nonlinear Controller for Maximum Power-Point Tracking with Artificial Neural Network in Grid-Connected Photovoltaic Systems. *Energies* **2016**, *9*, 1005. [CrossRef]

12. Liberos, M.; González-Medina, R.; Garcerá, G.; Figueres, E. Modelling and Control of Parallel-Connected Transformerless Inverters for Large Photovoltaic Farms. *Energies* **2017**, *10*, 1242. [CrossRef]

13. Zhao, J.; Zhou, X.; Ma, Y.; Liu, Y. Analysis of Dynamic Characteristic for Solar Arrays in Series and Global Maximum Power Point Tracking Based on Optimal Initial Value Incremental Conductance Strategy under Partially Shaded Conditions. *Energies* **2017**, *10*, 120. [CrossRef]

14. Han, Y.; Chen, W.; Li, Q. Energy Management Strategy Based on Multiple Operating States for a Photovoltaic/Fuel Cell/Energy Storage DC Microgrid. *Energies* **2017**, *10*, 136. [CrossRef]

15. Rezk, H.; Tyukhov, I.; Raupov, A. Experimental implementation of meteorological data and photovoltaic solar radiation monitoring system. *Int. Trans. Electr. Energy Syst.* **2015**, *25*, 3573–3585. [CrossRef]

16. Tyukhov, I.I.; Rezk, H.; Vasant, P. Modern Optimization Algorithms and Applications in Solar Photovoltaic Engineering. In *Sustaining Power Resources through Energy Optimization and Engineering Pandian Vasant and Nikolai Voropai*; IGI Global: Hershey, PA, USA, 2016; Chapter 16; pp. 390–445, ISBN 978-1466697553.

17. Grimaccia, F.; Leva, S.; Dolara, A. Survey on PV Modules' Common Faults After an O&M Flight Extensive Campaign Over Different Plants in Italy. *IEEE J. Photovolt.* **2017**, *7*, 810–816.

18. Villarini, M.; Cesarotti, V.; Alfonsi, L.; Introna, V. Optimization of photovoltaic maintenance plan by means of a FMEA approach based on real data. *Energy Convers. Manag.* **2017**, *152*, 1–12. [CrossRef]

19. Energy Monitoring—Solar Results to the Web. Available online: http://www.rmeter.com/ (accessed on 10 May 2018).

20. UO Solar Radiation Monitoring Laboratory. Available online: http://solardat.uoregon.edu/AboutSrml.html and http://solardat.uoregon.edu/Photovoltaics.html (accessed on 26 April 2018).

21. Dubard, J.; Filtz, J.-R.; Cassagne, V.; Legrain, P. Photovoltaic module performance measurements traceability: Uncertainties survey. *Measurement* **2014**, *51*, 451–456. [CrossRef]

22. Chao, K.H.; Chen, C.T. A remote supervision fault diagnosis meter for photovoltaic power generation systems. *Measurement* **2017**, *104*, 93–104. [CrossRef]

23. Beránek, V.; Libra, M. Monitoring system for photovoltaic plants. In Proceedings of the 4th International Conference TAE 2010, Prague, Czech Republic, 7–10 September 2010; pp. 668–675, ISBN 978-80-213-2088-8.

24. Poulek, V.; Khudysh, A.; Libra, M. Self powered solar tracker for Low Concentration PV (LCPV) systems. *Sol. Energy* **2016**, *127*, 109–112. [CrossRef]

25. Reatti, A.; Kazimierczuk, M.K.; Catelani, M.; Ciani, L. Monitoring and field data acquisition system for hybrid static concentrator plant. *Measurement* **2017**, *98*, 384–392. [CrossRef]

26. Poulek, V.; Libra, M. New solar tracker. *Sol. Energy Mater. Solar Cells* **1998**, *51*, 113–120. [CrossRef]

27. Poulek, V.; Khudysh, A.; Libra, M. Innovative low concentration PV systems with bifacial solar modules. *Sol. Energy* **2015**, *120*, 113–116. [CrossRef]

28. Libra, M.; Beránek, V.; Sedláček, J.; Poulek, V.; Tyukhov, I.I. Roof photovoltaic power plant operation during the solar eclipse. *Sol. Energy* **2016**, *140*, 109–112. [CrossRef]

29. Poulek, V.; Strebkov, D.S.; Persic, I.S.; Libra, M. Towards 50 years lifetime of PV solar modules laminated with silicone gel technology. *Sol. Energy* **2012**, *86*, 3103–3108. [CrossRef]

30. Matuska, T.; Sourek, B.; Jirka, V.; Pokorny, N. Glazed PVT collector with polysiloxane encapsulation of PV cells: Performance and economic analysis. *Int. J. Photoenergy* **2015**, *2015*, 718316. [CrossRef]

31. Shemelin, V.; Matuška, T. Detailed Modeling of Flat Plate Solar Collector with Vacuum Glazing. *Int. J. Photoenergy* **2017**, *2017*, 1587592. [CrossRef]

32. Correa-Betanzo, C.; Calleja, H.; Lizarraga, A. Photovoltaic system assessment considering temperature and overcast conditions: Light load efficiency enhancement technique. *Sol. Energy* **2016**, *137*, 148–157. [CrossRef]

33. Chamberlin, C.E.; Rocheleau, M.A.; Marshall, M.W.; Reis, A.M.; Coleman, N.T.; Lehman, P.A. Comparison of PV Module Performance before and after 11 and 20 Years of Field Exposure. In Proceedings of the 37th IEEE Photovoltaic Specialists Conference, Seattle, DC, USA, 19–24 June 2011; pp. 000101–000105, ISBN 978-142449965-6.

34. PVGIS. Photovoltaic Geographical Information System-Interactive Maps. Available online: http://re.jrc.ec.europa.eu/pvgis/apps4/pvest.php# (accessed on 4 June 2018).

35. Kaplani, E. Detection of Degradation Effects in Field-Aged c-Si Solar Cells through IR Thermography and Digital Image Processing. *Int. J. Photoenergy* **2012**, *2012*, 396792. [CrossRef]

36. Olšan, T.; Libra, M.; Poulek, V.; Chalupa, B.; Sedláček, J. Combination of Three Methods of Photovoltaic Solar modules Damage Evaluation. *Sci. Agric. Bohem.* **2017**, *48*, 98–101.

Article

Occupancy-Based HVAC Control with Short-Term Occupancy Prediction Algorithms for Energy-Efficient Buildings

Jin Dong [1,*,†], **Christopher Winstead** [2,†], **James Nutaro** [2,†] and **Teja Kuruganti** [2,†]

[1] Energy and Transportation Science Division, Oak Ridge National Laboratory, Oak Ridge, TN 37831, USA
[2] Computational Sciences and Engineering Division, Oak Ridge National Laboratory, Oak Ridge, TN 37831, USA; winsteadcj@ornl.gov (C.W.); nutarojj@ornl.gov (J.N.); kurugantipv@ornl.gov (T.K.)
* Correspondence: dongj@ornl.gov; Tel.: +1-865-576-6742
† These authors contributed equally to this work.

Received: 24 August 2018; Accepted: 6 September 2018; Published: 13 September 2018

Abstract: This study aims to develop a concrete occupancy prediction as well as an optimal occupancy-based control solution for improving the efficiency of Heating, Ventilation, and Air-Conditioning (HVAC) systems. Accurate occupancy prediction is a key enabler for demand-based HVAC control so as to ensure HVAC is not run needlessly when when a room/zone is unoccupied. In this paper, we propose simple yet effective algorithms to predict occupancy alongside an algorithm for automatically assigning temperature set-points. Utilizing past occupancy observations, we introduce three different techniques for occupancy prediction. Firstly, we propose an identification-based approach, which identifies the model via Expectation Maximization (EM) algorithm. Secondly, we study a novel finite state automata (FSA) which can be reconstructed by a general systems problem solver (GSPS). Thirdly, we introduce an alternative stochastic model based on uncertain basis functions. The results show that all the proposed occupancy prediction techniques could achieve around 70% accuracy. Then, we have proposed a scheme to adaptively adjust the temperature set-points according to a novel temperature set algorithm with customers' different discomfort tolerance indexes. By cooperating with the temperature set algorithm, our occupancy-based HVAC control shows 20% energy saving while still maintaining building comfort requirements.

Keywords: occupancy model; occupancy-based control; model predictive control; energy efficiency; building climate control

1. Introduction

1.1. Background of Research

More than 30% of building energy is consumed by HVAC systems, which usually operate on a fixed schedule predefined by building owners or operation managers. Currently, most existing building control systems still condition rooms with a set-point assuming maximum occupancy from early morning until late evening during weekdays. As a result, rooms are often needlessly over-conditioned, which may lead to a significant waste in energy consumption. Occupancy based controls, as a promising remedy for the aforementioned issue, can achieve significant energy savings by temporally matching the building energy consumption and building usage. This has the potential to reduce up to a third of HVAC energy consumption. Moreover, accurate and reliable occupancy detection is becoming a key enabler for demand-response HVAC control, which requires the capturing of occupancy changes in real time [1]. By taking advantage of occupancy information, we can reduce building energy consumption via optimized scheduling of HVAC [2], as well as shading blinds and natural ventilation to make effective use of available natural resources [3–5].

Even in the rare cases where occupancy information is integrated into the HVAC operation, only binary decisions (occupied or not) are made, with the actual number of occupants in the building is ignored. However, even under this binary case, [6] has discovered that there exists potential annual energy savings of 10–42% if actual occupancy information has been properly utilized. In actuality, the energy consumption of that building is dominated by the occupancy and related activities [7]. It follows that there exists optimal control parameters, based on the instantaneous number of humans in a building and their associated behaviors, with great energy savings potential.

In fact, occupancy in a building is stochastic both in time and space, which greatly affects actual power consumption for an individual zone or building. Consequently, this will not only affect our decisions for improving energy efficiency but also in implementing the advanced demand response (Typically peak-shaving applications in modern energy management systems [8,9]). The authors of [10,11] discovered that average occupancy level for commercial buildings is at most a third of its maximum designed-for occupancy. Thus, accurate occupancy-sensing data provides significant insight for an online adaptive HVAC control strategy utilizing to the exact number of occupants in a building over a certain time period [12–14]. Moreover, occupant behavior is well recognized as a dominant source of the discrepency between predicted and actual building performance, and developing accurate short-term occupancy prediction will greatly enhance implementation of realistic building energy modeling and control.

1.2. Literature Review

1.2.1. Occupancy Models

Despite a plethora of potential application scenarios, buildings' occupancy modeling remains a cumbersome, error-prone and expensive process [15]. A through literature review for real-time occupancy detection and modeling in commercial buildings has been delivered in [16]. For occupancy modeling, various occupant behavior models have been developed in [17–20]. Moreover, such occupancy models have been integrated with operable windows, blinds, and lighting in EnergyPlus [19]. More recently, occupancy information has also been applied in Home Energy Management System (HEMS) using Markov-chain algorithms [21] or machine learning algorithms [22].

As previously mentioned, the uncertain occupancy information plays a central role in developing demand-driven HVAC control strategy. Due to the stochastic nature of occupancy, short-term prediction of it for individual rooms remains a challenging task. Previous occupancy modeling studies have focused on representing different detailedness of occupants' behavior, such as binary data (i.e., presence and absence) [23], accurate discrete values (i.e., the number of occupants) [24], or continuous probability distributions [25]. All these models achieved a balanced trade-off between model accuracy and complexity, depending on the actual application scenario. Consequently, an appropriate modeling complexity must be chosen for any specific case.

1.2.2. Occupancy-Based Control

When provided accurate occupancy models, demand-driven control can utilize such information to coordinate real-time HVAC usage, reducing energy use and maintaining indoor thermal comfort in buildings [13,26–29]. It has been reported in [30] that a 75% energy savings can be achieved by using a robust design which is less sensitive to occupant variation. Further, when integrated with model-based control strategies, 42% energy savings have been achieved by using real-time occupancy data [1].

The main task of a traditional HVAC control system is to maintain temperature and indoor air quality within a desired comfort range while minimizing energy use. Current mainstream HVAC control practice depends on the choice of predefined dead-band values, which involves a significant amount of tedious tuning. In fact, this tuning has become increasingly challenging with the rising complexity of modern HVAC systems, particularly with regard to the uncertain characteristics of occupancy [31].

An alternative approach is to use the well-known model predictive control (MPC) approach, which takes into account weather and occupancy forecasts (as shown in Figure 1). At each sampling time, MPC minimizes the energy use by optimizing a plan for future HVAC operation based on predictions of the weather and occupancy for a future time horizon [31]. MPC has been widely applied in building climate control systems and has demonstrated promising energy savings as studies in [32,33] show.

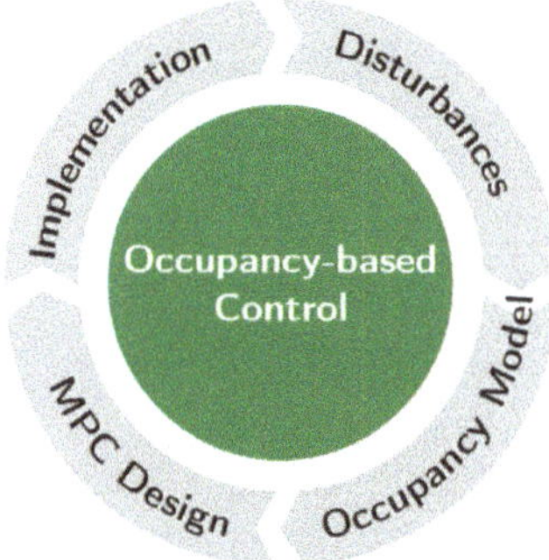

Figure 1. Scheme of occupancy-based control.

1.3. Main Idea and Outline

We note here that most of the aforementioned techniques require an off-line training process using a large amount of collected measurement. However, the focus of our paper is to provide an alternative on-line strategy for short-term occupancy prediction. The proposed occupancy-based control framework aims to minimize the total HVAC energy consumption while maintaining a comfortable indoor environment in buildings by utilizing a non-fixed temperature setpoint for the HVAC controller. To accomplish this, we recall one efficient algorithm [34] for optimally assigning temperature set-point based on both real-time forecasts of occupancy information of the building. To determine the temperature set-points for the planning horizon, a novel temperature setpoint algorithm is introduced, where a discomfort tolerance index is also included. After determining the optimal future temperature set-points, it is further integrated with an MPC framework to complete the occupancy-based control strategy.

Statement of contributions:
Our contribution in this paper is threefold: Firstly, we design a suitable utility function alongside a temperature set algorithm to capture the trade-off between occupancy comfort and energy consumption. Secondly, we propose three different occupancy estimation algorithms that enable short-term stochastic modeling of occupancy in buildings. Finally, we analyze and validate energy-saving performance of the proposed techniques. Detailed comparisons are provided for energy consumptions both with various occupancy estimation algorithms, and without any occupancy information. As mentioned in [35], very few implementations of occupancy models in building simulation are reported. To the best knowledge of the authors, there exist few available guidelines or analysis utilizing predicted occupancy information for occupancy-based HVAC control techniques. This paper bridges the gap between reliable stochastic occupancy modeling and energy efficiency building simulation.

It is an extension of authors' previous work [34], where a more powerful prediction algorithm—Uncertain Basis has been brought into the picture. Most importantly, we present novel quantified analysis for true energy savings with demand-based HVAC control strategy in this paper.

Finally, some discussions are provided to better illustrate true energy-saving benefit by using the proposed demand-based control strategy.

The paper is structured as follows: Section 2 defines the mathematical problem under consideration and HVAC model we use. We also set up a temperature setpoint algorithm which could adaptively tune the temperature setpoint of the room based on the real occupancy information. Section 3 contains details about the occupancy prediction algorithms. Three different prediction techniques, especially the uncertain basis technique is introduced to predict the occupancy for the first time. Section 4 presents the simulation results regarding the estimation performance of all the aforementioned estimation algorithms. Finally, Section 5 draws the conclusions and ideas for future directions of development.

2. Problem Formulation

In order to use occupancy for demand-based HVAC control, we first must illustrate our HVAC model settings and control strategy. MPC is a class of algorithms designed to exploit building models and forecasts of interior and exterior disturbance signals. An MPC algorithm then computes open-loop optimal control actions by optimizing a cost function over a finite time horizon.

The reference temperature set point will serve as a bridge to relate occupancy prediction with MPC control strategy via the temperature set algorithm proposed in [34].

2.1. Building Thermal Model

In this section, we describe the typical one-dimensional resistance-capacitance (RC) model used in MPC design. The model stems from a physics-based continuous-time model, which captures the dynamics of indoor temperature, interior-wall surface temperature, as well as exterior-wall core temperature. This building thermal model has been widely applied in dozens of researches [32,36,37] for simulating residential and commercial buildings. It is described by

$$\dot{x}_1 = \frac{1}{C_1}\left[(K_1 + K_2)(x_2 - x_1) + K_5(x_3 - x_1) + K_3(\delta_1 - x_1) + u_1 + u_2 + \delta_2 + \delta_3\right]$$
$$\dot{x}_2 = \frac{1}{C_2}\left[(K_1 + K_2)(x_1 - x_2) + \delta_2\right]$$
$$\dot{x}_3 = \frac{1}{C_3}\left[K_5(x_1 - x_3) + K_4(\delta_1 - x_3)\right]$$

where the variables are defined in Table 1, and the parameter values are provided in Table 2.

Table 1. Building parameter definition.

Variables	Definition
x_1	Indoor air temperature (°C)
x_2	Interior-wall temperature (°C)
x_3	Exterior-wall core temperature (°C)
u_1	Cooling power (≤ 0) (kW)
u_2	Heating power (≥ 0) (kW)
δ_1	Ambient temperature (°C)
δ_2	Solar radiation (kW/m^2)
δ_3	Internal heat gain (kW)

Table 2. Building parameter values.

Building Parameter Values	Unit
$C_1 = 9.356 \times 10^5$	kJ/°C
$C_2 = 2.970 \times 10^6$	kJ/°C
$C_w = 6.695 \times 10^5$	kJ/°C
$K_1 = 16.48$	kJ/°C
$K_2 = 108.5$	kJ/°C
$K_3 = 5$	kJ/°C
$K_4 = 30.5$	kJ/°C
$K_5 = 23.04$	kJ/°C

The system states are x_1, x_2, and x_3. The model inputs consist of control decision variables and exterior disturbance signals. The control decision variables are the demands sent to HVAC systems (with u_1 represents the cooling power, while u_2 corresponds to the heating power). The disturbances are δ_1, δ_2, and δ_3.

To translate the model into an MPC-friendly model, we must define the state vector x, the control signal vector u, and the environment stochastic disturbance vector ω as:

$$x := \begin{bmatrix} x_1 \\ x_2 \\ x_3 \end{bmatrix}, \quad u := \begin{bmatrix} u_1 \\ u_2 \end{bmatrix}, \quad \omega := \begin{bmatrix} \delta_1 \\ \delta_2 \\ \delta_3 \end{bmatrix}.$$

The continuous-time state-space model can then be described compactly as:

$$\dot{x} = A_c x + B_c u + C_c \omega \tag{1}$$

where

$$A_c := \begin{bmatrix} -\frac{1}{C_1}(K_1 + K_2 + K_3 + K_5) & \frac{1}{C_1}(K_1 + K_2) & \frac{K_5}{C_1} \\ \frac{K_1 + K_2}{C_2} & -\frac{(K_1 + K_2)}{C_2} & 0 \\ \frac{K_5}{C_3} & 0 & -\frac{(K_4 + K_5)}{C_3} \end{bmatrix}$$

$$B_c := \begin{bmatrix} \frac{1}{C_1} & \frac{1}{C_1} \\ 0 & 0 \\ 0 & 0 \end{bmatrix}, \quad C_c := \begin{bmatrix} \frac{K_3}{C_1} & \frac{1}{C_1} & \frac{1}{C_1} \\ 0 & \frac{1}{C_2} & 0 \\ \frac{K_4}{C_3} & 0 & 0 \end{bmatrix}. \tag{2}$$

We then consider the discrete-time (sampled) version of Equation (1) described by

$$x_{k+1} = A_d x_k + B_d u_k + C_d \omega_k \tag{3}$$

where k is the discrete-time index, $x_k = [x_{1,k}\, x_{2,k}\, x_{3,k}]^T$ and the parameters $[A_d, B_d, C_d]$ are computed from the continuous-time model parameters in Equation (2).

It should be mentioned that such state-space matrices A, B, G can be easily generated for any given buildings through either physics-based or data-driven modeling techniques.

2.2. Baseline Control Strategy (or RBC)

The performance of the proposed adaptive control scheme will be compared with a baseline rule-based on/off control (RBC) scheme commonly used by thermostats in residential homes. These RBC algorithms represent the core logic behind the most popular mechanical and digital controls of thermostats in residential homes. Figure 2a,b describe the overall schemes for summer and winter cases, respectively. Basically, the RBC rules compare the indoor temperature T with a given reference temperature T_{ref}, which is allowed to drift by a cooling/heating dead band ΔT_c or ΔT_h, respectively.

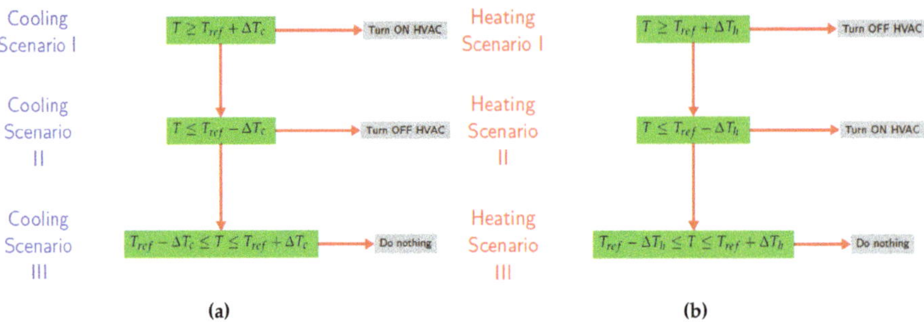

Figure 2. Overall scheme of baseline control (or RBC): (**a**) Overall scheme of baseline line control (or RBC) in summer cooling case; (**b**) Overall scheme of baseline line control (or RBC) in winter heating case.

2.3. System Model

The system states are the room air temperature t_1, interior wall surface temperature t_2, and exterior-wall core temperature t_3. u represents the control input, which is heating power S_h in this scenario (S_h comes as electrical power kW, since power conversion coefficients from heating load are absorbed in the model), and v is the outside disturbance including: the outdoor dry-bulb temperature $T_{outdoor}\,^{\circ}C$, the approximated sky temperature $T_{sky}\,^{\circ}C$, the internal load of space $Q_{internal}$ [W], and the solar radiation on the nodes Q_{solar} [W]. All variables with subscripts H correspond to the HVAC.

The thermal model for any given building can then be described as:

$$\dot{X} = AX + BU + GV \tag{4}$$

where

- *States* $X = [t_1, t_2, t_3], \quad$ *Inputs* $U = [S_h]$,
- *Disturbance* $V = \left[T_{outdoor}, T_{sky}, Q_{internal}, Q_{solar}\right]$.

We assume the following constraints are imposed on the temperature and control inputs (for winter heating):

$$23.3\,^{\circ}C(74\,^{\circ}F) \leq t_{1,k} \leq 25.5\,^{\circ}C(78\,^{\circ}F), \quad 0 \leq U_k \leq 1\,\text{kW}, \tag{5}$$

where $U_k \leq 0$ means cooling (we can consider similar heating case when U_k positive). It should be mentioned that sub-index k represents the kth time step. Additionally, we consider a variable speed HVAC system, where 0 represents HVAC totally OFF, and 1 (-1) means working at the maximum heating (cooling) power. It should be mentioned that while the temperature comfort interval has been chosen by ASHRAE, it can be adjusted to any other values based on user's preference.

2.4. Cost Function

In this MPC problem, we desire to minimize both the temperature deviation from the reference setpoints and the energy consumption while simultaneously enforcing a performance guarantee that ensures the indoor temperature always falls in a pre-defined comfort zone. We can assign set-points for all three temperature states, where x_r represents the indoor temperature set-points (dominated by current and future occupancy information). Ultimately, our objective is to find the finite-horizon control sequences which minimize the following finite horizon objective function:

$$V_N\left(e_0, U, V\right) := \tfrac{1}{2}\left[(x_N - x_r)^T P(x_N - x_r) + \sum_{k=1}^{N-1} e_k^T Q e_k + \sum_{k=0}^{N-1} u_k^T R u_k\right], \tag{6}$$

where $P \geq 0$, $Q \geq 0$ (i.e., semi-definite positive matrices), $R > 0$ (i.e., positive definite matrix), N is the prediction horizon, and

$$X := [x_0^T, ..., x_N^T]^T, \quad U := [u_0^T, ..., u_{N-1}^T]^T, \quad V := [\omega_0^T, ..., \omega_{N-1}^T]^T.$$

Thus, once we have accurate occupancy predictions available provided in Section 3, we can adaptively adjust temperature set-points according to the novel temperature set algorithm proposed in [34]. From this, we will be able to achieve occupancy-based optimal control (as shown in Figure 3) to improve energy efficiency of the buildings.

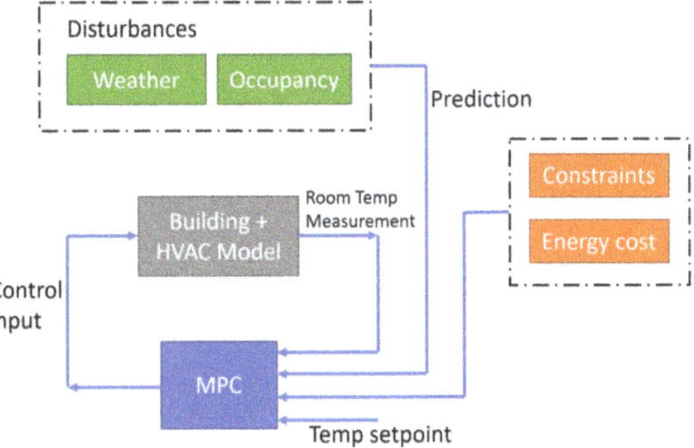

Figure 3. The proposed occupancy-based control setup for a building HVAC system.

2.5. Temperature Set Algorithm

In conventional practice, the HVAC operates under a fixed dead-band (chosen by the users) for indoor temperature. Currently, most temperature set-points are predefined by the building owner or administrator, and do not change frequently during the regular operation periods.

We recall our own simple yet effective algorithm (shown in Algorithm 1) to assign reference temperature set-points for each half hour of the day based on the real-time occupancy [34]. Inside this temperature set algorithm, we need first define the maximum ($\max(O_h)$) and minimum($\min(O_h)$) occupancy values (based on previous field measurement or survey) during normal operation. Similarly, we also must define the comfort band chosen by the customers. Obviously, the larger the band is, the more energy savings will be achieved. The beauty of Algorithm 1 is its ability to identify a temperature set-point depending on the occupancy information. The temperature set-points are then assigned to each half hour of the day based on the range in which the occupancy number of corresponding half hours fall in.

Following [38], a discomfort tolerance index α is defined to characterize building users' different choices/tolerance on thermal comfort (discomfort). Discomfort tolerance is considered high when $\alpha > 0$, and low when $\alpha < 0$.

Algorithm 1 Temperature Setting Algorithm [34]

1: **Step 1:**
2:     Initialize α
3: **Step 2:**
4:     $n \leftarrow \frac{T_{max} - T_{min}}{k} + 1$
5: **for all** hour $h = 1$ to 48 **do**
6:     $Range \leftarrow max(O_h) - min(O_h)$
7:     $r_0 \leftarrow min(O_h)$
8: **end for**
9: **if** $\alpha = 0$ **then**
10:     Go to step 3
11: **else**
12:     Go to step 4
13: **end if**
14: **Step 3:**
15: **for all** set-point j $(j = 1$ to $n)$ **do**
16:     $r_j \leftarrow r_{j-1} + \frac{Range}{n}$
17:     Go to Step 5
18: **end for**
19: **Step 4:**
20: **for all** set-point j $(j = 1$ to $n)$ **do**
21:     $r_j \leftarrow r_{j-1} + Range * \frac{2^{\alpha(j-1)}(1-2^{\alpha})}{(1-2^{\alpha n})}$
22: **end for**
23: **Step 5:**
24: **for all** hour $h = 1$ to 48 **do**
25:     $T_h^{set} \leftarrow k[argmin\{j : O_h \leq r_j\} - 1] + T_{min}$
26: **end for**

---

3. Occupancy Prediction Algorithms

In this section, we introduce our algorithm for the detailed estimation of future occupancy, and we show how we could predict the number of occupants in the room. An overview of all the developed algorithms is summarized in Figure 4.

In order to study a practical occupancy model, we collected real occupancy data from an occupancy sensor for a room. We randomly pick a segment of data dated "13 October 2010–5 April 2011", i.e., 174 days. The sampling interval is 30 min, so each individual sensor collects 48 occupancy samples each day. i.e., we have 8352 samples for each sensor. For simplicity, we consider the 8352 samples from a single sensor.

Natural questions which arise are: what is the probability for this room to be occupied; and how many people will be in the room?

To answer the first question, we could compute the probability for the room to be occupied by observing historic data. This has been reported in our earlier paper [34], so we skip this model here due to space constraints.

For the second problem, we need to apply more intelligent techniques (proposed in Sections 3.1–3.5) to predict the number of people in the room.

Due to occupancy's stochastic characteristic, it is not realistic to expect the real occupancy of the room to exactly follow the given schedule. Therefore, we should desire to accurately predict the occupancy information based on the most recent measurement. Moreover, detailed occupancy estimation considers not only whether the building is occupied or not, but also takes into account the number of occupants in the building. Here we will introduce some background and several alternative techniques that may be adequate for occupancy estimation.

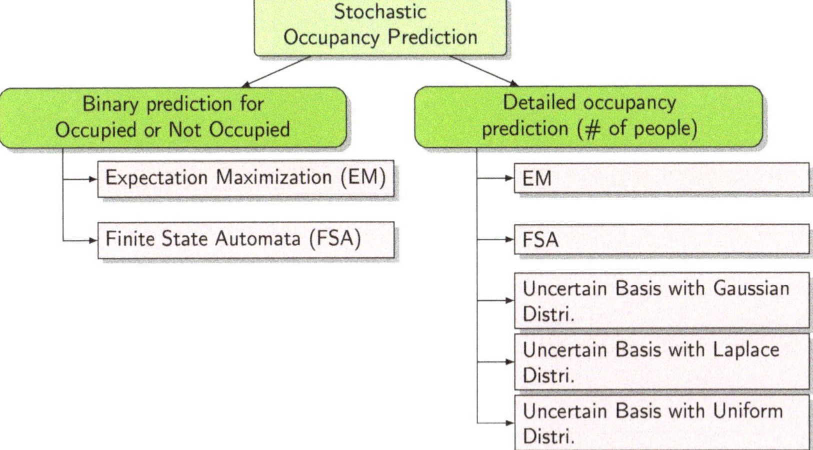

Figure 4. Overview of the occupancy prediction algorithms.

3.1. Expectation Maximization (EM)

The first occupancy modeling algorithm relies on the state-space model, which has been very popular in both societies of control systems and signal processing due to its advantage of on-line recursive implementation. The EM algorithm is a data-driven approach that builds and updates the estimated occupancy relying purely on collected measurements. Its core mechanism consists of two simple equations , i.e., a state x_k Equation (7) and a measurement y_k Equation (8).

A standard EM model in discrete time can be written as:

$$x_{k+1} = A_k x_k + B_k w_k \tag{7}$$
$$y_k = C_k x_k + D_k v_k \tag{8}$$

where $x_{k+1} \in \mathbb{R}^{n \times 1}$ ($\mathbb{R}^{n \times 1}$ denotes the space of real vectors with dimension $n \times 1$) is the state that characterizes the occupancy; it is a variable of the time series $\{x_k\}$ determined by the previous state x_k and the noise term $w_k \in \mathbb{R}^{m \times 1}$ introduced at each k. $A_k \in \mathbb{R}^{n \times n}$ and $B_k \in \mathbb{R}^{n \times m}$ are corresponding coefficients.

The beauty of the EM algorithm is the time varying nature inherited in the state-space model, which enables the model to adapt dynamically to the most recent occupancy model. Moreover, it takes into account the noise terms w_k and v_k that capture small perturbations or uncertainties introduced at each time k. This greatly alleviates the challenges associated with occupancy uncertainties in the model.

We can estimate the unknown system parameters $\beta_k = \{A_k, B_k, C_k, D_k\}$ and states $\{x_k\}$ through a finite set of received signal measurement data $Y = \{y_1, y_2, ...\}$ following [39]. Finally, we can achieve a one-step prediction of the occupancy by [40]:

$$\hat{y}_{k+1} = \hat{C}_k(\hat{A}_k \hat{x}_{k/k} + \hat{K}_k(y_k - \hat{C}_k \hat{x}_{k/k})) \tag{9}$$

where $\hat{y}_{k+1}$ denotes the predicted occupancy at $k+1$ and $\hat{K}_k$ is the Kalman gain.

3.2. Finite State Automata (FSA)

Probabilistic FSA [41] have previously been introduced to describe distributions over strings. FSA has been used quite successfully to address several complex sequential pattern recognition problems, such as continuous speech recognition, computational biology [42] and linguistics [43]. The general Systems Problem Solver (GSPS) proposed in [44], provides a novel and highly effective method for reconstructing the input/output behavior of FSA. The detailed algorithm and system formulation can be found in [45].

Given a system and a sequence of length n generated by that system, we may posit a relation between its variables. This relation takes the form of a function f that maps observations of some variables at times $n, n-1, \ldots, k$ with $k \geq 1$ to observations of other variables at time n.

In this occupancy problem, we want to know the exact number of people in the room. So we assume different numbers of occupants as different states in the FSA. As long as finite states are involved, the general form of the rule and the methods for constructing and forecasting are the same.

Such a relation is called a *rule*. Typically this purpose is forecasting; i.e., to predict future observations from past and current observations. This relation takes the form of a function f that maps observations of some variables at times $t, t-1, \ldots, t-n$ to observations of other variables at time t. Rules are posited by the observer for some particular purpose.

Continuing the example above, consider strings comprised of the letters a, b, c. Take these strings to be our system and the variable v_1 to be a single character in a string. Observations of this variable are indexed by position in the string ordered from left to right so that the first character in the string is $v_1(1)$, the second is $v_1(2)$, and so forth. One example of a sequence for this system is *abcabcabc*.

A possible rule that predicts the next letter in this string is $v_1(n) = f(v_1(n-1))$. This rule relates the characters in subsequences of length two such that the leftmost character predicts the rightmost character. Interpreting the sequence *abcabcabc* with this rule yields the following function: $f(a) = b$ which occurs three times, $f(b) = c$ which occurs three times, and $f(c) = a$ which occurs two times.

3.3. Simplified Binary States FSA

In general, the length of the "look back depth" used is decided by actual problem. For our specific occupancy prediction problem, the system of interest has a single variable with two possible values: occupied (b) or not occupied (a). After tail and error check, we posit a rule that looks back three steps and also considers the time of day, i.e., $v_1(n) = f(v_1(n-1), v_1(n-2), v_1(n-3), t(n))$. The time of day $t(n) = t_n$ is used to characterize the different rules for different time period during a day; for the available data use there are 48 times that can be considered, as the sampling interval of the sensor is 30 min. Given the data and a rule, we can build a model for forecasting with the simple procedure described by Klir [44]. More details about working mechanisms for FSA can be found in [34].

To illustrate this procedure, consider the mask and sequence in Table 3. The system that generated this sequence has time variable t_n and a single variable v that can take the value b or a. The rule for this mask built with 3000 data points is shown in Table 4. It should be noted that in order to simplify the explaination, we decouple the time variable from the rule. To be specific, Table 4 represents the rule for 3:00 p.m. using 3000 data points. Considering the sampling interval is every 30 min, we could generate 48 such rule tables for the whole day (24 h).

Table 3. A sequence and mask for a system with uncertain output. Circles are input data for the function f and the square is the output data.

Round	Time (v_1)	Occupancy (v_2)
11	t_{11} (circled)	a (squared)
10	t_{10}	ⓑ
9	t_9	ⓐ
8	t_8	ⓑ
7	t_7	a
6	t_6	a
5	t_5	b
4	t_4	a
3	t_3	b
2	t_2	b
1	t_1	a

Table 4. Rule for the sequence and mask using a subset of 3000 points restricted to 3:00 p.m. (*a*-unoccupied, *b*-occupied).

Input	Output	Count	Likelihood
aaa	a	47	0.959
	b	2	0.041
aab	a	0	0
	b	1	1
aba	a	1	1
	b	0	0
abb	a	0	0
	b	1	1
baa	a	1	0.5
	b	1	0.5
bab	a	1	0.33
	b	2	0.67
bba	a	0	0
	b	1	1
bbb	a	0	0
	b	4	1

To illustrate how this model is applied for forecasting, suppose we begin with the latest observation sequence as *bab* at 3:00 p.m. The next output is *a* with probability 0.33 or *b* with probability 0.67. If we were at a different time step other than 3:00 p.m., then the output is selected according to the corresponding rule table at that time step. Once the time step is fixed, then the second output is *a* with some probability p or *b* with probability $1 - p$. Continuing in this fashion, we can construct a tree of possible futures and use these possible futures to inform the control system.

3.4. Estimating Number of Occupants

We consider three methods for anticipating the actual number of occupants within a room.

FSA with 3 or More Input/Output Values

Though the method described in Section 3.3 only involves two states, it is readily extended to estimate the exact number of occupancy by using the number of occupants as variable rather than the binary occupied/unoccupied.

In the following simulation results, we show that the discussion for the binary model applies directly to the model with multiple states.

3.5. Basis Function

Finally, we introduce the third powerful algorithm, i.e., the uncertain basis functions. If we consider y_k as the current occupancy measurements detected by occupancy sensor, it can be well represented by the following basis functions:

$$y_k = \sum_{i=1}^{p} A_i \phi_{i,k} \quad k = 1, 2, ..., N, \tag{10}$$

where $\{\phi_{1,k}, ..., \phi_{p,k}\}$ are the basis functions, and $p < N$, and A_i is the corresponding coefficient of each basis function [46,47].

However, this trivial representation may not be able to capture all the dynamics of occupancy due to the limitation of fixed basis functions, which eliminates the uncertainty nature. An alternative way to cast this formulation is to assume that each basic function depends on some unknown parameter vectors θ_i, where only some statistics of the distribution of θ_i are known. Then, the corresponding coefficients can be estimated by minimizing an expected cost function following the technique developed in [48].

Following [48], we assume that each θ_i is independent. The basic functions are further represented by $\phi_{i,k}(\theta_i)$. The main objective here is to find the best coefficients to minimize the expected cost function $\hat{J}(A)$ defined as:

$$\hat{J}(A) = \mathbb{E}_\theta \left[\sum_{k=1}^{N} |y_k - \sum_{i=1}^{p} A_i \phi_{i,k}(\theta_i)|^2 \right] \tag{11}$$

where $\mathbb{E}_\theta$ is the expectation with respect to θ_i. The measured values are real, so we could estimate the coefficients A following steps presented in [39].

We focus on predicting future occupancy using the basis function:

$$\hat{y}_k = \sum_{i=1}^{p} \hat{A}_i \mathbb{E}_\theta[\phi_{i,k}(\theta_i)] \tag{12}$$

This means that we can predict occupancy values with p random basis functions. Similarly as in [39], we assume the basis functions to be governed by three different distributions, i.e., *Gaussian, Laplace and Uniform*. Consequently, we are able to compute occupancy predictions under each distribution.

4. Case Studies

To illustrate the effectiveness of the occupancy prediction techniques proposed in the last section, we assess their performance using the aforementioned occupancy measurement (at the beginning of Section 3). In the first part, performance of three occupancy prediction techniques are examined, and corresponding accuracy comparison are provided. In the second part, different temperature set trajectories s are obtained using our temperature set algorithm. The algorithm is employed to assign reference temperature set-points for each hour of the day. Lastly, these reference temperature set-points are utilized in the standard HVAC control strategies. In essence, different energy saving benefits are studied for traditional ON/OFF control and advanced MPC control, respectively.

4.1. Definition of the Performance Indexes

We define the estimation accuracy as the total number of correct predictions divided by the total number of predictions. The Root Mean Squared Error (RMSE), which characterizes the absolute estimation errors.

More formally, if we let ACC represent the accuracy, then using indicator functions, we obtain the accuracy of the estimator:

$$ACC(\hat{O}) := \frac{N - \sum_{k=1}^{N} \mathbb{1}(|O(k) - \hat{O}(k)|)}{N}. \tag{13}$$

where O and $\hat{O}$ are true and estimated occupancy, respectively. Let us denote the characteristic function of estimation error $\mathbb{1}(EO(k))$ as:

$$\mathbb{1}(EO(k)) := \mathbb{1}(|O(k) - \hat{O}(k)|) = \begin{cases} 1 & \text{if } O(k) > 0 \\ 0 & \text{otherwise,} \end{cases} \tag{14}$$

Here, the RMSE is defined as the square root of the mean square error:

$$MSE(\hat{O}) := \frac{1}{N} \sum_{k=1}^{N} (\hat{O}(k) - O(k))^2. \tag{15}$$

Therefore, RMSE will be the square root of (15).

4.2. Occupancy Prediction Performance

4.2.1. GSPS Model

In this GSPS-based scenario, we assume we have access to a large enough historical data set. We trained the model using the last 3000 and 5000 data points, respectively. The prediction results are shown in Figure 5a,b. As expected, the more data involved in training the FSA, the better the prediction results. This is also confirmed by the estimation error comparison shown in Table 5.

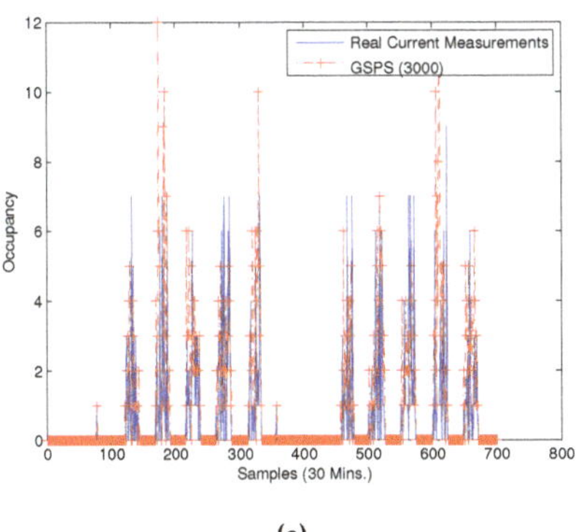
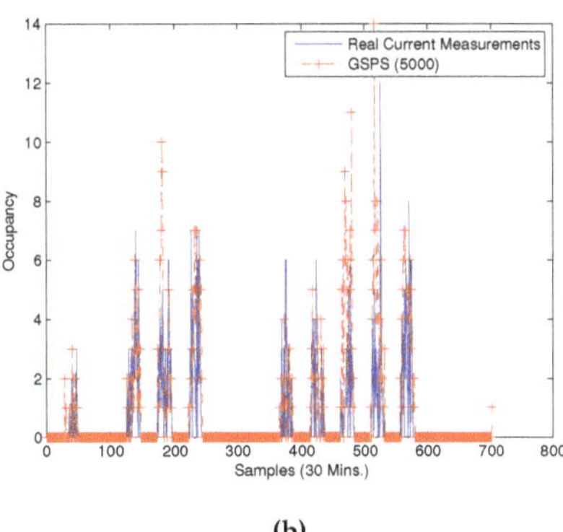

(a)   (b)

Figure 5. Occupancy estimation using GSPS model: (**a**) Occupancy estimation using GSPS model 3000 points; (**b**) Occupancy estimation using GSPS model 5000 points.

4.2.2. EM Method

Next, the elegant EM algorithm is applied to occupancy prediction. The performance is depicted in Figure 6, where only the last 20 sample points are used to predict the occupancy information at the very next time step.

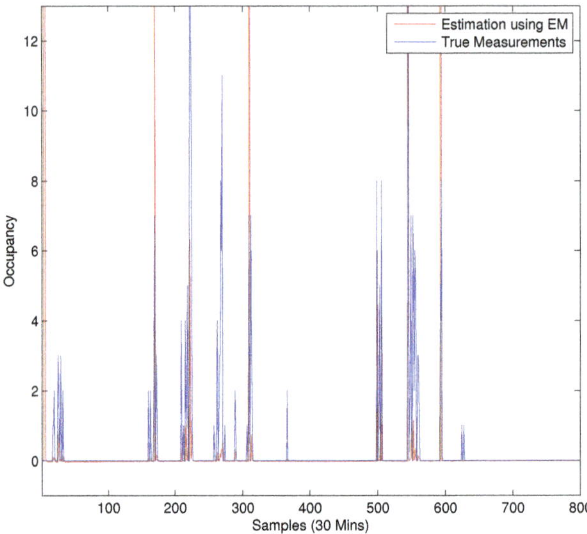

Figure 6. Occupancy estimation using EM algorithm.

4.2.3. Uncertain Basis Functions

In order to the show robustness of the proposed uncertain basis technique, we tested it with three different distributions for θ_i. It should be mentioned that only 10 last sampling points are used to build the optimal basis for each distribution. Figure 7 shows prediction comparison results using three different distributions.

Table 5. Performance comparison of occupancy prediction algorithms.

Methods	Estimation RMSE	Accuracy
GSPS (3000)	3.078	70.0%
GSPS (5000)	2.646	71.5%
EM	3.715	61.5%
Basis-Gaussian	3.211	68.4%
Basis-Laplace	2.946	70.9%
Basis-Uniform	2.571	72.6%

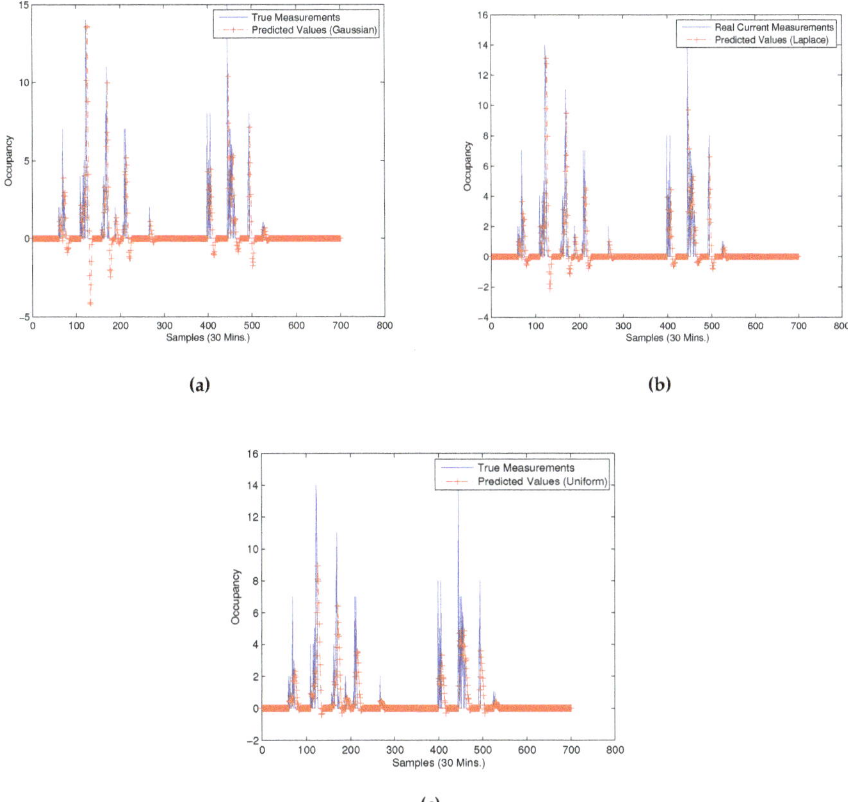

(a)

(b)

(c)

Figure 7. Occupancy estimation using three different basis functions: (**a**) Gaussian basis functions; (**b**) Laplace basis functions; (**c**) Uniform basis functions.

4.3. Temperature Set Points

Through the above simulation results, we achieve the desired occupancy prediction. However, our goal is to design temperature set-points based on these occupancy sequences. In order to further determine the effectiveness of these occupancy prediction results, we integrate the occupancy prediction results into the temperature set algorithms. The one corresponding to basic functions using *uniform distribution* is presented in Figure 8.

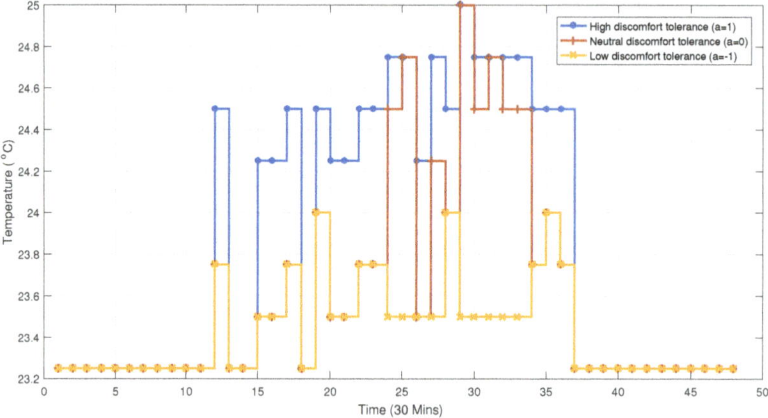

Figure 8. Temperature set point for uncertain basis method.

4.4. Occupancy-Based Control

To show the energy-saving performance using the proposed stochastic models for occupancy and temperature set algorithm, we insert the obtained temperature reference into simple ON/OFF switching control framework. A fixed reference temperature 23 °C is chosen for the baseline scenario. An occupancy-dependent reference temperature generated via our temperature set algorithm replaces original fixed schedule. This simple step will adaptively tune the set-point of HVAC systems according to current occupancy information, avoiding wasting energy for empty rooms.

In order to quantify the performance using different algorithms, we utilize the energy cost using traditional ON/OFF control strategy without any occupancy information as a benchmark. It should be mentioned that energy cost is defined as 2-norm, i.e., $U_{total} = \sqrt{\sum_{k=1}^{T} U_k^2}$.

Next, a comparison between different algorithms is made by changing only the occupancy information. Detailed numerical result is given in the right end column of Table 5. We are expected to save approximately 13% and 20% energy consumptions for traditional ON/OFF and MPC control strategies, respectively.

Table 6. Comparison of energy saving.

Methods	Control Cost (U_{total})	Energy Saving
Basic Control (No Occupancy info)	5.43	0%
Basis-Gaussian (Basic Control)	4.77	13%
Basis-Gaussian (MPC)	4.38	20%

4.5. Summary of the Results

In this section we compare four occupancy prediction algorithms, all trained using the same training set described at the beginning of Section 3. Figure 8 shows the realizations of occupancy predictions can be applied to the corresponding test set. Ideally, we can increase temperature set point when less occupants will be present in the room. As expected, we notice higher temperature set points are achieved corresponding to a larger tolerance index, as we are studying a summer cooling case in this paper. Table 5 summarizes the achieved numerical performance and accuracy comparison for the three algorithms. Generally speaking, both EM and Uncertain basis methods can provide reliable predictions with just a few historical data points. The GSPS method meanwhile, requires many more points to build a reliable model. Additionally, GSPS and Uncertain basis methods achieve a higher accuracy, while the EM method provides a degraded prediction result. Each method may find its own suitable application scenarios depending on the accuracy requirement and data structure.

It should be remarked that, although some mismatches exist for non-zero jumps in Figures 5a–7c, all prediction algorithms track the 0 base line (non-occupied status) perfectly. Therefore, all prediction techniques are effective for identifying empty rooms, with an over 90 percent accuracy rate. Moreover, the accuracy conditions we set are extremely restrictive. In other words, the accuracy is said to be satisfied only when the estimated number of occupants is exactly the same as the real measurement. So in this case, the accuracy is void if the estimated number is 13 while actual number is 12.

Furthermore, the obtained occupancy status is successfully applied into the temperature set algorithm which dominates energy consumption in building climate systems. In this final experiment, we applied the designed temperature setpoints into two different algorithms - basic control and MPC (designed in Section 2), and compare their energy consumption. The building thermal zone model we picked has also been introduced in Section 2. The detailed energy consumption data of the simulation has been given in Table 6. These control tests complete our occupancy-based control framework, which showcases up to 20% energy saving benefits via the proposed corresponding control framework.

5. Conclusions and Future Work

In this paper, we propose three different occupancy prediction methods for demand-based HVAC control. All three proposed short-term stochastic modeling methods, GSPS, EM and uncertain basis, achieved more than 70% accuracy in the experimental studies. Furthermore, we have designed a novel temperature set algorithm to correctly assign temperature set points based on the instantaneous occupancy information. To complement the occupancy-based framework, we have integrated the temperature set point naturally into the MPC algorithm. Finally, detailed comparisons are provided for energy consumptions with various occupancy estimation algorithms and without any occupancy information. This paper delivered a novel end-to-end solution, which connects a reliable stochastic occupancy modeling study with the occupancy-based control design. Consequently, we have seen up to 20% energy saving by integrating the proposed technique into two standard HVAC control strategies.

A great number of increasingly complex models are being developed for HVAC systems. However, the limited number of implementations of such models in demand-based control and the lack of occupants' effects limits their ability to improve efficiency while guaranteeing a comfortable temperature environment in buildings. Our near future work will involve detailed internal heat gain subject to different occupancy situations and various application scenarios, particularly the hot topic of building-to-grid integration. Another interested direction is to perform the sensitivity analysis for changing the set point. Basically, to answer the question, when is the best time to change the set-point; and how long/much will the electricity consumption reflect the change. This knowledge is critical for doing demand-response using buildings.

Author Contributions: Conceptualization, J.D., T.K. and J.N.; Methodology, J.D., T.K. and J.N.; Software, J.D. and C.W.; Validation, J.D. and C.W.; Formal Analysis, J.D. and T.K.; Investigation, J.D. and T.K.; Writing—Original Draft Preparation, J.D.; Writing—Review & Editing, J.N. and J.D.

Acknowledgments: This manuscript has been authored by UT-Battelle, LLC under Contract No. DE-AC05-00OR22725 with the U.S. Department of Energy. This material is based upon work supported by the U.S. Department of Energy. The United States Government retains and the publisher, by accepting the article for publication, acknowledges that the United States Government retains a non-exclusive, paid-up, irrevocable, world-wide license to publish or reproduce the published form of this manuscript, or allow others to do so, for United States Government purposes. The Department of Energy will provide public access to these results of federally sponsored research in accordance with the DOE Public Access Plan (http://energy.gov/downloads/doe-public-access-plan).

Conflicts of Interest: The authors declare no conflict of interest.

References

1. Erickson, V.L.; Carreira-Perpiñán, M.Á.; Cerpa, A.E. Occupancy modeling and prediction for building energy management. *ACM Trans. Sens. Net.* **2014**, *10*, 42. [CrossRef]
2. Sun, B.; Luh, P.B.; Jia, Q.S.; Jiang, Z.; Wang, F.; Song, C. Building energy management: Integrated control of active and passive heating, cooling, lighting, shading, and ventilation systems. *IEEE Trans. Automation Sci. Eng.* **2013**, *10*, 588–602.
3. Tzempelikos, A.; Athienitis, A.K. The impact of shading design and control on building cooling and lighting demand. *Solar Energy* **2007**, *81*, 369–382. [CrossRef]
4. Van Moeseke, G.; Bruyère, I.; De Herde, A. Impact of control rules on the efficiency of shading devices and free cooling for office buildings. *Build. Environ.* **2007**, *42*, 784–793. [CrossRef]
5. Valdiserri, P.; Biserni, C.; Garai, M. Energy performance of a ventilation system for an apartment according to the Italian regulation. *Inter. J. Energy Environ. Engin.* **2016**, *7*, 353–359. [CrossRef]
6. Erickson, V.L.; Carreira-Perpiñán, M.Á.; Cerpa, A.E. OBSERVE: Occupancy-based system for efficient reduction of HVAC energy. In Proceedings of the 10th ACM/IEEE International Conference on Information Processing in Sensor Networks, Chicago, IL, USA, 12–14 April 2011; pp. 258–269.
7. Shih, H.C. A robust occupancy detection and tracking algorithm for the automatic monitoring and commissioning of a building. *Energy Build.* **2014**, *77*, 270–280. [CrossRef]
8. Sharma, I.; Dong, J.; Malikopoulos, A.A.; Street, M.; Ostrowski, J.; Kuruganti, T.; Jackson, R. A modeling framework for optimal energy management of a residential building. *Energy Build.* **2016**, *130*, 55–63. [CrossRef]
9. Dong, J.; Olama, M.M.; Kuruganti, T.; Nutaro, J.; Xue, Y.; Sharma, I.; Djouadi, S.M. Adaptive building load control to enable high penetration of solar photovoltaic generation. In Proceedings of the Power & Energy Society General Meeting IEEE, Chicago, IL, USA, 16–20 July 2017; pp. 1–5.
10. Rafsanjani, H.N.; Ahn, C.R.; Alahmad, M. A review of approaches for sensing, understanding, and improving occupancy-related energy-use behaviors in commercial buildings. *Energies* **2015**, *8*, 10996–11029. [CrossRef]
11. Brandemuehl, M.J.; Braun, J.E. The impact of demand-controlled and economizer ventilation strategies on energy use in buildings. *ASHRAE Trans.* **1999**, *105*, 39.
12. Harle, R.K.; Hopper, A. The potential for location-aware power management. In Proceedings of the 10th international conference on Ubiquitous computing, Seoul, Korea, 21–24 September 2008; pp. 302–311.
13. Erickson, V.L.; Cerpa, A.E. Occupancy based demand response HVAC control strategy. In Proceedings of the 2nd ACM Workshop on Embedded Sensing Systems for Energy-Efficiency in Building, Zurich, Switzerland, 3–5 November 2010; pp. 7–12.
14. Garg, V.; Bansal, N. Smart occupancy sensors to reduce energy consumption. *Energy Build.* **2000**, *32*, 81–87. [CrossRef]
15. Nguyen, T.A.; Aiello, M. Energy intelligent buildings based on user activity: A survey. *Energy Build.* **2013**, *56*, 244–257. [CrossRef]
16. Labeodan, T.; Zeiler, W.; Boxem, G.; Zhao, Y. Occupancy measurement in commercial office buildings for demand-driven control applications—A survey and detection system evaluation. *Energy Build.* **2015**, *93*, 303–314. [CrossRef]
17. Gunay, H.B.; O'Brien, W.; Beausoleil-Morrison, I. A critical review of observation studies, modeling, and simulation of adaptive occupant behaviors in offices. *Build. Environ.* **2013**, *70*, 31–47. [CrossRef]
18. Parys, W.; Saelens, D.; Hens, H. Coupling of dynamic building simulation with stochastic modelling of occupant behaviour in offices–a review-based integrated methodology. *J. Build. Perform. Simul.* **2011**, *4*, 339–358. [CrossRef]
19. Gunay, H.B.; O'Brien, W.; Beausoleil-Morrison, I. Implementation and comparison of existing occupant behaviour models in EnergyPlus. *J. Build. Perform. Simul.* **2016**, *9*, 567–588. [CrossRef]
20. Yang, J.; Santamouris, M.; Lee, S.E. Review of occupancy sensing systems and occupancy modeling methodologies for the application in institutional buildings. *Energy Build.* **2016**, *121*, 344–349. [CrossRef]
21. Louis, J.N.; Caló, A.; Leiviskä, K.; Pongrácz, E. Modelling home electricity management for sustainability: The impact of response levels, technological deployment & occupancy. *Energy Build.* **2016**, *119*, 218–232.
22. Peng, Y.; Rysanek, A.; Nagy, Z.; Schlüter, A. Using machine learning techniques for occupancy-prediction-based cooling control in office buildings. *Appl. Energy* **2018**, *211*, 1343–1358. [CrossRef]

23. Chaney, J.; Owens, E.H.; Peacock, A.D. An evidence based approach to determining residential occupancy and its role in demand response management. *Energy Build.* **2016**, *125*, 254–266. [CrossRef]

24. Wang, W.; Chen, J.; Huang, G.; Lu, Y. Energy efficient HVAC control for an IPS-enabled large space in commercial buildings through dynamic spatial occupancy distribution. *Appl. Energy* **2017**, *207*, 305–323. [CrossRef]

25. D'Oca, S.; Hong, T. Occupancy schedules learning process through a data mining framework. *Energy Build.* **2015**, *88*, 395–408. [CrossRef]

26. Scott, J.; Bernheim Brush, A.; Krumm, J.; Meyers, B.; Hazas, M.; Hodges, S.; Villar, N. PreHeat: controlling home heating using occupancy prediction. In Proceedings of the 13th international conference on Ubiquitous computing, Beijing, China, 17–21 September 2011; pp. 281–290.

27. Lu, J.; Sookoor, T.; Srinivasan, V.; Gao, G.; Holben, B.; Stankovic, J.; Field, E.; Whitehouse, K. The smart thermostat: using occupancy sensors to save energy in homes. In Proceedings of the 8th ACM Conference on Embedded Networked Sensor Systems, Zurich, Switzerland, 3–5 November 2010; pp. 211–224.

28. Gunay, H.B.; O'Brien, W.; Beausoleil-Morrison, I. Development of an occupancy learning algorithm for terminal heating and cooling units. *Build. Environ.* **2015**, *93*, 71–85. [CrossRef]

29. Peng, Y.; Rysanek, A.; Nagy, Z.; Schlüter, A. Occupancy learning-based demand-driven cooling control for office spaces. *Build. Environ.* **2017**, *122*, 145–160. [CrossRef]

30. Karjalainen, S. Should we design buildings that are less sensitive to occupant behaviour? A simulation study of effects of behaviour and design on office energy consumption. *Energy Efficiency* **2016**, *9*, 1257–1270. [CrossRef]

31. Oldewurtel, F.; Sturzenegger, D.; Morari, M. Importance of occupancy information for building climate control. *Appl. Energy* **2013**, *101*, 521–532. [CrossRef]

32. Oldewurtel, F.; Parisio, A.; Jones, C.N.; Morari, M.; Gyalistras, D.; Gwerder, M.; Stauch, V.; Lehmann, B.; Wirth, K. Energy efficient building climate control using stochastic model predictive control and weather predictions. In Proceedings of the 2010 American Control Conference, Baltimore, MD, USA, 30 June–2 July 2010; pp. 5100–5105.

33. Oldewurtel, F.; Parisio, A.; Jones, C.N.; Gyalistras, D.; Gwerder, M.; Stauch, V.; Lehmann, B.; Morari, M. Use of model predictive control and weather forecasts for energy efficient building climate control. *Energy Build.* **2012**, *45*, 15–27. [CrossRef]

34. Dong, J.; Winstead, C.; Djouadi, S.M.; Nutaro, J.J.; Kuruganti, T. Stochastic Modeling of Short-term Occupancy for Energy Efficient Buildings. In Proceedings of the 4th International High Performance Buildings Conference at Purdue, West Lafayette, IN, USA, 11–14 July 2016.

35. Gaetani, I.; Hoes, P.J.; Hensen, J.L. Occupant behavior in building energy simulation: Towards a fit-for-purpose modeling strategy. *Energy Build.* **2016**, *121*, 188–204. [CrossRef]

36. Gwerder, M.; Tödtli, J. Predictive control for integrated room automation. In Proceedings of the 8th REHVA World Congress Clima, Lausanne, Switzerland, 9–12 October 2005.

37. Olama, M.M.; Kuruganti, T.; Nutaro, J.J.; Dong, J. Coordination and Control of Building HVAC Systems to Provide Frequency Regulation to the Electric Grid. *Energies* **2018**, *11*, 1852. [CrossRef]

38. Avci, M.; Erkoc, M.; Asfour, S.S. Residential HVAC load control strategy in real-time electricity pricing environment. In Proceedings of the 2012 IEEE Energytech, Cleveland, OH, USA, 29–31 May 2012; pp. 1–6.

39. Dong, J.; Ma, X.; Djouadi, S.; Li, H.; Liu, Y. Frequency Prediction of Power Systems in FNET Based on State-Space Approach and Uncertain Basis Functions. *IEEE Trans. Power Syst.* **2014**, *29*, 2602–2612. [CrossRef]

40. Verhaegen, M.; Verdult, V. *Filtering and System identification: a Least Squares Approach*; Cambridge University Press: Cambridgeshire, UK, 2007.

41. Paz, A. *Introduction to Probabilistic Automata*; Academic Press, Manhattan, NY, USA, 2014.

42. Lyngso, R.B.; Pedersen, C.; Nielsen, H. Metrics and similarity measures for hidden Markov models. In Proceedings of the 7th International Conference on Intelligent Systems for Molecular Biology, Heidelberg, Germany, 6–10 August 1999; pp. 178–186.

43. Mohri, M. Finite-state transducers in language and speech processing. *Comput. Linguist.* **1997**, *23*, 269–311.

44. Klir, G. *An Approach to General Systems Theory*; Van Nostrand Reinhold: New York, NY, USA, 1969.

45. Klir, G. *Architecture of Systems Problem Solving*; Springer Science & Business Media, Berlin, Germany, 2013.

46. Dong, J.; Ma, X.; Djouadi, S.M.; Li, H.; Liu, Y. Frequency prediction of power systems in FNET based on state-space approach and uncertain basis functions. *IEEE Trans. Power Syst.* **2014**, *29*, 2602–2612. [CrossRef]

47. Dong, J.; Kuruganti, T.; Djouadi, S.M. Very short-term photovoltaic power forecasting using uncertain basis function. In Proceedings of the Information Sciences and Systems (CISS) 51st Annual Conference on IEEE, Baltimore, MD, USA, 22–24 March 2017; pp. 1–6.

48. Kay, S. Signal Fitting With Uncertain Basis Functions. *Signal Process. Lett. IEEE* **2011**, *18*, 383–386. [CrossRef]

Article

Analysis of Winding Vibration Characteristics of Power Transformers Based on the Finite-Element Method

Xiaomu Duan [1], Tong Zhao [1,*], Jinxin Liu [2], Li Zhang [1] and Liang Zou [1]

[1] School of Electrical Engineering, Shandong University, Jinan 250061, China; 201613055@mail.sdu.edu.cn (X.D.); zhleee@sdu.edu.cn (L.Z.); zouliang@sdu.edu.cn (L.Z.)
[2] State Grid, Jining Power Supply Company, Jining 272000, China; ljxwillingheart@gmail.com
* Correspondence: zhaotong@sdu.edu.cn; Tel.: +86-531-8169-6129

Received: 25 August 2018; Accepted: 7 September 2018; Published: 11 September 2018

Abstract: The winding is the core component of a transformer, and the technology used to diagnose its current state directly affects the operation and maintenance of the transformer. The mechanical vibration characteristics of a dry-type transformer winding are studied in this paper. A short-circuit test was performed on an SCB10-1000/10 dry-type transformer, and the vibration signal at the surface was measured. Based on actual experimental conditions, a vibration-simulation model of the transformer was established using COMSOL Multiphysics software. A multiphysics coupling simulation of the circuit, magnetic field, and solid mechanics of the transformer was performed on this model. The simulation results were compared with measured data to verify the validity of the simulation model. The simulation model for a transformer operating under normal conditions was then used to develop simulation models of transformer-winding looseness, winding deformation, and winding-insulation failure, and the winding fault vibration characteristics were analyzed. The results provide a basis for detecting and analyzing the mechanical state of transformer windings.

Keywords: power transformer winding; vibration characteristics; multiphysical field analysis; short-circuit experiment; winding-fault characteristics

---

1. Introduction

The safety and reliability of power transformers, which are the core pieces of equipment in a power grid, are important for the reliable operation of the entire power grid [1]. Foreign statistics show that approximately 2% of transformers that run continuously for more than four years will experience accidents of varying degrees [2]. The high failure rate of transformers has always affected the safe and stable operation of the power grid [3], and it is not difficult to find that mechanical faults in transformer are often due to latent issues upon reviewing historical cases of transformer accidents [4]. Transformer faults generally involve the failure of main components and accessories, with the primary source of these faults being due to windings and core failures. In China, faults have occurred in 18 transformers in and above the 110 kV class [5]. Of these faults, 10 (55.6%) were caused by winding issues. From 2006 to 2010, the State Grid Corporation of China (SGCC) compiled statistics on the causes of faults in 46 transformers, of which 26 (56.5%) were caused by winding deformation [6]. In 2013, there were five accidents in transformers of the 110 kV class and above belonging to the Guangxi Power Supply Company of the Southern China Power Grid [7]. Of these, four cases (80%) were caused by winding deformation.

The study of electromagnetic vibration in transformers began in the 1920s, mainly by large power-transformer manufacturers and related research institutions. However, that work was limited by the technology available at that time, when winding fault simulations were not ideal [8]. Fahnoe H.

studied the forced vibration of a transformer's vibrating iron core under magnetostriction and resonance at the harmonic frequency [9]. A substantial amount of simulation of the transformer was carried out. The modal-resonance frequency of the transformer was analyzed such that the transformer avoided resonance, but the simulation results were not verified via experimentation [10]. Foster S. L. and others used finite element numerical analysis to calculate the electromagnetic field and structural force field of large transformers, and obtained the vibration of the transformer core [11]. By combining electromagnetic-field theory with the theory of structural mechanics, Yang Qingxin and other scholars in China established a mathematical model of the electromagnetic vibration of the iron core of a power transformer [12–14]. The model was used to simulate magnetostriction of silicon steel sheets. On this basis, the distribution of the sound field around the core was analyzed. These researchers paid considerable attention to the vibration of the iron core, but the vibration of the windings at various working conditions was less important to them. Liu Dichen and other scholars established an electromagnetic mechanical sound field finite-element model of a transformer core and its winding [15]. In ANSYS Workbench, a finite-element model of the transformer winding, iron core, and oil tank were established. Transient electromagnetic-field analysis was used to obtain the alternating electromagnetic force of the transformer core and winding under the effect of alternating currents. Noise distribution was analyzed, but little attention was paid to the spectrum analysis of the windings under various fault conditions. Ji Shengchang and other scholars discussed in detail the relationship between the vibration of the winding, the iron core, the load current, and the no-load voltage, and proposed a method for extracting the characteristics of the vibration signal of the transformer based on wavelet analysis [16,17]. Through simulation and experimentation, Yu Xiaohui and others discussed the interaction between the tightening force and the natural frequency of the winding and concluded that the pretension of the windings can change their natural frequency [18]. A comprehensive analysis of the research conducted by experts around the world reveals that, although many effective diagnostic methods based on vibration signals have been proposed, there still exist problems, such as incomplete simulations of the various types of winding faults and poor diagnostic accuracy.

In recent years, various nondestructive testing methods for transformer-winding deformation have been developed, such as the frequency-response analysis method for comparing transformer frequency-response changes, and vibration analysis method for judging winding state based on the transformer-vibration signal. The principle of frequency-response analysis is to detect the amplitude-frequency response characteristics of each winding of the transformer, and compare the detection results horizontally or vertically. According to the difference of amplitude-frequency response characteristics, winding faults that may occur in the transformer are comprehensively judged. In recent years, scholars have paid more and more attention to vibration-detection transformer research. The vibration-analysis method discriminates the winding state of the transformer by detecting the vibration signal transmitted to the body surface [19,20]. The principle is to reflect the winding states by detecting a change in the mechanical characteristics of the winding. The frequency-response method has many factors that affect the test results, such as the position of the signal source, the length of test leads, the length of the test instrument grounding wire, the position of the transformer tap changer, and connection mode. Compared with the frequency-response method, the vibration-analysis method has fewer factors affecting the test results. The noise of the transformer cooling system will pollute the vibration signal. When collecting the signal, it should be as far away from the fan group as possible, or you should take noise-reduction measures. The vibration test results of transformer-winding deformation are affected by the vibration of the core. Power frequency 150 Hz and 250 Hz components appear in the frequency spectrum of transformer-vibration signals under a three-phase asymmetric operation. The severe overvoltage generated in the asymmetric phase increases the amplitude of the resonant frequency of the core, which interferes with the test results of winding. When a short-circuit fault occurs, the vibration of the iron core is far less than that of winding, and it can be approximated that the detected vibration contains only vibration signals of the winding. The frequency-response method is blackout detection, while the vibration-detection method is live detection [21,22]. It can continuously monitor transformer-winding deformation and reflect the decline trend of transformer short-circuit resistance after repeated short-circuit shocks, which reduces the difficulty of online monitoring and fault diagnosis of the power transformer.

Because the vibration of the transformer is a complicated process, the interaction between the magnetic field and the load current, as well as between the magnetostriction of the silicon steel sheet and the structural change in the transformer, produce changes in the vibration signal in both the time and frequency domains, thus increasing the difficulty in fault monitoring and diagnosis. In this paper, the SCB10-1000/10 dry-type transformer is studied in detail from the perspective of simulation modeling, fault simulation, and feature analysis to obtain state diagnostics on transformer winding via vibration analysis. Based on the mechanical-vibration characteristics of dry-type transformer windings, a short-circuit experiment was performed on the SCB10-1000/10 transformer, and the vibration signals at its surface were measured. A vibration-simulation model of the SCB10-1000/10 transformer was established using COMSOL Multiphysics 5.3, and the coupling calculations were performed with regard to the circuit, magnetic field, and solid mechanics of the transformer, among other areas of physics. By comparing the simulated data to the actual data of the transformer, the accuracy of the model was proven. Using this model, faults like loosening, deformation, and loss of insulation from the transformer windings were simulated, and the vibration characteristics of the winding fault were subsequently analyzed. The model utilizes multiphysical field-coupling simulation of the electromagnetic solid mechanics of dry transformer windings, which can provide a new basis for the state simulation and fault diagnosis of transformer windings.

2. Study of the Mechanical Vibration Characteristics of Transformer Windings

2.1. Vibration-Signal Conduction Process and Winding Electrodynamic Analysis of Dry-Type Transformers

This paper focuses on dry-type transformers. To understand the mechanism behind the mechanical vibration of transformer windings, a short-circuit experiment of the SCB10-1000/10 dry-type transformer was conducted, and the vibration signal at the surface was measured. The vibration of power transformers during operation is complicated and influenced by many factors, but there are two main phenomena: the vibration caused by the electric force on the winding and the vibration caused by the Lorentz force and the magnetostrictive force on the silicon steel sheet [23]. Figure 1 shows the conduction process of a vibration wave for a dry-type transformer. The vibration caused by the winding and the iron core is transferred to the surface of the fixed clamp of the transformer through the rigid component that connects the two. A dry-type transformer consists of layer-type windings, which cause vibration from the effect of the electrical power. These windings pass through the rigid connecting component to the fixed-clamp surface. The iron core of a dry-type transformer is subjected to magnetostrictive force and the action of the Lorentz force, which is transmitted to the surface of the fixed clamp of the transformer by a support unit, such as the cushion block or the fastening bolt [24].

When the load current of the dry-type transformer is loaded, leakage of the magnetic field occurs in its vicinity, which produces electrical power and causes mechanical vibration of the transformer winding. This vibration is transferred through the connecting component to the surface of the transformer clamp. When the transformer is in a steady state, the load current inside the winding can be found as follows:

$$i_t = I \cos \omega t \tag{1}$$

In Equation (1), the current effective value is presented, where ω represents the current angular frequency.

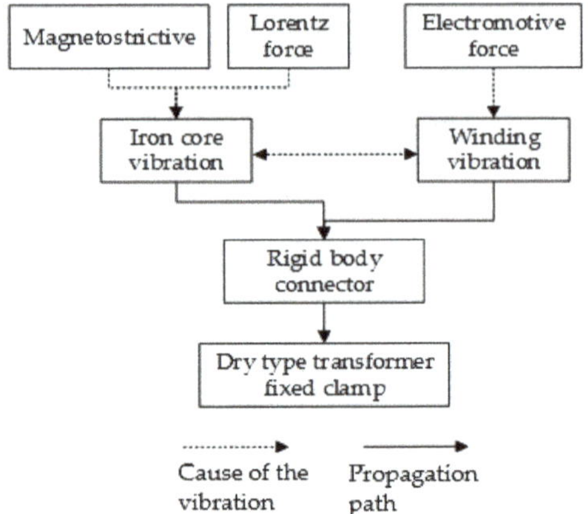

Figure 1. Dry-type transformer-vibration transmission route.

The vibration of the transformer body is mainly caused by core vibration, which, in turn, is caused by the magnetostriction of the silicon steel sheet and winding vibration resulting from load current. The vibration of the core is caused by the magnetostriction of the silicon steel sheet in a strong magnetic field [25]. The amplitude of the vibration is directly proportional to the square of the excitation voltage, and the fundamental frequency is two times greater than the voltage frequency. The vibration of the winding is caused by the electromagnetic force produced by the current in the winding. The amplitude of the vibration is proportional to the square of the winding current, and the basic frequency is two times greater than the current frequency. In the short-circuit test of the transformer winding, due to low excitation voltage, the vibration of the winding is far greater than the vibration of the core. Therefore, the detected vibration signals can be approximated as containing only the vibration signals of the winding.

The leakage of the magnetic field around the winding of the transformer is a function that changes with time. When the winding generates a change in position, the distribution of the leakage of the magnetic field around the winding also changes [26]. To calculate the force on a single conductor, the discrete magnetic field value is fitted to a continuous distribution function. By the Biot–Savart Law, magnetic flux density B at a certain point on the windings can be expressed as follows:

$$\vec{B_t} = \frac{u_0}{4\pi} i_t \int_{l'} \frac{dl' \times r^0}{r^2} \tag{2}$$

At a given point in space, all quantities except i_t are constant. Thus, in the calculation of the static electromagnetic field, the magnetic-flux leakage density B_t of the winding can be simplified to the following:

$$\vec{B_t} = \vec{k} I \cos \omega t \tag{3}$$

where $\vec{k}$ is the proportionality constant between magnetic-flux density and load current.

The axial magnetic-field leakage induced by the load current flowing through the transformer winding is B_{zt}, and radial electromagnetic force F_x is induced by the action of the load current. Similarly, the magnetic leakage field B_{xt}, induced by radial induction, can induce axial electromagnetic force F_z through the load current. The axial force and radial force of the conductor can be calculated from the electric-force equation as follows:

$$F_x = i_t B_{zt} 2\pi R$$
$$F_z = i_t B_{xt} 2\pi R \tag{4}$$

By vector calculation, the axial and radial electric forces are integrated to simplify the electric power of the winding:

$$\begin{aligned}
F &= \sqrt{F_x^2 + F_z^2} = i_t B_t 2\pi R \\
&= I \cos \omega t \cdot kI \cos \omega t \cdot 2\pi R \\
&= 2\pi R k I^2 \left(\tfrac{1}{2} + \tfrac{1}{2} \cos 2\omega t\right)
\end{aligned} \tag{5}$$

where i_t is the load current in the winding, ω represents the angular frequency of the power, and R represents the transformer-winding radius. Equation (5) shows that the magnitude of the electric force on the transformer winding is proportional to the square of the load current flowing through the transformer winding, and the fundamental frequency of the vibration signal is twice the power frequency of the power grid.

2.2. Transformer Short-Circuit Experiment

Based on the analysis above, when the secondary winding of the transformer is short-circuited, the vibration of the body is mainly caused by winding vibration. To eliminate the disturbance caused by vibration of the iron core and analyze only the vibration characteristics of the transformer winding, a short-circuit experiment was performed on an SCB10-1000/10 transformer, which is a 10 kV SCB11 epoxy resin-cast dry-type transformer.

The parts with strong vibration signal are more likely to have fault accidents occur [27]. High signal-to-noise ratio can also be obtained by selecting a strong vibration area of the transformer winding [28]. According to the model studied in this paper, the vibration signal near the winding is strong. In order to understand the vibration condition of the winding, the inner side of the upper winding is the first choice, that is, the position selected in this experiment. During the experiment, the intensity of the vibration signal is tested in different areas. The results show that the vibration signal near the winding is tested. Considering the difficulty of installing the vibration sensor and the intensity of the vibration signal, the test point near the B-phase winding, the test point near the B-phase winding, and the test point on the inner and upper side of the C-phase winding are selected. Figure 2 shows the location of the vibration-acceleration sensor relative to the transformer for the field experiment. The key technical parameters are shown in Table 1.

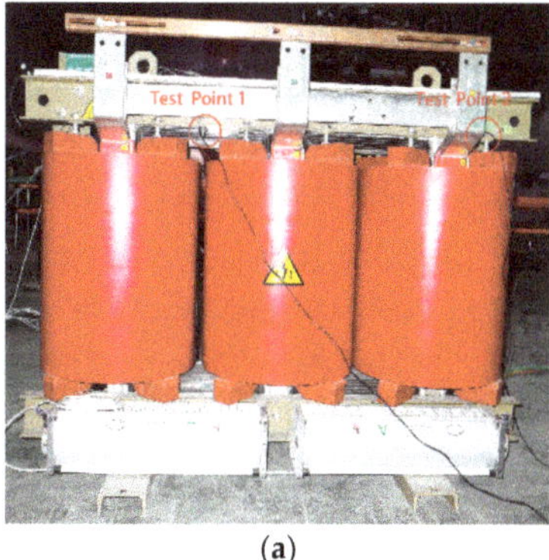

(a)

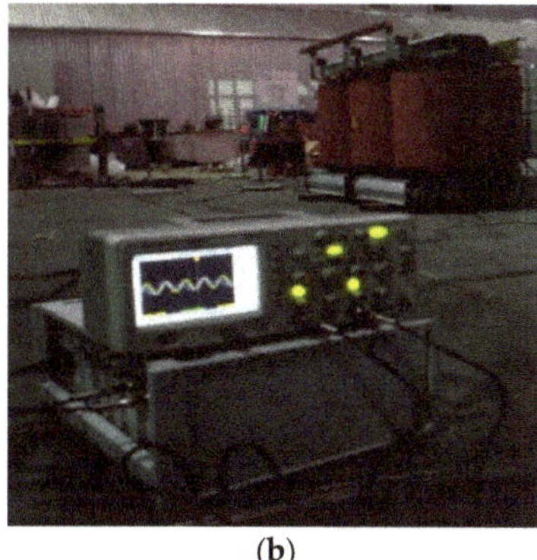

(b)

Figure 2. (**a**) Position of vibration-acceleration sensor; (**b**) field experiment.

When the transformer is short-circuit tested, the secondary low-voltage side is short-circuited, and a three-phase voltage is applied to the primary high-voltage side such that the load current in the winding attains its rated value. The vibration signals were measured using a vibration acceleration sensor (YD70C) (Xieli Science and Technology, Qinhuangdao, China), a charge amplifier (DHF-10), and a Tek oscilloscope.

Table 1. Main technical parameters of the SCB10-1000/10 dry-type transformer.

Main Technical Indicators	Parameter
Phase number	Three-phase
Rated frequency	50 Hz
Rated capacity	1000 kVA
Rated voltage	10,000/400 V
Rated current	57.74/1443.38
Connection mode	Dyn11
Cooling mode	AN
Short-circuit impedance (%)	5.91

Figure 3 shows a set of typical mechanical-vibration signals measured during the experiment. The ordinate axis in this picture represents vibration acceleration a. The relationship between vibration acceleration a and output voltage U of the charge amplifier is as follows:

$$U = a \times S_1 \times S_2 \tag{6}$$

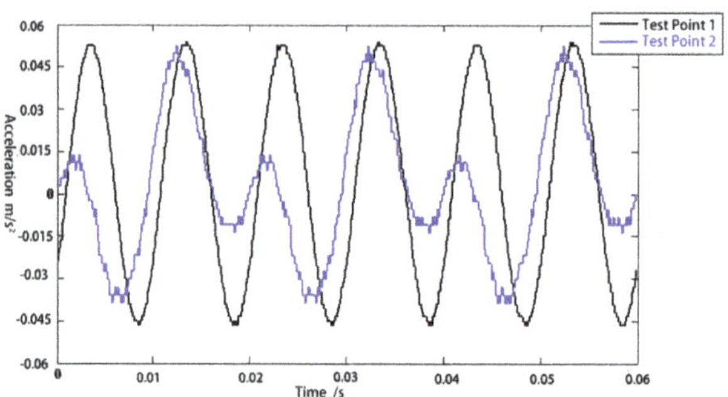

Figure 3. A set of typical mechanical-vibration signals.

The sensitivity S_1 of the vibration acceleration sensor was $2\ pC/ms^{-2}$, and the sensitivity S_2 of the charge amplifier was $656\ V/pC$. The mechanical-vibration-signal diagram in Figure 3 can be obtained via transformation.

3. Multiphysical Field-Coupling Model for Winding Vibration of Dry-Type Transformer

In this paper, COMSOL Multiphysics software was used to simulate the vibration of the SCB10-1000/10 dry-type transformer winding, and coupling simulation analysis of the magnetic field and the solid mechanics was performed.

3.1. Establishment and Mesh Generation of a Vibration-Simulation Model for a Dry-Type Transformer

The structure of the transformer is complex and includes a variety of components such as the winding, iron core, and cooling device. For the purpose of the simulation, the transformer is simplified and treated as an ideal model in which the internal cooling device and supporting fastening components are neglected. The finite-element geometric model was set up based on the actual structural parameters of the SCB10-1000/10 dry-type transformer (Xuzhou Debon Electric Equipment Co., Ltd., Xuzhou, China), and a fixed constraint was applied at both ends of the winding to simulate clamping. To simulate the electrical insulation of the actual transformer, three-phase high- and low-voltage winding turns and layers were used, which allowed the electromagnetic field to be solved. The solid mechanics model was set up as a fully coupled solution to understand the connection between the differential equation of motion and the differential equation of the electromagnetic field.

The simulation model is shown in Figure 4a. To improve efficiency, the geometric transformer model was simplified by considering it symmetric. Figure 4b illustrates the simplified geometric model of the body.

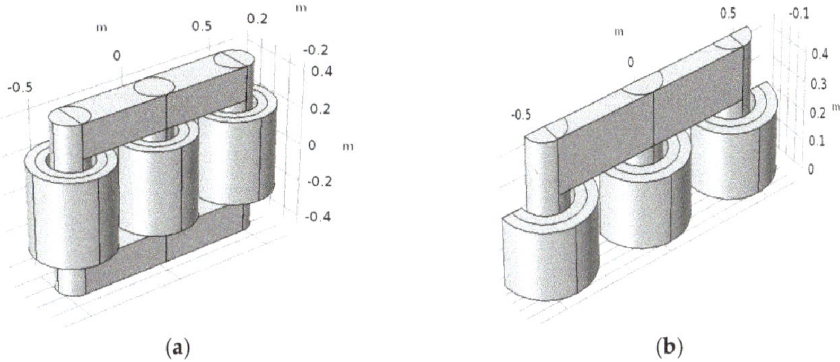

Figure 4. (**a**) Integral model of transformer; (**b**) simplified model of transformer.

The geometric model of the transformer is meshed by means of a free tetrahedron network, and the mesh model shown in Figure 5a can be obtained. The mesh quality is shown in Figure 5b, and the closer the value is to 1, the higher the mesh quality is. The winding and core structural parameters of the SCB10-1000/10 dry-type transformer are shown in Tables 2 and 3.

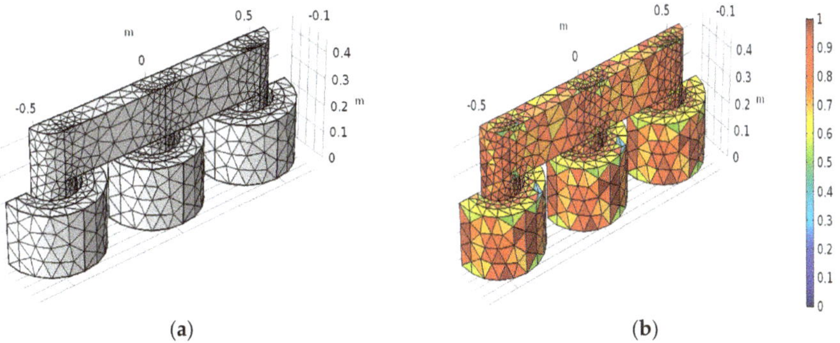

Figure 5. (**a**) Mesh-generation results of geometric model; (**b**) quality diagram of grid division.

Table 2. Main parameters of the SCB10-1000/10 dry-type transformer coil structure.

Coil	Parameter Type	Size (mm)	Coil	Parameter Type	Size (mm)
	Internal diameter	286		Internal diameter	252
	External diameter	369		External diameter	280
High pressure	Height	405	Low pressure	Height	446
	Turn number	1125		Turn number	45
	Type	Layer type		Type	Layer type

Table 3. Main parameters of the transformer core structure.

Parameter Type	Parameter	Parameter Type	Size (mm)
Structure	Three-phase three-column	General length	990
Joint method of side column	Standard oblique connection	Total height	900
Silicon steel sheet material	35Q165	Thickness	196
Core column radius (mm)	90	Upper- and lower-yoke height	180

3.2. Simulation Model of Electromagnetic Field

Figure 6 shows the circuit diagram of the transformer. In the diagram, an AC voltage source with a 50 Hz frequency was applied to the three-phase high-voltage side, and the low-voltage side was short-circuited. A 50 Hz AC voltage to the rated current was applied to the high-voltage winding by an external voltage source. The three-phase induction current of the low-voltage winding was obtained by electromagnetic-coupling calculation, as shown in Figure 7. As a result, the amplitude of the three-phase induction current of the low-voltage winding was 2041 A, and its effective value was 1443.38 A. The simulation results are consistent with the rated current of the transformer, as shown in Table 1.

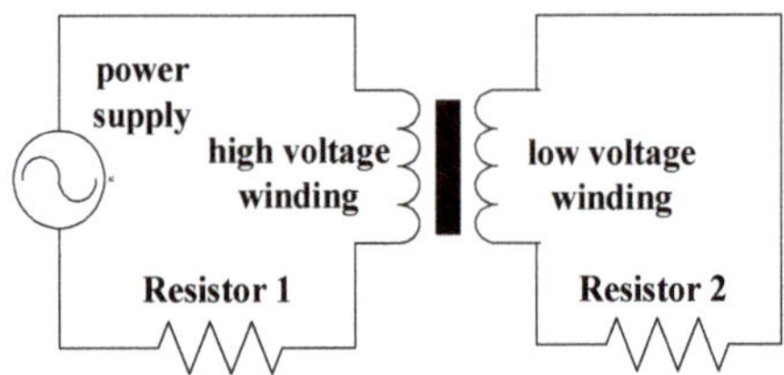

Figure 6. External-circuit equivalent diagram.

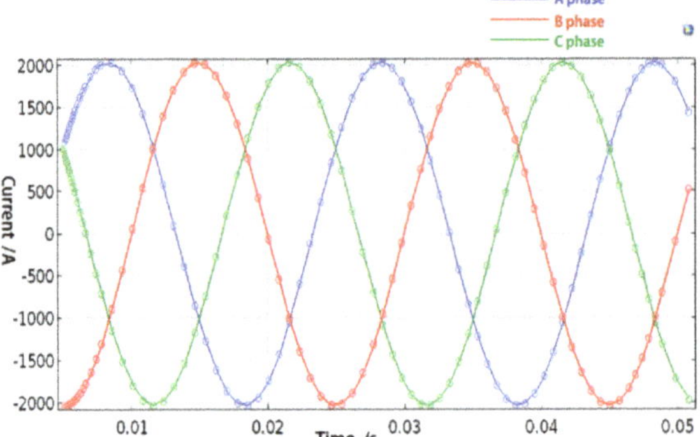

Figure 7. Transformer low-voltage coil three-phase induction current.

The current obtained from the circuit module is used as the excitation source of the magnetic-field model, and current density J was brought into the magnetic-field differential equation as follows:

$$\mu_0^{-1}\mu_r^{-1}\nabla^2 A = J^e \tag{7}$$

In Equation (7), μ_0 represents the permeability of free space and has a value of $4\pi *10^{-7} H/m$, R is the relative permeability, A is the vector magnetic potential, and J^e is the current density. Additionally, the following relationship exists in the electromagnetic-field model:

$$B = \mu_0\mu_r H = \nabla \times A \tag{8}$$

where B represents the magnetic-flux density, and H represents the magnetic-field intensity. Figure 8 shows the H–B curve of the transformer when it is in operation. The curve in Figure 8 also provides information on the material properties of the dry-type transformer core, which agrees with the actual material properties of the transformer.

The magnetic-flux density distribution around the winding and the core of the dry-type transformer during a short-circuit test was obtained by coupling the electromagnetic-field model.

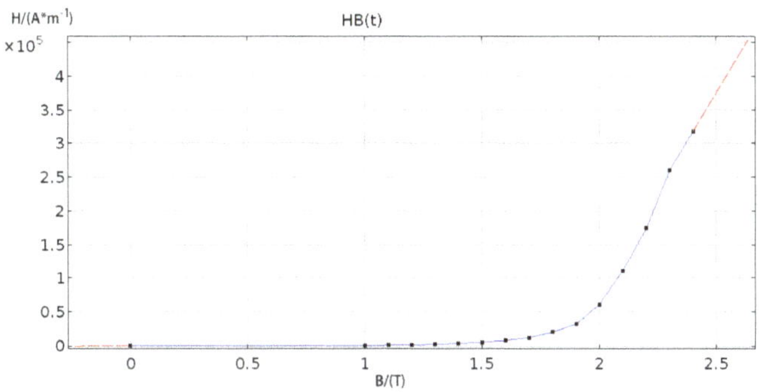

Figure 8. H–B curve.

As seen from Figure 9, the maximum flux-leakage density of the dry-type transformer winding was 0.08 T, and the maximum flux density of the core was 1.75 T. The results of the simulation are in line with the output parameters of the SCB10-1000/10 dry-type transformer. The results verify the correctness of the model, demonstrating that the model can be further used for the simulation of various faults in the windings.

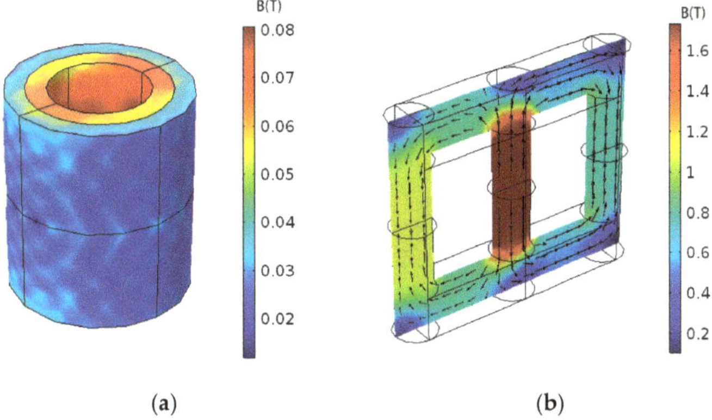

    (**a**)   (**b**)

Figure 9. (**a**) Flux-density distribution of the transformer winding; (**b**) Flux-density distribution of the transformer iron core.

3.3. Modeling of Solid Mechanics and Analysis of Winding Vibration

In the solid-mechanics model, the vibration characteristics of the dry-type transformer winding must be coupled. The mass inertia, elasticity, and damping of the winding are the key factors affecting its vibration. Equation (9) provides the differential equations of motion for the solid mechanics.

$$M_i \frac{d^2 z}{dt^2} + C_i \frac{dz}{dt} + k_i z = f(t) \tag{9}$$

In Equation (9), M_i represents the mass matrix; C_i represents the damping coefficient matrix, $k_i[0,1]$ represents the stiffness coefficient matrix, z represents the deformation (displacement) of the winding, $\frac{dz}{dt}$ represents the deformation velocity of the winding, $\frac{d^2 z}{dt^2}$ represents the deformation acceleration of the winding, and $f(t)$ represents the magnitude of the force on the winding. Equation (5) for electrodynamic force is incorporated into the differential equation of Motion (9).

$$lM_i\frac{d^2z}{dt^2} + C_i\frac{dz}{dt} + k_iz \\ = \frac{1}{2}BI^2 + \frac{1}{2}BI^2\cos 2\omega t + M_ig \tag{10}$$

To converge to a solution, it is necessary to impose fixed constraints on both ends of the transformer winding.

In Equation (10), mass matrix M_i, damping coefficient C_i, and stiffness coefficient k_i are all constant, which makes it a constant coefficient differential equation. The solution of the equation is composed of the general solution and the special solution. For the homogeneous part,

$$M_i\frac{d^2z}{dt^2} + C_i\frac{dz}{dt} + k_iz = 0 \tag{11}$$

the general solution is as follows:

$$z_0 = Ye^{-\frac{C_it}{2M_i}}\sin(\omega_0 t + \theta) \tag{12}$$

In Equation (12), Y and θ are constants whose values are determined by the initial conditions, and ω_0 represents the natural frequency of the vibration of the transformer winding, which is expressed as follows:

$$\omega_0 = \sqrt{\frac{K_i}{M_i} - \left(\frac{C_i}{2M_i}\right)^2} \tag{13}$$

There are two special solutions for Equation (10):

$$z_1 = \frac{0.5BI^2 + M_ig}{K_i} = D \tag{14}$$

$$z_2 = G\cos(2\omega t + \psi) \tag{15}$$

In the equation: $\quad G = \dfrac{BI^2}{\sqrt{(K_i - 4M_i\omega^2)^2 + 4C_i^2\omega^2}};$
$\quad\quad\quad\quad\quad\quad \tan\psi = -\dfrac{2C_i\omega}{K_i - 4M_i\omega^2}$.

The total displacement of the transformer winding at any time can be expressed as follows in Equation (16):

$$z = z_0 + z_1 + z_2 = Ye^{-\frac{C_it}{2M_i}}\sin(\omega_0 t + \theta) + D + G\cos(2\omega t + \psi) \tag{16}$$

where t, the total displacement of the winding, and the displacement velocity are zero. Integral constants Y and θ can then be obtained.

By quadratic derivation of the total displacement equation of the transformer winding, the vibration acceleration a of the transformer winding at any time t can be obtained as in Equation (17):

$$a = -\omega_0^2Ye^{-\frac{C_it}{2M}}\sin(\omega_0 t + \theta) \\ -4\omega^2G\sin(2\omega t + \psi) \tag{17}$$

From the solution above, it can be seen that the vibration characteristics of a given transformer's windings are mainly related to the elastic coefficient and the winding geometry, that is, when elastic coefficient K and the geometric structure of the transformer winding are changed, the vibration acceleration of the transformer-winding surface changes accordingly. Table 4 shows the main material properties of the winding and the core in the solid mechanics model [29].

Table 4. The main material properties of the solid mechanics model.

Structure	Modulus of Elasticity (Pa)	Density (g/cm^3)	Poisson's Coefficient
Winding	1.16×1011	3.2	0.32
Iron core	2×1011	7.6	0.24

The amplitude and axial forces on the transformer winding are obtained through a simulation of the solid mechanics, as demonstrated in Figure 10. The legend shows the magnitude of stress, and the red arrows indicate the direction of force. As seen in Figure 10a, the external high-voltage windings receive outward traction, while the inner low-voltage windings are pushed inward; the force of the outer winding is obviously lower than that of the inner winding. Figure 10b shows that the two ends of the transformer winding are subjected to inward extrusion pressure. Therefore, the above simulation results are in line with the actual force on the transformer windings.

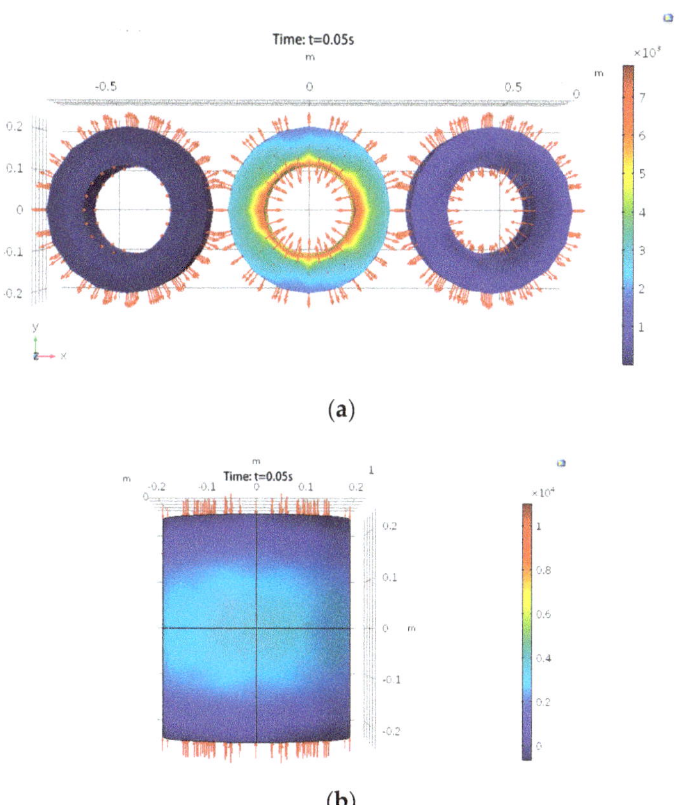

(a)

(b)

Figure 10. (a) Radial-stress distribution diagram of transformer windings; (b) axial-stress distribution diagram of transformer winding.

To further verify the accuracy of the above model, the vibration signals measured by the transformer short-circuit test were compared with the vibration signals from the simulation. Figure 11 shows the layout of the actual vibration-signal measurement points and the simulation-model measurement points. The locations in the simulation model are consistent with the vibration signal measured in the experiment.

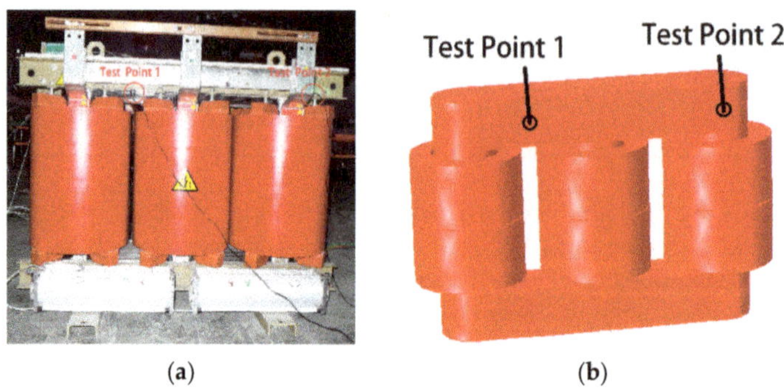

(**a**)   (**b**)

Figure 11. (**a**) Experiment locations of measurement points; (**b**) Simulation locations of measurement points.

Figure 12 shows the calculated results of vibration at the placement of the acceleration vibration sensor and the measured results. Because the current passing through the winding in the short-circuit experiment is rated current, the simulation-calculation conditions are close to the actual measurement conditions. As shown in Figure 12, the simulation results are in good agreement with the measured results, which further illustrates the validity of the finite-element model and the correctness of the calculation results. In addition, the vibration waveform of the test point appears in the shape of the top cusp, which may be due to the material characteristics [30].

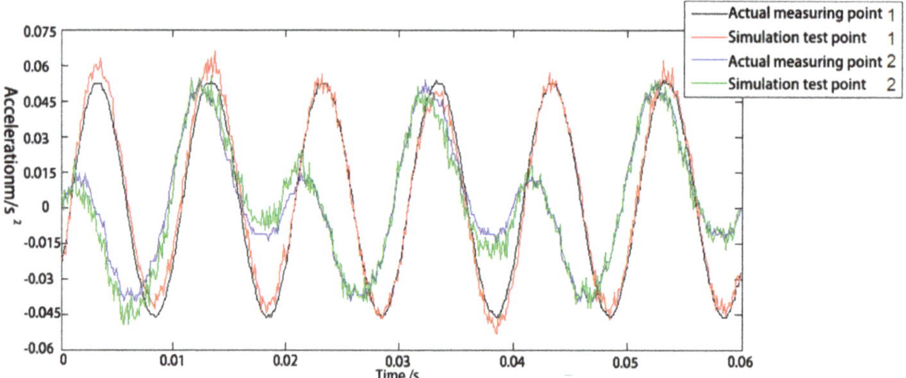

Figure 12. Comparison of vibration signals at each test point.

4. Simulation and Analysis of the Mechanical Faults of Transformer Windings

4.1. Simulation Geometric Model of the Mechanical Faults of Transformer Winding

Models were simulated for winding-insulation loss, winding looseness, and winding deformation, with the number of coil turns held constant. The nonlinear characteristics of insulating pads and clamps in transformer windings cause their elastic modulus to vary with pretightening force. That is to say, the elastic coefficient of insulating pads on both sides of windings is related to the degree of tightening after assembly [31]. In this paper, as shown in Figure 13a, the core and clip were reasonably simplified. Pretightening force was reduced by changing the elastic coefficient of the B-phase winding material, and the loosening of the B-phase winding was simulated by slightly extending the length of the B-phase winding. Transformer-winding deformation is usually caused by the electric force formed by the short-circuit current and axial magnetic field. For the winding-deformation faults simulated in Figure 13b, the high-voltage windings were subjected to radial electrodynamic force of outward expansion, resulting in the windings bulging. The low-voltage winding was twisted by the internal radial electric force. In the model studied in this paper, in order to study the axial and radial displacement of winding after a winding-deformation fault more intuitively, circular winding was changed into an octagonal cylinder with obvious edges and angles to simulate the bulging condition after winding deformation. As shown in Figure 13c, the effect of the winding insulation was simulated by reducing the number of B-phase winding coils. During the simulation, the other conditions were not changed.

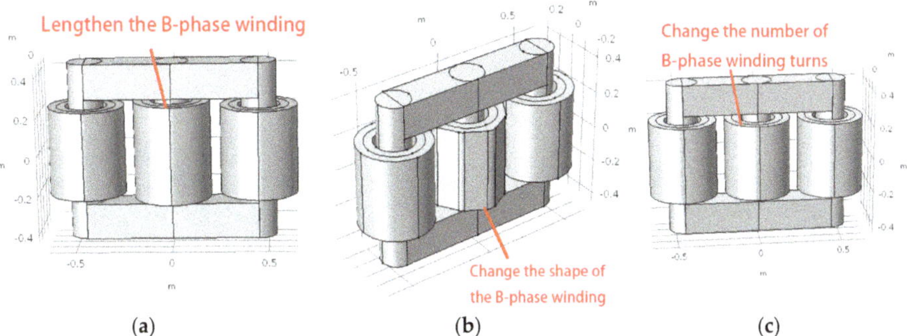

Figure 13. (**a**) Simulation model of winding loosening; (**b**) simulation model of winding-insulation failure; (**c**) simulation model of winding deformation.

4.2. Analysis of the Simulation Results of Three Typical Faults

As shown in Figures 14 and 15, the total displacement of normal windings and the total displacement of windings under the effect of insulation shedding, winding loosening, and winding deformation were compared for a time period of 0.005 s.

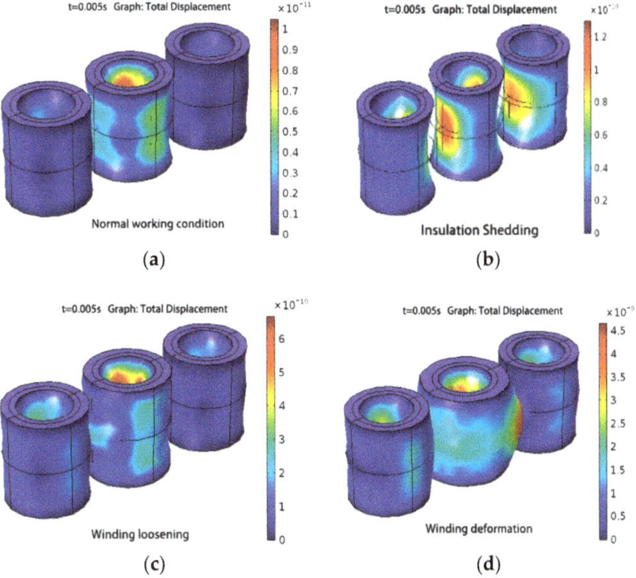

Figure 14. Axial diagram of the total displacement of windings under various working conditions. (**a**) Normai working; (**b**) Insulation Shedding; (**c**) Winding loosing and (**d**) Winding deformation.

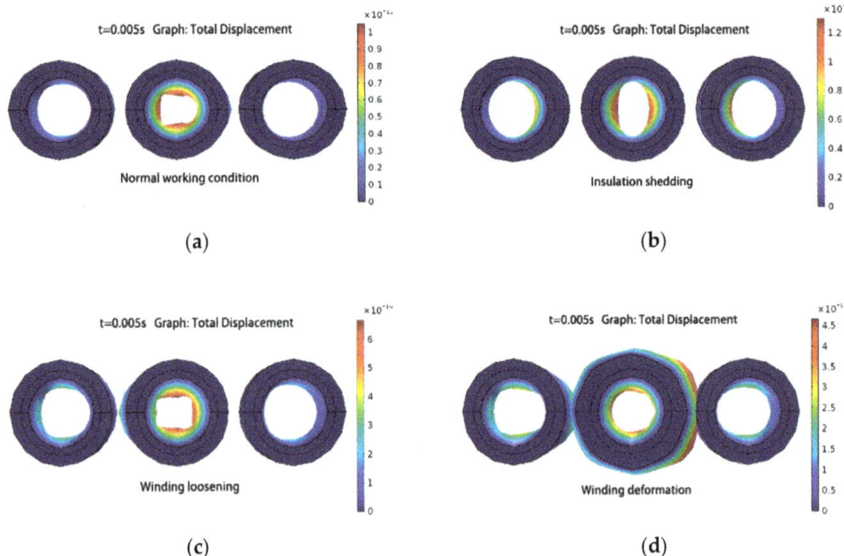

Figure 15. Amplitude direction diagram of winding force displacement under various conditions.
(**a**) Normal working; (**b**) Insulation Shedding; (**c**) Winding loosening and (**d**) Winding deformation.

Figure 14 shows that the total displacement of the transformer windings was smallest under normal conditions, while the winding was more prone to displacement in the event of a fault. Total displacement was greatest in the case of winding deformation. The results validate the cumulative characteristics of winding mechanical faults, that is, small mechanical faults increase force on the winding, which makes it relatively easy for greater mechanical faults to occur. Therefore, considerable attention should be paid to the mechanical state of the winding. If a small mechanical failure occurs, it should be dealt with in a timely manner; if it is not, it becomes relatively easy for a serious accident to occur.

Figure 15 shows that the total displacement of the low-voltage winding in the inner part of the transformer was greater than the displacement of the lateral high-voltage winding, which indicates that the force of the low-voltage coil was greater than that of the high-voltage coil. This result is due to the opposing directions of current flow along the windings between the high- and low-voltage sides of the transformer, as well as the mutual exclusion of the electromagnetic force in the radial direction of the two windings, which is in accordance with the actual force. This result verifies the accuracy of the model.

Figure 16 shows a time-domain diagram of the vibration-acceleration signal at measurement point 1 under normal working conditions, as well as in the cases of insulation shedding, winding loosening, and winding deformation.

As seen from Figure 16, the vibration signal at measurement point 1 under normal conditions was more stable than the vibration signal at the time of failure, and the average amplitude of the vibration signal was approximately 0.06 m/s^2. When insulation shedding occurred, the maximum amplitude of the vibration-acceleration signal at measurement point 1 was 0.08 m/s^2. When the winding was loose, the maximum amplitude of the vibration acceleration signal at measurement point 1 was 0.28 m/s^2. When winding deformation occurred, the maximum amplitude of the vibration-acceleration signal at measurement point 1 was 0.25 m/s^2. When a winding fault occurred, the amplitude of the vibration signal was nonstationary. Moreover, when the winding was loose, and the winding deformation failed, the change was more intense. When the transformer windings failed, vibration signals would obviously change. The amplitude of the vibration signal in the fault was higher than the amplitude of the normal vibration signal, and the fluctuation of the signal was intense.

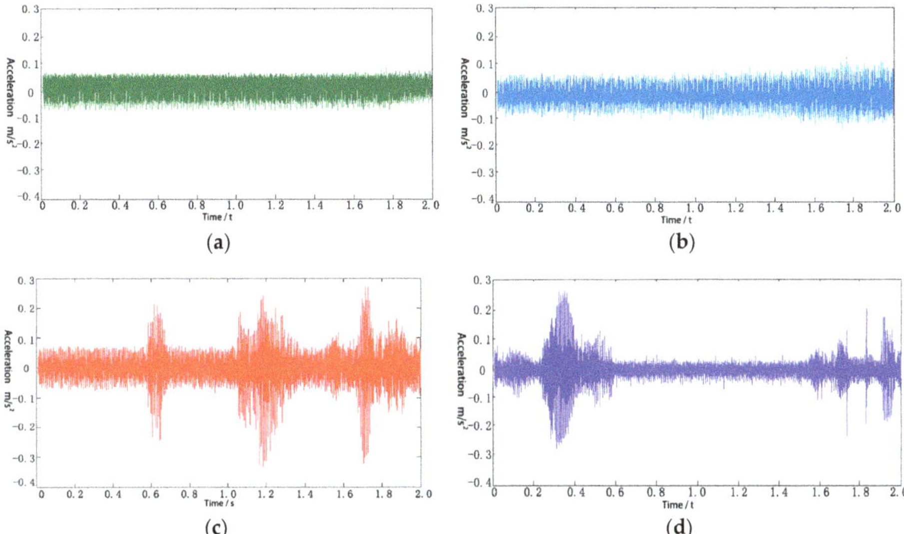

Figure 16. Time-domain diagram of vibration-acceleration signals at measurement point 1 under various working conditions: (**a**) normal working conditions; (**b**) insulation shedding; (**c**) winding loosening; (**d**) winding deformation.

Figure 17 presents a spectrum-analysis diagram of the vibration signal at measurement point 1, when the transformer was working under normal conditions, as well as for the cases of insulation shedding, winding loosening, and winding deformation.

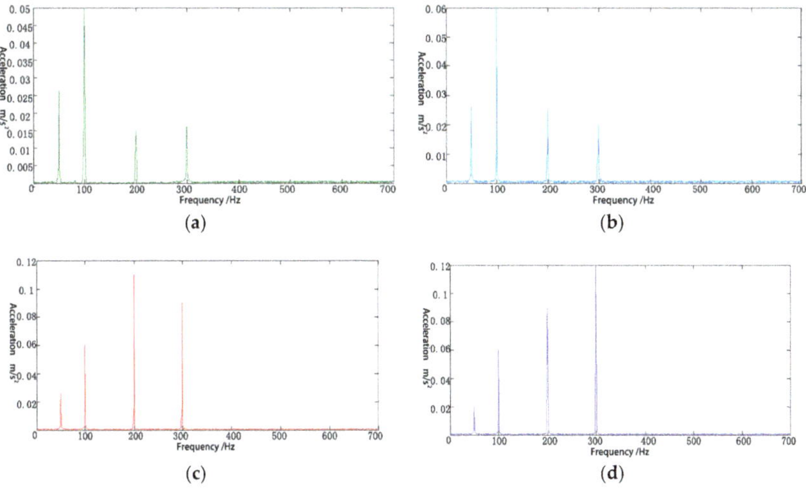

Figure 17. Spectrum diagram of vibration signal at measurement point 1 under different working conditions: (**a**) normal condition; (**b**) insulation shedding; (**c**) winding loosening; (**d**) winding deformation.

As shown in Figure 17, the maximum frequency attained by the vibration-acceleration signal at transformer measurement point 1 was 100 Hz under normal operating conditions, at which point the amplitude was 0.05. Signals with frequencies of 50 Hz, 200 Hz, and 300 Hz can also be seen in this diagram. When insulation was shed, the amplitudes of the 100 Hz, 200 Hz, and 300 Hz signals increased. When the windings were loose, the maximum amplitude of the signal in the frequency domain occurred at 200 Hz, and the amplitude of the signal of 300 Hz increased apparently as well. When winding deformation occurred, the maximum amplitude of the vibration signal occurred at 300 Hz, and the amplitude of the signal spectrum at 200 Hz was greater than the amplitude at 100 Hz. These results are consistent with the experimental results of transformer faults in Reference [32], which verifies the accuracy of the model.

By studying the vibration characteristics of transformer windings, it is known that the corresponding mechanical vibration will change when the mechanical state of the windings changes. In order to make the change more intuitive and distinguish the characteristics of different mechanical states, based on the analysis of transformer-winding vibration characteristics, some artificial-intelligence algorithms can be used to identify fault features, such as Improved Variational Mode Decomposition (IVMD)-Weight Divergence, which is a feature-extraction method presented for vibration signals of transformer windings. In the process of feature extraction, the mechanical vibration signals of transformer windings in different states are decomposed into a series of finite-bandwidth intrinsic-mode functions (IMFs) by means of Improved Variational Mode Decomposition; K-L divergence (K-L) between the IMF component and the original vibration signal is calculated, and the weighting coefficients are multiplied. Weight divergence is obtained to represent the time-frequency domain complexity of mechanical-vibration signals of transformer windings in different states. It can effectively extract the characteristics of a mechanical-vibration signal of transformer windings [33].

Many scholars have studied the vibration state of the transformer. Gu Hongxia of Kunming University of Technology, through the finite-element analysis of winding-loosening faults, showed that the natural frequency of the four orders in winding-loosening faults and the natural frequency of the normal state of the winding transfer to a low frequency [34]. The simulation experiment shows that it can be used. Natural frequency is used to judge the working condition of power-transformer windings. However, this paper only focuses on the loosening fault of windings, and other faults such as winding deformation and insulation shedding are not studied. From the point of view of acoustic measurement, C. Bartoletti extracts the weighted values and proportions of the middle-, low-, and high-frequency signals from the noise signals for transformer-fault diagnosis [35]. In this paper, the operation state of the transformer is diagnosed by the sound signal. But the vibration signal is transmitted through air, and noise reduction is needed to obtain the available vibration signal. The monitoring effect is not as accurate as the vibration of the transformer itself. Hyun-Mo Ahn has more analysis on the vibration and stress of short-circuit faults, but less research on the vibration caused by load changes in normal operation [36]. Ji Shengchang of Xi'an Jiaotong University has done long-term work on the vibration characteristics of the transformer. The radial vibration of the transformer winding, axial-vibration characteristics of the pressure plate, and influencing factors were studied [37]. The vibration law and propagation characteristics of transformer windings were revealed. There is no theoretical calculation of the radial vibration of the winding, and no research on the change rule of the vibration characteristics of the typical winding mechanical-fault state, which makes it difficult to judge the actual fault.

By setting winding looseness, winding deformation, and insulation shedding, and by means of simulation and experiment, the frequency-spectrum characteristics of winding-vibration signals under different conditions were analyzed. From the research, it can be seen that the transformer-vibration model provides a reliable basis for transformer-winding state simulation and fault diagnosis, and can be used to easily calculate the distribution of the magnetic field and the winding characteristics of the transformer. Moreover, this model can be used to analyze vibration control during the design of a transformer, thereby shortening the design cycle. It also provides a theoretical basis and method of calculation for further analysis and verification of new methods of reducing electromagnetic noise in transformers.

5. Conclusions

In this paper, the mechanical-vibration characteristics of transformer windings were studied theoretically. Changes in the mechanical state of the windings were reflected in their vibration signals. A simulation model of vibration was established for an SCB10-1000/10 dry-type transformer winding using COMSOL software, and a multifield-coupling simulation of the circuit, magnetic field, and solid mechanics was performed. The following conclusions were obtained through simulation and experimental analysis. By changing the geometry model and power parameters, similar conclusions can be obtained for other types of dry-type transformers, which can be extended to the study of winding-vibration characteristics of various types of dry-type power transformers. For the oil-immersed transformer, the simulation model was greatly changed, so it was necessary to add an oil-tank wall to the transformer model, replace the medium with insulation oil, and consider the vibration-signal propagation process in the insulation oil. At the same time, we also studied the occurrence and propagation of the winding vibration of oil-immersed transformers. Because the oil-immersed transformer was closed, the vibration-signal sensor could only be mounted on the wall of the oil tank. When analyzing the signal received by the vibration-signal sensor, we needed to consider the properties of the different insulating oil. Therefore, the vibration analysis in this paper was limited to dry-type transformers.

(1) A short-circuit experiment was performed on an SCB10-1000/10 transformer, which is a dry-type transformer. Information on the vibration acceleration at the surface of the transformer was obtained using a vibration-acceleration sensor (YD70C), a charge amplifier (DHF-10), and a Tek oscilloscope. Vibration-signal analysis showed that, when the secondary winding of the transformer was short-circuited, the signal was mainly composed of a 100 Hz component and contained relatively small components at 50 Hz, 200 Hz, and 300 Hz.

(2) Based on the vibration data obtained from the transformer short-circuit experiment, a multiphysical field-coupling vibration-simulation model of the dry-type transformer winding is established using the parameters of the actual transformer. The vibration-acceleration signal was obtained from actual measurement points, which was then compared to the corresponding points in the simulation. The similarity between the two exceeded 80%. Therefore, the model can be used to investigate the vibration of transformer windings and possesses high value in engineering applications.

(3) Based on the simulation model, for normal working conditions, fault simulations of the transformer windings were carried out. Specifically, winding loosening, deformation, and insulation shedding were simulated, and the vibration characteristics of these winding faults were analyzed. When insulation was shed, the amplitude in the frequency domain at 100 Hz, 200 Hz, and 300 Hz increased. When the winding was loose, the maximum amplitude in the frequency domain appeared at 200 Hz, and the amplitude at 300 Hz was also relatively larger. When winding deformation occurred, the maximum amplitude of the vibration signal occurred at 300 Hz, and the amplitude at 200 Hz exceeded the amplitude at 100 Hz. The simulation results are consistent with the experimental results, which verify the accuracy of the fault model. This method can provide a new basis for simulating and diagnosing transformer-winding faults.

Author Contributions: X.D. and J.L. designed the multiphysics field-coupling model for transformer-winding vibration, performed the data analysis, and prepared the manuscript. L.Z. (Li Zhang) and L.Z. (Liang Zou) assisted the project and managed to obtain the transformer-vibration signal. T.Z. led the project and research. All authors discussed the results and approved the publication.

Funding: This research was supposed by the Key Research and Development project of Shandong Province (2018GGX104009).

Conflicts of Interest: The authors declare no conflict of interest.

References

1. Zhang, F.; Ji, S.; Shi, Y.; Lu, W.; Li, D.; Chen, L. Research on transformer winding vibration and propagation characteristics. *Trans. China Electrotech. Soc.* **2018**, *38*, 2790–2798.
2. Ghoneim, S.S.M.; Taha, I.B.M. A New approach of DGA interpretation technique for transformer fault diagnosis. *Int. J. Electr. Power Energy Syst.* **2016**, *81*, 265–274. [CrossRef]
3. Wang, J.; Li, D. Accidents and defects analysis of 110 kV and above voltage transformers. *Guangxi Electr. Power* **2014**, *37*, 63–64.
4. Wang, Z. Cause analysis to deteriorate of 220 kV power transformer and its counter measure. *Inner Mong. Electr. Power* **2016**, *59*, 34–41. [CrossRef]
5. Wang, M. 220 kV power transformer damage analysis and countermeasures 2005 year 110 (66) kV and above transformer accident and defect statistical analysis. *Power Equip.* **2006**, *7*, 99–102.
6. Deng, Y. *Typical Examples of Transformer Equipment Failure (2006–2010 Years)*; China Electric Power Press: Beijing, China, 2012; pp. 16–23.
7. Wang, M. Statistic analysis of transformer's faults and defects at voltage 110 kV and above. *Distrib. Util.* **2007**, *24*, 1–5.
8. Report, A.C. Bibliography on transformer noise. *Power Appar. Syst. Part III* **1954**, *73*, 1760–1762. [CrossRef]
9. Fahnoe, H. A Study of sound levels of transformers. *Electr. Eng.* **2013**, *60*, 277–282. [CrossRef]
10. Masti, R.S.; Desmet, W.; Heylen, W.; Leuven, K. On the Influence of Core Laminations upon Power Transformer Noise. In Proceedings of the 2004 International Conference on Noise and Vibration Engineering, Heverlee, Belgium, 20–22 September 2004.
11. Foster, S.L.; Reiplinger, E. Charcteristics and control of transformer sound. *IEEE Trans. Power Appar. Syst.* **1981**, *PAS-100*, 1072–1077. [CrossRef]
12. Xu, L.; Liu, X. Study on the Three dimension attenuated model and the algorithm of environmental noise in substations. *Proc. CSEE* **2012**, *32*, I0024.
13. Wu, G.; Cheng, S.G.; Huang, L.; Huang, L.; Gao, F. Prediction on noise of 220 kV outdoor substation to environmental infection. *Noise Vib. Control* **2007**, *27*, 135–137.
14. Zhu, L.; Yang, Q.; Yan, R.; Zhan, X. Research on vibration and noise of power transformer cores including magnetostriction effects. *Trans. China Electrotech. Soc.* **2013**, *28*, 1–6.
15. Hu, J.; Liu, D.; Liao, Q.; Yan, Y.; Liang, S. Analysis of transformer electromagnetic vibration noise based on finite element method. *Trans. China Electrotech. Soc.* **2016**, *31*, 81–88.
16. Ji, S.; Li, Y.; Fu, C. Application of on-load current method in monitoring the condition of transformer's core based on the vibration analysis method. *Proc. CSEE* **2003**, *2*, 154–158.
17. Ji, S.C. *Vibration Characteristics of Transformer Winding and Core and Its Application in Fault Monitoring*; Xi'an Jiaotong University: Xi'an, China, 2003.
18. Yu, X.; Li, Y.; Jing, Y.; Li, H. Calculation and analysis of natural frequency of winding model of transformer. *Transformer* **2010**, *47*, 5–8.
19. Zhang, B.; Xu, J.Y.; Chen, J.B.; Li, H.; Lin, X.; Zang, Z. Diagnosis method of winding deformation based on power transformer vibration information. *High Volt. Eng.* **2015**, *41*, 2341–2349.
20. Cheng, J.; Li, Y.; Ji, S.; Hao, H. Application of vibration method on monitoring the winding and core condition of transformer. *High Volt. Eng.* **2005**, *31*, 43–45.
21. Malewski, R.; Poulin, B. Impulse testing of power transformers using the transfer function method. *IEEE Trans. Power Deliv.* **1988**, *3*, 476–489. [CrossRef]
22. Zhu, Y.; Ji, S.; Zhang, F.; Liu, Y.; Dong, H.; Cui, Z.; Wu, W. Vibration mechanism and influence factors in power transformers. *J. Xi'an Jiaotong Univ.* **2015**, *49*, 115–125.
23. Zhang, B. *The Study on Multi-Information Diagnosis Method of Power Transformer Winding Mechanical State*; Shenyang University of Technology: Shenyang, China, 2015.
24. Xie, P.; Rao, Z.; Zhu, Z. Finite element modeling and analysis on transformer windings. *J. Vib. Shock* **2006**, *25*, 134–137.
25. Xu, J.; Shao, Y.; Jin, Z.; Rao, Z.; Jiang, Y.; Zhou, K. The detection of transformer windings' deformation based on the method of frequency response analysis. *Noise Vib. Control* **2009**, *29*, 26–29.
26. Wan, D.F. *Magnetic Theory and Its Application*; Huazhong University of Science and Technology Press: Wuhan, China, 1996; pp. 56–62.

27. Cheng, J.; Ji, S.; Liu, J.; Li, Y. Analysis of the measuring position for on-line monitoring on vibration signal of the winding. *High Volt. Eng.* **2004**, *30*, 46–48.
28. Zhang, L.; Zhang, W.; Liu, J.; Zhao, T.; Zou, L.; Wang, X. A New Prediction Model for Transformer Winding Hotspot Temperature Fluctuation Based on Fuzzy Information Granulation and an Optimized Wavelet Neural Network. *Energies* **2017**, *10*, 1998. [CrossRef]
29. Nagata, T.; Hirai, K.; Iwasaki, S.; Ebisawa, Y. Estimation on On-load Vibration of Transformer Windings. In Proceedings of the 2002 IEEE International Conference on Power Engineering Society Winter, New York, NY, USA, 27–31 January 2002; pp. 1378–1382.
30. Wang, F.; Duan, R.; Geng, C.; Qian, G.; Lu, Y. Research of Vibration Characteristics of Power Transformer Winding Based on Magnetic-mechanical Coupling Field Theory. *Proc. CSEE* **2016**, *36*, 2555–2562.
31. Wang, X.; Hao, N.; Zhang, G. Simulation study on detection of transformer winding deformation by frequency response considering phase characteristic. *Transformer* **2018**, *55*, 25–31.
32. Wang, H.; Wang, N.; Li, T. Axial Nonlinear Vibration of Large Power Transformer Winding. *Power Syst. Technol.* **2000**, *24*, 42–45.
33. Liu, J.; Wang, G.; Zhao, T.; Zhang, L. Fault diagnosis of on-load tap-changer based on variational mode decomposition and relevance vector machine. *Energies* **2017**, *10*, 946. [CrossRef]
34. Gu, H. *Research on the Relationship between Winding Deformation and Vibration Signal of Power Transformer*; Kunming University of Technology: Kunming, China, 2017.
35. Bartoletti, C.; Desiderio, M.; Carlo, D.D.; Fazio, G. Vibro-acoustic techniques to diagnose power transformers. *IEEE Trans. Power Deliv.* **2004**, *19*, 221–229. [CrossRef]
36. Ahn, H.M.; Oh, Y.H.; Kim, J.K.; Song, J.S.; Hahn, S.C. Experimental verification and finite element analysis of short-circuit electromagnetic force for dry-type transformer. *IEEE Trans. Magn.* **2012**, *48*, 819–822. [CrossRef]
37. Ji, S.; Wang, S.; Li, Q.; Li, Y.; Sun, Q. The application of vibration method in monitoring the condition of transformer winding. *High Volt. Eng.* **2002**, *28*, 13–15.

Article

A Hybrid Electric Vehicle Dynamic Optimization Energy Management Strategy Based on a Compound-Structured Permanent-Magnet Motor

Qiwei Xu [1,*], Yunqi Mao [1], Meng Zhao [1] and Shumei Cui [2]

[1] State Key Laboratory of Power Transmission Equipment & System Security and New Technology, Chongqing University, Chongqing 400044, China; myq2016@cqu.edu.cn (Y.M.); zhaom2015@cqu.edu.cn (M.Z.)

[2] Department of Electrical Engineering, Harbin Institute of Technology, Harbin 150080, China; cuism@hit.edu.cn

* Correspondence: xuqw@cqu.edu.cn; Tel.: +86-185-2327-8964

Received: 15 July 2018; Accepted: 21 August 2018; Published: 23 August 2018

Abstract: A dynamic optimization energy management strategy called Hybrid Electric Vehicle Based on Compound Structured Permanent-Magnet Motor (CSPM-HEV) is investigated in this paper. CSPM-HEV has obvious advantages in power density, heat dissipation efficiency, torque performance and energy transmission efficiency. This paper describes the topology and working principle of the CSPM-HEV, and analyzes its operating mode and corresponding energy flow laws. On this basis, the relationship about the power loss of the vehicle, the CSPM transmission ratio i_{CSPM} and the CSPM-HEV power distribution coefficient f_1 were derived. According to the optimal combination of (i_{CSPM}, f_1), the engine power and speed which minimize the power loss of the vehicle, were calculated, thus realizing the instantaneous optimal control of the vehicle. In addition, in order to improve the instantaneously optimized control processing speed, a neural network controller was established. The drive axle demand power, speed and battery State of Charge (SOC), were taken as input variables. Then, the engine power and speed were taken as output variables. The simulation results show that the average speed of the instantaneous optimization strategy after BP neural network optimization is increased by 98.1%, the control effect is significant, and it has high application value.

Keywords: hybrid electric vehicle; compound structured permanent-magnet motor; energy management strategy; instantaneous optimization minimum power loss; back propagation (BP) neural network

---

1. Introduction

The compound structure permanent-magnet motor (CSPM) is a new electric transmission device. Its basic structure is shown in Figure 1. It consists of an inner rotor, an outer rotor and a stator. The outer rotor is composed of two layers of inner and outer layers. Ignoring the influence of magnetic field coupling, the motor can be regarded as a combination of two independent motors inside and outside. The internal motor EM1 is composed of an inner permanent magnet and an inner rotor, and the external motor EM2 is composed of an outer permanent magnet and a stator. The CSPM has two mechanical ports and two electrical ports, which realize efficient energy transmission, so the CSPM has broad application prospects in wind power generation, ship propulsion systems and hybrid electric vehicles [1].

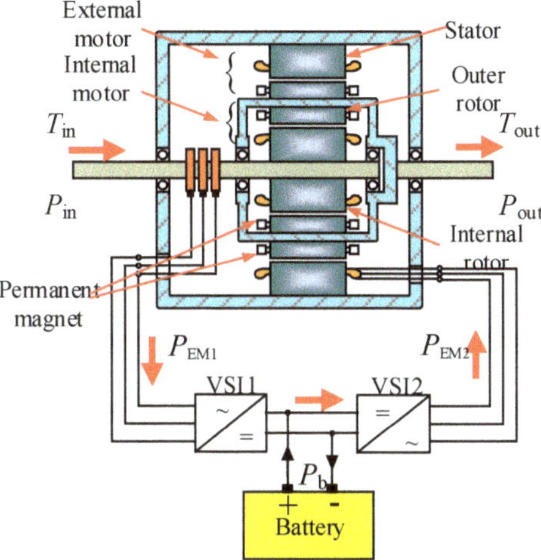

Figure 1. Structure of compound structure permanent-magnet motor.

A correlative research study demonstrates that a suitable Energy Management Strategy (EMS) can reduce HEV fuel consumption [2–19], and it is also beneficial to emissions reduction. In [2], dynamic programming was investigated for reducing the fuel consumption of vehicles when the future driving conditions are acquired. However, due to the computational complexity, the results obtained from dynamic programming cannot be executed directly. To solve this question, approximated dynamic programming [3] is proposed as an alternative solution. Since analytical optimization methods employ a mathematical equation to achieve the solution, their computation speed is faster than that of simple numerical methods. In this category, as an optimal control method, EMS based on Pontryagin's minimum principle (PMP) is proposed [4]. Unfortunately, this approach only works under the condition that the future driving conditions are known in advance. That is to say, vehicles must be able to communicate with road traffic systems [5]. In practice, currently this is difficult to achieve on a large scale.

For online implementation, in [6,7], the authors have developed a generic framework of online EMS for HEVs, where an evolutionary algorithm is used for online optimization of the power-split and battery SOC management. Due to the merit that the rule-based energy management algorithms are easy to implement in real time, this control strategy have been broadly applied in practical HEV EMS. In [8–10], rule-based EMS are used to split the power demand between the internal combustion engine (ICE) and the battery. In [11,12] the authors adopted a rule-based EMS and they introduced a proper supervisory environment for a complex structure control. However, to create these rules, extensive engineering experience and extensive experimental data are needed. Moreover, this cannot improve the fuel economy significantly. To overcome this problem, [13,14] propose a rule-based strategy combine with a fuzzy inference system. As a fuzzy system, due to its simple logic, it can be implemented in real time applications. In addition, it easy to model nonlinearity and uncertainty. However, the drawback of fuzzy systems is that they cannot achieve large modifications of the system modelling. This matter can be settled by the use of genetic algorithms (GAs) [15], which optimize the membership functions of the fuzzy controller for several matters while using appropriate fitness function. The trouble with GAs is random convergence of the solutions. Moreover, the operational speed of the optimization algorithm impedes the widespread application of GAs. To overcome this problem, the Artificial Neural Network (ANN) concept was introduced in [16–18]. ANNs are especially good at self-learning, adaptive ability and parallel distributed processing. Therefore, ANNs can be used to save the time consumed for optimization. They are suitable for improvement of instantaneous optimization strategies [19].

There are a variety of power source output couplings in HEVs with new topologies. The energy management strategy in traditional HEVs cannot achieve efficient power output distribution. Therefore, an optimized energy management strategy is proposed to improve the power, energy transmission efficiency and fuel economy of the new HEV. Through the optimized energy management strategy of this paper, the power, comfort, energy saving and environmental protection of HEVs are satisfied under different working conditions. The research of this paper promotes the development of electromagnetic continuously variable transmission. This paper develops a new theoretical foundation for the practical application of CSPM in HEVs.

Section 2 of this paper analyzes the operation mode of CSPM-HEV. Moreover, in order to facilitate the simulation analysis, the simulation model of CSPM-HEV was established and the simulation parameters of each component were determined in Section 3. In Section 4, based on the CSPM-HEV energy transfer characteristics, an instantaneous optimization energy management strategy based on a BP neural network was proposed, and the corresponding controller was designed in Section 5.

2. CSPM-HEV Operating Mode Analysis

After the structure optimization of CSPM, the magnetic field coupling, between internal and external motors in a CSPM can be neglected, so the CSPM can be seen as two independent permanent-magnet motors. Figure 2 is its equivalent schematic diagram. For simplification of the analysis, it was based on the equivalent permanent-magnet motor structure shown in Figure 2.

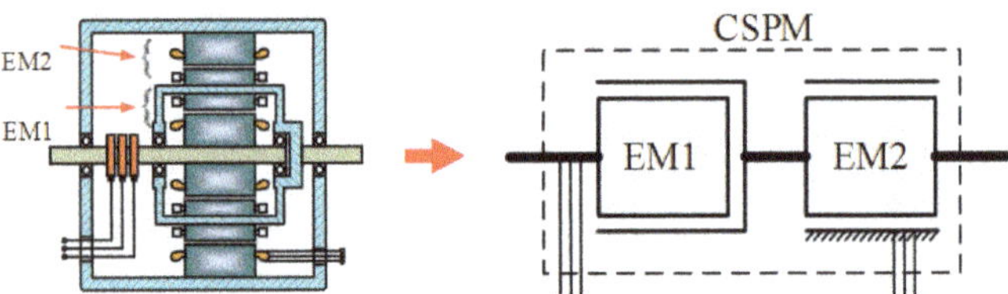

Figure 2. Equivalent diagram of compound-structure permanent-magnet motor.

According to the function of the CSPM in the HEV, the operating modes can be mainly divided into hybrid mode, starter mode, regenerative braking mode, generator mode and motor mode.

(1) Hybrid mode

When the vehicle demand torque is large or the battery SOC is low, the engine and the battery will jointly provide energy for vehicle. The CSPM works in the hybrid mode at this time, and its power flow is shown in Figure 3.

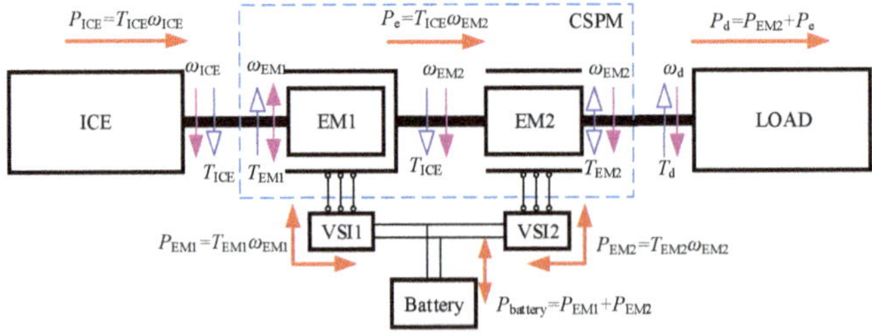

Figure 3. Hybrid mode and the related power flow diagram.

(2) Starter mode

In a conventional vehicle, the engine needs an external force to start. This process is completed by an electric motor. However, the CSPM-HEV avoids this process, and the inner motor EM1 can realize this function. Its power flow is shown in Figure 4.

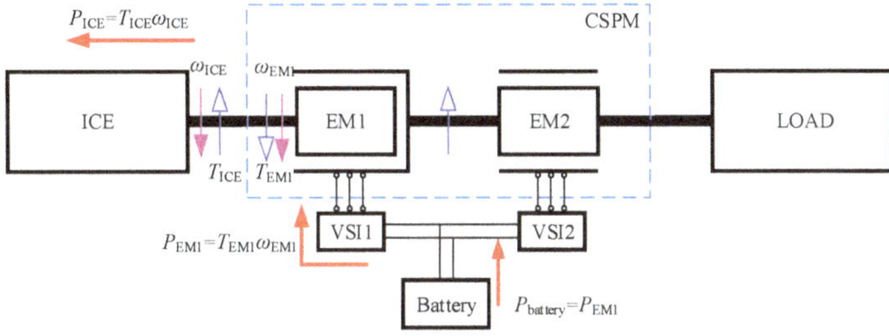

Figure 4. Starter mode and the related power flow diagram.

(3) Generator mode

If a vehicle is static for a short time and the battery SOC is low, the external motor of the CSPM stops working, and the engine drives the inner motor to operate as generator. Therefore, the battery is charged. Its power flow is shown in Figure 5.

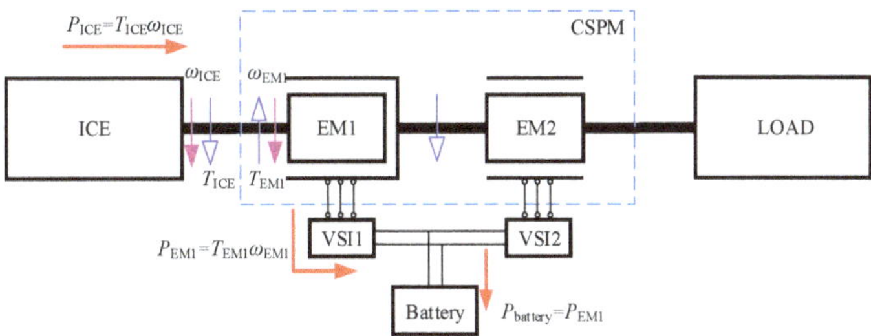

Figure 5. Generating mode and the related power flow diagram.

(4) Motor mode

When the vehicle runs at low speed and the battery SOC is high, the CSPM-HEV works in the pure electric mode. At this time, the inner motor EM1 and the engine do not work, and the outer motor EM2 works as a motor to directly drive the vehicle, and the system energy is all provided by the battery. Its power flow is shown in Figure 6.

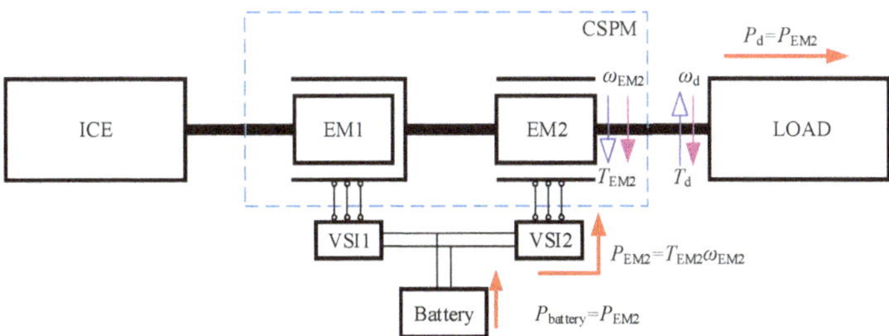

Figure 6. Motor mode and the related power flow diagram.

(5) Regenerative braking mode

When the vehicle brakes, the EM2 works in the power generation state, and the recovered energy is stored in the battery. The power flow is shown in Figure 7.

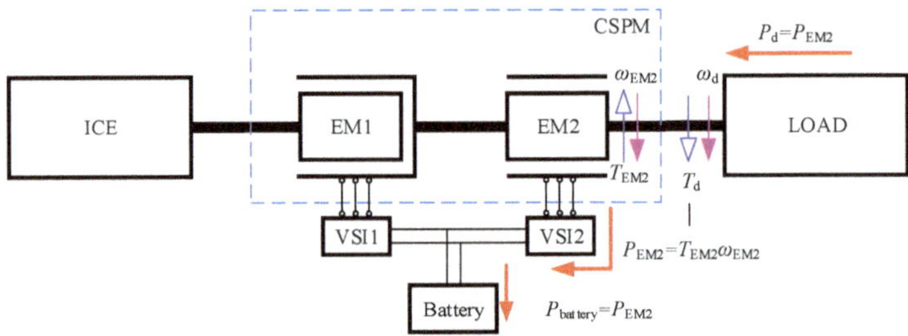

Figure 7. Regenerative braking mode and the related power flow diagram.

3. CSPM-HEV Model and Parameters

According to the mathematical model of the ICE, CSPM, battery and other components, the simulation models were built in the MATLAB/Simulink (MATLAB 2014a, MathWorks, Natick, MA, USA) environment [20,21]. Afterwards, due to the connection relationship of each component, a vehicle model of the CSPM was established based on the ADVISOR (ADVISOR 2002, AVL List GmbH, Graz, Austria) simulation platform. The entire model is shown in Figure 8. Simulation parameters and main component parameters of CSPM-HEV are shown in Tables 1 and 2 respectively.

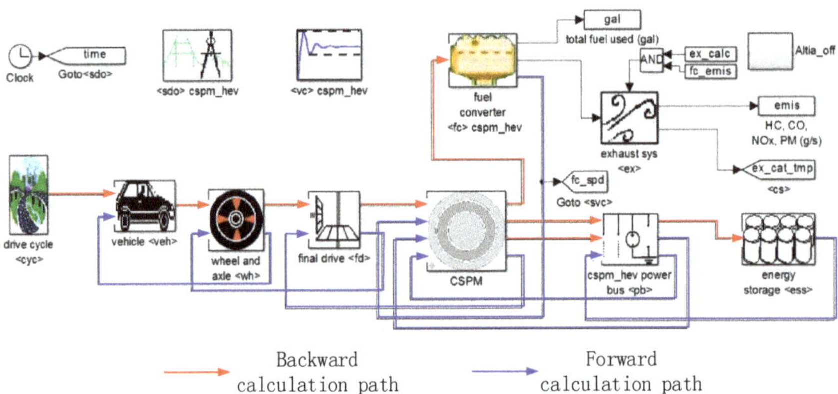

Figure 8. Simulation model of CSPM-HEV.

Table 1. Simulation parameters of CSPM-HEV.

Parameter	Data	Parameter	Data
Curb weight	1360 kg	Main reducer efficiency	0.95
Frontal area	1.746 m^2	Rolling resistance coefficient	0.01
Air resistance coefficient	0.3	Tire Rolling radius	0.2928 m
Final drive ratios	3.905		

Table 2. The main component parameters of CSPM-HEV.

Component	Parameter	Data
ICE	Power	43 kW
	Torque	100 Nm
CSPM	Type	Double magnet type
	External motor power	30 kW
	Internal motor power	20 kW
	Torque peak of external motor	305 Nm
	Torque peak of internal motor	100 Nm
Battery	Rated voltage	300 V

4. Instantaneous Energy Management Strategy Based on BP Neural Network

To some extent, the instantaneous optimization energy management strategy (IO-EMS) compensates for the shortcomings of energy management strategies based on rules and global optimization. Due to the independence from expert experience, the optimized control can be achieved at each moment. Compared with global optimization, IO-EMS can adapt to different working conditions. However, due to the massive calculations required, IO-EMS cannot guarantee the real-time performance, which hinders its application.

Artificial Neural Networks (ANNs) are a computing system formed by connecting a number of simple processing units (neurons). It is also called "neural network", which mimics the operating mechanism of the human brain and enables the machine to have some features of the human brain. The control system utilized the continuous learning, profound memory and adaptability of the neural network, so that the control system can realize the non-linear mapping between input and output. Neural networks are suitable for modeling and the control of systems with many uncertainties [22].

This section develops an IO-EMS based on the principle of "minimum power loss", and it proposes a real-time energy management strategy based on a BP neural network (BP-EMS). BP-EMS improves the poor real-time performance of IO-EMS. The optimal solution at each moment in each operating condition was collected as a training sample of the neural network. Then, the BP neural network was trained. Afterwards, the BP neural network controller was built.

The optimization based on the "minimum power loss" principle is to find a set of values, which minimizes power loss among all the power flow combinations, and these values are used to calculate the torque and speed of engine demand. Thus, the minimum power loss of the system is achieved, and the optimal vehicle efficiency is achieved eventually.

The speed difference between the two shafts is continuously variable, which is similar to the traditional continuously variable transmission, so that the transmission ratio i_{CSPM} can be defined according to the concept in the mechanical transmission:

$$i_{CSPM} = \frac{\omega_{ICE\_G}}{\omega_{EM2}} \tag{1}$$

where ω_{ICE_G} is output speed of the engine after the gear.

When the CSPM-HEV runs in hybrid mode, the power between the EM1 and the engine has the following relationship:

$$P_{EM1} = T_{EM1}(\omega_{EM2} - \omega_{ICE\_G}) = P_{ICE}(1 - \frac{1}{i_{CSPM}}) \tag{2}$$

From Equation (2), we can get that if $i_{CSPM} > 1$, EM1 works in the power generation state; if $i_{CSPM} < 1$, EM1 works in the electric state.

When the CSPM-HEV runs, the required power of the drive axle is definite. For simplification of power flow analysis, the engine and the EM1 can be regarded as a whole, and they are defined as the engine block, as shown in Figure 9.

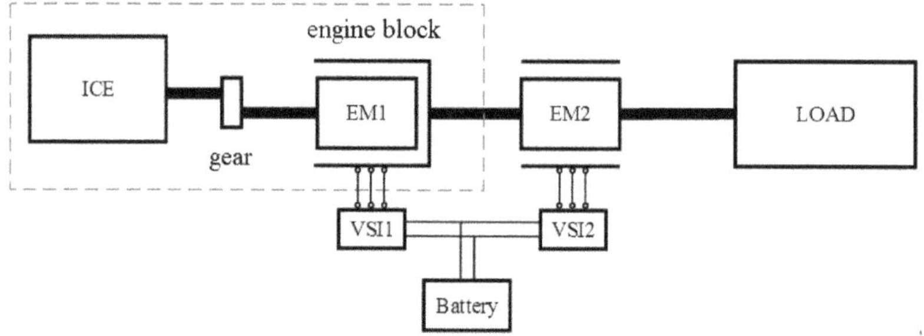

Figure 9. ICE group block diagram.

Since the external motor EM2 is independent with the engine, the power distribution coefficient f_1 can be defined. At this time, the output power of the engine group and the EM2 are:

$$P_{ICE} + P_{EM1} = f_1 \times P_{req} \tag{3}$$

$$P_{EM2} = (1 - f_1) \times P_{req} \tag{4}$$

In the equation, P_{req} is the required power of the axle.

Combining Equation (2) with Equation (3), the power of the engine and internal motor EM1 can be expressed as:

$$P_{ICE} = \frac{i_{CSPM}}{2i_{CSPM} - 1} \times f_1 \times P_{req} \tag{5}$$

$$P_{EM1} = \frac{i_{CSPM} - 1}{2i_{CSPM} - 1} \times f_1 \times P_{req} \tag{6}$$

Due to the inevitable power loss during the powertrain transmission and the mechanical transmission, the power loss of the drive axle is mainly concentrated in four parts: engine, internal motor EM1 and its controller, external motor EM2 and its controller, battery, respectively. The total system power loss P_{loss_real} can be expressed as:

$$
\begin{aligned}
P_{loss\_real} = P_{ICE} \times \frac{1-\eta_{ICE}}{\eta_{ICE}} + |P_{EM1}| \times \frac{1-\eta_{EM1}}{\eta_{EM1}} \\
+ |P_{EM2}| \times \frac{1-\eta_{EM2}}{\eta_{EM2}} + |P_{bat}| \times \frac{1-\eta_{bat}}{\eta_{bat}}
\end{aligned}
\tag{7}
$$

$$P_{bat} = P_{EM1} \times \frac{1}{\eta_{EM1}} + P_{EM2} \times \frac{1}{\eta_{EM2}} \tag{8}$$

In the above equations, η_{ICE} is the engine efficiency; η_{EM1} is the EM1 and its controller efficiency; η_{EM2} is the EM2 and its controller efficiency; and η_{bat} is the battery efficiency. They have the following relationship:

$$
\begin{cases}
0 \leq \eta_{ICE} \leq 1 \\
0 \leq \eta_{EM1} \leq 1 \\
0 \leq \eta_{EM2} \leq 1 \\
0 \leq \eta_{bat} \leq 1
\end{cases}
\tag{9}
$$

Substituting the output power of engine and EM1/EM2 into Equation (7):

$$P_{loss\_real} = P_{req} \times \left[\frac{i_{CSPM}}{2i_{CSPM}-1} \times f_1 \times \frac{1-\eta_{ICE}}{\eta_{ICE}} \right.$$
$$+ \left| \frac{i_{CSPM}-1}{2i_{CSPM}-1} \right| \times f_1 \times \frac{1-\eta_{EM1}}{\eta_{EM1}} + |1-f_1| \times \frac{1-\eta_{EM2}}{\eta_{EM2}}$$
$$+ \left. \left| (1-f_1) \times \frac{1}{\eta_{EM2}} + \frac{i_{CSPM}-1}{2i_{CSPM}-1} \times f_1 \times \frac{1}{\eta_{EM1}} \right| \times \frac{1-\eta_{bat}}{\eta_{bat}} \right] \tag{10}$$

In the CSPM-HEV, the energy in the battery comes from two parts: one is the energy recovered by regenerative braking, and the other is converted from the mechanical energy of the engine. The latter accounts for a large proportion. Since the battery will wear out during charging and discharging, when analyzing the system loss, the loss of the battery should be considered. Therefore, the battery loss compensation factor is introduced, which can be expressed as:

$$K = \frac{1}{\overline{\eta}_{ICE} \times \overline{\eta}_{CSPM} \times \overline{\eta}_{bat}} - 1 \tag{11}$$

In the equation, $\overline{\eta}_{ICE}$, $\overline{\eta}_{CSPM}$, $\overline{\eta}_{bat}$ represent the average efficiency of the engine, CSPM, and battery, respectively.

When the battery is discharged, the system power loss can be expressed as:

$$P_{loss} = P_{req} \times \left[\frac{i_{CSPM}}{2i_{CSPM}-1} \times f_1 \times \frac{1-\eta_{ICE}}{\eta_{ICE}} \right.$$
$$+ \left| \frac{i_{CSPM}-1}{2i_{CSPM}-1} \right| \times f_1 \times \frac{1-\eta_{EM1}}{\eta_{EM1}} + |1-f_1| \times \frac{1-\eta_{EM2}}{\eta_{EM2}}$$
$$+ \left. \left| (1-f_1) \times \frac{1}{\eta_{EM2}} + \frac{i_{CSPM}-1}{2i_{CSPM}-1} \times f_1 \times \frac{1}{\eta_{EM1}} \right| \times \frac{1-\eta_{bat}+K}{\eta_{bat}} \right] \tag{12}$$

When the battery is charging, the system power loss P_{loss} is equal to P_{loss_real}:

$$P_{loss} = P_{loss\_real} \tag{13}$$

From Equation (10), it can be seen that the efficiency of each system component can be obtained by looking it up in the table, according to the corresponding i_{CSPM} and f_1. Consequently, the power loss of the vehicle will be only related to these two variables. Therefore, when the power distribution coefficient f_1 is a definite value, and Equation (10) is a linear equation with one unknown. It can be known from MATLAB calculation that the power loss becomes small firstly and then becomes large as the gear ratio i_{CSPM} increases. In this process, there is a minimum value of the power loss. If within the value range of the power distribution coefficient f_1, all minimum values form a set, then, the minimum value in this set is the minimum power loss under the specific operating condition. Similarly, if the transmission ratio is determined first, then the relationship between the power distribution coefficient and the power loss is same as the above conclusion.

At each moment under the IO-EMS, there are multiple combinations of (i_{CSPM}, f_1), and the power loss of each combination will be compared with each other, to find the combination that minimize the power loss:

$$J^* = \min \left\{ P_{loss}(i_{CSPM}, f_1) \right\} \tag{14}$$

According to the (i_{CSPM}, f_1) combination at each moment, the required power, speed and torque of the engine, EM1 and EM2 can be calculated under the current combination:

$$\begin{cases} P_{ICE} = \frac{i_{CSPM}}{2i_{CSPM}-1} \times f_1 \times P_{req} \\ \omega_{ICE} = \frac{i_{CSPM} \times \omega_{EM2}}{i_{gear}} \\ T_{ICE} = \frac{P_{ICE}}{\omega_{ICE}} \end{cases} \tag{15}$$

$$\begin{cases} \omega_{EM1} = (i_{CSPM} - 1) \times \omega_{EM2} \\ T_{EM1} = -T_{ICE} \end{cases} \quad (16)$$

$$\begin{cases} T_{EM2} = \frac{(1-f_1) \times P_{req}}{\omega_{EM2}} \\ \omega_{EM2} = \omega_{req} \end{cases} \quad (17)$$

In the equation, i_{gear} is the transmission ratio of the gear, ω_{req} is the required speed of the axle.

In the process of solving the IO-EMS algorithm, the power, speed, and torque of the engine, EM1, and EM2 must satisfy the following constraints:

$$\begin{cases} T_{ICE\_min} \leq T_{ICE} \leq T_{ICE\_max} \\ \omega_{ICE\_min} \leq \omega_{ICE} \leq \omega_{ICE\_max} \\ P_{ICE\_min} \leq P_{ICE} \leq P_{ICE\_max} \\ T_{EM1,2\_min} \leq T_{EM1,2} \leq T_{EM1,2\_max} \\ \omega_{EM1,2\_min} \leq \omega_{EM1,2} \leq \omega_{EM1,2\_max} \\ P_{EM1,2\_min} \leq P_{EM1,2} \leq P_{EM1,2\_max} \\ SOC_{min} \leq SOC \leq SOC_{max} \end{cases} \quad (18)$$

The implementation of the IO-EMS based on the minimum power loss is as follows: At every moment when the CSPM-HEV operates in the hybrid mode, all possible combinations (i_{CSPM}, f_1) are listed firstly, then, these combinations are adopted to calculate the power, torque, and speed required for the engine, EM1, and EM2 by Equations (15)–(17). The calculation results must satisfy the constraints shown in Equation (18). Those combinations that do not meet the constraints will be discarded. Afterwards, according to the required power, torque and speed of the engine, EM1 and EM2, the efficiency of the component is obtained by looking up the table. After this, the power loss, which satisfy the condition, is calculated by Equations (12) and (13). When the loss P_{loss} is minimum, the corresponding (i_{CSPM}, f_1) combination is obtained. Next, according to Equations (15)–(17), the power, speed, and torque of the engine, EM1, and EM2 corresponding to the combination are calculated. Eventually, these data are used to control the system operation, and the minimum vehicle power loss is realized.

5. Neural Network Controller Design

In order to realize the Back-Propagation Neural Network Energy Management Strategy (BP-EMS), this paper designed a neural network controller, structure of which is shown in the Figure 10. According to the three input variables in the IO-EMS, three neurons were designed on the input layer, they are axle power demand, axle speed demand and battery SOC, respectively. Afterwards, two neurons were designed on the output layer, these two neurons represented the engine power and speed respectively. By the engine speed and torque, the operating point of the internal and external motors can be calculated, so that realizing the efficient energy management of the system.

For the three-layer BP neural network, the number of input and output layer neurons is determined, so the hidden layer neurons determine the network structure and have an important impact on network performance. If the number of neurons in the hidden layer is too small, the network is difficult to train and its performance will be poor; if there are too many hidden layer neurons, the training time of the network will be increased greatly, and the training will easily fall into a local minimum without the optimal advantage. Moreover, the excessive neurons will also bring difficulties for hardware and software implementations. According to the principle of simplifying the network structure as much as possible, this paper determined the number of hidden layer neurons by continuous experimentation.

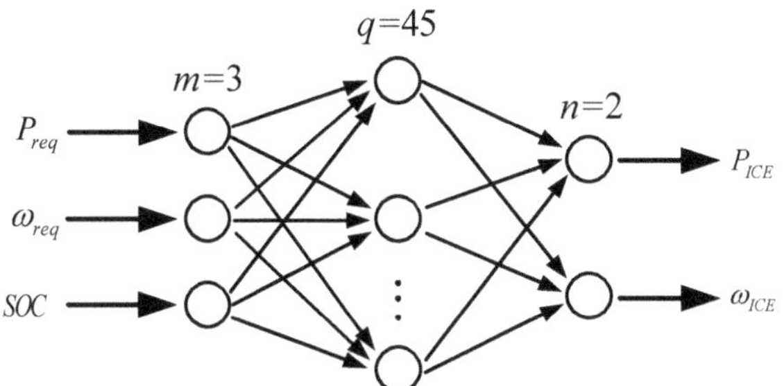

Figure 10. BP neural network energy management controller structure diagram.

5.1. BP Neural Network Training

There is an important relationship between the selection of training samples and their performance in neural networks, so the following aspects should be noted when learning the sample selection of neural network controllers:

(1) The sample should be widely representative and reflect the working characteristics of all possible operating conditions of the HEV.
(2) The neural network after the sample learning has good generalization ability.
(3) Do not have too many samples, otherwise it may lead to over-fit of the network.

In the BP neural network controller of this paper, the input and output variables such as vehicle power, speed, and battery SOC vary greatly in magnitude, and they need to be normalized before training, that is, all the data were converted into (0, 1), to eliminate the impact of different orders of magnitude on the network. The linear function method was used here, as shown in the following equation:

$$\overline{x_i} = \frac{x_i - x_{\min}}{x_{\max} - x_{\min}} \tag{19}$$

In the equation, x_i represents the sample; $x_{\max}$ and $x_{\min}$ represent the sample maximum and minimum values respectively; $\overline{x_i}$ is the standardized sample.

In order to overcome the drawbacks of the standard BP algorithm, such as the difficulty in adjusting initial network weight, learning rate and momentum coefficient, and the long training time and slow convergence rate, the neural network controller adopted Levenberg-Marquardt algorithm for sample learning. The flow chart is shown in the Figure 11, and the training steps are as follows:

Step 1: The required power, speed, and battery SOC are putted in the training sample to the network and the error between the output and the target value $e_i(W)(i = 1, \ldots, N_1, N_1$ as the total number of training samples) is calculated. Here, the error indicator was defined:

$$E_k(W) = \frac{1}{2} \sum_{i=1}^{N_1} e_i^2(W) \tag{20}$$

In the equation, W is the vector group formed by the weights and thresholds in the network, k is the number of iterations of network learning.

Step 2: If the requirements are satisfied. If yes, training will be saved and ended. If not, go to the next step.

Step 3: The Jacobian matrix is calculated:

$$J(W) = \begin{bmatrix} \dfrac{\partial e_1(W)}{\partial W_1} & \dfrac{\partial e_1(W)}{\partial W_2} & \cdots & \dfrac{\partial e_1(W)}{\partial W_n} \\[2mm] \dfrac{\partial e_2(W)}{\partial W_1} & \dfrac{\partial e_2(W)}{\partial W_2} & \cdots & \dfrac{\partial e_2(W)}{\partial W_n} \\[2mm] \vdots & \vdots & \ddots & \vdots \\[2mm] \dfrac{\partial e_{N_1}(W)}{\partial W_1} & \dfrac{\partial e_{N_1}(W)}{\partial W_2} & \cdots & \dfrac{\partial e_{N_1}(W)}{\partial W_n} \end{bmatrix} \tag{21}$$

Step 4: The adjustment rate is solved through the following equation:

$$\Delta W = -[\mu I + J(W)^{\mathrm{T}} J(W)]^{-1} J(W)^{\mathrm{T}} e(W) \tag{22}$$

In the equation, μ is a non-negative number that indicates the speed of network learning.

$$e(W) = [e_1(W), e_2(W), \ldots, e_{N_1}(W)]^{\mathrm{T}} \tag{23}$$

Step 5: $W + \Delta W$ is calculated by Equation (20). If $E_{k+1}(W) < E_k(W)$, then $\mu = \mu/2$, $W_{k+1} = W_k + \Delta W$, skip to Step 1; otherwise, go to Step 4.

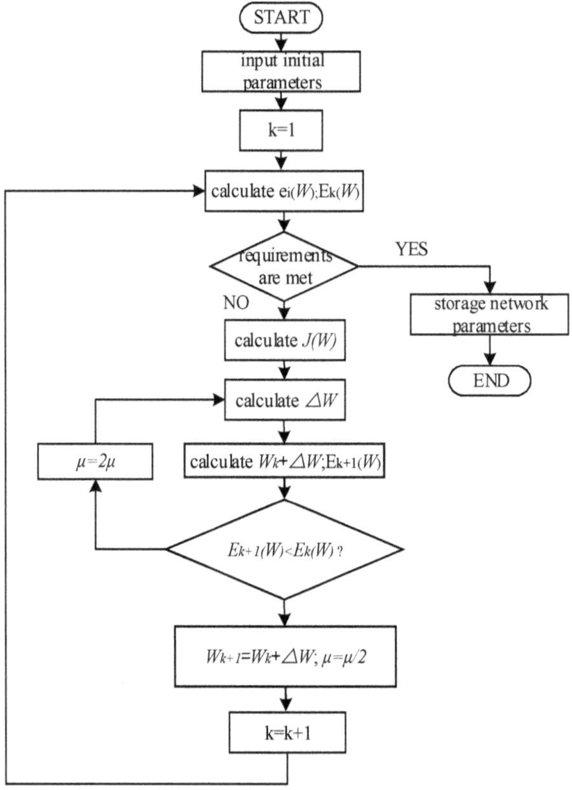

Figure 11. Levenberg-Marquardt algorithm flowchart.

5.2. Analysis of Simulation Results

In order to verify the control effect of the BP-EMS, the Urban Dynamometer Driving Schedule (UDDS), US06 Highway Cycle (US06 HWY), and New European Driving Cycle (NEDC) were simulated and the results were compared with the corresponding results of IO-EMS. Figure 12 shows the engine output power and torque command value based on the two control strategies under UDDS conditions. It can be seen from this figure, if engine power and torque are negative values, the invalid command values of the engine in the closed state are removed, the BP neural network controller will give a good effect on the sample learning. It realizes the nonlinear mapping between vehicle demand power and speed, battery SOC and engine power and torque command value.

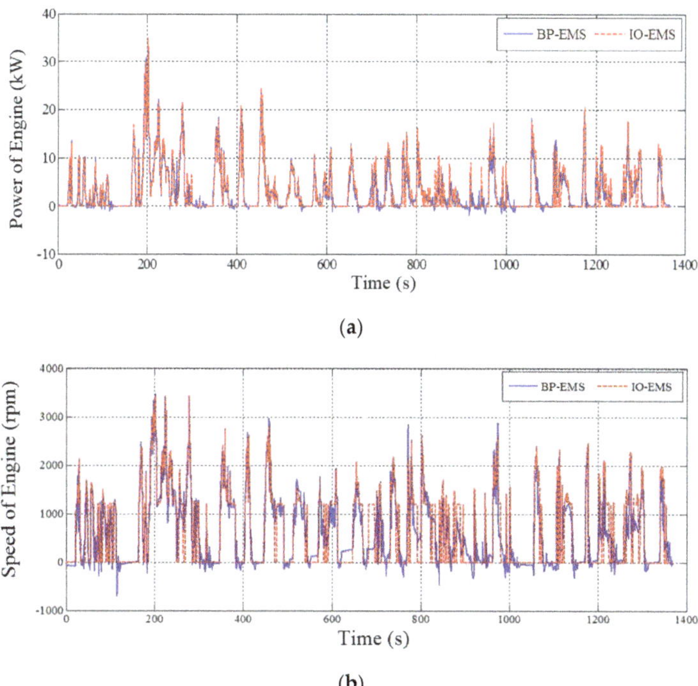

(**a**)

(**b**)

Figure 12. Power (**a**) and speed (**b**) of ICE during UDDS driving cycle.

From simulation results of US06 HWY and NEDC driving cycle, it can get the same analysis results as UDDS driving cycle. Simulation results of US06 HWY and NEDC are shown in Figures 13 and 14 respectively.

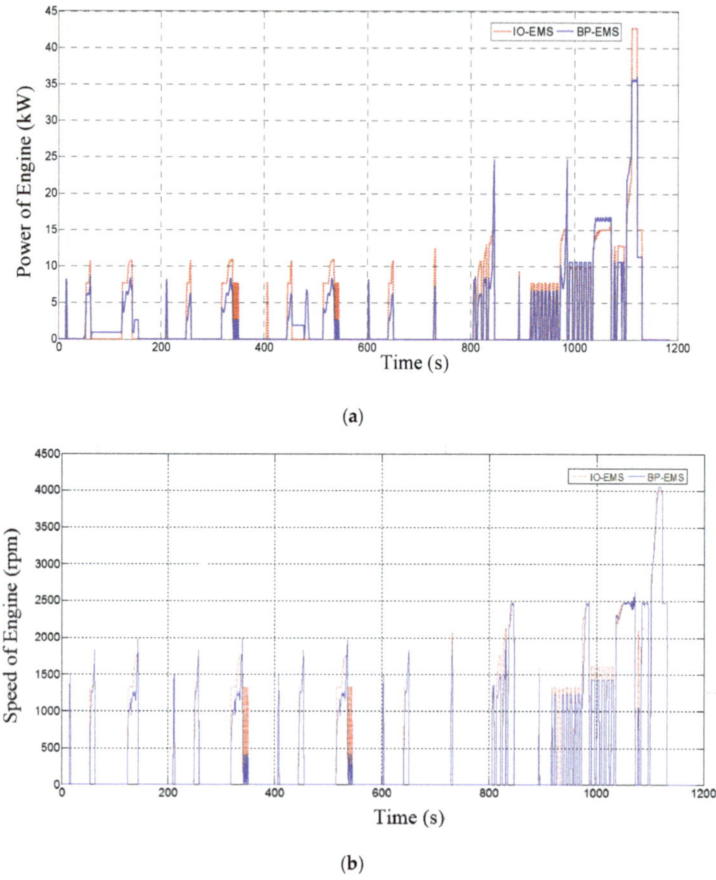

(**a**)

(**b**)

Figure 13. Power (**a**) and speed (**b**) of ICE during NEDC driving cycle.

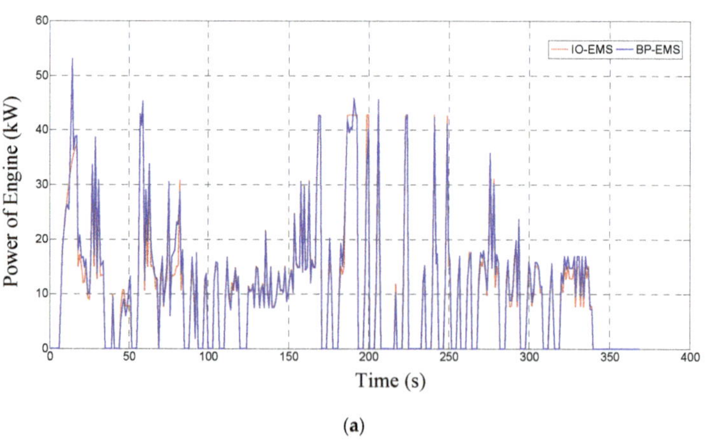

(**a**)

Figure 14. *Cont.*

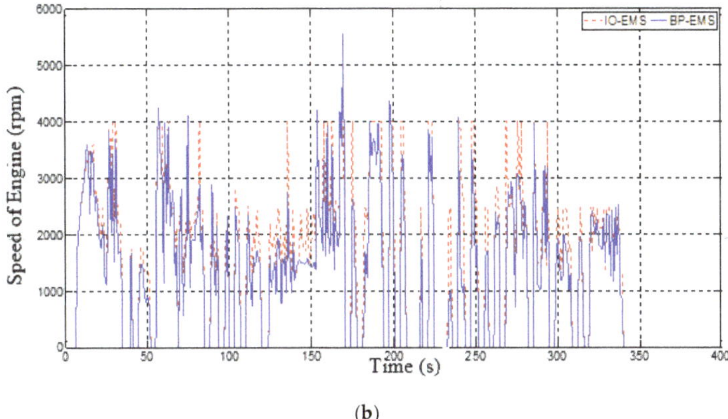

(b)

Figure 14. Power (**a**) and speed (**b**) of ICE during US06 driving cycle.

In order to make the simulation results more convincing, this paper uses Mean Square Error (MSE) to process the data obtained from the two control strategies in different working conditions. The results are shown in the Table 3. For the simplification of calculation, the unit of speed is a thousand revolutions per minute. The calculation results show that the control effect of the two control strategies is similar, and the introduction of neural network does not decrease the control performance of instantaneous optimization.

Table 3. Mean square error of obtained data from two control strategies.

Cycle Condition	Parameter	Mean Square Error
UDDS	Power of Engine	3.6
	Speed of Engine	0.063
US06	Power of Engine	5.5
	Speed of Engine	0.135
NEDC	Power of Engine	9.8
	Speed of Engine	0.536

Figure 15 shows the battery SOC variation based on two control strategies under UDDS conditions. It can be seen that the battery SOC is stable at the end, and the changes are similar, indicating the effectiveness of the real-time EMS based on a BP neural network.

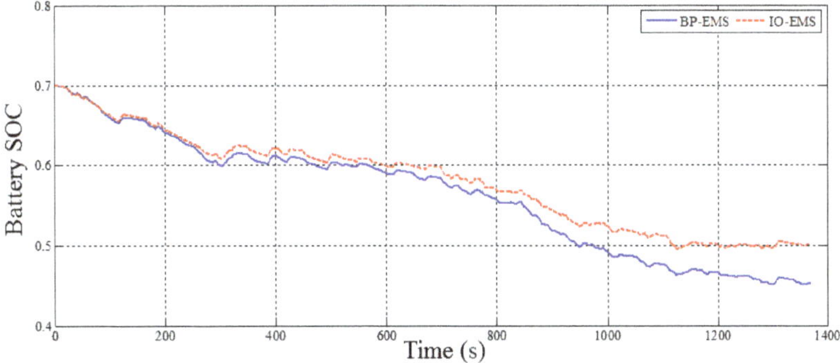

Figure 15. State of charge of battery during UDDS driving cycle.

It can be seen from Table 4 that the BP-EMS can simulate the control rules of the IO-EMS well, and the control effect is similar to that of the IO-EMS. The average fuel consumption difference between these two strategies under various operating conditions is 1.2%, while the average operation speed of the BP-EMS is improved effectively (the improvement degree is 98.1%), which shows that the improved control strategy has a high application value.

Table 4. Fuel consumption and operation time in two strategies.

Cycle Condition	BP-EMS		IO-EMS	
	Fuel Consumption (L/100 km)	Operation Time (s)	Fuel Consumption (L/100 km)	Operation Time (s)
UDDS	3.71	20	3.70	1230
NEDC	3.69	18	3.64	1059
US06 HWY	4.39	11	4.31	356

6. Conclusions

In this paper, an instantaneous optimization strategy based on the principle of "minimum power loss" is developed for CSPM-HEV. To solve the problem of complex algorithms and poor real-time performance, in addition, acBP neural network is employed to realize the real-time energy management of CSPM-HEV. The results may be summarized as follows:

(1) The transmission ratio of the CSPM i_{CSPM} was defined with reference to the traditional mechanical transmission, furthermore, the CSPM-HEV power distribution coefficient was raised for analyzing the system. Afterwards, the relationship about the power loss of the vehicle and (i_{CSPM}, f_1) was derived, as a result, and the instantaneous optimization energy management based on the principle of "minimum power loss" is established. The strategy calculates the engine power and speed at the current time according to the optimal combination of i_{CSPM} and f_1, so that achieve the instantaneous optimal control of the vehicle.

(2) According to the simulation results of the instantaneous optimization strategy under various working conditions, the learning samples were made. The vehicle power, speed and battery SOC were input variables, the engine power and speed were output variables, afterwards, the neural network controller was established, so the real-time energy management strategy based on a BP neural network is fulfilled.

(3) The real-time energy management strategy based on a BP neural network was simulated and compared with the results of a traditional instantaneous optimization strategy. The results show that the BP-EMS can greatly improve the running speed and optimize the control effect, and it also realizes the nonlinear mapping between engine output and drive axle demand power, speed and battery SOC, as a result, the instantaneous optimal control of CSPM-HEV is completed. In spite of benefits of BP-EMS, since the training sample of the BP neural network controller is the result of the instantaneous optimization strategy in the paper, it is unable to reach the global optimization. In the future, the authors will train samples based on the results of the global optimization algorithm, and employ a hybrid vehicle working condition recognition technology, which is expected to further improve the fuel economy of CSPM-HEV.

Author Contributions: Conceptualization, Q.X. and Y.M.; Methodology, Q.X.; Software, M.Z.; Validation, Q.X., Y.M. and M.Z.; Formal Analysis, Y.M.; Investigation, Y.M.; Resources, Q.X.; Data Curation, Q.X.; Writing-Original Draft Preparation, Q.X.; Writing-Review & Editing, Q.X. and Y.M.; Supervision, Q.X., Project Administration, S.C.

Funding: This research was funded by (National Natural Science Foundation of China) grant number (51507021), Chongqing Science and Technology Commission of China under Project No. cstc2013jcyjA60001, graduate research and innovation foundation of Chongqing, China under Project No. CYS17008 and The State Key Laboratory of Power Transmission Equipment & System Security and New Technology in Chongqing University of China under Project No. 2007DA10512716303.

Conflicts of Interest: The authors declare no conflict of interest.

References

1. Hoeijmakers, M.J.; Ferreira, J.A. The electrical variable transmission. In Proceedings of the 39th IAS Annual Meeting, Seattle, WA, USA, 3–7 October 2004; Institute of Electrical and Electronics Engineers Inc.: Piscataway, NJ, USA, 2004.
2. Abdrakhmanov, R.; Adouane, L. Dynamic programming resolution and database knowledge for online predictive energy management of hybrid vehicles. In Proceedings of the 14th International Conference on Informatics in Control, Automation and Robotics (ICINCO 2017), Madrid, Spain, 26–28 July 2017; Volume 1, pp. 132–143.
3. Johannesson, L.; Asbogard, M.; Egardt, B. Assessing the potential of predictive control for hybrid vehicle powertrains using stochastic dynamic programming. *IEEE Trans. Intell. Transp. Syst.* **2007**, *8*, 71–83. [CrossRef]
4. Ouddah, N.; Adouane, L.; Abdrakhamanov, R.; Kamal, E. Optimal energy management strategy of plug-in hybrid electric bus in urban conditions. In Proceedings of the 14th International Conference on Informatics in Control, Automation and Robotics (ICINCO 2017), Madrid, Spain, 26–28 July 2017; Volume 1, pp. 304–311.
5. Amini, M.H.; Karabasoglu, O. Optimal operation of interdependent power systems and electrified transportation networks. *Energies* **2018**, *11*, 196. [CrossRef]
6. Qi, X.; Wu, G.; Boriboonsomsin, K.; Barth, M.J. An on-line energy management strategy for plug-in hybrid electric vehicles using an estimation distribution algorithm. In Proceedings of the 17th International IEEE Conference on Intelligent Transportation Systems (ITSC), Qingdao, China, 8–11 October 2014; pp. 2480–2485.
7. Qi, X.; Wu, G.; Boriboonsomsin, K.; Barth, M.J. Development and evaluation of an evolutionary algorithm-based online energy management system for plug-in hybrid electric vehicles. *IEEE Trans. Intell. Transp. Syst.* **2017**, *18*, 2181–2191. [CrossRef]
8. Peng, J.; Fan, H.; He, H.; Pan, D. A rule-based energy management strategy for a plug-in hybrid school bus based on a controller area network bus. *Energies* **2015**, *8*, 5122–5142. [CrossRef]
9. Hofman, T.; van Druten, R.M.; Serrarens, A.F.A.; Steinbuch, M. Rule-based energy management strategies for hybrid vehicles. *Int. J. Elect. Hybrid Veh.* **2007**, *1*, 71–94. [CrossRef]
10. Hofman, T.; Steinbuch, M.; van Druten, R.M.; Serrarens, A.F.A. Rule-based energy management strategies for hybrid vehicle drivetrains: A fundamental approach in reducing computation time. In Proceedings of the 4th IFAC Symposium on Mechatronic Systems, Heidelberg, Germany, 12–14 September 2006; pp. 1–6.
11. Adel, B.; Youtong, Z.; Shua, S. Parallel HEV hybrid controller modeling for power management. *World Electr. Veh. J.* **2010**, *4*, 190–196. [CrossRef]
12. Hajizadeh, A.; Golkar, M.A. Intelligent power management strategy of hybrid distributed generation system. *Electr. Power Energy Syst.* **2007**, *29*, 783–795. [CrossRef]
13. Lihao, Y.; Youjun, W.; Congmin, Z. Study on fuzzy energy management strategy of parallel hybrid vehicle based on quantum PSO algorithm. *Int. J. Multimedia Ubiquitous Eng.* **2016**, *11*, 147–158. [CrossRef]
14. Denis, N.; Dubois, M.R.; Desrochers, A. Fuzzy-based blended control for the energy management of a parallel plug-in hybrid electric vehicle. *Intell. Transp. Syst.* **2015**, *9*, 30–37. [CrossRef]
15. Martnez, C.M.; Hu, X.; Cao, D.; Velenis, E.; Gao, B.; Weller, M. Energy management in plug-in hybrid electric vehicles: Recent progress and a connected vehicles perspective. *IEEE Trans. Veh. Technol.* **2017**, *66*, 4534–4549. [CrossRef]
16. Dai, X.; Li, C.K.; Rad, A.B. An approach to tune fuzzy controllers based on reinforcement learning for autonomous vehicle control. *IEEE Trans. Intell. Transp. Syst.* **2005**, *6*, 285–293. [CrossRef]
17. Qi, X.; Wu, G.; Boriboonsomsin, K.; Barth, M.J.; Gonder, J. Data driven reinforcement learning-based real-time energy management system for plug-in hybrid electric vehicles. *Transp. Res. Rec. J. Transp. Res. Board* **2016**, *2572*, 1–8. [CrossRef]
18. Qi, X.; Luo, Y.; Wu, G.; Boriboonsomsin, K.; Barth, M.J. Deep reinforcement learning-based vehicle energy efficiency autonomous learning system. In Proceedings of the 2017 IEEE Intelligent Vehicles Symposium (IV), Los Angeles, CA, USA, 11–14 June 2017; pp. 1228–1233.
19. Zhang, X.; Liu, Y.; Zhang, J. A Fuzzy Neural Network Energy Management Strategy for Parallel Hybrid Electric Vehicle. In Proceedings of the 9th International Conference on Modelling, Identification and Control (ICMIC 2017), Kunming, China, 10–12 July 2017.

20. Xu, Q.; Cui, S.; Song, L.; Zhang, Q. Research on the Power Management Strategy of Hybrid Electric Vehicles Based on Electric Variable Transmissions. *Energies* **2014**, *7*, 934–960. [CrossRef]
21. Xu, Q.; Sun, J.; Luo, L.; Cui, S.; Zhang, Q. A Study on Magnetic Decoupling of Compound-Structure Permanent-Magnet Motor for HEVs Application. *Energies* **2016**, *9*, 819. [CrossRef]
22. Wang, X. *Introduction of Neural Networks*; Science Press of China: Beijing, China, 2017.

Article

Low Cost Position Controller for Exhaust Gas Recirculation Valve System

Habib Bhuiyan and Jung-Hyo Lee *

Department of Electrical Engineering, Kunsan National University, Gunsan 54150, Korea; bhuiyanmdhabib@gmail.com
* Correspondence: jhlee82@kunsan.ac.kr; Tel.: +82-634-694-707

Received: 30 July 2018; Accepted: 17 August 2018; Published: 20 August 2018

Abstract: This paper proposes a position control method for a low-cost exhaust gas recirculation (EGR) valve system for automotive applications. Generally, position control systems used in automotive applications have many restrictions, such as cost and space. The mechanical structure of the actuator causes high friction and large differences between static friction and coulomb friction. When this large friction difference occurs, the position control vibrates when the controller uses a conventional linear controller such as the P or PI controller. In this paper, we introduce an inexpensive position control method that can be applied under the high-difference-friction mechanical systems. The proposed method is verified through the use of experiments by comparing it with the results obtained when using a conventional control system.

Keywords: position control; static friction; exhaust gas recirculation (EGR) valve system; automotive application

---

1. Introduction

Recently, many mechanical components used in vehicles have been replaced by electrical components to increase efficiency. These components are not only found in hybrid electric vehicles or electric vehicles but they have also been applied to gasoline and diesel vehicles such as Motor Driven Power Steering (MDPS) and Integrated Starter and Generator (ISG). These electric automotive components increase drive efficiency and reduce fossil fuel consumption. These changes are being applied to the transmission system and the engine valve system. Among these changes, the exhaust gas recirculation (EGR) valve is the mechanical component being targeted to replace the small DC motor [1–5]. However, in general, the mechanical systems using the EGR valve have a low acceptable cost and a narrow space for implementation; therefore, the electrical system including the actuator should be cost-effective and small. To achieve this, the mechanical actuating system cannot avoid being roughly designed, which implies high friction forces. Also, the difference between coulomb friction and static friction is very large, so obtaining a correct and a fast response in terms of position control is almost impossible using a conventional linear control system such as P, PI, or PID.

To achieve position control given this friction torque, some research has been proposed [6–10]. In [6], H infinite control and impulse control were combined for a fast control response. Robust control was achieved using a disturbance observer [7]. A fuzzy controller [8] and a neural network controller [9] were proposed to overcome this problem. In [10], an adaptive control method for friction compensation was proposed. These methods can dramatically reduce the effect of friction; however, a large number of parameters have to be set and the processing burden for realization is also complex in a low-cost drive system.

This paper proposes a position control method for a low-cost system. The general position control method for this low-cost system is a P-PI control method, as described in [11]. As mentioned above, correct and fast control cannot be achieved with this linear controller in mechanical systems that have this friction condition. Generally, in this case, feedforward compensation is adopted for improving the control performance [11–13]. However, these feedforward data are incorrect because of the aging of the mechanical system and environmental changes, such as temperature and humidity. Moreover, feedforward compensation can improve the dynamics of the controller; however, it cannot be the solution for unstable control performance that is caused by the difference in static and coulomb friction torque. In this paper, to achieve the stable position control, we first analyze the EGR valve mechanical model, define the cause of the vibration. Then, a proposed novel and simple algorithm that may be adapted to low-cost system to solve this problem are illustrated. Finally, we compare the performance of the conventional method to our proposed method to verify its superiority using experiments.

2. Mechanical Model of EGR Valve and Torque Measurement

2.1. Model Analysis of EGR Valve

Figure 1 shows the mechanical composition of an EGR valve. In general, an EGR valve is composed of a spring for recovering the initial position of the valve, joint and gear for transforming the power from rotation to translation, a DC motor, and a throttle valve, which is the source and the actuator of the mechanical system, respectively.

Figure 1. Mechanical composition of exhaust gas recirculation (EGR) valve.

First, the motor operating this system is a DC motor. Therefore, the generated torque from the motor is:

$$T_e = k_t i_a \tag{1}$$

where k_t is the torque constant and i_a is the armature current of DC motor.

The mechanical equation of the valve system shown in Figure 1 can be described as:

$$T_e = J\frac{d^2\theta_r}{dt^2} + T_{fric} + T_{spring} + T_L \tag{2}$$

where J is inertia, θ_r is the rotating angle, T_{fric} is the friction torque, T_{spring} is the spring torque, and T_L is the load torque.

This rotating angle is transferred to a linear position by the mechanical joint and gear. The linear position x can be expressed as:

$$x = r\{\cos(\theta_{L0}) - \cos(\theta_L + \theta_{L0})\} \tag{3}$$

$$\theta_L = \frac{\theta_r}{n} \tag{4}$$

where n is the ratio of the gear, r is the joint distance, θ_L is the joint angle, and θ_{L0} is the initial joint angle.

The spring force according to the linear position is described as:

$$F_{spring} = k_{spr}(x + x_0) \tag{5}$$

where k_{spr} is the spring character constant and x_0 is the initial linear position.

This spring force can be transferred to the torque on the load side:

$$T_{springL} = \frac{rk_{spr}}{n}(x + x_0)\sin(\theta_L + \theta_{L0}) \tag{6}$$

Then, by transferring the spring torque on the load side to the motor side, Equation (6) can be changed to:

$$T_{spring} = \frac{rk_{spr}}{n}\sin(\theta_L + \theta_{L0})[r\{\cos(\theta_{L0}) - \cos(\theta_L + \theta_{L0})\} + x_0] \tag{7}$$

Equation (7) indicates that the spring torque is only affected by the spring position. However, in practice, the spring torque is not only affected by the position but also by the speed direction. To apply this to Equation (7), we defined the spring coulomb friction torque as:

$$f_{spr\_c} = F_{spr\_col}\mathrm{sgn}(\frac{dx}{dt}) \tag{8}$$

As shown in Equation (8), the spring coulomb friction is negative when the motor speed is in reverse. As a result, spring torque can be modelled as:

$$T_{spr\_col} = \frac{rF_{spr\_col}}{n} \tag{9}$$

$$\begin{aligned}
&T_{spring} \\
&= \frac{rk_{spr}}{n}\sin(\theta_L + \theta_{L0})[r\{\cos(\theta_{L0}) - \cos(\theta_L + \theta_{L0})\} + x_0] + T_{spr_col}\sin(\theta_L + \theta_{L0})\mathrm{sgn}(\omega_r)
\end{aligned} \tag{10}$$

The EGR valve mechanical system is not only affected by the spring but also by the joint and the gear. The low-cost gear and the joint causing friction like a lead-screw emphasize the nonlinear static friction. In this paper, the LuGre friction model described by Yao et al. [10] is derived:

$$T_{fric} = [T_{ge\_col} + (T_{ge\_sta} - T_{ge\_col})e^{-(\omega_r/\omega_s)^2}]\mathrm{sgn}(\omega_r) \tag{11}$$

where T_{ge_col} is the coulomb friction torque on the gear and joint, T_{ge_sta} is the static friction torque on the gear and the joint, and ω_s is the Stribeck velocity.

In this paper, these modeled load torques were measured experimentally to implement a feedforward controller as previously reported [12]. This feedforward compensation can reduce the burden on the feedback controller and can help to enhance the control performance when nonlinear load has to be controlled by a linear controller.

2.2. Measurement Procedures of Spring and Friction Torque

Figures 2 and 3 show the measured spring torque and friction torque of the tested EGR valve, respectively. First of all, the electric torque from the motor is proportional to the DC motor current based on Equation (1). Therefore, we assumed that the current waveform can indirectly describe the generated torque. To measure the spring torque, we followed the steps below for identifying them.

1. Perform speed control on the initial EGR valve position.
2. Apply the speed reference from 10 rpm to 300 rpm.

3. Measure the averaged current. The speed at which the lowest averaged current is observed is the Stribeck velocity. Repeat the experiment as necessary for gathering data.

4. Control the motor using Stribeck velocity. The measured instantaneous current on steady state is the spring torque, with the assumption that the friction torque at Stribeck velocity can be ignored.

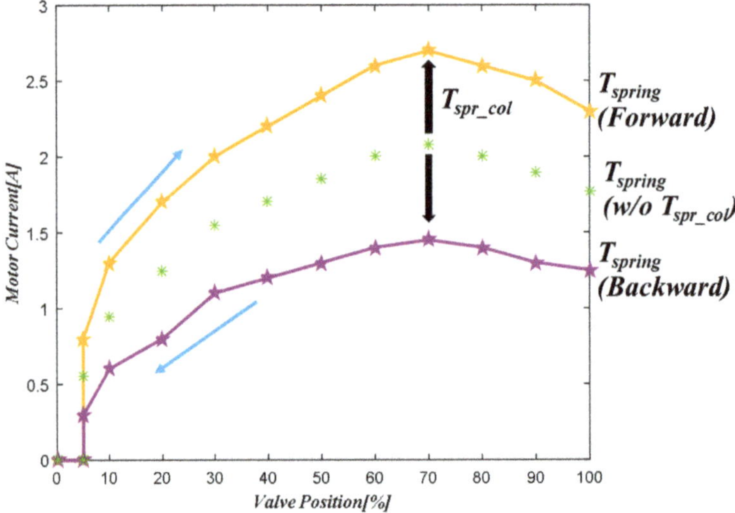

Figure 2. Measured spring torque.

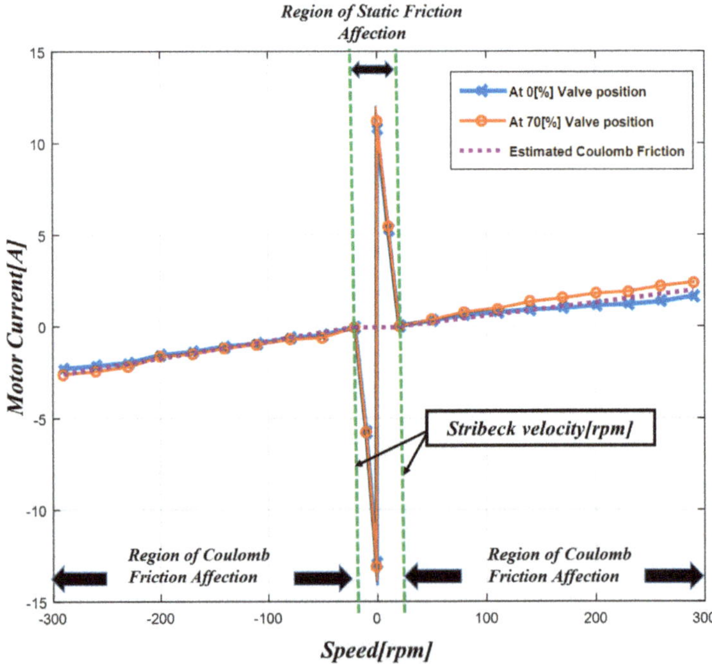

Figure 3. Measured friction torque.

With the obtained spring torque, we measured the static friction torque with the following steps.

1. Perform speed control started on each EGR valve position.
2. Set up the small gain on the current controller in order to apply the ramp increasing current reference.
3. Sudden current changes occur due to the position of the movement; measure the peak current point.
4. Subtract the spring torque amount from the measurement in Step 3. The remaining value is the static friction torque.

The coulomb friction torque can be obtained with the following steps.

1. Perform speed control started on the initial EGR valve position.
2. Apply the speed reference from 100 rpm and 300 rpm.
3. Measure the instantaneous current on the steady state of each speed.
4. Subtract the instantaneous current at 100 rpm from the current at 300 rpm.
5. Divide 200 rpm from the result of Step 4 for removing the spring torque component.
6. Multiply the speed from the result obtained in Step 5. This is the coulomb friction torque.

As shown in Figure 2, different spring torques occurred according to the valve position direction. If the valve position direction was to open the valve, the spring torque increased due to the coulomb friction in the spring torque, which is depicted in Equation (8). Reversely, if the valve position direction was to close the valve, the spring torque decreased. Figure 3 indicates that the friction torque at each position had almost the same static friction torque. Also, the static friction in the reverse direction had different values from the positive direction value. Coulomb friction torque calculation is based on a simple principle. First, the torque equation at 300 rpm can be described as:

$$T_e(\omega_{r2}) = J\frac{d^2\theta_r}{dt^2} + T_{fric}(\omega_{r2}) + T_{spring}(\omega_{r2}) \tag{12}$$

where ω_{r2} is the angular speed of 300 rpm.

If the steady state condition is only effective for identifying the coulomb friction torque, the inertia term can be neglected. With Equation (12), the torque difference that represents Step 4 can be calculated as:

$$\begin{aligned} T_e(\omega_{r2}) &- T_e(\omega_{r1}) \\ &= (T_{fric}(\omega_{r2}) + T_{spring}(\omega_{r2})) - (T_{fric}(\omega_{r1}) + T_{spring}(\omega_{r1})) \end{aligned} \tag{13}$$

where ω_{r1} is the angular speed of 100 rpm.

As described in Equation (5), if the positions coincide, the spring torque is not affected by the speed. Static friction torque does not interfere during the constant speed operation, so Equation (13) can be simply described as:

$$T_e(\omega_{r2}) - T_e(\omega_{r1}) = T_{ge\_col}(\omega_{r2}) - T_{ge\_col}(\omega_{r1}) \tag{14}$$

If the coulomb friction is proportional to the speed, it can be expressed by the coulomb friction gain and the speed. The assumption that this gain is almost the same all over the position, to simplify the coulomb friction, Equation (14) can be transformed into:

$$T_e(\omega_{r2}) - T_e(\omega_{r1}) = B(\omega_{r2} - \omega_{r1}) \tag{15}$$

where B is the coulomb friction gain.

In this paper, we assumed that the coulomb friction occurs over the Stribeck velocity. Estimated coulomb friction gain is 0.0082 rpm/A for the forward direction and 0.0089 rpm/A for the reverse direction, respectively. Stribeck velocity of forward direction is 21 rpm and reverse direction is 17 rpm, respectively.

Figure 4 illustrates the measurement of the static friction torque. As shown in the figure, the current is increased to overcome the static friction force. However, the valve position does not move. If the current reaches the point described in the figure, the valve position starts to move due to the generated motor torque being greater than the static torque. At this time, speed increases radically when the static friction torque and the coulomb friction torque has a large difference. Note that the controlled speed is 20 rpm, which is the Stribeck velocity. In this case, the coulomb friction torque current is 0.8 A. However, the static friction torque is 10.4 A, which means that the static friction torque is over the 10 times that of the coulomb friction torque.

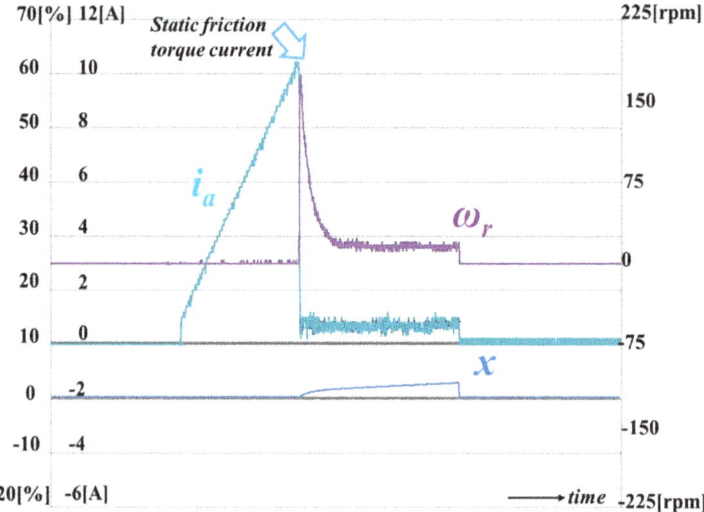

Figure 4. Static friction torque measurement at zero valve position (green: current, purple: speed, blue: position).

3. Proposed Position Controller

Figure 5 shows the conventional P-PI controller and the proposed control system [11]. As shown in the figure, the proposed control system does not use a speed controller. The main reason for this is that the motor position detection sensor is absent in the actual products to reduce the cost. Although the speed information can be obtained from the linear position sensor used for detecting the valve position, the sensing dynamics of the linear position sensor is insufficient to calculate the motor speed. Moreover, the speed information is the derivative component of the position information, so it is essential to use a filter to mitigate the noise. This worsens the restrictions on the bandwidth of the controller, which is already restricted due to the slow response of the linear position sensor.

For the same reason, the D controller cannot be used because the effective derivative component of the position error is difficult to obtain. Moreover, it can amplify the noise of the position information signal. Therefore, the PI controller was selected as the position controller in this system. Essentially, this position control system has performance problems.

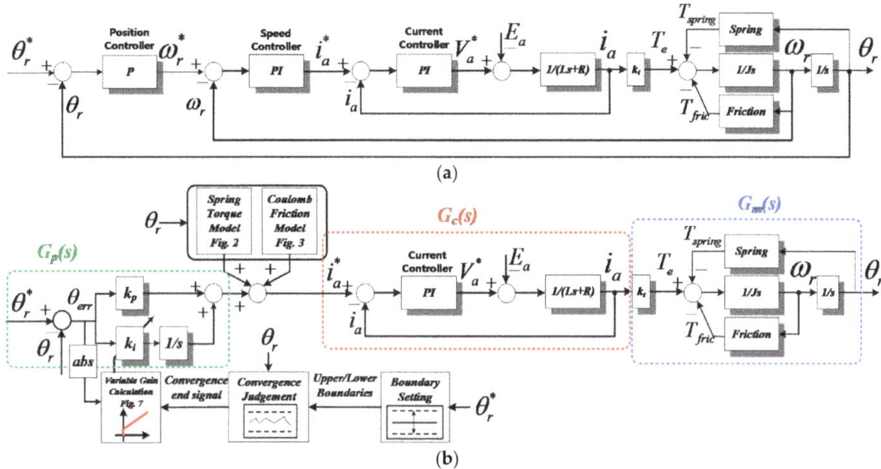

Figure 5. Position control method (**a**) P-PI controller [11] and the (**b**) proposed position controller.

First, proposed position control transfer function shown in Figure 5 can be described as:

$$\frac{\theta_r}{\theta_r^*} = \frac{G_p(s)G_c(s)G_m(s)}{1 + G_p(s)G_c(s)G_m(s)} \tag{16}$$

The control dynamic of the current controller is much faster than the position controller. The transfer function of the current controller $G_c(s)$ can be approximated as the unity in the position control view. Assuming that the spring load torque is fully compensated by the feedforward path, the transfer function can be changed to:

$$\frac{\theta_r}{\theta_r^*} = \frac{k_p k_m s + k_i k_m}{s^3 + k_p k_m s + k_i k_m} \tag{17}$$

where k_m is k_t/J.

Insert this transfer function into the final value theorem. The error of the step response can be obtained as:

$$e_\infty = \lim_{s \to 0} \frac{1}{1 + G_p(s)G_m(s)} = 1 \tag{18}$$

From the above equation, the PI controller for position control has an error in the steady state. To solve this problem, the proposed control method was derived from the hysteresis control. The proposed control sets the allowable boundary to perceive that the practical position follows the reference. If the sensed linear position is inside the boundary, the timer is activated to observe that the controlled position is stably located in the boundary or it is just during a transient operation. In this paper, the time to perceive the controlled position to be in the steady state was 200 ms.

Figure 6 shows the problems experienced by the conventional PI controller. If the sensed position gradually reaches the reference position, the controller output is also reduced. It also reduces the generated current and the motor speed. In advance, if the speed is reduced below the Stribeck velocity, the static friction torque majorly affects the entire load torque. As a result, as indicated by the figure, the motor is stopped when the motor current does not overcome the static friction torque. Next, the I controller integrates the position error when the sensed position does not exactly follow the reference. This integrated error gradually increases or decreases the current reference. If the specific current value reached by the generated motor torque is above the static friction torque, it causes sudden speed variation shown in Figure 4, which creates the position vibration.

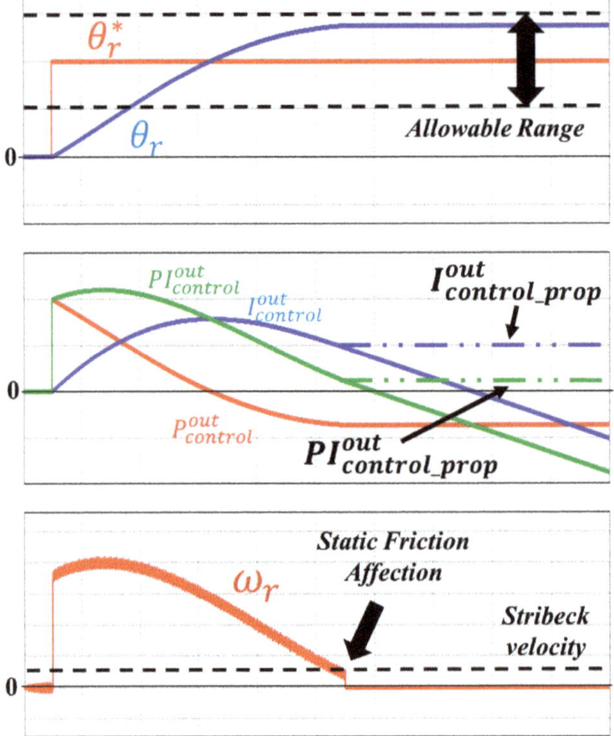

Figure 6. The operation of conventional and proposed PI controller.

To solve this problem, the variable I controller gain was adopted according to the position error. When the position error reached the boundary, I controller gain was reduced to the minimum accordance with the position error, as shown in Figure 7. As mentioned above, the I controller is the root of the position control vibration, so this controller was inactivated, since the valve position was located inside the allowable range. In this case, only the P controller affects the current reference generation. As a result, position vibration did not occur with the proposed position control method. In this paper, this allowable range was 5% of the position reference.

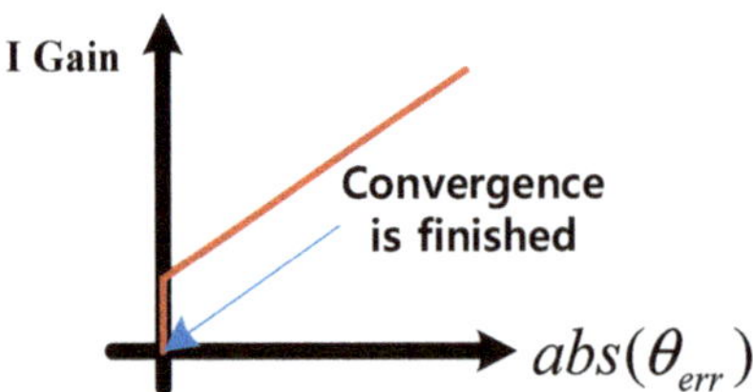

Figure 7. Variable I controller gain adaption according to the position error.

4. Experimental Results

Figure 8 shows the experimental test setup. To compare our method with the conventional control method, a high performance DSP TMS320F28335 from TI was used. A speed sensor was also instantaneously implemented in the mechanical system. The sampling frequency of the current controller and the switching frequency were the same: 20 kHz. The position control frequency was 2 kHz. The motor parameters are shown in Table 1.

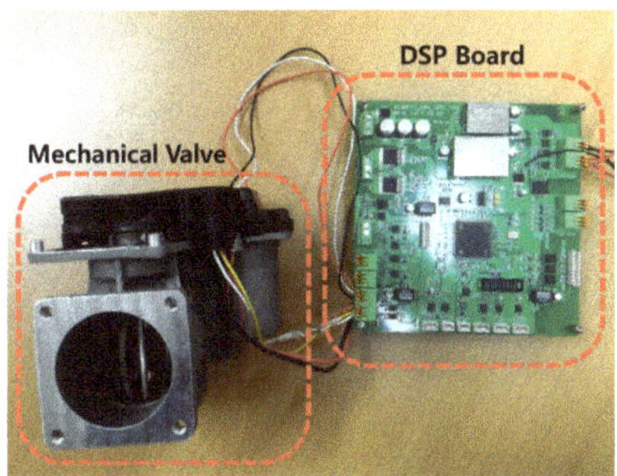

Figure 8. Experiment setup.

Table 1. Motor parameters.

Parameter	Value	Unit
Rated power	200	W
Input voltage	12	V
Max. current	20	A
Rated speed	500	rpm

Figure 9 shows the position dynamic responses of 10% and 100% of the valve reference. Due to the high static friction torque, if the position error was small, the I controller needed some time to generate the output for the suitable torque against the static friction torque. Therefore, the gain had to be tuned considering the maximum allowable control response time when the smallest position reference was applied. As shown in the figure, when the 10% reference was applied, the control response time was much longer than the result obtained when the 100% reference was applied because of the static friction torque. This control gain cannot be increased infinitely because of the overshoot restriction. Therefore, control gain tuning involves a trade-off by considering two aspects: response time and overshoot.

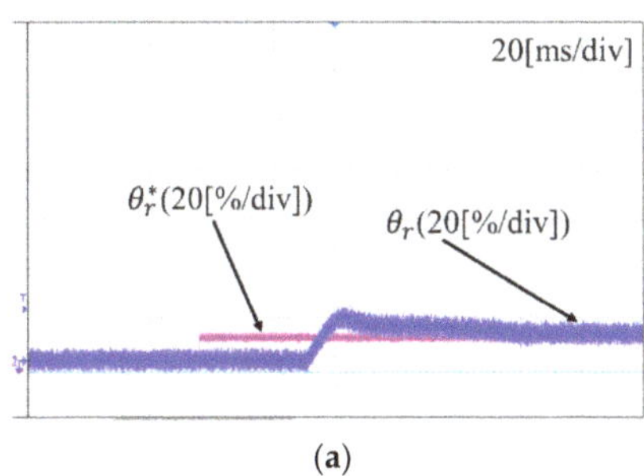

(a)

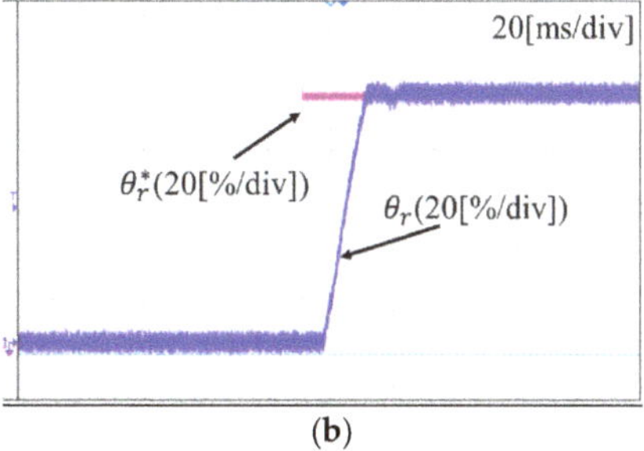

(b)

Figure 9. Position dynamic response when a (**a**) 10% and (**b**) 100% valve reference were applied.

Figure 10 compares the experimental results obtained using both a conventional P-PI controller [11] and the proposed controller. As shown in the figure, the position controlled by the conventional method vibrated due to the large difference between static and coulomb friction torque.

When the current reaches 2.5 A for the forward direction, the position movement is limited because the static friction torque resists the movement, however, since the current is above the 10 A, the position is radically moved forward because of the sudden friction change to coulomb friction torque. Due to the I controller affection, the current reaches to almost 17 A, and then, the position is over the reference. When the current reaches 17 A, the conventional controller starts to operate to move the position to the reverse direction. However, the position is almost stuck because of the high static friction torque, resulting in a slow decrease of the current to 2.5 A due to a small error integration of I controller. Subsequently, the controller starts to operate, and the position moves forward repeatedly. This operation caused the vibration as shown in Figure 10a. Note that the repeated current and the position are almost the same as shown in the figure, which meant that the static and coulomb friction torques at each position were hardly changed.

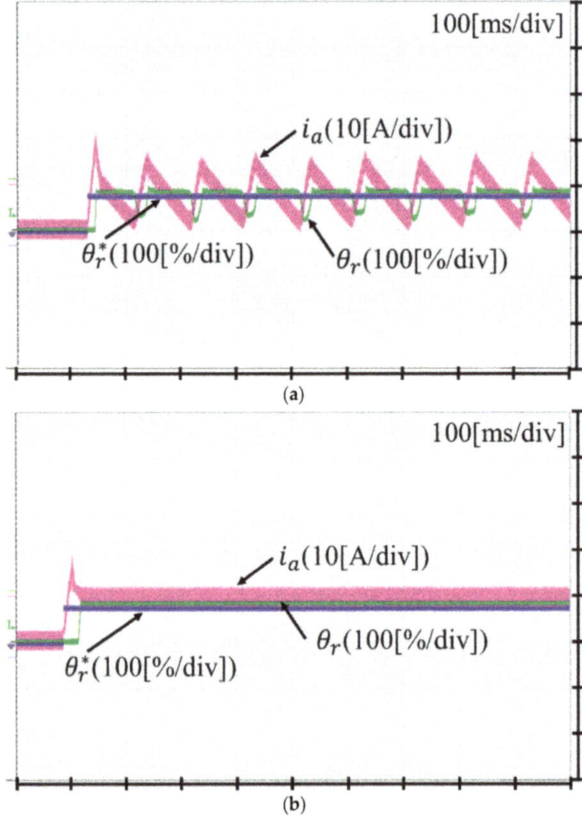

Figure 10. Comparison of the results between (**a**) conventional and (**b**) proposed position control (80% reference).

5. Conclusions

This paper proposed a position control method for a cost effective and a fast response time, which could be used in vehicle valve systems. Because the low-cost mechanical system has large differences in static friction and coulomb friction, the position and the current vibrations occur with the conventional P-PI linear controller. To solve this problem, this paper analyzed the EGR valve mechanical system and illustrated the procedure of extracting the parameters based on the predefined mathematical analysis. From this analysis, the proposed control method entailed acceptable boundary and selectable operation of the I controller. As a result, it could achieve the proper control performance which has an acceptable position error. The proposed method was verified by comparing our method with the conventional method in an experiment.

Author Contributions: J.-H.L. designed the experiment; J.-H.L. performed the experiment; J.-H.L. analyzed the theory. J.-H.L. wrote the manuscript. J.-H.L. and H.B. participated in research plan development and revised the manuscript. All authors have contributed to the manuscript.

Acknowledgments: This work is supported the Human Resources Development Program (Grant No. 20174010201350) by the Korea Institute of Energy Technology Evaluation and Planning (KETEP) grants. This research was supported by Korea Electric Power Corporation (Grant number: R18XA04).

Conflicts of Interest: The authors declare no conflict of interest.

References

1. Murtaza, G.; Butt, Y.A.; Bhatti, A.I. Higher Order Sliding Mode Based Control Scheme for Air Path of Diesel Engine. In Proceedings of the 2016 International Conference on Emerging Technologies (ICET), Islamabad, Pakistan, 18–19 October 2016.
2. Chen, S.K.; Yanakiev, O. *Transient NOx Emission Reduction Using Exhaust Oxygen Concentration Based Control for a Diesel Engine*; SAE Technical Paper No. 2005-01-0372; Ford Motor Company: Dearborn, MI, USA, 2005.
3. Heywood, J.B. *Internal, Combustion Engine Fundamentals*; McGraw-Hill Book Co.: New York, NY, USA, 1998.
4. Zheng, M.; Reader, G.T.; Hawley, J.G. Diesel engine exhaust gas recirculation—A review on advanced and novel concepts. *Energy Convers. Manag.* **2004**, *45*, 883–900. [CrossRef]
5. Seo, E.-S.; Shin, H.-B. Modeling of EGR Valve Actuator. *Trans. KIPE* **2017**, *22*, 390–396.
6. Kim, D.T.; Zhang, Z.J. Position Control of a Pneumatic Cylinder Considering Friction Compensation. *J. Drive Control* **2013**, *10*, 1–6. [CrossRef]
7. Byun, J.H. A Study on the Position Control of a Motor Cylinder with Nonlineal Friction. *J. Korean Soc. Power Syst. Eng.* **2008**, *12*, 80–86.
8. Chopra, S.; Mitra, R.; Kumar, V. Reduction of Fuzzy Rules and Membership Functions and Its Application to Fuzzy PI and PD Type Controllers. *Int. J. Control Autom. Syst.* **2006**, *4*, 438–447.
9. Ni, Z.; Wang, M. Research on the fuzzy neural network PID control of load simulator based on friction torque compensation. In Proceedings of the Sixth International Conference on Intelligent Human-Machine Systems and Cybernetics, Hangzhou, China, 26–27 August 2014; pp. 292–295.
10. Yao, J.; Deng, W.; Jiao, Z. Adaptive Control of Hydraulic Actuators with LuGre Model-Based Friction Compensation. *IEEE Trans. Ind. Electron.* **2015**, *62*, 6469–6477. [CrossRef]
11. Kim, H.J.; Park, H.S.; Heo, H.J.; Kim, J.M. Improvement of Position Control Performance of EGR Valve System with Low Control Frequency. In Proceedings of the 2017 IEEE 3rd International Future Energy Electronics Conference and ECCE Asia (IFEEC 2017—ECCE Asia), Kaohsiung, Taiwan, 3–7 June 2017; pp. 394–399.
12. Oh, B.G.; Lee, M.K.; Park, Y.S.; Lee, K.Y.; Sunwoo, M.H.; Nam, K.H.; Cho, S.H. Feedforward EGR Control of a Passenger Car Diesel Engine Equipped with a DC Motor Type EGR Valve. *Trans. KSAE* **2011**, *19*, 14–21.
13. Lee, H.S.; Tomizuka, M. Robust Motion Controller Design for High Accuracy Position Systems. *IEEE Trans. Ind. Electron.* **1996**, *43*, 48–55.

Article

Coordination and Control of Building HVAC Systems to Provide Frequency Regulation to the Electric Grid

Mohammed M. Olama [1,†], **Teja Kuruganti** [1,†], **James Nutaro** [1,†] and **Jin Dong** [2,*,†]

1   Computational Sciences and Engineering Division, Oak Ridge National Laboratory, Oak Ridge, TN 37831, USA; olamahussemm@ornl.gov (M.M.O.); kurugantipv@ornl.gov (T.K.); nutarojj@ornl.gov (J.N.)
2   Energy and Transportation Science Division, Oak Ridge National Laboratory, Oak Ridge, TN 37831, USA
*   Correspondence: dongj@ornl.gov; Tel.: +1-865-576-6742
†   These authors contributed equally to this work.

Received: 11 June 2018; Accepted: 9 July 2018; Published: 16 July 2018

Abstract: Buildings consume 73% of electricity produced in the United States and, currently, they are largely passive participants in the electric grid. However, the flexibility in building loads can be exploited to provide ancillary services to enhance the grid reliability. In this paper, we investigate two control strategies that allow Heating, Ventilation and Air-Conditioning (HVAC) systems in commercial and residential buildings to provide frequency regulation services to the grid while maintaining occupants comfort. The first optimal control strategy is based on model predictive control acting on a variable air volume HVAC system (continuously variable HVAC load), which is available in large commercial buildings. The second strategy is rule-based control acting on an aggregate of on/off HVAC systems, which are available in residential buildings in addition to many small to medium size commercial buildings. Hardware constraints that include limiting the switching between the different states for on/off HVAC units to maintain their lifetimes are considered. Simulations illustrate that the proposed control strategies provide frequency regulation to the grid, without affecting the indoor climate significantly.

Keywords: ancillary service; frequency regulation; demand response; commercial/residential buildings; HVAC systems; model predictive control; rule-based control.

---

1. Introduction

To ensure the functionality and reliability of a power grid, supply and demand must be balanced instantaneously and continuously. Balancing generation and load at all time scales, given the randomness in dynamics of generation and demand, is challenging. Correcting the mismatch requires ancillary services such as regulation and load following. The Federal Energy Regulatory Commission (FERC) has defined such services as those "necessary to support the transmission of electric power from seller to purchaser given the obligations of control areas and transmitting utilities within those control areas to maintain reliable operation of the interconnected transmission system." This quotation highlights the importance of ancillary services for both bulk-power reliability and support of commercial transactions [1]. Furthermore, a large amount of ancillary service will be required in the future if a large fraction of the energy needs is to be met from renewable energy sources with their associated unpredictability and volatility.

The traditional electric grid is load-following and is based on centralized generation assets that are controlled to compensate for demand changes in order to maintain a stable and reliable grid. Higher penetration of renewables and distributed energy resources, with their uncontrollable generation variability, imposes significant grid stability and control challenges. Demand-side control techniques are expected to address these challenges by increasing reliability and stability, reducing reserve margins, reducing peak demand, and improving energy efficiency. The inherent flexibility of many electric loads, when harnessed without impacting consumer comfort, can be an inexpensive source of ancillary service. Although employing loads for system services raise several challenges, several key advantages can be achieved: (i) reducing overall grid emissions by using loads to provide system services [2]; (ii) instantaneous response of loads to operator requests, versus slow response of generators to make significant output changes [3]; and (iii) less variability associated to a very large number of small loads with respect to that of a small number of large generators [3]. The key technical impediments to reliable utilization of loads for system services are the development of deployable control strategies over wide-area and the development of inexpensive and scalable sensing, communication, and control infrastructure [4].

Buildings account for 73% of total electricity consumption in the United States and therefore will play a crucial role in the future of the national electric power system. Total annual US energy consumptions are roughly equal for residential and commercial buildings with Heating, Ventilation and Air-Conditioning (HVAC) loads that account for about half of their energy use. Nonetheless, only about 10% of all commercial buildings use automation systems to control their energy usage, and an insignificant percentage of these buildings provide ancillary services to power system operators [5]. The unrealized potential to incorporate buildings into the grid to provide ancillary services is therefore very large and will help mitigate the global challenge of providing reliable, cost effective, and clean energy.

The use of commercial building HVAC systems for providing ancillary services ia examined in [6–18]. In particular, the works in [6–8] address the usage of commercial buildings for demand response programs, which typically involve reduction of peak power in emergency situations. The works in [9–12] illustrate the potential promise of model predictive control (MPC) for energy efficiency in buildings and for integrating time-of-use rates for shifting loads. The work in [13] shows that the supply fans in air handling units (AHUs) with variable frequency drives (VFDs) can provide high frequency ancillary service of about 70% of the regulation reserves in the time scale of 8 s to 3 min. In [14], the time scale of ancillary service from commercial building HVAC systems is extended to the range of 3 min to 1 h by using the flexibility in the power demand from chillers. Recent research on residential loads [19,20] such as HVAC and refrigerators has shown that such loads can provide ancillary service with the help of appropriate control algorithms. One drawback of residential loads is that they are largely on/off control, which greatly reduces the flexibility of control strategies that can be applied. Such a drawback will be tackled for the first time in this paper by considering a coordination and control of an aggregate of on/off loads to provide desired grid-response. In particular, the proposed controller will manipulate the total power consumption of available HVAC loads according to requested change in power from the grid side, represented by a regulation signal, to enhance the grid reliability and stability.

In this paper, we consider HVAC loads as an ancillary service for providing frequency regulation to the grid. We investigate two control strategies of HVAC systems to provide such services. In the first strategy, we consider an optimal strategy based on MPC acting on a continuously variable HVAC system, which is available in most large commercial buildings. MPC has been widely employed in energy efficiency in buildings, but only a few in control strategies for ancillary services to the grid. In the second strategy, we consider rule-based coordination and control of an aggregate of on/off HVAC systems, which are widely-spread in residential buildings in addition to many small to medium commercial buildings. This strategy is based on priority control of room temperatures of multiple on/off HVAC systems. Numerical results show that it is feasible to use a small portion (less than 20%) of the total HVAC power in residential/commercial buildings for regulation services to the grid, with little change in their indoor environments.

2. HVAC Thermal Dynamical Model

In this section, a simple, yet realistic building thermal model is introduced to represent the building with a HVAC system. It is a continuous-time linear time-invariant (LTI) system model based on the dynamics of the room temperature, interior-wall surface temperature, and exterior-wall core temperature. This physics-based lumped thermal model is initially proposed in [21] and employed in [22–24] for simulating residential and commercial buildings. It is described by

$$
\begin{aligned}
\dot{x}_1 &= \tfrac{1}{C_1}[(K_1 + K_2)(x_2 - x_1) + K_5(x_3 - x_1) + K_3(\delta_1 - x_1) + u_1 \\
&\quad + u_2 + \delta_2 + \delta_3] \\
\dot{x}_2 &= \tfrac{1}{C_2}[(K_1 + K_2)(x_1 - x_2) + \delta_2] \\
\dot{x}_3 &= \tfrac{1}{C_3}[K_5(x_1 - x_3) + K_4(\delta_1 - x_3)]
\end{aligned}
$$

where the variables used in the above model are defined in Table 1, and the parameter values are provided in Table 2.

Table 1. Building parameter definition.

Variables	Definition
x_1	room air temperature (°C)
x_2	interior-wall surface temperature (°C)
x_3	exterior-wall core temperature (°C)
u_1	cooling power (≤ 0) (kW)
u_2	heating power (≥ 0) (kW)
δ_1	outside air temperature (°C)
δ_2	solar radiation (kW/m^2)
δ_3	internal heat sources (kW)

Table 2. Building parameter values.

$C_1 = 9.356 \times 10^5$	kJ/°C	$C_2 = 2.970 \times 10^6$	kJ/°C
$C_w = 6.695 \times 10^5$	kJ/°C	$K_1 = 16.48$	kW/°C
$K_2 = 108.5$	kW/°C	$K_3 = 5$	kW/°C
$K_4 = 30.5$	kW/°C	$K_5 = 23.04$	kW/°C

The system states are x_1, x_2, and x_3. The model inputs are divided into manipulated variables and disturbance inputs. The manipulated variables are the cooling power u_1 and the heating power u_2, and they can be combined to one variable $u = u_1 + u_2$. Without loss of generality, we assume cooling and heating are not functioning simultaneously in our study, which is usually the case for small residential buildings. The disturbances are δ_1, δ_2, and δ_3.

Define the state vector x, the control signal vector u, and the environment stochastic disturbance vector ω as:

$$
x := \begin{bmatrix} x_1 \\ x_2 \\ x_3 \end{bmatrix}, \quad
u := \begin{bmatrix} u_1 \\ u_2 \end{bmatrix}, \quad
\omega := \begin{bmatrix} \delta_1 \\ \delta_2 \\ \delta_3 \end{bmatrix}.
$$

The continuous-time state-space model can then be described compactly as:

$$
\dot{x} = A_c x + B_c u + C_c \omega \tag{1}
$$

where

$$
A_c := \begin{bmatrix}
-\tfrac{1}{C_1}(K_1 + K_2 + K_3 + K_5) & \tfrac{1}{C_1}(K_1 + K_2) & \tfrac{K_5}{C_1} \\
\tfrac{K_1 + K_2}{C_2} & -\tfrac{(K_1 + K_2)}{C_2} & 0 \\
\tfrac{K_5}{C_3} & 0 & -\tfrac{(K_4 + K_5)}{C_3}
\end{bmatrix}
$$

$$B_c := \begin{bmatrix} \frac{1}{C_1} & \frac{1}{C_1} \\ 0 & 0 \\ 0 & 0 \end{bmatrix}, \qquad C_c := \begin{bmatrix} \frac{K_3}{C_1} & \frac{1}{C_1} & \frac{1}{C_1} \\ 0 & \frac{1}{C_2} & 0 \\ \frac{K_4}{C_3} & 0 & 0 \end{bmatrix}. \tag{2}$$

We consider the discrete-time (sampled) version of Equation (1) described by

$$x_{k+1} = A_d x_k + B_d u_k + C_d \omega_k \tag{3}$$

where k is the discrete-time index, $x_k = [x_{1,k}\, x_{2,k}\, x_{3,k}]^T$ and the parameters $[A_d, B_d, C_d]$ are computed from the continuous-time model parameters in (2).

The control input u is the critical actuator yielding its own working properties and conditions. It takes continuous values within a certain bound for a given HVAC system in large commercial buildings, where the variable frequency drive (VFD), which is available in variable air volume (VAV) HVAC systems and allows continuous input power, is responsible for changing the air handling unit fan speed. On the other hand, the control input u takes discrete values, usually two to three states, in the on/off HVAC systems that are available in residential buildings in addition to many small to medium commercial buildings. In this paper, both the VAV and on/off HVAC systems are investigated to provide ancillary services to the grid.

3. Control Design for a VAV HVAC Unit

In this section, a controller is designed for a large commercial building having a VAV HVAC units with continuous control variables. The intended controller should control the operation of the HVAC unit such that: (i) the *change* in the power consumed by the HVAC unit is as close as reasonably possible to the requested *change* in power in the regulation signal for that building; and (ii) the reported thermostat temperature for the HVAC unit is as close as possible to its set point. To achieve these two objectives, we have designed the feedback control system shown in Figure 1. It is an MPC strategy that is widely-used in the industry and displays its main strength when applied to problems with constraints imposed on both the manipulated and controlled variables. The designed controller has three input signals and one output signal. The input signals are the room temperature set point r_k, the regulation signal u_k^{Reg}, and the current reported room (thermostat) temperature y_k (y_k represents $x_{1,k}$ of the thermal model in Equation (3), the other model states are assumed to be non-measurable). The output signal of the controller is a scaled version of the input power to the HVAC unit, u_k. The plant represents the discrete-time thermal model for the HVAC and building system described in Equation (3).

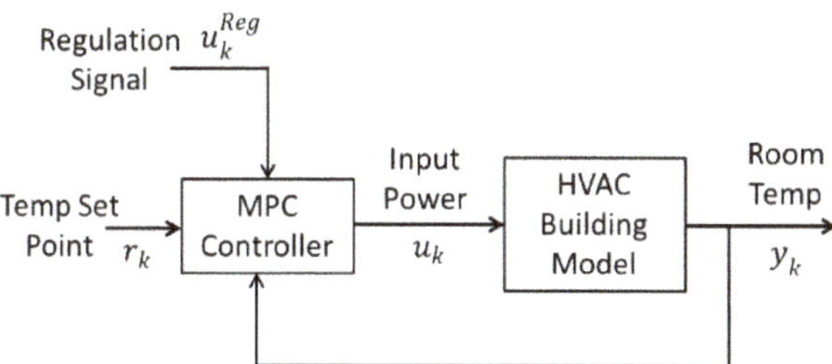

Figure 1. The proposed control architecture for a building with a VAV HVAC system.

The controller minimizes the cost function described by

$$J(z_k) = J_y(z_k) + J_u(z_k)$$
$$= \sum_{i=1}^{p} \left\{ \omega_i^y \left[r_{k+i|k} - y_{k+i|k} \right] \right\}^2 + \left\{ \omega_i^u \left[u_{k+i|k}^{Target} - u_{k+i|k} \right] \right\}^2 \tag{4}$$

subject to

$$y_i^{min} \leq y_{k+i|k} \leq y_i^{max}, \, i = 1 : p$$
$$u_i^{min} \leq u_{k+i|k} \leq u_i^{max}, \, i = 1 : p \tag{5}$$

where k is the current control interval, p is the prediction horizon, $y_{k+i|k}$ is the predicted value of room temperature at ith prediction horizon step, $r_{k+i|k}$ is the thermostat set point at ith prediction horizon step, ω_i^y is the tuning weight for the room temperature at ith prediction horizon step, $u_{k+i|k}$ is the estimated value (to be computed) of the input power to the HVAC unit (manipulated variable) at ith prediction horizon step, $u_{k+i|k}^{Target}$ is the target value for the manipulated variable at ith prediction horizon step, ω_i^u is the tuning weight for the manipulated variable at ith prediction horizon step, and $z_k = \left[u_{k|k}, u_{k+1|k}, u_{k+p-1|k} \right]$.

The values p, ω_i^y, and ω_i^u are controller specifications, and are constants. The controller receives $r_{k+i|k}$ and $u_{k+i|k}^{Target}$ values for the entire horizon and uses the state observer to predict the plant outputs. At instant k, the controller state estimates are available, thus J is a function of z_k only. Since the regulation signal represents the change in the power consumed by the HVAC unit, $u_{k+i|k}^{Target}$ is described by

$$u_{k+i|k}^{Target} = u_{k+i|k}^{Baseline} + u_{k+i|k}^{Reg} \tag{6}$$

where $u_{k+i|k}^{Baseline}$ is the optimal value for the manipulated variable at ith prediction horizon step without considering the impact of the regulation signal, i.e. the optimal control signal such that the room air temperature is as close as possible to its set point.

The regulation signal used in our analysis is taken from the PJM dynamic (D) regulation control signal, which is used for fast-responding resources and constructed from the area control error (ACE) that measures the amount of negative or positive power needed in the power system [25]. Figure 2 shows the normalized regulation signal for a specific day that is sampled every 1 min. A positive ACE value represents the case where an increase in the power consumption is requested and a negative value represents the case where a decrease in the power consumption is requested. The regulation signal can simply be distributed among multiple VAV HVAC units by maintaining its shape and scaling it down to proper levels to maintain occupant comfort. Thanks to the continuous input power ability of VAV HVAC systems that allows this simple distribution of the regulation signal among multiple HVAC units. Thus, without loss of generality, we only consider one VAV HVAC unit in the analysis presented next.

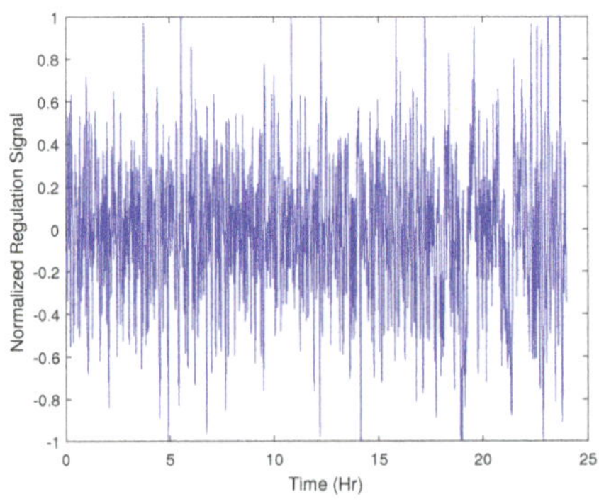

Interconnection LLC (PJM).

Figure 2. The normalized PJM dynamic regulation signal.

To illustrate the performance of the control scheme in Equations (4)–(6), simulations are conducted for a typical summer day with the following setup: simulation time-step is selected to be one minute to satisfy the variations in the regulation signal, initial room temperature $T_0 = 26\,^\circ$C, room temperature set point $r_k = 21\,^\circ$C, $20.5\,^\circ$C $\le y_k \le 21.5\,^\circ$C, $0 \ge u_k \ge -6$ kW (cooling scenario), $p = 1$, and the outside air temperature (profile 1) demonstrated in Figure 3 is considered. The random noise that exists in the temperature profile represents changes in solar irradiance due to temporary cloud cover.

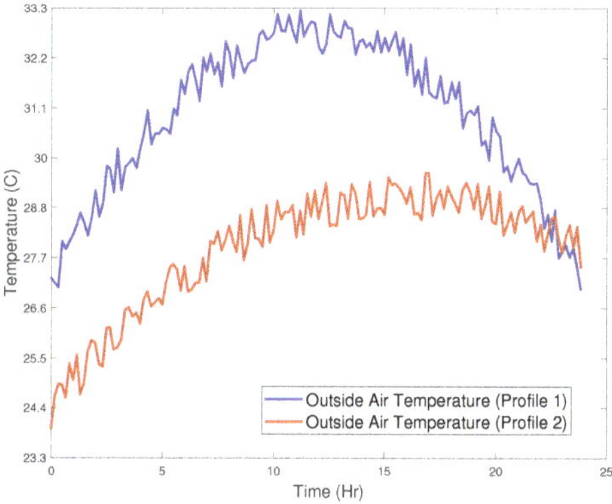

Figure 3. Two outside air temperature profiles for typical summer days.

The simulation results for this scenario are demonstrated in Figure 4, where the lower graph shows that the change in the power consumed by the HVAC unit closely follows the regulation signal, while the room temperature is close enough to the temperature set point (within $\pm 0.5\,^\circ$C) , as shown in the upper graph. Note that, in this example, the maximum power in the regulation signal is 1 kW, which is about 17% of the maximum power consumed by an HVAC unit (6 kW).

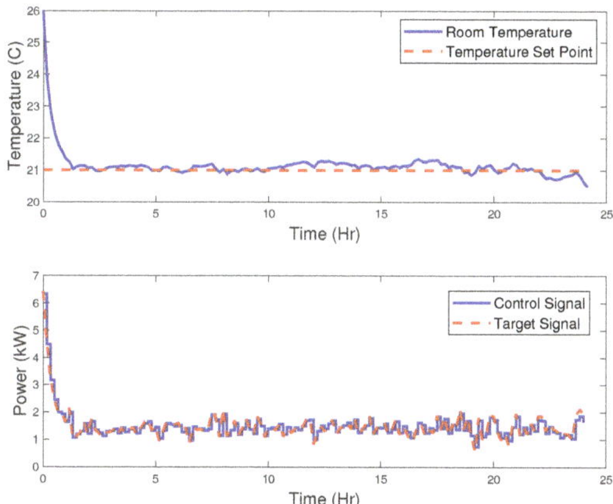

Figure 4. Simulation results illustrating the performance of the designed MPC controller for a building with a VAV HVAC system and a 1 kW maximum regulation power.

Table 3 illustrates the performance of the designed MPC controller for this scenario over different regulation power levels. We observe that the requested change in power (maximum regulation signal magnitude) should be less than 17% of the total power consumed by the HVAC unit to maintain occupant comfort (room temperature is within $\pm 0.5\,^{\circ}$C from its set point). The root-mean-square-error (RMSE) of the room temperature from its set point is also presented in Table 3.

In the next section, we introduce the second control strategy that considers coordination and control of an aggregate of on/off HVAC systems.

Table 3. The performance of the designed MPC controller for a building with a VAV HVAC system over different regulation power levels.

Max Regulation (kW)	Ratio of Max Reg. to Max Power Consumption (%)	Max Room Temp. Dev. from Set Point ($^{\circ}$C)	RMSE ($^{\circ}$C)
0.3	5	0.17	0.08
0.6	10	0.32	0.11
0.9	15	0.47	0.14
1	17	0.51	0.15
1.2	20	0.62	0.18
1.5	25	0.76	0.21
1.8	30	0.91	0.25

4. Control Design for an Aggregate of on/off HVAC Units

Now, we address the more challenging on/off HVAC system, which is widely used in residential buildings. It is obvious that an on/off HVAC system does not provide the required flexibility for tracking a regulation signal as in the VAV HVAC case because of the limited number of states in the on/off HVAC, usually two states (on and off). However, we show in this section that, by proper control and coordination of a fleet of on/off HVAC systems (available in one or many building(s)), the required flexibility for tracking a regulation signal can be reached. Thus, the objective in this case is to design a controller (strategy) for an aggregate of on/off HVAC systems that takes into consideration the regulation signal provided by the power utility such that its impact should be minimal on room temperatures of the buildings. The proposed controller should coordinate the operation of the HVAC units across all buildings such that: (i) the total *change* in the power consumed by *all* HVAC units is as close as reasonably possible to the requested *change* in power in the regulation signal; and (ii) the reported thermostat temperature for each HVAC unit is as close as possible to its set point. In addition, the following constraints should be satisfied:

1. The on/off HVAC unit has the following three states:

    (a) Off (no power consumed).
    (b) Stage 1 cooling (power consumed is 3 kW).
    (c) Stage 2 cooling (power consumed is 6 kW).

2. Switching between the different states for an HVAC unit is not allowed before certain time duration from the last switch (we assume it is 10 min in our analysis). This is required to maintain the lifetime of the HVAC units. More frequent switching of an HVAC unit may cause some damage to the unit and shorten its lifetime.
3. Thermostat temperature control (dead band) for each building is $\pm 0.5\,^{\circ}$C from its set point.

The proposed control architecture for this strategy is illustrated in Figure 5, where the feedback message from each HVAC unit to the central controller contains its current room temperature and thermostat set point, i.e. $m_i = \{y_i, r_i\}, i = 1, \cdots, M$, and M is the total number of HVAC units. Note that the manipulated variables, $u_1, \cdots, u_M$, take only three values, as indicated in Constraint 1 above.

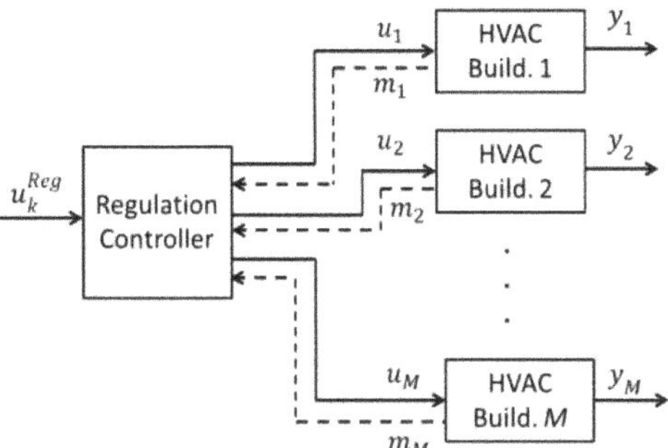

Figure 5. The proposed rule-based control architecture for a fleet of on/off HVAC units.

Without loss of generality, by considering the cooling case in HVAC units and assuming all buildings have the same temperature set points, the centralized rule-based control strategy is designed as follows:

1. If the regulation signal is *positive*, this indicates an increase in the power consumption is requested (more cooling). In this case, the controller should prioritize room temperatures with the highest temperatures to cool them first. Thus, the controller should *increase* the power provided to HVAC units with the *highest* temperature zones or buildings.
2. If the regulation signal is *negative*, this indicates a decrease in the power consumption is requested (less cooling). In this case, the controller should prioritize room temperatures with the lowest temperatures to not cool them first. Thus, the controller should *decrease* the power provided to HVAC units with the *lowest* temperature zones or buildings.

For the case where buildings have different temperature set points, the controller prioritizes room temperature deviations from their set points with the highest positive temperature deviations to cool them first when the regulation signal is positive and with the highest negative temperature deviations to not cool them first when the regulation signal is negative. It should be remarked that the home owner could very easily shift the set point if he feels cold or hot. This priority control strategy is described in Algorithm 1. Moreover, the heating case in HVAC units is designed in an opposite way, for instance, when the regulation signal is positive, the controller increases the power provided to HVAC units with the lowest temperature zones or buildings. Note that this priority control strategy is chosen based on its simplicity to implement, as it does not require solving an optimization problem such as the MPC strategy in the previous section. Thus, it is computationally efficient and much easier to implement in practice. Lastly, it is hardly fair to directly compare these two control strategies because each strategy is used for different type of HVAC loads. Using MPC for on/off HVAC loads requires solving more complicated mixed integer linear programming problem, which is out of the scope of this manuscript and is left for future work.

In the next section, numerical results are presented to illustrate the performance of the rule-based control strategy.

Algorithm 1: The rule-based control strategy for a fleet of on/off HVAC systems (cooling case).

Inputs : Current room temperatures, thermostat set points, and power consumptions for all buildings, i.e. $\{y_i, r_i, u_i\}, i = 1, \cdots, M$, in addition to the regulation signal u^{Reg}.

Output: Updated input powers to the on/off HVAC units (manipulated variables), i.e. $u_i, i = 1, \cdots, M$.

foreach *time step* **do**

    Compute the room temperature deviations from their set points

    Sort the computed deviations in descending order

    Compute the number of HVAC units, N, that is required to satisfy the regulation signal

    if *the regulation signal is positive* **then**

        foreach *HVAC unit starting from the highest room temperature deviation in descending order until reaching the Nth change in power consumption of HVAC units* **do**

            Check whether the constraints are satisfied (previous state lasted for more than 10 min, and HVAC current power consumption is not Stage 2 cooling)

            Increase the input power to the HVAC unit (from 0 to 3 kW or from 3 kW to 6 kW)

    if *the regulation signal is negative* **then**

        foreach *HVAC unit starting from the lowest room temperature deviation in ascending order until reaching the Nth change in power consumption of HVAC units* **do**

            Check whether the constraints are satisfied (previous state lasted for more than 10 min and HVAC current state is not OFF)

            Reduce the input power to the HVAC unit (from 6 kW to 3 kW or from 3 kW to 0)

    Distribute the updated HVAC power consumptions

    Repeat for the next time step

---

5. Numerical Results

We consider 50 on/off HVAC units (buildings) in our analysis with the same PJM dynamic regulation signal used in Section 3, but it is now scaled up to 50 kW to satisfy the ratio of the maximum regulation to the maximum power consumed by all HVAC units (300 kW) as described in Section 3. In addition, initial room temperatures are assumed to be normally distributed around their set points with unit variance and all HVAC units are initially at off state. In addition, the same outside temperature profile and building parameters as in Section 3 are used here. The simulation time-step is 1 min and the 10-min switching constraint of on/off HVAC units is maintained in the control strategy.

The performance of the designed rule-based control strategy has been investigated for a 24-h period and compared with the baseline case where no regulation is considered (the only objective is to satisfy the thermostat temperature for each HVAC unit). Figure 6 shows the room temperature and the corresponding power consumed by the HVAC unit for one building selected at random (Building 39). Figure 7 shows the baseline case for the same scenario in Figure 6 for comparison. Notice the variations in the room temperature under the regulation case (Figure 6) as compared with the one without regulation (Figure 7). These variations are due to the impact of satisfying the regulation condition. Despite these variations, the room temperature remained most of the time within the allowed band, which is $\pm0.5\ °C$ from the set point. In addition, it is observed that the impact of the regulation on the HVAC total power consumption and duty cycle (fraction of time in which HVAC unit is ON) is minimal (see their values on top of the figures). In addition, notice that this HVAC unit did not switch to Stage 2 cooling because of either there are higher room temperatures at that time than the one at Building 39 that have higher priorities, or the regulation signal at that time was not high enough to turn on this unit to Stage 2 cooling after taking care of the other buildings with higher room temperatures. Similar behaviors as in Building 39 are observed in the remaining buildings, except that Stage 2 cooling is observed for short time durations in few HVAC units.

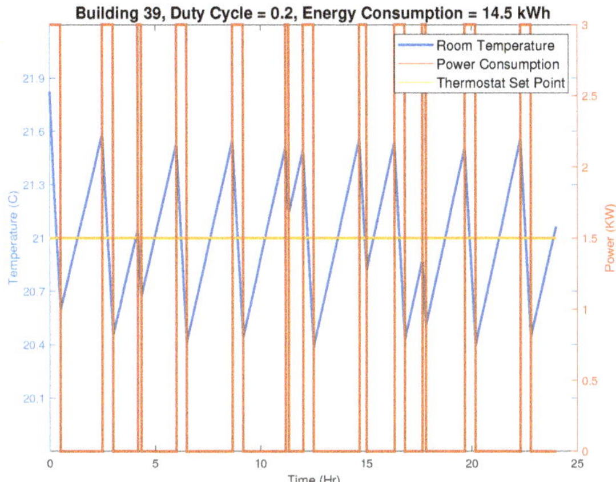

Figure 6. Simulation results showing room temperature and power consumption of Building 39 for the designed *rule-based* controller

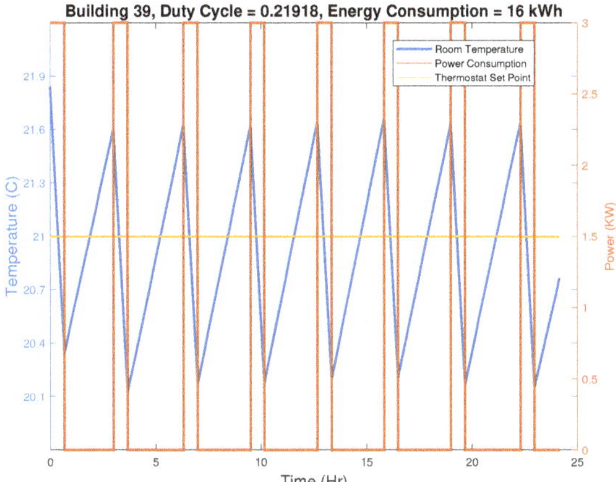

Figure 7. Simulation results showing room temperature and power consumption of Building 39 for the *baseline* controller (no regulation is considered).

Figure 8 demonstrates room temperatures for all 50 buildings. Observe that the deviations from the set points are less than 0.5 °C most of the time. This is due to the appropriate selection of the total number of buildings (and their corresponding maximum power consumption) relative to the maximum regulation power.

Figure 9 shows the HVAC duty cycles (DCs) for the 50 buildings using the rule-based and baseline controllers. Notice that about 20% duty cycle is typical for a hot summer day. It can be observed that nearly all buildings provide the same amount of ancillary service. The average duty cycles for all buildings for the designed rule-based and baseline controllers are 0.2037 and 0.2034, respectively. These duty cycles correspond to total energy consumption by all 50 buildings for 24-h interval of an amount 741.5 kWh and 737 kWh, respectively. Notice that nearly the same duty cycle (power consumption) took place for both scenarios. Assuming the utility pays 10 cents per kWh for building owners subscribed to provide ancillary service to the grid, the owners of the 50 buildings in this scenario will receive credit of \$798 per month.

Figure 10 demonstrates that the total power consumed by all 50 buildings closely follows the regulation signal. This is due to the proper designed control strategy and the appropriate selection of the total number of buildings relative to the maximum regulation power.

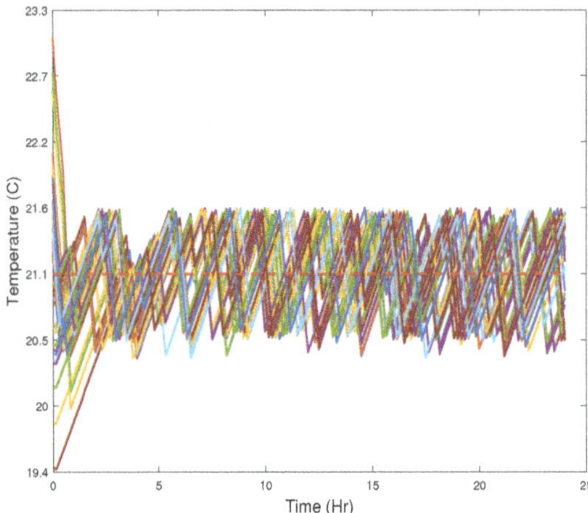

Figure 8. Simulation results showing room temperature deviations of all buildings from their set points for the designed *rule-based* controller.

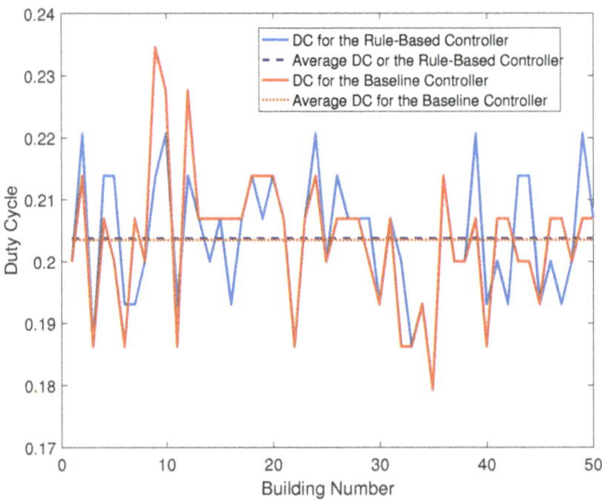

Figure 9. Simulation results showing the HVAC duty cycles for the 50 buildings for the designed rule-based and baseline controllers.

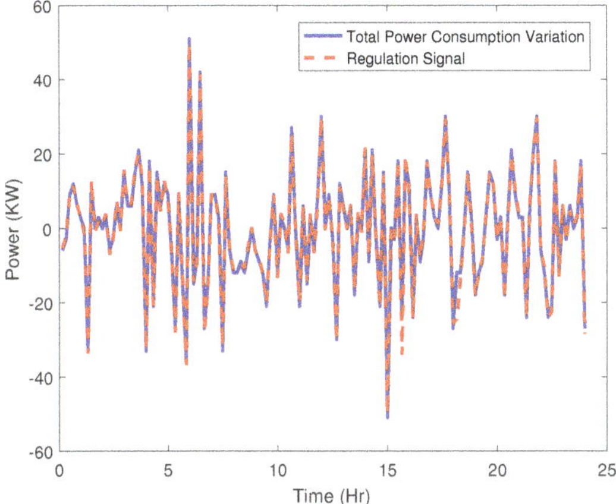

Figure 10. Simulation results showing the aggregate power consumption variation for all 50 buildings with on/off HVAC units and 50 kW maximum regulation power.

Sensitivity analysis has been conducted for the designed rule-based controller under different buildings parameters ($C_1, C_2, C_3, K_1, K_2, K_3, K_4$, and K_5 are within 20% tolerance from their nominal values) and outside air temperatures, where similar behaviors as in previous scenarios (Figures 6–10) are observed. For example, when different building parameters and outside air temperatures are used (Profile 2 in Figure 3), numerical results show that the room temperatures for all 50 buildings are within $\pm 0.5\,^{\circ}\mathrm{C}$ from their set points almost all of the time while satisfying the required regulation, as illustrated in Figure 11.

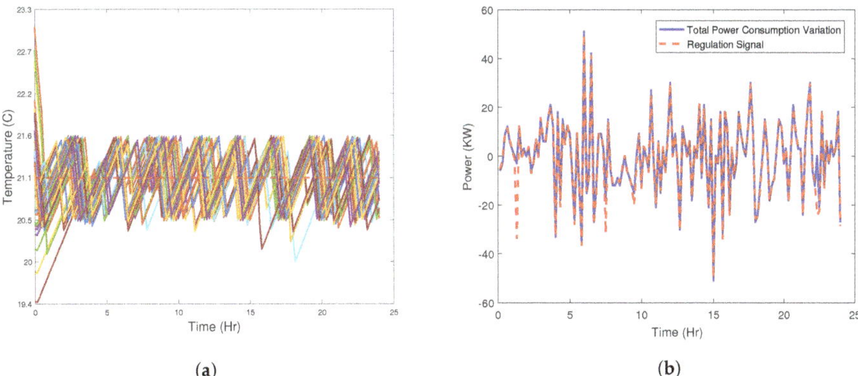

(a)   **(b)**

Figure 11. Simulation results showing the performance of the designed rule-based controller for 50 buildings with on/off HVAC units, 50 kW maximum regulation power, and outside air temperature profile 2: (**a**) room temperature deviations for all buildings from their set points; and (**b**) the aggregate power consumption variation and the regulation signal.

It is also interesting to investigate the impact of the total number of buildings on the performance of the rule-based control strategy. Figure 12 illustrates the room temperatures and aggregate power consumption variations when the total number of buildings is reduced to the half (25 buildings).

We can observe that the performance in this scenario is not as good as the 50 buildings scenario; quite few room temperatures went out of the ±0.5 °C band for some time and the aggregate power consumption variation for all building did not fully follow the regulation signal. This unsatisfactory performance is due to the fact that each HVAC unit contributed about 33% of its power capacity as an ancillary service to the grid, where our analysis for this particular example (in Section 3 and in this section as well) indicates that each HVAC unit should not contribute more than 17% of its power capacity as ancillary service to the grid to ensure satisfactory performance. It should be mentioned that we assume all the buildings have correct sizing of the HVAC system. In practice, it is also worth investigating the system sizing problem [26,27] before looking into the proposed control strategy.

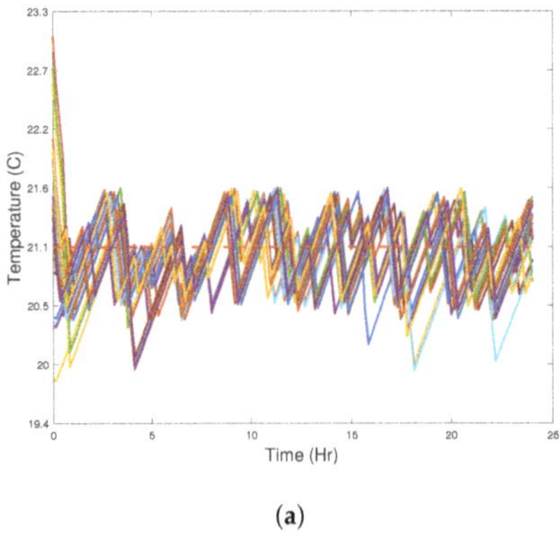
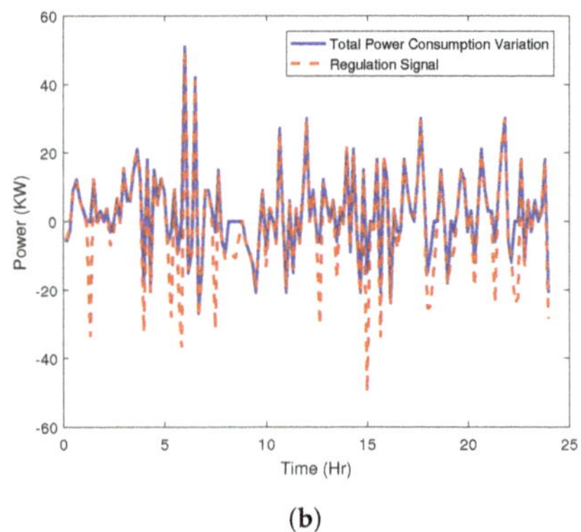

(a)
(b)

Figure 12. Simulation results showing the performance of the designed rule-based controller for 25 buildings with on/off HVAC units and 50 kW maximum regulation power: (**a**) room temperature deviations of all 25 buildings from their set points; and (**b**) the aggregate power consumption variation and the regulation signal.

Table 4 shows the maximum deviation and root-mean-square-error (RMSE) of the regulation signal and indoor temperature tracking for all buildings. Notice the improved performance with the increased number of recruited buildings, which provides additional flexibility for tracking the regulation signal and thus satisfying the grid requirements. Note that the 17% contribution limit of HVAC power capacity as ancillary service to the grid is specific to this particular example (specific to the assigned building parameters and disturbances) and should be verified/tested for different settings. Therefore, an initial testing/tuning is recommended before implementing the proposed rule-based controller to ensure satisfactory performance.

Table 4. The performance of the designed rule-based controller for a different number of buildings (HVAC units).

# of Buildings	Regulation Signal Tracking Error		Indoor Temperature Tracking Error	
	Max. Error (kW)	RMSE (kW)	Max. Error (°C)	RMSE (°C)
25	36.83	4.28	1.03	0.34
50	8.21	0.4	0.61	0.27
75	2.54	0.22	0.53	0.21

6. Conclusions and Future Directions

We have investigated two control strategies for HVAC units to provide ancillary services to the grid. In the first strategy, an optimal control based on MPC acting on a VAV HVAC unit is examined, while, in the second strategy, a rule-based control of a fleet of on/off HVAC units is examined. The presented analysis has validated our argument that coordination and control of a fleet of on/off HVAC units provide the required flexibility for ancillary services to the grid, with little impact on indoor environments. If we assume the regulation requirement of the United States is 1% of its peak load (similar to the one in PJM), the total regulation reserves that are potentially available from all residential/commercial buildings in the US are about three times of the total regulation capacity currently needed [28].

The novelty of the analysis is to use HVAC loads as ancillary service to the grid. In particular, the proposed controller should manipulate the total power consumption of available HVAC loads according to requested change in power from the grid side, represented by a regulation signal, to enhance the grid reliability and stability. The designed rule-based controller can be implemented in a centralized mode, where each building communicates its current state (room temperature, power consumption, and temperature set point) to a centralized aggregator that selects the next states of the HVAC units. Then, the selected strategies are communicated back to buildings (smart thermostats) to be applied. For large number of buildings, a cluster tree topology would be more feasible, as each building communicates its current state to a server node within its cluster, and the server nodes communicate their aggregate information to a centralized node where the control strategies are chosen and communicated back to the buildings through the cluster nodes. Another implementation option could be through the cloud, where the information is shared (via the Internet) and analyzed in utility or third-party data centers.

Future work is to conduct further studies to design a decentralized framework for the rule-based control strategy.

Author Contributions: Conceptualization, M.O., T.K. and J.N.; Methodology, M.O., T.K. and J.N.; Software, M.O. and J.D.; Validation, M.O. and J.D.; Formal Analysis, M.O. and T.K.; Investigation, M.O. and T.K.; Writing—Original Draft Preparation, M.O.; Writing—Review & Editing, J.N. and J.D.

Acknowledgments: This manuscript has been authored by UT-Battelle, LLC under Contract No. DE-AC05-00OR22725 with the U.S. Department of Energy. This material is based upon work supported by the U.S. Department of Energy, Office of Energy Efficiency and Renewable Energy. The United States Government retains and the publisher, by accepting the article for publication, acknowledges that the United States Government retains a non-exclusive, paid-up, irrevocable, world-wide license to publish or reproduce the published form of this manuscript, or allow others to do so, for United States Government purposes. The Department of Energy will provide public access to these results of federally sponsored research in accordance with the DOE Public Access Plan (http://energy.gov/downloads/doe-public-access-plan).

Conflicts of Interest: The authors declare no conflict of interest.

References

1. Kirby, B.J. *Frequency Regulation Basics and Trends*; United States Department of Energy: Washington, DC, USA, 2005.
2. Strbac, G. Demand side management: Benefits and challenges. *Energy Policy* **2008**, *36*, 4419–4426. [CrossRef]
3. Kirby, B.J. *Spinning Reserve from Responsive Loads*; United States Department of Energy: Washington, DC, USA, 2003.
4. Woo, C.K.; Kollman, E.; Orans, R.; Price, S.; Horii, B. Now that California has AMI, what can the state do with it? *Energy Policy* **2008**, *36*, 1366–1374. [CrossRef]
5. Kelso, J.D. *Buildings Energy Data Book*; United States Department of Energy: Washington, DC, USA, 2017.
6. Pavlak, G.S.; Henze, G.P.; Cushing, V.J. Optimizing commercial building participation in energy and ancillary service markets. *Energy Build.* **2014**, *81*, 115–126. [CrossRef]
7. Zhao, P.; Henze, G.P.; Plamp, S.; Cushing, V.J. Evaluation of commercial building HVAC systems as frequency regulation providers. *Energy Build.* **2013**, *67*, 225–235. [CrossRef]
8. Kiliccote, S.; Piette, M.A.; Hansen, D. *Advanced Controls and Communications for Demand Response and Energy Efficiency in Commercial Buildings*; Lawrence Berkeley National Laboratory: Berkeley, CA, USA, 2006.
9. Maasoumy, M.; Sanandaji, B.M.; Sangiovanni-Vincentelli, A.; Poolla, K. Model predictive control of regulation services from commercial buildings to the smart grid. In Proceedings of the 2014 American Control Conference, Portland, OR, USA, 4–6 June 2014; pp. 2226–2233.

10. Maasoumy, M.; Sangiovanni-Vincentelli, A. Total and peak energy consumption minimization of building hvac systems using model predictive control. *IEEE Des. Test Comput.* **2012**, *29*, 26–35. [CrossRef]

11. Oldewurtel, F.; Ulbig, A.; Parisio, A.; Andersson, G.; Morari, M. Reducing peak electricity demand in building climate control using real-time pricing and model predictive control. In Proceedings of the 49th IEEE Conference on Decision and Control, Atlanta, GA, USA, 15–17 December 2010; pp. 1927–1932.

12. Maasoumy, M.; Sangiovanni-Vincentelli, A. Optimal control of building hvac systems in the presence of imperfect predictions. In Proceedings of the Dynamics and Systems and Control Conference, Fort Lauderdale, FL, USA, 17–19 October 2012; pp. 257–266.

13. Hao, H.; Lin, Y.; Kowli, A.S.; Barooah, P.; Meyn, S. Ancillary service to the grid through control of fans in commercial building HVAC systems. *IEEE Trans. Smart Grid* **2014**, *5*, 2066–2074. [CrossRef]

14. Rui, X.; Liu, X.; Meng, J. Dynamic frequency regulation method based on thermostatically controlled appliances in the power system. *Energy Procedia* **2016**, *88*, 382–388. [CrossRef]

15. Vrettos, E.; Oldewurtel, F.; Zhu, F.; Andersson, G. Robust provision of frequency reserves by office building aggregations. *IFAC Proc. Vol.* **2014**, *47*, 12068–12073. [CrossRef]

16. Vrettos, E.; Kara, E.C.; MacDonald, J.; Andersson, G.; Callaway, D.S. Experimental demonstration of frequency regulation by commercial buildings—Part I: Modeling and hierarchical control design. *IEEE Trans. Smart Grid* **2016**, *9*. [CrossRef]

17. Vrettos, E.; Kara, E.C.; MacDonald, J.; Andersson, G.; Callaway, D.S. Experimental demonstration of frequency regulation by commercial buildings—Part II: results and performance evaluation. *IEEE Trans. Smart Grid* **2016**, *9*. [CrossRef]

18. Lin, Y.; Barooah, P.; Meyn, S.P. Low-frequency power-grid ancillary services from commercial building HVAC systems. In Proceedings of the IEEE International Conference on Smart Grid Communications, Vancouver, BC, Canada, 21–24 October 2013; pp. 169–174.

19. Koch, S.; Mathieu, J.L.; Callaway, D.S. Modeling and control of aggregated heterogeneous thermostatically controlled loads for ancillary services. *IEEE Trans. Power Syst.* **2011**, *31*, 1–7.

20. Short, J.A.; Infield, D.G.; Freris, L.L. Stabilization of grid frequency through dynamic demand control. *IEEE Trans. Power Syst.* **2007**, *22*, 1284–1293. [CrossRef]

21. Gwerder, M.; Tödtli, J. Predictive control for integrated room automation. In Proceedings of the 8th REHVA World Congress Clima, Lausanne, Switzerland, 9–12 October 2005.

22. Oldewurtel, F.; Parisio, A.; Jones, C.N.; Morari, M.; Gyalistras, D.; Gwerder, M.; Stauch, V.; Lehmann, B.; Wirth, K. Energy efficient building climate control using stochastic model predictive control and weather predictions. In Proceedings of the 2010 American Control Conference, Baltimore, MD, USA, 30 June–2 July 2010; pp. 5100–5105.

23. Lu, N. An evaluation of the HVAC load potential for providing load balancing service. *IEEE Trans. Smart Grid* **2012**, *3*, 1263–1270. [CrossRef]

24. Ma, X.; Dong, J.; Djouadi, S.M.; Nutaro, J.J.; Kuruganti, T. Stochastic control of energy efficient buildings: A semidefinite programming approach. In Proceedings of the IEEE International Conference on Smart Grid Communications, Miami, FL, USA, 2–5 November 2015; pp. 780–785.

25. PJM. *Ancillary Services*; Technical Report; PJM: Norristown, PA, USA. Available online: http://www.pjm.com/markets-and-operations/ancillary-services.aspx (accessed on 11 June 2018).

26. Zogou, O.; Stamatelos, A. Optimization of thermal performance of a building with ground source heat pump system. *Energy Convers. and Manag.* **2007**, *48*, 2853–2863. [CrossRef]

27. Zogou, O.; Stamatelos, A. *Application of Building Energy Simulation in Sizing and Design of an Office Building and Its HVAC Equipment (Chapter 11)*; Energy and Building Efficiency, Air Quality and Conservation; Nova Publishers, Inc.: Hauppauge, NY, USA, 2009; pp. 279–324.

28. U.S. Energy Information Administration (EIA). *How Much Electric Supply Capacity Is Needed to Keep U.S. Electricity Grids Reliable?* Technical Report; U.S. Energy Information Administration: Washington, DC, USA, 2013.

 *energies*

Article

Decision Tree-Based Preventive Control Applications to Enhance Fault Ride Through Capability of Doubly-Fed Induction Generator in Power Systems

Dione Vieira *, Marcus Nunes and Ubiratan Bezerra

Institute of Technology, Graduate Program in Electrical Engineering, Federal University of Para,
Belém-PA 66075-110, Brazil; mvan@ufpa.br (M.N.); bira@ufpa.br (U.B.)
* Correspondence: dionevieira@ufpa.br; Tel.: +55-091-3355-8000

Received: 2 June 2018; Accepted: 2 July 2018; Published: 4 July 2018

Abstract: The development of a preventive control methodology to increase the capacity of voltage sag recovery (Fault Ride Through Capability (FRTC)) of a doubly-fed induction generator (DFIG) connected in an electrical network is presented. This methodology, which is based on the decision trees (DT) technique, assists with monitoring and support for security and preventive control, ensuring that wind systems remain connected to the power system even after the occurrence of disturbances in the electric system. Based on offline studies, DT discovers inherent attributes of the FRTC scenario related to electrical system behavior and provides a quick prediction model for real-time applications. From the obtained results, it is possible to check that the DFIG is contributing to a system's operation security from the availability of power dispatch and participation in the voltage control. It is also noted that the use of DT, in addition to classifying the system's operational state with good accuracy, also significantly facilitates the operator's task, by directing him to monitor the most critical variables of the monitored operation state for a given system's topological configuration.

Keywords: decision tree; preventive control; Fault Ride Through Capability; doubly-fed induction generator

---

1. Introduction

Doubly-fed induction generators (DFIGs) have excellent control and high energy efficiency when compared to fixed speed wind power systems, which makes them the best choice for many wind farm installations worldwide, if economic aspects are also taken into account [1]. Vector control techniques, especially those of oriented fields, allow decoupling of a machine's active and reactive power control loops. Thus, DFIGs can independently control the active and reactive power, enabling the control of the machine's terminal voltage or power factor.

However, when compared to variable speed synchronous generators wind systems, DFIG is easily affected by disturbances, because its stator windings are directly connected to the electric network. In the event of network failures, for example, short circuits, the DFIG terminal voltage may be very low in relation to its nominal value, and currents in the stator and rotor windings may be very high, representing a threat to operational security which can lead to the burning of the generator and converter components [2]. Formerly, the DFIG was disconnected from the electrical network and returned to normal operation only when the system had recovered from a fault occurrence. As the integration of wind systems into the electrical grid has increased, it has been established that wind systems must remain connected to the power system during disturbances, since the disconnection of large wind farms could cause stability problems. In this regard, different regulatory agencies in many countries have established technical requirements of survival during voltage sags (Fault Ride Through Capability (FRTC)), aimed at increasing the operational security [3].

The main existing methods to increase FRTC fall into two categories, namely, control improvement and hardware modification. Some proposed methods reduce the overcurrent during network failures or increase the DFIG transient stability margins, modifying the control of static converters, for example, through demagnetizing current control and double control techniques [4,5]. The methods of control improvement may present lower costs; however, the DFIG behavior is not satisfactory during voltage sags due to the limitations of converters [6].

Several hardware modifications have been proposed to reduce this problem. The conventional crowbar and chopper scheme is still widely used to reduce rotor overcurrent and DC link overvoltage in order to improve wind turbine operational security [7]. However, this scheme is not able to avoid electromagnetic torque oscillation which may damage the gearbox in extreme circumstances [8].

Other solutions have been adopted for the integration of large-scale wind farms, such as the use of static synchronous compensators (STATCOM) connected in parallel with the wind farm transmission line which can inject reactive power to assist with voltage control during electrical network failures [9]; dynamic voltage restorers (DVRs) connected in series with the transmission line may offset the terminal voltage by a transformer connected to the electrical network [10]. In addition to these alternatives, another cheaper solution is the use of a fault current limiter (FCL) as the series dynamic braking resistor [11] and the bridge type fault current limiter [12]. When using a power interruption circuit to perform commutations between normal operation and faults, the FCL can enlarge the transmission line's equivalent impedance to reduce the overcurrent in both stators as rotor-sides during failures in the electric network.

In recent years, several hardware devices with new techniques have been proposed, such as the energy storage device and superconducting fault current limiter (SFCL). Combined control strategies are also used in these new devices, but they still have some limitations, as described in [13–16]. However, until now, no work has presented applications of preventive action based on the decision tree method for the increase in the DFIG Fault Ride Through capability. The use of automatic machine learning techniques provides a promising approach for defining the main control variables and their security limits in the DFIG operation, as will be presented in this article.

Traditionally, the data mining technique called decision tree (DT) has been widely applied in the area of energy systems for security evaluation and the application of preventive control [17–19]. The DT utilizes offline studies to discover intrinsic attributes of the electrical system. The knowledge obtained by DT can be directly used to aid the adoption of preventive actions in order to enlarge operational security in addition to providing a quick prediction model for real-time applications [20].

In addition, DT significantly reduces the set of options to be used in preventive control actions, allowing operators to remain more focused on the really critical security-related variables. Another significant aspect of DT is the fact that it presents a description of the critical variables that affect the system. This systemic characteristic is important because the set of critical variables for each network topological configuration can be distributed by various parts of the electrical system, often in places that would not be necessarily so apparent to the operator.

In addition to DFIG, other wind power technologies, such as the direct drive can also be adopted, as well as other forms of clean energy, such as photovoltaic generation and triboelectric nanogenerators. For this, it is important that these other forms of unconventional generation have good penetration in the electrical grid, because the larger their contributions to the generation of energy, the greater the contribution of their variables to preventive control based on the decision tree will be.

Thus, the main contribution of this work is the application of a preventive control method based on the decision tree method to enhance the DFIG FRTC. The innovative methodology has two methods of analysis: a local and a systemic one. The first chooses the attributes of the main control variables of the system and the second chooses only the DFIG controllable attributes which were selected by the DT during offline training as the attributes to be tuned during the preventive control process. The trained DT identifies the DFIG operational security limits for each topological configuration of the electrical system. The limits indicated are finally used as a guide to design preventive control strategies, thereby ensuring that the wind power system remains connected to the electrical network after a power system failure.

2. Methods and Materials

2.1. Doubly-Fed Inductcion Generator (DFIG)

DFIG is an induction generator that normally works in variable speed mode and is connected to the electrical network through static converters linked to the generator rotor. The most common configuration adopted by manufacturers is a schema with two static converters with Pulse-Width Modulation (PWM), connected to the rotor circuit and the power grid, respectively, as shown in Figure 1. This allows the generator to operate with shaft speeds above and below the synchronous speed, decoupling the system frequency from the generator rotation. DFIG is designed to make the most of the wind potential under different wind speeds and generator shaft rotation.

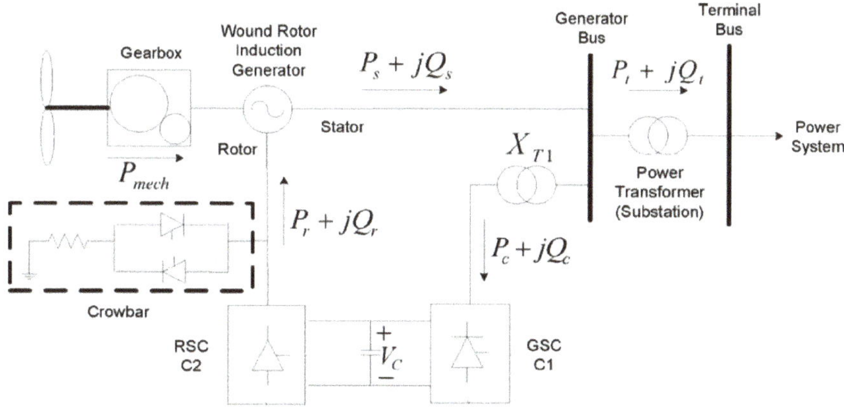

Figure 1. Schematic diagram of a doubly-fed induction generator.

The Rotor Side Converter (RSC) controls, from the rotor injected power and current, the active and reactive powers circulating through the stator. The first models of doubly-fed induction generators adopted constant power factor control, usually unity, providing the maximum active power. With the increased penetration of wind systems, DFIG went on to provide reactive power under conditions of power system failures [21]. However, many power system operators offer a financial compensation to variable speed generators when supplying reactive power to the grid (ancillary service) [22,23].

The DFIG provides reactive power through both the stator and the Grid Side Converter (GSC). However, the GSC generally operates with a unit power factor, and does not provide reactive power to the electrical grid, controlling only the DC link voltage. To provide greater support to voltage control and to increase the reactive power capacity [24], the GSC converter must operate with a power factor that is different from unity.

The generator dynamic model adopted in this study was the software ANATEM (version10.4.6, The Electrical Energy Research Center (Cepel), Rio de Janeiro, Brazil) default model and the adopted model of turbine controllers was the ALSTOM ECO74 provided by ONS (Brasília, Brazil) [25–27].

The FRTC technical requirement is defined as the ability of a generator to support network failures with resulting electrical voltage sags and remain connected after the occurrence of the disturbance [3]. For this, it is necessary that the generator terminal voltage remains above the defined FRTC curve and that the failure is eliminated during the time period defined by the same curve. In cases of voltage sags during one or more phases of wind generation at the connection point with the electrical network, the wind plant should continue operating if the voltage at its terminals remains above the curve shown in Figure 2; otherwise, the generator must be disconnected.

Figure 2 presents the terminal voltage tolerance limits of wind farms connected to the electrical network during the occurrence of power system disturbances.

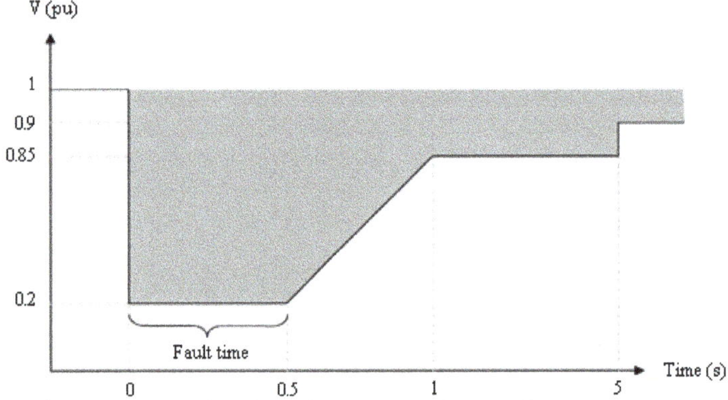

Figure 2. Voltage at wind generator terminals (Source: ONS).

2.2. Decision Tree (DT)

DT is a sorting algorithm belonging to the class of machine learning techniques that has the ability to learn through examples in order to sort records in a database. One of its most important features is the recursive partition of a dataset into several subsets until they contain only instances of a single class to allow better analysis of the problem. The DT so built presents results organized into a simple and easily interpretable form that can be used as a tool for decision making support.

A DT is essentially a series of "if–then" statements, and its creation was based on the hierarchical model, that is, from the root node to the leave nodes. The nodes correspond to the attributes' names, the nodes links represent the attributes' values, and the leaves represent the different existing classes. The classification occurs following the path from the root node to the leaves where the classes are assigned, as highlighted in Figure 3.

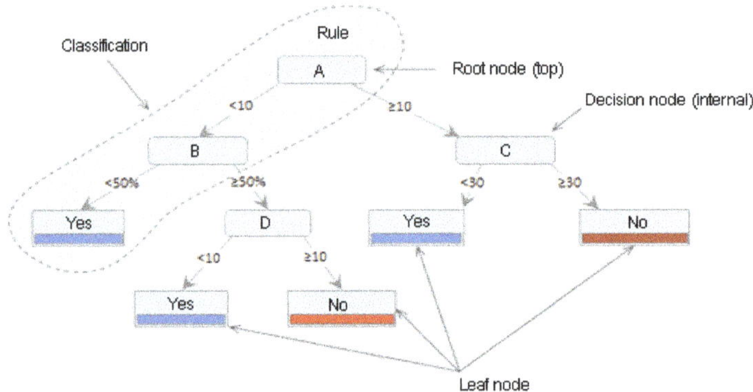

Figure 3. Example of a decision tree structure.

The first decision tree-based classifiers arose in the late 50, from Hunt's work, where several experiments were presented for the induction of rules. The "Classification and Regression Trees"

(CART) algorithm was subsequently developed by Friedman, and Quinlan developed the "Iterative Dichotomiser 3" algorithm (ID3), and as a new development, after these two algorithms, came the C4.5 algorithm [28].

The C4.5 algorithm creates decision trees from a database in a similar way to the ID3 algorithm—by using the concept of entropy. Entropy is a measure of the degree of impurity in an arbitrary sample set, that is, it is the measure of disorder or randomness. Given a class A attribute of a sample set (S), in which A can take *Vi* values of different classes, the entropy of A concerning this classification is defined with Equation (1):

$$Entropy(A) = -\sum_{i=1}^{m} p_i \log_2 p_i \tag{1}$$

where m is the total number of classes and $pi = p\ (A = Vi)$ is the probability of the class A attribute being equal to the class attribute whose index is (i), i.e., the ratio of the number of samples with a value of *Vi* in relation to the total number of samples of S.

The higher the entropy of an attribute, the more uniform the distribution of its values is. An entropy equal to zero means that one class in the data set has occurred, and it will be equal to 1 if the number of samples in each class are equal. An entropy near zero indicates that the classes are not uniform. Figure 4 represents the variation in entropy ($H(p)$) as a function of the probability (p).

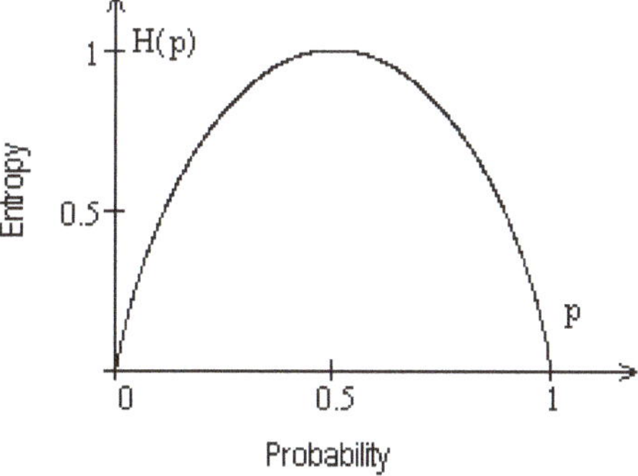

Figure 4. Entropy diagram.

During the process of creating a decision tree, the correct choice of attributes defines the success of the algorithm. Among the various criteria for choosing a candidate attribute to a node, the information gain is used. The information gain is based on entropy. The information gain is given by the sum of the individual entropies minus the joint entropy and is a measure of correlation between two variables.

Consider a sample set containing a class attribute (set as A) and one of the predictive attributes (set as B). The information gain (GI) of predictive attribute B is defined as the difference between the entropy of the class A attribute ($Entropy(A)$) and the conditional entropy of predictive attribute B, which is set as the value of the class A attribute ($Entropy(B \mid A)$). The information gain is given by Equation (2):

$$GI(B, A) = Entropy(A) - Entropy(B|A) \tag{2}$$

where the second term of Equation (2) is the conditional entropy, defined as the entropy of a predictive attribute, B, which was previously known the class A attribute. This is given by Equation (3):

$$Entropy(B|A) = -\sum_{i=1}^{m} p_i.Entropy(B|A = v_i) \tag{3}$$

where m is the total number of classes in the sample set, and B is the predictive attribute that is being considered. A is the class attribute, assuming value Vi. The term ($Entropy\,(B\,|\,A = Vi)$) is the entropy of predictive attribute B being given the value ($A = Vi$) of the class attribute.

$$Entropy(B|A = v_i) = -\sum_{i=1}^{m} p(B|A = v_i) \log_2 p(B|A - v_i) \tag{4}$$

where m is the number of classes that the class A attribute can assume, $p(B\,|\,A = Vi)$ is the conditional probability of attribute B, that is, the proportion given by the ratio between the number of examples of B with $A = Vi$ and the total number of samples in the class.

In the process of DT construction, the attribute that has the highest information gain should be placed as the root node to enable data to be sorted more quickly. The construction of the decision tree has three goals: to decrease the entropy, to be consistent with the data set, and to have the smallest number of nodes.

3. Methodology

The methodology of preventive control developed in this article can be applied in real-time operation and in very short-term planning studies focusing on the operational security assessment of wind systems integrated into the electrical grid. The methodology also enables the accomplishment of a preventive assessment of the future operating state with topological changes such as unplanned outages of transmission lines and generating units and the switching of capacitor banks and reactors among other contingencies. This feature allows operators to analyze preventive actions to be taken if these contingencies happen in future operations. Figure 5 shows the flowchart of offline procedures used to establish the database.

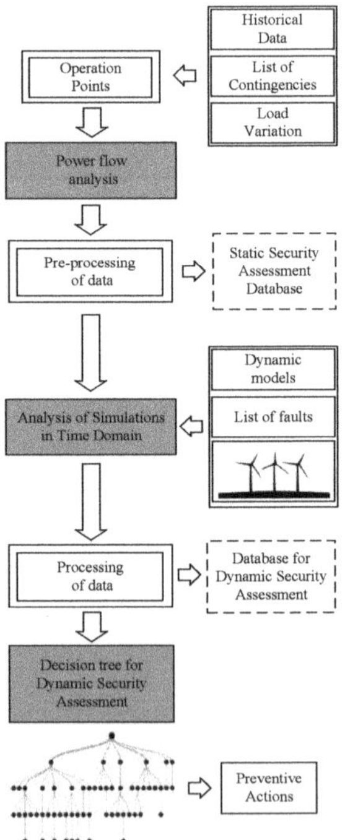

Figure 5. Offline steps for the decision tree creation process.

It is important to note that the proposed procedure involves an offline step and another step in real time. The database creation is performed offline, as the integration step of the methodology with the Supervisory Control and Data Acquisition/Energy Management System (SCADA/EMS). The database is formed from the information of each operation point provided by the state estimator for each topological configuration determined by the network configurator.

From the operation data, planning data, and load variations around each operation point, several cases are created for decision tree training.

From historical data and a list of the most critical contingencies or those with greater probability of occurrence, new operating scenarios are generated in order to simulate the load variation that may occur during a normal operation day. Using the computer program ANAREDE [25], which is adopted for power system steady state studies, load flow simulations are performed to obtain the initial conditions for each operation point of the system.

At this step, using the power flow study solutions, the database for static security assessment is created. The static database is required to obtain data to allow the initial conditions of the operation point to be analyzed by the dynamic simulation studies. Now, using the computational tool ANATEM [26], adopted for the analysis of electromechanical transients in electrical systems, time domain simulations are performed with dynamic models (synchronous generators, wind generators, and associated controls), forming a dynamic database to create decision trees to be used for dynamic security assessment.

Finally, using the data mining software Rapidminer [29], decision trees are constructed to assist with preventive control. The created decision trees evaluate dynamic security in real time, highlighting only those variables that have the largest influence on every topological configuration of the electrical system, thereby generating a smaller set of variables that deserve the attention of operators at the power system operation point. Each DT branch carries a rule that, when met, guarantees the security of the wind generator operation after network failure. Figure 6 shows this real-time operation scenario schematically.

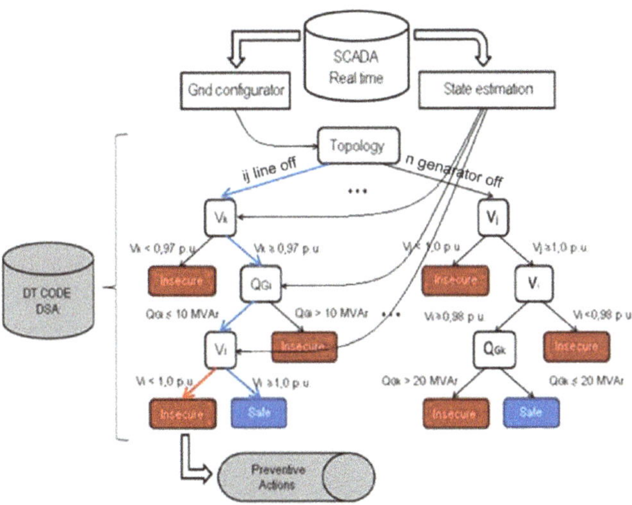

Figure 6. Real-time step scheme dynamic security assessment module based on decision tree.

After classifying the system's operational status, preventive actions can be performed if the system presents any operation limit violation, as indicated by the decision tree.

Preventive control actions can be taken to prevent the occurrence of serious consequences to the electrical system. In this case, the decision variables presented by the DT indicate the path to maintain DFIG dynamic security within each power system topology. FLUPOT software (version9.7.2, The Electrical Energy Research Center (Cepel), Rio de Janeiro, Brazil) [25], a computer program for power system optimization, is used to obtain optimal solutions for power system operation using the voltage security constraints indicated by the DT solution. Figure 7 shows the operation scheme involving preventive control actions.

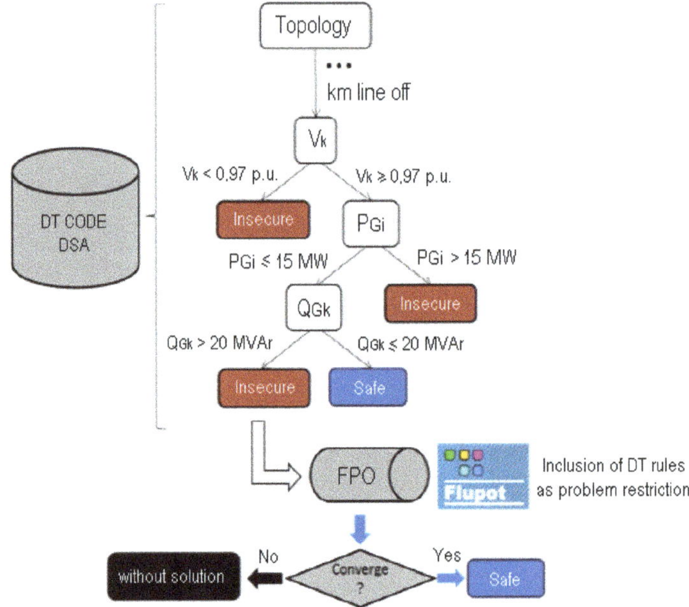

Figure 7. Real-time scheme decision tree-based preventive control module.

The methodology developed in this article may be applied to local and systemic scenarios, as illustrated in Figure 8. In systemic scenarios, electrical variables related to the electrical system as a whole are chosen as attributes, while in local scenarios, only the DFIG controllable attributes are chosen to participate in the preventive actions to be implemented by the DT.

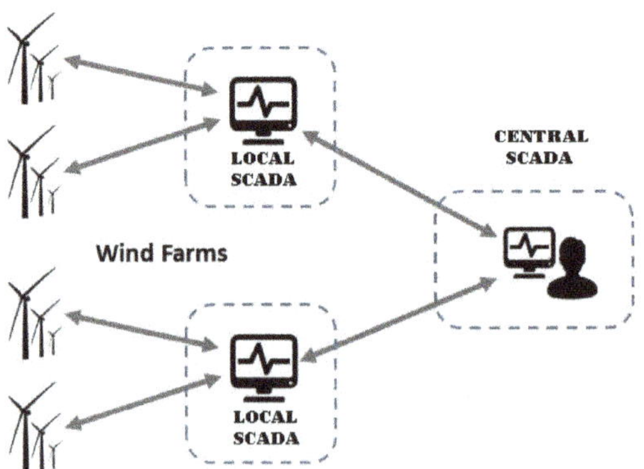

Figure 8. Application of a preventive control scheme for local and systemic actions.

To create the decision tree in systemic case studies of preventive control applications with a focus on DFIG FRTC, a large database must be created with a large number of objects. Each row represents a static pre-contingency condition of the power system, (every object is an attribute of a power flow solution with control variables available in the whole system), along with the results (disconnection and Fault Ride Through) of a dynamic simulation in the time domain. Table 1 presents the systemic database structure.

Table 1. Systemic database structure.

Topology	V1	V2	Vi	Pg1	Pg2	Pgi	Qg1	Qg2	Qgi	Atribute
Complete	V1	V2	Vi	Pg1	Pg2	Pgi	Qg1	Qg2	Qgi	Disconnection
Complete	V1	V2	Vi	Pg1	Pg2	Pgi	Qg1	Qg2	Qgi	Fault Ride Through
N-1	V1	V2	Vi	Pg1	Pg2	Pgi	Qg1	Qg2	Qgi	Disconnection
N-1	V1	V2	Vi	Pg1	Pg2	Pgi	Qg1	Qg2	Qgi	Disconnection
N-1	V1	V2	Vi	Pg1	Pg2	Pgi	Qg1	Qg2	Qgi	Fault Ride Through
N-1	V1	V2	Vi	Pg1	Pg2	Pgi	Qg1	Qg2	Qgi	Fault Ride Through
N-1	V1	V2	Vi	Pg1	Pg2	Pgi	Qg1	Qg2	Qgi	Disconnection
N-1	V1	V2	Vi	Pg1	Pg2	Pgi	Qg1	Qg2	Qgi	Fault Ride Through
N-1	V1	V2	Vi	Pg1	Pg2	Pgi	Qg1	Qg2	Qgi	Fault Ride Through
N-1	V1	V2	Vi	Pg1	Pg2	Pgi	Qg1	Qg2	Qgi	Disconnection

With respect to the decision tree creation applied to the local case study, each row represents a static pre-contingency condition of the power system (every object is an attribute of a power flow solution with control variables of the local wind system), along with the results (Disconnection and Fault Ride Through) of a dynamic simulation in the time domain. Table 2 presents the systemic database structure.

Table 2. Local database structure.

Topology	Vi	Pgi	Qgi	Atribute
Complete	Vi	Pgi	Qgi	Disconnection
Complete	Vi	Pgi	Qgi	Fault Ride Through
N-1	Vi	Pgi	Qgi	Disconnection
N-1	Vi	Pgi	Qgi	Disconnection
N-1	Vi	Pgi	Qgi	Fault Ride Through
N-1	Vi	Pgi	Qgi	Fault Ride Through
N-1	Vi	Pgi	Qgi	Disconnection
N-1	Vi	Pgi	Qgi	Fault Ride Through
N-1	Vi	Pgi	Qgi	Fault Ride Through
N-1	Vi	Pgi	Qgi	Disconnection

4. Results

For the purpose of testing and validating the proposed methodology for preventive control based on decision trees, case studies were carried out using the IEEE New England-39 bus test system. The results will be presented in the following text.

The New England test system is originally formed by 39 buses having 10 synchronous generators. Generator 1 is an equivalent model representing part of the electrical network over which there is no control, and generators 2 to 10 are controlled by automatic voltage regulators. Figure 9 illustrates the single-line diagram for the modified test system, considering the integration of variable speed wind generators at buses 40, 41, 42, 43, 44, and 45.

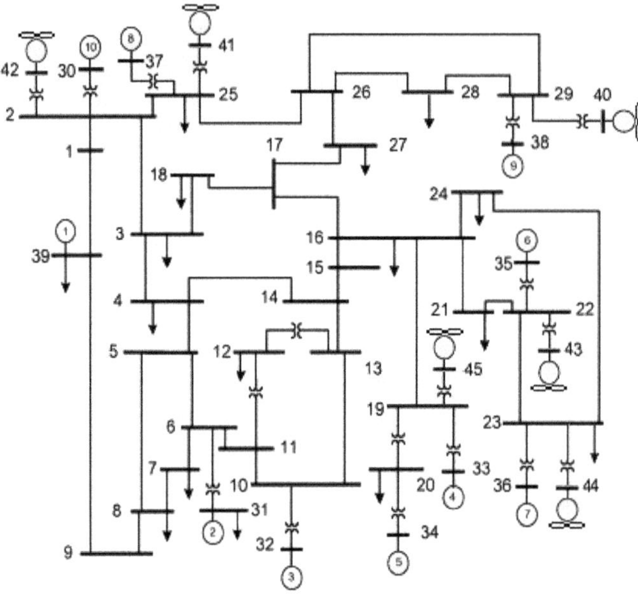

Figure 9. Application of a preventive control scheme for local and systemic actions.

The IEEE test system was modified by considering the integration of wind farms representing 30% of the total power generated by the system. Six new buses were inserted with wind farms. Each wind generating unit was given a nominal power of 1670 kW. Table 3 presents a summary of the wind generation given by the test system, detailing the installation bus number, the number of turbines installed, and the resulting maximum power generation.

Table 3. Wind turbine information.

Bus Number	Number of Wind Turbines	Maximum Power (MW)
40	300	500
41	180	300
42	90	150
43	210	350
44	180	300
45	156	260

4.1. Case Study of the Preventive Control Application Focusing on the DFIG FRTC—Training and Testing the Systemic Analysis Method

Firstly, the modified IEEE-39 bus test system was considered for simulation. From a defined base case, new scenarios were generated for different load variation conditions. Another four (N-1) type topological conditions representing line and generator outages were added to the database and were complemented with new operating scenarios. Then, these files were simulated using the power system analysis software ANAREDE to obtain the initial conditions of the bus voltages, transmission lines power flows, set points for control variables, and generation of active and reactive power values, forming the simulation database. For each topology, 200 simulations were carried out, representing a total of 1000 scenarios.

Later, with the purpose of carrying out time domain simulations, files of the initial conditions, representing the electric network, were added, comprising dynamic data describing synchronous generators, wind turbines, and associated controls to accomplish simulation studies to generate preventive control actions against disturbances that cause wind generators to be disconnected due to voltage sags. The critical event for testing was the application of a short circuit of 100 ms in transmission line LT 28–29. The dynamic simulations were carried out using the time domain dynamic analysis software ANATEM.

The database for dynamic security assessment was then generated with a symbolic attribute (the topological signature) and the numeric attributes, Vi (voltage magnitude), θi (phase angle), Pgi (generated active power), and Qgi (generated reactive power), which constitute the pre-fault conditions, as well as two target attributes, "Fault Ride Through" (if the voltage magnitude does not exceed the established limits) and "Disconnection" (if the voltage exceeds the predefined limits).

Finally, the data miner software Rapidminer was used for the creation of an intelligent system based on the decision tree method to assess the preventive control and dynamic security using, as a pattern, 70% of the data for training and 30% for testing. Figure 10 shows the decision tree model generated with topological orientation for dynamic security assessment and preventive control aid purposes, in order to ensure that the wind turbines will survive the voltage sags.

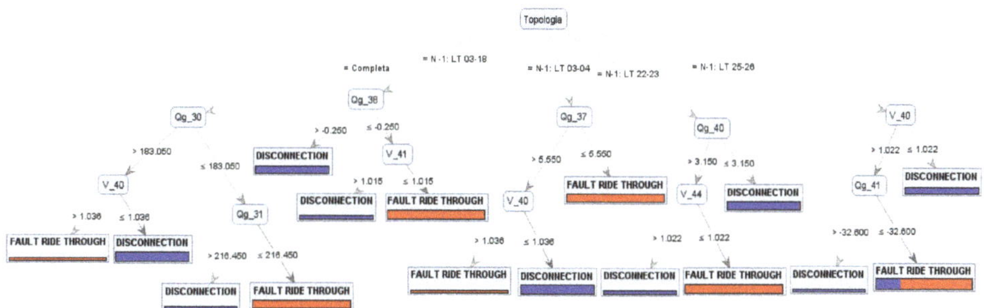

Figure 10. Decision tree for the application of preventive control during systemic actions.

Table 4 presents the confusion matrix corresponding to the created decision tree to evaluate the efficiency of the implemented model by determining whether the predicted value matches the final value.

Table 4. Confusion Matrix (Fault Ride Through Capability (FRTC)).

Accuracy: 98.67%		Real Class		
	-	Disconnection	FRTC	Class Precision
Predicted Class	Disconnection	143	01	99.31%
	FRTC	03	153	98.08%
Class Recall		97.95%	99.35%	-

It can be observed in the presented results, that the hit rate (accuracy) was 98.67%, and only one case predicted "Disconnection", when, in reality, it was "Fault Ride Through". The prediction of the "Disconnection" class reached a precision rate of 99.31%, while the prediction of "Fault Ride Through" class was accurate to 98.08%. Both classes presented good performances.

The selected attributes provided to Rapidminer were voltage magnitudes in all generation buses, active and reactive power values for PV buses, network topology, and the labels "Fault Ride Through" and "Disconnection". As can be seen in Figure 10, the decision tree root node is the system topological configuration.

When analyzing the model in Figure 11 provided by the decision tree, the first branch on the left represents the path that will ensure electrical power system security for the complete topology. The operator, observing the current system topology, needs only to follow the variables and guidelines established by the decision tree to ensure the wind farm will survive the voltage sag.

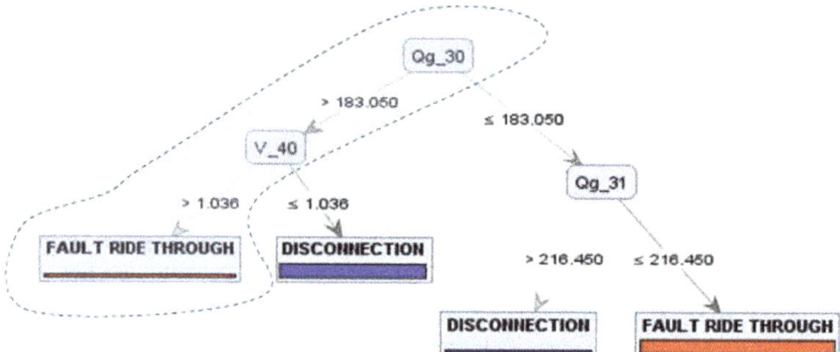

Figure 11. Decision tree branch for the complete topology.

It can be observed in Figure 11 that the decision tree rules circuled by the dotted line, if adopted, can ensure the wind turbines survive voltage sags under this current topology.

The rules found by the DT algorithm that have greater influence with regard to power system security are, in hierarchical order, as follows: If $Qg\_30 > 183.05$ MVAr and $V\_40 > 1.036$ pu, then the wind farm will remain connected, as noted in the presented results. The first variable of the first branch of the decision tree is the reactive power generated at bus 30 and the second variable is the wind turbine voltage magnitude at bus 40. The rules of this branch can be directly used to take preventive actions locally or remotely.

Figure 12 presents the dynamic behavior of the electrical system during wind generation when a short circuit lasting 100ms is applied in line LT 28–29. The result shows that due to this disturbance, the DFIG terminal voltage at bus 40 violates the limit of the voltage curve established by the standard which may lead to protection action, thereby disconnecting the wind farm.

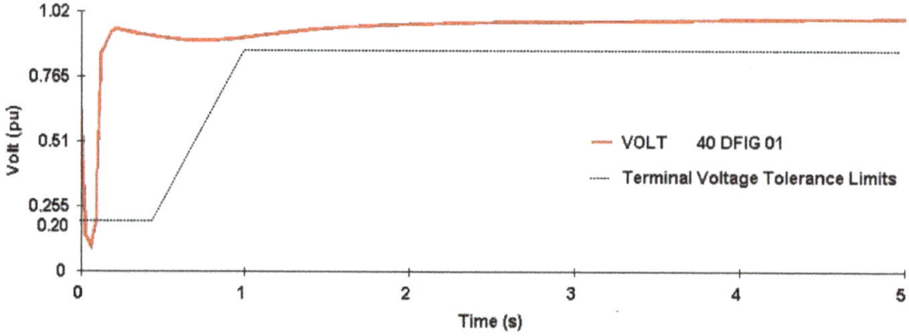

Figure 12. Doubly-fed induction generator (DFIG) terminal voltage at bus 40 after a short circuit in line LT 28–29.

In order to prevent the wind turbine being disconnected due to voltage sags, a case where voltage sag occurred at bus 40, with $Qg\_30 = 184.7$ MVAr and $V\_40 = 1.020$ pu was considered, and it appeared that part of the decision tree rules was not met as soon as the wind turbine voltage was below the limits set by the standard. Therefore, in this case, the wind farm would be disconnected.

So, as a solution, the optimization software FLUPOT was run, adopting an objective function involving voltage control and the decision tree rules. Thus, the voltage magnitude of bus 40 was changed from V_40 = 1.020 pu to V_40 = 1.041 pu, and the reactive power Qg_30 of 184.7 MVAr was changed to 184 MVAr. The obtained results are shown in Figure 13.

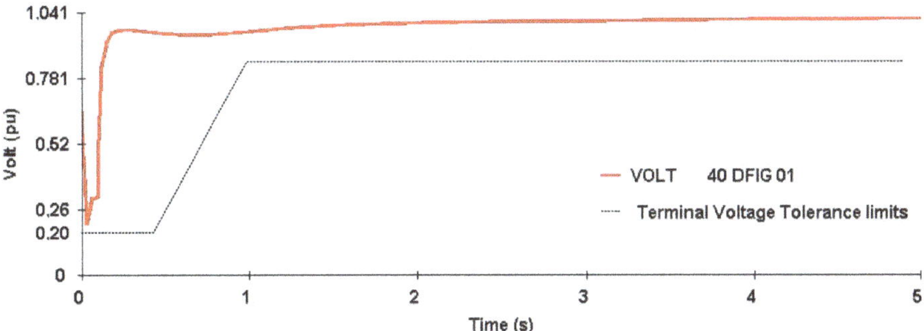

Figure 13. Terminal voltage of the DFIG at bus 40 after the short circuit with the decision tree (DT) rules and optimization of FLUPOT.

It can be observed in Figure 13 that the optimized adjustment considering the DT rules increased the wind turbine's survival following voltage sags at bus 40. From the obtained results, it is possible to verify that variable speed wind systems can also contribute to increase dynamic security through the availability of their control variables.

To ensure the electrical system's operation security, the operator observing the system topology under operation, needs to take into account the variables and rules established by the decision tree, taking preventive actions accordingly to ensure the continuity of the wind farm's operation after disturbances in the electrical system.

4.2. Case Study of the Preventive Control Application Focusing on the DFIG FRTC—Training and Testing of the Local Analysis

The preparation of the database for local analysis was similar the method used in the systemic study, as presented in Section 4.1, where the file corresponding to the IEEE 39-bus test system which was modified to include wind turbines generation representing 30% of the total electrical system generated power was used. Figure 14 shows the decision tree model generated with topological orientation for dynamic security assessment and preventive control aid purposes.

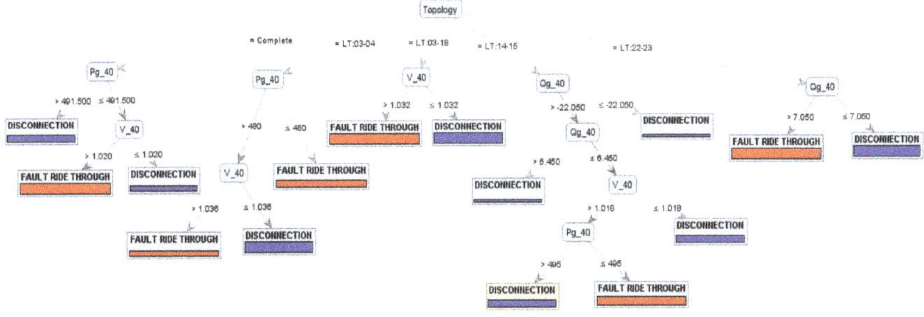

Figure 14. Decision tree for the application of preventive control during local actions.

Table 5 presents the confusion matrix corresponding to the created decision tree to evaluate the efficiency of the implemented model. It can be observed that the hit rate was 100%, with 100% also being the precision in predicting the "Disconnection" class.

Table 5. Confusion matrix.

Accuracy: 100%			Real Class		
	-	Disconnection	FRTC	Class Precision
Predicted Class	Disconnection	161	00	100%
	FRTC	0	139	100%
Class Recall		100%	100%	-

The prediction of the "Fault Ride Through" class also reached an accuracy of 100%. Both classes presented excellent performances.

The selected attributes provided to Rapidminer were voltage magnitudes, active and reactive generated power by the wind farm at bus 40, network topology, and the labels "Fault Ride Through" and "Disconnection". As can be seen in Figure 14, the decision tree root node was the system topological configuration.

When analyzing the model in Figure 15 provided by the decision tree, the first branch on the right represents the path that will ensure electrical power system security for the complete topology. The operator, observing the current system topology, needs only to follow the variables and guidelines established by the decision tree to ensure the wind farm will survive the voltage sag.

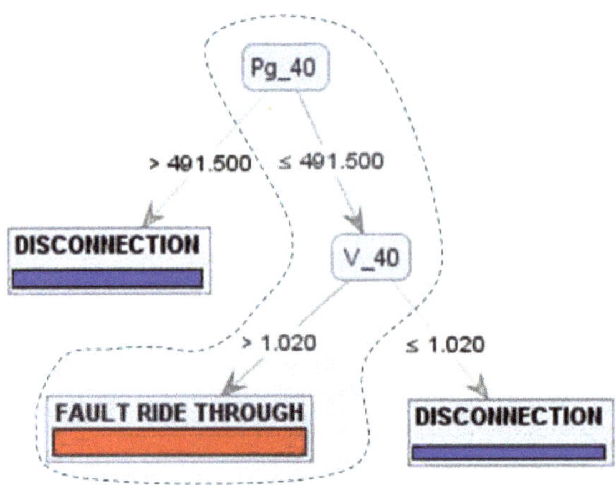

Figure 15. Decision tree branch for the complete topology.

It can be observed in Figure 14 that the decision tree rules circuled by dotted line, if adopted, can ensure the wind turbines survive voltage sags in bus 40. The rules found by the DT algorithm that have the greatest influence with regard to power system security are, in hierarchical order, the following: If Pg_40 ≤ 491.5 MW and V_40 > 1.020 pu, under these conditions, the wind farm will remain connected.

It is noted that this branch of the decision tree indicates a path whose rules must be met to ensure the wind system is maintained at a secure operating point.

This aspect is very important, because this new information regarding the system's full topology only provided by using the decision tree, will facilitate the operator's task significantly, which, in turn, will allow them pay attention to the monitoring of really critical variables.

Another important aspect to be highlighted is the much smaller number of variables indicated by the decision tree branch when compared to the number of attributes that belong to the database that was used by the Rapidminer software to create the decision tree. This is one main feature of decision trees, which is based on dimensionality reduction, due to the index which correlates the critical attributes to system security.

The intelligence contained in the rules of this decision tree branch can be directly used to aid in the system's dynamic security assessment as well as to take local preventive actions.

Figure 16 presents the dynamic behavior of the electrical system with wind generation when a short circuit lasting 100ms was applied in line LT 28–29. The result shows that due to this disturbance, the DFIG terminal voltage at bus 40 violated the limit of the voltage curve established by the standard which had the potential to lead to protection action, disconnecting the wind farm.

With the purpose of preventing against voltage sags, a case with an operating voltage sag at bus 40 with Pg_40 = 495 MW and V_40 = 1.030 pu was considered, and it was verified that part of the decision tree rules was not met once the wind turbine voltage was below the limits set by the standard, and therefore, the wind farm would be disconnected.

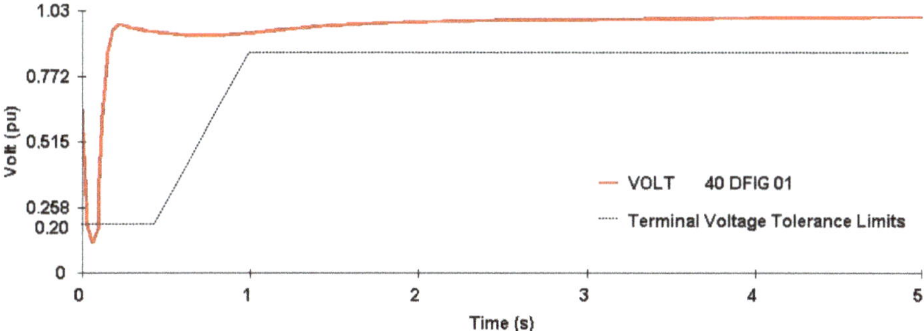

Figure 16. DFIG terminal voltage at bus 40 after a short circuit.

Using FLUPOT optimization software to adopt an objective function involving voltage control and the decision tree rules and changing the generated active power Pg_40 from 495 MW to 448 MW and also changing the voltage V_40 from 1.030 pu to 1.031 pu, the result shown in Figure 17 was obtained.

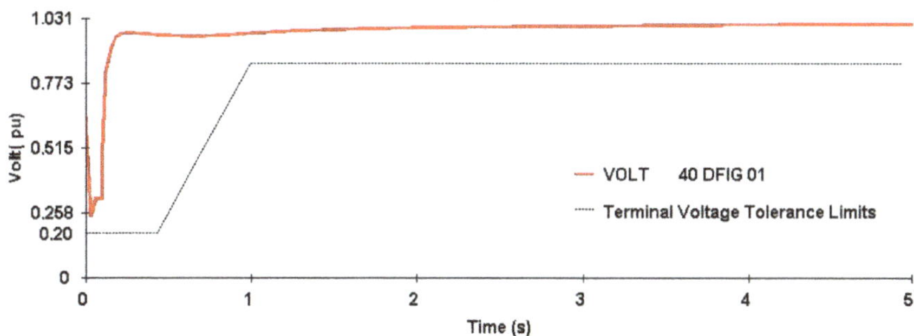

Figure 17. Terminal voltage of the DFIG in bus 40 after the short circuit with the DT rules and optimization of FLUPOT.

It can be observed in Figure 17 that the optimized adjustment considering the DT rules increased the wind turbine survival to voltage sags at bus 40. From the obtained results, it is possible to verify that variable speed wind systems can also contribute to an increase in dynamic security through the availability of their control variables.

5. Conclusions

This article presented the results of applying a proposed preventive control procedure based on the decision tree method to enhance the Fault Ride Through capability of variable speed wind turbines connected to a power system. The developed methodology was tested using the IEEE 39-bus system, which was modified by the insertion of doubly-fed induction generators. From the obtained results, it was possible to verify that the integration of DFIG wind turbines contributes to the enhancement of the power system operation security, considering the rules established criteria set out by the decision tree.

From the presented results obtained for the systemic case study, applying preventive control focused on DFIG FRTC, it was possible to verify, in the first branch of the decision tree, that the wind power system voltage at bus 40 and the reactive power of synchronous generator 30 contribute to the system's operation security and also to the continuity of electricity supply from a wind turbine after the occurrence of a disturbance in the electrical network. The systemic aspect is characterized by the contribution of all the control variables available in the electrical system; however, the decision tree selects only the variables of most relevance to operational security.

In relation to the results of the local case study, it was possible to verify, in the decision tree branch with full topology, that active power and voltage at bus 40 contribute to the continuity and lack of wind system shutdown. The local aspect is justified by the use of the control variables of the local wind system without the availability of remote control variables.

It was also found that the application of the decision tree, in addition to classifying the system's operational state with good accuracy has also indicated the way to maintain the electrical system dynamic security for each topology. Preventive control actions can be taken according to the DT rules to avoid dynamic security problems in the power system's operation. The use of an optimization tool, as presented in the article, may guarantee optimal operating conditions, using only the reduced set of variables indicated by the decision tree for this purpose, significantly reducing the operator's task in the operation monitoring and allowing him to pay more attention to the more critical variables in each operation topology.

Thus, this article has implemented a DT-based support tool which can be directly integrated into operation centers, ensuring a considerable confidence increase in operators' decision-making during electrical power system operation.

Author Contributions: D.V. conceived and designed the experiments; D.V. performed the experiments; D.V. and M.N. analyzed the data; D.V., M.N. and U.B. contributed materials/analysis tools; D.V., M.N. and U.B. wrote the paper.

Funding: This research was funded by CNPQ-UFPA.

Acknowledgments: This research was supported by CEAMAZON-Center of Excellence on Energy Efficiency in the Amazon.

References

1. Yang, L.; Yang, G.Y.; Xu, Z.; Dong, Z.Y.; Wong, K.P.; Ma, X. Optimal controller design of a doubly-fed induction generator wind turbine system for small signal stability enhancement. *IET Gener. Transm. Distrib.* **2010**, *4*, 579–597. [CrossRef]
2. Rashid, G.; Ali, M.H. Transient stability enhancement of doubly fed induction machine-based wind generator by bridge-type fault current limiter. *IEEE Trans. Energy Convers.* **2015**, *30*, 939–947. [CrossRef]
3. Tsili, M.; Papathanassiou, S. A review of grid code technical requirements for wind farms. *IET Renew. Power Gener.* **2009**, *3*, 308–332. [CrossRef]

4. Xiang, D.; Ran, L.; Tavner, P.J.; Yang, S. Control of a doubly fed induction generator in a wind turbine during grid fault ride-through. *IEEE Trans. Energy Convers.* **2006**, *21*, 652–662. [CrossRef]

5. Lopez, J.; Sanchis, P.; Gubia, E.; Ursúa, A.; Marroyo, L.; Roboam, X. Control of doubly fed induction generator under symmetrical voltage dips. In Proceedings of the 2008 IEEE International Symposium on Industrial Electronics, Cambridge, UK, 30 June–2 July 2008; pp. 2456–2462. [CrossRef]

6. Marques, G.D. Active stabilization method for the doubly-fed induction generator using a quadrature inner control loop. In Proceedings of the International Conference on Power Engineering, Energy and Electrical Drives, Setubal, Portugal, 12–14 April 2007; pp. 765–768. [CrossRef]

7. Abad, G.; López, J.; Rodríguez, M.; Marroyo, L.; Iwanski, G. *Doubly Fed Induction Machine: Modeling and Control for Wind Energy Generation*; Wiley-IEEE Press: Hoboken, NJ, USA, 2011; ISBN 978-0-470-76865-5.

8. Gounder, K.; Nanjundappan, D.; Boominathan, V. Enhancement of transient stability of distribution system with SCIG and DFIG based wind farms using STATCOM. *IET Renew. Power Gener.* **2016**, *10*, 1171–1180. [CrossRef]

9. Hossain, M.; Ali, H. Transient stability improvement of doubly fed induction generator based variable speed wind generator using DC resistive fault current limiter. *IET Renew. Power Gener.* **2016**, *10*, 150–157. [CrossRef]

10. Ramirez, D.; Martinez, S.; Platero, A.; Blazquez, F.; De Castro, R.M. Low-voltage ride-through capability for wind generators based on dynamic voltage restorers. *IEEE Trans. Energy Convers.* **2011**, *26*, 195–203. [CrossRef]

11. Okedu, E. Enhancing DFIG wind turbine during three-phase fault using parallel interleaved converters and dynamic resistor. *IET Renew. Power Gener.* **2016**, *10*, 1211–1219. [CrossRef]

12. Rashid, G.; Ali, H. Nonlinear control-based modified BFCL for LVRT capacity enhancement of DFIG based wind farm. *IEEE Trans. Energy Convers.* **2017**, *32*, 284–295. [CrossRef]

13. Shen, W.; Ke, P.; Sun, Z.; Kirschen, D.S.; Qiao, W.; Deng, X.T. Advanced auxiliary control of an energy storage device for transient voltage support of a doubly Fed induction generator. *IEEE Trans. Sustain. Energy* **2016**, *7*, 63–76. [CrossRef]

14. Yunus, A.S.; Abu-Siada, A.; Masoum, M.A. Application of SMES unit to improve DFIG power dispatch and dynamic performance during intermittent misfire and fire-through faults. *IEEE Trans. Appl. Supercond.* **2013**, *23*, 5701712. [CrossRef]

15. Ngamroo, I. Optimization of SMES-FCL for augmenting FRT performance and smoothing output power of grid-connected DFIG wind turbine. *IEEE Trans. Appl. Supercond.* **2016**, *26*, 1–5. [CrossRef]

16. Xiao, X.Y.; Yang, R.H.; Chen, X.Y.; Zheng, Z.X.; Li, C.S. Enhancing fault ride-through capability of DFIG with modified SMES-FCL and RSC control. *IET Gener. Transm. Distrib.* **2018**, *12*, 258–266. [CrossRef]

17. Diao, R.; Vittal, V.; Logic, N. Design of a Real-Time Security Assessment Tool for Situational Awareness Enhancement in Modern Power Systems. *IEEE Trans. Power Syst.* **2010**, *25*, 957–965. [CrossRef]

18. Genc, I.; Diao, R.; Vittal, V.; Kolluri, S.; Mandal, S. Decision Tree-Based Preventive and Corrective Control Applications for Dynamic Security Enhancement in Power Systems. *IEEE Trans. Power Syst.* **2010**, *25*, 1611–1619. [CrossRef]

19. Liu, C.; Sun, K.; Rather, Z.H.; Chen, Z.; Bak, C.L.; Thøgersen, P.; Lund, P. A Systematic Approach for Dynamic Security Assessment and the Corresponding Preventive Control Scheme Based on Decision Trees. *IEEE Trans. Power Syst.* **2014**, *29*, 717–730. [CrossRef]

20. Wehenkel, L.A. *Automatic Learning Techniques in Power Systems*; Kluwer: Norwell, MA, USA, 1998.

21. Ullah, N.R.; Thiringer, T.; Karlsson, D. Voltage and Transient Stability Support by Wind Farms Complying with the E.ON Netz Grid Code. *IEEE Trans. Power Syst.* **2007**, *22*, 1647–1656. [CrossRef]

22. Ullah, N.R.; Bhattacharya, K.; Thiringer, T. Wind Farms as Reactive Power Ancillary Service Providers—Technical and Economic Issues. *IEEE Trans. Energy Convers.* **2009**, *24*, 661–672. [CrossRef]

23. Braun, M. Reactive Power Supplied by Wind Energy Converters Cost-Benefit-Analysis. In Proceedings of the EWEC European Wind Energy Conference, Brussels, Belgium, 31 March–3 April 2008.

24. Engelhardt, S.; Erlich, I.; Feltes, C.; Kretschmann, J.; Shewarega, F. Reactive Power Capability of Wind Turbines Based on Doubly Fed Induction Generators. *IEEE Trans. Energy Convers.* **2011**, *25*, 364–372. [CrossRef]

25. CEPEL. *ELETROBRAS: R&D Center in Electrical Energy: ANAREDE-Electrical Networks Analysis Program*; User Guide, Version 9.7.2; CEPEL: Rio de Janeiro, Brasil, 2011.

26. CEPEL. *ELETROBRAS: R&D Center in Electrical Energy: ANATEM-Electromagnetic Transient Analysis Program*; User Guide, Version 10.4.6; CEPEL: Rio de Janeiro, Brasil, 2012.
27. ONS. Operador Nacional do Sistema Elétrico. Available online: http://www.ons.org.br (accessed on 20 January 2018).
28. Rokach, L.; Maimon, O. *Data Mining with Decision Tree: Theory and Applications*; World Scientific: River Edge, NJ, USA, 2008.
29. Hofmann, M.; Klinkenberg, R. *RapidMiner: Data Mining Use Cases and Business Analytics Applications*; Chapman & Hall/CRC: Boca Raton, FL, USA, 2013.

Article

Differential Evolution-Based Load Frequency Robust Control for Micro-Grids with Energy Storage Systems

Hongyue Li, Xihuai Wang * and Jianmei Xiao

Logistics Engineering College, Shanghai Maritime University, Shanghai 201306, China; 201540211019@stu.shmtu.edu.cn (H.L.); jmxiao@shmtu.edu.cn (J.X.)
* Correspondence: wxh@shmtu.edu.cn; Tel.: +86-021-38282637

Received: 28 May 2018; Accepted: 25 June 2018; Published: 27 June 2018

Abstract: In this paper, the secondary load frequency controller of the power systems with renewable energies is investigated by taking into account internal parameter perturbations and stochastic disturbances induced by the integration of renewable energies, and the power unbalance caused between the supply side and demand side. For this, the μ-synthesis robust approach based on structure singular value is researched to design the load frequency controller. In the proposed control scheme, in order to improve the power system stability, an ultracapacitor is introduced to the system to rapidly respond to any power changes. Firstly, the load frequency control model with uncertainties is established, and then, the robust controller is designed based on μ-synthesis theory. Furthermore, a novel method using integrated system performance indexes is proposed to select the weighting function during controller design process, and solved by a differential evolution algorithm. Finally, the controller robust stability and robust performance are verified via the calculation results, and the system dynamic performance is tested via numerical simulation. The results show the proposed method greatly improved the load frequency stability of a micro-grid power system.

Keywords: load frequency control; model uncertainty; μ-synthesis; differential evolution

---

1. Introduction

With the application of renewable energies (e.g., wind, solar, hydro) and the development of energy storage devices (e.g., battery, flywheel and ultra-capacitor), they are connected to the power systems to form micro-grids, have become an important way to reduce energy consumption and improve energy efficiency [1–4]. In the hybrid renewable energy system, due to the introduction of clean energy sources, some unstability factors are also introduced into the system, and we must considering that energy storage units have the ability to store energy from the system to improve the power system stability [5–8], new control strategy challenges are presented to ensure the power balance and frequency stable in power systems.

In order to enhance the power generation efficiency and improve the system stability, a number of scholars have devoted themselves to studying micro-grid power systems. In [9], in order to enhance the power system frequency stability, a novel intelligent methodology for battery energy storage system control and regulation is proposed. In [10], an actual model is proposed to guarantee the efficiency of energy storage in the micro-grid, and greatly improve the power system stability. This model is practical because the energy storage aging is considered. In [11–13], maximum power point tracking (MPPT) control, as a critical technology, is discussed and improved to enhance the stability of a micro-grid power system. In [14], in order to eliminate the micro-grid power system voltage imbalance and deviations, a novel hybrid bird-mating optimization approach is utilized for the connection decisions of distribution transformers. In [15,16], the fault analysis problem is mentioned, which is indispensable to guarantee the robustness of micro-grids. In [17], aimed at a high-voltage alternating current power system integrating an offshore wind farm and seashore wave power farm, in order to reduce the power fluctuations and keep the voltage stable, a novel intelligent damping controller for a static synchronous compensator is designed. Several key issues for the micro-grid system are discussed in the above studies, meanwhile, due to the power change between the generation side and the load side or system parameter perturbations, these changes will directly cause a power imbalance and lead to frequency fluctuation in the micro-grid. The frequency is one of the most important indicators of the power system, so it is required that the system have the ability to automatically take the frequency to the reference value when the power is changed. Thus, paying attention to studying the load frequency control is essential.

In a micro-grid the uncertainties that lead to frequency deviations are divided into two types in the frequency domain, one is the unstructured uncertainty with high frequency characteristics, e.g., external power disturbances, or the time delays of the control signals in the transmission process, and the other one is structured uncertainty with low frequency characteristics, e.g., the system parameter perturbations caused by equipment aging or electromagnetic interference. The two types of uncertainties may act on the system at the same time or separately, and lead to frequency fluctuations. In order to keep the frequency stable, lots of works have been done by relevant experts. The proportion integral derivative (PID) controllers, due to their simple structure and easy to implementation, were widely utilized to design load frequency controllers. In [18–21], controller parameter selections are posed as a multi-objective constraints problem and solved by optimization algorithms, but the robust stability and robust performance of PID controllers are not satisfactory. In [22,23], fuzzy methods are utilized to design a frequency controller to keep the micro-grid frequency stable, and this method has better performance robustness towards system parameter perturbations, but the control precision cannot be guaranteed because of the fuzziness of this method. In [24,25], the plug-in hybrid electric vehicle power control is utilized to compensate for the inadequate load frequency control capacity, and the mixed $H2/H\infty$ theory and the robust multivariable generalized predictive theory are researched, respectively. Some modern control theories were also researched and utilized to design secondary frequency controllers, such as model predictive control [26,27], the sliding mode control [28,29] and the active disturbance rejection control [30]. These methods show good robustness and good dynamic performance, but the calculation process is complex and the stability needs to be demonstrated in each case.

From the above analysis, considering that robust control theory has a strong stability and better performance, it is adopted to design the load frequency controller, and the $H\infty$ and μ-synthesis based on robust methods are proposed in [31–35]. In [31,32], the load frequency robust controller is designed based on $H\infty$ method to handle the time delay uncertainty in the micro-grid. In [33,34], the external disturbances caused by wind and solar power changes in a micro-grid are counteracted by a robust method. In [35], an integrated micro-grid composed of renewable energies and energy storage systems is introduced, and in this system, all of the power generations are modeled by a first-order inertial model, two types of uncertainties are considered in this model, and the load frequency controllers are discussed and compared based on the $H\infty$ and μ-synthesis methods, respectively. The results show the μ-synthesis with structured uncertainty gives better performance than $H\infty$ method. Both methods based on robust theory have excellent performance to deal with the unstructured uncertainties and structured uncertainties, and can guarantee a strong and robust stability and performance of the system.

The μ-synthesis robust method based on structural singular value theory, due to its excellent robust performance and low conservative nature, is of great interest to the scholars in the design of load frequency controllers. In the controller design process, the weighted functions play key roles and directly affect the control performance, as they not only determine the controller robust stability, but also determine the system's dynamic performance, so it is essential to choose an optimal weighted function coefficient. In [33,35], the load frequency controller based on μ-synthesis is introduced, but the selection process and selection principle of weighted functions are not given. As usual, the empirical method is adopted by the designers; this requires the designers be really good in frequency-domain control theory and have rich engineering experience. Some designers also find out the coefficients of weighted function by trial-and-error, but this is a massive task, and it is too hard for the two methods to get an optimum solution. For this reason, the application of μ-synthesis is greatly limited and the control performance is reduced.

From the above analysis, in this paper, an integrated micro-grid power system with renewable energy and energy storage unit is studied, and the corresponding load frequency controller is designed based on the μ-synthesis robust method. In order to find out the appropriate weighting functions to achieve a more robust performance and better dynamic performance, the weighting functions selection problem is transformed into a multi-objective problem, where the adaptive differential evolution algorithm is proposed to search the optimal solution.

The rest of this paper is organized as follows: in Section 2, the micro-grid attached with energy storage is described and the load frequency control model with uncertainties is established. In Section 3, the uncertain parameter model is built, the robustness index is given, and the DK iteration is introduced to solve the μ-synthesis controller problem. In Section 4, the weighting functions are selected via differential evolutionary algorithms and the controller is figured out. In Section 5, the robust stability and robust performance are demonstrated and analyzed. The results are tested and simulated in Section 6. Finally, the conclusions are presented in Section 7.

2. Model Description

In this section, the secondary load frequency control model of a micro-grid with energy storage is established and the model uncertainties with structured and unstructured uncertainty are described.

2.1. Description of Micro-Grid Power System

The micro-grid researched in this paper is shown in Figure 1. It is composed of a diesel generator, renewable energy and energy storage unit. Diesel generators, as the traditional power generation source, are connected to the ac bus by a transformer, and their working state should be as stable as possible in order to reduce the fuel consumption and exhaust emissions. The generators are driven by the diesel engine to generate electric power for the system, and the generators' speed determines the system frequency, which is completely dependent on the diesel engines' output torque, so the essence of load frequency control is to regulate the diesel engine output power. When a small disturbance acts on the power system and causes a small-range frequency deviation, the speed governor can suppress the deviation adequately, but for a larger disturbance, the primary frequency control is ineffective at bringing the deviation to zero. In this condition, the secondary load frequency control is essential to change the characteristics of the speed governor and finally take the frequency deviation to zero.

As mentioned, the energy storage is indispensable in the micro-grid power system. It has the ability to store the extra power generated by the renewable sources, while providing stored power to the system to maintain the power balance. Many kinds of energy storage forms are tested and utilized in micro-grid power systems, such as batteries, flywheels and ultracapacitors, among which, the battery is the most widely applied to improve the power system stability, but the charge/discharge rate is not satisfactory, while the flywheel has the ability to overcome this disadvantage, but its cost and maintenance are expensive [26]. Therefore, the ultra-capacitor is chosen as the energy storage unit in this paper because of its fast response and low cost. As described in Figure 1, the ultra-capacitor is connected to the ac bus by a dc/ac inverter, and the power in the inverter is bidirectional. If the generated power is much than the demand power, the ultra-capacitor is working in charging mode. Otherwise, the ultra-capacitor is working in discharging mode.

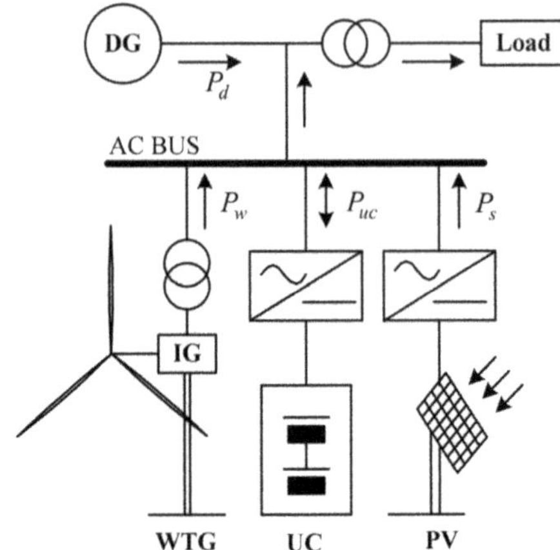

Figure 1. Micro-grid power system structure.

The photovoltaic panel and wind turbine generator are connected to the ac bus by a dc/ac inverter and ac/ac converter, respectively. Both transform the renewable energies into electric power. Due to the fact the power generated by wind and solar deeply depend on the weather conditions, this can be regarded as the unstable factor on the generation side. The power system working in stable state must satisfy the following equation:

$$\Delta P_d + \Delta P_{uc} + \Delta P_w + \Delta P_s + \Delta P_l = 0 \tag{1}$$

In the expression, ΔP_d is the diesel engine output power change, ΔP_{uc} is the ultra-capacitor output power change, ΔP_w is the wind turbine generator output power change, ΔP_s is the photovoltaic output power change, and ΔP_l is the load power change. Because the power in the ultra-capacitor is bidirectional, it is assumed that the power from the ultra-capacitor to the ac bus is positive, and the power from the ac bus to ultra-capacitor is negative.

2.2. Load Frequency Control Model

The model of micro-grid frequency control process used in the paper is shown in Figure 2.

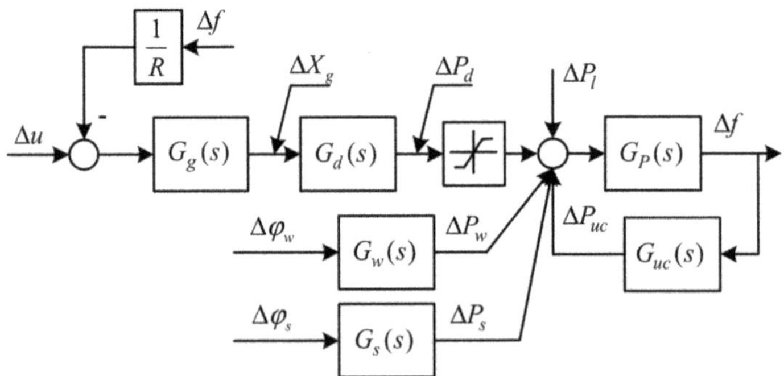

Figure 2. Load frequency control.

In the figure, the $G_g(s)$, $G_d(s)$, $G_s(s)$, $G_{uc}(s)$, $G_w(s)$, $G_s(s)$ are the transfer functions of governor, diesel engine, generator-load, ultra-capacitor, photovoltaic and wind turbine generator. There dynamic characteristics are as follows: the dynamics of the governor are expressed as [36–38]:

$$\Delta \dot{X}_g = -\frac{1}{T_g}\Delta X_g + \frac{1}{T_g}\left(\Delta u - \frac{1}{R}\Delta f\right) \tag{2}$$

where, ΔX_g is the governor output, T_g is the governor time constant, Δu is the control signal, R is the droop coefficient, Δf is the frequency deviation.

The dynamics of the prime mover are expressed as:

$$\Delta \dot{P}_d = -\frac{1}{T_d}\Delta P_d + \frac{1}{T_d}\Delta X_g \tag{3}$$

where, ΔP_d is the prime mover output power, and T_d is the prime mover time constant.

The power system dynamics are expressed as:

$$\Delta \dot{f} = -\frac{H}{M}\Delta f + \frac{1}{M}[\Delta P_d + \Delta P_w + \Delta P_s + \Delta P_{uc} + \Delta P_l] \tag{4}$$

where M is the inertia constant, H is the damping constant.

The dynamics of the ultra-capacitor model are expressed as:

$$\Delta \dot{P}_{uc} = -\frac{1}{T_{uc}}\Delta P_{uc} + \frac{1}{T_{uc}}\Delta f \tag{5}$$

where T_{uc} is the ultra-capacitor time constant.

The wind turbine generator and photovoltaic panel transform the renewable energy into electric power. The dynamic processes of the two are expressed as [39,40]:

$$\Delta \dot{P}_w = -\frac{1}{T_w}\Delta P_w + \frac{1}{T_w}\Delta \varphi_w \tag{6}$$

where ΔP_w is the wind turbine generator output power change, $\Delta \varphi_w$ is the wind power change, and T_w is the wind turbine generator time constant:

$$\Delta \dot{P}_s = -\frac{1}{T_s}\Delta P_s + \frac{1}{T_s}\Delta \varphi_s \tag{7}$$

where ΔP_s is the photovoltaic panel output power change, $\Delta \varphi_s$ is the solar power change, T_s is the photovoltaic panel time constant.

By simultaneously solving Equations (2)–(7), the micro-grid load frequency control state-space model can be written as:

$$\begin{aligned} \dot{x} &= Ax(t) + Bu(t) + Fw(t) \\ y &= Cx(t) \end{aligned} \tag{8}$$

In the model, $x = [\Delta X_g, \Delta P_g, \Delta P_b, \Delta P_W, \Delta P_S, \Delta f]^T$ is the state vector, $y = \Delta f$ is the measured output vector, $w = [\Delta \varphi_w, \Delta \varphi_s, \Delta P_l]^T$ is the disturbance vector, u is the input vector. $A \in \mathbb{R}^{n \times n}$ is the state matrix, $B \in \mathbb{R}^{n \times m}$ is the input matrix, $C \in \mathbb{R}^{v \times n}$ is the output matrix, $F \in \mathbb{R}^{n \times l}$ is the disturbance matrix: Align or replace as in the previous paragraphs as example

$$A = \begin{bmatrix} -\frac{1}{T_g} & 0 & 0 & 0 & 0 & -\frac{1}{T_g R} \\ \frac{1}{T_d} & -\frac{1}{T_d} & 0 & 0 & 0 & 0 \\ 0 & 0 & -\frac{1}{T_b} & 0 & 0 & 0 \\ 0 & 0 & 0 & -\frac{1}{T_w} & 0 & 0 \\ 0 & 0 & 0 & 0 & -\frac{1}{T_s} & 0 \\ 0 & \frac{1}{M} & \frac{1}{M} & \frac{1}{M} & \frac{1}{M} & -\frac{H}{M} \end{bmatrix}, \; B = \begin{bmatrix} \frac{1}{T_g} \\ 0 \\ \frac{1}{T_b} \\ 0 \\ 0 \\ 0 \end{bmatrix}, \; F = \begin{bmatrix} 0 & 0 & 0 \\ 0 & 0 & 0 \\ 0 & 0 & 0 \\ \frac{1}{T_w} & 0 & 0 \\ 0 & \frac{1}{T_s} & 0 \\ 0 & 0 & -\frac{1}{M} \end{bmatrix}, \; C = \begin{bmatrix} 0 \\ 0 \\ 0 \\ 0 \\ 0 \\ 1 \end{bmatrix}$$

The linear state space Equation (8) describes the load frequency control of a micro-grid power system. This is idealized without considered the parameter changes. Due to the environmental effects, such as temperature, electromagnetic interference, equipment aging, the parameters are inevitably perturbed and bring nonlinear factors to the system. With consideration of the parameter perturbation, Equation (8) can be rewritten as:

$$\dot{x} = (A + \Delta A)x(t) + (B + \Delta B)u(t) + (F + \Delta F)w(t)$$
$$y = Cx(t) \tag{9}$$

where, ΔA, ΔB, ΔF are the parameter uncertainties with block-diagonal structure, which have the same dimensions as A, B, F.

Separating the uncertainties from Equation (9), let:

$$f(x,t) = \Delta Ax(t) + \Delta Bu(t) + (F + \Delta F)w(t) \tag{10}$$

so, Equation (9) is rewritten as:

$$\dot{x} = Ax(t) + Bu(t) + f(x,t) \tag{11}$$

Without loss of generality, it is assumed that $\{A,B\}$ is controllable, $\{A,C\}$ is observable. $f(x,t)$ is norm bounded and meets the Lipschitz condition [29].

3. Controller Designed Based on μ-Synthesis

In this section, the load frequency control model considering parameter perturbation is established, and the robust stability and robust performance indexes are presented.

3.1. Uncertainty Model Establish

Assuming that in the load frequency control system, the parameters T_g, T_d, T_{uc}, T_w, T_s, M, H have perturbation errors, their values change within a certain interval, expressed as:

$$\begin{rcases} T_g = \overline{T}_g(1 + p_g \delta_g) \\ T_d = \overline{T}_d(1 + p_d \delta_d) \\ T_{uc} = \overline{T}_{uc}(1 + p_{uc}\delta_{uc}) \\ T_w = \overline{T}_w(1 + p_w \delta_w) \\ T_s = \overline{T}_s(1 + p_s \delta_s) \\ M = \overline{M}(1 + p_M \delta_M) \\ H = \overline{H}(1 + p_H \delta_H) \end{rcases} \tag{12}$$

In the expression, the parameters with over-lines denote the nominal value of the corresponding parameters. The corresponding p and δ in Equation (12) represent the possible perturbations of these seven parameters. In the present study, let $p = 0.3$, and $\delta \in [-1, 1]$. This shows that the parameters' uncertainties are perturbed within $\pm 30\%$. Considering the parameter perturbation, the load frequency control shown in Figure 2 can be restructured as shown in Figure 3.

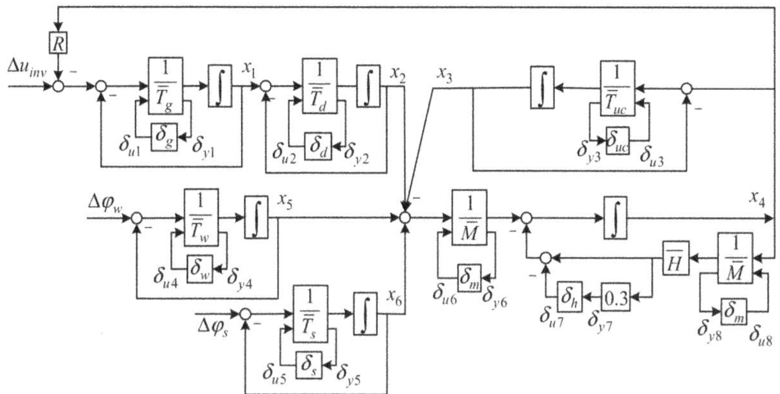

Figure 3. Load Frequency Control System with Uncertainty.

Extracting the uncertainties from the actual controlled system, the seven constant blocks in Figure 3 can be replaced by block diagrams in terms of $\overline{T}_g$, p_g, δ_g etc., in a unified approach. As introduced in model description section, all the parameters T_g, T_d, T_{uc}, T_w, T_s, M are in the denominator, so they can be represented as a low linear fractional transformation. Taking for example the case of T_g:

$$\frac{1}{T_g} = \frac{1}{\overline{T}_g(1+0.3\delta_g)} = \frac{1}{\overline{T}_g} - \frac{0.3}{\overline{T}_g}\delta_g\left(1+0.3\delta_g\right)^{-1}$$
$$= F(M_g, \delta_g) \tag{13}$$

with $M_g = \begin{bmatrix} \frac{1}{\overline{T}_g} & -\frac{0.3}{\overline{T}_g} \\ 1 & -0.3 \end{bmatrix}$. The parameters T_d, T_{uc}, T_w, T_s, M are similarly transformed using the same method. Especially, the parameter H is in the molecule, and can be represented as a upper linear fractional transformation:

$$H = \overline{H}(1+0.3\delta_h) = F_u(M_H, \delta_h) \tag{14}$$

with $M_H = \begin{bmatrix} \overline{H} & -1 \\ 0 & 0.3\overline{H} \end{bmatrix}$.

Through the above transformations and substitutions, the interconnection matrix is established; in the matrix, all of the system inputs, uncertainties input, and the system output uncertainties are contained:

$$\begin{bmatrix} \dot{x}_p \\ \delta_{yq} \end{bmatrix} = \widetilde{G} \begin{bmatrix} x_p \\ \delta_{uq} \end{bmatrix} \tag{15}$$

$$\delta_{uq}^T = diag[\delta_g, \ \delta_d, \ \delta_{uc}, \ \delta_w, \ \delta_s, \ \delta_m, \ \delta_h, \ \delta_m] \cdot \delta_{yq}^T \tag{16}$$

where $p = 1, 2, \ldots, 6$, $q = 1, 2, \ldots, 8$. $\widetilde{G}$ is the augmented matrix with nominal model and parameter uncertainty:

$$\tilde{G}=\begin{bmatrix}
-\frac{1}{T_g} & 0 & 0 & -\frac{1}{RT_g} & 0 & 0 & -\frac{0.3}{T_g} & 0 & 0 & 0 & 0 & 0 & 0 & 0 & \frac{1}{T_g} \\
\frac{1}{T_d} & -\frac{1}{T_d} & 0 & 0 & 0 & 0 & -\frac{0.3}{T_d} & 0 & 0 & 0 & 0 & 0 & 0 & 0 & 0 \\
0 & 0 & -\frac{1}{T_b} & \frac{1}{T_b} & 0 & 0 & 0 & 0 & -\frac{0.3}{T_b} & 0 & 0 & 0 & 0 & 0 & 0 \\
0 & 0 & 0 & 0 & -\frac{1}{T_w} & 0 & 0 & 0 & 0 & -\frac{0.3}{T_w} & 0 & 0 & 0 & 0 & 0 \\
0 & 0 & 0 & 0 & 0 & -\frac{1}{T_w} & 0 & 0 & 0 & 0 & -\frac{0.3}{T_s} & 0 & 0 & 0 & 0 \\
-1 & 0 & 0 & -\frac{1}{R} & 0 & 0 & 0.3 & 0 & 0 & 0 & 0 & 0 & 0 & 0 & 1 \\
1 & -1 & 0 & 0 & 0 & 0 & 0 & -0.3 & 0 & 0 & 0 & 0 & 0 & 0 & 0 \\
0 & 0 & -1 & 1 & 0 & 0 & 0 & 0 & -0.3 & 0 & 0 & 0 & 0 & 0 & 0 \\
0 & 0 & 0 & 0 & -1 & 0 & 0 & 0 & 0 & -0.3 & 0 & 0 & 0 & 0 & 0 \\
0 & 0 & 0 & 0 & 0 & -1 & 0 & 0 & 0 & 0 & -0.3 & 0 & 0 & 0 & 0 \\
0 & 0 & 0 & 0 & 0 & 0 & 0 & 0 & 0 & 0 & 0 & 0 & 0 & 0 & 0 \\
0 & 0 & -1 & 0 & 0 & 0 & 0 & 0 & 0 & 0 & 0 & 0 & -0.3 & 0 & 0 \\
0 & 0 & 0 & \frac{0.3\overline{D}}{M} & 0 & 0 & 0 & 0 & 0 & 0 & 0 & 0 & 0 & -\frac{0.09\overline{D}}{M} & 0 \\
0 & 0 & 0 & 1 & 0 & 0 & 0 & 0 & 0 & 0 & 0 & 0 & 0 & -0.3 & 0 \\
0 & 0 & 0 & 1 & 0 & 0 & 0 & 0 & 0 & 0 & 0 & 0 & 0 & 0 & 0
\end{bmatrix}$$

3.2. Robust Performance Analysis

The objective of load frequency control in the paper is to find a feedback control $u(s) = K(s)y(s)$, which makes sure that the closed loop system robust stability and robust performance demands are satisfied. The closed-loop system is shown in Figure 4.

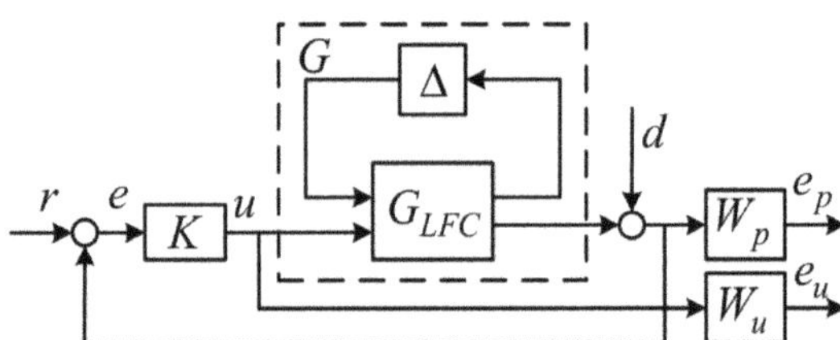

Figure 4. Closed-loop system structure.

In the figure, G_{LFC} is the nominal load frequency control model, Δ is the uncertainties with block-diagonal structure caused by parameter perturbations, G is transfer function matrix concluded from the nominal model and uncertainties model, $G = F_u(G_{LFC}, \Delta)$. W_p, W_u are weighting functions, reflecting the frequency characteristics of disturbances and the system performance index. In order to achieve robust stability, the closed-loop system should be internally stable, which satisfies the expression:

$$\left\| \begin{bmatrix} W_P(I+G_{LFC}K)^{-1} \\ W_u K(I+G_{LFC}K)^{-1} \end{bmatrix} \right\|_\infty < 1 \tag{17}$$

In addition to the robust stability, the system should also satisfy the robust performance need, for all the uncertainties:

$$\left\| \begin{bmatrix} W_P(I+GK)^{-1} \\ W_u K(I+GK)^{-1} \end{bmatrix} \right\|_\infty < 1 \tag{18}$$

3.3. DK Iteration

DK iteration is a common method to solve μ-synthesis controller. The steps are the following:

Step 1: Select the initial scale matrix D, generally set $D = I$;

Step 2: Hold D, and obtain the optimum solution for K via $H\infty$ optimization method. $K = \underset{K}{\arginf} \|F_l(P,K)\|_\infty$, P is the interconnected augmented matrix include the weighting function and the controlled object.

Step 3: Hold K to solve the convex optimization problem for D at the selected frequency domain and obtain the optimal estimation matrix, mark as $\widetilde{D}$. $\widetilde{D}(j\omega) = \underset{D}{\arginf}\overline{\sigma}[D(j\omega)F_l(P,K)D^{-1}(j\omega)]$.

Step 4: Let $D = \widetilde{D}$, return to Step 2, repeat steps 2 and 3, until the maximum iteration number is reached, or the constraint $\underset{\omega \in \mathbb{R}}{\sup}\underset{D}{\inf}\overline{\sigma}[D(j\omega)F_l(P,K)D^{-1}(j\omega)] < 1$ is satisfied.

Through the *DK* iterative process, the controller is solved.

4. Weighting Function Selection Based on Differential Evolution

In this section, the weighting functions in μ-synthesis are determined by a differential evolution method through solving the defined constraint conditions.

4.1. Parameters Setting

The weighting functions not only decide whether the system robust stability and robust performance demands are satisfied, but also play key roles in the *DK* iterative process. The weighting functions are listed as:

$$W_p = \frac{a_1 s + a_2}{a_3 s + a_4} \tag{19}$$

$$W_u = \frac{b_1 s + b_2}{b_3 s + b_4} \tag{20}$$

In the expressions, $a_1, \ldots, a_4, b_1, \ldots, b_4$ are undetermined coefficients, and the ranges of parameters are limited to $[0, 10^2]$. In order to obtain the optimal parameters, the differential evolution (DE) method is utilized because the method has less parameters to determine and does not easily fall into a local optimum, the DE has better performance than the genetic algorithm and particle swarm optimization [41].

4.2. Determination of Fitness Function

In order to get a better robust stability and robust performance, and also ensure that the closed-loop system has a satisfactory dynamic performance, the fitness function of the differential evolution method should consider all the factors. The selection of the weighting functions is constrained by a series of inequalities:

(1)   General indicator

$$\delta_1 = 20 \cdot \log[\overline{\sigma}[P(j\omega)]] \tag{21}$$

If $\delta_1 < 0$, let $\Phi_1 = \overline{\sigma}[P(j\omega)]$, if $\delta_1 > 0$, let $\Phi_1 = 1000$. P is the interconnected augmented matrix that includes the weighting function of the closed-loop system. $\overline{\sigma}$ is the upper bound of the maximum singular value.

(2)   Robust stability

$$\delta_2 = 20 \cdot \log[\overline{\sigma}[W_p(j\omega)(I + G_{LFC}(j\omega)K(j\omega))]] \tag{22}$$

$$\delta_3 = 20 \cdot \log[\overline{\sigma}[W_u(j\omega)K(j\omega)(I + G_{LFC}(j\omega)K(j\omega))]] \tag{23}$$

If $\delta_2 < 0$, let $\Phi_2 = \bar{\sigma}[W_p(j\omega)(I + G_{LFC}(j\omega)K(j\omega))]$, if $\delta_2 > 0$, $\Phi_2 = 1000$.
If $\delta_3 < 0$, let $\Phi_3 = \bar{\sigma}[W_u(j\omega)K(j\omega)(I + G_{LFC}(j\omega)K(j\omega))]$, if $\delta_3 > 0$, let $\Phi_3 = 1000$.

(3) Robust performance

$$\delta_4 = 20 \cdot \log[\bar{\sigma}[W_p(j\omega)(I + G(j\omega)K(j\omega))]] \tag{24}$$

$$\delta_5 = 20 \cdot \log[\bar{\sigma}[W_u(j\omega)K(j\omega)(I + G(j\omega)K(j\omega))]] \tag{25}$$

If $\delta_4 < 0$, let $\Phi_4 = \bar{\sigma}[W_p(j\omega)(I + G(j\omega)K(j\omega))]$, if $\delta_4 > 0$, $\Phi_4 = 1000$.
If $\delta_5 < 0$, let $\Phi_5 = \bar{\sigma}[W_u(j\omega)K(j\omega)(I + G(j\omega)K(j\omega))]$, if $\delta_5 > 0$, let $\Phi_5 = 1000$.

(4) Dynamic response performance

The dynamic characteristics and stability margin should be considered to ensure the better output performance as mentioned in [42]; let $\Phi_6 = 10t_r$, t_r be the rise time. $\Phi_7 = \int_0^{T_s} |e(t)|^2 dt$, $e(t)$ is the output error, and T_s is the simulation time.

(5) Stability margin

According to [42], for a control system with better stability margin, the amplitude margin should be around 6 dB and the phase margin should be around $45°$, thus, the stability margin is defined as follows:
If $6 < G_m < 20$, $\Phi_8 = 100/(G_m - 20)$, else $\Phi_8 = 1000$; If $30 < P_m < 60$, $\Phi_9 = 100/(P_m - 60)$, else $\Phi_9 = 1000$, where, G_m is the amplitude margin, P_m is the phase margin.

By the abovementioned method, the system performance index are expressed as the inequality constraints, and the objective function is designed as $f = \sum \Phi_i$, according to the minimum search principle, the fitness function is designed as $F = 1/f$. For this, the smaller of f, the bigger of F, and F is positive can be guaranteed.

4.3. Algorithm Steps

The design steps of micro-grid load frequency robust controller based on differential evolution method are as follows, and the flowchart is shown in Figure 5.

(1) Establish the load frequency control model which concluding the uncertainties.
(2) Initialize the differential evolution algorithm, and obtain the initial populations.
(3) Take the population parameters into the system and going DK iteration process. After iterations, the controller is obtained.
(4) Computing the system robust stability, robust performance and output dynamic performance, and verifying whether the performances are satisfied.
(5) If the performances are not satisfied, then executing the second differential evolution process, and repeat the step 3 and step 4.

The process is done until the desired robust stability and robust performance are achieved and dynamic performance is satisfactory, or the DE iteration is finished.

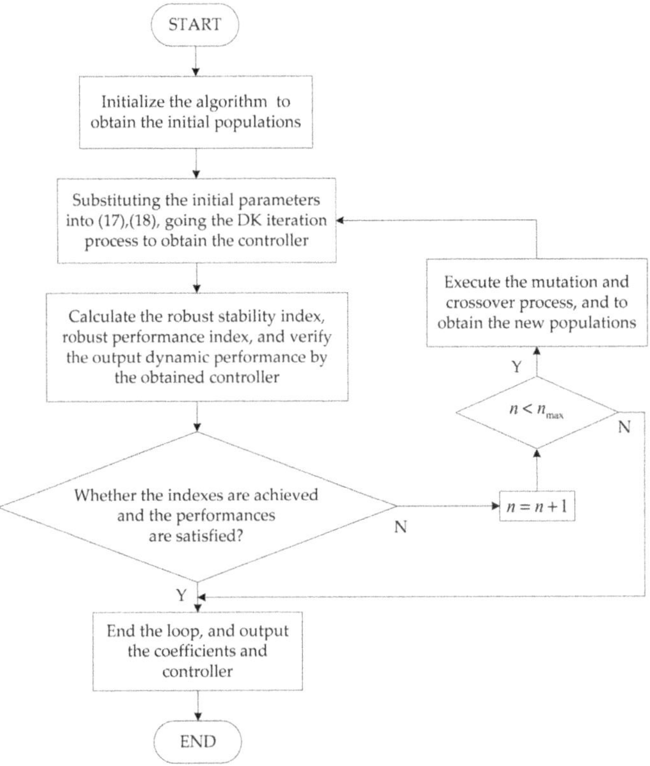

Figure 5. Flowchart of the proposed method.

5. Robust Stability and Robust Performance Analysis

In this section, the robust controller is solved based on the differential evolution algorithm. The algorithm parameters are setting as: number of parameters $D_d = 8$, population size $M_G = 50$, iteration number $G_T = 40$, variation factor $F_r = 0.6$, cross factor $CR = 0.3$. The parameters of micro-grid system are listed in Table 1, the parameters units are expressed as per-units (p.u.).

Table 1. System parameters.

Parameter Name	Value
Rated Frequency (Hz)	50
Rated power (MW)	2
Governor Time Constant T_g (s)	0.008
Diesel Time Constant T_d (s)	0.3
Ultracapacitor Time Constant T_{uc} (s)	0.1
wind turbine generator time constant T_w (s)	1.5
photovoltaic panel time constant T_s (s)	1.8
Inertia coefficient M (p.u./s)	0.15
Damping coefficient H (p.u./Hz)	0.008
Droop coefficient R (p.u./Hz)	2.4

After the iteration calculation, the weighting function coefficients are obtained, and the structure of weighting functions are shown as:

$$W_p(s) = \frac{s + 30}{2s + 0.5} \tag{26}$$

$$W_e(s) = \frac{s + 0.2}{0.01s + 50} \tag{27}$$

Taking the designed weighting functions into the closed loop system model, we then execute the *DK* iteration process to figure out the μ-synthesis controller.

The iteration result is listed in Table 2.

Table 2. *DK* iteration results.

Iterations	K Order	D Order	γ Value	μ Value	μ-RS	μ-RP
1	6	0	8.777	1.498	1.172	1.3516
2	8	2	0.744	0.520	0.549	0.817
3	12	6	0.439	0.436	0.255	0.505
4	20	14	0.390	0.389	0.255	0.505
5	20	14	0.369	0.369	0.255	0.505
6	30	24	0.360	0.359	0.255	0.505
7	30	24	0.353	0.352	0.255	0.505
8	32	26	0.349	0.349	0.255	0.505
9	32	26	0.348	0.348	0.255	0.505
10	34	28	0.347	0.347	0.255	0.505

In the table, after the first iteration, the value of μ and γ are larger than 1, which means that the robust stability (RS) and robust performance (RP) are not achieved. After the second iteration, the value of μ and γ are reduced to less than 1, indicating the system robust stability and robust performance have been reached. After the third iteration, the values are further reduced and the system robust stability and robust performance are further improved. As the iterations increase, the values of γ and μ continue to decrease, but the variation is small enough to disregard, and the indexes of robust stability and robust performance are no longer changed. From the table, we also see that the more iterations, the higher the controller orders. In other words, too many iteration calculations will lead to more conservativeness of the controller, and this is unnecessary. Properly, the fifth iteration result is adopted to design the controller to guarantee a compromise between the controller order and robust index.

From Table 2 and Figure 6, the robust stability index is 0.255, what indicates that the system stability is guaranteed for $\|\Delta\|_\infty < 1/0.255$. The robust performance index is 0.505, what indicates that the closed-loop system with the designed controller has achieved both the nominal performance and robust performance since Equation (26), for every diagonal Δ, $\|\Delta\|_\infty < 1$:

$$\left\| \begin{bmatrix} W_p(I + F(G,\Delta)K)^{-1} \\ W_u K(I + F(G,\Delta)K)^{-1} \end{bmatrix} \right\|_\infty < 0.505 \tag{28}$$

However, the order of controller is still much higher than the order of the plant transfer function. This is means that the implementation of the high order controller requires more hardware (equipment) and this brings more maintenance problems, which greatly increase the cost of the controller and reduce the reliability of the controller, so it is necessary to reduce the controller order to obtain a low order controller. The Hankel-norm approximation method is adopted in the paper to implement the order reduction [43]. After five iterations, the order of controller K_μ is 20, and reduced to 5 by Hankel-norm approximation. The controller transfer function is expressed as:

$$K_\mu = \frac{3.402 \times 10^{-8}s^5 + 10.26s^4 + 1.611 \times 10^4 s^3 + 3.879 \times 10^5 s^2 + 2.729 \times 10^6 s^1 + 5.409 \times 10^6}{s^5 + 1575s^4 + 5.053 \times 10^4 s^3 + 4.256 \times 10^5 s^2 + 1.251 \times 10^6 s^1 + 1.422 \times 10^6} \tag{29}$$

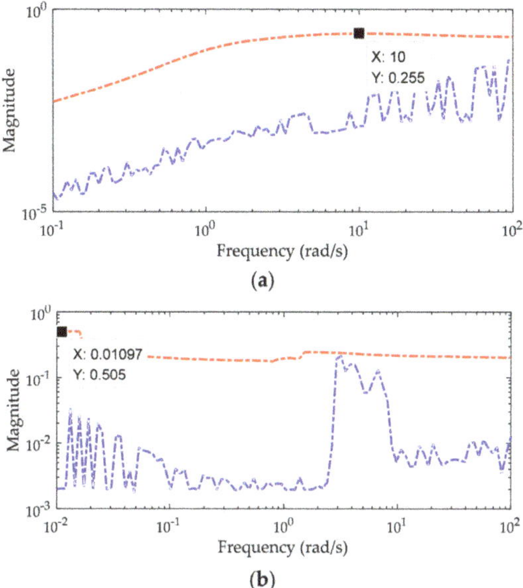

Figure 6. The index of robust stability and robust performance, (**a**) the robust stability of controller, upper bound (red) and lower bound (blue); (**b**) the robust performance of perturbed system, upper bound (red) and lower bound (blue).

The comparison of the amplitude-phase frequency characteristic curves is shown in Figure 7, which shows that the full-order controller and the reduced-order control have almost the same frequency characteristics. Thus, compared with the full-order controller, the reduced-order controller is much easier to implement and without performance degradation.

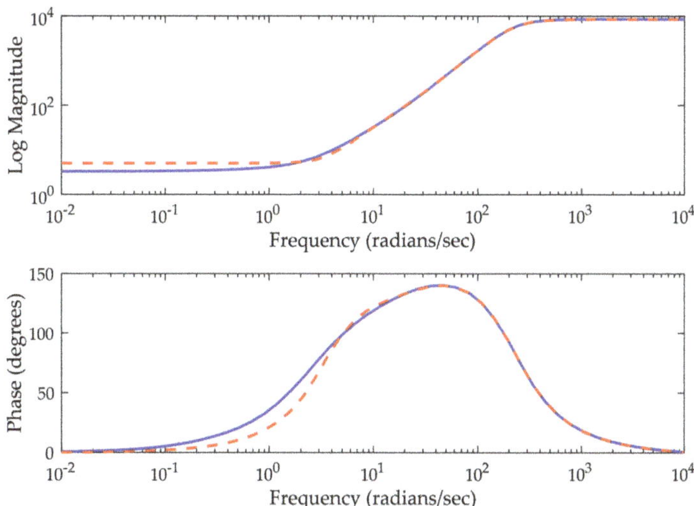

Figure 7. Bode diagram of the full-order controller (red) and reduced−order controller (blue).

6. Numerical Simulation

In this section, the controller designed by μ-synthesis is simulated and tested. In order to show the effectiveness of the proposed controller, the disturbance power and parameters perturbation are considered in the test. Further to show the robust stability and robust performance, the classic PID controller and $H\infty$ controller [32] are compared with the proposed controller in the simulation.

Firstly, in order to demonstrate the ultracapacitor performance, the micro-grids power system without energy storage unit and with battery unit are simulated respectively. We assumed that the system is disturbed by a step power $\Delta P_l = 0.01$p.u., shown as Figure 8a, and the frequency responses are shown as Figure 8b.

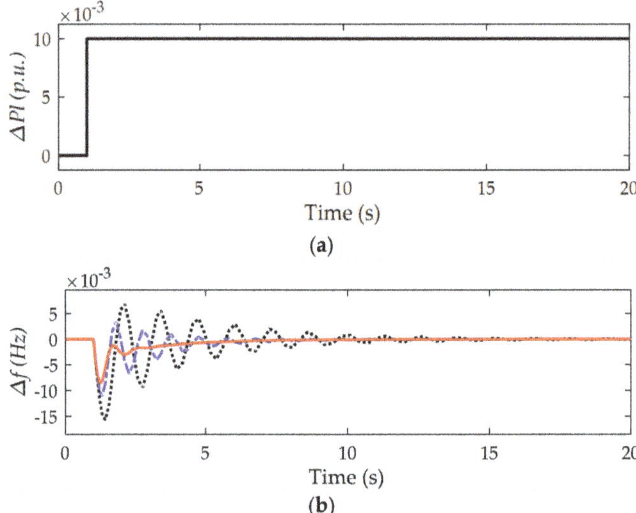

Figure 8. The power system frequency deviation disturbed by the step signal, without energy storage unit (black-dotted), with battery (blue-dashed), and with ultra-capacitor (red-solid), (**a**) the step signal; (**b**) the frequency deviation.

From the figures, when the disturbance occurred, the power system without energy storage restored to steady state in about 15 s, and the maximum frequency response is about 0.015 Hz. The stabilization time of the system with battery unit is 7 s, and the overshoot is 0.01 Hz, while for the ultracapacitor unit, the stabilization time is 4 s and the overshoot is 0.008 Hz. The results illustrate that the energy storage units can improve the stability of the micro-grid significantly, and have the ability to reduce the power change. Moreover, due to the fact the ultracapacitor has much faster charge/discharge speed than battery, it is more applicable to absorb the disturbance power with high-frequency and high- power features in the renewable power system.

Secondly, we assume that a step signal, which has the same magnitude as in Figure 8a, is given to the renewable energy power system with ultracapacitor. The response results are shown in Figure 9. From the figure, both the $H\infty$ controller and the μ-synthesis control have a faster setting time and smaller overshoot than the optimized PID controller, meanwhile, the μ-synthesis controller has much better performance than the $H\infty$ controller.

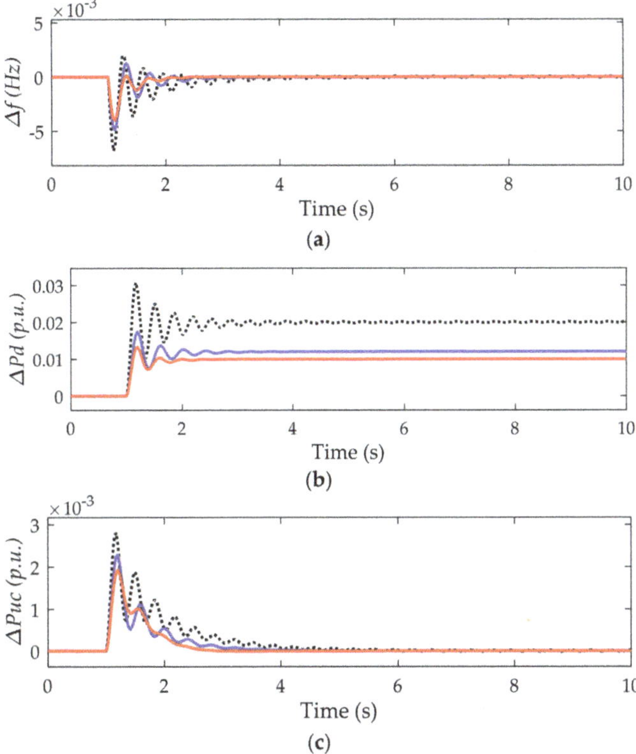

Figure 9. The system outputs with step signal, proportion integral derivative (PID) control (black), $H\infty$ control (blue) and μ-synthesis control (red), (**a**) frequency deviation; (**b**) diesel engine power deviation; (**c**) ultra-capacitor power deviation.

Thirdly, considering that due to the fact weather factors lead to power changes in the renewable energies' output, it is assumed that the sum power deviation of wind and solar change with a five second period in thirty seconds, then the magnitude is as shown in Figure 10a, and the system output responses are as shown in Figure 10. From the figure, the PID controller has a long setting-time and larger overshoot, an even becomes uncontrollable when the change power is bigger. The μ-synthesis controller has better performance than the $H\infty$ no matter how the renewable power changes. Figures 9 and 10 illustrate that the μ-synthesis method has much better robust stability than the $H\infty$ controller.

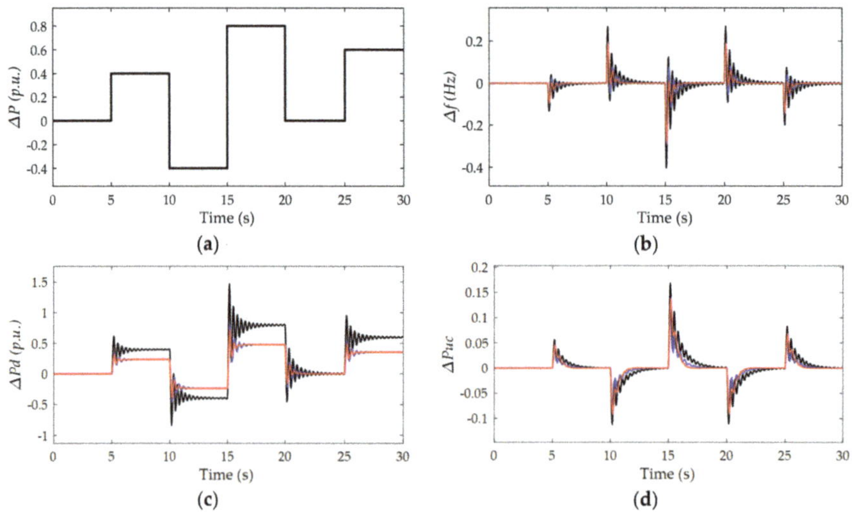

Figure 10. Outputs response with renewable energies power change by PID control (black), $H\infty$ control (blue) and μ-synthesis control (red), (**a**) renewable energies power change; (**b**) frequency deviation; (**c**) diesel engine power deviation; (**d**) ultra-capacitor power deviation.

Fourthly, in order to verify the system robust performance, it is assumed that a random external disturbance is applied to the system, and the frequency responses are as shown in Figure 11. From the figure, when the parameters are perturbed within 10%, the fluctuation becomes large, and further when the parameters are perturbed within 30%, the PID controller has become worse, but the $H\infty$ controller and μ-synthesis controller still limit the errors to a tolerable range. The results show that due to the fact the μ-synthesis controller has considered both the structured uncertainty and unstructured uncertainty when designed, it has better robust performance and nominal performance than the $H\infty$ controller, and can greatly improve the load frequency stability of the renewable energy power system.

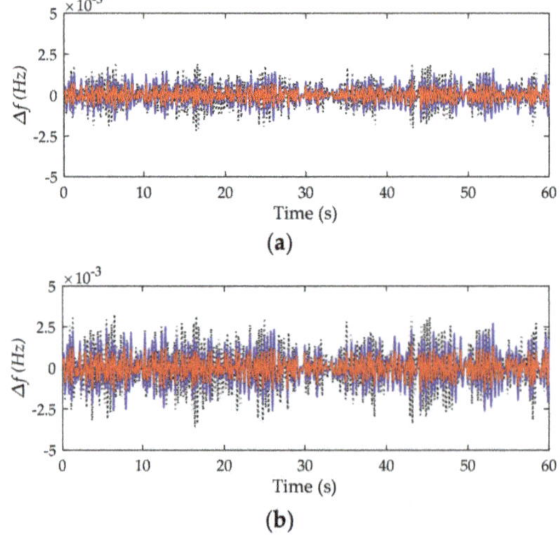

Figure 11. *Cont.*

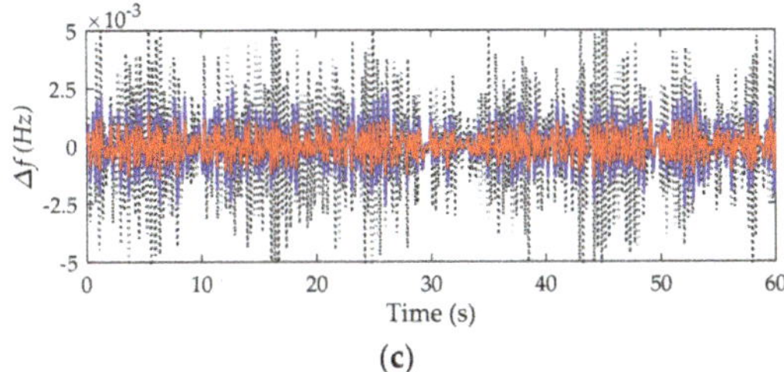

(c)

Figure 11. Frequency deviation with parameters perturbation, PID control (black), $H\infty$ control (blue) and μ-synthesis control (red), (**a**) without parameters perturbation; (**b**) parameters perturbed within 10%; (**c**) parameters perturbed within 30%.

7. Conclusions

In this paper, the robust μ-synthesis approach is used for load frequency control in a micro-grid power system. The load frequency control state space model with uncertainty is established. The μ-synthesis controller based on structure singular value is designed and solved by the *DK* iteration method. The controller performances are verified and tested in comparison with the PID controller and $H\infty$ controller. The results show that ultracapacitors can enhance the frequency stability of micro-grid power systems. Due to the fact the μ-synthesis controller has considered both the structured uncertainty and unstructured uncertainty when designed, it has more robust performance and better nominal performance than the $H\infty$ controller, and can greatly improve the load frequency stability of a micro-grid power system.

Author Contributions: Data curation, H.L.; Investigation, H.L., X.W. and J.X.; Methodology, H.L., X.W. and J.X.; Software, H.L.; Supervision, X.W. and J.X.; Writing—original draft, H.L.; Writing—review & editing, H.L.

Funding: This research was funded by the National Natural Science Foundation of China, grant number 61573241 and the Doctorial Innovation Foundation of Shanghai Maritime University, grant number 2016YCX067.

Acknowledgments: The authors gratefully acknowledge the financial support.

Conflicts of Interest: The authors declare no conflict of interest.

References

1. Meng, L.; Sanseverino, E.R.; Luna, A.; Dragicevic, T.; Vasquez, J.C.; Guerrero, J.M. Microgrid supervisory controllers and energy management systems: A literature review. *Renew. Sustain. Energy Rev.* **2016**, *60*, 1263–1273. [CrossRef]
2. Wang, C.S.; Liu, Y.X.; Li, X.L.; Guo, L.; Qiao, L.; Lu, H. Energy management system for stand-alone diesel-wind-biomass microgrid with energy storage system. *Energy* **2016**, *97*, 90–104. [CrossRef]
3. Sharma, S.; Bhattacharjee, S.; Bhattacharya, A. Grey wolf optimisation for optimal sizing of battery energy storage device to minimize operation cost of microgrid. *IET Gen. Transm. Distrib.* **2016**, *10*, 625–637. [CrossRef]
4. Zhang, Y.; Li, Y.W. Energy management strategy for supercapacitor in droop-controlled DC microgrid using virtual impedance. *IEEE Trans. Power Electron.* **2017**, *32*, 2704–2716. [CrossRef]
5. Sharma, R.K.; Mishra, S. Dynamic power management and control of PV PEM fuel cell based standalone AC/DC microgrid using hybrid energy storage. *IEEE Trans. Ind. Appl.* **2017**, *54*, 526–538. [CrossRef]
6. Ippolito, M.G.; Telaretti, E.; Zizzo, G.; Graditi, G. A new device for the control and the connection to the grid of combined RES-based generators and electric storage systems. In Proceedings of the International Conference on Clean Electrical Power, Alghero, Italy, 29 August 2013; pp. 262–267.
7. Pandey, S.K.; Mohanty, S.R.; Kishor, N. A literature survey on load–frequency control for conventional and distribution generation power systems. *Renew. Sustain. Energy Rev.* **2013**, *25*, 318–334. [CrossRef]

8. Mahmoud, M.S.; Hussain, S.A.; Abido, M.A. Modeling and control of microgrid: An overview. *J. Frankl. Inst.* **2014**, *351*, 2822–2859. [CrossRef]
9. Graditi, G.; Ciavarella, R.; Valenti, M. An innovative BESS management for dynamic frequency restoration. In Proceedings of the 2017 IEEE International Conference on Environment and Electrical Engineering and 2017 IEEE Industrial and Commercial Power Systems Europe, Milan, Italy, 6–9 June 2017; pp. 1–5.
10. Graditi, G.; Ciavarella, R.; Valenti, M.; Ferruzzi, G.; Zizzo, G. Frequency stability in microgrid: Control strategies and analysis of BESS aging effects. In Proceedings of the IEEE International Symposium on Power Electronics, Electrical Drives, Automation and Motion, Anacapri, Italy, 22–24 June 2016; pp. 295–299.
11. Ou, T.C.; Hong, C.M. Dynamic operation and control of microgrid hybrid power systems. *Energy* **2014**, *66*, 314–323. [CrossRef]
12. Hong, C.M.; Ou, T.C.; Lu, K.H. Development of intelligent MPPT (maximum power point tracking) control for a grid-connected hybrid power generation system. *Energy* **2013**, *50*, 270–279. [CrossRef]
13. Adhikari, S.; Li, F. Coordinated V-F and P-Q control of solar photovoltaic generators with MPPT and battery storage in microgrids. *IEEE Trans. Smart Grid* **2014**, *5*, 1270–1281. [CrossRef]
14. Ou, T.C.; Su, W.F.; Liu, X.Z. A modified bird-mating optimization with hill-climbing for connection decisions of transformers. *Energies* **2016**, *9*, 671. [CrossRef]
15. Ou, T.C. A novel unsymmetrical faults analysis for microgrid distribution systems. *Int. J. Electr. Power Energy Syst.* **2012**, *43*, 1017–1024. [CrossRef]
16. Ou, T.C. Ground fault current analysis with a direct building algorithm for microgrid distribution. *Int. J. Electr. Power Energy Syst.* **2013**, *53*, 867–875. [CrossRef]
17. Ou, T.C.; Lu, K.H.; Huang, C.J. Improvement of Transient Stability in a Hybrid Power Multi-System Using a Designed NIDC (Novel Intelligent Damping Controller). *Energies* **2017**, *10*, 488. [CrossRef]
18. Sathya, M.R.; Ansari, M.T. Load frequency control using Bat inspired algorithm based dual mode gain scheduling of PI controllers for interconnected power system. *Int. J. Electr. Power Energy Syst.* **2015**, *64*, 365–374. [CrossRef]
19. Sönmez, S.; Ayasun, S. Stability region in the parameter space of PI controller for a single-area load frequency control system with time delay. *IEEE Trans. Power Syst.* **2015**, *31*, 829–830. [CrossRef]
20. Zamani, A.; Barakati, S.M.; Yousofidarmian, S. Design of a fractional order PID controller using GBMO algorithm for load-frequency control with governor saturation consideration. *ISA Trans.* **2016**, *64*, 56–66. [CrossRef] [PubMed]
21. Pradhan, P.C.; Sahu, R.K.; Panda, S. Firefly algorithm optimized fuzzy PID controller for AGC of multi-area multi-source power systems with UPFC and SMES. *Eng. Sci. Technol.* **2016**, *19*, 338–354. [CrossRef]
22. Zhang, S.; Mishra, Y.; Shahidehpour, M. Fuzzy-logic based frequency controller for wind farms augmented with energy storage systems. *IEEE Trans. Power Syst.* **2016**, *31*, 1595–1603. [CrossRef]
23. Mahto, T.; Mukherjee, V. A novel scaling factor based fuzzy logic controller for frequency control of an isolated hybrid power system. *Energy* **2017**, *130*, 339–350. [CrossRef]
24. Vachirasricirikul, S.; Ngamroo, I. Robust LFC in a smart grid with wind power penetration by coordinated V2G control and frequency controller. *IEEE Trans. Smart Grid* **2017**, *5*, 371–380. [CrossRef]
25. Yang, J.; Zeng, Z.; Tang, Y.; Yan, J.; He, H.; Wu, Y. Load frequency control in isolated micro-grid with electrical vehicle based on multivariable generalized predictive theory. *Energies* **2015**, *8*, 2145–2164. [CrossRef]
26. Shankar, R.; Chatterjee, K.; Bhushan, R. Impact of energy storage system on load frequency control for diverse sources of interconnected power system in deregulated power environment. *Int. J. Electr. Power Energy Syst.* **2016**, *79*, 11–26. [CrossRef]
27. Ma, M.; Zhang, C.; Liu, X.; Chen, H. Distributed model predictive load frequency control of the multi-area power system after deregulation. *IEEE Trans. Ind. Electron.* **2017**, *64*, 5129–5139. [CrossRef]
28. Rinaldi, G.; Cucuzzella, M.; Ferrara, A. Third order sliding mode observer-based approach for distributed optimal load frequency control. *IEEE Control Syst. Lett.* **2017**, *1*, 215–220. [CrossRef]
29. Mu, C.; Tang, Y.; He, H. Improved sliding mode design for load frequency control of power system integrated an adaptive learning strategy. *IEEE Trans. Ind. Electron.* **2017**, *64*, 6742–6751. [CrossRef]
30. Fu, C.; Tan, W. Decentralized load frequency control for power systems with communication delays via active disturbance rejection. *IET Gen. Transm. Distrib.* **2017**, *12*, 1397–1403.
31. Zhang, C.K.; Jiang, L.; Wu, Q.H.; He, Y.; Wu, M. Delay-dependent robust load frequency control for time delay power systems. *IEEE Trans. Power Syst.* **2013**, *28*, 2192–2201. [CrossRef]

32. Ning, C. Robust $H\infty$ load-frequency control in interconnected power systems. *IET Control Theory Appl.* **2016**, *10*, 67–75.

33. Han, Y.; Young, P.M.; Jain, A.; Daniel, Z. Robust control for microgrid frequency deviation reduction with attached storage system. *IEEE Trans. Smart Grid* **2015**, *6*, 557–565. [CrossRef]

34. Zhao, J.; Xue, L.; Fu, Y.; Hu, X.G.; Li, F.X. Coordinated microgrid frequency regulation based on DFIG variable coefficient using virtual inertia and primary frequency control. *IEEE Trans. Energy Convers.* **2016**, *31*, 833–845. [CrossRef]

35. Bevrani, H.; Feizi, M.R.; Ataee, S. Robust Frequency Control in an Islanded Microgrid: $H\infty$ and μ-Synthesis Approaches. *IEEE Trans. Smart Grid* **2016**, *7*, 706–717. [CrossRef]

36. Zhu, D.; Hug-Glanzmann, G. Coordination of storage and generation in power system frequency control using an $H\infty$ approach. *IET Gen. Transm. Distrib.* **2013**, *7*, 1263–1271. [CrossRef]

37. Pandey, S.K.; Mohanty, S.R.; Kishor, N.; Catalão, J.P.S. Frequency regulation in hybrid power systems using particle swarm optimization and linear matrix inequalities based robust controller design. *Int. J. Electr. Power Energy Syst.* **2014**, *63*, 887–900. [CrossRef]

38. Sasaki, T.; Kadoya, T.; Enomoto, K. Study on load frequency control using redox flow batteries. *IEEE Trans. Power Syst.* **2004**, *19*, 660–667. [CrossRef]

39. Jayalakshmi, N.S.; Gaonkar, D.N. Performance study of isolated hybrid power system with multiple generation and energy storage units. In Proceedings of the 2011 International Conference on Power and Energy Systems, Chennai, India, 22–24 December 2011; pp. 1–5.

40. Mohanty, S.R.; Kishor, N.; Ray, P.K. Robust H-infinite loop shaping controller based on hybrid PSO and harmonic search for frequency regulation in hybrid distributed generation system. *Int. J. Electr. Power Energy Syst.* **2014**, *60*, 302–316. [CrossRef]

41. Das, S.; Suganthan, P.N. Differential Evolution: A Survey of the State-of-the-Art. *IEEE Trans. Evolut. Comput.* **2011**, *15*, 4–31. [CrossRef]

42. Shousong, H. *Automatic Control Theory*, 4th ed.; Science: Beijing, China, 2001; pp. 201–213.

43. Dawei, G.; Petkov, P.H.; Konstantinov, M.M. *Robust Control Design with MATLAB*; Springer: New York, NY, USA, 2005; pp. 158–161.

Article

Detection Method for Soft Internal Short Circuit in Lithium-Ion Battery Pack by Extracting Open Circuit Voltage of Faulted Cell

Minhwan Seo [1,†], **Taedong Goh** [2,†], **Minjun Park** [1,†] **and Sang Woo Kim** [1,*]

1 Department of Electrical Engineering, Pohang University of Science and Technology, Pohang 37673, Korea; mhseo09@postech.edu (M.S.); parkmj@postech.edu (M.P.)
2 Department of Creative-IT Engineering, Pohang University of Science and Technology, Pohang 37673, Korea; ehd1116@postech.edu
* Correspondence: swkim@postech.edu; Tel.: +82-54-279-2237
† These authors contributed equally to this work.

Received: 8 June 2018; Accepted: 25 June 2018; Published: 27 June 2018

Abstract: Early detection of internal short circuit which is main cause of thermal runaway in a lithium-ion battery is necessary to ensure battery safety for users. As a promising fault index, internal short circuit resistance can directly represent degree of the fault because it describes self-discharge phenomenon caused by the internal short circuit clearly. However, when voltages of individual cells in a lithium-ion battery pack are not provided, the effect of internal short circuit in the battery pack is not readily observed in whole terminal voltage of the pack, leading to difficulty in estimating accurate internal short circuit resistance. In this paper, estimating the resistance with the whole terminal voltages and the load currents of the pack, a detection method for the soft internal short circuit in the pack is proposed. Open circuit voltage of a faulted cell in the pack is extracted to reflect the self-discharge phenomenon obviously; this process yields accurate estimates of the resistance. The proposed method is verified with various soft short conditions in both simulations and experiments. The error of estimated resistance does not exceed 31.2% in the experiment, thereby enabling the battery management system to detect the internal short circuit early.

Keywords: lithium-ion battery pack; soft internal short circuit; model-based fault detection; battery safety; internal short circuit resistance

---

1. Introduction

Lithium-ion batteries are widely used as a power source in electric devices and electric vehicles [1,2], due to their high power density, high energy efficiency and excellent cycle stability [3,4]. The demand for them is expected to rise continuously in the coming years [5,6]. However, safety concerns related to the lithium-ion batteries still remain [7–9] because hazardous incidents such as fire accidents in the Boeing 787-8 aircraft [10] and battery failures in the Samsung Note7 [11] have been frequently reported by media [12]. The main cause of these two events is internal short circuit (ISCr) in the lithium-ion batteries. The ISCr can be caused by manufacturing defects [13,14] and various types of abuse such as overcharge [15,16] and overdischarge [17]. Furthermore, when a magnitude of ISCr resistance (R_{ISCr}) is lower than a particular value [18], a temperature of the battery exceeds a certain point due to the ISCr [19–21]. Then, the battery may experience thermal runaway with a fire and can even explode [22–25]. Therefore, the detection of soft ISCr, which has a large magnitude of the R_{ISCr}, is more necessary than the detection of hard ISCr with a small magnitude of the R_{ISCr} for user safety to prevent the lithium-ion battery from causing hazardous events such as the thermal runaway.

Recently, for these reasons, methods for detecting the ISCr have been suggested [26,27]. When the ISCr happens in the battery cell, terminal voltage of the cell decreases rapidly and then temperature of the cell increases drastically; these two characteristics can be used as pre-determined thresholds for detecting the ISCr [26]. However, to obtain the threshold values, prior ISCr-test with the batteries were required. Besides the threshold-based method, a detection algorithm based on an equivalent circuit model of the battery has been introduced [27]. Using variation of estimated parameters in both the equivalent circuit model and the energy balance equation, the method for detecting the ISCr was verified with large magnitude of the R_{ISCr}, which was larger than 10 Ω [27]; the 10 Ω is considered to be the minimum value which must be detected early before occurrence of the thermal runaway [23,27]. However, this algorithm was verified with only one type of current profile and the similar parameter variation may not be obtained depending on various other current profiles due to absence of the R_{ISCr} in the equivalent circuit model.

The R_{ISCr} is regarded as a promising fault index for detecting the ISCr because it can describe self-discharge phenomenon caused by the ISCr in the equivalent circuit model with R_{ISCr} [28,29] and represent heat generated by the ISCr [30]. Therefore, the ISCr detection methods, which directly use the R_{ISCr} as the fault index, have been introduced [29,31–33]. With measurement data of the self-discharge current and the terminal voltage of the battery with ISCr in the particular experiment configuration, the R_{ISCr} can be calculated correctly [29,31]. However, these experimental methods cannot be used as on-board ISCr detection when the restricted experiment system is not configured in actual application environment. Therefore, after analyzing the self-discharge phenomenon caused by the ISCr, the equation for calculating the R_{ISCr} was derived and then the R_{ISCr} (1~20 Ω) was estimated to detect the ISCr [32]. However, the accuracy of the R_{ISCr} estimated from two different load current profiles was low, because the R_{ISCr} in the equivalent circuit model was not used to estimate open circuit voltage (OCV) of the battery with ISCr. To overcome this error, the previously estimated R_{ISCr} in the model was used to update the model iteratively and to estimate the OCV, and then the next estimated R_{ISCr} (5~50 Ω) became accurate enough to detect the ISCr [33].

For a large capacity and a high power, a lithium-ion battery pack, where many battery cells are connected, is used in actual applications such as electric vehicles and energy storage system for the grid [34,35]. Studies for detecting the ISCr in the battery pack have been suggested [36–38]. The ISCr of the battery pack was detected based on the correlation coefficient of terminal voltages of individual cells [36], and the deviations of both state of charge (SOC) and heat generation power were used to detect the ISCr in the pack [37]. However, these two methods were verified with only the hard ISCr having a magnitude of R_{ISCr} (0.36 Ω for [36] and 0.35~2.4 Ω for [37], respectively), which is small enough to cause the dangerous incidents such as the thermal runaway in the cell [30]. Especially, temperature data of the individual cells were needed to detect the hard ISCr correctly [37]. Magnitude, differential value and fluctuation of estimated model parameters were acquired to detect the ISCr with a wide range of R_{ISCr} (1 ~100 Ω) after estimating the parameters in the mean-difference model of the battery pack [38]. These three detection methods for the battery pack have a common constraint: all terminal voltages of the individual cells in the battery pack must be provided.

If many battery cells are connected in series in the battery pack, many channels with high accuracy and high sample rate for measuring all the individual cell voltages increase the cost of battery management system (BMS). In addition, a data storage unit, needed to save and monitor the measurement data, can lead to increasing the cost of BMS [39]. Furthermore, depending on the applications of the battery pack and the BMS with various topologies, the individual cell voltages may not be provided with high precision and high sample rate and may not be saved due to the cost of the BMS [40,41]. When the data of individual cell voltages are not provided from the BMS, the ISCr detection methods for the pack [36–38] have trouble in deriving the properties of the faulted battery from the battery pack, resulting in problem of detecting the ISCr of the cell in the pack. Therefore, an algorithm for detecting the soft ISCr with load currents and whole terminal voltages of the battery pack is necessary definitely. Moreover, considering data computation, using the whole terminal voltage for diagnosing the ISCr in the battery pack is more efficient than using all individual cell voltages.

This paper proposes a method for detecting the soft ISCr in the lithium-ion battery pack, where normal batteries and a battery with ISCr are connected in series, with the load currents and the whole terminal voltages of the pack. To reflect the effect of ISCr in the battery pack clearly, the OCV of the faulted cell is extracted from the pack with ISCr. The proposed algorithm estimates the SOC of the battery pack with the extended Kalman filter (EKF) to increase accuracy of the pack SOC. Then, the SOC of the normal cell is obtained with the Coulomb counting method and a stable initial value, which is determined from the estimated SOC of the pack. Using the SOC estimates of the pack and the normal cell, the OCV of the battery cell with ISCr in the pack can be obtained; this shows the self-discharge phenomenon caused by the ISCr obviously. As a result, the R_{ISCr} can be estimated accurately, and the soft ISCr in the battery pack can be detected. To verify the proposed method, various soft ISCr cases were configured for simulation and experiment, and two load current profiles: dynamic stress test (DST) and urban dynamometer driving schedules (UDDS) were used.

The remainder of this paper is organized as follows: the proposed algorithm is carried out in Section 2; the configurations of simulation and experiment are introduced in Section 3; the results of the proposed algorithm and the discussions are presented in Section 4; the conclusions of this study and the outline of future work are summarized in Section 5.

2. Method Description

2.1. Overall Scheme for ISCr Detection Algorithm

To detect the soft ISCr in the battery pack, the proposed method estimates the R_{ISCr} of the faulted cell using the whole terminal voltages and the load currents of the pack. As a fault index, the estimated R_{ISCr} can directly inform the user of degree of the ISCr in the pack. If the soft ISCr having a large magnitude of the R_{ISCr} is detected, the BMS can give enough time to cope with the ISCr fault. The overall scheme of the proposed algorithm is depicted in Figure 1, which comprises of four estimation steps: estimating pack SOC (SOC_p); normal cell SOC (SOC_n); faulted cell SOC (SOC_f) and R_{ISCr}.

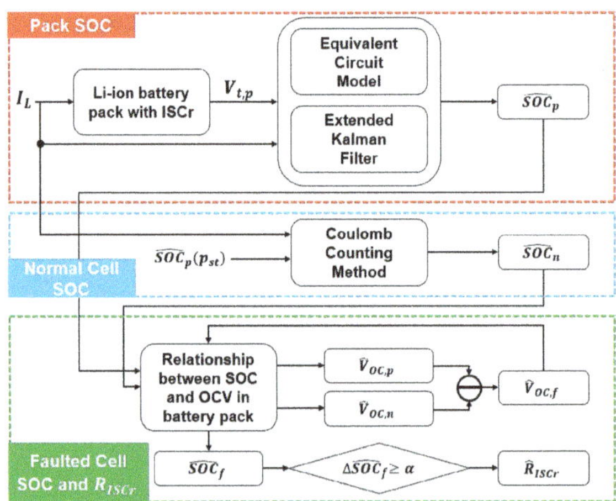

Figure 1. The scheme of the proposed algorithm.

When the ISCr occurs in a single battery cell, abnormal properties such as the decrease in terminal voltage and the increase in battery temperature are easily observed [42,43]. Hence, measured terminal voltage, current and temperature of the battery cell can be used to detect the ISCr. However, when an

ISCr occurs in the battery pack, which only provides the load currents and the whole terminal voltages of the battery pack, the decrease in the whole terminal voltage caused by the ISCr in the battery pack is not obeserved conspicuously. Thus, additional decrease in the OCV_f, induced by the self-discharge current in the faulted cell, should be extracted to ensure the high accuracy of R_{ISCr} estimates. First of all, using the equivalent circuit model of the battery pack with ISCr and the EKF algorithm, the SOC_p is estimated. As a second step, the SOC_n can be calculated by the Coulomb counting method with a initial value of the SOC_n which is obtained with the estimated SOC_p. Then, using the estimated SOCs of both the battery pack and the normal cells, the OCV of the faulted cell OCV_f and the SOC_f are obtained at the third and fourth estimation steps. Once the difference between the initial and present SOC_f estimates is more than a certain value, which is determined as 0.1 (10% of the total capacity of the cell) and discussed in Section 4.6, the R_{ISCr} of the faulted cell in the battery pack can be estimated and used to detect the soft ISCr in the pack as the fault index.

2.2. Equivalent Circuit Model of Battery Pack with ISCr

In Figure 2, the battery pack consists of several normal battery cells and one faulted battery cell, and the whole terminal voltage $V_{t,p}$ and the load current I_L are described, where m is the number of cells connected in series in the pack. The normal cell is represented by a simple equivalent circuit model [29] composed of the OCV (V_{OC}) and an internal resistance R. The sum of V_{OC}s and Rs of normal cells described in Equations (1) and (2) is used to express the model of normal cells with discretization step in Equation (3), and the sum of the terminal voltages of the normal cells $V_{t,n}$ can be induced by Ohm's law in Equation (3) [44].

$$V_{OC,n}(k) = \sum_{j=1}^{m-1} V_{OC,j}(k), \tag{1}$$

$$R_n = \sum_{j=1}^{m-1} R_j, \tag{2}$$

$$V_{t,n}(k) = V_{OC,n}(k) + R_n I_L(k). \tag{3}$$

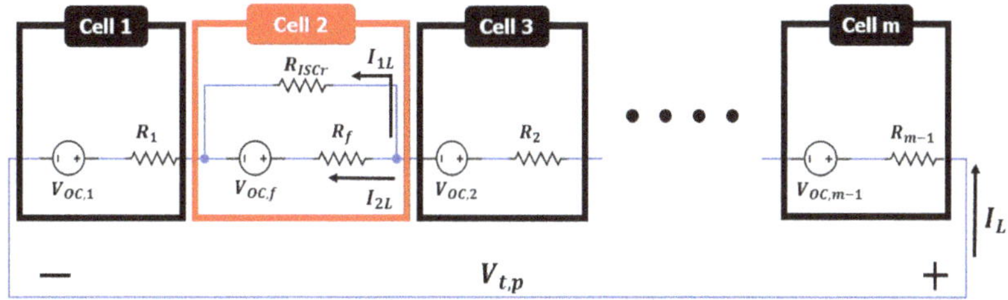

Figure 2. Equivalent circuit model of the lithium-ion battery pack with internal short circuit (ISCr).

The cell with ISCr is represented by the simple model with the R_{ISCr}, which is connected with the model of the normal cell in parallel [29,38]. The subscript f is used particularly in parameters associated with the faulted cell. The I_L is divided into two currents which are the self-discharge current I_{1L} flowing through the R_{ISCr} and residual current I_{2L} (Equation (4)), and $V_{t,f}$ is the terminal voltage of the faulted cell. The faulted cell model is described in Equations (5) and (6) [33,38]. To represent the model of the battery pack with ISCr, the $V_{t,p}$ is obtained by adding the $V_{t,n}$ and the $V_{t,f}$, as shown in Equation (7).

$$I_L(k) = I_{1L}(k) + I_{2L}(k), \tag{4}$$

$$V_{t,f}(k) = V_{OC,f}(k) + R_f I_{2L}(k), \tag{5}$$

$$V_{t,f}(k) = \frac{R_{ISCr}}{R_f + R_{ISCr}} V_{OC,f}(k) + \frac{R_{ISCr}R_f}{R_f + R_{ISCr}} I_L(k), \tag{6}$$

$$V_{t,p}(k) = V_{OC,n}(k) + \frac{R_{ISCr}}{R_f + R_{ISCr}} V_{OC,f}(k) + (R_n + \frac{R_{ISCr}R_f}{R_f + R_{ISCr}}) I_L(k). \tag{7}$$

2.3. Estimation of Pack SOC Using EKF

The EKF algorithm is a common method to estimate accurate SOC because the estimates are not affected by measurement noise dominantly due to properties of the battery system reflected in the state space model [45–47]. In this paper, the EKF algorithm is used to estimate the SOC_p correctly. After assuming that $R_{ISCr} \gg R_f$ in the model of battery pack with ISCr (Equation (7)), the simplified $V_{t,p}$ can be expressed with battery pack OCV ($V_{OC,p}$) and whole internal resistance R_p which is the sum of R_n and R_f in Equation (8). The recursive least squares (RLS) algorithm is used to identify the parameter of the normal battery model [48]. Using the RLS algorithm, the R_p is obtained from the $V_{t,p}$ and the I_L of the normal battery pack. Then, the estimated R_p is used to configure the state space model of battery pack with ISCr in the EKF algorithm. When the I_L is positive during battery pack charging and negative during discharging, in recursive discrete-time form, the SOC_p is calculated with the Coulomb counting method [44] in Equation (9), where η is the charging and discharging efficiency, Δt is the sample period and C_n is the nominal capacity of the normal battery pack. In this study, the charging and discharging efficiency is defined as 1.

$$V_{t,p}(k) = V_{OC,p}(k) + R_p I_L(k), \tag{8}$$

$$SOC_p(k+1) = SOC_p(k) + \frac{\eta \Delta t}{C_n} I_L(k). \tag{9}$$

To estimate the SOC_p using the EKF algorithm, the corresponding equations are listed in Table 1, where x_k is the SOC_p, y_k is the $V_{t,p}$, u_k is the I_L, k is the sample index, and w_k and v_k are the zero mean Gaussian noise with covariance of Q and T.

Table 1. Essential equations for estimating the SOC_p in the extended Kalman filter (EKF) [45–47].

Description	Equation	Step
State space model	$x_{k+1} = f(x_k, u_k) + w_k = x_k + \frac{\Delta t}{C_n} u_k + w_k$ $y_k = g(x_{k+1}, u_k) + v_k = V_{OC,p}(x_{k+1}) + R_p u_k + v_k$	
State transition matrix	$A_k = \left. \frac{\partial f}{\partial x} \right|_{x=x_k} = 1$	
Observation matrix	$C_k = \left. \frac{\partial g}{\partial x} \right|_{x=x_k} = \left. \frac{dV_{OC,p}}{dSOC_p} \right|_{SOC_p = \widehat{SOC}_{p_k}}$	
Initial assumed values	$\hat{x}_0^+ = E[x_0] = x_0$ $P_0^+ = E(x_0 - \hat{x}_0^+)(x_0 - \hat{x}_0^+)^T = P_0$ $Q = Q_0,\ T = T_0$	For $k = 0$
Error covariance matrix Kalman gain	$\hat{x}_k^- = f(\hat{x}_{k-1}^+, u_k)$ $P_k^- = A_{k-1}P_{k-1}^+ A_{k-1}^T + Q$ $L_k = P_k^-(C_k)^T[C_k P_k^- C_k^T + T]^{-1}$ $\hat{x}_k^+ = \hat{x}_k^- + L_k[y_k - g(\hat{x}_k^-, u_k)]$ $P_k^+ = (I - L_k C_k)P_k^-$	For $k = 1, 2, 3, \cdots, \infty$

2.4. Estimation of Normal Cell SOC

After starting to estimate the SOC_p, the imprecise SOC_p can be obtained for several seconds because of the incorrectly assumed initial value of the SOC_p. Thus, the estimated SOC_p obtained from the stable point p_{st}, where P_k is lower than a certain small value, is used to estimate the SOC_n, and the p_{st} is the sample index. By assuming that the $SOC_p(p_{st})$ equals to the $SOC_n(p_{st})$ at the stable point, the next SOC_n can be calculated with the I_L and the Coulomb counting method described in Equation (9).

2.5. Estimation of OCV and SOC for Faulted Cell

The relationship between OCV and SOC of the normal battery pack (Figure 3) obtained from the prior test [49] is essential for conducting the proposed algorithm. The fully charged battery pack is rested for 3600 s to obtain the OCV, which is equal to the terminal voltage, at 100% SOC. Then the battery pack is discharged with 0.5C (1 A) for 720 s to set 90% SOC, where the C (C-rate) is defined as the charge and discharge current of the battery, and then rested for 3600 s to get the OCV at 90% SOC. By repeating the process, the relationship between OCV and SOC can be obtain. Using the relationship, the OCV_p and the OCV_n can be acquired and are expressed in Equations (10) and (11), where h is the function representing the relationship. By subtracting OCV_ns from the OCV_p, the OCV_f is calculated. In sequence, the SOC_f is obtained from the inverse function h^{-1} of h (Equation (12)).

$$\hat{V}_{OC,p} = h(\widehat{SOC}_p), \tag{10}$$

$$\hat{V}_{OC,n} = h(\widehat{SOC}_n), \tag{11}$$

$$\widehat{SOC}_f = h^{-1}(\hat{V}_{OC,f}). \tag{12}$$

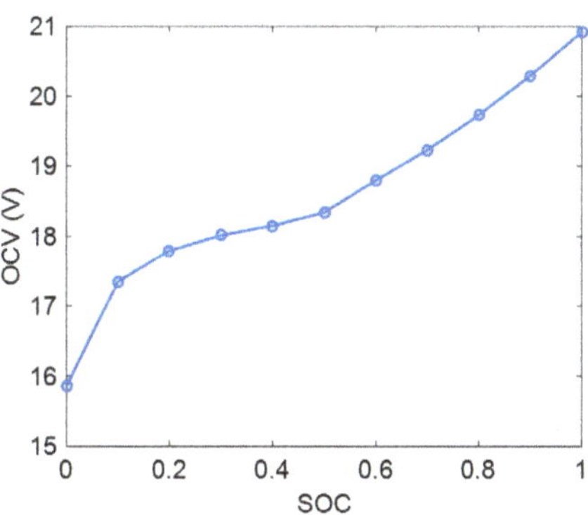

Figure 3. Relationship between open circuit voltage (OCV) and state of charge (SOC) of the normal battery pack.

In the process of using the h^{-1} to obtain the SOC_f, the accuracy of SOC_f is dependent on the slope from the relationship between OCV and SOC; i.e., in the range from 100% SOC to 50% SOC with the steep slope from the relationship a small SOC_f error is caused by error of the $V_{OC,f}$, while a large SOC_f error can be induced by the same error of the $V_{OC,f}$ in the range from 50% SOC to 10% SOC with the gradual slope. To avoid this problem, the data of $V_{t,p}$ and I_L were used to conduct the proposed algorithm until the estimated $SOC_f \leq 0.55$.

2.6. Calculation of R_{ISCr}

Using the self-discharge phenomenon [32,33], the R_{ISCr} can be estimated with the estimated SOC_f. To formulate Equation (16) for calculating the R_{ISCr}, the Coulomb counting method is used with respect to the SOC_f. $I_L - I_{1L}$ is used in Equation (13), because the faulted cell is discharged from the I_{2L} instead of the I_L.

$$SOC_f(k) = SOC_f(0) + \frac{\Delta t}{C_n} \sum_{n=1}^{k} [I_L(n) - I_{1L}(n)] \tag{13}$$

To delete the unknown terms, $SOC_f(0)$ and I_{1L}, the $k-1$th equation in Equation (13) is subtracted from the kth equation in Equation (13), and then the I_{1L} is replaced with $\frac{V_{t,f}}{R_{ISCr}}$. In Equation (14), $\frac{V_{t,p}}{m}$ is used instead of the $V_{t,f}$ because the $V_{t,f}$ is not provided measurement data from the battery pack with ISCr.

$$SOC_f(k) - SOC_f(k-1) = \frac{\Delta t}{C_n} I_L(k) - \frac{\Delta t}{C_n} \frac{V_{t,p}(k)/m}{R_{ISCr}} \tag{14}$$

The estimated R_{ISCr} from Equation (14) using a short interval between k and $k-1$ is vulnerable to errors of the estimated SOC_f because the slight variation of the estimated SOC_f in a short interval cannot reflect dominant self-discharge phenomenon from the ISCr. Therefore, the interval must be extended by adding the $k-1, k-2, k-3, \cdots, p_{st}+1$th equations in Equation (14) to the kth equations in Equation (14) ($p_{st}+1 < k$); i.e., the R_{ISCr} is estimated once the difference between the initial estimated SOC ($SOC_f(p_{st})$) and present estimated SOC ($SOC_f(k)$) is more than 0.1, which is 10% of the total capacity of the cell.

$$SOC_f(k) - SOC_f(p_{st}) = \frac{\Delta t}{C_n} \sum_{n=p_{st}+1}^{k} I_L(n) - \frac{\Delta t}{C_n} \frac{1}{mR_{ISCr}} \sum_{n=p_{st}+1}^{k} V_{t,p}(n) \tag{15}$$

The self-discharge phenomenon from the ISCr can be explained with the last term of Equation (15). For the normal battery cell with $R_{ISCr} \cong \infty$, the last term can be approximated to zero. However, when the ISCr occurs in the cell, the non-zero last term represents an additional decline in the SOC of the faulted cell due to the self-discharge current. Consequently, the estimated R_{ISCr} ($\widehat{R}_{ISCr}$) can be obtained with Equation (16) after Equation (15) is rearranged.

$$\widehat{R}_{ISCr} = \frac{\frac{\Delta t}{C_n} \frac{1}{m} \sum_{n=p_{st}+1}^{k} V_{t,p}(n)}{[\widehat{SOC}_f(p_{st}) - \widehat{SOC}_f(k)] + \frac{\Delta t}{C_n} \sum_{n=p_{st}+1}^{k} I_L(n)} \tag{16}$$

2.7. Parallel Processing of Proposed Algorithm

Once the $SOC_f(p_{st})$ is obtained, the $SOC_f(p_{st})$ is used in Equation (16) to calculate the R_{ISCr} continuously. If error exists in the $SOC_f(p_{st})$, the error can affect the accuracy of R_{ISCr}. Therefore, it is necessary to estimate R_{ISCr}s from the various p_{st} positions and apply these estimated R_{ISCr}s to the ISCr fault index. Based on the estimates of $SOC_f(p_{st})$, if the difference between $SOC_f(p_{st})$ and $SOC_f(k)$ is more than or equal to $0.01, 0.02, 0.03, \cdots$, the proposed methods are carried out sequentially and implemented in parallel with previously executed method to diversify the stable point of estimated SOC of the faulted cell. In this paper, four proposed methods were executed additionally in parallel.

3. Simulation and Experiment

3.1. Simulation Configuration

The simulation model of the battery pack with ISCr was configured with MATLAB/Simulink (MATLAB R2017b, MathWorks, Natick, MA, USA) [50]. In both simulation and experiment configuration, the battery pack was composed of four normal battery cells and a faulted battery cell with ISCr. A first-order RC model [51], where the RC represents a parallel resistor-capacitor sub-circuit, was utilized to describe the normal cell in the battery pack. The parameters for the first-order RC model were estimated with the RLS algorithm [48] and the experimental data of terminal voltage and load current, measured when the normal cell was discharged by load current profiles. The normal cell is same with the cell in the battery pack of the experimental environment. In this study, two load current profiles such as the DST and the UDDS were used to verify the proposed method in simulation and experiment. Prior characteristic tests for obtaining the capacity and the relationship between OCV and SOC of the normal cell were conducted, and these two data were also used for the simulation configuration. The simulation model of the ISCr-faulted cell was represented by connecting the R_{ISCr} with the normal model in parallel. In this study, various resistance values such as 50 Ω, 30 Ω, 20 Ω, 10 Ω and 5 Ω were used to represent diverse ISCr fault conditions.

3.2. Experimental Configuration

Figure 4 shows the experimental set-up for the ISCr tests with the battery packs. Two identical battery packs, A and B, were used to acquire the experimental data. The battery packs were tested in a thermal chamber, and the ambient temperature was maintained at 25 $\pm$ 1 °C. The important specifications of the cell are shown in Table 2. The prior tests for battery packs were conducted to obtain the capacities and the relationship between OCV and SOC. Based on the nominal capacity in Table 2, for capacity test, the packs were charged with the constant-current constant-voltage (CC-CV) protocol. For all experiments including the prior tests and the ISCr fault tests, when the batteries were charged, the CC-CV protocol was used. Charge-current was 0.5C (1 A) with upper cutoff-voltage 4.2 V in the CC mode, and cutoff-current was 0.05C (0.1 A) in the CV mode. Then, the packs were discharged with 0.5C as CC discharging [52]. The discharged capacities were regarded as true capacities. The measured true capacities of the battery packs, 2.1974 Ah and 2.1949 Ah for pack A and B, respectively, were used for acquiring the correct relationship between OCV and SOC. To check the distribution of capacities of cells in the pack, the capacity test was conducted for individual cells. As a result, the mean and standard deviation are 2.1944 Ah and 0.0062, respectively, for pack A; and 2.1959 Ah and 0.0071, respectively, for pack B.

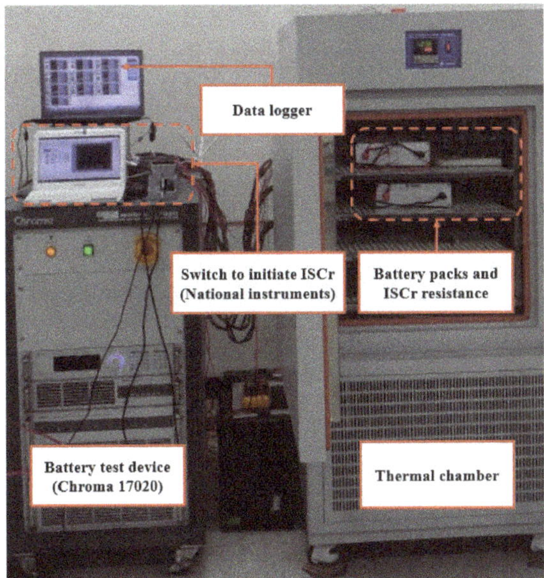

Figure 4. Experimental configuration for ISCr tests.

The experiments of the battery packs were conducted using a battery test device (Regenerative Battery Pack Test System 17020, Chroma, Taoyuan, Taiwan) with the sample period of 0.1 s. The five 10 Ω resistances, which have $\pm$5% tolerance, were used to make various true R_{ISCr}s. Their measured true values were 49.91 Ω, 29.98 Ω, 20.00 Ω, 10.02 Ω and 5.00 Ω, and were used to calculate relative errors of the fault index. These resistances were connected with one of the cells in the pack in parallel to represent the ISCr, and a switch was used to initiate the ISCr faults in the packs when the load current profiles were applied to the packs. For the ISCr experiments, the initial SOC of the pack was set to 90% after the pack was charged and rested for 3600 s. Subsequently, the DST current profile was used to discharge the pack A until its SOC reaches 10% of its total capacity to prevent the pack from being over-discharged, while the UDDS current profile was used for the pack B.

Table 2. Tested battery.

Specification Parameters	Values
Model	INR 18650-20R
Type	LiNiCoMnO$_2$
Dimension	$\varnothing$18.33 $\times$ 64.85 mm
Mass	45.0 g
Operating temperature	$-20\sim+75\,^\circ$C
Nominal voltage	3.6 V
Charge cut-off voltage	4.2 V
Discharge cut-off voltage	2.5 V
Nominal capacity	2.0 Ah

4. Results and Discussions

4.1. Terminal Voltages of Pack

Figure 5 depicts the terminal voltages of battery pack A and B in the experiments depending on the magnitudes of R_{ISCr}. The terminal voltages of the packs were measured until the pack SOC reached 10%. As the magnitude of R_{ISCr} is small, the terminal voltages decreased rapidly compared with the voltages of the normal battery pack, leading to rapid termination of experiments for the battery packs with ISCr. The additional decline in terminal voltages caused by the self-discharge phenomenon of the ISCr was not observed clearly. In addition, because the terminal voltages of the packs were affected by waveforms of load current profiles in common, the terminal voltages fluctuated and the slight difference of voltages between the normal pack and the pack with ISCr was not monitored readily. Therefore, it was difficult to detect the soft ISCr directly with only the measurment data of terminal voltages of the packs.

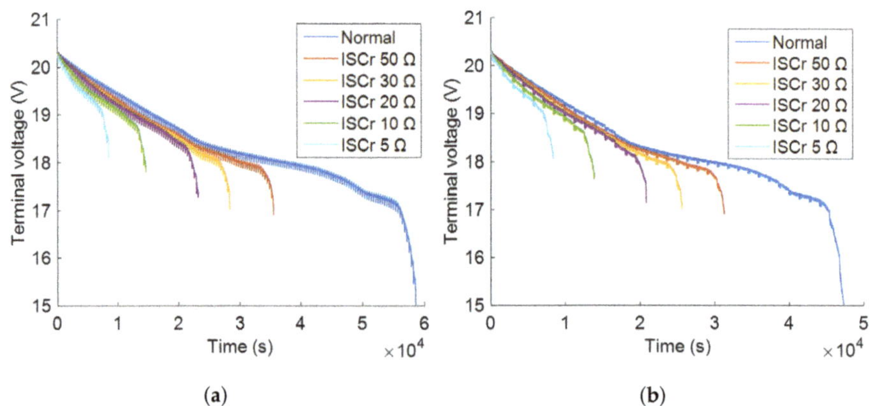

Figure 5. Terminal voltages of battery packs with different R_{ISCr}s in experiment with (**a**) dynamic stress test (DST) current profile, and (**b**) urban dynamometer driving schedules (UDDS) current profile.

4.2. Estimation Results of SOCs for Pack and Faulted Cell

As illustrated in Figure 6a, the initial values of SOC_p estimates were 0.5 due to the initially assumed value of the x_0, and the estimated SOC_ps of the battery pack with ISCr decreased faster than that of the normal battery pack. However, the difference between the normal SOC_p and others with various ISCr fault cases did not reflect significantly the effect of self-discharge from the ISCr in the battery pack, consequently leading to inaccurate estimates of the R_{ISCr}. Thus, deriving the properties of the faulted cell from the pack with ISCr was necessary to observe the effect of ISCr clearly. The estimated SOC_f is described in Figure 6b with the specific SOC_f range from 90% SOC to 55% SOC. There were no estimated values of SOC_f before the p_{st} was reached, because the SOC_n was calculated after the p_{st} was reached. Please note that by extracting the $V_{OC,f}$ with these $V_{OC,n}$ and $V_{OC,p}$ from the pack with ISCr, the effect of the self-discharge phenomenon, which is not dominantly observed in the SOC_p (Figure 6a), becomes enlarged noticeably in the SOC_f (Figure 6b).

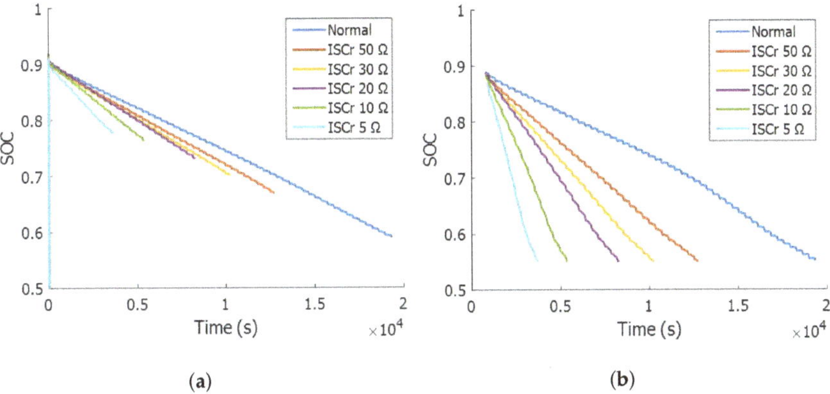

Figure 6. Estimated SOCs with different R_{ISCr}s in experiment with DST current profile: (**a**) battery pack and (**b**) faulted battery cell.

4.3. Estimated R_{ISCr}s from Parallel Processing

When the battery pack A was connected with true R_{ISCr} (49.91 Ω) and discharged with the DST current profile, the R_{ISCr}s were estimated from different four stable points of estimated SOC of the faulted cell ($SOC_f(p_{st})$) using the parallel processing (Figure 7). Although the four proposed methods were executed additionally in parallel with the firstly implemented method to diversify the $SOC_f(p_{st})$s, the number of $SOC_f(p_{st})$s can be different in accordance with the ISCr fault conditions in the pack; i.e., even though the four proposed algorithms are added sequentially in all ISCr fault cases, new $SOC_f(p_{st})$s may not be extracted because of the condition, where the $SOC_f(p_{st})$ was obtained if the P_k was lower than the certain small value (1.4×10^{-6}) in Section 2.4. In the case of the experiment with the true R_{ISCr} (49.91 Ω), the 1st stable point were obtained from the firstly executed method, while the three stable points were extracted from the the the four added algorithms. Figure 7 shows the slightly different $\widehat{R}_{ISCr}$s for the true R_{ISCr} (49.91 Ω). Although the $\widehat{R}_{ISCr,4}$ obtained from the 4th stable point was most accurate among them in this case, the order of accuracy of $\widehat{R}_{ISCr}$s from the diverse stable points was changed depending on various ISCr fault cases. Therefore, to reflect all $\widehat{R}_{ISCr}$s in an ISCr fault condition, mean value ($\overline{R}_{ISCr}$) of them was used as the fault index in simulation and experiment results.

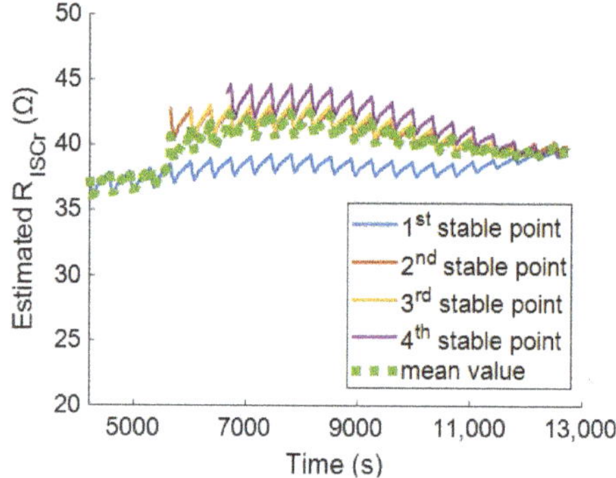

Figure 7. Estimated R_{ISCr}s from different stable points in experiment with true R_{ISCr} 49.91 Ω and DST current profile.

4.4. Estimation Results of R_{ISCr} in Simulation

Figure 8a shows the $\overline{R}_{ISCr}$s with different ISCr fault conditions when the DST current profile was applied to the pack with ISCr, and Figure 8b depicts the $\overline{R}_{ISCr}$s obtained from the UDDS current profile. The $\overline{R}_{ISCr}$s in the specific SOC_f range from 90% SOC to 55% SOC are described in Figure 8. Depending on the various ISCr fault cases, the $SOC_f(p_{st})$ and the 1st stable point, where the $\widehat{R}_{ISCr,1}$ starts to be obtained, were different. Thus, the different start points for estimating the R_{ISCr}s are aligned to zero in Figure 8, and the time of x-axis represents the total time used to obtain the $\overline{R}_{ISCr}$ with the $\widehat{R}_{ISCr}$s. Estimation results from two different current profiles were similar, and the values of $\overline{R}_{ISCr}$ slightly fluctuated because they were affected by waveforms of the current profiles.

The R_{ISCr} in Equation (16) is estimated with difference between the $SOC_f(p_{st})$ and the $SOC_f(k)$, which reflects the self-discharge phenomenon caused by the ISCr. At the same time, to discharge the pack, the difference between the $SOC_f(p_{st})$ and the $SOC_f(k)$ becomes large as the magnitude of R_{ISCr} is small (Figure 6b). In addition, the large difference can be obtained from long discharge time. The large difference between the $SOC_f(p_{st})$ and the $SOC_f(k)$ is insensitive to errors of estimated SOC_f, leading to accurate estimates of the R_{ISCr}. In cases of ISCr 50 Ω, 30 Ω and 20 Ω, compared to cases of ISCr 10 Ω and 5 Ω, large errors of the $\overline{R}_{ISCr}$, caused by errors of estimated SOC_f, occur in the early stage for estimating the R_{ISCr} because of subtle difference between the $SOC_f(p_{st})$ and the $SOC_f(k)$ in short discharge time. However, the $\overline{R}_{ISCr}$s in cases of ISCr 50 Ω, 30 Ω and 20 Ω gradually approach true R_{ISCr}s because the difference between the $SOC_f(p_{st})$ and the $SOC_f(k)$ becomes large gradually as the pack is discharged, resulting in the dominant effect of the self-discharge phenomenon compared with that of the estimation errors of SOC_f. It is also the reason that the $\overline{R}_{ISCr}$ is used as the fault index, which was calculated with the $\widehat{R}_{ISCr}$s for the different stable points in the parallel processing considering errors caused by the position of $SOC_f(p_{st})$. Although the large difference was obtained for each cases ISCr 10 Ω and 5 Ω, the $\overline{R}_{ISCr}$ still had errors, because using the $\frac{V_{t,p}}{m}$ instead of the $V_{t,f}$ in Equation (16) greatly affected the R_{ISCr} estimates compared to the cases of ISCr 50 Ω, 30 Ω and 20 Ω; i.e., while the $\frac{V_{t,p}}{m}$ was similar with the $V_{t,f}$ for ISCr 50 Ω, 30 Ω and 20 Ω, the difference between the $\frac{V_{t,p}}{m}$ and the $V_{t,f}$ became large in cases of ISCr 10 Ω and 5 Ω, leading to large errors of the $\overline{R}_{ISCr}$.

To evaluate the accuracy of the $\overline{R}_{ISCr}$, relative errors of the estimates were calculated with Equation (17). The relative errors of the final values in $\overline{R}_{ISCr}$s (final relative error) with the various ISCr fault conditions are shown in Table 3. It should be noted that the final relative errors are less than or equal to 10% except for the ISCr 5 Ω case. Even though the final relative errors of ISCr 5 Ω in the DST and the UDDS were about 26%, the ISCr fault (5 Ω) can be detected with the $\overline{R}_{ISCr}$. In addition, although the magnitude of R_{ISCr} (10 Ω) is regarded as the minimum value which must be detected early [23,27], the R_{ISCr} (5 Ω) in the cell cannot sufficiently increase the temperature of the battery which is too low to cause the thermal runaway [30,37]. The reason that the magnitude of the R_{ISCr} is 5 Ω and more for verification is to show that the proposed algorithm can be applied to detection of the soft ISCr. If the proposed method detects various soft ISCr faults, we can conclude that there is sufficient time to provide against the thermal runaway in the battery pack.

$$\text{Relative error} = \frac{\mid R_{ISCr} - \overline{R}_{ISCr} \mid}{R_{ISCr}} \times 100\% \tag{17}$$

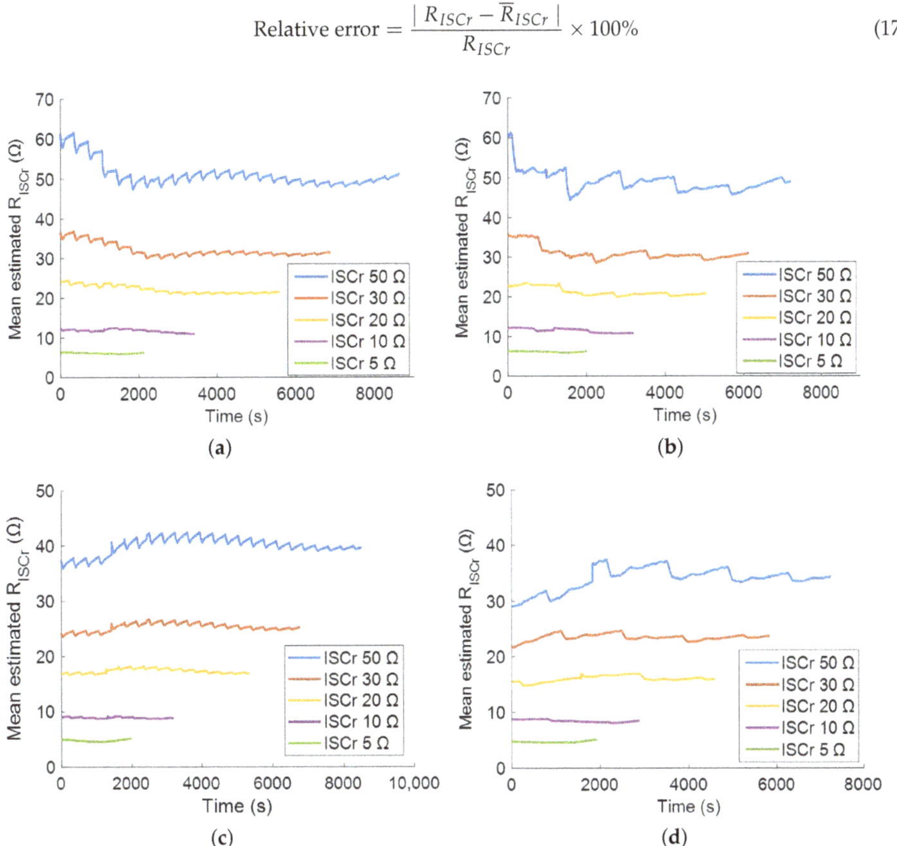

Figure 8. Estimated $\overline{R}_{ISCr}$s from the various ISCr fault cases in simulations: (**a**) DST current profile and (**b**) UDDS current profile and in experiments: (**c**) DST current profile and (**d**) UDDS current profile.

Table 3. Final relative errors (%) of $\overline{R}_{ISCr}$ in simulation depending on the ISCr faults.

Discharge Condition	True ISCr Resistance				
	5 Ω	10 Ω	20 Ω	30 Ω	50 Ω
DST	26.2	9.9	7.9	4.1	2.1
UDDS	25.4	9.0	4.4	2.9	1.7

4.5. Estimation Results of R_{ISCr} in Experiment

Figure 8c,d show the estimation results of the $\overline{R}_{ISCr}$ in the experiment with the DST and the UDDS current profiles. The tendency of the obtained $\overline{R}_{ISCr}$ in experimental results was similar to that of the simulation due to the reasons explained in Section 4.4. In addition, the simplified model of the battery pack with ISCr, which was induced by assuming $R_{ISCr} \gg R_f$ (Equation (8)), was validated, because all estimated R_{ISCr}s were indeed much greater than the R_f in both simulation and experiment.

In particular, the main difference between simulation results and experimental results was that the relative errors in the experiment increased (Table 4) because in all ISCr faults cases, the $\overline{R}_{ISCr}$s in the experiment were generally under-estimated compared to those of the simulation. Contrary to the configuration of the simulation, the characteristics of individual cells in the experiment, such as capacity, internal resistance, and relationship between OCV and SOC, were not identical. Due to both the model difference and measurement noise, the errors of SOC_p estimates of the experiment increased compared to that of the simulation, leading to large errors of the estimated SOC_f in the experiment. This large errors of SOC_f estimates increased the values of the denominator in Equation (16); this was main cause to incur the under-estimation of $\overline{R}_{ISCr}$ in the experiment. Meanwhile, in case of ISCr 5 Ω, the final relative errors in the experiments decreased compared to the results of simulation, because the $\overline{R}_{ISCr}$s for ISCr 5 Ω with large relative errors in the simulation became close to the true value of R_{ISCr} by the under-estimation in the experiment. Although the relative errors of the experiment increased compared to those of the simulation, the obtained $\overline{R}_{ISCr}$ was accurate enough to be used as the fault index to detect the soft ISCr before the thermal runaway occurs in the battery pack, and to classify the various ISCr fault conditions.

Table 4. Final relative errors (%) of $\overline{R}_{ISCr}$ in experiment depending on the ISCr faults.

Discharge Condition	True ISCr Resistance				
	5.0 Ω	10.02 Ω	20.0 Ω	29.98 Ω	49.91 Ω
DST	3.3	11.8	15.1	15.8	20.7
UDDS	2.2	15.4	20.1	20.7	31.2

4.6. Other Discussions

For the normal battery pack, the self-discharge current I_{1L} cannot flow through R_{ISCr} and the faulted cell is discharged by the load current I_L. Therefore, the SOC_f is represented as Equation (18) with the I_L, and the equation for obtaining the $\widehat{R}_{ISCr}$ for the normal battery pack ($\widehat{R}_{ISCr,n}$) can simply be expressed with both the measured $V_{t,p}$ and the ϵ (Equation (19)) which is the difference value between the estimated SOC_f and the true SOC_f.

$$SOC_f(k) = SOC_f(p_{st}) + \frac{\Delta t}{C_n} \sum_{n=p_{st}+1}^{k} I_L(n) \tag{18}$$

$$\widehat{R}_{ISCr,n} = \frac{\frac{\Delta t}{C_n} \frac{1}{m} \sum_{n=p_{st}+1}^{k} V_{t,p}(n)}{\epsilon} \tag{19}$$

When the SOC_f is estimated with the DST load currents and the whole terminal voltages of the normal battery pack in the specific region, the maximum value and relative error of the ϵ were 0.0197 and 3.6%, respectively, for the simulation; and were 0.0502 and 8.4%, respectively, for the experiment. Actually, in Section 4.5 as the reason for the under-estimation of $\overline{R}_{ISCr}$ in the experiment, it was checked that the error of estimated SOC_f for the experiment was larger than it for the simulation. These two maximum errors were used to calculate the $\widehat{R}_{ISCr,n}$ for the normal cases in both simulation and experiment. Due to the non-zero value of ϵ, the obtained $\widehat{R}_{ISCr,n}$ monotonically increased. In Figure 9, the dotted lines show $\widehat{R}_{ISCr,n}$s of the normal battery pack in the specific region and represent reliable maximum values of $\overline{R}_{ISCr}$s for all ISCr fault cases. If the obtained $\overline{R}_{ISCr}$ exists in the region above the dotted line, the $\overline{R}_{ISCr}$ are determined as unreliable estimation values, while region under the dotted line is defined as the reliable estimation region of the $\overline{R}_{ISCr}$.

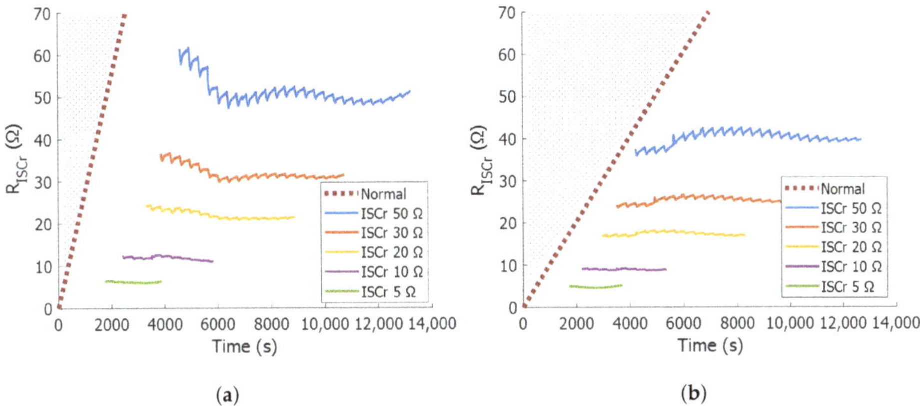

Figure 9. Reliable maximum $\widehat{R}_{ISCr,n}$ in simulation and experiment with normal battery pack and DST current profile: (**a**) simulation and (**b**) experiment.

From the stable point p_{st}, sufficient time to obtain reliable $\widehat{R}_{ISCr}$ is necessary. If the $\widehat{R}_{ISCr}$ is obtained when the k is very close to the p_{st} in Equation (16), the $\widehat{R}_{ISCr}$ can be located in the region above the dotted line, because the values of the dotted line are small in the early stage shown in Figure 9 and the obtained $\widehat{R}_{ISCr}$ for ISCr fault cases can be affected by both the estimation errors of SOC_f and measurement noise in the $V_{t,p}$ and the I_L. Thus, in the proposed method, the R_{ISCr} was estimated once the difference between the $SOC_f(p_{st})$ and the $SOC_f(k)$ was more than or equal to 0.1; this condition guaranteed that the $\widehat{R}_{ISCr}$s in various ISCr faults of simulation and experiment exist in the reliable estimation region.

When a battery pack is manufactured with used lithium-ion cells, a variation in the characteristics of individual cells in the pack becomes large [41], resulting in large error of R_{ISCr} estimates. However, the battery pack made with used cells can be operated with balanced voltage and SOC due to the proper screening process in configuring the battery pack [53]. Therefore, the proposed method can be applied in both fresh and reused battery pack.

If an ambient temperature in operation environment of the battery varies and the battery model which does not reflect thermal properties of the battery is used to estimate the SOC, errors of estimated SOC become large [54–57], leading to large error of R_{ISCr} estimates. Although in this study the proposed method focused on detecting the soft ISCr at constant temperature, depending on real applications the ambient temperature can be changed [58]. Therefore, detection of ISCr in the battery under varing ambient temperature is a maningful and interesting subject of reaserch as a future work.

5. Conclusions

In this paper, a method for detecting the ISCr early in the lithium-ion battery pack was introduced. The battery pack with ISCr was represented with the equivalent circuit model with the R_{ISCr} and the EKF algorithm was used to estimate SOC_p accurately. The OCV of the faulted cell was derived from the battery pack with ISCr to reflect the self-discharge phenomenon caused by the ISCr in the battery pack clearly, because the effect of ISCr in the battery pack was not observed in both the $V_{t,p}$ and the SOC_p obviously. Using the Coulomb counting method and the stable initial value of the SOC_n, obtained from the estimated SOC_p, the SOC_n was calculated. The OCV_p and the OCV_n were acquired from the relationship between OCV and SOC of the normal battery pack, and then the OCV_f was calculated with these two OCV values. Subsequently, the R_{ISCr} (5~50 Ω) of the battery pack with ISCr was estimated accurately using the self-discharge phenomenon in the SOC_f. The proposed algorithm was verified for various soft ISCr fault conditions such as diverse magnitudes of true R_{ISCr} and two load current profiles in both the simulation and the experiment. In addition, through estimating the

R_{ISCr} from the normal battery pack and analyzing it, it was checked that estimated R_{ISCr}s in the various fault cases were reliable. Using the proposed algorithm, the R_{ISCr} was estimated with high accuracy, and the soft ISCr in the battery pack can be detected using the $\overline{R}_{ISCr}$ as the fault index. Our future research will focus on increasing the accuracy of the R_{ISCr} estimates and extending the availability of our proposed algorithm to both the aged battery pack and the battery under varing ambient temperature.

Author Contributions: M.S. suggested the main idea of this paper and constructed the proposed detection technique. T.G. contributed the mathematical modeling and thoroughly reviewed the paper. M.P. conducted the experiments and thoroughly reviewed the paper. S.W.K. supervised the work and finally reviewed the paper.

Funding: This research was funded by the National Research Foundation of Korea (NRF) grant funded by the Korea government (MSIT) grant number [2018R1A2B6005522], MSIT: Ministry of Science and ICT.

Acknowledgments: This work was supported by the National Research Foundation of Korea (NRF) grant funded by the Korea government (MSIT), MSIT: Ministry of Science and ICT (2018R1A2B6005522).

Conflicts of Interest: The authors declare no conflict of interest.

Nomenclature

Symbols

R_{ISCr}	ISCr resistance, Ω
V_t	Terminal voltage, V
I_L	Load current, A
V_{OC}	OCV, V
R	Internal resistance, Ω

Symbols

I_{1L}	Self-discharge current, A
I_{2L}	Residual current, A
η	Charging and discharging efficiency
Δt	Sample period
C_n	Nominal capacity
h	Function of relation between OCV ans SOC
h^{-1}	Inverse function of h
ϵ	Difference error
x	EKF state variable
y	EKF output variable
u	EKF input variable
w, v	EKF process/measurement errors
Q, T	EKF covariances of Gaussian noise
A	EKF state transition matrix
C	EKF observation matrix
P	EKF error covariance matrix
L	EKF Kalman gain
f	EKF state update function
g	EKF output update function
p_{st}	Stable point, sample index

Subscripts

p	battery pack
f	faulted cell
n	normal cell
k	iterration index
j	cell index
m	number of cells in the pack
$1, 2, 3, 4$	1st, 2nd, 3rd, 4th stable points

Abbreviations

ISCr	Internal short circuit
OCV	Open circuit voltage
SOC	State of charge
BMS	Battery management system
EKF	Extended Kalman filter
DST	Dynamic stress test
UDDS	Urban dynamometer driving schedule
CC-CV	Constant-current constant-voltage

References

1. Capasso, C.; Veneri, O.T. Experimental analysis on the performance of lithium based batteries for road full electric and hybrid vehicles. *Appl. Energy* **2014**, *136*, 921–930. [CrossRef]
2. Panchal, S.; Rashid, M.; Long, F.; Mathew, M.; Fraser, R.; Fowler, M. *Degradation Testing and Modeling of 200 Ah LiFePO₄ Battery for EV*; SAE Technical Paper; SAE: Warrendale, PA, USA, 2018.
3. Tarascon, J.M.; Armand, M. Issues and challenges facing rechargeable lithium batteries. In *Materials for Sustainable Energy: A Collection of Peer-Reviewed Research and Review Articles from Nature Publishing Group*; World Scientific: Hackensack, NJ, USA, 2011; pp. 171–179.
4. Deng, D.; Kim, M.G.; Lee, J.Y.; Cho, J. Green energy storage materials: Nanostructured TiO_2 and Sn-based anodes for lithium-ion batteries. *Energy Environ. Sci.* **2009**, *2*, 818–837. [CrossRef]
5. Nykvist, B.; Nilsson, M. Rapidly falling costs of battery packs for electric vehicles. *Nat. Clim. Chang.* **2015**, *5*, 329. [CrossRef]
6. Pillot, C. Battery market development for consumer electronics, automotive, and industrial: Materials requirements and trends. In Proceedings of the 5th Israeli Power Sources Conference, Herzelia, Israel, 21 May 2015; pp. 1–40.
7. Tsujikawa, T.; Yabuta, K.; Arakawa, M.; Hayashi, K. Safety of large-capacity lithium-ion battery and evaluation of battery system for telecommunications. *J. Power Sources* **2013**, *244*, 11–16. [CrossRef]
8. Abada, S.; Marlair, G.; Lecocq, A.; Petit, M.; Sauvant-Moynot, V.; Huet, F. Safety focused modeling of lithium-ion batteries: A review. *J. Power Sources* **2016**, *306*, 178–192. [CrossRef]
9. Gao, Z.; Chin, C.S.; Chiew, J.H.K.; Jia, J.; Zhang, C. Design and Implementation of a Smart Lithium-Ion Battery System with Real-Time Fault Diagnosis Capability for Electric Vehicles. *Energies* **2017**, *10*, 1503. [CrossRef]
10. Kolly, J.M.; Panagiotou, J.; Czech, B.A. *The Investigation of a Lithium-Ion Battery Fire Onboard a Boeing 787 by the US National Transportation Safety Board*; Safety Research Corporation of America: Dothan, AL, USA, 2013; pp. 1–18.
11. Samsung Investigation Reveals New Details about Note7 Battery Failures. Available online: https://consumerist. com/2017/01/22/samsung-investigation-reveals-new-details-about-note7-battery-failures/ (accessed on 12 June 2018).
12. Lisbona, D.; Snee, T. A review of hazards associated with primary lithium and lithium-ion batteries. *Process Saf. Environ. Prot.* **2011**, *89*, 434–442. [CrossRef]
13. Cai, W.; Wang, H.; Maleki, H.; Howard, J.; Lara-Curzio, E. Experimental simulation of internal short circuit in Li-ion and Li-ion-polymer cells. *J. Power Sources* **2011**, *196*, 7779–7783. [CrossRef]

14. Wu, Y.; Saxena, S.; Xing, Y.; Wang, Y.; Li, C.; Yung, W.K.; Pecht, M. Analysis of Manufacturing-Induced Defects and Structural Deformations in Lithium-Ion Batteries Using Computed Tomography. *Energies* **2018**, *11*, 925. [CrossRef]

15. Maleki, H.; Howard, J.N. Internal short circuit in Li-ion cells. *J. Power Sources* **2009**, *191*, 568–574. [CrossRef]

16. Leising, R.A.; Palazzo, M.J.; Takeuchi, E.S.; Takeuchi, K.J. A study of the overcharge reaction of lithium-ion batteries. *J. Power Sources* **2001**, *97*, 681–683. [CrossRef]

17. Maleki, H.; Howard, J.N. Effects of overdischarge on performance and thermal stability of a Li-ion cell. *J. Power Sources* **2006**, *160*, 1395–1402. [CrossRef]

18. Kriston, A.; Pfrang, A.; Döring, H.; Fritsch, B.; Ruiz, V.; Adanouj, I.; Boon-Brett, L. External short circuit performance of Graphite-LiNi$_{1/3}$Co$_{1/3}$Mn$_{1/3}$O$_2$ and Graphite-LiNi$_{0.8}$Co$_{0.15}$Al$_{0.05}$O$_2$ cells at different external resistances. *J. Power Sources* **2017**, *361*, 170–181. [CrossRef]

19. Santhanagopalan, S.; Ramadass, P.; Zhang, J.Z. Analysis of internal short-circuit in a lithium ion cell. *J. Power Sources* **2009**, *194*, 550–557. [CrossRef]

20. Spotnitz, R.; Muller, R.P. Simulation of abuse behavior of lithium-ion batteries. *Electrochem. Soc. Interface* **2012**, *21*, 57–60. [CrossRef]

21. Xu, J.; Wu, Y.; Yin, S. Investigation of effects of design parameters on the internal short-circuit in cylindrical lithium-ion batteries. *RSC Adv.* **2017**, *7*, 14360–14371. [CrossRef]

22. Wang, Q.; Ping, P.; Zhao, X.; Chu, G.; Sun, J.; Chen, C. Thermal runaway caused fire and explosion of lithium ion battery. *J. Power Sources* **2012**, *208*, 210–224. [CrossRef]

23. Feng, X.; Fang, M.; He, X.; Ouyang, M.; Lu, L.; Wang, H.; Zhang, M. Thermal runaway features of large format prismatic lithium ion battery using extended volume accelerating rate calorimetry. *J. Power Sources* **2014**, *255*, 294–301. [CrossRef]

24. Jhu, C.Y.; Wang, Y.W.; Wen, C.Y.; Shu, C.M. Thermal runaway potential of LiCoO$_2$ and Li(Ni$_{1/3}$Co$_{1/3}$Mn$_{1/3}$)O$_2$ batteries determined with adiabatic calorimetry methodology. *Appl. Energy* **2012**, *100*, 127–131. [CrossRef]

25. Zavalis, T.G.; Behm, M.; Lindbergh, G. Investigation of short-circuit scenarios in a lithium-ion battery cell. *J. Electrochem. Soc.* **2012**, *159*, A848–A859. [CrossRef]

26. Xia, B.; Mi, C.; Chen, Z.; Robert, B. Multiple cell lithium-ion battery system electric fault online diagnostics. In Proceedings of the Transportation Electrification Conference and Expo (ITEC), Dearborn, MI, USA, 14–17 June 2015; pp. 1–7.

27. Feng, X.; Weng, C.; Ouyang, M.; Sun, J. Online internal short circuit detection for a large format lithium ion battery. *Appl. Energy* **2016**, *161*, 168–180. [CrossRef]

28. Kim, G.H.; Smith, K.; Ireland, J.; Pesaran, A. Fail-safe design for large capacity lithium-ion battery systems. *J. Power Sources* **2012**, *210*, 243–253. [CrossRef]

29. Guo, R.; Lu, L.; Ouyang, M.; Feng, X. Mechanism of the entire overdischarge process and overdischarge-induced internal short circuit in lithium-ion batteries. *Sci. Rep.* **2016**, *6*, 30248. [CrossRef] [PubMed]

30. Feng, X.; He, X.; Lu, L.; Ouyang, M. Analysis on the Fault Features for Internal Short Circuit Detection Using an Electrochemical-Thermal Coupled Model. *J. Electrochem. Soc.* **2018**, *165*, A155–A167. [CrossRef]

31. Sazhin, S. V.; Dufek, E. J.; Gering, K. L. Enhancing Li-Ion Battery Safety by Early Detection of Nascent Internal Shorts. *J. Electrochem. Soc.* **2017**, *164*, A6281–A6287. [CrossRef]

32. Seo, M.; Goh, T.; Koo, G.; Park, M.; Kim, S.W. Detection of internal short circuit in Li-ion battery by estimating its resistance. In Proceedings of the 4th IIAE International Conference on Intelligent Systems and Image Processing (ICISIP2016), Kyoto, Japan, 8–12 September 2016; pp. 212–217.

33. Seo, M.; Goh, T.; Park, M.; Koo, G.; Kim, S.W. Detection of Internal Short Circuit in Lithium Ion Battery Using Model-Based Switching Model Method. *Energies* **2017**, *10*, 76. [CrossRef]

34. Chen, H.; Cong, T.N.; Yang, W.; Tan, C.; Li, Y.; Ding, Y. Progress in electrical energy storage system: A critical review. *Prog. Nat. Sci.* **2009**, *19*, 291–312. [CrossRef]

35. Hesse, H.C.; Schimpe, M.; Kucevic, D.; Jossen, A. Lithium-Ion Battery Storage for the Grid—A Review of Stationary Battery Storage System Design Tailored for Applications in Modern Power Grids. *Energies* **2017**, *10*, 2107. [CrossRef]

36. Xia, B.; Shang, Y.; Nguyen, T.; Mi, C. A correlation based fault detection method for short circuits in battery packs. *J. Power Sources* **2017**, *337*, 1–10. [CrossRef]

37. Feng, X.; Pan, Y.; He, X.; Wang, L.; Ouyang, M. Detecting the internal short circuit in large-format lithium-ion battery using model-based fault-diagnosis algorithm. *J. Energy Storage* **2018**, *18*, 26–39. [CrossRef]

38. Ouyang, M.; Zhang, M.; Feng, X.; Lu, L.; Li, J.; He, X.; Zheng, Y. Internal short circuit detection for battery pack using equivalent parameter and consistency method. *J. Power Sources* **2015**, *294*, 272–283. [CrossRef]

39. Lu, L.; Han, X.; Li, J.; Hua, J.; Ouyang, M. A review on the key issues for lithium-ion battery management in electric vehicles. *J. Power Sources* **2013**, *226*, 272–288. [CrossRef]

40. Andrea, D. *Battery Management Systems for Large Lithium-Ion Battery Packs*; Artech House: Norwood, MA, USA, 2010.

41. Väyrynen, A.; Salminen, J. Lithium ion battery production. *J. Chem. Thermodyn.* **2012**, *46*, 80–85. [CrossRef]

42. Zhao, W.; Luo, G.; Wang, C.Y. Modeling nail penetration process in large-format Li-ion cells. *J. Electrochem. Soc.* **2015**, *162*, A207–A217. [CrossRef]

43. Zhao, R.; Liu, J.; Gu, J. Simulation and experimental study on lithium ion battery short circuit. *Appl. Energy* **2016**, *173*, 29–39. [CrossRef]

44. Plett, G.L. Extended Kalman filtering for battery management systems of LiPB-based HEV battery packs: Part 3. State and parameter estimation. *J. Power Sources* **2004**, *134*, 277–292. [CrossRef]

45. Tong, S.; Klein, M.P.; Park, J.W. On-line optimization of battery open circuit voltage for improved state-of-charge and state-of-health estimation. *J. Power Sources* **2015**, *293*, 416–428. [CrossRef]

46. Hussein, A.A.H.; Batarseh, I. State-of-charge estimation for a single Lithium battery cell using Extended Kalman Filter. In Proceedings of the Power and Energy Society General Meeting, Detroit, MI, USA, 24–29 July 2011; pp. 1–5.

47. Wang, D.; Bao, Y.; Shi, J. Online Lithium-Ion Battery Internal Resistance Measurement Application in State-of-Charge Estimation Using the Extended Kalman Filter. *Energies* **2017**, *10*, 1284. [CrossRef]

48. He, H.; Zhang, X.; Xiong, R.; Xu, Y.; Guo, H. Online model-based estimation of state-of-charge and open-circuit voltage of lithium-ion batteries in electric vehicles. *Energy* **2012**, *39*, 310–318. [CrossRef]

49. Xing, Y.; He, W.; Pecht, M.; Tsui, K.L. State of charge estimation of lithium-ion batteries using the open-circuit voltage at various ambient temperatures. *Appl. Energy* **2014**, *113*, 106–115. [CrossRef]

50. Yao, L.W.; Aziz, J.A.; Kong, P.Y.; Idris, N.R.N. Modeling of lithium-ion battery using MATLAB/simulink. In Proceedings of the 2013 39th Annual Conference of the IEEE Industrial Electronics Society, Vienna, Austria, 10–13 November 2013; pp. 1729–1734.

51. Liaw, B.Y.; Nagasubramanian, G.; Jungst, R.G.; Doughty, D.H. Modeling of lithium ion cells—A simple equivalent-circuit model approach. *Solid State Ionics* **2004**, *175*, 835–839.

52. Wang, J.; Purewal, J.; Liu, P.; Hicks-Garner, J.; Soukazian, S.; Sherman, E.; Verbrugge, M.W. Degradation of lithium ion batteries employing graphite negatives and nickel–cobalt–manganese oxide+ spinel manganese oxide positives: Part 1, aging mechanisms and life estimation. *J. Power Sources* **2014**, *269*, 937–948. [CrossRef]

53. Kim, J.; Shin, J.; Chun, C.; Cho, B.H. Stable configuration of a Li-ion series battery pack based on a screening process for improved voltage/SOC balancing. *IEEE Trans. Power Electron.* **2012**, *27*, 411–424. [CrossRef]

54. Madani, S.S.; Schaltz, E.; Knudsen Kær, S. Review of Parameter Determination for Thermal Modeling of Lithium Ion Batteries. *Batteries* **2018**, *4*, 20. [CrossRef]

55. Chin, C.S.; Gao, Z. State-of-Charge Estimation of Battery Pack under Varying Ambient Temperature Using an Adaptive Sequential Extreme Learning Machine. *Energies* **2018**, *11*, 711. [CrossRef]

56. Gao, Z.; Chin, C.S.; Woo, W.L.; Jia, J. Integrated equivalent circuit and thermal model for simulation of temperature-dependent LiFePO4 battery in actual embedded application. *Energies* **2017**, *10*, 85. [CrossRef]

57. Gao, Z.C.; Chin, C.S.; Toh, W.D.; Chiew, J.; Jia, J. State-of-Charge Estimation and Active Cell Pack Balancing Design of Lithium Battery Power System for Smart Electric Vehicle. *J. Adv. Transp.* **2017**, *2017*. [CrossRef]

58. Panchal, S.; Mcgrory, J.; Kong, J.; Fraser, R.; Fowler, M.; Dincer, I.; Agelin-Chaab, M. Cycling degradation testing and analysis of a LiFePO4 battery at actual conditions. *Int. J. Energy Res.* **2017**, *41*, 2565–2575. [CrossRef]

MDPI

St. Alban-Anlage 66

4052 Basel

Switzerland

Tel. +41 61 683 77 34

Fax +41 61 302 89 18

www.mdpi.com

Energies Editorial Office

E-mail: energies@mdpi.com

www.mdpi.com/journal/energies